T0363685

Austral Ark
The State of Wildlife in Australia and New Zealand

Australia and New Zealand are home to a remarkable and unique assemblage of flora and fauna. Sadly, by virtue of their long isolation and a naïve and vulnerable biota, both countries have suffered substantial losses to biodiversity since European contact.

Bringing together the contributions of leading conservation biologists, *Austral Ark* presents the special features and historical context of Austral biota, and explains what is being conserved and why. The threatening processes occurring worldwide are discussed, along with the unique conservation problems faced at regional level. At the same time, the book highlights many examples of conservation success resulting from the innovative solutions that have been developed to safeguard native species and habitats in both Australia and New Zealand.

Austral Ark fills an important gap regarding wildlife gains and declines, and how best to take conservation forward to keep this extraordinary area of the world thriving.

Adam Stow is an Associate Professor in the Department of Biological Sciences at Macquarie University in Australia. His research focusses on three broad themes: conservation genetics, evolutionary processes and molecular ecology. He is Head of the Conservation Genetics Group which carries out research on animals, both vertebrate and invertebrate, in both marine and terrestrial environments. He is a nominated scientific expert to the state and federal governments for shark conservation and an academic editor for the open access scientific journal *PLoS ONE*. Regarded as a talented science communicator, he promotes wildlife conservation to a range of audiences.

Norman Maclean is Emeritus Professor of Genetics at Southampton University. He is the Molecular Editor of *Journal of Fish Biology*, has authored and edited a dozen books on genetics and has published about 150 papers. He is a member of RSPB, BTO, Hampshire and Isle of Wight Wildlife Trust and Southampton Natural History Society, and has studied wildlife (as an amateur) in over 50 countries. In 2010 he edited *Silent Summer: The State of Wildlife in Britain and Ireland*, published by Cambridge University Press.

Gregory I. Holwell is a Senior Lecturer in the School of Biological Sciences at the University of Auckland. His research focuses on the behaviour, ecology, evolution and conservation of insects and other invertebrates throughout Australia, New Zealand and South East Asia. Much of his research has centred on praying mantises, most recently investigating the impact of an invasive South African praying mantis on New Zealand's only native mantis. He is a passionate naturalist and advocate for the conservation of our little-known invertebrate fauna.

Austral Ark

The State of Wildlife in Australia and New Zealand

Edited by

Adam Stow

Macquarie University, Ryde, Sydney, Australia

Norman Maclean

University of Southampton, U.K.

Gregory I. Holwell

University of Auckland, New Zealand

CAMBRIDGE
UNIVERSITY PRESS

Shaftesbury Road, Cambridge CB2 8EA, United Kingdom

One Liberty Plaza, 20th Floor, New York, NY 10006, USA

477 Williamstown Road, Port Melbourne, VIC 3207, Australia

314–321, 3rd Floor, Plot 3, Splendor Forum, Jasola District Centre, New Delhi – 110025, India

103 Penang Road, #05–06/07, Visioncrest Commercial, Singapore 238467

Cambridge University Press is part of Cambridge University Press & Assessment, a department of the University of Cambridge.

We share the University's mission to contribute to society through the pursuit of education, learning and research at the highest international levels of excellence.

www.cambridge.org
Information on this title: www.cambridge.org/9781107033542

First published 2015 (version 2, September 2022)

Printed in the United Kingdom by TJ Books Limited, Padstow Cornwall

A catalogue record for this publication is available from the British Library

Library of Congress Cataloguing in Publication data
Austral ark : the state of wildlife in Australia and New Zealand / edited by Adam Stow, Macquarie University, Ryde, Sydney, Australia, Norman Maclean, University of Southampton, U. K., Gregory I. Holwell, University of Auckland, New Zealand.
  pages   cm
Includes index.
ISBN 978-1-107-03354-2
1. Wildlife conservation – Australia.   2. Wildlife conservation – New Zealand.
3. Biodiversity – Australia.   4. Biodiversity – New Zealand.   5. Endangered species – Australia.
6. Endangered species – New Zealand.   I. Stow, Adam, 1971 – editor of compilation.
II. Maclean, Norman, 1932- editor of compilation.   III. Holwell, Gregory I., editor of compilation.   IV. Title: State of wildlife in Australia and New Zealand.
QL84.7.A1A86   2014
333.95′40993–dc23

2014013701

ISBN 978-1-107-03354-2 Hardback

Additional resources for this publication at www.cambridge.org/9781107033542

CONTENTS

Colour plate section between pages 304 and 305.

CONTRIBUTORS

Vanessa M. Adams Research Institute for the Environment and Livelihoods, Charles Darwin University, Darwin, NT 0909, Australia and Australian Research Council Centre of Excellence for Coral Reef Studies, James Cook University, Townsville, QLD 4811, Australia

Nigel R. Andrew Centre for Behavioural and Physiological Ecology, Zoology, University of New England, Armidale, NSW, Australia, 2351

Maria G. Asmyhr Macquarie University, Department of Biological Sciences, NSW, Australia, 2109

Megan Barnes University of Queensland, Environmental Decisions Group, St Lucia, QLD, Australia, 4072

Imogen E. Bassett Centre for Biodiversity and Biosecurity School of Biological Sciences, Tâmaki Campus, The University of Auckland, Private Bag 92019, Auckland Mail Centre, Auckland, New Zealand, 1142. *Currently* Auckland Council, Private Bag 92300, Auckland 1142.

Linda J. Beaumont Macquarie University, Department of Biological Sciences, North Ryde, NSW, Australia, 2109

Luciano B. Beheregaray Flinders University, School of Biological Sciences, Adelaide SA, Australia, 5001

Phillip J. Bishop Department of Zoology, University of Otago, P.O. Box 56, Dunedin, New Zealand, 9054

Allan H. Burbidge Department of Parks and Wildlife, Wildlife Place, Woodvale WA, Australia, 6026

Abigail Cabrelli Macquarie University, Department of Biological Sciences, North Ryde, NSW, Australia, 2109

Louise Chilvers Institute of Vet, Animal and Biomedical Sciences, Massey University, Palmerston North, New Zealand, 4410

Sarah Comer Department of Parks and Wildlife, South Coast Region, 120 Albany Highway, Albany, WA, Australia, 6330

Steven J. B. Cooper Australian Centre for Evolutionary Biology and Biodiversity, School of Earth and Environmental Sciences, The University of Adelaide, Adelaide, SA 5005, Australia and Evolutionary Biology Unit, South Australian Museum, North Terrace, Adelaide, SA, Australia, 5000

Ian D. Craigie Australian Research Council Centre of Excellence for Coral Reef Studies, James Cook University, Townsville, QLD, Australia, 4811

Saul Cunningham, CSIRO, Black Mountain laboratories, Box 1700, Canberra, ACT, Australia, 2601

Rebecca Dunlop Cetacean Ecology and Acoustics Laboratory, School of Veterinary Science, University of Queensland, Qld, Australia, 4343

Graham J. Edgar University of Tasmania, Private Bag 49, HOBART TAS, Nubeena Crescent, Taroona Sandy Bay, Australia, 7001

Mark D. B. Eldridge Terrestrial Vertebrates, Australian Museum, 6 College Street, Sydney, New South Wales, Australia, 2010

Robert M. Ewers Imperial College London, Silwood Park Campus, Ascot SL5 7PY, UK

Peter Fairweather Flinders University, Sturt Road, Bedford Park, South Australia, 5042

Leanne Faulks Department of Ecology and Genetics, Evolutionary Biology Centre, Uppsala University, Norbyvägen 18 D, 752 36 Uppsala, Sweden

Donald C. Franklin Charles Darwin University, Darwin NT, Australia, 0909

Poppy Lakeman Fraser Imperial College London, Silwood Park Campus, Ascot SL5 7PY, UK

Josie A. Galbraith Centre for Biodiversity and Biosecurity School of Biological Sciences, Tâmaki Campus, The University of Auckland Private Bag, 92019, Auckland Mail Centre, Auckland, New Zealand, 1142

Rachael V. Gallagher Department of Biological Sciences, Macquarie University, North Ryde, NSW, Australia, 2109

Stephen T. Garnett Charles Darwin University, Darwin NT, Australia, 0909

Dean Gilligan Fisheries and Ecosystems Research, Industry & Investment NSW, Level 1 Braysyth Building, Cnr Beach Rd & Orient St, P.O. Box 17, Batemans Bay NSW, Australia, 2536

Michael Gillings Department of Biological Sciences, Macquarie University, Sydney, NSW, Australia, 2109

Al S. Glen Landcare Research P.O. Box 69040, Lincoln, New Zealand, 7640

Alana Grech Department of Environment and Geography, Macquarie University, Building E7A Room 602 & 603, Herring Road, North Ryde, NSW, Australia, 2109

Robert Harcourt Department of Biological Sciences, Macquarie University, Sydney, NSW, Australia, 2109

Dan Harley Threatened Species Biologist, Healesville Sanctuary, P.O. Box 248, Healesville, Victoria, Australia, 3777

Peter S. Harlow Herpetofauna Division, Taronga Conservation Society Australia, Mosman, NSW, Australia, 2088

Catherine A. Herbert Faculty of Veterinary Science, The University of Sydney, Camperdown, New South Wales, Australia, 2006

Jean-Marc Hero Environmental Futures Research Institute, School of Environment, Griffith University, Gold Coast Campus, QLD, Australia, 4222

Rod Hitchmough Department of Conservation – Te Papa Atawhai, National Office, P.O. Box 10, New Zealand, 420

Marc Hockings School of Geography Planning and Environmental Management, University of Queensland, St Lucia, QLD, Australia, 4072

Gregory I. Holwell School of Biological Sciences, University of Auckland, Private Bag 92019, Auckland, New Zealand, 1142

Grant C. Hose Department of Biological Sciences, Macquarie University, NSW, Australia, 2109

Conrad J. Hoskin School of Marine & Tropical Biology, James Cook University, Townsville, Queensland, Australia, 4811

Lesley Hughes Department of Biological Sciences, Macquarie University, North Ryde, NSW, Australia, 2109

William F. Humphreys Western Australian Museum, Locked Bag 49, Welshpool DC, Western Australia, Australia, 6986

Vanessa Flora Jaiteh Centre for Fish, Fisheries and Aquatic Ecosystems Research/Asia Research Centre, Murdoch University, Perth, Western Australia, Australia and Lembaga Ilmu Penelitian Indonesia (LIPI), Ambon, Indonesia

Mike J. Joy Institute of Agriculture & Environment, College of Sciences, Massey University, Private Bag 11 222, Palmerston North, New Zealand, 4442

Cheryl R. Krull Centre for Biodiversity and Biosecurity School of Biological Sciences, Tâmaki Campus, The University of Auckland Private Bag, 92019, Auckland Mail Centre, Auckland, New Zealand, 1142

Daphne E. Lee Department of Geology, University of Otago, P.O. Box 56, Dunedin, New Zealand

William G. Lee Landcare Research, Private Bag 1930, Dunedin, New Zealand and School of Biological Sciences, University of Auckland, New Zealand, 1010

Sarah Legge Suite 5, 280 Hay St, Subiaco, Western Australia, 6008

Carlos A. Lehnebach Museum of New Zealand Te Papa Tongarewa, 55 Cable Street, P.O. Box 467, Wellington, New Zealand, 6011

Michelle R. Leishman Department of Biological Sciences, Macquarie University, North Ryde, NSW, Australia, 2109

David B. Lindenmayer The Australian National University, Canberra, ACT, Australia, 0200

Katrin Lowe Environmental Futures Research Institute, School of Environment, Griffith University, Gold Coast Campus, QLD, Australia, 4222

Helene Marsh School of Earth and Environmental Science, James Cook University, Townsville, Qld, Australia, 4811

Paolo Momigliano Department of Biological Sciences, Macquarie University, Sydney, New South Wales, Australia, 2109

Jo M. Monks Department of Conservation – Te Papa Atawhai, Private Bag 4715, Christchurch, New Zealand, 8140

Edward J. Narayan Environmental Futures Research Institute, School of Environment, Griffith University, Gold Coast Campus, QLD, Australia, 4222

Helen W. Nathan Centre for Biodiversity and Biosecurity School of Biological Sciences, Tâmaki Campus, The University of Auckland Private Bag, 92019, Auckland Mail Centre, Auckland, New Zealand, 1142

Nicola J. Nelson Allan Wilson Centre for Molecular Ecology and Evolution, School of Biological Sciences, Victoria University of Wellington, P.O. Box 600, Wellington, New Zealand, 6140

David A. Nipperess Department of Biological Sciences, Macquarie University, Sydney, NSW, Australia, 2109

Mike Noad Cetacean Ecology and Acoustics Laboratory, School of Veterinary Science, University of Queensland, Qld, Australia, 4343

Thalie B. Partridge Charles Darwin University, Darwin, Northern Territory, Australia, 0909

Benjamin L. Phillips The University of Melbourne, Parkville, Vic, Australia, 3010, 4810

David A. Pike School of Marine and Tropical Biology and Centre for Tropical Environmental & Sustainability Science, James Cook University, Queensland, Australia, 4811

Robert L. Pressey Australian Research Council Centre of Excellence for Coral Reef Studies, James Cook University, Townsville, QLD, Australia, 4811

J. Dale Roberts School of Animal Biology, Centre for Evolutionary Biology, and Centre of Excellence in Natural Resource Management, University of Western Australia, P.O. Box 5771, Albany WA, Australia, 6330

Maurizio Rossetto National Herbarium of NSW, Mrs Macquaries Road, Sydney, NSW, Australia, 2000

Nick Shears Leigh Marine Laboratory, University of Auckland, PO Box 349, Warkworth, New Zealand

Richard Shine School of Biological Sciences A08, University of Sydney, NSW, Australia, 2006

David Slip Taronga Conservation Society, Bradley's Head Road, Mosman, NSW, Australia, 2088

Conrad Speed Science Division, Department of Environment and Conservation, Marine Science Program, Kensington, Western Australia, Australia

Margaret C. Stanley Centre for Biodiversity and Biosecurity School of Biological Sciences, Tâmaki Campus, The University of Auckland Private Bag, 92019, Auckland Mail Centre, Auckland, New Zealand, 1142

Adam Stow Department of Biology, Macquarie University, Ryde, Sydney, Australia, 2109

Judit K. Szabo Charles Darwin University, Darwin NT, Australia, 0909

Martin Taylor Protected Areas and Conservation Science Manager, WWF-Australia, 1/17 Burnett Lane, Brisbane, Australia, 4000

Hannah L. Thomas Marine Programme, UNEP World Conservation Monitoring Centre, Cambridge, United Kingdom

Darren F. Ward New Zealand Arthropod Collection, Landcare Research, Private Bag 92170, Auckland, New Zealand

Trevor J. Ward University of Technology, Sydney, P.O. Box 123, Broadway NSW, Australia, 2007

Jonathan K. Webb School of the Environment, University of Technology Sydney, PO Box 123, Broadway NSW, Australia, 2007

Sarah Withers School of Biological Sciences, University of Auckland, Private Bag 92019, Auckland 1142, New Zealand

John C. Z. Woinarski Charles Darwin University, Darwin, Northern Territory, Australia, 0909

FOREWORD

Australia and New Zealand can be thought of as *arks*. For tens of millions of years they have drifted, isolated, across the surface of the Earth, carrying unique plants and animals. Indeed it is the isolation of their wildlife that gives these southern lands their uniqueness, as well as inherent vulnerability. Long separated, many of their creatures are naive to the defences, competitive techniques or hunting tactics of wildlife introduced from elsewhere. Sadly, it's clear that much of the wildlife of these southern lands is in crisis. As the new epoch dominated by human destruction, the *anthropocene*, continues to bite, both solutions and action are needed to slow down the damage.

The issues facing wildlife on the lands and surrounding waters of Australia and New Zealand are immense, and while it's impossible to cover all threats in a single book, the editors have selected comment from leading researchers on key threats including introduced plants and animals, pollution, habitat fragmentation and climate change. To evaluate these threats, and prioritise our response to them, we need up-to-date information on areas currently protected from human activity and the conservation status of particular groups. *Austral Ark* serves this purpose and one can hope that this information will raise awareness, inspire action and, perhaps with excessive optimism, that these descriptions on the state of the wildlife will provide a reference point in the future that shows how bad things were in the early twenty-first century.

Professor Tim Flannery

INTRODUCTION

Both Australia and New Zealand are countries blessed with an extraordinary diversity of fascinating animals and plants, and the biodiversity of our region is a major source of pleasure to residents and visiting tourists alike. However, all is not well with our wildlife and conservation of our precious native species is proving to be an increasingly uphill struggle. Most people realise that past major extinction events have been a characteristic of the history of life on planet Earth, such as those of the Permian era some 250 million years ago (MYA), and the end of the Cretaceous (65 MYA), which marked the demise of the dinosaurs. However, the notion that we are now living amidst a current 'great extinction' may come as a surprise to many. There is one important respect in which the present great extinction differs from the preceding ones, namely that most of the available evidence suggests that it is anthropogenic in origin; it is 'man made'. Even if we already accept this picture of a current man-made extinction event, there is a tendency to assume that it is mainly happening elsewhere.

This book is an attempt to take stock of the state of wildlife in Australia and New Zealand, and to draw attention to the severity of the numerous species declines and extinctions. However, a wake-up call is all very well, but without careful documentation of the present state of play, it is easy for us all to remain complacent. There is no way in which one or two people (in our case three) can assemble this kind of information on their own: the knowledge required is just too detailed. So we, the co-editors, have sought out those people who have the necessary expertise, and persuaded them to write the chapters of this book. It is a measure of their concern for the biodiversity of the region that these authors have taken the time and trouble to do so.

The choice to make two separate countries the target for this wildlife audit perhaps needs some justification. The slow drift of tectonic plates has separated Australia and New Zealand and surrounded each by large expanses of ocean, and consequently, through the process of evolution, wildlife has diverged to become exceptionally different by world standards (see Chapters 1 and 2). Australia and New Zealand have many historical and contemporary environmental differences, which over time have given rise to each landmass possessing a strikingly different wildlife, and this allows for interesting contrasts. Nonetheless, there are many commonalities in the wildlife, and for fauna at least New Zealand and Australia can be grouped together into their own exclusive zoogeographic region (Holt *et al.*, 2013).

There are also many conservation issues that are shared by Australia and New Zealand. This is likely to reflect the similar timing of European colonisation in both countries. As a result, both areas have been subject to recent and major environmental changes, and high rates of extinction, despite population sizes and densities of both countries being among the lowest anywhere. In Australia the average number of people per square kilometre is 2 to 3, and 15 in New Zealand, compared with 243 in the United Kingdom. In some ways, the long-term isolation that shaped the unique assemblage of organisms in the region has also rendered the wildlife naive, and more vulnerable to invaders. In the arid interior of Australia, it can be especially haunting when, many hundreds of kilometres from the nearest person, there are still telltale signs of animals recently lost from this expansive and seemingly untouched environment. In essence it's a case of 'out of sight' and 'out of mind'. Being island nations there are extensive coast lines, and the central highlands of New Zealand, and arid interior of Australia, have resulted in the highest concentrations of people being on the coasts. However, the marine environment by its very nature is subject to the same issue of being largely out of sight for the majority of people. Nevertheless, some relatively pristine examples of temperate coastal ecosystems are preserved, especially in New Zealand on its many small islands. In Australia, there is a very extensive coastline and approximately 5000 km is in the tropics, which unlike many tropical regions elsewhere, has very few people and astounding diversity represented in features such as coral reefs. There is an impressively unique wildlife in the Australian and New Zealand region, a fact that for the most part is widely appreciated. The extent to which this is in peril is generally less appreciated.

There is also cause for optimism. This book highlights many examples of conservation success, and the people actively engaged in researching our threatened species and habitats, and informing management decisions, are well-represented among the authors of the following chapters. In this sense, *Austral Ark* represents a unique synthesis of the work and views of Australia and New Zealand's leading conservation biologists. Australia represents perhaps the most diverse landscape managed by a single government, spanning the tropics and deserts through to sub-Antarctic and Antarctic territories. New Zealand is also ecologically diverse and nearly 30% of the country is under some degree of protection and under public ownership. Thus the potential for optimistic planning for the future of the region's remarkable species is great.

By considering the causes of extinctions and declines such as climate change, over-fishing, pollution, invasive species and habitat loss, together with other chapters devoted to specific faunal and floral groups, we have endeavoured to make the book reasonably comprehensive without letting it become too large or too expensive. The book therefore endeavours to tell a story as well as to organise and detail the evidence and relevant case histories. In addition to considering the history of our wildlife in the two countries and discussing the present state of play, we also try to indicate, where possible, what the future of our wildlife is likely to be over the next half century or so (that is within the lifetime of many of our readers) and what new measures could be introduced to help remedy the situation.

REFERENCES

Holt, B. G., Lessard, J. P., Borregaard, M. K., *et al.* (2013). An update of Wallace's Zoogeographic Regions of the World, *Science*, **339** (6115), pp. 74–78. doi:10.1126/science.1228282.

CHAPTER 1

A separate creation: diversity, distinctiveness and conservation of Australian wildlife

David A. Nipperess

Summary

Australia is biologically diverse, with around 150 000 described species, representing perhaps 25% of the total number present. However, this biota is more notable for its endemism than its richness (e.g. 94% of Australian frog species are found nowhere else). Australia is distinctive, not only in terms of endemism, but also in terms of evolutionary adaptations (e.g. large hopping mammals) and ecological processes (e.g. nutrient cycling by fire). Distinctiveness is attributed to three principal factors: (1) a long period of geographic isolation; (2) the preponderance of ancient soils low in key nutrients; and (3) an increasingly arid and inherently unpredictable climate. Australia is also unfortunately distinctive in the scale of biodiversity loss since European settlement with 98 species and subspecies listed as extinct, and a further 1700 threatened with extinction. Both for historical extinctions and currently threatened species, habitat loss and introduced species are the key threats, while climate change is the emerging and possibly most significant threat of the twenty-first century. In the face of these perils, Australia's distinctive wildlife needs special attention because it makes such a large contribution to the biodiversity and cumulative evolutionary history of the planet.

1.1 Introduction

Australia is a biologically unusual continent. This is easily shown by a few examples such as the presence of large hopping marsupials, the prevalence of fire-adapted vegetation, and the sheer diversity of arid-zone lizards. Entire groups of organisms are found nowhere else. Many more are largely confined to the Australian continent, with only a

Austral Ark: The State of Wildlife in Australia and New Zealand, eds. A. Stow, N. Maclean and G. I. Holwell. Published by Cambridge University Press. © Cambridge University Press 2015.

few representatives on nearby islands, such as New Guinea. While visiting Australia and pondering the unusual Australian animals, Charles Darwin wrote in his diary: *"An unbeliever in everything beyond his own reason, might exclaim 'Surely two distinct creators must have been at work ...'"* (p. 402 of Darwin, 2001). The Australian fauna was so different from that found in Europe, Asia or the Americas, it was as though it was created completely separately from that elsewhere. Of course, Darwin was essentially correct in that Australian wildlife, to a large extent, have been 'created' separately. This separateness, however, was not the work of a separate supernatural entity, but rather the result of a long period of independent evolution on an isolated continent subjected to significant and unusual environmental change.

Precisely because of this distinctiveness, the conservation of Australian wildlife is essential to the conservation of global biodiversity. Loss of Australian species would result in the loss of large parts of the global tree of life, as many are both geographically and evolutionarily distinct. Unfortunately, many Australian species have already become extinct and more will likely follow in the coming decades. At the same time, many species that have evolved elsewhere have arrived, changing the structure and dynamics of Australian ecosystems, which, coupled with extinctions, has diminished the distinctiveness that the long years of separate evolution have wrought.

This chapter aims to provide a broad overview of Australian wildlife, focusing on diversity, distinctiveness and conservation of the biota. The term 'wildlife' is here used in the broad sense of naturally occurring, native populations (whether animals, plants or 'other') and should not be read as meaning terrestrial vertebrates in particular. Nevertheless, examples will often be drawn from terrestrial vertebrates. This is because data on vertebrates and terrestrial systems are more complete and more familiar to the author. When referring to 'Australia', this comprises the national territory of the Commonwealth of Australia (primarily the area of the Australian continental shelf including Tasmania but excluding New Guinea) rather than a discrete biogeographical or geological unit.

1.2 Australian biodiversity

1.2.1 Richness

The total number of species in Australia, or indeed any part of the world, is uncertain. The primary reason for this is that many species, especially in hyper-diverse groups such as the insects, remain undiscovered by science and thus undescribed (Purvis and Hector, 2000). This gulf between the known and unknown has been termed the *Linnean Shortfall* (Whittaker *et al.*, 2005). The sheer size of the biota will ensure that the shortfall will remain for some time to come. A recent global estimate (Mora *et al.*, 2011) suggests that there are 8.7 million species of eucaryotes, on land and at sea, of which approximately 1.2 million have been described, leaving more than 80% still to be given a scientific name.

In the case of Australia, the Linnean Shortfall is likely especially pronounced. Although formal scientific description began shortly after the modern system of taxonomy was founded in 1758 (with the publication of the 10th edition of Linnaeus' *Systema Naturae*), rates of description have been slower than many parts of the world (Plate 4). This is likely because of the large size of the continent, the small human population, and the concentration of available expertise in a few cities far from many of Australia's biodiversity hotspots. Even for the well-known mammalian fauna, the rate of discovery more closely resembles that of Brazil than either the United States or Canada (Plate 4). Indeed the curve

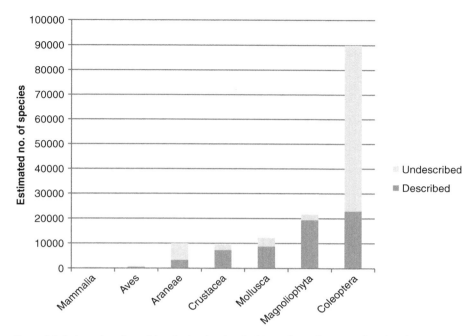

Figure 1.1 Estimated numbers of described and undescribed Australian native species for selected groups. Data taken from Chapman (2009).

may not have yet reached an asymptote, suggesting further new mammal species await discovery in the twenty-first century (Ceballos and Ehrlich, 2009). Of course, what applies to mammals applies much more so to insects and other less-studied taxa.

Considering all described taxa, Chapman (2009) estimates that Australia is home to 147 579 native species. Note that this is an estimate with an unknown margin of error because: (1) historically, species names have not been entered into a central repository but rather are scattered across numerous scientific journals (Minelli, 2003); (2) a large proportion of published names are now considered synonyms (Godfray, 2002); and (3) related to the last point, there is uncertainty and disagreements about species delineation (Vane-Wright, 2003) and thus the number of valid species names. There has, however, been progress in the compilation of published species names in a central, publically available repository. At the time of writing, the Atlas of Living Australia (www.ala.org.au) has a national list of 152 080 accepted species names – quite similar to the estimate of Chapman (2009). Much more difficult, of course, is to estimate how many species remain to be described. Chapman (2009), by adding up individual estimates for all taxonomic groups, estimates that Australia is home to 556 398 species (described and undescribed). If this is a reasonable figure, then about three-quarters of Australia's species await formal scientific recognition. However, the ratio of described to undescribed varies considerably among groups – mammals (Mammalia) are estimated to be 99% described while beetles (Coleoptera) may be only 25% (Figure 1.1).

The diversity of the Australian biota compared to the rest of the world can be assessed only for well-known groups such as birds and mammals. Overall, Australia appears neither especially species-rich nor particularly species-poor when compared to regions of similar area (Figure 1.2). A significant source of variation among countries is the

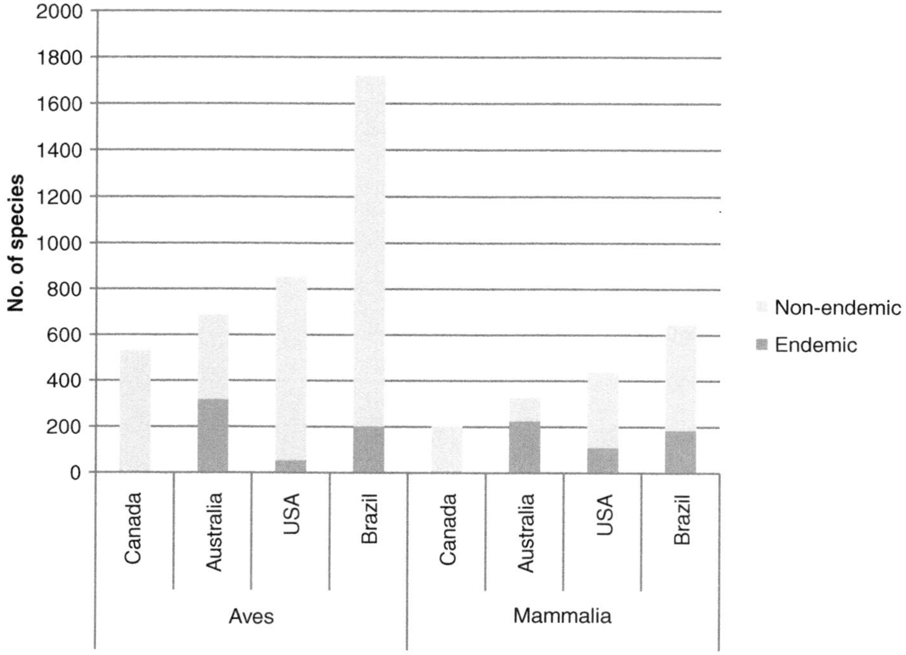

Figure 1.2 Species richness and endemism of extant birds (Aves) and mammals (Mammalia) in Australia compared to other countries of similar area. Data sourced from IUCN Red List (www.iucnredlist.org) for mammals and BirdLife International (www.birdlife.org) for birds.

latitudinal gradient they encompass as species richness is highest at the tropics, due at least in part to higher primary productivity (Gaston, 2000). However, the United States has more mammal and bird species than Australia despite being further from the equator. Clearly other factors are also important, such as Australia's geographic isolation, low topographic relief, extensive aridity and low soil fertility (discussed further below).

1.2.2 Endemism

While Australia may not be exceptional in terms of species richness, this is not the case when it comes to endemism. When compared to other countries of similar area, Australia has proportionally many more endemic species (i.e. species found nowhere else) of birds (46%) and mammals (69% including marine species) (Figure 1.2). High levels of endemism are found throughout the biota. Chapman (2009) estimates that 94% of species of amphibians, 93% of flowering plants, and 93% of reptiles are endemic to Australia. In the case of arthropods, endemic species are likely to makeup about 90% of the fauna, with specific groups, such as cicadas (Cicadoidea), being even higher (Raven and Yeates, 2007). Of those species native to, but not strictly endemic to, Australia, many are found only in areas in the immediate vicinity such as New Guinea.

Endemism can be thought of in a temporal as well as a spatial sense – the length of time a particular lineage has been restricted to an area contributes to the degree of endemism. This can be assessed by noting endemic higher taxa such as genera, tribes, subfamilies and families (Table 1.1). In general, these taxa can be described as *palaeoendemics* as they each represent ancient lineages with a long period of association with the Australian

Table 1.1   Examples of Australian palaeoendemics. Those taxa marked with an asterisk are monotypic (represented by a single species).

Taxon	Taxonomic rank
Wollemia (Wollemi Pine)*	Genus
Austrobaileyaceae (Flowering plant)*	Family
Xanthorrhoeoideae (Grass-trees)	Subfamily
Euastacus (Spiny Crayfish)	Genus
Hemiphlibiidae (Ancient Greenling – Damselfly)*	Family
Lamingtoniidae (Beetle)*	Family
Myraboliidae (Beetles)	Family
Tettigarctidae (Hairy Cicadas)	Family
Hypertrophidae (Moths)	Family
Ceratodontidae (Queensland Lungfish)*	Family
Pseudemydura (Western Swamp Turtle)*	Genus
Carphodactylidae (Geckoes)	Family
Ornithorhynchidae (Platypus)*	Family
Potoroidae (Potoroos and Bettongs)	Family
Atrichornithidae (Scrub-birds)	Family
Pardalotidae (Pardalotes)	Family

continent. Many are relictual, with one or a few surviving species (e.g. Wollemi Pine), while others represent substantial indigenous radiations (e.g. Spiny Crayfish).

Because of the many palaeoendemics, Australia has a biota that is evolutionarily distinct (see Box 1.1). Whole groups of organisms are found nowhere else (Table 1.1), while others, such as the Eucalypts (Eucalyptae) are very much centred on the Australian continent. The cause of this distinctiveness, is of course, a long period of independent evolution due to isolation but environmental factors may have also played a role. From a conservation perspective, high distinctiveness means high complementarity (Margules and Pressey, 2000), with Australian sites making large and unique contributions to global biodiversity (Box 1.1).

1.2.3 Provincialism

Starting with Buffon (1761), it has long been noted that the ranges of organisms are not distributed randomly across the globe but are spatially clustered, producing distinctive regions, often with relatively sharp boundaries (Lomolino *et al.*, 2010). Consequently, numerous schemes have been proposed to divide the Earth's terrestrial surface into biogeographically distinct regions or provinces based on the distributions of various taxa (Sclater, 1858; Wallace, 1876; Udvardy, 1975; Holt *et al.*, 2013). Almost invariably, the Australian continent, including Tasmania, is seen as a discrete biogeographic unit, with affinity with the nearby landmasses of New Guinea, Eastern Indonesia, Melanesia and New Zealand. In the still commonly used classification of Wallace (1876), Australia plus the above-mentioned regions form the Australian (or Australasian) Realm, with its north-western boundary marked by Wallace's Line (Plate 1). Marine biogeographic

Box 1.1 Evolutionary distinctiveness of mammalian faunas

The distinctiveness of the Australian biota has long been recognised but rarely quantified (Westoby, 1993). The idea of distinctiveness is related to endemism because a biota with many endemic taxa is clearly distinctive by any reasonable measure. Distinctiveness also implies a temporal (evolutionary) component in that the proportion of palaeoendemics should contribute more heavily to distinctiveness than more recent arrivals (neoendemics). Distinctiveness is important because it indicates the *complementarity* of Australia to the rest of the world as a conservation hotspot. Complementarity is a measure of the contribution of a location to overall biodiversity (Margules and Pressey, 2000) and, conversely, the extent of the loss to biodiversity should that location be lost to conservation.

Plate 2 is an attempt to map evolutionary distinctiveness for mammals for the world's terrestrial ecoregions. Distinctiveness was measured as the average dissimilarity in Phylogenetic Diversity (PD-dissimilarity) between each ecoregion and all other ecoregions. PD-dissimilarity measures the extent to which the branches in a phylogenetic tree are *not* shared between a pair of regions (Nipperess *et al.*, 2010). The Simpson variant of PD-dissimilarity was used, which ignores differences in species richness (Leprieur *et al.*, 2012), and has the following formula:

$$dissimilarity = \frac{\min\{b,c\}}{a + \min\{b,c\}},$$

where a is the total branch length of a phylogenetic tree that is shared between two ecoregions, b is the total branch length occurring in the first ecoregion but not the second, and c is the total branch length occurring in the second ecoregion but not the first. The phylogenetic tree used was the supertree of mammals published by Fritz *et al.* (2009).

It is clear that Australia, and the southern continents generally, make substantial contributions to global mammalian phylogenetic diversity. New Zealand, with only two living species of terrestrial mammal, is much less distinct, although one of those two species, the endemic Lesser Short-tailed Bat (*Mystacina tuberculata*), is the only surviving member of its family (Mystacinidae). A long period of geographic isolation, both from the northern continents and from each other, has engendered significant indigenous radiations of mammals in South America, Africa, Madagascar and Australia (Meredith *et al.*, 2011). In contrast, North America and Eurasia, have had intermittent land connections since the breakup of Pangaea in the mid Jurassic (~180 Ma), allowing greater dispersal and thus less evolutionary distinctiveness (Lomolino *et al.*, 2010). The pattern shown here for mammals is also seen in other vertebrate groups (Holt *et al.*, 2013), indicating that the geological history of land connections is a fundamental factor contributing to the evolutionary distinctiveness, and thus the complementarity, of continental biotas.

classifications are less common because distributional data have historically been sparse, and many marine taxa have large distributions making boundaries less distinct. However, in a recent and comprehensive classification of the world's coastal waters (Spalding *et al.*, 2007), Australian national territory largely spans two marine realms

(Central Indo-Pacific, Temperate Australasia), with Australia's sub-Antarctic territories (e.g. Macquarie Island) falling within the Southern Ocean realm (Plate 1).

Provincialism, the clustering of species distributions, is the result of two principal factors: limits to dispersal and environmental gradients (Lomolino *et al.*, 2010). Dispersal barriers are obviously important in explaining the general distinctiveness of the Australian continent compared to surrounding areas due to the intervening ocean. Within the continent, there is limited relief and thus few effective topographic barriers to dispersal compared to other continents. However, Australia experiences substantial environmental gradients. Mean annual temperatures range from less than 6 °C in upland Tasmania to more than 28 °C in parts of the Northern Territory and Western Australia, while mean annual rainfall varies from less than 200 mm in central Australia to over 3000 mm in Far North Queensland (data from Bureau of Meteorology, Australian Government). There is also a significant gradient in the seasonality of rainfall, with strong monsoonal summer rainfall in the north and winter rainfall in the south.

Australia has several distinct terrestrial biomes, each characterised by a particular vegetation structure, and corresponding to major climatic zones of temperature and rainfall (Plate 1). The scheme followed here is that of Olson *et al.* (2001) but many others have been suggested – see Archer and Fox (1984) for a dated but still relevant review. All classifications agree on the fundamental distinction between the arid centre, dominated by grasslands and Acacia shrublands, and the wetter (mesic) fringe, dominated by Eucalypt woodlands and open forests (Augee and Fox, 2000). Within the mesic fringe, there is a key division between the tropical north, with strongly seasonal summer rainfall, and the temperate south, with less seasonal winter rainfall (Bowman *et al.*, 2010; Byrne *et al.*, 2011). This north–south division is also reflected in the marine realms of Spalding *et al.* (2007). Because it is relatively flat, Australia has a very small montane biome in the Australian Alps. Rainforest is restricted to isolated pockets along the east coast with the largest extent occurring in Far North Queensland (classified as moist broadleaf forest in Olson *et al.*, 2001). Yet finer biogeographical divisions can be made. Olson *et al.* (2001) recognises 37 distinct terrestrial ecoregions on the Australian continent and Tasmania, and Spalding *et al.* (2007) describes an additional 17 marine ecoregions along the coast (Plate 1). These units reflect variation in climate and topography and consequently show significant turnover in species composition.

1.3 Key factors shaping the Australian biota
1.3.1 Isolation

Splendid isolation (to misappropriate a phrase) is the principal defining characteristic of Australian wildlife. Geographic isolation is not a recent phenomenon but extends deep into the past and has had a profound effect upon the composition and evolution of Australia's biota. So sudden is the change in the terrestrial fauna, as one passes from west to east through the Indo-Malayan archipelago, that it motivated Wallace (1860) to draw a line to mark where the recognisably Asian fauna ended and the distinctly and peculiarly Australasian fauna began (Plate 1). Unknown to Wallace, his line corresponded almost exactly to the Asian continental shelf and thus to the limit of historical land connections between the islands of the archipelago and the Asian mainland.

The relative proximity of Australia to Asia is, geologically speaking, a recent phenomenon. In the early Cretaceous (97–140 Ma), Australia was part of the supercontinent of

Gondwana (also including Africa, Arabia, South America, India, Madagascar, Antarctica, New Zealand and some smaller fragments). Australia was located at high latitudes, straddling 60° S at the centre, and experienced a generally cool, wet climate (Rich *et al.*, 1988; White, 2006). Gondwana had already begun to break up by this time (see the review by Upchurch, 2008), although the Australian plate would not separate from Antarctica until the Eocene (about 45 Ma) (White, 2006). By the Oligocene (23–33 Ma), Australia was completely isolated from other landmasses, floating northwards on the convection currents of the Earth's mantle. As Australia drifted, the islands to the north, up to Wallace's Line, were formed from a complex amalgam of Gondwanan microcontinental fragments and new land driven up from the seafloor as the northern edge of the Australian plate collided with Asia (Michaux, 2010).

Because of this geological history, Australia's wildlife can be divided into a Gondwanan component (Table 1.2), reflecting ancient (>45 Ma) dispersal routes among the southern continents, and a post-Gondwanan component, reflecting more recent (<45 Ma) dispersals, especially from Asia. The Gondwanan component has, of course, a long period of association with the continent and includes all the palaeoendemic taxa (Table 1.1). Gondwanan taxa often have disjunct distributions (Table 1.2), being found on two or more southern continents, now widely separated, and their evolutionary relationships generally reflect this by having major branching events that correspond to ancient tectonic rifts (Upchurch, 2008). However, not all taxa with Gondwanan origins have remained restricted to these landmasses. Possibly the most spectacular example is the passerine birds (Passeriformes). Phylogenetic and fossil evidence point to the origin of the group in East Gondwana in the late Cretaceous (~80 Ma), a subsequent radiation in Australia, followed by dispersal to the rest of the world, where they now comprise around half of all bird species (Edwards and Boles, 2002).

Table 1.2   Examples of Gondwanan taxa occurring in Australia with disjunct distributions on other southern continents. Key: + = taxon is currently present; * = taxon was recently (<10 ka) present; SA = South America; NZ = New Zealand; NC = New Caledonia; In = India; Ma = Madagascar; Af = Africa.

Taxon	SA	NZ	NC	In	Ma	Af
Podocarpaceae (Podocarp Conifers)	+	+	+	+	+	+
Proteaceae (Proteas)	+	+	+	+	+	+
Nothofagaceae (Southern Beech)	+	+	+			
Rytididae (Land Snails)	+	+			+	+
Peripatidae (Velvet Worms)	+	+				+
Peloridiidae (Moss Bugs)	+	+	+			
Palaephatidae (Gondwanaland Moths)	+					+
Galaxiidae (Galaxiid Fish)	+	+	+			+
Dipnoi (Lungfish)	+					+
Australobatrachia (Frogs)	+					
Pleurodira (Side-necked Turtles)	+				+	+
Diplodactyloidea (Geckoes)		+	+			
Marsupialia (Marsupials)	+					
Palaeognathae (Ratites)	+	+			*	+
Psittaciformes (Parrots)	+	+	+	+	+	+

Beginning at least as early as the Oligocene (23–33 Ma), Australia has received immigrants along non-Gondwanan dispersal routes. These dispersals have mostly been from the north but long-distance oceanic dispersals have also occurred (Schwarz *et al.*, 2006). The influence of northern dispersal has of course become stronger over time. As a result, the proportion of endemic genera of plants is considerably less in the tropical north (14%) compared to the temperate south (47%) (Augee and Fox, 2000). Interestingly, much of Australia's diverse reptile fauna has been founded from post-Gondwanan immigrants (Hugall *et al.*, 2008; Sanders *et al.*, 2008; Vidal *et al.*, 2012), with perhaps the most intriguing example being that of the Elapid snakes (Elapidae). Australian Elapids share a common ancestor with the sea snakes (Hydrophiinae), suggesting a marine origin, and despite their current diversity, probably entered Australia only in the last ten million years (Sanders *et al.*, 2008). Despite Australia's reputation as a marsupial haven, native placental mammals currently make up almost 47% of terrestrial mammal species (Van Dyck and Strahan, 2008), all of these being either bats (Chiroptera) or Old World Rats and Mice (Murinae). Bats have a long association with the continent (ancestors of the modern bat fauna were already present around 25 Ma; Hand, 2006) while Murines arrived much more recently (probably 5 Ma; Aplin, 2006) but have since become a numerically and ecologically important group.

Isolation has created a great natural experiment in that Australian wildlife has evolved separately from the rest of the world. We see many instances of convergence of ecological roles between Australian and non-Australian taxa, despite considerable phylogenetic divergences, indicating that common evolutionary pressures can lead to similar, independently derived solutions. Probably the most frequently cited examples are comparisons between North American placental mammals and Australian marsupials, such as the Thylacine (*Thylacinus cynocephalus*) and the Wolf (*Canis lupus*). These comparisons should not be taken too literally as there are definite differences in form and function – thylacines were both less agile and less social than wolves (Van Dyck and Strahan, 2008). More convincing examples can be found in Australian deserts. Marsupial Moles (*Notoryctes* spp.) bear an incredible likeness to the Golden Moles (Chrysochloridae) of southern Africa, including the unusual dentition (Archer *et al.*, 2011). Hopping Mice (*Notomys* spp.) have evolved the same set of morphological, ecological, physiological and behavioural characteristics as other unrelated hopping rodents found in the arid zones of Africa, central Asia and North America (Mares, 1993).

1.3.2 Nutrient poverty

Unlike New Zealand, Australia has experienced very little orogeny (mountain-building), glaciation or volcanism since separating from Gondwana (White, 2006). This has resulted in a land surface that is generally very old and a topography that is mostly flat. Except for alluvial systems and a few sites in Eastern Australia affected by hotspot volcanism, soils are ancient and largely leached of nutrients (oligotrophic) due to an extended period of weathering with little opportunity for replenishment (Augee and Fox, 2000). This nutrient poverty has been a huge challenge for Australian wildlife and has played a substantive role in determining the form and function of species and ecosystems.

Australian vegetation exhibits an unusual combination of features related to nutrient poverty. Sclerophylly (the characteristic of having long-lived, hardened leaves) is associated worldwide with Mediterranean climates (Beadle, 1966) but, in Australia, it is common in all biomes except rainforest. Although often considered an adaptation to drought, the occurrence of sclerophylly seems most reliably correlated with nutrient limitation (Beadle, 1966; Westoby, 1993). Sclerophyllous leaves are generally resistant to damage, well-protected by chemical defences, and low in the essential nutrients nitrogen and phosphorous (Orians and Milewski, 2007; Morton *et al.*, 2011). Nitrogen-fixing species, such as Wattles (*Acacia*), are common, ensuring that phosphorous is the key limiting nutrient in Australian ecosystems (Augee and Fox, 2000; Orians and Milewski, 2007). High light availability means that Australian plants can generally photosynthesise year-round, producing large amounts of carbon-rich, but nutrient-poor, tissues such as wood, oils, exudates and nectar (Orians and Milewski, 2007; Morton *et al.*, 2011).

The unusual features of the Australian flora have direct effects on the ecology of the fauna. The generally low quality of plant production has likely constrained the diversity and dietary habits of herbivores (Orians and Milewski, 2007). For example, despite being much smaller in area, New Guinea has nearly twice the number of arboreal mammalian herbivores (Flannery, 1995), with Australian species being generally less specialised for folivory (the Koala, *Phascolarctos cinereus*, is an obvious exception). Feeding on nectar and other plant exudates is, in contrast to folivory, a relatively common adaptation. Australia has a large radiation of nectar-feeding birds (mostly in the family Meliphagidae), which attain a large average body size compared to avian nectarivores elsewhere (Orians and Milewski, 2007). The Honey Possum (*Tarsipes rostratus*) is the only mammal in the world to subsist exclusively on nectar and pollen (Van Dyck and Strahan, 2008). Australia has a large diversity of sap-sucking insects with the plant lice (Psylloidea) representing more than 10% of the world's species (Austin *et al.*, 2004). Low-quality plant production, coupled with aridity, is also thought to constrain the activity of detritivores (Morton *et al.*, 2011). The likely exception are termites (Termitoidea), which can occur at high densities in Australia, and are uniquely adapted to deal with the breakdown of nutrient-poor woody plant material (Orians and Milewski, 2007).

Nutrient poverty is thought to be the principal reason for the importance of fire in Australian ecosystems. Constraints on herbivory and detritivory allow a build-up of carbon-rich plant material, which, when combined with hot, dry conditions, leads to fire. In effect, fire largely replaces detritivore and generalised herbivore activity as a recycler of nutrients in Australian ecosystems (Morton *et al.*, 2011). As a result, Australian plants have evolved a remarkable array of adaptations to survive fire or to exploit conditions after fire. Many species can resprout after fire from lignotubers or epicormic buds, while others re-establish from seed that can germinate only when exposed to fire (Augee and Fox, 2000). Interestingly, resprouting from epicormic buds may have arisen in Eucalypts (Eucalypteae) around 60 Ma, suggesting that fire has long been a feature of Australian ecosystems (Crisp *et al.*, 2011).

1.3.3 Aridity and climatic variability

Since the early Cretaceous, Australia has drifted more than 30 degrees of latitude resulting in profound climate change (White, 2006). Moving towards the equator has of course increased temperatures, but only relative to global averages for the time, which have generally been decreasing for the past 50 Ma (Zachos *et al.*, 2008). Probably more important has been increased insolation and decreased seasonality, allowing for year-round photosynthesis (Orians and Milewski, 2007). The most significant change, beginning in the early Miocene (~20 Ma), has been the onset of aridity (Byrne *et al.*, 2008). This was triggered by a northward shift to the mid-latitudes, where descending cool air absorbs much of the available moisture (Lomolino *et al.*, 2010). Following a brief period of wetter conditions in the early Pliocene (~ 5 Ma), the trend towards increasing aridity accelerated, in association with a rapid global cooling trend (Martin, 2006; Zachos *et al.*, 2008). Deserts and xeric shrublands now constitute the largest biome in Australia (Plate 1), covering approximately 70% of the land surface (Byrne *et al.*, 2008).

Widespread aridity has had a strong effect on Australian vegetation. Rainforest covered much of the continent in the Eocene, when Australia broke from Antarctica, but is now limited to a few pockets along the east coast, and has been replaced by more open vegetation types dominated by sclerophyllous taxa (Martin, 1990, 2006; Byrne *et al.*, 2011). Coincident with the rise of sclerophylly and the decline of rainforest has been an increase in the frequency and intensity of fire, as evinced by the charcoal record (Martin, 1990). Indeed the occurrence of rainforest today seems to be better predicted by the long absence of fire rather than either nutrient availability or rainfall (Bowman, 1998), although all three are inter-related.

With increasing aridity, Australian lineages have had the options of contracting to mesic refugia, adapting (and even thriving) under the changing conditions, or becoming extinct. Thus, the wetter parts of Australia, and especially rainforest, might be expected to be home to many relictual taxa representing ancient radiations, while the arid zone should be characterised by relatively recent (20 Ma or less) radiations of groups that have adapted to the new situation (Byrne *et al.*, 2011). Among plants for example, Southern Beech (*Nothofagus*) and other Gondwanan taxa have contracted to the few remaining cool, wet locations in southeastern Australia (especially Tasmania), while Wattles (*Acacia*), Banksias (*Banksia*), Eucalypts (*Eucalyptus*) and other sclerophyllous taxa have undergone rapid recent radiations (Crisp *et al.*, 2004; Martin, 2006). The rise of sclerophyllous plants has been accompanied by a parallel radiation of insects such as the Gum Treehoppers (Eurymelinae) and Gall-inducing Thrips (Phlaeothripidae), among many others (Austin *et al.*, 2004). Birds of the arid zone are, on average, more closely related to one another than they are in wetter regions, indicating a relatively recent radiation of a subset of taxa of mesic origin adapting to arid conditions (Hawkins *et al.*, 2005). Similarly, among Australian Skinks (Scincidae), almost all of the substantial arid zone diversity has been derived from rapid recent radiations from two genera: *Lerista* and *Ctenotus* (Rabosky *et al.*, 2007).

In addition to increasing aridity, Australian wildlife has had to contend with strong climatic variability. Since at least the late Pliocene (2.5 Ma), oscillations in the Earth's orbit have produced a pattern of glacial and interglacial cycles responsible for the rapid growth and decline of the northern ice-sheets (Zachos, 2001). In Australia, glacial periods corresponded to depressions in temperature of as much as 10 °C (leading to severe frosts but very little glaciation), major drops in sea level (exceeding 120 m), and extended and severe drought (Byrne *et al.*, 2008). By the time of the last glacial maximum around 21 ka, most of Australia's inland lakes had dried up and western Tasmania, now heavily forested, was an open grassy steppe (Colhoun, 2000; Byrne *et al.*, 2008). For much of Australia's biota, glacial–interglacial cycles corresponded to repeated range contractions and expansions centred on climate refugia (Byrne, 2009).

Since at least the Pleistocene, the El Niño Southern Oscillation (ENSO) has provided an additional source of climatic variability. ENSO is a climatic phenomenon referring to irregular warming (El Niño) and cooling (La Niña) phases in the surface temperature of the eastern Pacific Ocean (Nichols, 1991). Few landmasses are as strongly and comprehensively affected by ENSO as Australia, where El Niño phases bring continent-scale droughts while La Niña phases bring cooler, wetter conditions (Nichols, 1991). Effectively, ENSO increases both the variability and irregularity of rainfall in Australia. It is unclear how long ENSO has been a feature of the Australian climate, although palaeoenvironmental data indicates a strong influence dating back at least 45 ka (Turney *et al.*, 2004).

While there are many adaptations to aridity and climatic variability among Australian animals, the conservation of energy seems to be a common theme, at least among vertebrates (Orians and Milewski, 2007). Animals with low energy requirements are better able to survive extended periods of irregular rainfall and the consequent limits to food availability. Nutrient poverty also favours energy conservation because low energy requirements allow persistence on low-quality food (Orians and Milewski, 2007). Reptiles, in particular, have an advantage over other vertebrates in their combination of low metabolic rate and high resistance to evaporative water loss (Dawson and Dawson, 2006). Snakes and lizards show remarkable diversity in the arid centre of Australia and this success is likely a result of these adaptations (Morton and James, 1988). The success of reptiles, relative to mammals, in large predatory niches is likely also a result of the selective advantage of energy conservation (Flannery, 1991, but see Wroe *et al.*, 1999, for a dissenting view). The rapid radiation of kangaroos and wallabies (Macropodidae) since the onset of aridity (Meredith *et al.*, 2008) is also attributable, at least in part, to adaptations for energy conservation. Hopping, a considerably more energy-efficient means of locomotion than running, has evolved several times in desert mammals, but only in Australia do hopping mammals achieve large body size and so effectively dominate grazing niches (Webster and Dawson, 2004).

1.4 Conservation issues

1.4.1 Prehistoric extinctions

During the late Pleistocene, Australia lost at least 21 genera of vertebrates, although the exact timing and ultimate cause (human impact versus climate change) of these extinctions continues to be debated (Barnosky *et al.*, 2004). The timing of these events is important to determine the relative likelihood of extreme climatic conditions or interaction with humans as the primary cause. Through careful selection of reliable radiometric dates and new methods of analysis, there is growing evidence that most of the extinctions occurred over a relatively short period around 46 ka (Roberts *et al.*, 2001; Gillespie *et al.*, 2006; Grün *et al.*, 2010; Prideaux *et al.*, 2010). This estimate does not coincide with extreme climatic conditions (the last glacial maximum was at 21 ka) but does coincide with the earliest evidence of humans (Gillespie, 2006) with a relatively brief overlap in time of about 4000 years (Gillespie *et al.*, 2006). It seems therefore that humans are the likely agent of the Pleistocene extinctions in Australia by either hunting or habitat alteration or both. While the evidence in support of this interpretation is growing, acceptance of this scenario is certainly not universal (Wroe *et al.*, 2013).

What is not in dispute is that these prehistoric extinctions had a definite body size bias – few species weighing over 40 kg survived (Johnson and Prideaux, 2004). The extinct species are therefore collectively referred to as the *megafauna*, although it is important to realise that not all megafaunal species became extinct (the Red Kangaroo, *Macropus rufus*, for example). Body size is a reliable correlate of extinction risk in modern mammals (Cardillo *et al.*, 2005), with large-bodied species having limited capacity for population growth and thus high susceptibility to environmental change (whether human-caused or not). Population modelling has shown that megafaunal species were likely to have succumbed relatively quickly if actively hunted by humans (Brook and Bowman, 2004), leading to a *blitzkrieg* scenario of rapid extinction within a few millennia of initial human contact (Burney and Flannery, 2005).

The removal of large herbivores from Australian ecosystems in the late Pleistocene would likely have led to profound and rapid changes in vegetation. Large herbivores have a substantial influence on the structure and composition of vegetation by creating disturbance gradients; dispersing seeds; enhancing nutrient cycling (through urine and faeces); and slowing the build-up of dry plant biomass, thus reducing the frequency and intensity of fires (Owen-Smith, 1987). Evidence from sediment cores from Lynch's Crater, Far North Queensland, shows wholesale replacement of rainforest with sclerophyll vegetation, and a concomitant increase in the prevalence of fire, beginning around 43 ka (Rule *et al.*, 2012). Owing to the timing, such a shift cannot be explained by a climatic transition but coincides with the emerging consensus on the timing of the megafaunal extinction. This interpretation is further confirmed by the coincident and rapid loss of spores of *Sporormiella*, a fungus that lives on the dung of large herbivores (Rule *et al.*, 2012). While this evidence is compelling, there are currently limited data to determine if ecosystem transformation was truly a continent-wide phenomenon (Johnson, 2009). Analysis of charcoal records on a continental scale do not show any evidence for an increase in the prevalence of fire for this period, suggesting that neither the demise of the megafauna nor the initiation of human sources of ignition had a large-scale effect on fire regimes (Mooney *et al.*, 2012).

In the period between the megafaunal extinction event and European settlement, there is little evidence for extinction, whether human-mediated or not, even during the harsh conditions of the last glacial maximum around 21 ka. Among mammals, only the Nullarbor Dwarf Bettong (*Bettongia pusilla*) may have been become extinct before 1788 (McNamara, 1997). Some species did, however, experience significant range contractions during the relatively benign conditions of the Holocene. The best-known examples are the Thylacine or Tasmanian Tiger (*Thylacinus cynocephalus*), the Tasmanian Devil (*Sarcophilus harrisii*) and the Tasmanian Native Hen (*Gallinula mortierii*), which were present on the Australian mainland at the beginning of the Holocene (~11.5 ka) but subsequently became restricted to Tasmania (Johnson and Wroe, 2003). The demise of these mainland populations coincides broadly with a period (~5 ka) of increasing human population size and activity known as *intensification* (Lourandos, 1983; Kohen, 1995; Johnson and Brook, 2011). Intensification is associated with the establishment of new technologies (e.g. stone-tipped spears) and the introduction of the Dingo (*Canis lupus dingo*) as a human commensal (Johnson and Wroe, 2003) reflecting a period of cultural exchange perhaps triggered by the expansion of Austronesian speakers into the southwest Pacific (Gray *et al.*, 2009). The best evidence for the intensive use of fire by people also dates to this period (Kershaw *et al.*, 2002). The implication is that intensification is the cause of these mammal range contractions – a conclusion made more compelling by the fact that there is no evidence for intensification in Tasmania, including no introduction of the Dingo (Johnson and Wroe, 2003).

1.4.2 Historic extinctions

There have been a depressingly large number of extinctions in Australia since European settlement began in 1788. The Environment Protection and Biodiversity Conservation (EPBC) Act (1999) lists 98 species and subspecies with no remaining wild populations known in Australia (Table 1.3). Most of these are endemics, meaning no remaining wild populations anywhere, and, except for the Pedder Galaxias (*Galaxias pedderensis*), no captive populations either. The majority are therefore global extinctions with no chance of recovery. Compared to countries of similar landmass, Australia has suffered a relatively large number of historic extinctions. The International Union for the Conservation of Nature (IUCN) lists 35 extinct species of vertebrate for Australia, compared to five for Brazil, six for Canada, and 49 for the United States (of which 20 were endemic to the Hawaiian Islands).

At least for vertebrates, it is clear that the extinctions were non-random. Species that were endemic to small offshore islands comprise nearly half the extinct vertebrate species and subspecies on the EPBC list. Insular species are especially vulnerable to extinction both because limited island area constrains population size (a well-known correlate of extinction risk – see O'Grady *et al.*, 2004) and because island endemics are generally ecologically naïve. Island species have evolved in isolation with few or no competitors or predators and thus, when faced with a novel species (including humans), fare poorly because of a lack of appropriate physical or behavioural adaptations (Clavero *et al.*, 2009). Accounts of early visitors to Lord Howe Island, for example, record the extreme tameness of the birds whose naivety was such that visiting sailors were able to harvest them in large numbers, strongly contributing to the early extinction of the White Gallinule (*Porphyrio albus*) and several other species (Hindwood, 1940). At least among mammals, body size was clearly also a factor in historically recorded extinctions. Species with an intermediate body size (the so-called *Critical Weight Range* (CWR) of between 35 g and 5.5 kg) have more frequently become extinct than either smaller or larger species (Burbidge and McKenzie, 1989; Cardillo and Bromham, 2001; Johnson and Isaac, 2009), especially if they also lived in arid environments and/or sheltered on the ground (McKenzie *et al.*, 2007).

A cumulative plot of last known records (Figure 1.3) for the EPBC-listed species and subspecies suggests at least three phases to Australian historical extinctions. These phases correspond approximately to those identified by Johnson (2006) for mammals. The first phase began almost immediately after European settlement in 1788 and ends around 1880. This period saw the extinction of several island bird species as well as small mammals from the southeastern and southwestern coastal fringe. The second phase (marked in grey on Figure 1.3) from around 1880 to 1940 marks a period of more rapid extinction, which included several more species of island bird as well as many CWR mammals. The final phase from 1940 sees a slower rate of extinction and the loss of several mammal species from the arid inland, some montane frog species, and yet more island birds. Worryingly, rates of extinction seem to be increasing towards the present but have not yet reached the levels seen in the dark times of 1880–1940.

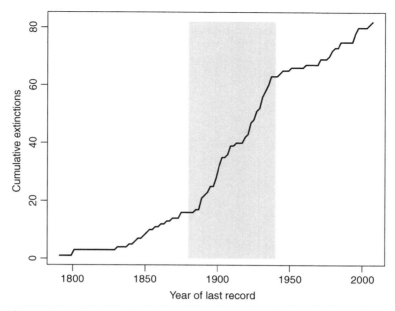

Figure 1.3 Cumulative historic extinctions of Australian species and subspecies. Dates of extinctions are inferred as the last known record of a wild specimen. Species and subspecies are those listed as extinct under the Environment Protection and Biodiversity Conservation (EPBC) Act 1999. Taxa without reliable dates for last records are omitted. Grey box delineates period of accelerated extinctions from about 1880–1940. Data sourced from the Species Profile and Threats (SPRAT) database (http://www.environment.gov.au/cgi-bin/sprat/public/sprat.pl).

The causes of historical extinctions are varied and include habitat loss and alteration, over-exploitation and outright persecution. In many cases, records are extremely scant, often because a species disappeared shortly after it was first described, and thus it is very difficult to attribute causes with any certainty. However, there is a strong case for the role of introduced predators, and especially the Red Fox (*Vulpes vulpes*), in the extinction or decline of many of Australia's terrestrial vertebrates. The fox was first deliberately introduced to the wild in Victoria around 1870 (Van Dyck and Strahan, 2008), shortly before the main phase of historical extinctions. Foxes spread rapidly across the continent, occupying all areas except the northern tropics and Tasmania by the 1930s (Johnson, 2006). Data on bounties paid for skins in New South Wales show that many mammal species declined rapidly after the first appearance of foxes (Short, 1998). Species in the Critical Weight Range, and especially those that sheltered on the ground, were particularly vulnerable because they are the most likely prey for a medium-sized predator like the fox (Johnson and Isaac, 2009). Similar to island birds, native mammals show predator naivety, having not co-evolved with foxes, and thus commonly have poor physical or behavioural defences (Short *et al.*, 2002).

1.4.3 Threatened species and threatening processes

Discounting extinct species (for which we can do nothing), there are nearly 1700 species and subspecies listed as threatened under the EPBC Act (1999). The list consists largely of vertebrates and flowering plants (Table 1.3). While larger species are more likely to be threatened (Cardillo *et al.*, 2005) due to, for example, lower rates of

Table 1.3   Count of species and subspecies listed under the Environment Protection and Biodiversity Conservation (EPBC) Act (1999) by threat category and major taxonomic grouping. Data sourced from the Species Profile and Threats (SPRAT) database (http://www.environment.gov.au/cgi-bin/sprat/public/sprat.pl).

Phylum / Division	Extinct	Critically endangered	Endangered	Vulnerable	Total
Annelida	1	0	0	1	2
Arthropoda	0	7	12	8	27
Bryophyta	0	0	0	1	1
Chordata	55	23	125	195	398
Cycadophyta	0	0	6	10	16
Echinodermata	0	1	0	0	1
Lycopodiophyta	2	0	4	4	10
Magnoliophyta	33	118	506	623	1280
Mollusca	0	12	4	0	16
Onychophora	0	0	1	0	1
Pinophyta	0	0	3	1	4
Polypodiophyta	5	2	12	10	29
Psilophyta	1	0	0	1	2
Rhodophyta	1	0	0	0	1
Streptophyta	0	0	1	0	1
Total	98	163	674	854	1789

reproduction, there is no doubt that threatened species lists such as the EPBC Act and the IUCN Red List are heavily biased towards those groups for which we have the most information. There are likely many species in poorly known groups that are either threatened with extinction or have actually become extinct without anyone having noticed. It is suggested for example that many species of Dung Beetle (Scarabaeidae) may have become extinct or suffered serious declines with the extinction of the CWR mammal species on which they depended (Coggan, 2012). Mesibov (2001) makes the pertinent point that it is simply impractical to methodically list invertebrate species for conservation action when there are so many and so little is known about their taxonomy, let alone their ecology.

Many threatening processes have been identified as important for heightening extinction risk for Australian terrestrial wildlife, of which landscape transformation and introduced species have historically been the most important. Landscape transformation (including outright loss of habitat) is identified as a key threat to almost all of the threatened species and subspecies on the EPBC list. Since 1788, Australia has lost nearly 40% of its forests (Bradshaw, 2012) with most clearance concentrated in the temperate and Mediterranean forest biomes of the southeast and southwest (Plate 1). However, no biome has escaped transformation of some kind, including soil compaction by grazing, altered fire regimes, erosion and salinisation. Landscape transformation and ecosystem disruption have likely assisted the spread of introduced species, which have proven especially successful in Australia (Low, 1999). Owing to Australia's long isolation, exotic species (which include infectious diseases) can have an advantage over native species because of ecological naivety of native prey or competitors (Clavero *et al.*, 2009); absence of parasites, pathogens or predators (Harvey *et al.*, 2012); or a lack of competitors for particular resources (Neumann, 1979). In combination with landscape transformation, introduced species have triggered the formation of *novel ecosystems* with combinations of species, both native and introduced, not seen before (Hobbs *et al.*, 2006).

The rapidly emerging threat for Australian wildlife in the twenty-first century is climate change. While Australian species have experienced considerable fluctuation in climate since 2.5 Ma, anthropogenic climate change is especially threatening because of the speed of its onset, and its synergistic effects with habitat transformation and introduced species (Root and Schneider, 2006). Much like previous periods of climatic shift, species must either adapt to the changing conditions or move (or contract) to places where preferred conditions still exist (Byrne *et al.*, 2011). If climate change is either too extreme or too rapid, then neither option will be viable. If temperatures rise more than 2 °C, it is predicted that many species of vertebrate in the rainforests of Far North Queensland will become extinct because suitable habitat will no longer be available (Williams *et al.*, 2003). Removal and fragmentation of habitat makes migration difficult, emphasising the need to improve connectivity between reserves and identify (and, if necessary, restore) refugia where climatic conditions will remain most stable (Shoo *et al.*, 2010).

1.5   Concluding remarks

Australia makes a unique contribution to global biodiversity. The long years of isolation have ensured a high degree of endemism and evolutionary distinctiveness, while nutrient poverty and climatic variability have led to the development of truly peculiar evolutionary adaptations and ecological processes. Despite this, Australian ecosystems are impoverished – much that was distinctive, incredible and beautiful has been lost. Although the challenges are many, we must conserve what is truly great about this island nation – its extraordinary wildlife.

REFERENCES

Aplin, K. P. (2006). Ten million years of rodent evolution in Australasia: phylogenetic evidence and a speculative historical biogeography. In *Evolution and Biogeography of Australasian Vertebrates*. Oatlands: Auscipub, pp. 707–744.

Archer, M. and Fox, B. (1984). Background to vertebrate zoogeography in Australia. In *Vertebrate Zoogeography and Evolution in Australasia*. Carlisle: Hesperian Press, pp. 1–15.

Archer, M., Beck, R., Gott, M., *et al.* (2011). Australia's first fossil marsupial mole (Notoryctemorphia) resolves controversies about their evolution and palaeoenvironmental origins. *Proceedings Of The Royal Society Of London Series B – Biological Sciences*, **278**, pp. 1498–1506.

Augee, M. L. and Fox, M. D. (2000). *Biology of Australia and New Zealand*. Frenchs Forest: Benjamin Cummings.

Austin, A., Yeates, D., Cassis, G., *et al.* (2004). Insects 'Down Under' – diversity, endemism and evolution of the Australian insect fauna: examples from select orders. *Australian Journal Of Entomology*, **43**, pp. 216–234.

Barnosky, A. D., Matzke, N., Tomiya, S., *et al.* (2004). Assessing the causes of Late Pleistocene extinctions on the continents. *Science*, **306**, pp. 70–75.

Beadle, N. (1966). Soil phosphate and its role in molding segments of the Australian flora and vegetation, with special reference to xeromorphy and sclerophylly. *Ecology*, **47**, pp. 992–1007.

Bowman, D. M. J. S. (1998). The impact of aboriginal landscape burning on the Australian biota. *New Phytologist*, **140**, pp. 385–410.

Bowman, D. M. J. S., Brown, G. K., Braby, M. F., *et al.* (2010). Biogeography of the Australian monsoon tropics. *Journal of Biogeography*, **37**, pp. 201–216.

Bradshaw, C. J. A. (2012). Little left to lose: deforestation and forest degradation in Australia since European colonization. *Journal of Plant Ecology*, **5**, pp. 109–120.

Brook, B. W. and Bowman, D. M. J. S. (2004). The uncertain blitzkrieg of Pleistocene megafauna. *Journal of Biogeography*, **31**, pp. 517–523.

Buffon, G. L. L. (1761). *Histoire Naturelle, Generale et Particuliere*, vol. **9**. Paris: Imprimerie Royale.

Burbidge, A. A. and McKenzie, N. L. (1989). Patterns in the modern decline of Western Australia's vertebrate fauna: causes and conservation implications. *Biological Conservation*, **50**, pp. 143–198.

Burney, D. A. and Flannery, T. F. (2005). Fifty millennia of catastrophic extinctions after human contact. *Trends in Ecology and Evolution*, **20**, pp. 395–401.

Byrne, M. (2009). Evidence for multiple refugia at different time scales during Pleistocene climatic oscillations in southern Australia inferred from phylogeography. *Quaternary Science Reviews*, **27**, pp. 2576–2585.

Byrne, M., Steane, D. A., Joseph, L., *et al.* (2008). Birth of a biome: insights into the assembly and maintenance of the Australian arid zone biota. *Molecular Ecology*, **17**, pp. 4398–4417.

Byrne, M., Yeates, D. K., Joseph, L., *et al.* (2011). Decline of a biome: evolution, contraction, fragmentation, extinction and invasion of the Australian mesic zone biota. *Journal of Biogeography*, **38**, pp. 1635–1656.

Cardillo, M. and Bromham, L. (2001). Body size and risk of extinction in Australian mammals. *Conservation Biology*, **15**, pp. 1435–1440.

Cardillo, M., Mace, G. M., Jones, K. E., *et al.* (2005). Multiple causes of high extinction risk in large mammal species. *Science*, **309**, pp. 1239–1241.

Ceballos, G. and Ehrlich, P. R. (2009). Discoveries of new mammal species and their implications for conservation and ecosystem services. *Proceedings of the National Academy of Sciences*, **106**, pp. 3841–3846.

Chapman, A. D. (2009). *Numbers of Living Species in Australia and the World*. 2nd edn. Canberra: Australian Biological Resources Study.

Clavero, M., Brotons, L., Pons, P. and Sol, D. (2009). Prominent role of invasive species in avian biodiversity loss. *Biological Conservation*, **142**, pp. 2043–2049.

Coggan, N. (2012). Are native dung beetle species following mammals in the critical weight range towards extinction? *Proceedings of the Linnaean Society of New South Wales*, **134**, pp. A5–A9.

Colhoun, E. A. (2000). Vegetation and climate change during the Last Interglacial–Glacial cycle in western Tasmania, Australia. *Palaeogeography, Palaeoclimatology, Palaeoecology*, **155**, pp. 195–209.

Crisp, M. D., Burrows, G. E., Cook, L. G., Thornill, A. H. and Bowman, D. M. J. S. (2011). Flammable biomes dominated by eucalypts originated at the Cretaceous–Palaeogene boundary. *Nature Communications*, **2**, pp. 193.

Crisp, M. D., Cook, L. G. and Steane, D. A. (2004). Radiation of the Australian flora: what can comparisons of molecular phylogenies across multiple taxa tell us about the evolution of diversity in present-day communities? *Philosophical Transactions of the Royal Society of London: Biological Sciences*, **359**, pp. 1551–1571.

Darwin, C. 2001. *Charles Darwin's Beagle Diary*. Cambridge: Cambridge University Press.

Dawson, T. J. and Dawson, L. (2006). Evolution of arid Australia and consequences for vertebrates. In *Evolution and Biogeography of Australasian Vertebrates*. Oatlands: Auscipub, pp. 51–70.

Edwards, S. and Boles, W. E. (2002). Out of Gondwana: the origin of passerine birds. *Trends in Ecology and Evolution*, **17**, pp. 347–349.

Flannery, T. F. (1991). The mystery of the meganesian meat-eaters. *Australian Natural History*, **23**, pp. 722–729.

Flannery, T. F. (1995). *Mammals of New Guinea*. Revised and updated edn. Chatswood: Reed Books.

Fritz, S. A., Bininda-Emonds, O. R. P. and Purvis, A. (2009). Geographical variation in predictors of mammalian extinction risk: big is bad, but only in the tropics. *Ecology Letters*, **12**, pp. 538–549.

Gaston, K. J. (2000). Global patterns in biodiversity. *Nature*, **405**, pp. 220–227.

Gillespie, R. (2006). Dating the first Australians. *Radiocarbon*, **44**, pp. 455–472.

Gillespie, R., Brook, B. W. and Baynes, A. (2006). Short overlap of humans and megafauna in Pleistocene Australia. *Alcheringa*, **30** (Suppl. 1), pp. 163–186.

Godfray, H. C. J. (2002). Challenges for taxonomy. *Nature*, **417**, pp. 17–19.

Gray, R. D., Drummond, A. J. and Greenhill, S. J. (2009). Language phylogenies reveal expansion pulses and pauses in Pacific settlement. *Science*, **323**, pp. 479–483.

Grün, R., Eggins, S., Aubert, M., *et al*. (2010). ESR and U-series analyses of faunal material from Cuddie Springs, NSW, Australia: implications for the timing of the extinction of the Australian megafauna. *Quaternary Science Reviews*, **29**, pp. 596–610.

Hand, S. J. (2006). Bat beginnings and biogeography: the Australasian record. In *Evolution and Biogeography of Australasian Vertebrates*. Oatlands: Auscipub, pp. 673–705.

Harvey, K. J., Nipperess, D. A., Britton, D. R. and Hughes, L. (2012). Australian family ties: does a lack of relatives help invasive plants escape natural enemies? *Biological Invasions*, **14**, pp. 2423–2434.

Hawkins, B. A., Diniz-Filho, J. A. F. and Soeller, S. A. (2005). Water links the historical and contemporary components of the Australian bird diversity gradient. *Journal of Biogeography*, **32**, pp. 1035–1042.

Hindwood, K. A. 1940. The birds of Lord Howe Island. *Emu*, **40**, pp. 1–86.

Hobbs, R. J., Arico, S., Aronson, J., *et al.* (2006). Novel ecosystems: theoretical and management aspects of the new ecological world order. *Global Ecology and Biogeography*, **15**, pp. 1–7.

Holt, B. G., Lessard, J-P, Borregaard, M. K., *et al.* (2013). An update of Wallace's zoogeographic regions of the world. *Science*, **339**, pp. 74–78.

Hugall, A. F., Foster, R., Hutchinson, M. and Lee, M. S. Y. (2008). Phylogeny of Australasian agamid lizards based on nuclear and mitochondrial genes: implications for morphological evolution and biogeography. *Biological Journal Of The Linnean Society*, **93**, pp. 343–358.

Johnson, C. (2006). *Australia's Mammal Extinctions: a 50,000 Year History*. Port Melbourne: Cambridge University Press.

Johnson, C. N. (2009). Ecological consequences of Late Quaternary extinctions of megafauna. *Proceedings Of The Royal Society Of London Series B – Biological Sciences*, **276**, pp. 2509–2519.

Johnson, C. N. and Brook, B. W. (2011). Reconstructing the dynamics of ancient human populations from radiocarbon dates: 10 000 years of population growth in Australia. *Proceedings Of The Royal Society Of London Series B – Biological Sciences*, **278**, pp. 3748–3754.

Johnson, C. N. and Isaac, J. L. (2009). Body mass and extinction risk in Australian marsupials: The 'Critical Weight Range' revisited. *Austral Ecology*, **34**, pp. 35–40.

Johnson, C. N. and Prideaux, G. J. (2004). Extinctions of herbivorous mammals in the late Pleistocene of Australia in relation to their feeding ecology: no evidence for environmental change as cause of extinction. *Austral Ecology*, **29**, pp. 553–557.

Johnson, C. N. and Wroe, S. (2003). Causes of extinction of vertebrates during the Holocene of mainland Australia: arrival of the dingo, or human impact? *The Holocene*, **13**, pp.941–948.

Kershaw, A. P., Clark, J. S., Gill, A. M. and D'Costa, D. M. (2002). A history of fire in Australia. In *Flammable Australia: the Fire Regimes and Biodiversity of a Continent*. Cambridge: Cambridge University Press, pp. 3–25.

Kohen, J. (1995). *Aboriginal Environmental Impacts*. Sydney: Universityof New South Wales Press.

Leprieur, F., Albouy, C., De Bortoli, J., *et al.* (2012). Quantifying phylogenetic beta diversity: distinguishing between 'true' turnover of lineages and phylogenetic diversity gradients. *PLoS ONE*, **7**, e42760.

Linnaeus, C. (1758). *Systema naturae per regna tria naturæ, secundum classes, ordines, genera, species, cum characteribus, differentiis, synonymis, locis.* 10th edn. Stockholm: Holmiae (Salvius).

Lomolino, M. V., Riddle, B. R., Whittaker, R. J. and Brown, J. H. (2010). *Biogeography*. 4th edn. Sunderland: Sinauer Associates.

Lourandos, H. (1983). Intensification: a late Pleistocene–Holocene archaeological sequence from southwestern Victoria. *Archaeology in Oceania*, **18**, pp. 81–94.

Low, T. (1999). *Feral Future: the Untold Story of Australia's Exotic Invaders*. Melbourne: Penguin.

Mares, M. A. (1993). Desert rodents, seed consumption, and convergence. *BioScience*, **43**, pp. 372–379.

Margules, C. and Pressey, R. (2000). Systematic conservation planning. *Nature*, **405**, pp. 243–253.

Martin, H. A. (1990). Tertiary climate and phytogeography in southeastern Australia. *Review of Palaeobotany and Palynology*, **65**, pp. 47–55.

Martin, H. A. (2006). Cenozoic climatic change and the development of the arid vegetation in Australia. *Journal of Arid Environments*, **66**, pp. 533–563.

McKenzie, N. L., Burbidge, A. A., Baynes, A., *et al.* (2007). Analysis of factors implicated in the recent decline of Australia's mammal fauna. *Journal of Biogeography*, **34**, pp. 597–611.

McNamara, J. A. (1997). Some smaller Macropod fossils of South Australia. *Proceedings of the Linnaean Society of New South Wales*, **117**, pp. 97–105.

Meredith, R. W., Janecka, J. E., Gatesy, J., *et al.* (2011). Impacts of the Cretaceous Terrestrial Revolution and KPg extinction on mammal diversification. *Science*, **334**, pp. 521–524.

Meredith, R. W., Westerman, M., Case, J. A. and Springer, M. S. (2008). A phylogeny and timescale for marsupial evolution based on sequences for five nuclear genes. *Journal of Mammalian Evolution*, **15**, pp. 1–36.

Mesibov, R. (2001). The Milabeena Marvel, or why single-species conservation is inappropriate for cryptic invertebrates. *The Tasmanian Naturalist*, **123**, pp. 16–23.

Michaux, B. (2010). Biogeology of Wallacea: geotectonic models, areas of endemism, and natural biogeographical units. *Biological Journal Of The Linnean Society*, **101**, pp. 193–212.

Minelli, A. (2003). The status of taxonomic literature. *Trends in Ecology and Evolution*, **18**, pp. 75–76.

Mooney, S. D., Harrison, S. P., Bartlein, P. J. and Stevenson, J. (2012). The prehistory of fire in Australasia. In *Flammable Australia: Fire Regimes, Biodiversity and Ecosystems in a Changing World*. Collingwood: CSIRO Publishing, pp. 3–25.

Mora, C., Tittensor, D. P., Adl, S., Simpson, A. G. B. and Worm, B. (2011). How many species are there on Earth and in the ocean? *PLoS Biology*, **9**, e1001127.

Morton, S. R. and James, C. D. (1988). The diversity and abundance of lizards in arid Australia: a new hypothesis. *The American Naturalist*, **132**, pp. 237–256.

Morton, S. R., Stafford Smith, D. M., Dickman, C. R., *et al.* (2011). A fresh framework for the ecology of arid Australia. *Journal of Arid Environments*, **75**, pp. 313–329.

Neumann, F. G. (1979). Insect pest management in Australian Radiata Pine plantations. *Australian Forestry*, **42**, pp. 30–38.

Nichols, N. (1991). The El Nino/Southern Oscillation and Australian vegetation. *Vegetatio*, **91**, pp. 23–36.

Nipperess, D. A., Faith, D. P. and Barton, K. (2010). Resemblance in phylogenetic diversity among ecological assemblages. *Journal of Vegetation Science*, **21**, pp. 809–820.

O'Grady, J., Reed, D., Brook, B. W. and Frankham, R. (2004). What are the best correlates of predicted extinction risk? *Biological Conservation*, **118**, pp. 513–520.

Olson, D. M., Dinerstein, E., Wikramanayake, E. D., *et al.* (2001). Terrestrial ecoregions of the world: a new map of life on Earth. *BioScience*, **51**, pp. 933–938.

Orians, G. H. and Milewski, A. V. (2007). Ecology of Australia: the effects of nutrient-poor soils and intense fires. *Biological Reviews*, **82**, pp. 393–423.

Owen-Smith, N. (1987). Pleistocene extinctions: the pivotal role of megaherbivores. *Paleobiology*, **13**, pp. 351–362.

Prideaux, G. J., Gully, G. A., Couzens, A. M. C., *et al.* (2010). Timing and dynamics of Late Pleistocene mammal extinctions in southwestern Australia. *Proceedings of the National Academy of Sciences*, **107**, pp. 22 157–22 162.

Purvis, A. and Hector, A. (2000). Getting the measure of biodiversity. *Nature*, **405**, pp. 212–219.

Rabosky, D. L., Donnellan, S. C., Talaba, A. L. and Lovette, I. J. (2007). Exceptional among-lineage variation in diversification rates during the radiation of Australia's most diverse vertebrate clade. *Proceedings Of The Royal Society Of London Series B – Biological Sciences*, **274**, pp. 2915–2923.

Raven, P. H. and Yeates, D. K. (2007). Australian biodiversity: threats for the present, opportunities for the future. *Australian Journal Of Entomology*, **46**, pp. 177–187.

Rich, P. V., Rich, T. H., Wagstaff, B. E., *et al.* (1988). Evidence for low temperatures and biologic diversity in Cretaceous high latitudes of Australia. *Science*, **242**, pp. 1403–1406.

Roberts, R. G., Flannery, T. F., Ayliffe, L. K., *et al.* (2001). New ages for the last Australian megafauna: continent-wide extinction about 46,000 years ago. *Science*, **292**, pp. 1888–1892.

Root, T. L. and Schneider, S. H. (2006). Conservation and climate change: the challenges ahead. *Conservation Biology*, **20**, pp. 706–708.

Rule, S., Brook, B. W., Haberle, S. G., *et al.* (2012). The aftermath of megafaunal extinction: ecosystem transformation in Pleistocene Australia. *Science*, **335**, pp. 1483–1486.

Sanders, K. L., Lee, M. S. Y., Leys, R., Foster, R. and Keogh, J. S. (2008). Molecular phylogeny and divergence dates for Australasian elapids and sea snakes (hydrophiinae): evidence from seven genes for rapid evolutionary radiations. *Journal of Evolutionary Biology*, **21**, pp. 682–695.

Schwarz, M., Fuller, S., Tierney, S. and Cooper, S. J. (2006). Molecular phylogenetics of the Exoneurine Allodapine bees reveal an ancient and puzzling dispersal from Africa to Australia. *Systematic Biology*, **55**, pp. 31–45.

Sclater, P. L. (1858). On the general geographical distribution of the members of the class Aves. *Journal of the Linnean Society, Zoology*, **2**, pp. 130–145.

Shoo, L. P., Storlie, C., VanDerWal, J., Little, J. and Williams, S. E. (2010). Targeted protection and restoration to conserve tropical biodiversity in a warming world. *Global Change Biology*, **17**, pp. 186–193.

Short, J. (1998). The extinction of rat-kangaroos (Marsupialia: Potoroidae) in New South Wales, Australia. *Biological Conservation*, **86**, pp. 365–377.

Short, J., Kinnear, J. E. and Robley, A. (2002). Surplus killing by introduced predators in Australia – evidence for ineffective anti-predator adaptations in native prey species? *Biological Conservation*, **103**, pp. 283–301.

Spalding, M. D., Fox, H. E., Allen, G. R., *et al.* (2007). Marine ecoregions of the world: a bioregionalization of coastal and shelf areas. *BioScience*, **57**, pp. 573–583.

Turney, C. S. M., Kershaw, A. P., Clemens, S. C., *et al.* (2004). Millennial and orbital variations of El Nino/Southern Oscillation and high-latitude climate in the last glacial period. *Nature*, **428**, pp. 306–310.

Udvardy, M. D. F. (1975). A classification of the biogeographical provinces of the world. *IUCN Occasional Paper*, **18**, pp. 1–50.

Upchurch, P. (2008). Gondwanan break-up: legacies of a lost world? *Trends in Ecology and Evolution*, **23**, pp. 229–236.

Van Dyck, S. and Strahan, R. (2008). *The Mammals of Australia*. 3rd edn. Chatswood: New Holland Publishers.

Vane-Wright, R. (2003). Indifferent Philosophy versus Almighty Authority: on consistency, consensus and unitary taxonomy. *Systematics and Biodiversity*, **1**, pp. 3–11.

Vidal, N., Marin, J., Sassi, J., *et al.* (2012). Molecular evidence for an Asian origin of monitor lizards followed by Tertiary dispersals to Africa and Australasia. *Biology Letters*, **8**, pp. 853–855.

Wallace, A. R. (1860). On the zoological geography of the Malay Archipelago. *Journal of the Proceedings of the Linnean Society*, **4**, pp. 172–184.

Wallace, A. R. (1876). *The Geographic Distribution of Animals*. London: Macmillan.

Webster, K. N. and Dawson, T. J. (2004). Is the energetics of mammalian hopping locomotion advantageous in arid environments? *Australian Mammalogy*, **26**, pp. 153–160.

Westoby, M. (1993). Biodiversity in Australia compared with other continents. In *Species Diversity in Ecological Communities: Historical and Geographical Perspectives*. Chicago: University of Chicago Press, pp. 170–177.

White, M. E. (2006). Environments of the geological past. In *Evolution and Biogeography of Australasian Vertebrates*. Oatlands: Auscipub, pp. 17–50.

Whittaker, R., Araujo, M., Jepson, P., *et al.* (2005). Conservation biogeography: assessment and prospect. *Diversity and Distributions*, **11**, pp. 3–23.

Williams, S. E., Bolitho, E. E. and Fox, S. (2003). Climate change in Australian tropical rainforests: an impending environmental catastrophe. *Proceedings Of The Royal Society Of London Series B – Biological Sciences*, **270**, pp. 1887–1892.

Wroe, S., Field, J. H., Archer, M., *et al.* (2013). Climate change frames debate over the extinction of megafauna in Sahul (Pleistocene Australia–New Guinea). *Proceedings Of The National Academy of Sciences*, **110**, pp. 8777–8781.

Wroe, S., Myers, T. J., Wells, R. T. and Gillespie, A. (1999). Estimating the weight of the Pleistocene marsupial lion, *Thylacoleo carnifex* (Thylacoleonidae: Marsupialia): implications for the ecomorphology of a marsupial super-predator and hypotheses of impoverishment of Australian marsupial carnivore faunas. *Australian Journal of Zoology*, **47**, pp. 489–498.

Zachos, J. (2001). Trends, rhythms, and aberrations in global climate 65 Ma to present. *Science*, **292**, pp. 686–693.

Zachos, J. C., Dickens, G. R. and Zeebe, R. E. (2008). An early Cenozoic perspective on greenhouse warming and carbon-cycle dynamics. *Nature*, **451**, pp. 279–283.

CHAPTER 2

New Zealand – a land apart

William G. Lee and Daphne E. Lee

Summary

New Zealand is well known for its distinctive biota and high proportion of endemic taxa. A recent checklist of fossil and living plants and animals in terrestrial and marine ecosystems highlights spectacular local radiations coupled with unusual ecological niches and an amazing array of cryptic diversity across most habitats. Although New Zealand and Australia share common biotic antecedents, for the past 20 million years the New Zealand terrestrial biota has occupied constantly mesic habitats but has undergone major extinctions as climate cooled. On land, plants and animals have diversified in non-forest ecosystems, particularly shrublands and grasslands. New Zealand faces major challenges in protecting biodiversity and ecosystem services, in part because the biota is adjusting to relatively recent human occupation, and land use and marine harvesting are intensifying. Models for effective conservation in terrestrial environments are dependent on enduring predator control, legislative protection of remaining indigenous habitats and threatened species, and novel approaches that compensate land owners for contributing towards national biodiversity outcomes. Marine protected areas are minimal in extent although managed fish quota systems appear to be sustaining some species. Indigenous biodiversity in New Zealand remains vulnerable especially in ecosystems dominated by human activities.

2.1 Introduction

New Zealand is often compared to an Ark, laden with a unique Gondwanan biota. While this emphasises remoteness and antiquity as major drivers of our biotic distinctiveness, it less adequately captures the composite elements of the biota or the complex history of immigration, extinction and speciation, and a changing geography and climate during the Cenozoic in both marine and terrestrial ecosystems. All of these factors have contributed to the evolution of an overwhelmingly endemic idiosyncratic biota that continues to adjust to natural environmental change and more recent anthropogenic impacts.

Austral Ark: The State of Wildlife in Australia and New Zealand, eds. A. Stow, N. Maclean and G. I. Holwell. Published by Cambridge University Press. © Cambridge University Press 2015.

New Zealand is a global biodiversity hotspot, a status reflecting both the high concentration of endemic species and their vulnerability to threats from past and present rates of habitat loss (Myers *et al.* 2000). Late human arrival (*c.* 1350 AD) makes the original biota and ecology of great significance for evaluating the effects of settlement (Wilmshurst *et al.* 2008). Many trees in New Zealand forests pre-date human arrival, but overall ecosystems are still adjusting to the dramatic loss of large indigenous vertebrates and imposition of many stresses including frequent fire, pests and weeds. It is this relatively brief interaction between humans and the biota that makes New Zealand particularly interesting and unusual. In New Zealand, no equilibrium has yet been reached between human activities and biodiversity, and species and habitats continue to respond and generally decline. In addition, extinction debts from human actions centuries ago are still unfolding.

It is appropriate that New Zealand biodiversity has its own chapter in this particular book as over recent years some have begun to think of our terrestrial ecosystems as little more than Australian biotic dependencies formed from flotsam and jetsam over the past few million years. Although a component of our biota is recently (past few Mya) derived from Australia, and interactions with the western continent have been important, this by no means includes the majority of the flora and fauna inhabiting New Zealand. Parochialism aside, New Zealand and Australia provide two fascinating examples of extraordinary evolutionary responses arising from an initially similar biota to quite different changing environments and therefore selection factors during the past 80 million years.

In this chapter we outline the geological history of New Zealand in relation to Australia before giving a profile of the biota and discussing the key factors shaping its character over evolutionary and ecological timescales. We then examine the extinctions and conservation status of key groups, highlighting past and ongoing threats to the persistence of biodiversity. Finally, we outline the conservation approaches used in New Zealand and the challenges ahead.

2.2 Australasia – divergent paths

Until the late Cretaceous (*c.* 83 Mya) the continent Zealandia (the area currently defined by the continental shelf including New Zealand) was a region of the eastern Gondwana supercontinent that shared with Australia a contemporaneous Gondwanan biota that included *Nothofagus*, dinosaurs and tuatara on land, and mosasaurs and ammonites in the oceans. Some of these animals, including dinosaurs, marine reptiles, and ammonites, later became extinct around 65 Mya during the global K/T boundary extinction event.

Following gradual opening of the Tasman Sea, which began in the south at about 83 Mya, Zealandia drifted slowly east and north, reaching its current latitudinal position at about 40 Mya, by which time any land connections between Australia and Zealandia were severed. This may have precluded the arrival of mammals and snakes in New Zealand, although it would not be surprising if they had been present. New Zealand's current isolated position 2000 km east of Australia has been maintained for the past 30 million years, although at times there have been intermittent connections to the northwest via New Caledonia and various volcanic island chains (Lee *et al.* 2001).

After initial rifting, the Zealandia continent began to subside slowly due to crustal thinning. By about 40 million years ago, it was still a substantial, fairly low-lying landmass, about three times its present size. During the late Oligocene, about 25 Mya, the land area was reduced to a series of large low-lying islands occupying an area of between 50 000 to 20 000 square kilometres (c.f. the current land areas of Tasmania and New Caledonia respectively), but still supporting a diverse biota (Ferguson *et al.* 2010; Jordan *et al.* 2011; Worthy *et al.* 2011). Around 25 Mya a new plate boundary that developed between the Australian and Pacific plates initiated uplift and the land area increased to about its present size by the early Miocene. Today, the three large, and hundreds of smaller islands, that make up modern New Zealand comprise about 10% of the original land area of the Zealandia continent (Graham 2008).

From 50 to 10 Mya, New Zealand's climate was at least marginally subtropical and supported what are nowadays considered quintessential Australian groups such as crocodiles (Molnar and Pole 1997) and Proteaceae (Carpenter *et al.* 2012). Following the Miocene climatic optimum about 16 Mya, the seas around New Zealand gradually cooled and for the first time in *c.* 40 million years the climate became warm to cool temperate as it is today. Tectonic activity associated with uplift along the Alpine Fault produced the first mountains over 1000 m elevation, including the Southern Alps, beginning about 5 Mya. The full range of alpine habitats in New Zealand of the type that we are familiar with today developed once the Southern Alps had reached their current elevation of >3000 m around 2 Mya (Lee *et al.* 2001; Heenan and McGlone 2013).

In contrast to Australia, New Zealand has always had a wet, humid and oceanic climate with only marginal sporadic aridity in localised rain-shadow areas in the central and eastern South Island since Pliocene times. Currently, annual rainfall can vary from 300 mm in Central Otago to over 7000 mm along the Southern Alps, but in contrast to truly semi-arid environments, precipitation can and does occur at any time of the year (Walker and Wilson 2002).

Intermittent local volcanism has occurred in New Zealand for the past 80 million years and continues. Since the Pliocene intense disturbances associated with New Zealand's position across the boundary between two tectonic plates has caused huge increases in disturbances such as landslides, accelerated erosion, and flooding, all of which continually act to rejuvenate soils and landscapes.

Another difference is that New Zealand has lacked terrestrial environments at lower latitudes, causing extinction of the subtropical elements of the biota, which in Australia were able to move north during cooler periods (Lee *et al.* 2001). The impacts of cooling climate were exacerbated by major shifts in the geographic extent and configuration of the New Zealand archipelago, associated with sea-level changes and strong temperature fluctuations during glacial/interglacial cycles.

The biotic dissimilarities that we see today between the two countries are thus a result of different environmental trajectories, geographic extent, and extinction and speciation processes. In contrast to Australia, arid and tropical biomes have not featured as factors in the diversification of the New Zealand biota.

2.3 Profile of the New Zealand biota

Only in the past decade have we have been able to gain anything approaching an adequate overview of the past and present biota of the New Zealand region, thanks to the recently published authoritative *New Zealand Inventory of Biodiversity* series edited by Dennis Gordon (Gordon 2009, 2010, 2012), which has been used extensively to compile the following account.

Plate 3 shows that New Zealand's current land area (268 680 sq. km) comprises a mere 6.5% of the total Exclusive Economic Zone (4 083 744 sq. km) associated with our extensive continental shelf. Combining marine and terrestrial zones, New Zealand currently supports at least *c.* 56 500 taxa, 70% in terrestrial ecosystems (61% on land and 9% in fresh water) and 30% in marine environments. These figures include estimates of those known taxonomic entities yet to be described but not those systematists consider still likely to be found. Because of limits to exploration in the marine environment and inadequate taxonomic scrutiny everywhere, these will invariably represent considerable underestimates.

Levels of endemism What stands out in Tables 2.1 and 2.2 is how much our perception of biodiversity is skewed towards the minority of species that are large, obvious, and on land. For example, familiar vascular plant communities and species represent, on a numerical basis, much less than 5% of New Zealand's known biodiversity, microbes excluded, although they dominate most biodiversity assessments and conservation planning.

In general, the New Zealand terrestrial and, to some extent, the marine biota display remarkably high levels of endemism, coupled with the absence or very low representation of many groups that are common or ubiquitous elsewhere in the Southern Hemisphere. The most striking difference is the almost complete lack of terrestrial mammals, apart from a few species of bats. In contrast, other groups, such as terrestrial snails are highly diverse in relation to the land area. Similarly, many birds and invertebrates have taken over the functional roles of the missing mammals. Among birds, many are (or were, until human colonisers arrived) flightless, large, ground dwelling, and nocturnal (Holdaway 1989).

Tables 2.1 and 2.2 give the relative proportions of endemics in selected groups. Amongst seed plants, 82% are endemic to New Zealand. In bryophytes and fungi with tiny propagules the proportion is lower (<40%). However, New Zealand is the most important hot-spot of liverwort diversity in the world, with more species (>600 species) and a higher degree of endemism (50%) than any Northern Hemisphere liverwort flora, and double the levels found in the Australian flora. There are two monotypic endemic families (Jubulopsidaceae, Verdoorniaceae) and 18 endemic genera (Engel and Glenny 2008).

Almost all terrestrial vertebrates are endemic including at high taxonomic levels. For example there are endemic orders (e.g. Sphenodontia – tuatara), families (e.g. Leiopelmatida – frogs; Acanthisittidae – wrens; Mystacinidae – short-tailed bat), and dozens of endemic genera (e.g. *Apteryx* – kiwi). Notably, all New Zealand parrots and wrens are the sister groups to all other parrots (Wright *et al.* 2008) and passerines (Ericson *et al.* 2002) respectively.

New discoveries, based in part on molecular evidence for many cryptic species, are showing how little is known about the diversity of even some larger groups of vertebrates. For example, indigenous lizards are remarkably diverse: there are 43 species of endemic gecko and 50 species of endemic skink in the New Zealand fauna, making the lizard diversity equal in richness to the number of terrestrial birds (*c.* 90).

In terrestrial invertebrates, endemism is also high at both specific and generic level. For example all of the 34 species of Onychophora (*Peripatus*) and 51 species of mayfly (Ephemeroptera) are endemic to New Zealand. Overall, 75% of Arthropoda are endemic.

Even among predominantly marine animal groups, endemism levels above 30% are exhibited by phyla as different as Porifera, Platyhelminthes, Brachiopoda, Bryozoa,

Table 2.1   Species diversity and endemicity for selected New Zealand animal groups derived from Gordon (2009, 2010, 2012). The number of taxa in marine, terrestrial and freshwater environments includes adventives, i.e. those species introduced from other countries now sustaining populations in New Zealand.

Taxon	Ecosystems			Total number of taxa	% Native species endemic	Number adventive taxa
	Marine	Terrestrial	Freshwater			
Porifera sponges	724	0	5	729	46	0
Ctenophora comb jellies	19	0	0	19	21	0
Cnidaria e.g. corals	1112	0	14	1102	21	24
Platyhelminthes e.g. flatworms	324	137	78	504	39	35
Gnathifera e.g. rotifers	44	2	484	530	4	0
Mollusca	**3593**	**906**	**89**	**4538**	**86**	**50**
Polyplacophora	64	0	0	60	85	0
Bivalvia	673	0	5	674	85	4
Scaphopoda	45	0	0	45	96	0
Gastropoda	2671	905	84	3660	87	46
Prosobranchia	2170	49	66	2285	82	1
Heterobranchia	206	0	0	206	83	0
Opisthobranchia	272	0	0	272	50	6
Pulmonata	23	857	18	898	97	39
Brachiopoda	38	0	0	38	39	0
Bryozoa	953	0	8	961	61	24
Annelida	792	207	55	1054	41	80
Nemertea	29	6	4	39	72	0
Echinodermata	623	0	0	623	38	0
Hemichordata	7	0	0	7	14	0
Tunicata	192	0	0	192	65	11
Chordata	**1427**	**264**	**96**	**1787**	**26**	**96**
Pisces	1112	0	58	1170	19	22
Chondrichthyes Cartilaginous fishes	106	0	0	106	28	0
Actinopterygii Bony fishes	1006	0	58	1064	21	22
Amphibia	0	8	2	10	100	3
Reptilia	7	84	0	91	83	1
Aves	122	137	35	294	56	38
Mammalia	48	3	0	51	32	32
Tardigrada	5	0	83	88	38	0
Onychophora	0	34	0	34	100	0

Table 2.1   (*cont.*)

Taxon	Ecosystems			Total number of taxa	% Native species endemic	Number adventive taxa
	Marine	Terrestrial	Freshwater			
Arthropoda	**2926**	**17 609**	**1242**	**21 784**	**81**	**1773**
Chelicerata e.g. spiders	115	3273	158	3546	78	114
Myriapoda e.g. millipedes	0	295	0	295	99	28
Crustacea e.g. crabs	2614	123	236	2973	38	46
Hexapoda e.g. insects	197	14 000	848	15 045	90	1585
Nematoda	211	540	9	760	15	152

Table 2.2   Species diversity and endemicity for selected New Zealand plant, fungal and lichen groups derived from Gordon (2012). The number of taxa in marine, terrestrial and freshwater environments includes adventives, i.e. those species introduced from other countries now sustaining populations in New Zealand.

Taxon	Ecosystems			Total number of taxa	% Native species endemic	Number adventive taxa
	Marine	Terrestrial	Freshwater			
Algae						
Rhodophyta	541	0	20	561	35	15
Chlorophyta	156	35	394	585	5	10
Charophyta	0	1	534	535	8	1
Bryophyta						
mosses, liverworts	0	1086	35	1122	36	36
Tracheophyta Higher plants	**5**	**4642**	**108**	**4744**	**75**	**2523**
Ferns & fern allies	0	226	5	231	45	49
Pinopsida Conifers	0	60	0	20	100	40
Magnoliopsida Angiosperms	5	4341	0	4346	85	2430
Fungi						
Chytridiomycota	6	106	39	151	9	13
Ascomycota	77	5210	128	5415	20	1485
Basidiomycota	3	2771	4	2778	37	481
Zygomycota	0	111	17	128	8	19
Lichens	0	1767	0	1767	30	N/A

Annelida, Nemertea, Echinodermata, and Tunicata. The molluscan fauna of New Zealand is unusual globally in many respects. Endemicity (above 85% at species level) is extraordinarily high, and the diversity, size and ecological niche in many groups are remarkably different from molluscs found elsewhere.

Patterns of endemism Last century, many biogeographic studies attempted to find regional patterns of endemism, usually within a restricted group of plants or animals

(Wardle 1963; Diamond 1984; Heads 1997; Trewick *et al.* 2011). What is now emerging is a complex pattern of speciation across most of New Zealand, except where volcanism and glacial activity are too recent for differentiation to have occurred. Late Miocene, Pliocene and Pleistocene climatic cooling and the absence of warm-climate refuges at low latitudes caused widespread extinction amongst the woody flora, with many formerly more diverse families and genera nowadays being represented by a few relictual taxa (Jordan *et al.* 2011; Bannister *et al.* 2012). In contrast, these same environmental changes, coupled with factors such as cyclic fall and rise of sea-level, and concomitant formation and loss of islands, glacial/interglacial cycles and severe climates, and the emergence of novel habitats (e.g. alpine zone, landscape-level alluvial disturbances), periodically fragmented populations. Over the past 20 million years, these provided new opportunities for increased speciation in both resident and immigrant taxa, in groups as diverse as lizards (Chapple *et al.* 2009; Nielsen *et al.* 2011) and vascular plants (Wagstaff *et al.* 2002).

Endemism within terrestrial ecosystems is related in a complex manner to habitat history, particularly disturbance. For example, New Zealand has diverse freshwater ecosystems but most of our faunal diversity is associated with rivers and streams that have evolved in forested landscapes. Lakes notably lack the specialised faunas found elsewhere, perhaps reflecting the transience, relative youthfulness and frigidity of our predominantly glacial and volcanic lakes. Stream invertebrates, largely feeders on fine detritus, are highly endemic and have strong Gondwanan affinities. Ephemeroptera (mayflies) for example have 19 endemic genera (Winterbourn 2004). Our freshwater fish fauna of *c.* 38 species is also predominantly confined to riverine habitats and is dominated by Galaxiidae (56%), a Southern Hemisphere group with a centre of diversity in New Zealand. These include diadromous *Galaxias* species that migrate to sea for much of their lives, occupy multiple river systems and have broad geographic ranges, and *Neochanna* (mudfish) that aestivate over summer locally in drier parts of the country. For vascular plants the patterns are reversed with submerged macrophytes more common in lakes, although in this group, endemism is relatively low (29%).

Currently we can provide only a very approximate view of biodiversity in New Zealand ecosystems because of our inability to measure many entities. This has been highlighted in a mountain beech (*Nothofagus solandri* var. *cliffortioides*) forest in eastern South Island, containing very few vascular plant species. However, a study of saprobic fungi in decomposing beech logs revealed at least 151 taxa of mainly corticiod fungi and mushrooms (Allen *et al.* 2000), identifying significant cryptic biodiversity that is frequently overlooked.

Antiquity of the New Zealand biota The direct Gondwanan links of the indigenous biota have been challenged over the last decade in debates about whether or not Zealandia, the now mostly submerged continent, supported any terrestrial environments at all during the peak marine transgression in the late Oligocene (Campbell and Hutching 2007; Biffen *et al.* 2010). However, pre-Oligocene arrival dates, based on fossils, or molecular calibrations independent of the rifting of New Zealand from Australia, for extant trees such as *Knightia* and *Agathis*, and animal lineages such as frogs, tuatara, wrens, kakapo and landsnails, argue for continuous land throughout the Cenozoic. The absence of any signal in the fossil record of major biotic turnover during this period (*c.* 25 Mya) also supports the persistence of land (Lee *et al.* 2001). In general this debate has established when different components of the biota arrived, immigrants establishing when land connections were available or during major extinction phases, or after the appearance of new habitats such as the alpine zone less than 5 Mya ago. Overall, results have indicated that New Zealand is much more than merely a recent collection of Australian biotic vagrants. These discussions have also entered the conservation realm, with some implying that geological antiquity and conservation value are directly linked, downplaying the value of putative recent arrivals. Lineages with long histories in New Zealand are more likely to have higher endemism at genus, family or higher taxonomic levels that will influence their conservation significance. However, extant species in both ancient and recent lineages in New Zealand (and globally) extend back only a few million years at most. They all represent lineages that have survived and evolved in New Zealand for differing periods, but have been shaped by a unique suite of environmental factors.

2.4 What is so amazing about the indigenous biota?

In this section we provide some brief highlights from subsets of the terrestrial, freshwater and marine floras and faunas that are merely the tip of the New Zealand 'bioberg'.

Plant habit Among the interesting vascular plant forms in New Zealand, perhaps the most unusual and fascinating are the grey shrubs and heteroblastic trees that are super-ficially difficult to discriminate. The development of this wire-plant architecture, commonly known as divarication, occurs in over 50 endemic woody plant species, representing 16 families of angiosperms. The habit is characterised by an outer canopy of dense, inter-tangled, fine, and extremely strong branches that protect small leaves, and has long puzzled ecologists. Adaptive explanations for this curious habit vacillated between climate (e.g. drought, frost, wind) (Howell *et al.* 2002) and anti-herbivore defences against large ratite birds that roamed New Zealand (Bond *et al.* 2004). The controversy continues and can be viewed as part of the universal debate among ecologists over the relative strength of abiotic versus biotic factors in the evolution of biota and particular traits.

Terrestrial snails New Zealand's land snail fauna (Order Pulmonata) is globally distinctive and extraordinarily diverse for a relatively small land area. There is low family diversity (only 11 out of *c.* 80 families known globally) but in striking contrast there are about 65 endemic genera, and of the 1400 species, all but five are endemic. The fauna is dominated by species with flattish shells that vary from a few mm to 10 cm in diameter. In some forest sites, communities of between 30 and 70 species co-exist in areas of only a few square metres (Barker and Mayhill 1999). They occupy a wide range of habitats from beach dune systems and lowland forest to alpine grasslands. Many are nocturnal, and some are long-lived (up to 30 years or more). The majority are detritivores living in leaf litter on the forest floor where they feed on microorganisms associated with plant decay. Others such as *Powelliphanta*, *Wainuia* and *Rhytida* are carnivores which feed on earthworms, amphipods, and other snails, respectively. Few feed primarily on green plant tissues in contrast to snails in many continental areas.

Orthopterans Giant weta (family Anostostomatidae), which weigh up to 70 g and have bodies over 10 cm in length, represent some of the largest insects in the world (Gibbs 2001). The three main groups are ground-weta, tusked-weta and tree-weta, which comprise over 60 inadequately understood endemic taxa. Both arboreal and ground-dwelling species are nocturnal and usually feed on plants and other insects across varying habitats, including caves. Remarkably, one species of tree-weta (*Hemideina maori*) occupies the alpine zone due to an ability to tolerate sub-zero temperatures by partial freezing. In this species males have elaborate weaponry that is used to defend large harems (Koning and Jamieson 2001). In New Zealand weta have been able to expand their ecological niche considerably in the absence of competition and predation from small native mammals.

Bats Currently represented by a single species in the endemic genus *Mystacina* (a second species became extinct last century), the adult short-tailed bats are tiny animals less than 10 cm in length and weighing *c.* 25 g. They form labile colonies, usually roosting during the day in natural cavities, except during breeding when males form individual roosts and attract females with distinctive scent and ultrasound. Although manoeuvrable fliers, short-tailed bats frequently forage on the ground. This uncommon behaviour is assisted by their foldable membranous wings and elaborate claws that aid climbing across uneven terrain. They predominantly feed on forest-floor arthropods and pollen and nectar. Recently they have been shown to be the main pollinator of *Dactylanthus taylorii*, a remarkable root parasite in the family Balanophoraceae whose rarity (and conservation status) has been linked to declines in bat populations (Ecroyd 1995). In addition to its other peculiar features, *Mystacina* is also a participant in a symbiotic relationship with the New Zealand batfly, a small, wingless dipterid fly that is placed in its own endemic genus *Mystacinobia* and family Mystacinobiidae.

Reef fish New Zealand does not have a highly diverse marine fish fauna, but shallow-water sub-tidal environments in New Zealand support an amazing abundance and diversity of triplefin fishes (family Tripterygiidae), which at the endemic level include 12 genera and 26 species, representing 40% and 15% respectively of the global total. As Clements (2006) has noted, this level of endemism is at least tenfold that of the New Zealand marine fish fauna as a whole. Most New Zealand triplefins feed on a variety of benthic invertebrates, and there is considerable dietary overlap between species with specialists and more generalist taxa occurring somewhat randomly across the lineages in New Zealand. Species distributions mostly overlap geographically, but species have diverged greatly in terms of habitat (Wellenreuther *et al.* 2007), especially in terms of water depth and exposure. Uniquely in the world, New Zealand triplefins have diversified to inhabit estuaries and deep water down to 500 m, and there is also a schooling, planktivorous species. The group appears to have radiated in New Zealand in response to the unusually low diversity of similar fishes including blennies and gobies.

Seabirds New Zealand is well known for the paucity of terrestrial mammals and the dominance of birds as predators, browsers, scavengers, pollinators and seed dispersers. Birds fill many of the functional roles occupied by mammals elsewhere, but it is only in recent years that we have discovered the remarkable diversity and functional significance of seabirds in island and coastal ecosystems. Approximately a quarter (86) of the world's seabirds regularly breed in the New Zealand archipelago, transferring vast amounts of nutrients from marine ecosystems while they nest and feed juveniles, and creating special habitats for several endemic plant species (Bellingham *et al.* 2010). By using GPS tracking systems we now know that these birds use New Zealand as a breeding station, in between travelling thousands of kilometres across oceans every year (Shaffer *et al.* 2006). The rediscovery of seabirds as ecosystem engineers is providing a new approach and impetus to restoring island ecosystems.

2.5 Evolutionary context

Soils In comparison with Australia, New Zealand soils are perceived as nutrient-rich, with high levels of phosphorus in young, alluvially and colluvially derived soils (Wright *et al.* 2010). On older more stable land surfaces in New Zealand, nitrogen levels are high, particularly when augmented by avian activity which was almost certainly more widespread in the past. Paradoxically, although strong edaphic differences structure habitats and are associated with species differentiation in major plant radiations (Pirie *et al.* 2010), universally slow relative growth rates, long-lived leaves, and conservative patterns of nutrient recycling amongst the vascular flora suggest long-term adaptations to nutrient-poor conditions (Craine *et al.* 2003). This may reflect the paucity of low nutrients during the early (50–5 million years) evolution of plants when New Zealand was a low-lying archipelago, prior to the emergence of mountains, but it is puzzling that more recent arrivals over the past 5 million years also seem to exhibit these characteristics.

Disturbances Major disturbances connected with volcanoes, earthquakes, cyclones/wind-throw, flooding and landslides have been regular occurrences in New Zealand, and are responsible for diversity at spatial scales ranging from several hectares to those observed across large regions. They have contributed to sustaining those elements of the biota generally adapted to early successional communities in somewhat more stable landscapes. However, the evolution of distinctive forms and strategies to cope with these disturbances seem to be rare. For example, we lack specialised or endemic plant taxa on volcanic soils even though they have been a regular though localised part of the New Zealand landscape for millions of years. In contrast some endemic plant and animal taxa are associated with colluvial screes that appeared in the past few million years as mountains rose, valley sides were over-steepened by glaciers and landslide frequency increased.

Fire initiated by volcanism, lightning strikes and self-combusting coal seams has always been present in the New Zealand landscape, and there is evidence for burning from paleo-charcoal (Ogden *et al.* 1998). However, unlike Australia, broad syndromes and adaptions to fire are very rare in the current biota (e.g. induced mast flowering in *Chionochloa rigida*, serotonous capsules in populations of *Leptospermum scoparium*). Moreover, much of the biota is extremely fire-sensitive, as indicated by the losses suffered during regular burning since human arrival (McGlone 2001).

2.6 Extinctions

An inconvenient biota When humans first arrived in New Zealand they discovered a relatively benign climate but a very inconvenient terrestrial biota for sustaining a human population. Once the largest birds (especially moa) were consumed and much of the eastern forest on South Island transformed into tussock grassland and scrubland by fire (McGlone 1989), Maori became increasingly dependent on freshwater and near-shore marine ecosystems for food, supplemented, after contact with European explorers and whalers, with introduced pigs (*Sus scrofa*) and goats (*Capra hircus*) and crops such as potatoes (*Solanum tuberosum*).

As Europeans settled throughout the nineteenth century they found few indigenous species of immediate use and economic benefit. After the taller, slow-growing conifers were harvested for timber, few, if any, plants suitable for arable crops or pastoral and forestry species could be found among the indigenous biota. However, they were very familiar with and had access to many potentially useful domestic and wild species in their original homelands. And so began the largest and most rapid importation of non-native species to any country in historical times. As species were introduced from around the globe, large-scale clearance of forest and drainage of wetlands was undertaken to provide for settlements and pastoral development, initiating major ecological transformation of the New Zealand landscape. The aim was to provide food for a growing population and crucial exports to sustain an economy remote from most markets. To achieve this goal, we imported fast-growing grasses, legumes and plantation trees and numerous crops and animals (Allen and Lee 2006).

Maori have continued to harvest a range of freshwater and coastal species but indigenous species overall have played variable roles in the economic growth of New Zealand. Seals and whales were important in the nineteenth century and marine species of fish (e.g. hoki) and crustaceans (e.g. crayfish) have developed into major industries in the twentieth century. More recently, an aquacultural industry has developed based on cultivation of the endemic green-lipped mussel (*Perna canaliculus*). Juvenile *Galaxias* (whitebait) in the lower reaches of rivers and 'game birds' including a few native species are both hunted seasonally under licence.

It is hard to overestimate the significance of the absence of a biota adaptable to human requirements and capable of sustaining growing populations in New Zealand. Many countries regularly utilise native species for timber, crops and pasture but in New Zealand virtually all of these species have had to be imported and native biodiversity displaced to provide areas primarily focused on production.

Inviting trouble In our search for an alternative biota, New Zealand embarked on an unprecedented massive global introduction programme. This activity was fuelled by Acclimatisation Societies and provincial governments supported by recreational hunting, forestry, horticultural and agricultural industries, and encouraged by a public keen to develop spectacular gardens in New Zealand's temperate maritime climate. The scale of these introductions, both accidental and deliberate, almost defies belief; for over 150 years New Zealand has welcomed plants and animals to the extent that we are now one of the most invaded countries in the world (Allen and Lee 2006). Currently over 90% of mammals, 45% of freshwater fish, 50% of vascular plants, and at least 30% of amphibians and bird species are introduced. Invertebrate groups with over 100 naturalised species include Chelicerata, Hemiptera, Coleoptera, Hymenoptera, Lepidoptera and Nematoda, although these were mostly accidental introductions associated with trade (Tables 2.1 and 2.2).

Plants, birds and mammals provide exemplary models of the processes associated with invasions in New Zealand. Over 23 000 vascular plant species have been introduced, largely confined to residential gardens. However, the naturalised adventive flora (*c.* 2430) now outnumbers the comparable indigenous flora. Approximately 50% of the naturalisations are from the families Gramineae, Asteraceae, Fabaceae and Rosaceae (Williams and Cameron 2006). Naturalisation is primarily associated with widespread plantings by humans. From this pool over 200 species have become major environmental weeds although a lesser number significantly impact native species and habitats.

At least 120 bird species have been brought to New Zealand with 34 successfully establishing, nearly half from Fringillidae (finches), Phasianidae (galliforms) and

Anatidae (ducks, geese and swans). Successful species establishment was more closely related to introduction effort, large natural geographic ranges and similarities of home climate to New Zealand than life history or demographic traits (Duncan *et al.* 2006). Except for agricultural pests, avian introductions are generally accepted and valued as game birds. In wetlands, for example, they may replace the functional role of extinct native bird species.

In addition to domestic stock, mammalian introductions (King 2005) include rodents, mustelids, lagomorphs, ungulates and the brush-tailed possum (*Trichosurus vulpecula*), all groups that were not previously present in New Zealand. Apart from rodents, all were deliberately established to provide recreational game and new industries, while mustelids were imported as a biocontrol for rabbits. All have impacted indigenous biota although debates continue about their roles in the decline of biodiversity.

Many of these introductions support a very successful primary agricultural economy among other industries but some have also had considerable ongoing costs and biodiversity impacts. Some introductions, such as feral deer, are favoured by different sectors and their control or eradication is often a contentious issue, especially if poisons are used.

The marine environment is also susceptible to invasion but it is only in recent decades that we have become aware of significant impacts from introduced species, reflecting the problems of monitoring habitats that are difficult to access. Notable examples include the cord grass, *Spartina alterniflora*, previously widely planted to stabilise estuaries, and Japanese kelp, *Undaria pinnatifida*, likely introduced in ballast water, which is now establishing in coastal waters (Clayton and Edwards 2006). Once established, eradication or even range control is proving both expensive and impractical.

Because New Zealand lacked sport fish, freshwater ecosystems have, since the commencement of European settlement in the 1860s, been deliberately stocked with salmonids, notably brown (*Salmo trutta*) and rainbow (*Oncorhynchus mykiss*) trout, which form the basis for a world renowned and economically important recreational fishing industry. For a century, the impacts of these fish were unknown but studies have now shown that they predate *Galaxias*, native crayfish (*Paranephrops zealandicus*) and mayfly nymphs (*Nesameletus* and *Deleatidium* spp.), which can foster periphyton algal growth (Townsend and Simon 2006). There are now at least 20 introduced fish (McIntosh and McDowell 2004), mostly in warmer regions, accompanied by introduced macrophytes. Over 70 freshwater aquatic plants are naturalised, and at least three (*Lagarosiphon major*, *Hydrilla verticillata*, *Ceratophyllum demersum*) have become problem weeds displacing native plant species and altering food webs (Clayton and Edwards 2006). In 2004, the freshwater diatom *Didymosphenia geminata* was discovered in southern New Zealand, initiating a national biosecurity alert and strict movement control and hygiene regulations for river users, largely driven by concern for the trout fishery.

New Zealand has excelled at deliberately introducing inappropriate biota which has had long-term consequences for biodiversity. Meanwhile, biodiversity protection groups have become experts at managing invasive species impacts and national agencies have developed world-leading legislation and risk assessment protocols that are attempting to reduce the risk of further unwanted organisms crossing the border.

Extinctions and conservation status This assessment of extinction and conservation status is based on the New Zealand Threat Classification System (Hitchmough 2013) Table 2.3 summarises the Nationally Critical taxa, the highest threat category, which is based on a combination of small population size (usually less than 250 individuals) and

Table 2.3   Threatened and At Risk taxa for New Zealand derived from Hitchmough (2013).

		Number Nationally Critical	% Threatened	% At Risk	% Threatened or At Risk	Total assessed
Invertebrates	Beetles	35	10.2	60.4	70.6	442
	Terrestrial snails	28	10.3	56.6	66.8	458
	Powelliphanta	8	68.1	30.4	98.6	69
	Spiders	3	0.4	13.6	14.0	1138
	Flies	0	0.1	16.9	17.0	860
	Hemiptera	9	6.0	32.9	38.9	149
	Hymenoptera	2	2.8	27.8	30.6	72
	Lepidoptera	13	26.2	36.9	63.1	187
	Freshwater invertebrates	11	9.8	69.9	79.7	143
	Marine invertebrates	10	10.7	81.8	92.5	307
Vertebrates	Marine fish	0	0.0	24.8	24.8	218
	Marine mammals	13	11.2	20.7	31.8	242
	Bats	1	57.1	14.3	71.4	7
	Birds 2008	24	18.0	21.7	39.7	428
	Reptiles 2008	6	15.6	46.8	62.4	109
	Frogs	1	27.3	9.1	36.4	11
	Freshwater fish	4	18.7	26.7	45.3	75
Plants	Vascular plants	155	11.2	29.0	40.2	2580
	Algae	1	1.6	60.7	62.3	61
	Bryophytes	31	12.6	35.0	47.6	357
Others	Fungi	62	5.5	1.4	6.9	1605
	Lichens	4	0.6	9.8	10.5	1799

high ongoing or predicted decline, and gives the total of all taxa included on the official list (i.e. considered Threatened or At Risk).

Birds Losses are best known and understood for New Zealand's renowned avifauna. In the past 800 years, 58 (26%) of the original breeding bird species in New Zealand have become extinct, including over 40% of all endemic taxa. For the North Island and South Island the proportion of breeding birds lost is approximately 50%. These figures, however, disguise the biological and ecological significance of the eliminations, which included all nine species of moa (Dinornithiformes), four of the six species of wrens (Acanthisittidae), our largest wattlebird (*Heteralocha acutirostris*), most of the aerial predators including the world's largest eagle (*Harpagornis moorei*), and numerous wetland species (Tennyson and Martinson 2006). Such carnage was caused by a mix of human hunting and predation by the Pacific rat (*Rattus exulans*), but since the 1800s, other rodents (Norway rats – *Rattus norvegicus*; Ship rats – *Rattus rattus* and probably mice – *Mus musculus*) and mustelids (stoat – *Mustela eriminea*; ferret – *Mustela furo*; weasel – *Mustelia nivalis vulgaris*) have been the main culprits. Recognising the primary role of predator guilds in the extinction of native bird species in New Zealand has taken a long time. Debates have centred on identification of the most critical predators, the role of competitors (e.g. mammalian herbivores) and the importance of other changes (e.g. habitat loss, climate change).

What is apparent is that the losses have been disproportionately high amongst large-bodied flightless taxa in endemic clades (Duncan and Blackburn 2004). More species would have been lost but for the decades-long delay of predators reaching some remote sites, and the fortuitous presence of numerous offshore islands to act as refugia. Nonetheless, 24 species remain Nationally Critical, and a total of 170 taxa are either Threatened or At Risk. Management is focussing on predator control but effective methods for species requiring landscape-level security is as yet unattainable.

Other terrestrial animals The greater short-tailed bat, *Mystacina robusta*, became extinct during the 1960s, largely as a result of predation by rats. All other native bat species are considered either Threatened or At Risk, with small and fragmented populations (O'Donnell *et al*. 2010). Among reptiles and amphibians, the giant gecko *Hoplodactylus delcourti* and three frogs (all *Leiopelma* species) represent notable losses of globally unusual and/or ancient groups. Reptiles are highly over-represented in our threatened fauna, with 68 taxa either Threatened or At Risk. Among other terrestrial animals, extinctions are known with less certainty. One freshwater fish (the grayling – *Prototroctes oxyrhynchus*) four beetles, and three lice associated with extinct birds are officially listed as extinct. The threat status across these groups is rising rapidly, reflecting impacts of predation and habitat modifications, and improved taxonomic analysis. All native frogs are threatened. In freshwater systems, 34 fish and 114 invertebrates are considered either Threatened or At Risk. Major terrestrial invertebrate groups (Coleoptera, Diptera, Hemiptera, Hymenoptera, Leidoptera, Orthoptera and Nematodes), previously unrecognised for conservation purposes, now contribute over 700 taxa to the endangered species list for New Zealand. Frogs, weta and some carnivorous snails now have dedicated recovery programmes.

There have been no reported extinctions in the marine environment around New Zealand, although 54 fish and 284 invertebrates are considered either Threatened or At Risk. Population fluctuations of harvested fish species are apparent but these are now managed within set sustainable limits. Marine mammals, including seals and whales, are slowly recovering from over-exploitation in centuries past, but dolphin species continue to be threatened by coastal net fishing.

Plants At least six vascular plant species have become extinct following human settlement (de Lange *et al*. 2010), comprising a mistletoe found only in northern regions, and five small habitat-specialist herbs. Extinction is much less likely for plants because of the availability of numerous spatial and temporal refuges. Many are currently confined to cliffs or other inaccessible habitats but a growing number of taxa have few populations and appear functionally extinct, lacking significant regeneration in natural ecosystems. Presently 974 taxa are considered Threatened (243) or At Risk (731), representing over 40% of the total native vascular flora; of these 142 are Nationally Critical with small and declining populations. Habitat loss, particularly of naturally rare ecosystems and lowland nutrient-rich environments, is the main local factor currently threatening plant species. Mammalian canopy and ground herbivory, however, is also reducing abundance of palatable species. Equally important, introduced grasses and other weeds tolerant of grazing may effectively eliminate plant regeneration. The conservation of these threatened species *in situ* is very difficult unless the competitors can be suppressed without damaging the native plants. Recently, threatened plants such as *Clianthus puniceus, Hebe armstrongii* and *Leonohebe cupressoides* have become popular as garden plants, which protects the species but not their natural habitat.

Ecosystems New Zealand has begun to assess biodiversity loss at a range of scales based on the need to ensure representation across all major physical environments. The threatened environment classification provides national information on how much indigenous vegetation remains within land environments and its legal protection status, and shows how past vegetation loss and legal protection are distributed across New Zealand's landscape (Walker *et al.* 2006). Currently 157 of New Zealand's 500 land environments are classified in the category of highest threat (Category 1, <10% indigenous cover remaining), and 67% are included in one of five 'threatened environment' categories based on low levels of indigenous cover remaining (<30%) and/or low levels of protection (<20%).

In New Zealand, significant rare and endemic biodiversity is often located in small, relatively uncommon ecosystems. A typology has recently been developed that recognises 72 naturally uncommon ecosystems nationally (Wiser *et al.* 2013), mostly associated with more extreme non-forested environments. Applying the criteria for the IUCN's Ecosystem Red List to New Zealand's naturally uncommon ecosystems has identified 18 naturally uncommon ecosystems being ranked as critically endangered, 17 as endangered, and ten as vulnerable.

A similar hierarchical multivariate approach classifying segments of freshwater rivers and streams has been undertaken to encapsulate both environmental and biological similarity (Leathwick *et al.* 2011) for resource management and conservation planning. There is also a classification of marine environments (Snelder *et al.* 2006) around New Zealand using variables driving spatial variation in major biologic patterns. These new techniques are identifying major inadequacies in our protection network but do provide spatial tools for assessing representation across systems for the first time.

New threats At high latitudes in the southern oceans, climate change predictions for New Zealand appear relatively muted compared with changes forecast for Australia. New Zealand is significantly buffered by the surrounding ocean. Nevertheless, a moderate rise in temperature is predicted and winters have been warming over the past 100 years. Extreme wet/dry episodes and cyclonic events also appear to be increasing. In terrestrial ecosystems nesting dates for some birds, the timing of eel migrations, and increased seeding in *Nothofagus* are thought to be responses to warming conditions although temperature-sensitive ecotones such as upper treeline appear to be stable (McGlone and Walker 2011). The combined effect of warmer winters and increased seeding could pulse mammalian predator numbers and exacerbate impacts on threatened birds and insects. However, climate change impacts on the availability of marine food for seabirds are causing significant declines in albatross, petrels and gulls, although the precise mechanism varies with species. These changes appear unrelated to harvesting or by-catch issues, and most likely reflect temperature changes and movements in ocean currents around New Zealand (Mills *et al.* 2008).

2.7 Conservation

Approaches Conservation in New Zealand has traditionally focussed on threatened birds and forest ecosystems, a perspective which is embodied in the name of our oldest conservation NGO, the Royal Forest and Bird Protection Society. Nearly 50% of New Zealand still supports indigenous vegetation, mostly forest (70%), and approximately

66% of this is under some form of legal protection. However, in recent years New Zealand has also expanded the indigenous grassland protection network to include 15% of that present in 1840, mostly at higher altitudes and often to secure water flows for downstream use.

In the twenty-first century, our biodiversity perspective has continued to broaden to encompass a greater array of species, including invertebrates, lizards and marine mammals, alongside an expanded range of ecosystems, including primary and secondary indigenous grasslands, scrublands and naturally rare ecosystems. Community imagination is being captured increasingly by the possibility of having accessible predator-free areas on the mainland within fenced areas to protect local biodiversity and return those components that have become regionally extinct. The response of the biodiversity to near-zero pest levels has been astounding (Innes *et al.* 2004; Burns *et al.* 2011), and already species are emerging that were previously considered long gone from the area. There are now moves to implement this approach nationally to work towards a predator-free (i.e. carnivorous mammals-free) New Zealand.

Current New Zealand conservation is challenged in many ways. The legislation protecting threatened species and habitats is very weak, and is generally only one consideration among many in planning consent processes. We have become world-leaders at pest control in confined places, initially on islands and more recently in intensively managed areas, often within predator-proof fences, or using toxins and traps. However, we are less successful at dealing with the ongoing attenuated, pest-induced declines of species. The lack of representation of lowland ecosystems, nutrient pollution and water abstraction in many streams and rivers, and over-fishing in marine environments all present formidable problems, especially during a global recession when prosperity is defined solely in economic terms.

More recently, the Department of Conservation has shifted from species to ecosystem management and has developed an increasing focus on prioritisation approaches that identify the most important or vulnerable species or sites for intensive management, in part to match the shrinking budget available for biodiversity protection. Attempts to integrate biodiversity persistence with primary industries are underway but successful outcomes for biodiversity are unlikely because of the understandable tendency for agribusiness to monopolise land and ecosystem services, driven by strong commodity prices. Agricultural intensification is further marginalising indigenous biodiversity that nowadays is restricted to higher elevations and on steeper slopes.

To sustain or gain additional biodiversity benefits, generally via pest control, biodiversity offsets have been introduced as part of environmental compensation packages associated with major developments under the Resource Management Act. It is premature to evaluate their success, but in other countries the biodiversity restoration promised is rarely enduring and frequently falls far short of delivering the goal of no net loss.

Biosecurity legislation and practice in New Zealand is one of the strongest globally, arising from the need to maintain pristine condition with respect to several major animal and crop diseases, and also to protect our indigenous biodiversity. The procedures are not bullet-proof and have been thwarted deliberately and accidentally in recent years but with some notable successes (e.g. eradication of early incursions of the Australian painted apple moth *Orgyia anartoides*) and failures (e.g. Argentine ant *Linepithema humile*).

2.8 Concluding remarks

The New Zealand biota is one of the gems of global biodiversity based on the diversity, endemism and distinctiveness of the species and ecosystems present. We continue to discover new species and are obtaining a greater understanding of how ecosystems function and the services they provide. The indigenous biodiversity of today represents the outcome of millions of years of environmental change, biotic interchange and turnover, and natural selection in a dynamic environment. Although vast tracts of alpine ecosystems and high-rainfall forest remain, many of the larger birds and invertebrates in these systems have long gone, and a pervasive silence extends across the landscape. The future of many iconic species remains uncertain outside of predator-free islands and mainland predator-free sanctuaries, and habitat loss continues unabated. However, there are signs of hope. Some of our largest and most active NGOs are involved in conservation and we now have protected areas and new national parks that extend into areas formerly used for forestry and pastoral farming. Many urban centres are attempting to foster a return of native species and habitats and native plant nurseries now sell thousands of plants annually to support these activities. Areas that are predator-free, though only a small proportion of the total area available, are increasing on islands and the mainland, and environmental concerns about water quality and quantity are increasingly driving the management of our streams and rivers. Near-shore marine reserves are being created, albeit at glacial speed and at micro-proportions, and constraints on by-catch and quotas are being widely applied within New Zealand's exclusive marine economic zone. We are also seeing some fish stocks increasing in response to prudent management.

Biodiversity changes over a human lifespan are frequently barely discernible and yet we are aware of both gains and losses over the 40 years we have been privileged to study native ecosystems in southern New Zealand. In Fiordland, an eerie silence has descended on the forests reflecting declines in native birds, and the lowland and montane landscapes of Otago and Southland have become increasingly monopolised by intensive agriculture, with residual native habitats destroyed in gullies and wetlands that previously persisted. Yet in and around some of our major cities, groups are actively managing patches of native vegetation to promote native animal populations.

However, overall the losses and declines in native biodiversity are considerable and many species and habitats will not persist in the face of current land-use changes and economic drivers. Hopefully this book will contribute towards the growing awareness of the multiple values of what could potentially be lost.

Acknowledgements

We thank Matt McGlone, David Nipperess, and Kendall Clemens for helpful comments on the manuscript, Anne Austin for editorial assistance, Luke Easterbrook for providing the diagram, and Susan Walker for compiling the threatened species information.

REFERENCES

Allen, R. B. and Lee, W. G. (eds.) (2006). *Biological Invasions of New Zealand*. Berlin, Heidelberg, Germany, Springer.

Allen, R. B., Buchanan, P. K., Clinton, P. W. and Cone, A. J. (2000). Composition and diversity of fungi on decaying logs in a New Zealand temperate beech (*Nothofagus*) forest. *Canadian Journal for Forestry Research* **30**, 1025–1033.

Bannister, J. M., Conran, J. G. and Lee, D. E. (2012). Lauraceae from rainforest surrounding an early Miocene maar lake, Otago, southern New Zealand. *Review of Palaeobotany and Palynology* **178**, 13–34.

Barker, G. M. and Mayhill, P. C. (1999). Patterns of diversity and habitat relationships in terrestrial mollusc communities of the Pukeamaru Ecological District, northeastern New Zealand. *Journal of Biogeography* **26**, 215–238.

Bellingham, P., Towns, D., Cameron, E., *et al.* (2010). New Zealand island restoration: seabirds, predators, and the importance of history. *New Zealand Journal of Ecology* **34**, 115–136.

Biffin, E., Hill, R. S. and Lowe, A. J. (2010). Did kauri (*Agathis*: Araucariaceae) really survive the Oligocene drowning of New Zealand? *Systematic Biology* **59**, 594–602.

Bond, W. J., Lee, W. G. and Craine, J. M. (2004). Plant structural defenses against bird browsers: a legacy of New Zealand's extinct moas. *Oikos* **104**, 500–508.

Burns, B., Innes, J. and Day, T. D. (2011). The use and potential of pest-proof fencing for ecosystem restoration and fauna reintroduction in New Zealand. In **Somers, M. J. and Hayward, M. W.** (eds.) *Fencing for Conservation*. New York, Springer-US, pp. 65–90.

Campbell H. J. and Hutching, G. (2007). *In Search of Ancient New Zealand*. Auckland, Penguin; and Lower Hutt, New Zealand, GNS Science. 240 pp.

Carpenter, R. J., Bannister, J. M., Jordan, G. J. and Lee, D. E. (2012). Proteaceae leaf fossils from the Oligo-Miocene of New Zealand: new species and evidence of biome and trait conservatism. *Australian Systematic Botany* **25**, 375–389.

Chapple, D. G., Ritchie, P. A. and Daugherty, C. H. (2009). Origin, diversification, and systematics of the New Zealand skink fauna (Reptilia: Scincidae). *Molecular Phylogenetics and Evolution* **52**, 470–487.

Claton, J. and Edwards, T. (2006). Aquatic plants as environmental indicators of ecological condition in New Zealand lakes. *Hydrobiologia* **570**, 147–151.

Clements, K. (2006). Triplefins. *New Zealand Geographic* **078**, 94–105.

Craine, J. M. and Lee, W. G. (2003). Covariation in leaf and root traits for native and non-native grasses along an altitudinal gradient in New Zealand. *Oecologia* **134**, 471–478.

De Lange, P., Heenan, P., Norton, D., *et al.* (2010). *Threatened Plants of New Zealand*. Christchurch, New Zealand, Canterbury University Press.

Diamond, J. (1984). Distributions of New Zealand birds on real and virtual islands. *New Zealand Journal of Ecology* **7**, 37–55.

Duncan, R. P. and Blackburn, T. M. (2004). Extinction and endemism in the New Zealand avifauna. *Global Ecology and Biogeography* **13**, 509–517.

Duncan, R. P., Blackburn, T. M. and Cassey, P. (2006). Factors affecting the release, establishment and spread of introduced birds in New Zealand. In **Allen, R. B. and Lee, W. G.** (eds.) *Biological Invasions of New Zealand*. Berlin, Heidelberg, Springer, pp. 137–154.

Ecroyd, C. E. (1995). *Dactylanthus* and the bats: the link between two unique endangered NewZealand species and the role of the community in their survival. In

Saunders, D. A., Craig, J. L. and Mattiske, E. M. (eds.) *Nature Conservation 4: The Role of Networks*. Sydney, Australia, Surrey Beatty & Sons, pp. 78–87.

Engel, J. J. and Glenny, D. (2008). *A Flora of the Liverworts and Hornworts of New Zealand Volume 1*. St Louis, MO, Missouri Botanical Garden Press.

Ericson, P. G. P., Christidis, L., Cooper, A. *et al.* (2002). A Gondwanan origin of passerine birds supported by DNA. *Proceedings of the Royal Society B* **269**, 235–241.

Ferguson, D. K., Lee, D. E., Bannister, J. M. *et al.* (2010). The taphonomy of a remarkable leaf bed assemblage from the Late Oligocene–Early Miocene Gore Lignite Measures, southern New Zealand. *International Journal of Coal Geology* **83**: 173–181.

Gibbs, G. W. (2001). Habitats and biogeography of New Zealand's deinacridine and tusked weta species. In **Field, L. H.** (ed.) *The Biology of Wetas, King Crickets and Their Allies*. Wallingford, UK, CABI, pp. 35–56.

Gordon, D. P. (ed.) (2009). *New Zealand Inventory of Biodiversity. Volume One: Kingdom Animalia: Radiata, Lopotrochozoa, Deuterostomia*. Christchurch, New Zealand, Canterbury University Press. 648 pp.

Gordon, D. P. (ed.) (2010). *New Zealand Inventory of Biodiversity. Volume Two: Kingdom Animalia: Chaetognatha, Ecdysozoa, Ichnofossils*. Christchurch, New Zealand, Canterbury University Press. 528 pp.

Gordon, D. P. (ed.) (2012). *New Zealand Inventory of Biodiversity. Volume Three: Kingdoms Bacteria, Protozoa, Chromista, Plantae, Fungi*. Christchurch, New Zealand, Canterbury University Press, 616 pp.

Graham, I. J. (ed.) (2008). A continent on the move: New Zealand geoscience into the 21st century. *Geological Society of New Zealand Miscellaneous Publication 124*. 377 pp.

Heads, M. J. (1997). Regional patterns of biodiversity in New Zealand: one degree grid analysis of plant and animal distributions. *Journal of the Royal Society of New Zealand* **27**, 337–354.

Heenan, P. B. and McGlone, M. S. (2013). Evolution of New Zealand alpine and open-habitat plant species during the late Cenozoic. *New Zealand Journal of Ecology* **37**, 105–113.

Hitchmough, R. (2013). Summary of changes to the conservation status of taxa in 2008–11 New Zealand Threat Classification System listing cycle. *New Zealand Threat Classification Series 1*. Wellington, New Zealand, Department of Conservation.

Holdaway, R. N. (1989). New Zealand's pre-human avifauna and its vulnerability. *New Zealand Journal of Ecology* **12**, 11–25.

Howell, C. J., Kelly, D. and Turnbull, M. H. (2002). Moa ghosts exorcised? New Zealand's divaricate shrubs avoid photoinhibition. *Functional Ecology* **16**, 232–240.

Innes, J., Nugent, G., Prime, K. and Spurr, E. (2004). Responses of kukupa (*Hemiphaga novaeseelandiae*) and other birds to mammal pest control at Motatau, Northland. *New Zealand Journal of Ecology* **28**, 73–81.

Jordan, G. J., Carpenter, R. J., Bannister, J. M., *et al.* (2011). High conifer diversity in Oligo-Miocene New Zealand. *Australian Systematic Botany* **24**, 121–136.

King, C. M. (ed.) (2005). *The Handbook of New Zealand Mammals*. 2nd edn. Melbourne, Oxford University Press.

Koning, J. W. and Jamieson, I. G. (2001). Variation in size of male weaponry in a harem-defence polygynous insect, the mountain stone weta *Hemideina maori* (Orthoptera: Anostostomatidae). *New Zealand Journal of Zoology* **28**, 109–117.

Leathwick, J. R., Snelder, T., Chadderton, W. L., *et al.* (2011). Use of generalised dissimilarity modelling to improve the biological discrimination of river and stream classifications. *Freshwater Biology* **56**, 21–38.

Lee, D. E., Lee, W. G. and Mortimer, N. (2001). Where and why have all the flowers gone? Depletion and turnover in the New Zealand Cenozoic angiosperm flora in relation to palaeogeography and climate. *Australian Journal of Botany* **49**, 341–356.

McGlone, M. S. (1989). The Polynesian settlement of New Zealand in relation to environmental and biotic changes. *New Zealand Journal of Ecology* **12**, 115–129.

McGlone, M. S. (2001). The origin of the indigenous grasslands of southeastern South Island in relation to pre-human woody ecosystems. *New Zealand Journal of Ecology* **25**, 1–15.

McGlone, M. S. and Walker, S. (2011). Potential effects of climate change on New Zealand's terrestrial biodiversity and policy recommendations for mitigation, adaptation and research. *Science for Conservation 312*. Wellington, New Zealand, Department of Conservation.

McIntosh, A. and McDowell, R. (2004). Fish communities in rivers and streams. In Harding, J., Mosley, P., Pearson, C. and Sorrell, B. (eds.) *Freshwaters of New Zealand*. Christchurch, New Zealand, NZ Hydrological & Limnological Societies.

Mills, J. A., Yarrall, J. W., Bradford-Grieve, J. M., *et al.* (2008). The impact of climate fluctuation on food availability and reproductive performance of the planktivorous red-billed gull *Larus novaehollandiae scopulinus*. *Journal of Animal Ecology* **77**, 1129–1142.

Molnar, R. E. and Pole, M. (1997). A Miocene crocodilian from New Zealand. *Alcheringa* **21**, 65–70.

Myers, N., Mittermeier, R. A., Mittermeier, C. G., *et al.* (2000). Biodiversity hotspots for conservation priorities. *Nature* **403**, 853–858.

Nielsen, S. V., Bauer, A. M., Jackman, T. R. *et al.* (2011). New Zealand geckos (Diplodactylidae): cryptic diversity in a post-Gondwanan lineage with trans-Tasman affinities. *Molecular Phylogenetics and Evolution* **59**, 1–22.

O'Donnell, C. F. J., Christie, J. E., Hitchmough, R. A. *et al.* (2010). The conservation status of New Zealand bats, 2009. *New Zealand Journal of Zoology* **37**, 297–311.

Ogden, J., Basher, L. and McGlone, M. S. (1998). Fire, forest regeneration and links with early human habitation: evidence from New Zealand. *Annals of Botany* **81**, 687–696.

Pirie, M., Lloyd, K., Lee, W. G. and Linder, P. (2010). Diversification of *Chionochloa* (Poaceae) and biogeography of the New Zealand Southern Alps. *Journal of Biogeography* **37**, 379–392.

Shaffer, S. A., Tremblay, Y., Weimerskirch, H., *et al.* (2006). Migratory shearwaters integrate oceanic resources across the Pacific Ocean in an endless summer. *Proceedings of the National Academy of Sciences, USA* **103**, 12 799–12 802.

Snelder, T. H., Leathwick, J. R., Dey, K. L., *et al.* (2006). Development of an ecologic marine classification in the New Zealand region. *Environmental Management* **39**, 12–29.

Tennyson, A. and Martinson, P. (2006). *Extinct Birds of New Zealand*. Wellington, New Zealand, Te Papa Press.

Townsend, C. R. and Simon, K. S. (2006). Consequences of brown trout invasion for stream ecosystems. In **Allen, R. B. and Lee, W. G.** (eds.) *Biological Invasions of New Zealand*. Berlin, Heidelberg, Springer, pp. 213–225.

Trewick, S. A., Wallis, G. P. and Morgan-Richards, M. (2011). The invertebrate life of New Zealand: a phylogeographic approach. *Insects* **2**, 297–325.

Wagstaff, S. J., Bayly, M. J., Garnock-Jones, P. J. and Albach, D. C. (2002). Classification, origin, and diversification of the New Zealand Hebes (Scrophulariaceae). *Annals of the Missouri Botanical Garden* **89**, 38–63.

Walker, S. and Wilson, J. B. (2002). Tests for nonequilibrium, instability, and stabilizing processes in semiarid plant communities. *Ecology* **83**, 809–822.

Walker, S., Price, R., Rutledge, D., *et al.* (2006). Continuing indigenous cover loss in New Zealand. *New Zealand Journal of Ecology* **30**, 169–177.

Wardle, P. (1963). Evolution and distribution of the New Zealand flora as affected by Quaternary climates. *New Zealand Journal of Botany* **1**, 3–17.

Wellenreuther, M., Barrett, P. T. and Clements, K. D. (2007). Ecological diversification in habitat use by subtidal triplefin fishes (Tripterygiidae). *Marine Ecology Progress Series* **330**, 235–246.

Williams, P., and Cameron, E. K. (2006). Creating gardens: the diversity and progression of European plant introductions. In **Allen, R. B. and Lee, W. G.** (eds.) *Biological Invasions of New Zealand*. Berlin, Heidelberg, Springer, pp. 33–47.

Wilmshurst, J. M., Anderson, A. J., Higham, T. F. G. and Worthy, T. H. (2008). Dating the late prehistoric dispersal of Polynesians to New Zealand using the commensal Pacific rat. *Proceedings of the National Academy of Sciences USA* **105**, 7676–7680.

Winterbourn, M. (2004). Stream invertebrates. In **Harding, J., Mosley, P., Pearson, C. and Sorrell, B.** (eds.) *Freshwaters of New Zealand*. Christchurch, New Zealand, NZ Hydrological & Limnological Societies.

Wiser, S. K., Buxton, R. P., Clarkson, B. R., *et al.* (2013). New Zealand's naturally uncommon ecosystems. In **Dymond, J. R.** (ed.). *Ecosystem Services in New Zealand – Conditions and Trends*. Lincoln, New Zealand Manaaki Whenua Press.

Worthy, T. H., Tennyson, A. J. D. and Scofield, R. P. (2011). An Early Miocene diversity of parrots (Aves, Strigopidae, Nestorinae) from New Zealand. *Journal of Vertebrate Paleontology* **31**, 1102–1116.

Wright, D. M., Jordan, G. J., Lee, W. G., *et al.* (2010). Do leaves of plants on phosphorus-impoverished soils contain high concentrations of carbon-based defense compounds? *Functional Ecology* **24**, 52–61.

Wright, T. F., Schirtzinger, E. E., Matsumoto, T., *et al.* (2008). A multilocus molecular phylogeny of the parrots (Psittaciformes): support for a Gondwanan origin during the Cretaceous. *Molecular Biology and Evolution* **25**, 2141–2156.

CHAPTER 3

The ecological consequences of habitat loss and fragmentation in New Zealand and Australia

Poppy Lakeman Fraser, Robert M. Ewers and Saul Cunningham

Summary

Extensive loss and subdivision of Austral habitats has shaped the contemporary landscape and caused concomitant impacts on the region's flora and fauna. This chapter draws on the scientific literature to explore the ecological changes brought about by the loss and fragmentation of indigenous habitats in New Zealand and Australia. We explore what it means for a habitat to become fragmented; and investigate how the rate and pattern of this phenomenon sculpts these landscapes. Inevitably, habitat changes on this scale have a pronounced impact on the physical and biotic conditions within remaining fragments. These effects are experienced from the scale of the gene right up to the diversity of species; impacting not only the individual but the interaction between individuals; modifying not only the organism's immediate environment but their ability to disperse between patches; and affecting not only the organisms themselves but the ecological services they provide. Moreover, introduced species confound the impact of land-use change, and play an integral role influencing organisms in fragmented landscapes, so we go on to consider the interactions between these drivers. To conclude, this chapter reviews potential approaches for mitigating the impact of habitat fragmentation in order to conserve New Zealand and Australia's unique biodiversity.

3.1 Introduction

Habitat fragmentation is an umbrella term used to describe a variety of inter-correlated patterns and processes (Ewers and Didham, 2007). It is sometimes considered the 'lesser brother' of habitat loss, though it seldom occurs without destruction of natural habitats.

Austral Ark: The State of Wildlife in Australia and New Zealand, eds. A. Stow, N. Maclean and G. I. Holwell. Published by Cambridge University Press. © Cambridge University Press 2015.

Fragmentation refers to the division of continuous habitat into small disconnected patches that are often altered in shape and isolated from each other by a matrix of dissimilar habitat (Collinge, 1996). Worldwide, habitat fragmentation has become the single largest topic of research in conservation biology (Fazey et al., 2005), and habitat loss has been recognised as the most immediate threat to global biodiversity (Wilson, 1992). Empirical studies identify five key consequences of habitat fragmentation that impact biodiversity: patch area, edge effects, isolation, fragment shape and matrix influences (Ewers and Didham, 2006). These attributes directly and indirectly explain the biota that persists within habitat remnants. Here, we review literature on the ecological consequences of habitat loss and fragmentation that are relevant to New Zealand and Australia. This chapter is by no means an exhaustive review of habitat fragmentation studies in the two countries, but rather it provides an overview of how habitat fragmentation has impacted the ecology of this region.

3.2 Landscape change in New Zealand

Ecosystems in New Zealand and Australia have undergone vast transformation since human colonisation (Leathwick et al., 2004; Department for the Environment Water Heritage and the Arts, 2009). Habitat loss and fragmentation has occurred in both countries, although the time-periods involved are different. Human settlers arrived in Australia 60 000–70 000 years before present (Briscoe and Smith, 2002) altering the composition of flora and fauna. The development of 'fire stick farming' enabling hunting and passage is often considered to have been a leading process by which the environment was altered (Archer et al., 1991). In New Zealand, the arrival of the first people was as recent as about ~730 years ago (Wilmshurst et al., 2008), and again fire was primarily responsible for transforming the land (Kirch, 1982). Land use then changed again with the arrival of European settlers – in Australia in c. 1788 (Martin et al., 2012) and New Zealand in c. 1850 (Department of Conservation, 2000). Clearing the land for cropping and pasture has left much of the primary habitat reduced and divided. In New Zealand, at least three quarters of the natural environment has been significantly modified by human activities (Leathwick et al., 2004) and in Australia, of the ~25% of land hospitable to forests, 40% of the original forests have been lost (Bradshaw, 2012). In both countries this has left much of the remaining indigenous habitat highly fragmented (Bradshaw, 2012) and characterised by small remnant patches isolated by extensive pasture (Smale et al. 2008).

Information on the rates and patterns of land-use change in New Zealand and Australia has been documented across the literature (Ewers et al., 2006; Bradshaw, 2012) and outlined in Chapters 7 and 26. Here we provide an insight into the pattern of habitat loss and fragmentation over the past 1000 years in New Zealand. Deforestation occurred in relatively equal proportions on the two main islands (Ewers et al., 2006), but the fragmentation and spatial arrangement differs between the North and South Islands (Figure 3.1, Plate 24). Forest fragments are on average four times smaller in the North than South Island (61 vs. 244 ha respectively), and forest edge density is almost a third higher (0.93 vs. 0.72 km/km^2 respectively) (Ewers et al., 2006). The greater level of fragmentation on the North Island is partly due to a climate and landform that favoured the development of agriculture, with the associated growth in infrastructure dividing the remaining natural habitats. On the South Island, steep slopes and their unsuitability for

Figure 3.1 (Plate 24) The removal of native vegetation by people can leave remnant patches, often located along rivers or roads or difficult to develop areas such as steep hill sides. These habitat fragments are critical for the survival of native biodiversity, but altered biotic and abiotic conditions provide conservation challenges. A black and white version of this figure will appear in some formats. For the colour version, please refer to the plate section.

farming safeguarded large swathes of land along the Southern Alps, some of which were protected as water catchments in the 1890s (Park, 2000). The relative wealth of forest along the Southern Alps masks an underlying imbalance in the proportion of different forest types remaining on the South Island. While the Gondwanan gymnosperm forests on the West Coast are among the 'most extensive temperate rainforests on Earth' (Conservation International, 2007), forests in the drier and predominantly flat East Coast has been impacted by conversion to intensively cultivated land (Ewers *et al.*, 2006). Deforestation has however slowed considerably since the 1870s, with the nationwide deforestation rate being just 0.01% per year between 1997 and 2002 (Ewers *et al.*, 2006).

3.3 Ecological impacts of habitat loss and fragmentation
3.3.1 Abiotic impacts

First, we explore how the physical fragmentation of habitat alters the abiotic conditions of the environment, before reviewing the known impacts of fragmentation on gene flow, the abundance of individuals, species diversity, species interactions, the movement of species between patches and ecosystem services. We conclude by examining evidence for how habitat loss and fragmentation might interact with introduced species to exert combined impacts on native biodiversity.

Fragmentation can change the physical environment of fragments, with consequences for the organisms that live there. This impact can affect traits as diverse as the germination and early growth of the Kohekohe tree *Dysoxylum spectabile* (Young and Mitchell, 1994), composition of rainforest beetle assemblages (Grimbacher *et al.*, 2006) and the reproductive fitness of long-tailed bats, *Chalinolobus tuberculatus* (Sedgeley and O'Donnell, 2004).

Studies conducted in the podocarp–broadleaf forests of Northland (Young and Mitchell, 1994) and the Atherton Tableland in north-eastern Queensland (Grimbacher *et al.*, 2006) found generally consistent impacts of forest edges on microclimatic conditions. Forest edges are associated with higher air and soil temperatures, more photosynthetically active radiation and higher vapour pressure deficits than the forest interior. The microclimatic differences can be extreme, with light levels and wind speed reduced by 99% and 80% respectively in forest interiors compared with open conditions (Davies-Colley *et al.*, 2000). Microclimatic changes generally penetrate a relatively short distance into the forest, with elevated temperature and light gradients in North Island podocarp–broadleaf forests extending 50 m (Young and Mitchell, 1994). This distance, however, is influenced by the adjacent matrix habitat, with edge effects infiltrating forests five times further when adjacent to pasture rather than pine plantation (Denyer, 2000). Further, a complex effect of matrix type has been demonstrated in temperate forests of south-eastern Australia: areas adjacent to plantation matrix had higher daytime temperatures and vapour pressure deficit (VPD) than forest edges adjacent to agriculture, due to lower wind speeds at plantation edges (Wright *et al.*, 2010). The Wog Wog Habitat Fragmentation experiment set up by Margules (1992) in south eastern New South Wales demonstrated that pine stands can have impacts on the environmental conditions within indigenous *Eucalypt* forest fragments that last long after the plantation has been established (Farmilo *et al.*, 2013), with environmental differences still discernible after more than 20 years.

3.3.2 Biotic impacts

3.3.2.1 Abundance and population size

Habitat fragmentation can increase the abundance of some species, and drive others to low levels, or extinction. Amongst the five fragmentation processes, clear examples of edge effects (i.e. differences between the boundaries of two habitat types) and habitat quality (i.e. the value of an environment for organisms residing within) have been documented. Smaller and irregular-shaped patches are more influenced by edge effects, which can have positive, negative or neutral impacts on species abundance (Ries *et al.*, 2004). In New Zealand beetle communities, almost 90% of species had abundances that were significantly impacted by the presence of forest–grassland edges in the Lewis Pass region of the Southern Alps. Here, edge effects typically penetrated 20–250 m inside forest fragments, although one in eight of the common species had edge effects that extended at least 1 km into the interior of temperate beech (*Nothofagus* spp.) forest (Ewers and Didham, 2008). These effects do not, however, always result in reduced population sizes. For example, many indigenous species of beetle exhibited elevated abundance at the edge of Kahikatea–pasture boundaries in Waikato (Harris and Burns, 2000).

The quality of habitat in fragments and the surrounding matrix can influence species abundances. Much of the remnant vegetation in the upper south-east region of South Australia is confined to linear strips bordering transport infrastructure where vegetation can be highly degraded. In this instance, as habitat quality declined so did the abundance of the most commonly caught small mammal species, the introduced *Mus musculus* and the native *Cercartetus concinnus* (Carthew *et al.*, 2013). Although this indicates the value of these corridor habitats for certain species (particularly invasive species), the overall richness of mammals and reptiles was significantly higher in remnant fragments not bordering roads than at more isolated fragments along roadsides. Moreover, in north-east Queensland, Laurance (1991) found that the abundance of species in the matrix habitat surrounding forest remnants was a good predictor of vulnerability to habitat fragmentation. Those species of non-flying mammals which were abundant in the pasture matrix (such as the carnivorous marsupial *Antechinus flavipes* and the murid rodent *Hydromys chrysogaster*) maintained stable populations in fragmented landscapes, whereas those that avoided this vegetation type were reduced in abundance.

One approach to estimating the potential impact of habitat loss on population viability is to gauge the resources available to the organism. Although this method is not routinely undertaken, it is possible to estimate the number of breeding individuals required for a population to be self-supporting (Frankel and Soulé, 1981) and the subsequent amount of habitat needed to support these numbers (McLennan *et al.*, 1987). For example, McLennan *et al.* (1987) found that the North Island brown kiwi (*Apteryx australis mantelli*) in the Hawke's Bay region had an average range size of 30 ha and therefore there is a requirement for reserves to have 7500–15 000 ha of continuous forest to support a minimum viable population (calculated by Frankel and Soulé (1981) to be 500–1000 breeding individuals). Computer simulations of brown kiwi dispersal suggest similar minimum areas (10 000 ha). These figures are based on forested areas which are not only protected from further degradation, but have active predator management (Basse and McLennan, 2003), as kiwi chick survival is only 5% in areas without control of predators such as the stoat *Mustela erminea* (Basse *et al.*, 1999). In Hawke's Bay there are no large continuous patches of remnant vegetation (most forest patches are < 100 ha). Although this raises concern for kiwi populations, kiwis have been observed venturing over farmland outside the Paerata reserve on the North Island (Potter, 1990), suggesting that they may have the behaviour required to maintain population connectivity. This suggests that matrix habitats such as mixed native forest, scrub and plantation could be included in estimating areas suitable for self-sustaining kiwi populations.

3.3.2.2   Diversity and community composition

The Species Area Relationship (SAR) is commonly used to predict the number of species which a certain area of habitat can support and predicting species loss in diminished and isolated habitat patches (Tikkanen *et al.*, 2009). SARs are based on the concept that larger areas of habitat support more species, and many studies in New Zealand and Australia have detected this pattern: for plants (e.g. Williams, 1982), invertebrates (e.g. Derraik, 2009) and birds (e.g. East *et al.*, 1984). As such, it may appear that conserving large patches in fragmented landscapes may be preferable to conserving small patches. However, it is important to consider the exceptions that exist for these generalisations; for example, species that disperse well across the matrix habitat can effectively aggregate the resources available from multiple patches that are widely distributed across a landscape. This may suggest that small patches *can* support a diversity of species (particularly those species which are adapted to edgy environments) and can be extremely valuable for maintaining viable metapopulations (Lindenmayer and Fischer, 2006).

A number of studies have reported increased species diversity of plants and invertebrates at habitat edges (e.g. Ewers *et al.*, 2007). Edges may support higher diversity for a wide variety of reasons. These may include: the prevalence of pioneering species able to cope with these dynamic environments (Laurance *et al.*, 2001); an increase in generalist species that can use many environments (Webb, 1989); spill-over of species from adjacent habitats (Duelli *et al.*, 1990); or the presence of enhanced food availability such as rich invertebrate communities supporting rich bird communities (Helle and Muona, 1985).

Species exhibit a wide array of responses to the same changes in habitat that patterns of fragmentation bring about, and this variation in response may be due to species traits. For example, changes in the relative abundance of different feeding groups of beetle communities across forest edges has been reported; with detritivores, fungivores and saprophages having increased abundance in forest interiors, whereas herbivores were more prevalent in the surrounding pasture (Ewers and Didham, 2008). For honeyeater communities in mallee woodland remnants, those species which tolerated open disturbed habitats flourished and those which were woodland specialists declined with increasing fragmentation (Elliott *et al.*, 2012). Similarly, beetle species that were either rare, were specialists to native *Eucalypt* forest, or had higher trophic levels were more likely to decline in abundance in small isolated patches (Collinge, 2009).

3.3.2.3 Species interactions

A number of studies have found an altered frequency of species interactions at habitat edges. For example, the red mistletoe, *Peraxilla tetrapetala*, had higher fruit set at forest edges than in the forest interior as a result of higher bee visitation (Burgess *et al.*, 2006) and bird pollination (Montgomery *et al.*, 2003) (See Box 3.1 for more information on the impacts of fragmentation on pollination). Leaf herbivory by the common brushtail possum (*Trichosurus vulpecula*) on mistletoes was also significantly greater at edges, but this negative effect was partially offset by reduced flower predation by the native moth *Zelleria maculata* at edges (Bach and Kelly, 2004). Numbers of the box mistletoe, *Amyema miquelii*, in Victoria were enhanced in agricultural areas (MacRaild *et al.*, 2010), but once wooded vegetation fell below 15% of the total landscape the occurrence of this species declined rapidly (MacRaild *et al.*, 2010). This threshold effect was likely to be a consequence of the reduction in the woodland-dependent mistletoe bird (*Dicaeum hirundinaceum*), which is the primary seed disperser for *A. miquelii*.

Box 3.1 Pollination in fragmented landscapes: insights from the Austral Ark

Pollination is an intrinsically *spatial* process, requiring the movement of pollen from male to female flower parts. Although short-distance pollen movements (within individual plants or between near neighbours) are inevitably the most frequent, most plants have genetically controlled mechanisms to encourage outcrossing that mean pollen grains that have travelled over longer distances, and are therefore less likely to be closely related, are more likely to achieve fertilisation. These important long-distance dispersal events can occur over metres or kilometres. At these larger scales we expect habitat fragmentation to impact pollination. Fragmentation changes the density and arrangement of plants and therefore the spatial relationships among prospective mates. Most plants require an animal pollinator to disperse their pollen, and habitat fragmentation might also affect the abundance and behaviour of these pollinators.

These effects on pollinators can interact with effects on plant spatial arrangement leading to a wide spectrum of possible outcomes.

Extensive land clearing for agriculture has created extreme habitat fragmentation in parts of Australia, especially in the wheat–sheep belt (Hoekstra *et al.*, 2005), and research in these landscapes has been revealing. Studies commonly show reduced seed set in small remnant populations of plants (e.g. Lamont *et al.*, 1993). For some species we know that this depressed reproduction is caused by a shortfall in transfer of conspecific pollen grains (Cunningham, 2000a). For the bird-pollinated *Eremophila glabra* and the bee-pollinated *Dianella revoluta*, it seems that depressed pollination is not a result of insufficient pollinators or visits, but rather that pollinators in more fragmented patches deliver fewer suitable outcross pollen grains because the fragments have a lower effective pollen donor density (Cunningham 2000a,b; Duncan *et al.*, 2004; Bianchi and Cunningham, 2012).This phenomenon is to be expected in systems where pollinator movements are restricted to largely within-patch, but this does not seem to be the case for bird pollinators in this landscape. Honeyeaters fly many kilometres, can be seen flying in and out of patches frequently, and provide genetic connectivity over distances of 5 km and more (Byrne *et al.*, 2007; Llorens *et al.*, 2012; Elliott, 2009). Nevertheless, for *Eremophila glabra* this long-distance genetic connectivity, while helping to maintain seed production in fragments, does not prevent depression in the rate of seed set.

Plants with more specialised pollination relationships are vulnerable to loss of key pollinators, as appears to have been the case for *Freycinetia baueriana* in New Zealand (Lord, 1991), but most plants are open to a diversity of pollinators. Changes in the composition of pollinator communities will expose plants to different pollinator behaviours and changes in frequency of visits. For example, the Noisy Miner is an edge specialist that aggressively excludes other honeyeaters from the patches they occupy (Dow, 1977; Major *et al.*, 2001), with large impacts on the abundance of other bird species. These and other changes mean that in the mallee woodland fragments of central NSW isolated patches have more large-bodied honeyeaters (Elliott *et al.*, 2012). Native bee communities also respond to landscape-scale patchiness, such that while abundance of native bees is highest in flowering crops (such as canola) species richness is increased by the presence of woodland remnants and pasture trees within 100 m of the sampling point (Lentini *et al.*, 2012).

Habitat fragmentation studies have traditionally focussed on the conservation values of remnant patches, but the ecosystem services perspective draws attention to the impact of remnant-dependent pollinator fauna on crops in the agricultural land in the matrix of fragmented landscapes. In Australia's tropical Atherton tableland a number of horticultural crops show higher fruit set when insect pollinated. For Macadamia and Longan crops the insect pollination benefit to fruit set was greater for orchards nearer to rainforest habitat (Blanche *et al.*, 2006). Similarly, the beetle pollinators of Atemoya were more abundant and diverse in orchards nearer rainforest (Blanche and Cunningham, 2005). These studies are consistent with worldwide patterns of greater pollinator abundance and diversity in crops nearer to remnant vegetation (Ricketts *et al.*, 2008) and consequently better pollination outcomes

Box 3.1 (continued)

(Garibaldi *et al.*, 2011). Much of the benefit is associated with the diversity of native pollinators that use remnant patches (Garibaldi *et al.*, 2013), but exotic species can also be important crop pollinators (Rader *et al.*, 2012) and benefit from the habitats provided by native remnants in a heavily cleared landscape (Arthur *et al.*, 2010).

Introduced pollinators occur in fragmented landscapes worldwide, but they may be especially influential in Australia and New Zealand, interacting with crops and native species (Arthur *et al.*, 2010, Gross, 2001). Because of the region's distinctive biogeographic history much of the landscape has no native eusocial bees, and is missing two families that are important in much of the rest of the world (Andrenidae, Mellitdae), but instead has a high diversity of Colletid bees (Batley and Hogendoorn, 2009). Against this background invasive social bees have established including *Apis mellifera* and (in New Zealand and Tasmania) *Bombus terrestris*. The mild climate and abundance of nectar-rich plants (especially in the important families Myrtaceae and Proteacae) support strong populations of feral *Apis mellifera* (Oldroyd *et al.*, 1997). In Australia (though not New Zealand) these populations remain unaffected by the global *Varroa* mite epidemic (Cook *et al.*, 2007). Also distinctive is the richness of nectar-feeding vertebrates including two distinctive and widespread clades of nectar-feeding birds (Honeyeaters and Lorikeets), flower-visiting bats (Megachiroptera), and the world's only non-flying mammal flower specialist (*Tarsipes*: Wiens *et al.*, 1979). It may be that there is an evolutionary association between the abundance of these large-bodied nectar feeders and the abundance and diversity of plants with large nectar-rich flowers (Orians and Milewski, 2007). These distinctive features of the region's biogeographic history leave a heritage of plants and animals that are likely to show some distinctive patterns of response to the global phenomenon of habitat fragmentation.

Elevated predation rates at forest edges have been shown in many studies outside New Zealand and Australia, with most studies focussed on nest predation rates (e.g. Ries and Fagan, 2003). In the New Zealand context, Whyte *et al.* (2005) found that rates of attack on artificial wax eggs were highly variable across edge gradients and found no strong evidence for increased predation rates near habitat edges. In a similar experiment conducted in an Australian marine environment, a study by Smith *et al.* (2011) demonstrated a strong impact of edge effects on predation rates in seagrass communities. Potential prey species (such as King George whiting, *Sillaginodes punctate*) are found more commonly in the interior of grass patches whereas predatory fish (such as Australian salmon, *Arripis* sp.) frequented the adjacent sandy habitats. When King George whiting was tethered at positions across the edge gradient, more were removed by the salmon from the outer edges of the patch than the interior.

Isolation of habitat fragments can further disrupt species interactions in fragmented landscapes. An experimental manipulation of potted *Sporodanthus ferrugineus* (Restionaceae) plants in the Hauraki plains found that colonisation and damage rates by the Lepidopteran herbivore, *Batrachedra* spp. was reduced with increasing isolation of the plants (Watts and Didham, 2006). A moderate degree of isolation

(400 m) caused a near complete breakdown of the plant–insect interaction, although the interaction on restored 'plant islands' recovered within 196 and 308 weeks, showing that habitat restoration can reinstate interactions within a moderate time frame (Watts and Didham, 2006).

Fragmentation can also impact competitive interactions, as demonstrated by Braschler and Baur (2005) who investigated how grassland fragmentation alters competitive interactions in ant communities. Testing the hypothesis that generalist species are behaviourally dominant and are less affected by habitat fragmentation than specialists, these authors found that increased density of the generalist species, *Lasius paralienus* caused species richness and forager density to decline in the other species in habitat fragments but not in control plots. An increase in natural sugar resources was found in fragments suggesting that interspecific competition for sugar resources reduces with fragmentation (Braschler and Baur, 2005).

3.3.2.4 Dispersal patterns

Dispersal is essential for biota to persist in fragmented landscapes, as it supports gene flow (see Box 3.2) and the maintenance of healthy population sizes. The loss of forest cover and the spatial distribution of forest remnants are known to impact the dispersal patterns of several New Zealand and Australian species in different ways depending upon the capabilities of the species. For example, juvenile North Island robins (*Petroica long-pipes*) show a marked reluctance to fly over pasture. Most juvenile robins disperse among patches separated by less than 20 m (Wittern and Berggren, 2007), and were not observed crossing forest gaps greater than 110 m (Richard and Armstrong, 2010). By contrast, the brown kiwi preferentially followed forest corridors or 'stepping stones' of remnant forest fragments to reach destinations up to 1200 m from their original location (Potter, 1990). There are other species, such as the New Zealand pigeon, which are even stronger dispersers and are known to have flown more than 480 km in a 100 day period (Powlesland *et al.*, 2011). This strong flight ability suggests that this species is unlikely to be impacted by the isolation of forest patches.

Box 3.2 Genes in fragmented landscapes: insights from New Zealand case studies

Habitat loss leads to direct reductions in the size of populations, and fragmentation further isolates and subdivides those populations. A reduction in both genetic diversity and gene flow caused by fragmentation can lead to inbreeding and inbreeding depression. Inbreeding depression refers to a reduction in the reproductive ability of individuals and contributes to a higher risk of extinction (Keller and Waller, 2002; Jamieson *et al.*, 2008). Furthermore, reduced genetic diversity may limit the potential of populations to be resilient to new challenges (Vandergast *et al.*, 2007), such as changing climates or species invasions (Jamieson *et al.*, 2008).

There are relatively few studies from New Zealand that have shown changes in genetic diversity that are unequivocally caused by habitat loss and fragmentation. We do know that small populations of species such as the North Island kokako (*Callaeas cinerea wilsoni*) in Rotoehu forest and Mapara Wildlife Reserve had lower levels of

Box 3.2 (continued)

microsatellite variability than larger populations in the nearby Te Urewera National Park (Hudson *et al.*, 2000). However, although the initial decrease in kokako distributions was recognised as being caused by forest clearance, declines have also been recorded in extant forests (Rasch, 1992). The recent losses are thought to be due to the spread of introduced mammals over the past century (Rasch, 1992; Innes *et al.*, 1999).

Impacts of habitat loss and fragmentation have been detected in Grand skinks (*Oligosoma grande*), a species that has a naturally fragmented distribution restricted to rock outcrops. Berry and Gleeson (2005) presented mitochondrial and microsatellite DNA data that suggested there was probably always a strong genetic pattern of isolation-by-distance within this species that has very limited dispersal, but that this isolation has been exacerbated by habitat loss in the region (Berry and Gleeson, 2005; Berry *et al.*, 2005). At fine spatial scales, the loss of native habitat between rock outcrops has reduced dispersal rates (Berry *et al.*, 2005). This species exhibited no inbreeding avoidance in small populations (18% of matings occurred between full siblings), yet Berry (2006) found that this had no effect on the first-year survival of offspring and as such fragmented populations of this species may remain viable at low numbers.

For plants, genetic effects of habitat fragmentation can be mediated by changes to the pollinator community. This hypothesis was tested by Schmidt-Adam *et al.* (2000) who compared outcrossing rates in isolated stands of Pohutukawa (*Metrosideros excelsa*) on the mainland with those on islands. Mainland pollinator communities were dominated by introduced species whereas island pollinators were predominantly endemic species, but this difference did not result in any clear changes to outcrossing rates, which were very low in all populations (Schmidt-Adam *et al.*, 2000). In all populations the low level of outcrossing resulted in large numbers of inbred seedlings, but levels of gene fixation in the adults were much lower than in seedlings suggesting there is strong selection against the survival and reproduction of homozygote individuals (Schmidt-Adam *et al.*, 2000).

Reduced dispersal can result in micro-evolutionary changes to the genetics of isolated populations and have been demonstrated for a wide variety of taxa. For example, the ringtail possum (*Pseudocheirus peregrinus*) exhibited lower herterozygosity and allelic richness in small isolated patches than possums in larger patches, suggesting that the *Pinus radiata* plantation matrix surrounding native habitat fragments may have restricted the dispersal of this species more than had been previously suspected (Lancaster *et al.*, 2011). A similar result was found for Cunningham's skink (*Egernia cunninghami*), where patterns of pairwise relatedness demonstrated that both males and females of the rock-dwelling lizard had reduced dispersal in a deforested site than in a naturally vegetated reserve (Stow *et al.*, 2001). Finally, the very slow movement rates of snails can produce genetic differentiation across very small distances, such as the aquatic snail *Fonscochlea accepta*. Populations of *F. accepta* in central Australia separated by < 300 m had high genetic similarity that decreased rapidly with increasing distance (Worthington Wilmer *et al.*, 2008). The mode of dispersal used by an organism interacts with landscape structure to determine the rates at which genetic similarity declines across space. In the case of *F. accepta*, two dispersal modes were identified, with short-range dispersal of <300 m among springs by the movements of individual snails complemented by long-range dispersal using animal vectors crossing distances >3 km (Worthington Wilmer *et al.*, 2008).

3.4 Synergistic interactions between habitat fragmentation and introduced species

Understanding interactions between multiple drivers of global environmental change (GEC) – i.e. land-use change, climate change, nitrogen deposition and biotic exchange – is paramount if we are to accurately predict future anthropogenic-driven changes in biodiversity (Tylianakis *et al.*, 2008). Land-use change, including habitat loss and fragmentation, has been identified as being the greatest threat to biodiversity in the coming century (Pereira *et al.*, 2010), but an increasing body of research is illuminating the combined actions of multiple GEC drivers interacting to magnify the impacts of fragmentation (Ewers and Didham, 2006).

Biotic exchange, i.e. the introduction of exotic species, is widely considered to be the greatest contemporary threat to biodiversity in New Zealand (Department of Conservation, 2000) and, with habitat fragmentation, is jointly the most cited threat in Australia (Department for the Environment Water Heritage and the Arts, 2009). Didham *et al.* (2005) suggested that the prevalence of exotic species can be an indirect effect of habitat modification. For example, increasingly modified matrix habitats in the tussock grasslands of Banks Peninsula are associated with increased invasion of exotic grass species into native outcrop habitats (Wiser and Buxton, 2008), and habitat edges represent the focal point for weed invasion into forests (Wiser *et al.*, 1998). The invasion of weeds can alter the frequency of feeding interactions between herbivores and their food plants following a scrub invasion of grassland (Hobbs and Mooney, 1986), can indirectly alter the food avaiability for wetland bird species (Braithwaite *et al.*, 1989) and can affect the resource availability for native plant species when an exotic species creates a surrounding deficit of nutrients or moisture (Boswell and Espie, 1998). It has been shown, however, that lowland forest areas simultaneously impacted by fragmentation, introduced mammalian predators and livestock grazing can structurally recover when fencing and pest control management is applied (Dodd *et al.*, 2011).

Although most weeds are unable to invade intact forest (Craig *et al.*, 2000), habitat fragmentation can promote weed invasion in forest patches and lead to strong impacts on native species. Twenty-seven thousand invasive weed species have been introduced into Australia and the Assessment of Australia's Terrestrial Biodiversity report suggest that 'invasive species tend to be abundant and well established in highly modified landscapes' (National Land and Water Resource Audit, 2007). Small patches, and patches with high perimeter-to-edge ratios, are particularly vulnerable to weed invasion as demonstrated in Auckland reserves (Timmins and Williams, 1991), an alluvial forest remnant in the Bay of Plenty (Smale, 1984) and an urban remnant of alluvial forest in Hamilton (Whaley *et al.*, 1997). The South American invasive weed *Tradescantia fluminensis* is capable of smothering emerging native seedlings where it is found in fragmented patches in New Zealand (Timmins and Williams, 1991; Whaley *et al.*, 1997). It grows most vigorously in areas with high light levels and low canopy cover such as along the edges of native forest fragments. *T. fluminensis* radically alters native forest ecosystems by increasing litter decomposition rates (e.g. Standish *et al.*, 2004), is associated with reductions in the survival rates, abundance and species richness of native forest seedlings (Standish *et al.*, 2001), and alters the composition, and reduces the diversity, of invertebrate communities (e.g. Standish, 2004). In the northern coastal regions of Australia the introduced shrub *Mimosa pigra* has caused a decline in native herb species richness and led to populations of bird and reptile species falling in the affected areas (Braithwaite *et al.*, 1989).

In addition to weeds, more than 80 species of alien animal species have established in New Zealand (King, 1990), a number of which have known associations with habitat loss and fragmentation. Invasive Argentinian ants, for example, penetrate indigenous habitats up to 20 m into indigenous forests, 30 m into mangroves and 60 m into scrub (Ward and Harris, 2005), and the red fire ant (*Solenopsis geminate*) was widespread in isolated habitat patches in urban and agricultural areas in Queensland and the Northern Territory (Woodman *et al.*, 2008). Derraik (2009) found that, as fragment size decreased, the abundance of the native mosquito species (*Culex pervigilans)* steadily reduced as abundance of the exotic mosquito (*Aedes notoscriptus)* increased. Predatory mammals such as ferrets (*Mustela furo*), have been found to frequently use forest–pasture edges (Ragg and Moller, 2000), although this pattern did not translate into an increase in the predation rates of bird nests along forest edges (Whyte *et al.*, 2005).

Introduced species have the potential to confound investigations into the role of habitat fragmentation in determining biodiversity patterns in the present day in both New Zealand and Australia. For example, the spread of rabbits (*Oryctolagus cunicultus*) into Western Australia from the point of introduction in eastern Australia early in the nineteenth century (Stodart and Parer, 1988; and see Chapter 10) coincided with the beginning of native vegetation clearance to make space for agriculture (Hobbs, 2001). Consequently, Hobbs (2001) suggested that research which investigates impacts of habitat fragmentation alone, without considering habitat condition modified by invasive species, is 'unlikely to yield meaningful results'. In New Zealand, Brooks *et al.* (2002) suggested that SAR extinction models based on habitat loss underestimate the number of bird extinctions in New Zealand, and correlations with introduced species strongly suggest that invasive mammalian predators may explain the relatively high rates of avifaunal extinction (Blackburn, 2005). However, further analyses showed that invasion rates are closely correlated with rates of habitat modification across the Pacific Islands (Didham *et al.*, 2005), suggesting it is not possible to absolutely attribute extinctions to either driver alone.

3.5 Mitigating the effects of habitat fragmentation

The loss and fragmentation of habitats has had pervasive impacts on biodiversity in New Zealand and Australia, but the impacts themselves point towards potential methods to reduce the ecological damage arising from habitat fragmentation. For example, a potential mechanism to enhance the quality of habitat fragments is to surround them with a buffer zone of matrix habitats that are similar in structure to the habitat fragments themselves (Lindenmayer and Fischer, 2006). There is emerging evidence that corridors enhance the persistence of species in fragmented landscapes, and these may be created in future in 'greenway' concepts that use linear terrain features, either natural (ridge contours or rivers) or artificial (railways, canals and roads), to link patches of habitat to improve connectivity within landscapes. Such hands-on initiatives tackle directly the two greatest challenges for maintaining diversity in fragmented landscapes: protecting and enhancing the ecological quality of the remnant habitats, and promoting successful dispersal among those remnants.

Acknowledgements

We would like to acknowledge the Grantham Institute for Climate Change and the Gilchrist Educational Trust for funding this research.

REFERENCES

Archer, M., Hand, S.J., Godthelp, H. (1991) *Riversleigh: The Story of Animals in Ancient Rainforests of Inland Australia*. Reed Books, Sydney.

Arthur, A.D., Li, J., Henry, S., Cunningham, S.A. (2010) Influence of woody vegetation on pollinator densities in oilseed Brassica fields in an Australian temperate landscape. *Basic and Applied Ecology* **11**: 406–414.

Bach, C.E., Kelly, D. (2004) Effects of forest edges on herbivory in a New Zealand mistletoe, *Alepis flavida*. *New Zealand Journal of Ecology* **28**:195–205.

Basse, B., McLennan, J.A. (2003) Protected areas for kiwi in mainland forests of New Zealand: how large should they be? *New Zealand Journal of Ecology* **27**:95–105.

Basse, B., McLennan, J.A., Wake, G.C. (1999) Analysis of the impact of stoats, *Mustela erminea*, on northern brown kiwi, Apteryx mantelli, in New Zealand. *Wildlife Research* **26**:227–237.

Batley, M., Hogendoorn, K. (2009) Diversity and conservation status of native Australian bees. *Apidologie* **40**: 347–354.

Berry, O.F. (2006) Inbreeding and promiscuity in the endangered grand skink. *Conservation Genetics* **7**: 427–437.

Berry, O., Gleeson, D.M. (2005) Distinguishing historical fragmentation from a recent population decline – shrinking or pre-shrunk skink from New Zealand? *Biological Conservation* **123**: 197–210.

Berry, O., Tocher, M.D., Gleeson, D.M., *et al.*, (2005) Effect of vegetation matrix on animal dispersal: genetic evidence from a study of endangered skinks. *Conservation Biology* **19**: 855–864.

Bianchi, F.J.J.A., Cunningham, S.A. (2012) Unravelling the role of mate density and sex ratio in competition for pollen. *Oikos* **121**: 219–227.

Blackburn, T.M. (2005) Response to comment on "Avian extinction and mammalian introductions on oceanic islands." *Science* **307**:1412.

Blanche, R., Cunningham, S.A. (2005) Rain forest provides pollinating beetles for atemoya crops. *Journal of Economic Entomology* **98**: 1193–1201.

Blanche, K.R., Ludwig, J.A., Cunningham, S.A. (2006) Proximity to rainforest enhances pollination and fruit set in orchards. *Journal of Applied Ecology* **43**: 1182–1187.

Boswell, C. C., Espie, P. R. (1998) Uptake of moisture and nutrients by *Hieracium pilosella* and effects on soil in a dry sub-humid grassland. *New Zealand Journal of Agricultural Research* **41**:251–261.

Bradshaw, C. J. A. (2012) Little left to lose: deforestation and forest degradation in Australia since European colonization. *Journal of Plant Ecology* **5**:109–120.

Braithwaite, R. W., Lonsdale, W. M., Estbergs, J. A. (1989) Alien vegetation and native biota in tropical Australia: the impact of *Mimosa pigra*. *Biological Conservation* **48**:189–210.

Braschler, B., Baur, B. (2005) Experimental small-scale grassland fragmentation alters competitive interactions among ant species. *Oecologia* **143**:291–300.

Briscoe, G., Smith, L. (2002) The aboriginal population revisited: 70,000 years to the present. In: *Aboriginal History Monograph No. 10*. Canberra:ANU.

Brooks, T. M., Mittermeier, R. A., Mittermeier, C. G., *et al*. (2002) Habitat loss and extinction in the hotspots of biodiversity. *Conservation Biology* **16**:909–923.

Burgess, V. J., Kelly, D., Robertson, A. W., *et al*. (2006) Positive effects of forest edges on plant reproduction : literature review and a case study of bee visitation to flowers of *Peraxilla tetrapetala* (Loranthaceae). *New Zealand Journal of Ecology* **30**:179–190.

Byrne, M., Elliott, C. P., Yates, C., Coates, D. J. (2007) Extensive pollen dispersal in a bird-pollinated shrub, *Calothamnus quadrifidus*, in a fragmented landscape. *Molecular Ecology* **16**: 1303–1314.

Carthew, S. M., Garrett, L. A., Ruykys, L. (2013) Roadside vegetation can provide valuable habitat for small, terrestrial fauna in South Australia. *Biodiversity and Conservation* **22**:737–754.

Collinge, S. K. (1996) Ecological consequences of habitat fragmentation: implications for landscape architecture and planning. *Landscape and Urban Planning* **36**:59–77.

Collinge, S. K. (2009) *Ecology of Fragmented Landscapes*. The John Hopkins University Press, Baltimore.

Conservation International (2007) *Biodiversity Hotspots: New Zealand* (T. Brooks, N. De Silva, M. Foster, *et al*., eds.).

Cook, D. C., Thomas, M. B., Cunningham, S. A., Anderson, D. L., DeBarro, P. J. (2007) Predicting the economic impact of an invasive species on an ecosystem service. *Ecological Applications* **17**: 1832–1840.

Craig, J., Anderson, S., Clout, M., *et al*. (2000) Conservation issues in New Zealand. *Annual Review of Ecology and Systematics* **31**:61–78.

Cunningham, S. A. (2000a) Depressed pollination in habitat fragments causes low fruit set. *Proceedings of the Royal Society, Series B* **267**: 1149–1152.

Cunningham, S. A. (2000b) Effect of habitat fragmentation on the reproductive ecology of four mallee woodland species. *Conservation Biology* **14**: 758–768.

Davies-Colley, R. J., Payne, G. W., van Elswijk, M., *et al*. (2000) Microclimate gradients across a forest edge. *New Zealand Journal of Ecology* **24**:111–121.

Denyer, K. (2000) Maintaining biodiversity in a production matrix: the effects of adjacent landscape on indigenous forest fragments in the Waikato region. MSc:233.

Department for the Environment Water Heritage and the Arts (2009) Chapter 5 Threats to Australian biodiversity. In: Biotext Pty Ltd and Department of the Environment, Water, H. (ed.) *Assessment of Australia's Terrestrial Biodiversity 2008*, Report prepared by the Biodiversity Assessment Working Group of the National Land and Water Resources Audit for the Australian Government, Canberra, pp. 149–212.

Department of Conservation (2000) *The New Zealand Biodiversity Strategy*. Wellington, N.Z.

Derraik, J. G. B. (2009) Association between habitat size, brushtail possum density, and the mosquito fauna of native forests in the Auckland region, New Zealand. *Ecohealth* 6:229–238.

Didham, R. K., Tylianakis, J. M., Hutchison, M. A., *et al.* (2005) Are invasive species the drivers of ecological change? *Trends in Ecology & Evolution* 20:470–474.

Dodd, M., Barker, G., Burns, B., *et al.* (2011) Resilience of New Zealand indigenous forest fragments to impacts of livestock and pest mammals. *New Zealand Journal of Ecology* 35:83–95.

Dow, D. D. (1977) Indiscriminate interspecific aggression leading to almost sole occupancy of space by a single species of bird. *Emu* **77**: 115–112.

Duelli, P., Studer, M., Marchand, I., *et al.* (1990) Population movements of arthropods between natural and cultivated areas. *Biological Conservation* 54:193–207.

Duncan, D. H., Nicotra, A. B., Wood, J. T., Cunningham, S. A. (2004) Plant isolation reduces outcross pollen receipt in a partially self-compatible herb. *Journal of Ecology* **92**: 977–985.

East, R., Williams, G. R., Zealand, N. (1984) Island biogeography and the conservation of New Zealand's indigenous forest dwelling avifauna. *New Zealand Journal of Ecology* 7:27–35.

Elliott, C. P. (2009) Patterns and processes: ecology and genetic function of fragmented Emu Bush *(Eremophila glabra* ssp. *glabra)* populations. PhD Thesis, Austalian National University.

Elliott, C. P., Lindenmayer, D. B., Cunningham, S. A., *et al.* (2012) Landscape context affects honeyeater communities and their foraging behaviour in Australia: implications for plant pollination. *Landscape Ecology* 27:393–404.

Ewers, R. M., Didham, R. K. (2006) Confounding factors in the detection of species responses to habitat fragmentation. *Biological Reviews* 81:117–142.

Ewers, R. M., Didham, R. K. (2007) Habitat fragmentation: panchreston or paradigm? *Trends in Ecology & Evolution* 22:511.

Ewers, R. M., Didham, R. K. (2008) Pervasive impact of large-scale edge effects on a beetle community. *Proceedings of the National Academy of Sciences of the United States of America* 105:5426–5429.

Ewers, R. M., Kliskey, A. D., Walker, S., *et al.* (2006) Past and future trajectories of forest loss in New Zealand. *Biological Conservation* 133:312–325.

Ewers, R. M., Thorpe, S., Didham, R. K. (2007) Synergistic interactions between edge and area effects in a heavily fragmented landscape. *Ecology* 88:96–106.

Farmilo, B. J., Nimmo, D. G., Morgan, J. W. (2013) Pine plantations modify local conditions in forest fragments in southeastern Australia: insights from a fragmentation experiment. *Forest Ecology and Management* 305:264–272.

Fazey, I., Fischer, J., Lindenmayer, D. B. (2005) What do conservation biologists publish? *Biological Conservation* 124:63–73.

Frankel, O. H., Soulé, M. (1981) *Conservation and Evolution*. Cambridge University Press, Cambridge.

Garibaldi, L. A., Steffan-Dewenter, I., Kremen, C., *et al.* (2011) Stability of pollination services decreases with isolation from natural areas – a global synthesis. *Ecology Letters* 14: 1062–1072.

Garibaldi, L. A., Steffan-Dewenter, I., Winfree, R., *et al.* (2013) Wild pollinators enhance fruit set of crops regardless of honey bee abundance. *Science* 339: 1608–1611.

Grimbacher, P. S., Catterall, C. P., Kitching, R. L. (2006) Beetle species' responses suggest that microclimate mediates fragmentation effects in tropical Australian rainforest. *Austral Ecology* 31:458–470.

Gross, C. L. (2001) The effect of introduced honeybees on native bee visitation and fruit-set in *Dillwynia juniperina* (Fabaceae) in a fragmented ecosystem. *Biological Conservation* 102: 89–95.

Harris, R. J., Burns, B. R. (2000) Beetle assemblages of kahikatea forest fragments in a pasture-dominated landscape. *New Zealand Journal of Ecology* 24:57–67.

Helle, P., Muona, J. (1985) Invertebrate numbers in edges between clear-fellings and mature forest in northern Finland. *Silva Fennica* 19:281–294.

Hobbs, R. J. (2001) Synergisms among habitat fragmentation, livestock grazing, and biotic invasions in southwestern Australia. *Conservation Biology* 15:1522–1528.

Hobbs, R. J., Mooney, H. A. (1986) Community changes following shrub invasion of grassland. *Oecologia* 70:508–513.

Hoekstra, J. M., Boucher, T. M., Ricketts, T. H., Roberts, C. (2005) Confronting a biome crisis: global disparities of habitat loss and protection. *Ecology Letters* 8: 23–29.

Hudson, Q. J. J., Wilkins, R. J. J., Waas, J. R. R., *et al.* (2000) Low genetic variability in small populations of New Zealand kokako *Callaeas cinerea wilsoni*. *Biological Conservation* 96: 105–112.

Innes, J., Hay, R., Flux, I., *et al.* (1999) Successful recovery of North Island kokako *Callaeas cinerea wilsoni* populations, by adaptive management. *Biological Conservation* 87: 201–214.

Jamieson, I. G., Grueber, C. E., Waters, J. M., *et al.* (2008) Managing genetic diversity in threatened populations: a New Zealand perspective. *New Zealand Journal of Ecology* 32: 130–137.

Keller, L. F., Waller, D. M. (2002) Inbreeding effects in wild populations. *Trends in Ecology & Evolution* 17: 230–241.

King, C. M. (1990) *The Handbook of New Zealand Mammals*. Oxford University Press, Oxford.

Kirch, P. V. (1982) Ecology and the adaptation of Polynesian agricultural systems. *Archaeology in Oceania* 17:1–6.

Lamont, B. B., Klinkhamer, P. G. L., Witkowski, E. T. F. (1993) Population fragmentation may reduce frtility to zero in *Banksia goodii* – a demonstration of the allee effect. *Oecologia* **94**: 446–450.

Lancaster, M. L., Taylor, A. C., Cooper, S. J. B., *et al.* (2011) Limited ecological connectivity of an arboreal marsupial across a forest/plantation landscape despite apparent resilience to fragmentation. *Molecular Ecology* **20**:2258–2271.

Laurance, W. F. (1991) Ecological correlates of extinction proneness in Australian tropical rain forest mammals. *Conservation Biology* **5**:79–89.

Laurance, W. F., Perez-Salicrup, D., Delamonica, P., *et al.* (2001) Rain forest fragmentation and the structure of Amazonian liana communities. *Ecology* **82**:105–116.

Leathwick, J., McGlone, M., Walker, S. (2004) *New Zealand's Potential Vegetation Pattern.* Whenua Press, Lincoln.

Lentini, P. E., Martin, T. G., Gibbons, P., Fischer, J., Cunningham, S. A. (2012). Supporting wild pollination in a temperate agricultural landscape: maintaining mosaics of natural features and production. *Biological Conservation* **149**: 84–92.

Lindenmayer, D. B., Fischer, J. (2006) *Habitat Fragmentation and Landscape Change: an Ecological and Conservation Synthesis.* CSIRIO Publishing.

Llorens, T. M., Byrne, M., Yates, C. J., Nistelberger, H. M., Coates, D. J. (2012) Evaluation of the influence of different aspects of habitat fragmentation on mating patterns and pollen dispersal in the bird pollinated *Banksia sphaerocarpa* var. *caesia*. *Molecular Ecology* **21**: 314–328.

Lord, J. M. (1991) Pollination and seed dispersal in *Freycinetia baueriana*, a dioecious liane that has lost its bat pollinator. *New Zealand Journal of Botany* **29**: 83–86.

MacRaild, L. M., Radford, J. Q., Bennett, A. F. (2010) Non-linear effects of landscape properties on mistletoe parasitism in fragmented agricultural landscapes. *Landscape Ecology* **25**:395–406.

Major, R. E., Christie, F. J., Gowing, G. (2001) Influence of remnant and landscape attributes on Australian woodland bird communities. *Biological Conservation* **102**: 47–66.

Margules, C. R. (1992) The Wog Wog habitat fragmentation experiment. *Environmental Conservation* **19**:316–325.

Martin, T. G., Catterall, C. P., Manning, A. D., *et al.* (2012) Australian birds in a changing landscape : 220 years of European colonisation. In: Fuller, R. J. (ed.) *Birds and Habitat: Relationships in Changing Landscapes.* Cambridge University Press, Cambridge.

McLennan, J. A., Rudge, M. R., Potter, M. A. (1987) Range size and denning behaviour of brown kiwi, *Apteryx australis mantelli*, in Hawke's Bay, New Zealand. *New Zealand Journal of Ecology* **10**:97–107.

Montgomery, B. R., Kelly, D., Robertson, A. W., *et al.* (2003) Pollinator behaviour, not increased resources, boosts seed set on forest edges in a New Zealand Loranthaceous mistletoe. *New Zealand Journal of Botany* **41**:277–286.

National Land and Water Resource Audit (2007) *Weeds*.

Oldroyd, B. P., Thexton, E. G., Lawler, S. H., Crozier, R. H. (1997) Population demography of Australian feral bees (*Apis mellifera*). *Oecologia* **111**: 381–387.

Orians, G. H., Milewski, A. V. (2007) Ecology of Australia: the effects of nutrient-poor soils and intense fires. *Biological Review* **82**: 393–423.

Park, G. (2000) *New Zealand as Ecosystems: the Ecosystem Concept as a Tool for Environmental Management and Conservation.* Wellington, N.Z.

Pereira, H. M., Leadley, P. W., Proença, V., *et al.* (2010) Scenarios for global biodiversity in the 21st century. *Science* **330**:1496–1501.

Potter, M. A. (1990) Movement of North Island brown kiwi (*Apteryx australis mantelli*) between forest remnants. *New Zealand Journal of Ecology* **14**:17–24.

Powlesland, R. G., Moran, L. R., Wotton, D. M. (2011) Satellite tracking of kereru (*Hemiphaga novaeseelandiae*) in Southland, New Zealand: impacts, movements and home range. *New Zealand Journal of Ecology* **35**:229–235.

Rader, R., Howlett, B. G., Cunningham, S. A., Westcott, D. A., Edwards, W. (2012) Spatial and temporal variation in pollinator effectiveness: do unmanaged insects provide consistent pollination services to mass-flowering crops? *Journal of Applied Ecology* **49**: 126–134.

Ragg, J. R., Moller, H. (2000) Microhabitat selection by feral ferrets (*Mustela furo*) in a pastoral habitat, East Otago, New Zealand. *New Zealand Journal of Ecology* **24**:39–46.

Rasch, G. (1992) *Recovery Plan for North Island Kokako.* Wellington, New Zealand.

Richard, Y., Armstrong, D. P. (2010) Cost distance modelling of landscape connectivity and gap-crossing ability using radio-tracking data. *Journal of Applied Ecology* **47**:603–610.

Ricketts, T., Regetz J., Steffan-Dewenter, I., *et al.* (2008) Landscape effects on crop pollination services: are there general patterns? *Ecology Letters* **11**: 499–514.

Ries, L, Fagan, W. F. (2003) Habitat edges as a potential ecological trap for an insect predator. *Ecological Entomology* **28**:567–572.

Ries, L., Fletcher, R. J., Battin, J., *et al.* (2004) Ecological responses to habitat edges: mechanisms, models, and variability explained. *Annual Review of Ecology Evolution and Systematics* **35**:491–522.

Schmidt-Adam, G., Young, A. G., Murray, B. G. (2000) Low outcrossing rates and shift in pollinators in New Zealand pohutukawa (*Metrosideros excelsa*; Myrtaceae). *Botanical Society of America* **87**: 1265–1271.

Sedgeley, J. A., O'Donnell, C. F. J. (2004) Roost use by long-tailed bats in South Canterbury: examining predictions of roost-site selection in a highly fragmented landscape. *New Zealand Journal of Ecology* **28**:1–18.

Smale, M. C. (1984) White Pine Bush and an alluvial kahikatea (*Dacrycarpus dacrydioides*) forest remnant, eastern Bay of Plenty, New Zealand. *New Zealand Journal of Botany* **22**:201–206.

Smale, M. C., Dodd, M. B., Burns, B. R., *et al.* (2008) Long-term impacts of grazing on indigenous forest remnants on North Island hill country, New Zealand. *New Zealand Journal of Ecology* **32**:57–66.

Smith, T. M., Hindell, J. S., Jenkins, G. P., *et al.* (2011) Edge effects in patchy seagrass landscapes: the role of predation in determining fish distribution. *Journal of Experimental Marine Biology and Ecology* **399**:8–16.

Standish, R. J. (2004) Impact of an invasive clonal herb on epigaeic invertebrates in forest remnants in New Zealand. *Biological Conservation* **116**:49–58.

Standish, R. J., Robertson, A. W., Williams, P. A. (2001) The impact of an invasive weed *Tradescantia fluminensis* on native forest regeneration. *Journal of Applied Ecology* **38**:1253–1263.

Standish, R. J., Williams, P. A., Robertson, A. W., *et al.* (2004) Invasion by a perennial herb increases decomposition rate and alters nutrient availability in warm temperate lowland forest remnants. *Biological Invasions* **6**:71–81.

Stodart, E., Parer, I. (1988) *Colonisation of Australia by the Rabbit, Oryctolagus cunicultus.* Canberra, Australia.

Stow, A. J., Sunnucks, P., Briscoe, D. A., *et al.* (2001) The impact of habitat fragmentation on dispersal of Cunningham's skink (*Egernia cunninghami*): evidence from allelic and genotypic analyses of microsatellites. *Molecular Ecology* **10**:867–878.

Tikkanen, O. P., Punttila, P., Heikkila, R. (2009) Species–area relationships of red-listed species in old boreal forests: a large-scale data analysis. *Diversity and Distributions* **15**:852–862.

Timmins, S. M., Williams, P. A. (1991) Weed numbers in New Zealand's forest and scrub. *New Zealand Journal of Ecology* **15**:153–162.

Tylianakis, J. M., Didham, R. K., Bascompte, J., *et al.* (2008) Global change and species interactions in terrestrial ecosystems. *Ecology Letters* **11**:1351–1363.

Vandergast, A. G., Bohonak, A. J., Weissman, D. B. *et al.*, (2007) Understanding the genetic effects of recent habitat fragmentation in the context of evolutionary history: phylogeography and landscape genetics of a southern California endemic Jerusalem cricket (Orthopteta: Stenopelmatidae: Stenopelmatus). *Molecular Ecology* **16**: 977–992.

Ward, D. F., Harris, R. J. (2005) Invasibility of native habitats by Argentine ants, *Linepithema humile*, in New Zealand. *New Zealand Journal of Ecology* **29**:215–219.

Watts, C. H., Didham, R. K. (2006) Rapid recovery of an insect-plant interaction following habitat loss and experimental wetland restoration. *Oecologia* **148**:61–69.

Webb, N. R. (1989) Studies on the invertebrate fauna of fragmented heathland in Dorset, UK, and the implications for conservation. *Biological Conservation* **47**:153–165.

Whaley, P. T., Clarkson, B. D., Smale, M. C. (1997) Claudelands bush: ecology of an urban kahikatea (*Dacrycarpus dacrydioides*) forest remnant in Hamilton, New Zealand. *Tane* **36**:131–155.

Whyte, B. I., Didham, R. K., Briskie, J. V. (2005) The effects of forest edge and nest height on nest predation in two differing New Zealand forest habitats. *New Zealand Natural Sciences* **30**:19–34.

Williams, G. R. (1982) Species–area and similar relationships of insects and vascular plants on the southern out lying islands of New Zealand. *New Zealand Journal of Ecology* **5**:86–95.

Wilmshurst, J. M., Anderson, A. J., Higham, T. F. G., *et al.* (2008) Dating the late prehistoric dispersal of Polynesians to New Zealand using the commensal Pacific rat. *PNAS* **105**:7676–7680.

Wilson, E. O. (1992) *The Diversity of Life*. Belknap Press, Cambridge (MA).

Wines, D., Renfree, M., Wooller, R. O. (1979) Pollen loads of honey possums (*Tarsipes spenserae*) and nonflying mammal pollination in southwestern Australia. *Annals of the Missouri Botanical Garden* **66**: 830–838.

Wiser, S. K., Buxton, R. P. (2008) Context matters: matrix vegetation influences native and exotic species composition on habitat islands. *Ecology* **89**:380–391.

Wiser, S. K., Allen, R. B., Clinton, P. W., *et al.* (1998) Community structure and forest invasion by an exotic herb over 23 years. *Ecology* **79**:2071–2081.

Wittern, A. K., Berggren, Å. (2007) Natal dispersal in the North Island robin (*Petroica longipes*): the importance of connectivity in fragmented habitats. *Avian Conservation and Ecology* **2**.

Woodman, J. D., Baker, G. H., Evans, T. A., *et al.* (2008) *Soil Biodiversity and Ecology: Emphasising Earthworms, Termites and Ants as Key Macro-invertebrates*. Canberra.

Worthington Wilmer, J., Elkin, C., Wilcox, C., *et al.* (2008) The influence of multiple dispersal mechanisms and landscape structure on population clustering and connectivity in fragmented artesian spring snail populations. *Molecular Ecology* **17**:3733–3751.

Wright, T. E., Kasel, S., Tausz, M., *et al.* (2010) Edge microclimate of temperate woodlands as affected by adjoining land use. *Agricultural and Forest Meteorology* **150**:1138–1146.

Young, A., Mitchell, N. (1994) Microclimate and vegetation edge effects in a fragmented podocarp-broadleaf forest in New Zealand. *Biological Conservation* **67**:63–72.

CHAPTER 4

The impacts of climate change on Australian and New Zealand flora and fauna

Abigail Cabrelli, Linda Beaumont and Lesley Hughes

Summary

Over recent years, anthropogenic climate change has emerged as a considerable threat to the biota of Australia and New Zealand. Despite the relatively modest climatic changes that have occurred to date, species already appear to be responding by shifting their distributions, altering the timing of life-cycle events and modifying their behaviours. This chapter summarises the impacts of climate change on the species and ecosystems of Australia and New Zealand, describing the ways in which observed and projected responses differ from those occurring in the Northern Hemisphere due to the distinctiveness of our environment and biota. We also highlight the implications of these responses for species and ecosystem conservation.

4.1 Introduction

Nature's calendar is something we become familiar with from a young age. The flowering of golden wattles takes place in late winter, while the arrival of migratory birds and the drone of cicadas herald spring. As the season progresses Common Brown Butterflies emerge, webs are spun by St Andrew's Cross Spiders, and flame trees and jacarandas bloom. Sepal 'flowers' of the Christmas Bush turning red and Christmas Beetles littering the back porch signal the start of summer holidays. As the New Year dawns, By-the-Wind Sailor sea jellies can be found on our shores. Summer wanes, giving way to Autumn, leaves senesce and fall, while in the oceans Southern right and Humpback whales migrate to warmer waters.

But the timing of nature's calendar is changing. On all continents and in most oceans, diverse taxa – plants, birds, mammals, reptiles and insects, phytoplankton and coral – are demonstrating a globally coherent 'fingerprint' of climate change (Chambers *et al.*, 2013a; Parmesan and Yohe, 2003; Rosenzweig *et al.*, 2008). Earlier timing of spring

Austral Ark: The State of Wildlife in Australia and New Zealand, eds. A. Stow, N. Maclean and G. I. Holwell. Published by Cambridge University Press. © Cambridge University Press 2015.

life-cycle events, shifts in species range margins, behavioural, morphological and genetic changes, and population extinctions have been reported, indicating that species are responding to changes in climate, particularly temperature.

This chapter discusses ways in which flora and fauna in Australia and New Zealand have responded to anthropogenic climate change, what patterns might be expected to occur as the twenty-first century progresses, and what implications these responses have for species and ecosystem conservation.

4.2 Anthropogenic climate change

The Earth's climate has changed throughout history due to variation in its orbit around the Sun, changes in solar activity, and volcanic eruptions. As the Earth moved into and out of major ice ages, some species shifted their distributions to track the movement of climate zones (Willis and MacDonald, 2011). However, since the Industrial Revolution in the mid 1700s the Earth has undergone far more rapid change than at almost any time in its history. Moreover, human activity has substantially altered most landscapes, severely restricting the capacity of species to adapt to the changing climate by moving to more suitable habitats. Hence, while habitat loss is currently the predominant driver of ecosystem changes and species extinction (Hoekstra et al., 2005), as this century progresses the combined impacts of habitat loss and climate change on biodiversity will become increasingly apparent (Asner et al., 2010; Beaumont and Duursma, 2012). See also Box 4.1.

In line with global trends, climate in Australia and New Zealand has changed over the past century. Mean annual temperature has increased across both land masses, and precipitation patterns have altered (Table 4.1). These climatic changes are projected to increase in magnitude as the century progresses.

Box 4.1 Consequences of land-use change

The biosphere and climate system are tightly coupled, with interactions and feedbacks across a range of spatial and temporal scales. Forests and woodlands, in particular, have an important influence on local and regional climates by buffering extremes, recycling moisture and sequestering carbon, and through reflection of solar radiation (McAlpine et al., 2010). Large-scale land-use/land-cover change (LUCC) can significantly impact regional climate. Transformation of the Australian landscape since European settlement has contributed to regional climate changes (McAlpine et al., 2007), such as temperature and rainfall changes over eastern and western Australia (Narisma and Pitman, 2003), and has exacerbated droughts and climate extremes in eastern Australia (Deo et al., 2009). These impacts make it critical that land-use and climate change policies are strongly integrated, deforestation is significantly reduced, and investment in strategic reforestation occurs (McAlpine et al., 2010; Hughes, 2008).

Table 4.1   Summary of twentieth-century climate changes and projections for the twenty-first century.

Variable	Summary of twentieth-century climate changes	Projected future changes*
Land temperature	**Australia** T_{mean} increase of ~0.9 °C since 1910. Most warming has occurred since 1950. Rate of warming is not geographically uniform: largest increases in central Australia, minimum warming/slight cooling in NE Australia (CSIRO and BoM, 2007). **New Zealand** T_{mean} increased 0.91 °C (1909–2009), ranging from 1.58 °C in Auckland to 0.58 °C in Dunedin, with differences linked to spatial variation in SST (Mullan *et al.*, 2010).	**Australia** 2030: Median warming across Australia of 0.7–1.2 °C, depending on region. More warming inland than coastal regions. Less warming in winter. 2050: best estimate of 1.2–2.2 °C. 2070: 2.2–3.4 °C. **New Zealand** 2040: 0.9 °C; 2090: 2.1 °C, with less warming in spring.
Ocean temperature	**Australia** Increased 0.11–0.12 °C/decade (1950–2010) in northeast and northwest Australia; climate zones along northwest have shifted 100km south, and 200km south in the northeast (Lough, 2008). East Australian Current has strengthened, extending southwards. South Tasman SST has increased 0.23 °C/decade (Ridgway, 2007). **New Zealand** Increased 0.07 °C/decade (1909–2009); greater warming in N than S (Mullan *et al.*, 2010).	**Australia** 2030: Best estimate of 0.6–0.9 °C in southern Tasman Sea and northwest coast, 0.3–0.6 °C elsewhere. **New Zealand** Similar to land temperature increases.
Precipitation	**Australia** Increased in northwest since 1950s. East and southwest regions have become drier. Number of rain days increased in eastern central regions in spring, but decreased in NSW Tablelands (CSIRO and BoM, 2007). **New Zealand** Summer rainfall declines since late 1970s in North Island and western regions of South Island (Ummenhofer *et al.*, 2009).	**Australia** Varies spatially and seasonally. 2030 annual changes: −15% to +10% in north, −10% to little change in south. 2070 annual changes: −30% to +20% in central, eastern and northern areas, −30% to +5% in southwest. **New Zealand** Varies spatially and seasonally. Annual averages: increases in Tasman, West Coast, Otago, Southland and Chatham Islands; decreases in Northland, Auckland, Gisborne, Hawke's Bay.
Sea level rise	**Australia** 1920–2000: Mean increase 1.2 mm/year; 3 mm/year from 1993–2003 (CSIRO and BoM, 2007). **New Zealand** Mean increase 1.7 mm/year since early 1900s (Hannah and Bell, 2012).	**Australia** From 1990–2100: 18–59 cm. **New Zealand** From 1990–2100: 18–59 cm.

T_{mean} = annual mean temperature.
*Relative to 1980–1999. Australian projections from CSIRO and BoM (2007); NZ projections from MfE (2008).

4.3   Responses of species to climate change

To what extent have species and ecosystems already responded to anthropogenic climate change, and what will be the magnitude of future responses? Unfortunately, most meta-analyses of the biological impacts of climate change have been devoid of Southern Hemisphere examples. This is partly because Australia, and New Zealand particularly, suffer a lack of long-term datasets from which scientists can assemble clues to answer these questions. Whilst increasing research effort in Australia is highlighting the sensitivity of its biodiversity to climate change, our knowledge of responses by New Zealand's biota remains poor (Lundquist *et al.*, 2011).

To predict the responses and relative vulnerabilities of Australian and New Zealand species to climate change, it is important to consider the biogeographical histories of these two countries, and the environments in which their species have evolved. Both Australia and New Zealand have experienced a long period of geographical isolation following the break-up of Gondwanaland during the Cretaceous period, resulting in remarkable levels of endemism among their native plants and animals (Chapters 1 and 2). Many species have small geographic ranges spanning narrow climatic gradients, increasing their vulnerability to range contractions or displacement under climate change (e.g. Hughes *et al.*, 1996; Beaumont and Hughes, 2002). Many are also currently under threat from other stressors, such as habitat loss due to land clearing and invasive species, which are expected to compound the threat imposed by climate change.

The topography, soils and climates, however, differ markedly between these two countries. New Zealand has twice the vertical relief of Australia (Augee and Fox, 2000), suggesting that migration to higher elevations will be more feasible in New Zealand. The soils of New Zealand are also younger and richer in plant nutrients than those in Australia, increasing the likelihood that vegetation will be more responsive to rising CO_2 levels. New Zealand has a relatively stable, temperate climate (McGlone and Walker, 2011), whereas the climate of Australia is characterised by extremes and a high degree of variability. Many Australian species have therefore evolved to cope with high levels of environmental variation that will confer an advantage under climate change. Yet even in Australia, future climate change is expected to exceed the limits of natural variability to which species have become adapted, placing them under increasing pressure to respond by relocating to new areas, or staying and adapting *in situ* (Steffen *et al.*, 2009).

4.3.1 Relocation

Polewards or higher elevation shifts in species distributions are expected to be one of the principal responses to climate change (Chen *et al.*, 2011), with climate-driven range shifts already apparent amongst Australian marine species and terrestrial ecosystems. In the temperate seas off the coast of Tasmania, 45 fish species have undergone significant southerly range shifts since the late 1800s (Last *et al.*, 2010). Of 29 intertidal species, 16 have shifted southwards over the past 50 years, whilst only two species have been recorded at more northerly sites (Pitt *et al.*, 2010). The long-spined sea urchin *Centrostephanus rodgersii*, a voracious grazer of kelp beds, has extended its range pole-wards into Tasmanian coastal waters, in line with warming sea temperatures and a strengthening of the East Australian Current (Ling *et al.*, 2008; Banks *et al.*, 2010). Both native and feral animals in the Snowy Mountains now occur at higher elevations than in the past (Pickering *et al.*, 2004), while the ranges of 464 Australian bird species have moved an average of 12.7 km per decade since the 1960s (VanDerWal *et al.*, 2012).

Models simulating the potential for climate change to drive future shifts in species' distributions have received considerably more research attention than studies assessing twentieth-century range changes. This is partly due to the availability of generic tools for forecasting range shifts under climate change scenarios (Guisan and Zimmermann, 2000). These tools have been applied to a variety of Australian and New Zealand taxa, including mammals (Gibson *et al.*, 2010; Adams-Hosking *et al.*, 2011; Klamt *et al.*, 2011), reptiles (Penman *et al.*, 2010), amphibians (Fouquet *et al.*, 2010), fish (Bond *et al.*, 2011), birds (Reside *et al.*, 2012), plants (Fitzpatrick *et al.*, 2008), insects (Beaumont and Hughes, 2002) and invasive species (Gallagher *et al.*, 2010; Beaumont *et al.*, 2009; Bourdôt *et al.*, 2012). These studies consistently suggest that suitable habitat for many species may shift and/or contract as the climate changes. For example, 66%–89% of 243 Australian bird species assessed have been projected to undergo habitat contractions by 2080, the range of projections being due to differences in assumptions about dispersal capacity (Reside *et al.*, 2012). Among Western Australian Banksias (Proteaceae), 66% of species' ranges were projected to decline by 2080, and only 6% projected to expand or remain stable (Fitzpatrick *et al.*, 2008).

For many species in Australia, the flattest of all continents, tracking climate change will require long overland range shifts (Hughes, 2012) equivalent to 3–17 km/year, depending on the region (Dunlop and Brown, 2008). In contrast, the mountainous topography of New Zealand may provide greater scope for uphill movement. The abrupt *Nothofagus* tree line in the Southern Alps of New Zealand, however, has been unresponsive to recent climate change, potentially due to a scarcity of sites suitable for seedling establishment (Harsch *et al.*, 2012). Species likely to be more successful at shifting their ranges include those capable of flight, plants with seeds dispersed by vectors that can move long distances (Steffen *et al.*, 2009), and pelagic marine species, whose dispersal rates generally exceed those of terrestrial species (Grantham *et al.*, 2003). The limited dispersal capacity of other taxa, including non-flying vertebrates, many plants and most invertebrates, suggests they will be unlikely to match the isotherm shifts projected for the coming decades (Hughes, 2012).

Movement across landscapes will also be contingent upon the connectivity of suitable habitat and the permeability of the surrounding landscape. Geographic barriers, such as coasts and urban environments, and/or habitat fragmentation will restrict species' movements. Many species confined to habitats such as mountaintops, isolated lakes and small islands, and those whose ranges are already restricted by coasts, will simply have nowhere to go (Steffen *et al.*, 2009).

4.3.2 *In situ* changes

4.3.2.1 Phenological changes

Phenology is the study of the timing of biological events, such as flowering, breeding, and migration, and how these vary with climate. Temperature, precipitation and photoperiod are used by species as cues for different stages of their life-cycles, and flexible responses via changes in the timing of key events can be an important way of adapting to a changing environment. Globally, shifts in the timing of life-cycles provide strong evidence that species are already responding to climate change. While available data are more limited for Australia and New Zealand compared to the Northern Hemisphere (Chambers *et al.*, 2013a), most recorded phenological trends are consistent with global patterns. But some trends display important differences to those of the Northern Hemisphere, reflecting the distinctiveness of our environment and biota. For example, while phenological changes in the Northern Hemisphere are often correlated with temperature, precipitation (particularly the number of rain days) is frequently an important correlate of phenology changes in Australia, presumably because of its influence on resource availability (Chambers *et al.*, 2014).

Native plants from a range of families and bioregions demonstrate changes to the timing of flowering, with earlier flowering being almost three times more likely than delayed flowering (Chambers *et al.*, 2014) (Figure 4.1, Plate 50). Higher temperatures, declining soil water content, and changes in management practices have contributed to earlier ripening of wine grapes across southern Australia (Webb *et al.*, 2011). Migratory birds have advanced arrival by an average of 2.1 days/decade since the 1960s in Australia, although disparate shifts in temperature and precipitation across the continent has led to variation in phenological trends, most likely due to altered resource availability (Chambers *et al.*, 2013).

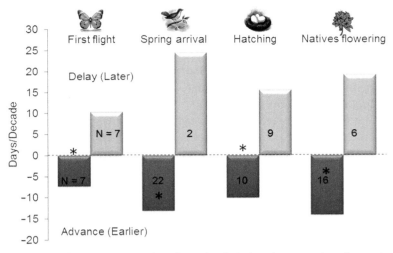

Figure 4.1 (Plate 50) Summary of significant phenological trends amongst Australian species (modified from data in Chambers *et al.*, 2014). Events are first flight date of butterflies, first arrival date of birds to breeding grounds, hatching date for birds, and flowering commencement date for native plants. Each graph shows the average *significant* ($p < 0.05$) delay or advance in phenology, with the asterisk indicating the average across all datasets. The number within each box indicates the number of datasets analysed. A black and white version of this figure will appear in some formats. For the colour version, please refer to the plate section.

The reproductive behaviour of sleepy lizards (*Tiliqua rugosa*) is also influenced by climate. Within this species, monogamous pairing takes place around eight weeks prior to mating. Data from the 1980s and 1990s show that pairing began earlier and lasted longer during later years of the study (Bull and Burzacott, 2006). Similarly, in alpine sites laying dates of the lizard *Acritoscincus duperreyi* may be occurring almost a month earlier compared with the late 1960s (Telemeco *et al.*, 2009). For species with temperature-dependent sex determination, temperature shifts can profoundly affect sex ratios, particularly if, like *A. duperreyi*, they cannot fully compensate for climate change impacts by adjusting nest depth, location or timing of ovipositioning (Telemeco *et al.*, 2009). Projections indicate that sex ratios among sea turtles at Ashmore Island and Bare Sand Island could be skewed towards all-female clutches by 2030 (Fuentes *et al.*, 2009).

Differences in the magnitude of trends across taxa may result in mismatches in the synchrony of life-cycles of interacting species, such as between predators and their prey and between host plants and their insect herbivores. First flight dates of some butterflies have become slightly later while spring migration and breeding of birds and flowering of plants has advanced (Figure 4.1, Plate 50). Earlier spring arrival of Richard's Pipit, *Anthus novaeseelandiae*, to the Australian alpine zone is associated with early snow melt (Green, 2010). The migratory Bogong Moth, *Agrotis infusa*, an important food source for the pipit and other insectivorous predators, is arriving to this region significantly later than in previous decades, and these contrasting responses may result in asynchrony of life-cycles (Green, 2010).

Box 4.2 Do a few degrees matter?

Temperature has profound effects on life. Metabolic and developmental rates in animals, and photosynthesis and respiration in plants is influenced by temperature and its interactions with moisture availability. Performance curves describe the relationship between body temperature (T_b) and physiological performance. These curves are characterised by an individual's thermal optima and thermal breadth. Performance is maximised at an optimal temperature (T_o), either side of which performance declines. If temperature reaches a critical level the impact is lethal. Thermal breadth defines the temperature range over which performance exceeds an arbitrary level.

Individuals can maintain homeostasis and reproduction across a given level of environmental variability. The impact of environmental temperature change depends on the individual's thermal breadth and T_b relative to T_o. If, prior to warming, T_b is below T_o higher temperature may be beneficial, increasing physiological performance towards the individual's maximum (Figure 4.2). Conversely, if T_b already exceeds T_o further warming increases stress. Therefore, individuals with wide thermal breadths (i.e. thermal generalists) are likely to be more tolerant of temperature increases than those with narrow thermal breadths (specialists). Similarly, populations at the species' range margin are more likely to respond to warming than those closer to the range core. As temperature increases, populations at the colder range margin may increase productivity (as T_b shifts towards T_o) or disperse to new climatically suitable areas, while populations at the warmer margin are likely to demonstrate negative responses.

As environmental temperature increases over time, the average temperature that an individual is exposed to changes as do the extremes. Cold conditions become

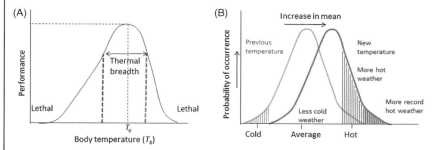

Figure 4.2 Performance curve for a hypothetical species. If body temperature (T_b) exceeds a threshold, the impact can be lethal. If T_b is below T_o, temperature increases may be beneficial. If T_b is above T_o higher temperature can reduce performance, and may be lethal. B. A small shift in mean temperature has a disproportionate increase (or decrease) in the amount of extreme high (or low) temperature an organism experiences.

warmer and less frequent while hot conditions become hotter still and occur more frequently. Extreme warming events have already affected Australian species, with examples including mortality of flying foxes (Welbergen *et al.*, 2008) and loss of marginal populations of the seaweed, *Scytothalia dorycarpa*, resulting in range contraction of ~100 km (Smale and Wernberg, 2013).

4.3.2.2   Physiological changes

All organisms live within a limited range of environmental temperatures (Box 4.2) and climate warming, particularly extreme heat events, can directly affect performance and survival. For example, reef-building corals, such as those in the Great Barrier Reef, exist within 1–2 °C of their upper thermal limit in summer, and extreme sea surface temperatures can lead to a physiological stress response in which corals expel their symbiotic zooxanthellae, leading to bleaching and subsequent mortality. There are no records of coral bleaching within the Great Barrier Reef prior to 1979, while at least seven bleaching events have occurred since. The most serious of these, in 1998 and 2002, affected up to 60% of individual reefs (Hughes, 2012). Bleaching in reefs off the coast of Western Australia, such as Ningaloo Reef, was observed for the first time in 2011 (Wernberg *et al.*, 2013).

Elevated levels of atmospheric CO_2 can result in a 'fertilisation effect': photosynthesis is increased, which influences growth rates, the allocation of biomass, reproduction, and the concentrations of defensive chemicals. Increasing CO_2 is likely to have already contributed to ecosystem-level changes in Australia such as the expansion of monsoon rainforests within the savanna matrix in northern Australia (Bowman *et al.*, 2010). Increasing density of woody plants in Australian savannas has also been attributed to elevated CO_2 combined with reduced burning (Fensham *et al.*, 2009). In many regions, however, increasing CO_2 is unlikely to result in large-scale changes in vegetation because plant growth is limited by factors such as water availability and soil nutrients. Rather, responses may primarily consist of alterations in water use, leaf chemistry and biomass allocation (Hovenden and Williams, 2010).

Physiologically based models have been used to estimate changes to forest productivity, and have shown that interactions between elevated CO_2, site fertility, soil depth and climate can result in a broad range of responses. For Australia, productivity in most major plantation regions is projected to increase, although decreases are likely in south-west Western Australia if predicted declines in precipitation occur (Battaglia *et al.*, 2009). In New Zealand, productivity of *Pinus radiata* is predicted to increase by ~19% by 2040, provided soil fertility is not limiting (Kirschbaum *et al.*, 2012).

4.3.2.3   Morphological changes

Morphological responses to climate change are considerably less well documented than other *in situ* changes. Yet evidence of declining body size, which is expected under climate change because higher surface area-to-volume ratios facilitate heat loss, is emerging. In south-eastern Australia, four out of eight passerine bird species have shown a significant decrease in body size over the past 100 years (Gardner *et al.*, 2009). Among southern populations body size is now similar to that of more northern populations pre-1950, and is equivalent to a 7° southward shift in latitude. In plants, increasing temperatures over recent decades have been linked to a 2 mm decline in leaf width among specimens of the shrub *Dodonaea viscosa* subsp. *angustissima* in South Australia, equivalent to a 3° southward latitudinal shift (Guerin *et al.*, 2012; Guerin and Lowe, 2013).

4.3.2.4 Behavioural changes

Behavioural changes may buffer some species against the impacts of rapid climate change, at least in the short term. Of particular importance are behaviours that enable organisms to modify their microclimate, such as burrowing, forming aggregations, or altering nest structure (Hassall *et al.*, 2007). For example, Australia's three-lined skinks (*Acritoscincus duperreyi*) are building deeper nests in response to warming, although in this instance, the behavioural change has been insufficient to achieve preferred nest temperatures for the duration of the nesting season (Telemeco *et al.*, 2009). As a consequence, mean incubation temperatures in natural nests have now crossed the threshold at which offspring gender is determined by temperature rather than by genotype.

4.3.2.5 Genetic changes

When exposed to novel conditions, strong selective pressures can rapidly promote genetic change in some species. In eastern Australia, a higher frequency of heat-tolerant genotypes has developed among southern populations of fruit flies (*Drosophila melanogaster*) since the early 1980s (Umina *et al.*, 2005). These populations now have the genetic constitution of more northerly populations, equivalent to a 4° southward shift in latitude.

However, not all species will have the capacity to adapt to climate change genetically. Little evidence of heritability for fundamental stress genes among the reef-building coral (*Acropora millepora*) on the Great Barrier Reef, for instance, has raised concerns about whether evolutionary adaptation in this species could occur fast enough to keep pace with environmental changes (Császár *et al.*, 2010). Other anthropogenic effects, such as overexploitation and habitat fragmentation, are expected to further limit evolutionary potential by decreasing genetic diversity and disrupting gene flow, respectively.

4.3.3 Changes in species interactions and the composition of communities

If a species' response to climate change renders it relatively advantaged or disadvantaged, flow-on effects to interacting species (such as predators, competitors, pollinators and parasites) are likely. In south-eastern Australia, the core range of the swamp rat (*Rattus lutreolus*) is projected to shift and overlap with that of the broad-toothed rat (*Mastacomys fuscus*), increasing competition (Green *et al.*, 2008). Rising sea surface temperatures off the coast of eastern Australia have indirectly reduced nesting and recruitment among populations of the Loggerhead sea turtle (*Caretta caretta*) via impacts on ocean productivity and prey abundance (Chaloupka *et al.*, 2008). As sea-level rises, the landward expansion of mangroves, saltmarshes, and intertidal habitats may be constrained by natural or man-made barriers, with the decline of these ecosystems negatively affecting species abundance and richness. This is particularly critical for migratory wading birds and significant Ramsar wetland sites, such as those at Farewell Spit and Firth of Thames in New Zealand (Lundquist *et al.*, 2011).

Given that plants form a foundation for food webs and provide habitat for myriad species, changes to plant structure and function will have cascading impacts through ecosystems. Plants grown at high concentrations of atmospheric CO_2, for example, tend to have lower concentrations of nitrogen in their tissues, as well as increased levels of carbon-based defensive chemicals such as phenols and tannins (Hovenden and Williams, 2010). Sclerophyllous vegetation, particularly eucalypts, already has foliage that is lower in nutrients and higher in secondary metabolites compared with other vegetation types, and further changes to leaf chemistry have significant implications for the diets of herbivores and granivores (Hovenden and Williams, 2010).

Many pathogens are also sensitive to environmental factors, and climate-driven changes in their distributions and population dynamics will affect hosts (Harvell *et al.*, 2002). The temperate climate of New Zealand currently precludes establishment of many pests from warmer regions, a situation that may change as temperatures increase (Kriticos, 2012). Existing diseases may also become more severe: Douglas-fir trees in the North Island of New Zealand are projected to be at greater risk of the foliage disease, Swiss needle cast, caused by the fungal pathogen *Phaeocryptopus gaeumanni* (Watt *et al.*, 2011). Conversely, survival rates of chytrid fungus (*Batrachochytrium dendrobaditis*) decrease with temperatures exceeding 28–30 °C (Fisher *et al.*, 2009; Kilpatrick *et al.*, 2010). This pathogen is linked to declines in amphibian populations globally (Kilpatrick *et al.*, 2010) including the northern and southern corroboree frog (*Pseudophryne pengilleyi* and *P. corroboree*, respectively) in the New South Wales Southern Tablelands (Hunter *et al.*, 2010).

As individual species respond to climate change, community composition will also shift. A recent dramatic example of this has been associated with the southerly range expansion of the long-spined sea urchin, *Centrostephanus rodgersii*, from the coast of NSW to Tasmania. The urchin feeds on kelp, creating 'barrens' of bare rock, causing loss of local habitat for at least 150 other species (Ling *et al.*, 2008). Such a dramatic impact of the addition or loss of a single species depends upon its ecological characteristics and its role in the community. In general, the loss of a top carnivore, a structurally dominant plant, or a species at the base of the food chain will have significant community-level consequences (Steffen *et al.*, 2009). For many species, these indirect biotic effects will likely be of greater importance than the direct impacts of changes in temperature and rainfall.

4.4   Vulnerable ecosystems

Over the next century, ecological communities and ecosystems as we know them today, are likely to be transformed as the impacts of climate change accelerate. The most vulnerable systems will be those that: are currently restricted, either in geographical extent or climatic niche; contain a high proportion of narrow-ranged endemic species; face high exposure to factors such as rising sea levels or changed fire regimes; comprise a high proportion of species that have a limited capacity to adapt; or are currently suffering from other threats such as fragmentation, over-allocation of water flows, or disturbance by invasive species (also see Box 4.3). The key drivers of transformation will be intensification of extreme events such as heatwaves, tropical cyclones and droughts, changes in fire regimes, and changes in hydrology (Steffen *et al.*, 2009; Lundquist *et al.*, 2011; Laurance *et al.*, 2011). Transformation of these communities will also have consequences for ecosystem services, including food provision, water security, nutrient cycling, pollination and pest control.

One of the most vulnerable Australian ecosystems is the alpine zone, which has already undergone significant losses of snow. The potential complete loss of snow cover by the end of the century (Hennessy *et al.*, 2007) will have dramatic effects on many species already considered rare and threatened, especially those mammals dependent on snow cover to protect them from predation (Pickering *et al.*, 2004). Expansion of woody species to higher elevations at the expense of herbaceous plants, increased competition from lower elevation species (both native and exotic), higher fire frequencies, and increasing mismatches in the timing of life-cycles, are likely impacts in alpine areas in both Australia and New Zealand (McGlone and Walker, 2011; Pickering *et al.*, 2004).

Box 4.3. Can tuatara survive climate change?

The tuatara is a reptile endemic to New Zealand (Figure 4.3, Plate 51). Although they resemble lizards, the two species (*Sphenodon punctatus* and the extinct *S. diversum*) are part of a distinct lineage, order Rhynchocephalia (Hay *et al.*, 2010). Once widespread throughout New Zealand, tuatara were extirpated from the mainland islands following repeated introductions of mammalian predators by humans (Towns and Daugherty, 1994). There remain 32 extant populations on offshore islands, 17 of which have fewer than 100 adults (Mitchell *et al.*, 2010).

Climate change is a significant threat to tuatara because their isolation on islands prevents migration to cooler areas. Long generation times (≈ 100 years), slow growth rates (sexual maturity at ≈ 15 years), small clutch sizes ≈ eight eggs), lengthy inter-clutch intervals and low genetic diversity result in an extremely low adaptive capacity.

Tuatara have temperature-dependent sex determination (TSD) in which offspring gender is determined by incubation temperature rather than by chromosomes. Although TSD is present in many reptiles, tuatara possess a particularly rare form (FM or type 1b) in which males are produced at higher incubation temperatures (Mitchell *et al.*, 2006). Thus, warming under climate change may bias populations towards males, increasing the potential for demographically driven extinctions

Figure 4.3 (Plate 51) A tuatara (*Sphenodon punctatus*). A black and white version of this figure will appear in some formats. For the colour version, please refer to the plate section.

Box 4.3. (continued)

(Mitchell *et al.*, 2008; Nelson *et al.*, 2004). For example, while the tuatara population inhabiting North Brothers Island would be expected to persist for at least 2000 years if hatchling sex ratios remain below 75% male, extinction would occur within 300 years if this ratio increases to 85% male (Mitchell *et al.*, 2010).

Tuatara may be able to buffer negative impacts of climate change through behavioural changes, such as nesting later in the season, or selecting shadier nest sites, thereby increasing the likelihood of females being produced (Mitchell *et al.*, 2008). If they are unable to compensate fully for climate change, human intervention by covering rookeries with shade is likely to help balance sex ratios. However, the greatest chance for the future survival of tuatara may depend on translocation to other islands, or protected mainland areas, where cooler nesting sites are available (Mitchell *et al.*, 2008).

The montane areas of the wet tropics region of North Queensland, an area of extremely high species richness and endemism, will be very vulnerable because many species are confined to cool mountaintop 'islands' surrounded by warmer rainforest (Williams *et al.*, 2003). Freshwater wetlands in some regions of both New Zealand and Australia will be at risk from reductions in rainfall, exacerbated by over-allocation of water resources to agriculture and urban settlements (Pittock and Finlayson, 2011). Low-lying coastal freshwater swamps in the Northern Territory have already suffered saltwater intrusion, partly attributed to rising sea levels; the iconic Kakadu National Park may be threatened by this process in the future (BMT WBM, 2010). Coral reefs are threatened by rising sea surface temperatures combined with ocean acidification and potential increases in damage due to higher intensity cyclones.

4.5 Implications for the conservation of Australian and New Zealand biodiversity

By the second half of the twenty-first century, the Earth's climate is likely to be hotter than it has been during the period over which most present-day species evolved, and it is increasingly likely that stabilisation of greenhouse gases will not occur within a time frame to 'allow ecosystems to adapt naturally to climate change' (UN General Assembly, 1994). Accelerating climate change is shifting the conservation goalposts. Many species currently 'protected' in reserves will need to migrate to cope with a changing climate and some will need active human intervention to do so. Climate change will almost certainly accelerate the high rates of species extinction and exacerbate the ecosystem degradation already evident from human activities. Managing the inevitable future transformation of landscapes, whilst reducing the risk of species extinctions, presents a formidable challenge. Increasing the natural adaptive capacity and resilience of our biota to the new threat of climate change by reducing existing threats, restoring cleared and degraded habitat, increasing habitat connectivity, and protecting climatic refugia will be critical. In addition, we will increasingly need to consider more active interventionist strategies to manage species distributions and engineer ecological communities, and this will present many challenges to traditional conservation practice.

REFERENCES

Adams-Hosking, C., Grantham, H. S., Rhodes, J. R., McAlpine, C. & Moss, P. T. 2011. Modelling climate-change-induced shifts in the distribution of the koala. *Wildlife Research*, **38**, 122–130.

Asner, G. P., Loarie, S. R. & Heyder, U. 2010. Combined effects of climate and land-use change on the future of humid tropical forests. *Conservation Letters*, **3**, 395–403.

Augee, M. & Fox, M. 2000. *Biology of Australia and New Zealand*. Sydney, Australia, Pearson Education.

Banks, S. C., Ling, S. D., Johnson, C. R. *et al.* 2010. Genetic structure of a recent climate change-driven range extension. *Molecular Ecology*, **19**, 2011–2024.

Battaglia, M., Bruck, J., Brack, C. & Baker, T. 2009. Climate change and Australia's plantation estate: analysis of vulnerability and preliminary investigation of adaptation options. *Forest and Wood Products, Australia*.

Beaumont, L. J. & Duursma, D. 2012. Global projections of 21st century land-use changes in regions adjacent to protected areas. *PLoS ONE*, **7**, e43714.

Beaumont, L. J. & Hughes, L. 2002. Potential changes in the distributions of latitudinally restricted Australian butterfly species in response to climate change. *Global Change Biology*, **8**, (954–971).

Beaumont, L. J., Gallagher, R. V., Downey, P. O., *et al.* 2009. Modelling the impact of *Hieracium* spp. on protected areas in Australia under future climates. *Ecography*, **32**, 757–764.

BMT WBM 2010. *Kakadu – Vulnerability to Climate Change Impacts*. A report to the Australian Government of Climate Change and Energy Efficiency.

Bond, N., Thomson, J., Reich, P. & Stein, J. 2011. Using species distribution models to infer potential climate change-induced range shifts of freshwater fish in south-eastern Australia. *Marine and Freshwater Research*, **62**, 1043–1061.

Bourdôt, G. W., Lamoureaux, S. L., Watt, M. S., Manning, L. K. & Kriticos, D. J. 2012. The potential global distribution of the invasive weed *Nassella neesiana* under current and future climates. *Biological Invasions*, **14**, 1545–1556.

Bowman, D., Murphy, B. P. & Banfai, D. S. 2010. Has global environmental change caused monsoon rainforests to expand in the Australian monsoon tropics? *Landscape Ecology*, **25**, 1247–1260.

Bull, C. M. & Burzacott, D. 2006. Changes in climate and in the timing of pairing of the Australian lizard, *Tiliqua rugosa*: a 15-year study. *Journal of Zoology*, **256**, 383–387.

Chaloupka, M., Kamezaki, N. & Limpus, C. 2008. Is climate change affecting the population dynamics of the endangered Pacific Loggerhead sea turtle? *Journal of Experimental Marine Biology and Ecology*, **356**, 136–143.

Chambers, L. E., Altwegg, R., Barbraud, R., *et al.* 2013. Phenological changes in the Southern Hemisphere. *PLoS ONE*, **8**(10), e75514.

Chambers, L. E., Beaumont, L. J. & Hudson, I. L. 2014. Continental scale analysis of bird migration timing: influences of climate and life history traits – a generalized mixture model clustering approach. *International Journal of Biometeorology*, **58**, 1147–1162.

Chen, I., Hill, J. K., Shiu, H. J. *et al.* 2011. Asymmetric boundary shifts of tropical montane Lepidoptera over four decades of climate warming. *Global Ecology and Biogeography*, **20**, 34–45.

Császár, N. B., Ralph, P. J., Frankham, R., Berkelmans, R. & Van Oppen, M. J. 2010. Estimating the potential for adaptation of corals to climate warming. *PLoS ONE*, **5**, e9751.

CSIRO & BOM 2007. *Climate Change in Australia*. Melbourne: CSIRO, Bureau of Meteorology.

Deo, R. C., Syktus, J. I., McAlpine, C. A., *et al.* 2009. Impact of historical land cover change on daily indices of climate extremes including droughts in eastern Australia. *Geophysical Research Letters*, **36**, 5.

Dunlop, M. & Brown, P. 2008. *Implications of Climate Change for Australia's National Reserve System: A Preliminary Assessment*, Report to the Department of Climate Change. Canberra, Australia: Department of Climate Change, Canberra, Australia.

Fensham, R. J., Fairfax, R. J. & Ward, D. P. 2009. Drought-induced tree death in savanna. *Global Change Biology*, **15**, 380–387.

Fisher, M. C., Garner, T. W. J. & Walker, S. F. 2009. Global emergence of *Batrachochytrium dendrobatidis* and *Amphibian Chytridiomycosis* in space, time, and host. *Annual Review of Microbiology*, **63**, 291–310.

Fitzpatrick, M. C., Gove, A. D., Sanders, N. J. & Dunn, R. R. 2008. Climate change, plant migration, and range collapse in a global biodiversity hotspot: the Banksia (Proteaceae) of Western Australia. *Global Change Biology*, **14**, 1337–1352.

Fouquet, A., Ficetola, G. F., Haigh, A. & Gemmell, N. 2010. Using ecological niche modelling to infer past, present and future environmental suitability for *Leiopelma hochstetteri*, an endangered New Zealand native frog. *Biological Conservation*, **143**, 1375–1384.

Fuentes, M., Maynard, J., Guinea, M., *et al.* 2009. Proxy indicators of sand temperature help project impacts of global warming on sea turtles in northern Australia. *Endangered Species Research*, **9**, 33–40.

Gallagher, R. V., Hughes, L., Leishman, M. R. & Wilson, P. D. 2010. Predicted impact of exotic vines on an endangered ecological community under future climate change. *Biological Invasions*, **12**, 4049–4063.

Gardner, J. L., Heinsohn, R. & Joseph, L. 2009. Shifting latitudinal clines in avian body size correlate with global warming in Australian passerines. *Proceedings of the Royal Society B – Biological Sciences*, **276**, 3845–3852.

Gibson, L., McNeill, A., De Tores, P., Wayne, A. & Yates, C. 2010. Will future climate change threaten a range restricted endemic species, the quokka (*Setonix brachyurus*), in south west Australia? *Biological Conservation*, **143**, 2453–2461.

Grantham, B. A., Eckert, G. L. & Shanks, A. L. 2003. Dispersal potential of marine invertebrates in diverse habitats. *Ecological Applications*, **13**, 108–116.

Green, K. 2010. Alpine taxa exhibit differing responses to climate warming in the Snowy Mountains of Australia. *Journal of Mountain Science*, **7**, 167–175.

Green, K., Stein, J. A. & Driessen, M. M. 2008. The projected distributions of *Mastacomys fuscus* and *Rattus lutreolus* in south-eastern Australia under a scenario of climate change: potential for increased competition? *Wildlife Research*, **35**, 113–119.

Guerin, G. R. & Lowe, A. J. 2013. Leaf morphology shift: new data and analysis support climate link. *Biology Letters*, **9**(1), 20120860.

Guerin, G. R., Wen, H. X. & Lowe, A. J. 2012. Leaf morphology shift linked to climate change. *Biology Letters*, **8**, 882–886.

Guisan, A. & Zimmermann, N. E. 2000. Predictive habitat distribution models in ecology. *Ecological Modelling*, **135**, 147–186.

Hannah, J. & Bell, R. G. 2012. Regional sea level trends in New Zealand. *Journal of Geophysical Research: Oceans*, **117**, C01004.

Harsch, M. A., Buxton, R., Duncan, R. P., *et al.* 2012 Causes of tree line stability: stem growth, recruitment and mortality rates over 15 years at New Zealand *Nothofagus* tree lines. *Journal of Biogeography*, **39**, 2061–2071.

Harvell, C. D., Mitchell, C. E., Ward, J. R., *et al.* 2002. Climate warming and disease risks for terrestrial and marine biota. *Science*, **296**, 2158–2162.

Hassall, C., Thompson, D., French, G. & Harvey, I. 2007. Historical changes in the phenology of British Odonata are related to climate. *Global Change Biology*, **13**, 1–9.

Hay, J., Sarre, S., Lambert, D., Allendorf, F. & Daugherty, C. 2010. Genetic diversity and taxonomy: a reassessment of species designation in tuatara (Sphenodon: Reptilia). *Conservation Genetics*, **11**, 1063–1081.

Hennessy, K., Whetton, P., Walsh, K., *et al.* 2007. Climate change effects on snow conditions in mainland Australia and adaptation at ski resorts through snowmaking. *Climate Research*, **35**, 255–270.

Hoekstra, J. M., Boucher, T. M., Ricketts, T. H. & Roberts, C. 2005. Confronting a biome crisis: global disparities of habitat loss and protection. *Ecology Letters*, **8**, 23–29.

Hovenden, M. J. & Williams, A. L. 2010. The impacts of rising CO_2 concentrations on Australian terrestrial species and ecosystems. *Austral Ecology*, **35**, 665–684.

Hughes, L. 2008. 10 things to do about climate change. In: **Lindenmayer, D., Dovers, S., Harriss Olsen, M. & Morton, S.** (eds.) *Ten Committments*. Canberra: CSIRO Publishing.

Hughes, L. 2012. Can Australian biodiversity adapt to climate change? In: **Lunney, D. & Hutchings, P.** (eds.) *Wildlife and Climate Change: Towards Robust Conservation Strategies for Australian Fauna*. Mosman, NSW, Australia: Royal Zoological Society of NSW.

Hughes, L., Cawsey, E. M. & Westoby, M. 1996. Climatic range sizes of *Eucalyptus* species in relation to future climate change. *Global Ecology and Biogeography Letters*, **5**, 23–29.

Hunter, D. A., Speare, R., Marantelli, G., *et al.* 2010. Presence of the amphibian chytrid fungus *Batrachochytrium dendrobatidis* in threatened corroboree frog populations in the Australian Alps. *Diseases of Aquatic Organisms*, **92**, 209–216.

Kilpatrick, A. M., Briggs, C. J. & Daszak, P. 2010. The ecology and impact of chytridiomycosis: an emerging disease of amphibians. *Trends in Ecology & Evolution*, **25**, 109–118.

Kirschbaum, M. U. F., Watt, M. S., Tait, A. & Ausseil, A.-G. E. 2012. Future wood productivity of *Pinus radiata* in New Zealand under expected climatic changes. *Global Change Biology*, **18**, 1342–1356.

Klamt, M., Thompson, R. & Davis, J. 2011. Early response of the platypus to climate warming. *Global Change Biology*, **17**, 3011–3018.

Kriticos, D. J. 2012. Regional climate-matching to estimate current and future sources of biodiversity threats. *Biological Invasions*, **14**, 1533–1544.

Last, P. R., White, W. T., Gledhill, D. C. *et al.* 2010. Long-term shifts in abundance and distribution of a temperate fish fauna: a response to climate change and fishing practices. *Global Ecology and Biogeography*, **20**, 58–72.

Laurance, W. F., Dell, B., Turton, S. M., *et al.* 2011. The 10 Australian ecosystems most vulnerable to tipping points. *Biological Conservation*, **144**, 1472–1480.

Ling, S., Johnson, C., Ridgway, K., Hobday, A. & Haddon, M. 2008. Climate-driven range extension of a sea urchin: inferring future trends by analysis of recent population dynamics. *Global Change Biology*, **15**, 719–731.

Lough, J. M. 2008. Shifting climate zones for Australia's tropical marine ecosystems. *Geophysical Research Letters*, **35**, L14708.

Lundquist, C. J., Ramsay, D., Bell, R., Swales, A. & Kerr, S. 2011. Predicted impacts of climate change on New Zealand's biodiversity. *Pacific Conservation Biology*, **17**, 179–191.

McAlpine, C. A., Syktus, J., Deo, R. C., *et al.* 2007. Modeling the impact of historical land cover change on Australia's regional climate. *Geophysical Research Letters*, **34**.

McAlpine, C. A., Ryan, J. G., Seabrook, L., *et al.* 2010. More than CO_2: a broader paradigm for managing climate change and variability to avoid ecosystem collapse. *Current Opinion in Environmental Sustainability*, **2**, 334–346.

McGlone, M. & Walker, S. 2011. *Potential Effects of Climate Change on New Zealand's Terrestrial Biodiversity and Policy Recommendations for Mitigation, Adaptation and Research*, Wellington, Department of Conservation.

MFE 2008. Climate change effects and impacts assessment: a guidance manual for local government in New Zealand. In: Mullan, B., Wratt, D., Dean, S., *et al.* (eds.) 2 edn. Wellington: Ministry for the Environment.

Mitchell, N. J., Allendorf, F. W., Keall, S. N., Daugherty, C. H. & Nelson, N. J. 2010. Demographic effects of temperature-dependent sex determination: will tuatara survive global warming? *Global Change Biology*, **16**, 60–72.

Mitchell, N. J., Kearney, M. R., Nelson, N. J. & Porter, W. P. 2008. Predicting the fate of a living fossil: how will global warming affect sex determination and hatching phenology in tuatara? *Proceedings of the Royal Society B – Biological Sciences*, **275**, 2185–2193.

Mitchell, N. J., Nelson, N. J., Cree, A., *et al.* 2006. Support for a rare pattern of temperature-dependent sex determination in archaic reptiles: evidence from two species of tuatara (Sphenodon). *Frontiers in Zoology*, **3**, doi:10.1186/1742-9994-3-9.

Mullan, A. B., Stuart, S. J., Hadfield, M. G. & Smith, M. J. 2010. Report on the Review of NIWA's 'Seven-Station' Temperature Series. *NIWA Information Series No. 78*. New Zealand.

Narisma, G. & Pitman, A. 2003. The impact of 200 years of land cover change on the Australian near-surface climate. *Journal of Hydrometeorology*, **4**, 424–436.

Nelson, N. J., Thompson, M. B., Pledger, S., Keall, S. N. & Daugherty, C. H. 2004. Do TSD, sex ratios, and nest characteristics influence the vulnerability of tuatara to global warming? *International Congress Series*, **1275**, 250–257.

Parmesan, C. & Yohe, G. 2003. A globally coherent fingerprint of climate change impacts across natural systems. *Nature*, **421**, 37–42.

Penman, T. D., Pike, D. A., Webb, J. K. & Shine, R. 2010. Predicting the impact of climate change on Australia's most endangered snake, *Hoplocephalus bungaroides*. *Diversity and Distributions*, **16**, 109–118.

Pickering, C., Good, R. & Green, K. 2004. *Potential Effects of Global Warming on the Biota of the Australian Alps*, Canberra, Technical Report. Australian Greenhouse Office.

Pitt, N. R., Poloczanska, E. S. & Hobday, A. J. 2010. Climate-driven range changes in Tasmanian intertidal fauna. *Marine and Freshwater Research*, **61**, 963–970.

Pittock, J. & Finlayson, C. M. 2011. Australia's Murray–Darling Basin: freshwater ecosystem conservation options in an era of climate change. *Marine and Freshwater Research*, **62**, 232–243.

Reside, A. E., VanDerWal, J., Kutt, A., Watson, I. & Williams, S. 2012. Fire regime shifts affect bird species distributions. *Diversity and Distributions*, **18**, 213–225.

Ridgway, K. R. 2007. Long-term trend and decadal variability of the southward penetration of the East Australian Current. *Geophysical Research Letters*, **34**, L13613.

Rosenzweig, C., Karoly, D., Vicarelli, M., *et al.* 2008. Attributing physical and biological impacts to anthropogenic climate change. *Nature*, **453**, 354–358.

Smale, D. A. & Wernberg, T. 2013. Extreme climatic event drives range contraction of a habitat-forming species. *Proceedings of the Royal Society B – Biological Sciences*, **280**, 20122829.

Steffen, W., Burbidge, A. A., Hughes, L., *et al.* 2009. *Australia's Biodiversity and Climate Change: A Strategic Assessment of the Vulnerability of Australia's Biodiversity to Climate Change*. A report to the Natural Resource Management Ministerial Council Commissioned by the Australian Government.

Telemeco, R. S., Elphick, M. J. & Shine, R. 2009. Nesting lizards (*Bassiana duperreyi*) compensate partly, but not completely, for climate change. *Ecology*, **90**, 17–22.

Towns, D. R. & Daugherty, C. H. 1994. Patterns of range contractions and extinctions in the New Zealand herpetofauna following human colonisation. *New Zealand Journal of Zoology*, **21**, 325–339.

Umina, P. A., Weeks, A. R., Kearney, M. R., McKechnie, S. W. & Hoffmann, A. A. 2005. A rapid shift in the classic clinal pattern in *Drosophila* reflecting climate change. *Science*, **308**, 691–693.

Ummenhofer, C. C., Sen Gupta, A. & England, M. H. 2009. Causes of late twentieth-century trends in New Zealand precipitation. *Journal of Climate*, **22**, 3–19.

UN General Assembly 1994. Article 2. *United Nations Framework Convention on Climate Change: resulition / adopted by the General Assembly*. Available at http://unfccc.int/ essential_background/convention/background/items/1353.php [Accessed 4th September 2013].

VanDerWal, J., Murphy, H. T., Kutt, A. S., *et al.* 2012. Focus on poleward shifts in species' distribution underestimates the fingerprint of climate change. *Nature Climate Change*, **3**, 239–243.

Watt, M. S., Stone, J. K., Hood, I. A. & Manning, L. K. 2011. Using a climatic niche model to predict the direct and indirect impacts of climate change on the distribution of Douglas-fir in New Zealand. *Global Change Biology*, **17**, 3608–3619.

Webb, L., Whetton, P. & Barlow, E. 2011. Observed trends in winegrape maturity in Australia. *Global Change Biology*, **17**, 2707–2719.

Welbergen, J. A., Klose, S. M., Markus, N. & Eby, P. 2008. Climate change and the effects of temperature extremes on Australian flying-foxes. *Proceedings of the Royal Society B – Biological Sciences*, **275**, 419–425.

Wernberg, T., Smale, D., Tuya, F., *et al.* 2013. An extreme climatic event alters marine ecosystem structure in a global biodiversity hotspot. *Nature Climate Change*, **3**, 78–82.

Williams, S. E., Bolitho, E. E. & Fox, S. 2003. Climate change in Australian tropical rainforests: an impending environmental catastrophe. *Proceedings of the Royal Society of London B*, **270**, 1887–1892.

Willis, K. & Macdonald, G. 2011. Long-term ecological records and their relevance to climate change predictions for a warmer world. *Annual Review of Ecology, Evolution, and Systematics*, **42**, 267–287.

CHAPTER 5

Unwelcome and unpredictable: the sorry saga of cane toads in Australia

Richard Shine and Benjamin L. Phillips

Summary

The history of biological invasions, and of attempts to combat them, is dominated by stories of futility. Especially in the case of invasive species that have a high public profile, like the cane toad (*Rhinella marina*) in Australia, the voices of scientists often have been drowned out in the roar of populist political debate. There have been many failures by science as well, beginning with the initial importation of toads for biocontrol, and extending through failed attempts to predict future patterns of toad colonisation and impact, or to devise practical means to ameliorate that impact. In our (highly biased) opinion, progress in understanding such issues and in developing effective means of toad control emerged only from detailed ecological research on toad biology, rather than enthusiastic but poorly informed attempts to find better ways to slaughter toads. The story of cane toad science in Australia is a cautionary tale for our ability to predict the future, because, to date, we have done an abysmal job of doing so.

'In the biological control of insect pests, . . . perhaps more than in any other [field], the path to successful achievement is strewn with the remains of optimistic attempts which have ended in abject failure. . . . Such a project is not to be embarked upon lightheartedly, but only after the most mature consideration, since a false step may have most disastrous . . . consequences through the upsetting of the whole biological balance.' R. W. Mungomery (1934, p. 5), written the year *before* he brought cane toads to Australia.

5.1 Introduction

Cane toads are the invaders that Australians love to hate. These large and 'warty' anurans, often likened to mobile cowpats in appearance (Figure 5.1, Plate 26), were brought to northeastern Australia in 1935 in a futile attempt to control beetles that were plaguing commercial sugarcane production (Lever, 2001; Kraus, 2009). The toads have since spread widely, and now occur across the northern parts of Queensland (QLD),

Austral Ark: The State of Wildlife in Australia and New Zealand, eds. A. Stow, N. Maclean and G. I. Holwell. Published by Cambridge University Press. © Cambridge University Press 2015.

83

Figure 5.1 (Plate 26) The cane toad (*Rhinella marina*) was brought from South America to Australia (via Puerto Rico and Hawaii), and released in northeastern Queensland in 1935. This is a male toad (note the orange colour, rugose dorsal skin, and dark nuptial pads on inner fingers) captured from the invasion front in 2004 (Mary River in Australia's Northern Territory). Photograph by Ben Phillips. A black and white version of this figure will appear in some formats. For the colour version, please refer to the plate section.

the Northern Territory (NT) and Western Australia (WA). These unwelcome anurans also have expanded their range southwards along the east coast, with established populations in northeastern New South Wales (NSW) and stowaway-founded satellite populations as far south as Sydney (White and Shine, 2009).

Australia is now home to many invasive species, but few have attracted as much attention from the general public as has the cane toad. Part of the reason is that cane toads are common in most places that they occur, and they are relatively easy to identify (but, see Somaweera *et al.*, 2010). They are also large, occasionally growing up to 230 mm long and weighing more than a kilogram. Their large body size, cumbersome manner and nocturnal habits may help to explain why so many Australians regard cane toads as repugnant. But there are probably also deep-rooted cultural reasons for the dislike of toads: like many invasive species, toads are seen as a legitimate target for xenophobic feelings (Simberloff, 2003). In our subjective view, some of the smaller native species of frogs would be unlikely to win a beauty contest if competing with cane toads, and yet toads attract an ire well beyond that directed to native frogs. Much of the public's hatred of toads is irrational. People often defend their dislike for the animal with nonsensical suggestions (e.g. that school-children will abandon their studies in favour of smoking toad-toxin cigarettes, as suggested by a newspaper in the city of Darwin) or opinions that are inconsistent with evidence – for example, that toad poisons will contaminate drinking water (see Beckmann and Shine, 2010), or that the toad invasion will devastate fisheries (Somaweera *et al.*, 2011). This mythology has strongly influenced decision-making by governments and land managers. Indeed, from the earliest decisions about the importation of toads, through to the contemporary efforts at toad control, most decisions have been influenced more by mythology of the toad rather than by factual evidence.

Regardless of its motivation, intense public concern about the cane toad invasion has stimulated extensive research over many decades, as well as a vast expenditure of time, money and volunteer effort on attempts to control cane toads. In this chapter, we look back on Australia's history with toads, and provide our own (admittedly biased) perspective on that history. Throughout, we examine the ability of researchers to predict the future, and to influence public policy. Thus, the chapter is structured in terms of a diverse suite of predictions (made by us as well as others) about toad invasion and impact, with an evaluation of the rationale for the prediction, and of the degree to which subsequent events have supported the initial suggestions. With an eye to irony (given the hopeless track record of predictions failing to be supported by subsequent research), we end with a new set of predictions about the future of cane toads in Australia.

5.2 Predictions about cane toad impact

5.2.1 Cane toads will help to control beetles in sugarcane

In the 1930s, when the decision was taken to import toads to Australia, scientists at the Bureau of Sugar Experimental Stations (BSES) had access to two kinds of evidence about the potential usefulness of toads for beetle control.

(a) A report (in the prestigious journal *Nature*, under the headline 'Toads save sugar crop': Anonymous, 1934) that the Puerto Rican sugarcane crop had been saved by introducing the cane toad. That report seems to have been based on a single year of higher-than-average sugar yields following toad introduction, an event that may well have resulted from local weather conditions.

(b) Dissection of toads from the Puerto Rican plantations, revealing that a high proportion of the insects consumed were 'harmful' pest species (Dexter, 1932).

This is shockingly weak evidence (Turvey, 2009). It is broadly equivalent to waking with a headache, eating a piece of chocolate and concluding, following the easing of the headache, that the chocolate saved you from an impending stroke because of the antioxidants it contained. So, why were toads brought to Australia? The answer lies more in circumstance and politics rather than science.

First, the decision to import cane toads into Australia was taken in the early 1930s, at a time when scientists were naïvely optimistic about the feasibility of solving ecological problems by introducing alien taxa. One of the great Australian success stories of biocontrol – *Cactoblastis* moths to kill prickly pear – had been played out only recently, and the future of biological control looked bright. Many translocations were being planned – for example, in 1935 (the same year that cane toads were brought to tropical Australia), European toads (*Bufo bufo*) were brought to Melbourne by CSIRO in the hope that they might be a useful biocontrol for insect pests in southern Australia (Lowe, 1999). Thankfully, the European toads were never released: if they had been, Australia would have faced an even greater toad problem.

Second, the decision to import cane toads came at a time when sugar producers were staggering under a string of bad seasons (Turvey, 2009). There was clearly a big problem, and the people to solve it would be lauded as heroes. The science, therefore, was secondary to the political agenda and never progressed beyond the 'proof of concept' stage. The reports of the toads' effectiveness in Puerto Rican and Hawaiian plantations were based on very weak data (i.e. no comparison of yields from areas with and without toads). And the dissection data – showing that toads consume harmful beetles (Dexter, 1932) – tells us nothing about the population-level impact of toads on beetle pests.

In hindsight, the decision to bring cane toads to Australia was a bad one. Although several publications have lampooned the decision, some authors have been unduly harsh. For example, the fact that beetles can fly whereas toads cannot (thus reducing the accessibility of beetles to foraging toads) was clearly understood by the scientists involved. Indeed, this issue was considered carefully by the man who ultimately brought cane toads to Australia. Reginald Mungomery, of the Bureau of Sugar Experimental Stations (BSES), speculated that toads may be of little use in southern cane-growing areas because the economically important beetle species in that region spent too little time on the ground to be accessible to foraging toads (N. Turvey, personal communication, based on original letters between Mungomery and his boss Arthur Bell). We should not forget that these were smart, passionate men – they were experts – but they were making predictions based on very little real evidence. Moreover, they badly wanted to find an answer to the problem, and biological control in the form of toads looked like a good possibility.

Nonetheless, the decision was clearly bad, and the process of evaluating potential effectiveness and impacts of the toad was lackadaisical – especially in comparison to efforts a few years earlier to control prickly pear (N. Turvey, personal communication). Did toads ever have an impact on cane beetles? Surprisingly, we still don't really know because appropriate experimental trials have never been undertaken. And they probably never will be: the relevance of toads for beetle control was obviated less than five years after their arrival in Australia, when the new technological fix – organochlorine pesticides – arrived on the scene. By then, of course, toads were here to stay. Recent newspaper reports suggest that cane-beetle grubs are still a major problem for the commercial sugarcane industry.

5.2.2   Cane toads will not affect other agricultural activities

As agricultural scientists, the sugarcane researchers responsible for introducing cane toads to Australia undoubtedly considered the possibility of collateral impact of toads on other agricultural industries. Complaints from poultry producers that toads would poison their chickens prompted Buzzacott (1938) to conduct experimental trials, leading him to conclude that domestic fowls can consume small toads without ill effect. Although the claim of negative impacts of toads on poultry continues to be repeated in popular discussion (such as in online forums), more rigorous research has supported Buzzacott's views (Beckmann and Shine, 2010). Apiarists also expressed concern about toad predation on their bees (Goodacre, 1947; Hewitt, 1956), a concern that is still echoed today; however, there is no evidence that toads have had an economically significant effect on commercial bee production.

The most important effect of toads on agriculture probably involves a segment of the agricultural community that didn't exist when toads were first introduced. Cane toads have impacted another (and far more successful) attempt at biological control: the introduction of dung beetles (in the 1960s through the 1980s) to break down piles of bovine faeces (cowpats) that were accumulating on pastureland and thus reducing productivity. Waterhouse's (1974) speculation that toad predation on dung beetles might reduce the effectiveness of these introduced insects in cowpat decomposition, has been supported by recent work (González-Bernal et al., 2013).

5.2.3   Cane toads will not have significant impacts on the native fauna

From the beginning, there were divergent views about the impact toads may have on native fauna. On one hand were the men from the BSES. Soon after toads were first released, Mungomery (1936) regarded the toad's introduction as a great success, with even more advantages than at first envisaged e.g. 'Flattering reports were also being received from some of the housewives, who asserted that the rather disquieting presence of cockroaches in their houses had in many cases been reduced through the activities of the toad' (1936, p. 68). The main critic of cane toad introduction to Australia in the 1930s was Walter W. Froggatt, an eminent entomologist (Froggatt, 1936). Froggatt argued that 'this great toad, immune from enemies, omnivorous in its habits, and breeding all the year round, may become as great a pest as the rabbit or the cactus' (1936, p. 164). He opposed the introduction of cane toads based on the general proposition that they might disturb the 'balance of Nature', but more specifically on his prediction that toads would prosper because their toxins rendered them invulnerable to predators (who, he assumed, would soon learn not to eat them) and that they would have major impacts as predators due to their fabled appetites for ground-dwelling fauna. The BSES scientists clearly resented Froggatt's criticisms, with Mungomery stating that 'we are mindful of certain adverse criticisms which have been directed against the importation of the toad colony, and certain innuendos concerning the ultimate result. It seems almost inevitable that all attempts at progress should meet with totally unwarranted opposition... [but] there appears to be no reason for the assumption that we have made an error in our judgement' (1936, pp. 73 and 74). In hindsight, of course, Froggatt was absolutely correct to sound a note of caution in a situation where that voice had been eerily silent. Froggatt's concern prompted a ban on the further release of toads in late 1936, a ban that was over-ruled less than six months later by word directly from the Prime Minister's office (Turvey, 2009). Again, evidence, science and the precautionary principle (espoused even by the proponents of the release of toads) played second fiddle to politics and public opinion.

Although Froggatt reached the right conclusion (introducing cane toads to Australia was indeed a bad idea), he did so on the wrong grounds. Predation by toads is not a major mechanism of impact (Greenlees *et al.*, 2007; Pearson *et al.*, 2009; Boland, 2004 reported toad predation on the nestlings of ground-nesting birds, but subsequent work challenges the generality of that effect: Beckmann and Shine, 2012; Table 5.1). Instead, the primary ecological impact of cane toads in Australia has been by fatally poisoning predators that attempt to ingest these toxic newcomers (Doody *et al.*, 2009; Shine, 2010; Figure 5.2 (Plate 27); Table 5.1). Many predators (such as snakes) are killed when toads arrive, and populations of large lizards (yellow-spotted monitors, bluetongue skinks) and marsupial quolls have declined precipitously (by >90%) after toad invasion (Burnett, 1997). In some areas, freshwater crocodile populations have plummeted also (Letnic *et al.*, 2008). Declines in these predator populations can have ramifying effects throughout a community (discussed further below).

In view of these dramatic effects, it is surprising to note that fatal toxic ingestion took a long time to be recognised as a mechanism of toad impact. Some early papers noted declines in frog-eating predators (such as snakes, frill-necked lizards, and goannas: Breeden, 1963; Rayward, 1974), but it wasn't until a seminal paper by Covacevich and Archer (1975) brought these various observations together that a clear picture began to emerge. The failure to *a priori* predict the impact of toads on predators (by Mungomery,

Table 5.1   A summary of published results on the predicted population-level ecological impact of cane toads in tropical Australia. The table deals only with direct effects of toads, so actual population-level effects *in situ* are often different from those predicted here.

Taxon	Mechanism of impact	Direction of impact	Intensity of impact	Reference
Invertebrates (most spp.)	Eaten by toads	Negative	Minor?	Greenlees *et al.* 2006
Meat ants	Eat toads	Positive	Unknown	Ward-Fear *et al.* 2009, 2010
Dytiscid beetles, etc.	Eat tadpoles	Positive	Unknown	Cabrera-Guzmàn *et al.* 2012
Fishes	Fatal ingestion of toad eggs	Negative	Minor	Greenlees and Shine 2011; Somaweera *et al.* 2011
Frogs	Compete for food	Negative	Minor	Greenlees *et al.* 2007; Shine *et al.* 2009; Shine 2014
	Eaten by toads	Negative	Minor	Greenlees *et al.* 2007, 2010
	Fatal ingestion of toads	Negative	Minor	Greenlees *et al.* 2010; Nelson *et al.* 2011
	Parasite transfer	Positive	Minor	Lettoof *et al.* 2013
	Parasite transfer	Negative	Minor	Pizzatto *et al.* 2010, 2012
Litoria splendida	Parasite transfer	Negative	Major?	Pizzatto and Shine 2011
Crocodiles	Fatal ingestion of toads	Negative	Variable	Letnic and Ward 2005; Letnic *et al.* 2008; Somaweera and Shine 2012
Turtles	Fatal ingestion of eggs	Negative	Minor	Greenlees and Shine 2011
Lizards (most species)	Fatal ingestion of toads	Negative	Minor	Shine 2010; Pearson *et al.*, 2014
Varanids	Fatal ingestion of toads	Negative	Major	Burnett 1997; Doody *et al.* 2009; Brown *et al.* 2013
Tiliquine skinks	Fatal ingestion of toads	Negative	Major	Price-Rees *et al.* 2010, 2012
Snakes (most species)	Fatal ingestion of toads	Negative	Minor	Phillips *et al.* 2003; Shine 2010
Tropidonophis	Eat toads	Positive	Minor	Llewelyn *et al.* 2010, 2014; Brown *et al.* 2011
Stegonotus	Eat toads	Positive	Minor	Brown *et al.* 2011
Birds	Fatal ingestion of toads	Negative	Minor	Beckmann and Shine 2009, 2012
	Eat toads	Positive	Minor	Beckmann and Shine 2009, 2010, 2012
Mammals (most spp.)	Fatal ingestion of toads	Negative	Minor	Webb *et al.* 2008; Kaemper *et al.* 2013
Dasyurus	Fatal ingestion of toads	Negative	Major	O'Donnell *et al.* 2010

Froggatt and contemporaries) may reflect their training (they were all entomologists), as well as a lack of evidence from other parts of the toads' introduced range. Within the toads' native range, for example, predators have adapted to deal with toads (either by not

Figure 5.2 (Plate 27) A radio-tracked death adder (*Acanthophis praelongus*) found dead in the field after it was fatally poisoned by a very large cane toad (*Rhinella marina*). Photograph by Ben Phillips. A black and white version of this figure will appear in some formats. For the colour version, please refer to the plate section.

eating them, or by evolving physiological resistance to bufotoxins: Llewelyn *et al.*, 2010; Shine, 2010; Ujvari *et al.*, 2013). In Hawaii, there were no large native predators to consume toads and die. Thus, the devastating impact of the toads' toxins on a native fauna was not seen until the toads were brought to a continent (Australia) that lacks native bufonids (toads) and thus, whose fauna included many species with no significant evolutionary history of exposure to the toads' distinctive toxins (Shine, 2010).

5.2.4 Cane toad invasion will have long-lasting impacts on most native species

Following Covacevich and Archer's (1975) report that cane toad toxins can kill native predators, and in the light of a growing public conservation ethos, the public and scientific debate about cane toad impacts took on a very different flavour. In strong contrast to the earlier complacency about minimal impact, recent discussions have argued that toad impacts have been substantial and long-lasting. In the most extreme example, some community groups have suggested that the toad invasion eliminates most native fauna. However, the available database has been scant, mostly consisting of subjective (albeit sometimes compelling) reports that abundance of specific taxa (notably quolls and goannas) declined precipitously at the time of toad arrival (e.g. Burnett, 1997). The first detailed study to examine the population-level impact of toads didn't happen until the late 1980s and was conducted in the remote wet–dry savanna around Boroloola in Australia's wet–dry tropics. Here, Bill Freeland and his team revealed a less pessimistic scenario. Based on surveys before and behind the advancing toad front, Freeland and his collaborators concluded that toads had little population-level effect on most native animal populations they examined (including frogs, a group identified by many people as a group at particular risk: Freeland and Kerin, 1988). Freeland (2004) also found that initial effects on large varanid lizards were devastating, but suggested that this impact may be a short-term one. Although the work was not conclusive, the possibility that toads were actually having fairly minor impacts attracted vociferous disagreement from at least some community-group leaders. Freeland's results were also ignored by much of the scientific community, perhaps because of a fear that a lack of quantifiable impact might not mean a real lack of impact.

Over most of this period, the toad front was expanding through relatively inaccessible areas where research was difficult. As the cane toad invasion front reached the vicinity of Darwin, where logistical impediments to research were reduced, studies on toad impact expanded enormously. For example, we conducted laboratory-based trials to assess the ability of potentially vulnerable native predators to survive toad ingestion (based upon toxin tolerance). Combined with data on dietary habits and geographic distribution, we predicted that about 30% of Australia's terrestrial snakes were potentially imperilled by cane toads (Phillips et al., 2003). A follow-up study (Smith and Phillips, 2006) reached similar conclusions for lizards and crocodiles. Finally, population monitoring revealed that toad invasion caused catastrophic population declines in varanid lizards (Doody et al., 2009) and perhaps quolls (O'Donnell et al., 2010; Table 5.1). Community groups were formed to combat the impending disaster, and (as noted above) some of those groups actively promulgated the idea that cane toads are an ecological catastrophe for *all* native wildlife.

The truth is somewhat less black-and-white. A partial resolution to the debate about the breadth and duration of cane toad impact has emerged from our long-term studies (collaboratively with others, notably Greg Brown) on toad impact at a site (Fogg Dam) on the Adelaide River floodplain, 60 km east of Darwin. For example, Greg Brown conducted nightly counts of snake numbers along standardised routes for seven years before toads arrived, and five years after they arrived (total of >3600 nights: Brown et al., 2011). Thus, we can now confidently identify which native species have declined since toad arrival, and which have not – at least at this single study site. Those data point to a dramatic impact of cane toads on a small number of species of large predators, notably yellow-spotted monitors (*Varanus panoptes*) and bluetongue skinks (*Tiliqua scincoides intermedia*: Price-Rees et al., 2010; Brown et al., 2011). However, most other native species have been relatively unaffected by toad invasion. Many potentially vulnerable predators (such as small marsupials, fishes and frogs) rapidly learnt to avoid the toxic newcomer (Webb et al., 2008; O'Donnell et al., 2010; Shine, 2010) (as originally predicted by Froggatt). Many invertebrates (ants, water-beetles, water-bugs) and vertebrates (rodents, birds) can tolerate the toad's toxins, and thus consume them without ill-effect (Beckmann and Shine, 2009; Ward-Fear et al., 2009; Cabrera-Guzmán et al., 2012). Even for most snakes – which are vulnerable to toad toxin, and seem to be slow to learn aversion – numbers often have increased rather than decreased since the arrival of toads (Brown et al., 2011). Similarly, the abundance of native frogs is largely unaffected by cane toad arrival (Catling et al., 1999; Grigg et al., 2006; Greenlees et al., 2010; Brown et al., 2011). Table 5.1 provides a summary of population-level impacts of cane toad invasion in tropical Australia.

Even populations of taxa that are severely affected by cane toad invasion may eventually recover. Although anecdotal, reports on abundances of varanid lizards and quolls suggest that these animals are once again common in some areas where toads have been present for several decades (Shine, 2010). Importantly, at least some of the toad-vulnerable predators in such areas recognise that toads are toxic, and refuse to consume them (Llewelyn et al., 2014).

Given the powerful toxins of cane toads, and the death of individuals of so many predator species that ingest toads (Table 5.1), why have population-level responses of most such species been so weak? At least three processes have been at work, none of which were fully appreciated in earlier research. First, many predators are capable of rapid aversion learning, so that they cease to consume the potentially deadly toads (Webb *et al.*, 2008; O'Donnell *et al.*, 2010). Second, evolutionary change can occur rapidly enough to modify the morphology, physiology and behaviour of vulnerable species, in ways that render them less likely to die if they seize a toad (Phillips and Shine, 2004, 2006b). Third, the indirect effects of removing top predators (such as goannas and quolls) may have major positive effects on their prey species – effects strong enough to outweigh occasional mortality due to ingestion of (or by) toads (Doody *et al.*, 2006; Shine, 2010; Brown *et al.*, 2011). More generally, it seems that such indirect effects of toads – that is, effects mediated through the effects of toads on other native species – may be strong enough to overwhelm direct effects in many cases. The decimation of varanid lizard ('goanna') populations may be the most important such process. For example, toad-induced mortality of varanids resulted in increased survival rates of the eggs of freshwater turtles (Doody *et al.*, 2006), crocodiles, and perhaps other species of lizards and snakes (Shine, 2010).

The apparent recovery of vulnerable predators a few decades after toad arrival remains puzzling. It might be due to adaptive changes, as have been documented in red-bellied blacksnakes *Pseudechis porphyriacus* (Phillips and Shine, 2004, 2006b). Alternatively, changes in toad abundance or age structure after initial colonisation might reduce toad impact (Shine, 2010). For example, invasion-front toad populations consist primarily of large adult animals, that possess so much toxin that ingesting them is fatal to a predator. In contrast, older-established populations have a broader age (and thus size) structure, with many juvenile toads that possess far less toxin than their larger conspecifics (Phillips and Shine, 2006a). Thus, predators in long-colonised areas may have a greater opportunity to encounter (and learn to avoid) juvenile toads, whereas that taste-aversion-learning opportunity is not available at the invasion front (Shine, 2010).

This more nuanced view of the long-term impact of cane toads has not, to date, penetrated the public's consciousness to any great degree. For example, community groups still claim devastating toad impacts on species that (based on actual data) appear to be unaffected (e.g. birds, fishes). Why has the new knowledge spread so slowly? Part of the reason likely is because that process takes time. Another factor is that the data paint a more complex picture, difficult to communicate in the short sound-bites that are the core of modern media. And third, public attitudes to toads are hard-boiled: people think that toads are 'bad', and thus are inclined to accept the idea that these feral pests <u>must</u> have bad effects on everything. Community groups have expended millions of dollars and thousands of person-hours attempting to stop toads entering Western Australia. A major justification for this expenditure was the impact of toads on native wildlife, even though the real extent of this impact has not been as bad as people feared.

5.3   Predictions about the rates and extent of range expansion by cane toads

5.3.1   Geographical extent of cane toad invasion

The early writing in favour of introducing toads into Australia was largely silent on the predicted final extent of the toads' distribution in Australia. This seems a staggering oversight given that toads now occupy more than 1.3 million square kilometres of

Australia (Urban *et al.*, 2007). The one exception again was Walter Froggatt, our 1936 voice of dissent who claimed there was 'no limit to [their] westward range' (p. 164).

Froggatt's prediction was vague and based on a hunch, but he was right: toads have certainly spread more than 2000 km westward from their initial release points around Gordonvale in north Queensland. Several researchers have attempted to predict the ultimate extent of cane toad colonisation in Australia (Phillips *et al.*, 2008), either by identifying areas where the climate resembles that in the toad's native range (e.g. Sutherst *et al.*, 1995), by extrapolating climates from the current localities inhabited by toads in Australia (van Beurden, 1981; Urban *et al.*, 2008), or by using laboratory-derived data on the effects of ambient temperature on toad growth and performance (e.g. Kearney *et al.*, 2008) to draw lines in the sand. The models have agreed in many places – for example, in predicting that toads will be able to thrive in a large coastal strip running from northern NSW to Broome in northern WA – but have differed widely in others. In particular, the western and southern edges of the toads' predicted range vary wildly between modelling efforts (Figure 5.3).

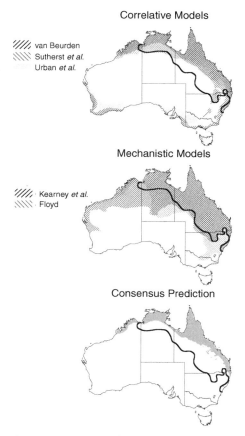

Figure 5.3 A comparison of predictions from mathematical models of the potential geographic distribution of cane toads in Australia. Correlative models use known toad localities to predict future range extent. Mechanistic models, on the other hand, use the toads' basic physiology to predict where they are unable to survive. The black line demarcates the extent of the toads' range as of 2008. (Figure courtesy of Michael Kearney. Reprinted from Phillips, B. L., Chipperfield, J. D., & Kearney, M. R. (2008) with permission of CSIRO Publishing.)

The increasing sophistication of analytical techniques, and the growing availability of empirical data, means that the more recent modelling exercises (e.g. Kearney *et al.*, 2008) are likely to be the most accurate. However, the task is a complex one, and none of the models will, ultimately, be perfectly correct. It is clear that toads have already expanded well beyond the bounds predicted by some of the earlier models (Phillips *et al.*, 2008). Models that draw empirical relationships between climate at a locality and the presence/absence of toads ('correlative models') have consistently under-predicted the extent of habitat available to toads. Extrapolating these known presence/climate associations into new climates results in a prediction that is often little better than an educated guess, and in the case of toads has tended to underestimate the toads' potential range in Australia. Models that use the toads' physiology to evaluate climatic suitability ('mechanistic models') have fared better, but inherently over-predict the extent of available habitat, so it is perhaps unsurprising that their predictions have not been over-run yet.

Nonetheless, all of these models will ultimately be wrong to some degree, it is just a matter of by how much. None of the current models, for example, capture the ability of toads to acclimate to new conditions (Kolbe *et al.*, 2010), and the inherent unpredictability of contemporary climate change makes future predictions of the toads' range even more uncertain. Other important difficulties in predicting toad distribution are the complex spatial and temporal heterogeneity of environmental conditions (e.g. 'heat islands' around cities), coupled with the toads' ability to stow away on trucks travelling to areas many hundreds of kilometres outside the main range (White and Shine, 2009). Additionally, rapid adaptive shifts may allow toads to survive in areas that seem unsuitable: for example, toads in arid Queensland display modifications to both their behavioural responses to desiccation and the permeability of their skin to water (Tingley *et al.*, 2012). As a result, toads may prove capable of building up populations even in areas that seem unsuitable for thermal or hydric reasons. One example of this is a population of cane toads that (at the time of writing) live in suburban Sydney, close to the lower thermal limit predicted by models (White and Shine, 2009). Many such populations may flourish only briefly, but might nonetheless inflict significant ecological damage during that period.

5.3.2   Rates of cane toad invasion

Given that there was little real comment on the likely extent of toads in Australia, there was obviously no comment on how rapidly toads might be expected to reach that final extent. The first prediction here came late and was based on solid empirical data. It was, nonetheless, very wrong. In 1985, Freeland and Martin predicted that toads would reach the Darwin area in 2027; in fact, they arrived in 2005 (Brown *et al.*, 2011): that is, in 20 not 40 years. Why this error of prediction, especially given that the spread of cane toads through Australia has been documented in detail? Surveys by several workers (e.g. Easteal *et al.*, 1985) established the times of arrival of toads at specific sites, allowing calculation of rates of spread. The problem with Freeland and Martin's (1985) prediction is that it did not recognise that the toad invasion was accelerating. Rates of frontal expansion have increased dramatically over the 77-year history of the toad's colonisation of tropical Australia (Phillips *et al.*, 2006). Toads expanded their range by about 10–15 km per annum in the years following their release, but this accelerated to about 50–60 km per annum by the time they reached the vicinity of Darwin. As a result, early predictions about the likely dates of arrival of toads in the western parts of Australia (by extrapolation from existing rates of spread) were in substantial error.

5.4 Predictions about toad control

Far more money has been spent on attempting to control toads than on understanding them. Broadly, the results of control efforts have been disappointing – for example, toads have continued to spread through tropical Australia at about the same rates in areas and at times where intensive 'toad-busting' efforts have been conducted, as in areas and at times when such efforts were lacking (Peacock, 2007). Some community groups believe that their collecting efforts have slowed the toad invasion, but cannot produce any evidence to support that claim. Mathematical models of the impacts of toad collection on population numbers conclude that such efforts will have minimal long-term effect (McCallum, 2006).

Although the prediction of toad-busting groups that 'we CAN stop the toads' has proven to be wrong, the removal of thousands of adult toads must have reduced local population densities in the short term. The only quantitative data on this effect come from the shores of Lake Argyle in Western Australia, where intensive collecting efforts reduced local toad abundances for a few months; however, a new wave of immigrants during the following wet-season brought overall toad densities up to higher levels than were present before the collecting effort (Somaweera and Shine, 2012).

Most research into landscape-scale control of cane toads has focussed on using pathogens – either viruses from the native range, or genetically engineered viruses to disrupt toad metamorphosis (Shanmuganathan *et al.*, 2010). Both of these programmes ran the risk of high collateral damage to native species, and have now been abandoned. More recent work following on from detailed study of the toads' dispersal ecology suggests that we can halt the spread of toads in key areas by removing artificial water-bodies (which provide a crucial dispersal corridor for toads in dry areas: Florance *et al.*, 2011; Tingley *et al.*, 2013; see Figure 5.4, Plate 19). In theory, we should be able to halt the spread of toads into Australia's Pilbara region, and work is currently underway to act on this possibility. Whether we achieve this aim will, ultimately, be a matter of political will and community goodwill.

Research on local-scale control methods also continues. In recent work, we have shown that the cannibalistic nature of toad tadpoles can be exploited by luring tadpoles into traps baited with pheromones derived from the toxins of adult toads (Crossland *et al.*, 2012; see Figure 5.5, Plate 43). Other larval pheromones also offer opportunities for control of toad populations at the tadpole stage of the life-cycle (Hagman and Shine, 2009; Crossland and Shine, 2012). Toad recruitment also might be reduced by manipulations of pondside vegetation (Hagman and Shine, 2006), increasing the numbers of competing native tadpoles or predatory aquatic invertebrates (Cabrera-Guzmán *et al.*, 2011, 2012), and/or by attracting carnivorous ants to sites of metamorph abundance (Ward-Fear *et al.*, 2009). These methods have yet to be deployed at spatial scales large enough to assess their effectiveness, but they seem likely to be useful in reducing toad presence in key areas, or for eradicating small, isolated populations of toads (e.g. those resulting from accidental transport to islands).

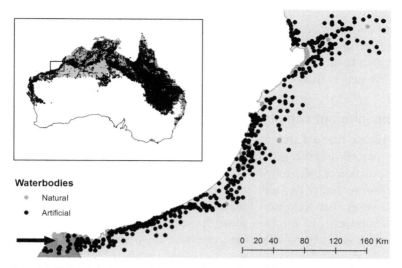

Figure 5.4 (Plate 19) In theory, excluding toads from a subset of these artificial waterbodies could prevent their natural spread into 268 000 square kilometres of potential habitat in Australia's Pilbara region. Can this be done? Can enough political will and community goodwill be mustered? Time will tell. (Image reproduced from Tingley, R., Phillips, B. L., Letnic, M., *et al.* (2013) with permission of the British Ecological Society.) A black and white version of this figure will appear in some formats. For the colour version, please refer to the plate section.

Figure 5.5 (Plate 43) One promising new approach to toad control involves curtailing recruitment, by trapping toad tadpoles. The toxin of adult cane toads is a powerful attractant for toad tadpoles, but repels native tadpoles. Thus, traps baited with toad toxin can eradicate toad tadpoles from natural waterbodies. Photograph by Michael Crossland. A black and white version of this figure will appear in some formats. For the colour version, please refer to the plate section.

5.5 Why have so many predictions failed?

Although some scientific predictions about cane toads have been validated by subsequent empirical data (e.g. the prediction by scientists that simply collecting adult toads would not curtail toad spread), our overall impression is of a litany of failure. In particular, why have so many attempts to predict toad impact and invasion dynamics been falsified by subsequent studies? Understanding the reasons for those failures may help us to improve our performance, or (at the least) encourage scientific humility.

We suggest that obstacles to successful prediction included at least three separate factors.

5.5.1   The inherent complexity of ecological communities

It is clear that a major mechanism of impact was initially ignored (i.e. the poisoning of predators); and that, when it was considered, the impact of individual deaths on the total population of predators was more complex than expected. For some native species, toad-driven mortality probably just replaced mortality that was previously driven by native predators that were now absent because of the toad. On top of this complexity of interaction, it is also now clear that the protagonists in these interactions changed rapidly, either by rapid learning, or evolved responses (Shine, 2012).

Intriguingly, the ecological impact of cane toads in Australia also has been modified by changes wrought by humans. Most obviously, toads relish the disturbed habitats and additional water sources created by humans, facilitating invasion success (Lever, 2001). Less obviously, prior biological invasions can modify the impact of subsequent arrivals. A perplexing pattern of geographic variation in tolerance to toad toxins by bluetongue skinks suggests that the introduction of a Madagascan weed with bufadienolide-like defensive chemicals may have preadapted eastern-Australian populations of these omnivorous lizards to the subsequent arrival of cane toads (Price-Rees et al., 2012).

Overall, recent research has revealed that cane toad impacts are far more complex than envisaged by earlier work. Attempting to predict anything in a system of such dynamic complexity is a truly formidable task. The magnitude of this obstacle was underestimated by both amateur and professional biologists alike.

5.5.2   A lack of empirical evidence

Despite the inherent difficulty (impossibility?) of prediction, many predictions were nonetheless made and these predictions were almost always based on very little empirical evidence, forcing authors to extrapolate apparent trends, and to rely upon untested assumptions. Given the intimidating complexity facing those making predictions, extrapolation and assumption was inevitable, but more information is always better. As one author puts it, 'knowing little . . . always makes it easier to fit what you know into a coherent pattern' (Kahneman, 2011). Very little information makes people more confident about their predictions than they should be; a well-known cognitive bias that affects everyone, not just experts. The failures to predict the eventual extent of the toads' range and the rate at which they would spread are failures based at least partly on a lack of good data to throw at the problem.

5.5.3   Discussions and predictions about cane toad invasion and impact occur within a political context

The iconic status of cane toads in Australian culture has resulted in intense popular concern about the toad's spread and impact. Because discussions about the likely consequences of toad invasion are a lightning rod for public debate, attitudes are influenced by political as well as scientific issues. For example, imagine a community group leader who publicly promulgated the notions that (1) cane toads have little effect on most native species, and (2) hand-collecting toads will have no long-term effect on their numbers. These conclusions (although consistent with scientific research) would impede the group's ability to gather resources and public support for their activities (Shine and Doody, 2011). Thus, political considerations encourage public disagreements about even straightforward results from empirical research.

5.6 Predicting the future

The dismal history of failed predictions about cane toad biology and impact in Australia should caution against further attempts. Given our criticisms of predictions made by earlier workers, however, it seems only fair to allow subsequent workers the opportunity to do the same to us. Some predictions seem clear-cut, although of course earlier workers felt the same way about their own guesses.

Cane toads will spread rapidly through tropical northwestern Australia, and perhaps into adjacent arid regions (at least after rainfall events) but will not establish major populations in the southern half of the continent. Although it is theoretically possible to prevent toads spreading to the Pilbara region, the success of that effort will depend on political will and community goodwill, as well as long-running vigilance. Toads will continue to arrive at extralimital areas through stowaway dispersal, and some of those newly founded populations will survive long enough to breed, and to affect local taxa. Cane toads will be a permanent component of Australian tropical ecosystems, although they likely will become less common overall through time. If some of the newly developed methods for toad control (e.g. Crossland *et al.*, 2012) work well on a landscape scale, removing cane toads will result in there being both winners and losers among the native fauna. Most obviously, a resurgence in abundance of varanid lizards would impose higher levels of predation on the many vertebrates and invertebrates consumed by these large predators (e.g. Doody *et al.*, 2006; Shine, 2010).

Lastly, what does our review identify as major gaps in our understanding? Although we now know a great deal more about toads than we did a decade ago, many uncertainties remain. Below, we list some of the most important.

(1) We know almost nothing about the biology of cane toads within their native range in the Americas.

(2) We know almost nothing about the rates of faunal recovery after toad invasion, and the reasons for such recovery.

(3) We do not understand the reasons for spatial heterogeneity in toad impact; in particular, why have some populations of freshwater crocodiles and death adders crashed whereas others have not (Letnic *et al.*, 2008; Brown *et al.*, 2011)?

(4) We do not understand the way the impacts of toads ramify throughout a community. Indirect effects are clearly important, but are very difficult to quantify.

(5) Do toads really become less common (and smaller and less toxic) in the decades after initial colonisation of an area? If so, why?

(6) Will we see the evolution of further host-switching between the parasites of toads and those of native frogs (Pizzatto *et al.*, 2012)?

These are big questions, and the list could go on for another several pages, but instead we finish with another, much smaller, unanswered question that provides a stark reminder of how little we really know about these troublesome anuran invaders. That is, what is the correct scientific name for the cane toad in Australia? In recent years, the long-used name *'Bufo marinus'* was replaced by *'Chaunus marinus'* (Frost *et al.*, 2006) and then, almost immediately, by *'Rhinella marina'* (Pramuk, 2007). Taxonomists continue to argue about the merits of these alternative names, but ultimately none of them may be correct. A phylogenetic analysis of these toads within their native range hinted that *'Rhinella marina'* may be a complex that contains three separate species-level taxa (Vallinoto *et al.*, 2009). We don't know which of these taxa was the one originally described by Linnaeus as *'Bufo marinus'*, raising the possibility that the cane toad – one of Australia's most high-profile and intensively studied pest species – may not, as yet, have even been formally named and described by scientists.

The fact that we may not even have the correct scientific name for the cane toad is symbolic of the speed with which current wisdom can be overturned. It is symbolic of how small scientific revolutions can play out on timescales that cause us to constantly revise our understanding, and which confound our ability to predict the future. The story of the toad is a story of failed predictions, the causes of which are universal. If we were to take a single lesson from the history of toad research, it would be to agree with Niels Bohr: 'prediction is very difficult, particularly about the future'.

Acknowledgements

We thank the Australian Research Council for funding our research on cane toads, and the members of Team Bufo for their insights, tolerance and enthusiasm.

REFERENCES

Anonymous. (1934). Toads save sugar crop. *Nature*, **134**, 877–887.

Beckmann, C., Shine, R. (2009). Are Australia's birds at risk due to the invasive cane toad? *Conservation Biology*, **23**, 1544–1549.

Beckmann, C., Shine, R. (2010). The power of myth: the (non) impact of invasive cane toads (*Bufo marinus*) on domestic chickens (*Gallus gallus*). *Animal Production Science*, **50**, 847–851.

Beckmann, C., Shine, R. (2012). How many of Australia's ground-nesting birds are likely to be at risk from the invasive cane toad (*Rhinella marina*)? *Emu*, **112**, 83–89.

Boland, C. R. J. (2004). Introduced cane toads are active nest predators and competitors of rainbow bee-eaters: observational and experimental evidence. *Biological Conservation*, **120**, 53–62.

Breeden, K. (1963). Cane toad (*Bufo marinus*). *Wildlife in Australia*, **1**, 31.

Brown, G. P., Phillips, B. L., Shine, R. (2011). The ecological impact of invasive cane toads on tropical snakes: field data do not support predictions from laboratory studies. *Ecology*, **92**, 422–431.

Brown, G. P., Ujvari, B., Madsen, T., Shine, R. (2013). Invader impact clarifies the roles of top-down and bottom-up effects on tropical snake populations. *Functional Ecology*, **27**, 351–361.

Burnett, S. (1997). Colonising cane toads cause population declines in native predators: reliable anecdotal information and management implications. *Pacific Conservation Biology*, **3**, 65–72.

Buzzacott, J. H. (1938). Giant toads will not poison fowls. *Cane Growers' Quarterly Bulletin*, July **1939**, 101.

Cabrera-Guzmán, E., Crossland, M. R., Shine, R. (2011). Can we use the tadpoles of Australian frogs to reduce recruitment of invasive cane toads? *Journal of Applied Ecology*, **48**, 462–470.

Cabrera-Guzmán, E., Crossland, M. R., Shine, R. (2012). Predation on the eggs and larvae of invasive cane toads (*Rhinella marina*) by native aquatic invertebrates in tropical Australia. *Biological Conservation*, **153**, 1–9.

Catling, P. C., Hertog, A., Burt, R. J., Wombey, J. C., Forrester, R. I. (1999). The short-term effect of cane toads (*Bufo marinus*) on native fauna in the Gulf Country of the Northern Territory. *Wildlife Research*, **26**, 161–185.

Covacevich, J., Archer, M. (1975). The distribution of the cane toad, *Bufo marinus*, in Australia and its effects on indigenous vertebrates. *Memoirs of the Queensland Museum*, **17**, 305–310.

Crossland, M. R., Shine, R. (2012). Embryonic exposure to conspecific chemicals suppresses cane toad growth and survival. *Biology Letters*, **8**, 226–229.

Crossland, M. R., Haramura, T., Salim, A. A., Capon, R. J., Shine, R. (2012). Exploiting intraspecific competitive mechanisms to control invasive cane toads (*Rhinella marina*). *Proceedings of the Royal Society B*, **279**, 3436–3442.

Dexter, R. (1932). The food habits of the imported toad *Bufo marinus* in the sugar cane sections of Puerto Rico. *Proceedings of the Fourth Congress of International Sugar Cane Technologists, San Juan*, **74**, 1–6.

Doody, J. S., Green, B., Sims, R., *et al.* (2006). Indirect impacts of invasive cane toads (*Bufo marinus*) on nest predation in pignosed turtles (*Carettochelys insculpta*). *Wildlife Research*, **33**, 49–54.

Doody, J. S., Green, B., Rhind, D., *et al.* (2009). Population-level declines in Australian predators caused by an invasive species. *Animal Conservation*, **12**, 46–53.

Easteal, S., van Beurden, E. K., Floyd, R. B., Sabath, M. D. (1985). Continuing geographical spread of *Bufo marinus* in Australia: range expansion between 1974 and 1980. *Journal of Herpetology*, **19**, 185–188.

Florance, D., Webb, J. K., Dempster, T., *et al.* (2011). Excluding access to invasion hubs can contain the spread of an invasive vertebrate. *Proceedings of the Royal Society B*, **278**, 2900–2908.

Freeland, W. J. (2004). *An Assessment of the Introduced Cane Toad's* (Bufo marinus Linnaeus) *Impacts on the Native Australian Fauna, with Particular Reference to the Eastern Kimberley Region*. Unpublished Report to the WA Department of Industry and Resources. Palmerston, NT: HLA-Envirosciences Pty Ltd.

Freeland, W. J., Kerin, S. H. (1988). Within habitat relationships between invading *Bufo marinus* and Australian species of frog during the tropical dry season. *Australian Wildlife Research*, **15**, 293–305.

Freeland, W. J., Martin, K. C. (1985). The rate of range expansion by *Bufo marinus* in northern Australia, 1980–84. *Australian Wildlife Research*, **12**, 550–559.

Froggatt, W. W. (1936). The introduction of the Great Mexican Toad *Bufo marinus* into Australia. *The Australian Naturalist*, **9**, 163–164.

Frost, D. R., Donnellan, S. C., Raxworthy, C., *et al.* (2006). The amphibian tree of life. *Bulletin of the American Museum of Natural History*, **297**, 1–370.

González-Bernal, E., Greenlees, M. J., Brown, G. P., Shine, R. (2013). Interacting biocontrol programs: invasive cane toads reduce rates of breakdown of cowpats by dung beetles. *Austral Ecology*, **38**, 891–895.

Goodacre, W. A. (1947). The giant toad (*Bufo marinus*) an enemy of bees. *Agricultural Gazette of NSW*, **58**, 374–375.

Greenlees, M. J., Shine, R. (2011). Impacts of eggs and tadpoles of the invasive cane toad (*Bufo marinus*) on aquatic predators in tropical Australia. *Austral Ecology*, **36**, 53–58.

Greenlees, M. J., Brown, G. P., Webb, J. K., Phillips, B. L., Shine, R. (2006). Effects of an invasive anuran (the cane toad, *Bufo marinus*) on the invertebrate fauna of a tropical Australian floodplain. *Animal Conservation*, **9**, 431–438.

Greenlees, M. J., Brown, G. P., Webb, J. K., Phillips, B. L., Shine, R. (2007). Do invasive cane toads (*Chaunus marinus*) compete with Australian frogs (*Cyclorana australis*)? *Austral Ecology*, **32**, 900–907.

Greenlees, M., Phillips, B. L., Shine, R. (2010). Adjusting to a toxic invader: native Australian frog learns not to prey on cane toads. *Behavioral Ecology*, **21**, 966–971.

Grigg, G., Taylor, A., McCallum, H., Fletcher, L. (2006). Monitoring the impact of cane toads (*Bufo marinus*) on Northern Territory frogs: a progress report. In *Science of Cane Toad Invasion and Control. Proceedings of the Invasive Animals CRC/CSIRO/Qld NRM&W Cane Toad Workshop, Brisbane, June 2006* (eds. K. L. Molloy and W. R. Henderson). Canberra, ACT: Invasive Animals Cooperative Research Centre, pp. 47–54.

Hagman, M., Shine, R. (2006). Spawning-site selection by feral cane toads (*Bufo marinus*) at an invasion front in tropical Australia. *Austral Ecology*, **31**, 551–558.

Hagman, M., Shine, R. (2009). Larval alarm pheromones as a potential control for invasive cane toads (*Bufo marinus*) in tropical Australia. *Chemoecology*, **19**, 211–217.

Hewitt, G. C. (1956). The giant American toad. *Walkabout*, **22**, 45.

Kaemper, W., Webb, J. K., Crowther, M. S., Greenlees, M. J., Shine, R. (2013). Behaviour and survivorship of a dasyurid predator (*Antechinus flavipes*) in response to encounters with the toxic and invasive cane toad (*Rhinella marina*). *Australian Mammalogy*, **35**, 136–143.

Kahneman, D. (2011). *Thinking, Fast and Slow*. New York: Farrar, Straus and Giroux.

Kearney, M. R., Phillips, B. L., Tracy, C. R., *et al.* (2008). Modelling species distributions without using species distributions: the cane toad in Australia under current and future climates. *Ecography*, **31**, 423–434.

Kolbe, J. J., Kearney, M., Shine, R. (2010). Modeling the consequences of thermal trait variation for the cane toad invasion of Australia. *Ecological Applications*, **20**, 2273–2285.

Kraus, F. (2009). *Alien Reptiles and Amphibians: A Scientific Compendium and Analysis*. Netherlands: Springer.

Letnic, M., Ward, S. (2005). Observations of freshwater crocodiles (*Crocodylus johnstoni*) preying upon cane toads (*Bufo marinus*) in the northern territory. *Herpetofauna*, **35**, 98–100.

Letnic, M., Webb, J. K., Shine, R. (2008). Invasive cane toads (*Bufo marinus*) cause mass mortality of freshwater crocodiles (*Crocodylus johnstoni*) in tropical Australia. *Biological Conservation*, **141**, 1773–1782.

Lettoof, D. C., Greenlees, M. J., Stockwell, M., Shine, R. (2013). Do invasive cane toads affect the parasite burdens of native Australian frogs? *International Journal for Parasitology: Parasites and Wildlife*, **2**, 155–164.

Lever, C. (2001). *The Cane Toad. The History and Ecology of a Successful Colonist*. Otley, West Yorkshire: Westbury Academic and Scientific Publishing.

Llewelyn, J., Schwarzkopf, L., Alford, R., Shine, R. (2010). Something different for dinner? Responses of a native Australian predator (the keelback snake) to an invasive prey species (the cane toad). *Biological Invasions*, **12**, 1045–1051.

Llewelyn, J., Schwarzkopf, L., Phillips, B. L., Shine, R. (2014). After the crash: how do predators adjust following the invasion of a novel toxic prey type? *Austral Ecology*, **39**, 190–197.

Lowe, T. (1999). *Feral Future: The Untold Story of Australia's Exotic Invaders*. Chicago, IL: University of Chicago Press.

McCallum, H. (2006). Modelling potential control strategies for cane toads. In *Science of Cane Toad Invasion and Control. Proceedings of the Invasive Animals CRC/CSIRO/Qld NRM&W Cane Toad Workshop, Brisbane, June 2006* (eds. K. L. Molloy and W. R. Henderson). Canberra, ACT: Invasive Animals Cooperative Research Centre, pp. 123–133.

Mungomery, R. (1934). The control of insect pests of sugar cane. *The Cane Growers' Quarterly Bulletin*, **2**, 1–8.

Mungomery, R. (1936). A survey of the feeding habits of the giant toad (*Bufo marinus* L.), and notes on its progress since its introduction to Queensland. *Proceedings of the Queensland Society of Sugar Cane Technologists*, **7**, 63–74.

Nelson, D. W. M., Crossland, M. R., Shine, R. (2011). Behavioural responses of native predators to an invasive toxic prey species. *Austral Ecology*, **36**, 605–611.

O'Donnell, S., Webb, J. K., Shine, R. (2010). Conditioned taste aversion enhances the survival of an endangered predator imperiled by a toxic invader. *Journal of Applied Ecology*, **47**, 558–565.

Peacock, T. (2007). *Community On-Ground Cane Toad Control in the Kimberley*. Review for Western Australia Department of Environment and Conservation. Canberra, ACT: Invasive Animals Cooperative Research Centre.

Pearson, D. J., Greenlees, M., Ward-Fear, G., Shine, R. (2009). Predicting the ecological impact of cane toads (*Bufo marinus*) on camaenid land snails in north-western Australia. *Wildlife Research*, **36**, 533–540.

Pearson, D. J., Greenlees, M. J., Phillips, B. L., *et al.* (2014). Behavioural responses of reptile predators to invasive cane toads in tropical Australia. *Austral Ecology*, **39**, 448–454.

Phillips, B. L., Shine, R. (2004). Adapting to an invasive species: toxic cane toads induce morphological change in Australian snakes. *Proceedings of the National Academy of Sciences of the United States of America*, **101**, 17150–17155.

Phillips, B. L., Shine, R. (2006a). Allometry and selection in a novel predator–prey system: Australian snakes and the invading cane toad. *Oikos*, **112**, 122–130.

Phillips, B. L., Shine, R. (2006b). An invasive species induces rapid adaptive change in a native predator: cane toads and black snakes in Australia. *Proceedings of the Royal Society B*, **273**, 1545–1550.

Phillips, B. L., Brown, G. P., Shine, R. (2003). Assessing the potential impact of cane toads *Bufo marinus* on Australian snakes. *Conservation Biology*, **17**, 1738–1747.

Phillips, B. L., Brown, G. P., Webb, J., Shine, R. (2006). Invasion and the evolution of speed in toads. *Nature*, **439**, 803

Phillips, B. L., Chipperfield, J. D., Kearney, M. R. (2008). The toad ahead: challenges of modelling the range and spread of an invasive species. *Wildlife Research*, **35**, 222–234.

Pizzatto, L., Shine, R. (2011). The effects of experimentally infecting Australian tree frogs with lungworms from invasive cane toads. *International Journal of Parasitology*, **41**, 943–949.

Pizzatto, L., Shilton, C. M., Shine, R. (2010). Infection dynamics of the lungworm *Rhabdias pseudosphaerocephala* in its natural host, the cane toad *Bufo marinus*, and in novel hosts (Australian frogs). *Journal of Wildlife Diseases*, **46**, 1152–1164.

Pizzatto, L., Kelehear, C., Dubey, S., Barton, D., Shine, R. (2012). Host-parasite relationships during a biologic invasion: 75 years post-invasion, cane toads and sympatric Australian frogs retain separate lungworm faunas. *Journal of Wildlife Diseases*, **48**, 951–961.

Pramuk, J. (2007). Phylogeny of South American *Bufo* (Anura: Bufonidae) inferred from combined evidence. *Zoological Journal of the Linnean Society*, **146**, 407–452.

Price-Rees, S. J., Brown, G. P., Shine, R. (2010). Predation on toxic cane toads (*Bufo marinus*) may imperil bluetongue lizards (*Tiliqua scincoides intermedia*, Scincidae) in tropical Australia. *Wildlife Research*, **37**, 166–173.

Price-Rees, S. J., Brown, G. P., Shine, R. (2012). Interacting impacts of invasive plants and invasive toads on native lizards. *American Naturalist*, **179**, 413–422.

Rayward, A. (1974). Giant toads: a threat to Australian wildlife. *Wildlife*, **17**, 506–507.

Shanmuganathan, T., Pallister, J., Doody, S., *et al.* (2010). Biological control of the cane toad in Australia: a review. *Animal Conservation*, **13** (Suppl. 1), 16–23.

Shine, R. (2010). The ecological impact of invasive cane toads (*Bufo marinus*) in Australia. *Quarterly Review of Biology*, **85**, 253–291.

Shine, R. (2012). Invasive species as drivers of evolutionary change: cane toads in Australia. *Evolutionary Applications*, **5**, 107–116.

Shine, R. (2014). A review of ecological interactions between native frogs and invasive cane toads in Australia. *Austral Ecology*, **39**, 1–16.

Shine, R., Doody, J. S. (2011). Invasive-species control: understanding conflicts between researchers and the general community. *Frontiers in Ecology and the Environment*, **9**, 400–406.

Shine, R., Greenlees, M., Crossland, M. R., Nelson, D. (2009). The myth of the toad-eating frog. *Frontiers in Ecology and the Environment*, **7**, 359–361.

Simberloff, D. (2003). Confronting introduced species: a form of xenophobia? *Biological Invasions*, **5**, 179–192.

Smith, J., Phillips, B. (2006). Toxic tucker: the potential impact of cane toads on Australian reptiles. *Pacific Conservation Biology*, **12**, 40–49.

Somaweera, R., Shine, R. (2012). The (non) impact of invasive cane toads on freshwater crocodiles at Lake Argyle in tropical Australia. *Animal Conservation*, **15**, 152–163.

Somaweera, R., Somaweera, N., Shine, R. (2010). Frogs under friendly fire: how accurately can the general public recognize invasive species? *Biological Conservation*, **143**, 1477–1484.

Somaweera, R., Crossland, M. R., Shine, R. (2011). Assessing the potential impact of invasive cane toads on a commercial freshwater fishery in tropical Australia. *Wildlife Research*, **38**, 380–385.

Sutherst, R. W., Floyd, R. B., Maywald, G. F. (1995). The potential geographical distribution of the cane toad, *Bufo marinus* L. in Australia. *Conservation Biology*, **9**, 294–299.

Tingley, R., Greenlees, M. J., Shine, R. (2012). Hydric balance and locomotor performance of an anuran (*Rhinella marina*) invading the Australian arid zone. *Oikos*, **121**, 1959–1965.

Tingley, R., Phillips, B. L., Letnic, M., *et al.* (2013). Identifying optimal barriers to halt the invasion of cane toads *Rhinella marina* in arid Australia. *Journal of Applied Ecology*, **50**, 129–137.

Turvey, N. (2009). The toad's tale. *Hot Topics from The Tropics*, **1**, 1–10.

Ujvari, B., Mun, H., Conigrave, A. D., *et al.* (2013). Isolation breeds naivety: Island living robs Australian varanid lizards of toad-toxin immunity via four-base-pair mutation. *Evolution*, **67**, 289–294.

Urban, M., Phillips, B. L., Skelly, D. K., Shine, R. (2007). The cane toad's (*Chaunus marinus*) increasing ability to invade Australia is revealed by a dynamically updated range model. *Proceedings of the Royal Society B*, **274**, 1413–1419.

Urban, M., Phillips, B. L., Skelly, D. K., Shine, R. (2008). A toad more traveled: the heterogeneous invasion dynamics of cane toads in Australia. *American Naturalist*, **171**, E134–E148.

Vallinoto, M., Sequeira, F., Sodre, D., *et al.* (2009). Phylogeny and biogeography of the *Rhinella marina* species complex (Amphibia, Bufonidae) revisited: implications for Neotropical diversification hypotheses. *Zoologica Scripta*, **39**, 128–140.

van Beurden, E. K. (1981). Bioclimatic limits to the spread of *Bufo marinus* in Australia: a baseline. *Proceedings of the Ecological Society of Australia*, **11**, 143–149.

Ward-Fear, G., Brown, G. P., Greenlees, M., Shine, R. (2009). Maladaptive traits in invasive species: in Australia, cane toads are more vulnerable to predatory ants than are native frogs. *Functional Ecology*, **23**, 559–568.

Ward-Fear, G., Brown, G. P., Shine, R. (2010). Using a native predator (the meat ant, *Iridomyrmex reburrus*) to reduce the abundance of an invasive species (the cane toad, *Bufo marinus*) in tropical Australia. *Journal of Applied Ecology*, **47**, 273–280.

Waterhouse, D. F. (1974). The biological control of dung. *Scientific American*, **230**, 100–109.

Webb, J. K., Brown, G. P., Child, T., *et al.* (2008). A native dasyurid predator (common planigale, *Planigale maculata*) rapidly learns to avoid toxic cane toads. *Austral Ecology*, **33**, 821–829.

White, A. W., Shine, R. (2009). The extra-limital spread of an invasive species via "stowaway" dispersal: toad to nowhere? *Animal Conservation*, **12**, 38–45.

CHAPTER 6

Invasive plants and invaded ecosystems in Australia: implications for biodiversity

Rachael V. Gallagher and Michelle R. Leishman

Summary

Exotic, invasive plants are a major cause of environmental degradation throughout Australia, affecting most ecosystems and vegetation types. Here, we investigate the origins of Australia's exotic plant flora, assess their economic and ecological impact and discuss the processes by which these species become successful invaders. Various management strategies and legislative measures designed to minimise the impact of weeds on native biodiversity are also examined. Most exotic invaders have been deliberately introduced to the Australian landscape via two main pathways – horticulture and agriculture – and many are taxonomically distinct, with no native Australian relatives. Following their introduction a proportion of these species have become naturalised, forming self-sustaining populations in the landscape. Approximately 10% of this naturalised pool of species have gone on to become serious invaders, capable of displacing native vegetation. We discuss the competing theories about how species make the transition along the 'invasion continuum', including the role of functional traits, introduction effort, propagule pressure, residence time and community susceptibility to invasion. We also consider the major functional types of exotic invaders (e.g. woody weeds, vines and scramblers, perennial grasses) and discuss the impacts and management challenges unique to each.

6.1 Introduction

6.1.1 Which exotic plants are in Australia and how did they get there?

Invasion by exotic plant species is recognised globally as one of the greatest threats to biodiversity (Millennium Ecosystem Assessment, 2005). This is equally true in Australia where exotic plants across a wide range of habitats significantly reduce native species abundance and diversity, contributing to extinction risk. Australia is

Austral Ark: The State of Wildlife in Australia and New Zealand, eds. A. Stow, N. Maclean and G. I. Holwell. Published by Cambridge University Press. © Cambridge University Press 2015.

particularly vulnerable to invasion by introduced plants due to its isolation and consequent unique biota, and few ecosystems in Australia are immune from invasion (Adair & Groves, 1998; Groves *et al.*, 2005).

The majority of exotic plant species have been introduced to Australia deliberately, primarily for agricultural production as fodder or crop species, or for horticultural acclimatisation. During Australia's early colonial history concerted efforts were made by societies to establish European flora, as well as to introduce productive species to feed the growing colony and their livestock. Aided by government initiatives such as the Commonwealth Plant Introduction scheme (Cook & Dias, 2006) and more recently the demand for horticultural stock, over 29 000 exotic plant species have now been introduced to Australia – some 8000 more than are currently recognised as native across the entire continent.

Scrutiny of which plants are introduced to Australia has now become standard and new importations are strictly controlled by weed risk assessment protocols and border protection measures, such as quarantine. Although this may have stemmed the flow of high-risk exotic species into Australia, new weed threats are likely to emerge from the large pool of introduced plants already present and being actively planted throughout the country. Many of these introduced species may be able to 'jump the garden fence' and form self-sustaining (naturalised) populations in natural and semi-natural ecosystems, posing a serious future threat to Australia's native biodiversity. Of the 29 000 exotic plants introduced to Australia, around 3000 have become naturalised (Groves *et al.*, 2005; Randall, 2007) and over 400 have been declared noxious or are under some form of legislative control (NRMMC, 2006). Thirty-two weeds, or suites of weeds, have been declared Weeds of National Significance (Figure 6.1, Plate 28) with the consequence that management is coordinated across state boundaries to try to limit their spread (AWC, 2012).

Australia's exotic plant species are taxonomically distinct from the native flora. For example, around one-third of the exotic plants introduced to Australia (37%) come from plant families with no native representatives (based on a comparison of native plant species listed in the Australian Plant Census (http://www.anbg.gov.au/chah/apc/about-APC.html) with introduced plant species listed in Randall (2007)). The introduction and establishment of plant lineages from other parts of the world can result in a reduction in the distinctiveness of regional floras, an effect known as biotic homogenisation (McKinney & Lockwood, 1999).

The vast majority of exotic plant species have been introduced to Australia for the horticultural trade and are widely planted (Table 6.1). An examination of those exotic plant species which have gone on to become serious invaders in Australia shows that 70% were introduced for horticulture or as aquatic ornamental species (Figure 6.2, Plate 53; *n* = 248 species). Agriculture is also an important source of invasive exotic species in Australia (17% of species) and almost 10% of invasive plants have been introduced accidentally. Over one-third (35%) of the invasive plants examined in Australia have been introduced from South America, whilst Africa and Asia also make relatively large contributions (22% and 17% respectively). Invasive plants in Australia encompass a wide range of growth forms; however, herbs and climbers are most common, making up just over half of all invaders examined (54% of 253 species).

Table 6.1 Reasons for introduction of Australia's ~29 000 exotic plant species

Reason for introduction	Approximate number of species introduced to Australia	% of species which have naturalised	% of species which have become weeds in native vegetation
Agriculture			
-food crops	220	38	25
-pasture & fodder	1100	32	23
Horticulture	25 750	7	5
Silviculture	630	24	16
Accidental	1300	80	60
TOTAL	29 000	10	7

Table adapted from Groves *et al.*, (2005).

Figure 6.1 (Plate 28) Australia's Weeds of National Significance (WONS). Thirty-two weeds, or suites of weeds, have been declared as WONS some of which are shown above. (a) Madeira vine (*Anredera cordifolia*) climbing toward the canopy of an established eucalypt; (b) Bitou bush (*Chrysanthemoides monilifera* subsp. *rotundata*) competing with the native coastal species *Banksia integrifolia* in Richmond River Nature Reserve, NSW; (c) showy flowers of *Lantana camara*; (d) thick mats of Ground asparagus (*Asparagus aethiopicus*) along a roadside in Stanwell Park, NSW; (e) Balloon vine (*Cardiospermum grandiflorum*) smothering riparian vegetation in south-east Queensland. Photographs (b), (c) and (d) courtesy of Mark Hamilton. A black and white version of this figure will appear in some formats. For the colour version, please refer to the plate section.

6.2 The impact of exotic plant species
6.2.1 Economic impacts

Exotic plant species constitute a significant economic burden the world over. In Australia, the costs of agricultural invaders have been estimated at over AUD$4 billion annually in lost productivity and the direct costs of control through management

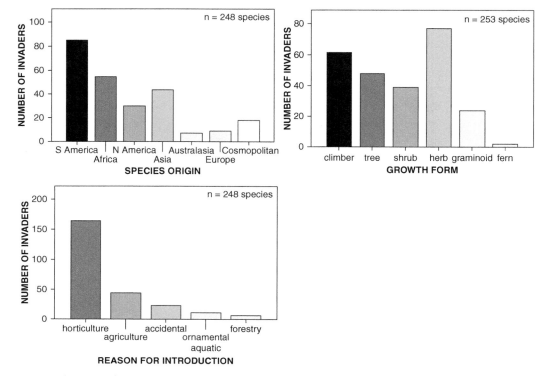

Figure 6.2 (Plate 53) Australia's invasive exotic flora: origins, reasons for introduction and growth forms. Data are for ~250 species of exotic plant which are listed as invasive in Randall (2007). A black and white version of this figure will appear in some formats. For the colour version, please refer to the plate section.

practices such as herbicide application, biocontrol and strategic planning, as well as prevention measures such as quarantine services (Sinden *et al.*, 2004). For example, it has been estimated that the potential impact of a single introduced grass species in Australia (Mexican feather grass, *Nassella tenuissima*) could be ~AUD$39 million over the next 60 years (Csurshes, 2008). Whilst the economic costs of weed control on agriculture and productivity are relatively easy to quantify, placing a dollar value on the impact of weeds on biodiversity is notoriously difficult. Biodiversity is more often viewed as having an intrinsic value, rather than being a specific, quantifiable commodity. Nevertheless, it is clear that considerable funding through all levels of government and natural resource management agencies is directed at removal of environmental weeds for ecosystem rehabilitation, through such programmes as Caring for our Country (http://www.nrm. gov.au/). Furthermore, the pricing of ecosystem services provided by natural systems, such as pollination and clean air and water, is becoming increasingly sophisticated (Bateman *et al.*, 2011) and offers hope that the ecological integrity and evolutionary legacy of natural systems will be viewed as integral to human well-being.

6.2.2 Biodiversity impacts

Invasion by exotic plant species can affect all levels of biological organisation, from individual species through to whole communities and ecosystems. Successful invasion can occur due to either changes in site environmental conditions that promote exotic plant species establishment and growth relative to that of native species (see Box 6.1), or

Box 6.1 Invasions on low fertility soils: the case of urban bushland remnants on Hawkesbury Sandstone soils in Sydney

Australia's soils are characterised generally as well-leached and infertile, reflecting low-nutrient parent material as well as the continent's great age and geological stability. Consequently the Australian flora is highly adapted to low nutrient conditions, with many species having a variety of sophisticated nutrient acquisition strategies, as well as long-lived sclerophyllous leaves that have low specific leaf area, low nutrient content and high nutrient withdrawal before senescence. Typically these low nutrient vegetation communities are highly resistant to invasion by exotic plants, even in urban areas where there are abundant exotic propagules (Lake & Leishman, 2004). The urban bushland remnants that occur on low nutrient Hawkesbury Sandstone-derived soils in the Sydney region provide an excellent example of the role of soil nutrients in facilitating exotic plant success. In these vegetation communities, soil total P values are typically less than 100 ppm. However in sites that receive nutrient-enriched storm-water run-off (derived from detergents from car washing, nutrient fertiliser applied to gardens, organic material and even the basalt-derived road surface), total P values are up to an order of magnitude greater, averaging around 600 ppm (Leishman, 1990). There is a very strong positive relationship between soil total P and percentage exotic cover in these urban remnants (Leishman *et al.*, 2004). Nutrient-enriched sites such as below stormwater outlets and roads, are heavily invaded by a suite of exotic species including privet (*Ligustrum sinense* and *Ligustrum lucidum*), asparagus fern (*Asparagus aethiopicus*) and tradescantia (*Tradescantia fluminensis*). Native species richness as well as percentage cover is greatly reduced in these nutrient-enriched sites compared to hillslopes that do not receive stormwater run-off (Lake & Leishman, 2004). Experimental studies have shown that invasive exotic species have better growth and survival under high nutrient conditions than native species, and that physical disturbance and additional water do not facilitate exotic species growth on these low nutrient soils (Leishman & Thomson, 2005). Finally, it has been shown that not only do additional nutrients favour exotic species, they also cause mortality of native species, most likely due to their inability to down-regulate phosphorus (Thomson & Leishman, 2004). Thus nutrient addition is clearly a driver for weed invasion and native species decline in this low nutrient system, and management actions are required to target nutrient input in these urban bushland remnants.

due to exotic species directly outcompeting natives for resources such as light, nutrients, space and water. Declines in the abundance and diversity of native plant species at sites where weeds have become dominant have been widely documented (Alvarez & Cushman, 2002; Lake & Leishman, 2004; Gaertner *et al.*, 2009; Hejda *et al.*, 2009; Vilà *et al.*, 2011) and largely result from changes in the dynamics of seedling recruitment and survivorship (Yurkonis & Meiners, 2004). This can lead to long-term or even permanent changes in the structure and function of vegetation communities and have flow-on effects to processes such as litter decomposition, rainfall interception, and fire regimes. These community-level changes can in turn affect landscape-level processes, such as catchment hydrology, fire regimes and nutrient cycling dynamics. Thus invasion of ecosystems by exotic plants can impact biodiversity either directly, or indirectly through

---

Box 6.2 Camphor laurel: a case study of an exotic plant's impacts at multiple levels

Native to Asia, Camphor laurel (*Cinnamonum camphorum*) was introduced to Australia as a garden plant in 1822 and has since become an aggressive invader of cleared rainforest and riparian habitats, particularly in northern NSW where it has spread across more than 91 000 hectares (Firth & Ensbey, 2009). Plant–animal interactions in these habitats are disrupted as camphor laurel fruits are particularly attractive to frugivores, with ripe berries eaten and transported by frugivorous birds which typically feed on the fruits of native rainforest plants (Neilan *et al.*, 2006; White & Vivian-Smith, 2011; Paul *et al.*, 2012).

Soil nutrient status and nitrogen cycling dynamics can also be modified by the presence of dense infestations of camphor laurel. The characteristics of camphor laurel leaf litter in monoculture stands differ markedly from the typical mixture of leaf material in comparable, non-invaded sites. Leaf litter from these Camphor laurel-dominated sites returns plant available nitrate to the soil at a slower rate than in replanted rainforest remnants, leading to altered soil nutrient properties (Paul *et al.*, 2010a). This in turn affects plant growth, with growth rates of native seedlings 25% lower in camphor-dominated sites compared to replanted and treated sites (Paul *et al.*, 2010b). It is likely that changes in litter quality and decomposition of this kind could also have flow-on effects on carbon and nutrient cycling across northern NSW given the vast extent of camphor laurel infestation in this region.

---

a multitude of pathways such as changes to vegetation structure, disturbance regimes and ecosystem processes (see Box 6.2).

The proliferation of invasive exotic plant species has led to increased extinction risk of populations of native plants and animals throughout Australia. For example, weed invasions have been implicated in the extinction of four native plant species (Groves & Willis, 1999) and have resulted in over 50 species being listed as endangered under the Commonwealth *Environment Protection Conservation and Biodiversity Act 1999*. In NSW, invasion by individual exotic plant species such as Scotch Broom (*Cytisus scoparius*), African Olive (*Olea europaea* subsp. *cuspidata*), Lantana (*Lantana camara*), Boneseed and Bitou Bush (*Chrysanthemoides monilifera*) and weed functional groups such as perennial grasses, and vines and scramblers have been listed as Key Threatening Processes to biodiversity under the *Threatened Species Conservation Act* (TSC Act). A survey of the impact of exotic plants on threatened biodiversity in NSW found that 279 plant species and 64 endangered ecological communities were listed as at risk of extinction under the NSW TSC Act due to invasive plants (Coutts-Smith & Downey, 2006). For example, the invasive vine Bridal Creeper (*Asparagus asparagoides*) smothers native ground-cover vegetation and threatens some of the remaining populations of the spiked rice flower (*Pimelea spicata*) in western Sydney with local extinction (Willis *et al.*, 2003). Bridal creeper has also been shown to dramatically reduce the richness and diversity of native plant communities in South and Western Australia, whilst having no discernible impact on recruitment patterns in co-occurring exotic species (Stephens *et al.*, 2008; Turner *et al.*, 2008).

Aside from directly affecting native plants and vegetation communities, weeds also pose a serious threat to the continued survival of a range of Australian native animals. In NSW, at least 62 animal species listed as at risk of extinction under the NSW TSC Act are threatened by exotic plants (Coutts-Smith & Downey, 2006). Australian fauna threatened by exotic plants include mammals (e.g. the Broad-toothed Rat – *Mastacomys fuscus*; Golden-tipped Bat – *Kerivoula paupuensis*), reptiles (e.g. the Five-clawed Worm Skink – *Anomalopus mackayi*; Blue Mountains Water Skink – *Eulamprus leuraensis*), invertebrates (e.g. Black Grass-dart Butterfly – *Ocybadistes knightorum*) and birds, in particular seabirds which nest on off-shore islands (e.g. the Little Penguin – *Eudyptula minor*). On Lord Howe Island, off Australia's east coast, dense infestations of Kikuyu (*Pennisetum clandestinum*) – an exotic South African grass – have reduced the number of breeding pairs of the wedge-tailed shearwater. This grass spread rapidly across the island after being deliberately introduced and now blocks the entrance to the burrows of a range of birds and, in some cases, strangles individuals which become entangled in impenetrable thickets (Hutton, 2003). Over AUD$4 million and thousands of volunteer hours have been invested in controlling infestations of Kikuyu and other weed species on Lord Howe Island over the past decade (DECC, 2007). Seabird habitat on Montague Island off the coast of NSW has been similarly affected by Kikuyu and there is now in place a comprehensive restoration programme to control this species and restore habitat for Little Penguins (http://www.environment.nsw.gov.au/parkmanagement/montagueislandshrp.htm).

6.3 Processes of invasion

Only a small proportion of the exotic plant species introduced to Australia have become serious invaders in the landscape. This pattern is seen throughout the globe; roughly one in ten of all species introduced to a region outside their native range become naturalised, and a further 10% of this pool go on to become invasive (known as 'The Ten's Rule'; Williamson & Fitter, 1996). Understanding the factors that allow this select group of species to successfully colonise, reproduce and displace native vegetation is a central theme in invasion biology and many schemes have been proposed to conceptualise the process of invasion. These schemes fall into two broad categories: those which emphasise the characteristics of invaders and the nature of their introduction history (species-based processes), and those which focus on the inherent invasibility of the plant community where the species is introduced (community-based processes).

6.3.1 The invasion continuum

All plant invasions begin in the same way – with the introduction of an exotic species to a novel region outside its native range. Following this initial introduction, species must pass through a series of abiotic and ecological barriers, each of which has the potential to impede its progression towards becoming invasive (Figure 6.3). This process is typically conceptualised to involve three distinct stages – introduction, naturalisation and invasion – which are known collectively as the 'invasion continuum' (Richardson *et al.*, 2000). Introduction is characterised by the deliberate or accidental movement of a species beyond its native, geographic range by humans. Introduced plants which have escaped cultivation, but are unable to form self-sustaining populations relying instead on repeated human introductions and husbandry for persistence, are termed casual aliens. By contrast, introduced species which are able to successfully reproduce and sustain populations over many generations in the absence of human intervention

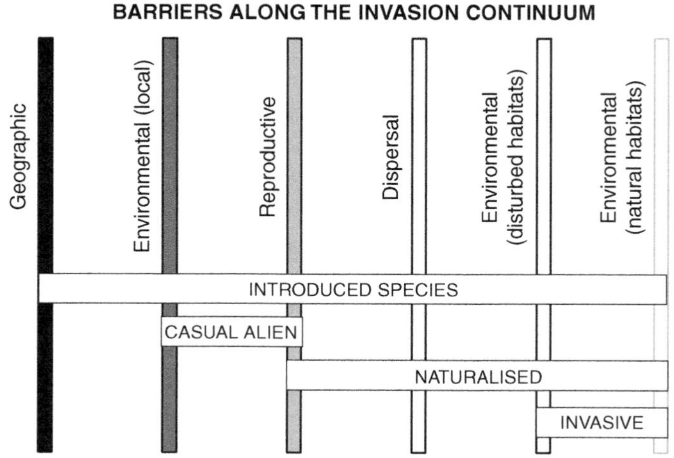

BARRIERS ALONG THE INVASION CONTINUUM

Figure 6.3 The invasion-continuum. A schematic representation of major barriers limiting the spread of introduced plants. Adapted from Richardson *et al.*, (2000).

are classified as naturalised. Naturalisation is accomplished when plants can overcome abiotic barriers such as unsuitable climatic or soil conditions, either through adaptation or through being introduced to regions where conditions fall inside their fundamental niche. Following the establishment of self-sustaining, naturalised populations in the landscape a species becomes invasive when it spreads beyond the original region of introduction to form new expanding populations and enters a period of exponential population growth. The time period between naturalisation and invasive spread varies widely between species, but may be decades.

It is important to recognise that many populations of an individual species may exist in the landscape, and that each of these populations may occupy different stages along the invasion continuum (Richardson & Pyšek, 2012). Some populations may be newly introduced to one region and yet progress to forming self-sustaining, naturalised populations, whilst being highly invasive in other regions. For example, Agapanthus (*Agapanthus praecox*) has a large number of introduced populations throughout Australia, some of which have formed naturalised populations in the Sydney region and a small fraction of these are now reported as invading bushland in the Blue Mountains region west of Sydney.

6.3.2   What makes a successful invasion?

There are three main factors that determine whether a species will successfully pass along the invasion continuum: the characteristics of the species, the introduction effort and the characteristics of the recipient community (Davis *et al.*, 2001). Residence time should also be considered.

6.3.2.1   Species characteristics

The characteristics, or functional traits, of a plant introduced to a new environment are important in determining invasion success, particularly those traits related to seed production and growth rates. However, only a few traits have been shown to be consistently associated with invasion success across a wide range of environments (e.g. specific leaf area, foliar nitrogen content; Leishman *et al.*, 2007; van Kleunen *et al.*, 2009). Thus, searching for a universal set of invasive traits without also taking into account the idiosyncrasies of individual invasions, such as introduction history or environmental context, is unlikely to produce clear answers.

6.3.2.2   Introduction effort and propagule pressure

It is clear that species which have been introduced to a novel region multiple times are those which are most likely to naturalise and go on to become serious invaders and that this process may take place independently of the functional traits the species possesses. When greater effort is invested in introducing an exotic plant to a novel region, the likelihood of the species encountering suitable conditions for establishment and spread increases. Introduction effort (the number of times a species is introduced) and propagule pressure (the amount of viable regenerative material in the landscape) can be particularly high for species used in agriculture, forestry or for rehabilitation purposes. For instance, the Australian native plant *Acacia saligna* was introduced multiple times across large areas of South Africa for fuelwood and site rehabilitation and has since gone on to become one of the worst invasive plants in this region (Griffin *et al.*, 2011; see Box 6.3). Similarly, Bitou bush (*Chrysanthemoides monilifera*) was planted widely along the east coast of Australia for sand dune stabilisation and is now considered one of Australia's worst environmental weeds, occurring along 80% of the NSW coastline and posing a significant threat to more than 150 native plant species (Winkler *et al.*, 2008).

Box 6.3 Australian native plants behaving badly

Even species considered native to a region or continent have the capacity to become serious invaders when introduced outside their known biogeographic range. In Australia, over 500 species of native plants have been introduced to areas outside their documented native range. These include species such as the iconic Western Australian kangaroo paw (*Anigozanthos* spp.) which has been widely planted throughout Australia, but only occurs naturally in the south-west corner of Western Australia. Kangaroo paws are prized horticultural species due to their showy floral displays, however they have now 'jumped the garden fence' to form self-sustaining populations which are rapidly spreading into bushland throughout South Australia and NSW (Le Roux *et al.*, 2010).

Species in the genus *Acacia* (wattles) offer the most compelling example of native plants behaving badly in Australia. Australia is home to over 1000 species of *Acacia* and over 40 of these have been introduced beyond the limits of their historic native range, largely for horticulture or for use in the rehabilitation of sand and mineral mining leases. Wattles indigenous to Western Australia have been introduced into the eastern states, and vice versa, creating a large-scale cross-continental transplant experiment. Ecological studies of these populations in their native and introduced ranges, and in common gardens, show that seed predation is generally lower and seed germination more successful in the introduced range (Harris *et al.*, unpublished data). In addition, some species may form different microbial associations in the native and introduced ranges (Birnbaum *et al.*, 2012) and, when grown in a common garden, seedlings from introduced range populations have higher allocation to above-ground biomass, greater specific leaf areas and relative growth rates, indicating adaptation to a competitive strategy in a novel environment (Harris *et al.*, 2012).

Box 6.3 (continued)

Acacias not only behave badly outside their native range within Australia, but have also become highly invasive in other regions of the world. Twenty-one Australian acacias are currently recognised as invasive in areas throughout Australasia and further afield (Portugal, South Africa, Mediterranean region). Australian acacias are particularly problematic in the Cape Floristic Region of South Africa, an area recognised as a biodiversity hotspot due to the very high percentage of endemic flora. Much of the iconic fynbos vegetation of the Cape has been invaded by Australian acacias (primarily, *A. cyclops, A. mearnsii, A. saligna* and *A. pycnantha*) since they were introduced as a forestry and fuel wood production species. A comparative study of the traits and range characteristics of invasive and non-invasive acacias found in this and other regions showed that, on average, the 21 invasive acacias have greater maximum heights, larger range sizes and exhibit a greater climatic niche breadth (a larger range of temperature and precipitation in their native Australian range) than do introduced, but non-invasive, acacia species (Gallagher *et al.*, 2011). Invasions by acacias are associated with a wide range of detrimental impacts on native vegetation. In South Africa, *A. cyclops* and *A. saligna* have been shown to disrupt dispersal mutualisms between native animals and plants, and to alter seed bank dynamics (Holmes & Cowling, 1997). The ability of acacias to fix nitrogen in the soil through mutualistic interactions with mycorrhizae has led to documented increases in soil fertility in invaded regions of Portugal (Rodríguez-Echeverría *et al.*, 2009). Increases in nitrogen content of this kind can substantially alter the recruitment environment and conditions for native plants adapted to low fertility soils, leading to permanent changes in the floristic diversity in invaded regions. See Figure 6.4 (Plate 22).

Figure 6.4 (Plate 22) Invasive Australian acacias (a) *Acacia saligna* seedlings colonising ground outside Stellenbosch, South Africa. This wattle is native to Western Australia but has spread to throughout the eastern states of Australia and is a serious invader in the Cape Floristic region of South Africa where it displaces native fynbos vegetation, (b) seeds of *Acacia longifolia* subsp. *longifolia* are produced in large numbers aiding in the spread of this species in its invasive range throughout the world, (c–d) ecologists harvest invasive acacias in the field for canopy biomass measurements and soil microbe analyses. A black and white version of this figure will appear in some formats. For the colour version, please refer to the plate section.

6.3.2.3   Community susceptibility

There are a number of key factors that are thought to influence a community's vulnerability to invasion by exotic plants. These are species richness (number of species), diversity (richness and abundance), and the availability of resources. The role of species richness in determining community invasibility has long been debated. Early studies observed high rates of plant invaders on islands and this was attributed to the relatively depauperate species assemblages of islands compared to mainland sites (Elton, 1958). Field-based experiments of model communities at local scales have confirmed this pattern (Tilman, 1999; Naeem *et al.*, 2000); however, regional-scale studies have shown the opposite i.e. areas of high native plant diversity support high numbers of exotic plant species (Lonsdale, 1999; Stohlgren *et al.*, 1999; Levine, 2000).

Resource availability is critical in determining invasion success. Limiting resources, such as soil nutrients, water, light or space may be increased by physical disturbance (i.e. processes such as floods, storms, landslides or animal diggings), which removes extant vegetation and thus reduces competition. Alternatively, resource availability may be directly increased by the addition of nutrients, for example due to run-off from agricultural fields or urban surfaces. Physical disturbance provides opportunities for plant colonisation and establishment, and many exotic plants are highly adapted to exploit these colonisation opportunities through seed dispersal and rapid growth. Consequently exotic plants are often associated with vegetation that has been cleared either for agricultural purposes or through natural disturbances, such as cyclones. Following disturbance, exotic plants typically establish along the edges of remnant vegetation, roads or clearings where resource availability is high. For instance, in tropical North Queensland a series of destructive cyclones during the past decade have increased colonisation opportunities for a range of invasive plants in rainforests, such as *Miconia calvescens* and *Turbina corymbosa*. It is also important to recognise that exotic plant invasions may act synergistically with other stressors, such as land-clearing and habitat fragmentation which results in a greater collective impact on native biodiversity than these threats alone.

6.3.2.4   Residence time

Several studies have shown the importance of considering residence time, or time since introduction, in order to understand patterns of exotic plant species invasion (e.g. Castro *et al.*, 2005; Hamilton *et al.*, 2005; Pysek & Jarosík, 2005). In general, the longer the time period since an exotic species was introduced, the greater the chance of it successfully naturalising and becoming invasive, due to increased likelihood of encountering a favourable stochastic event. Thus, for many species, population numbers build slowly over time to facilitate a steady increase in invasive potential. However, for some species the time from introduction to invasive spread may be rapid due to colonisation opportunities being opened up by abrupt changes in environmental conditions. For example, particularly wet years in the Adelaide River floodplain of the Northern Territory in the late 1970s transported the buoyant seed of the invasive shrub *Mimosa pigra* throughout this region, rapidly increasing its spread (Lonsdale, 1993). This example illustrates the importance of targeting eradication and control strategies for new invaders during the early stages of their establishment. The lag phase before exponential population growth takes hold is now widely recognised as the most cost-effective and successful stage to limit the impact of potentially invasive species on native biodiversity.

6.4 Functional types of weeds in Australia

The introduced flora of Australia can be grouped into a series of functional types, each of which possess a suite of characteristics adapted to colonising specific environments. Grouping species into functional types based on similarities in their physical traits, environmental tolerances and/or dispersal pathways aids in the design and implementation of broad-scale management plans, and helps to identify the ecological mechanisms which underpin invasions in different environments. For instance, functional types are commonly used to aggregate large numbers of exotic species, such as pasture grasses or vines and scramblers, into discrete groupings for the purposes of management and control under environmental legislation in Australia. Species may not belong exclusively to a single functional type; that is, some exotics may be placed into more than one functional grouping depending on the management action being deployed.

6.4.1 Six of the worst weed functional types
6.4.1.1 Bird-dispersed
Many of Australia's most problematic exotic plants are spread by frugivorous vertebrates, primarily birds. The offering of a nutritious reward – typically a fleshy fruit – in return for seed dispersal is a well-established evolutionary mechanism for moving plants around the landscape. However, in many instances exotic plants have exploited this mutualistic interaction between native birds and plants in order to spread into new regions (White & Vivian-Smith, 2011). Some of the most invasive plant species in Australia are bird-dispersed, including Bitou bush and Boneseed (two subspecies of *Chrysanthemoides monilifera*), Camphor laurel (*Cinnamomum camphora*), Lantana (*Lantana camara*; *L. montevidensis*) and the Privets (*Ligustrum lucidum* and *L. sinense*). In addition, a wide range of naturalised plants are being spread throughout the landscape by frugivores and emerging as the next generation of invaders (e.g. Passionfruits (*Passiflora* spp.); Wild tobacco (*Solanum mauritianum*); Ochna (*Ochna serrulata*) and African olive (*Olea europaea* subsp. *cuspidata*)).

Frugivorous birds generally have a wide diet breadth and feed on both native and exotic plants whilst foraging. As they seek out food sources, roosting sites and nesting areas in the landscape, the seed they carry and excrete may be spread to areas distant from the original source populations. This poses a serious challenge for weed management, particularly where seeds are deposited in isolated or remote areas. The ability to predict the spread of bird-dispersed weeds is highly dependent upon the availability of detailed data on both the attributes of dispersers (e.g. gape size for seed ingestion, time taken for seed to pass through the gut passage, home range size, roosting and foraging behaviour; Gosper *et al.*, 2005) and the traits of the fruit on offer (e.g. seed size and number, the provision of fleshy pulp and its sugar content, phenology, and defensive chemistry; Buckley *et al.*, 2006; Gosper & Vivian-Smith, 2010). The movements and behaviour of native rainforest frugivores have been extensively tracked in the Wet Tropics Bioregion of north Queensland and used to simulate the spread of the highly invasive species *Miconia calvescens* in rugged rainforest terrain (Murphy *et al.*, 2008; Hardesty *et al.*, 2011). This approach, whilst time-consuming, has proved a successful tool for designing surveys to locate bird-dispersed weeds in one of Australia's most iconic rainforest regions.

6.4.1.2 Woody weeds

Over the past two decades trees and woody shrubs have begun to emerge as serious invaders in the Australian landscape, as well as across the globe. A recent global survey of invasive trees and shrubs identified Australia as the biogeographic region with the largest number of woody invaders (*n* = 183; Richardson & Rejmánek, 2011). Invaders such as African boxthorn (*Lycium ferrocissimum*), Giant sensitive tree (*Mimosa pigra*), Prickly acacia (*Vachellia nilotica*), and Willow (*Salix* spp.) affect native plant recruitment, hydrology and river condition, and alter or reduce the types of habitat available for native wildlife (Lonsdale, 1992; Stokes & Cunnningham, 2006). As well as being the recipient of a large number of woody weeds, Australia's flora has also contributed some of the world's worst tree and shrub invaders. These include Broad-leaved paperbark (*Melaleuca quinquenervia*) which is highly invasive in the Florida Everglades (Serbesoff-King, 2003), Needlebush (*Hakea sericea*) which has colonised large areas of South African fynbos, and various species of acacia (see Box 6.3).

Woody species may take a longer time to progress along the invasion-continuum from introduced to naturalised to invasive due to the length of their life-cycle. Unlike annual species, which produce a seed crop every year allowing them to spread at a rapid rate, some woody species may take many years to reach reproductive maturity. For instance, the average time taken to produce the first seed crop in 21 invasive woody perennials in Australia is 2.7 years, with some species taking up to 15 years to produce a substantial seed crop (e.g. Radiata pine, *Pinus radiata*). Whilst woody weeds may take longer to produce seedlings, individual seed crops can be huge as a result of a large canopy area. An example is African olive (*Olea europaea* subsp. *cuspidata*) which can produce up to 950 seedlings/m^2 in areas where it has become established throughout the endangered Cumberland Plain woodland in western Sydney (Figure 6.5, Plate 29; Cuneo *et al.*, 2010; Cuneo & Leishman, 2012). Given their slow rate of naturalisation the next generation of woody weed problems may already have been introduced to Australia but are currently caught in a lag-phase and may emerge as serious invaders in the coming decades (Caley *et al.*, 2008).

6.4.1.3 Perennial grasses

The hunt for species that can flourish under dry conditions where water is scarce has led to the introduction of a range of perennial grasses for both agricultural and horticultural purposes to Australia over the past two centuries. Exotic perennial grasses such as Serrated tussock (*Nassella trichotoma*), Needle grasses (*N. neesiana, N. hyalina*) and Rat's tail grasses (*Sporobolus* spp.) have become invasive across a wide area of Australia, and some are yet to spread into the full complement of climatically suitable habitats within the continent (Gallagher *et al.*, 2012). In NSW, the invasion of

Figure 6.5 (Plate 29) African Olive invasion and control at the Mount Annan Botanic Gardens. (a) Dense hillside invasion of African olive in south-west Sydney. Characteristic seed dispersal by birds leads to visible 'halos' of established plants around the base of a mature eucalypt tree in the foreground, (b) small black African olive fruits (7–10 mm) are produced during the winter months and readily consumed by native and introduced birds, (c) control of African olive using 'drill and inject' method. Undiluted herbicide (glyphosate) is injected into drill holes, (d) mechanical chipping of dense African olive invasion in eucalypt woodland. Stumps are treated with herbicide and retained *in situ* for soil stability. Images provided by P. Cuneo. A black and white version of this figure will appear in some formats. For the colour version, please refer to the plate section.

perennial grasses into native vegetation has been listed as a Key Threatening Process under environmental legislation. This declaration provides formal recognition of the impacts of grass invasions on native vegetation and animals throughout the state. These impacts include the ability to outcompete native understorey species for essential resources, such as light and nutrients, increased fuel loads for seasonal fires, and in the case of the South African grass Kikuyu (*Pennisetum clandestinum*) the smothering of penguin nest sites on offshore islands (Weerheim *et al.*, 2003). Nationally, four of the 32 Weeds of National Significance are perennial grasses, some of which are capable of transforming the ecosystems they invade by modifying the grass-fire cycle (see Box 6.4).

Regrettably, only a small proportion of deliberately introduced exotic grasses have actually become important contributors to agricultural productivity in Australia; Lonsdale (1994) found that only 21 of the 463 species of exotic grass introduced to northern rangelands between 1945 and 1987 have proven useful. However, conflicts of interest may arise when devising management programmes to minimise the impact on biodiversity of those species considered as important productive grasses. For example, the widely planted pasture species Buffel grass (*Cenchrus ciliaris*) highlights the complex negotiations which result from the competing perceptions of some introduced species. Buffel grass was introduced throughout the rangelands of Australia in the 1860s in the saddles of inland camel traders and since this time has been deliberately seeded in vast areas of Australia (Jackson, 2005). This deep-rooted, drought-resistant perennial is capable of displacing native vegetation, in particular native grasses and ground-cover species, by monopolising soil nutrients and responding rapidly to rain

Box 6.4 Weeds that transform the ecosystems they invade

Transformer species are a distinct subset of invaders which alter the structure and function of the invaded vegetation community, as well as the availability of key resources to create favourable conditions for their own establishment and spread. For example, many invasive vine species in Australia are considered to be transformers due to their ability to both create and preferentially colonise canopy gaps. Exotic vine species generally possess a suite of traits which make them superior colonisers in high-light environments (i.e. wind-dispersed seeds, rapid growth rates), characteristic of edge and gap environments. Once established, vines over-top and smother host vegetation which can often result in the breakage of tree branches under the weight of vine biomass and, in some extreme cases, increased mortality of host plants. This process of overtopping and branch breakage in turn produces additional gaps within the host-tree canopy which provides new colonisation opportunities for the invading vine. This positive feedback loop (Figure 6.6, Plate 20) can permanently alter the composition and structure of native vegetation and is a particular problem in riparian areas where exotic vines, such as *Dolichondra unguis-cati* (Cat's claw creeper) and *Anredera cordifolia* (Madeira vine) reach high abundance (Batianoff & Butler, 2004; Downey & Turnbull, 2007). In recognition of the detrimental impacts of exotic vines and scramblers on the integrity of native vegetation, invasion by this group of transformers has been declared a Key Threatening Process under the NSW *Threatened Species Conservation Act 1995* and a suite of exotic vines have been declared Weeds of National Significance (WONS) throughout Australia.

In Australia's north, invasion by the exotic perennial grass *Andropogon gayanus* (Gamba grass) has profoundly transformed savanna vegetation by altering the frequency and intensity of seasonal fires (Figure 6.7, Plate 21). Savannas are charac-terised by a mixture of fire-tolerant grassy understorey species and trees and are found on approximately a quarter of Australia in seasonally droughted tropical regions (Hutley & Setterfield, 2008). In 1931, gamba grass was introduced to the

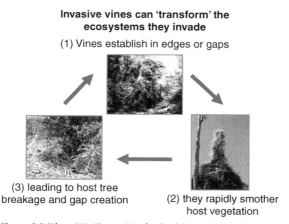

Invasive vines can 'transform' the ecosystems they invade

(1) Vines establish in edges or gaps

(3) leading to host tree breakage and gap creation

(2) they rapidly smother host vegetation

Figure 6.6 (Plate 20) The positive feedback loop which facilitates exotic vine invasions. A black and white version of this figure will appear in some formats. For the colour version, please refer to the plate section.

Box 6.4 (continued)

Northern Territory as a pasture grass from its native Africa, where it is found across a wide biogeographic area (Oram, 1990). Over the past 25 years dense infestations have developed across a 15 000 km^2 area in the mesic savannas of the Northern Territory (Setterfield *et al.*, 2010) and scattered populations now also occur in Queensland and Western Australia. Gamba grass can grow up to 4 m in height and produce up to 30 t of biomass per ha. As a result, fuel loads for fires can be up to seven times higher in gamba-infested areas compared to areas dominated by native grasses (Rossiter *et al.*, 2003). Gamba grass transforms savanna ecosystems by initiating a positive feedback cycle involving the burning of vast fuel loads in high intensity fires which in turn decreases native grass and tree recruitment and facilitates further invasion. This 'grass–fire cycle' (D'Antonio & Vitousek, 1992) is a potent driver of vegetation change in invaded regions in Australia, capable of permanently altering the composition and function of savanna ecosystems (Brooks *et al.*, 2010).

Figure 6.7 (Plate 21) Gamba grass invading ecosystems in Australia's north. (a) Tree death due to gamba grass invasion, (b) high intensity fire resulting from gamba infestation, (c) Gamba grass engulfing a vehicle in the Northern Territory. Photos courtesy of Samantha Setterfield, Charles Darwin University. A black and white version of this figure will appear in some formats. For the colour version, please refer to the plate section.

(Franks, 2002). However, Buffel grass remains highly prized for its utility as a fodder species and as a result continues to be actively planted by some agriculturalists. Therefore, efforts to limit the environmental impact of Buffel on native grass recruitment and rangeland biodiversity in Australia need to incorporate strategies that recognise the competing interests of conservationists and agriculturalists.

6.4.1.4 Garden escapes

Gardens are a major source of environmental weeds in Australia. Amongst the more than 25 000 species that have been introduced as ornamental garden plants since

European colonisation are a large number of species which have 'jumped the fence' to become major invaders. In fact, garden escapes are over-represented as a functional group amongst Australia's declared Weeds of National Significance (WONS). Twenty-three of the 32 declared WONS (72%) were originally introduced as ornamentals to either terrestrial or aquatic environments (AWC, 2012). In addition, species which were introduced for the horticultural trade made up 65% of all naturalised plants in Australia between 1971 and 1995 (Groves & Hosking, 1998). Despite the implementation of robust weed risk assessment protocols to screen out potentially problematic species from importation since this time, the horticultural trade is still the largest single contributor of environmental weeds to the Australian landscape (Groves *et al.*, 2005).

The dumping of garden waste into native bushland is a key factor; the planting of species such as Lantana, Mother-in-laws tongue (*Sansevieria trifasciata*), Privet (*Ligustrum* spp.), and Asparagus ferns (*Asparagus* spp.) in gardens has been widespread across Australia, and the inappropriate disposal of garden clippings, in addition to their capacity to produce large quantities of well-dispersed seeds, has seen these species successfully invade urban bushland remnants. On a global scale, the commercial trade in plants and seed material has gained pace in recent decades with the advent of the Internet and the globalisation of world trade. This has hastened the homogenisation of distinct floras across the world as species capable of thriving in a wide range of conditions are shifted to regions outside their native range.

6.4.1.5   Vines and scramblers

Climbing plants use the structural investment of their host species to place their leaves in adequate light for photosynthesis. In this way, climbers dispense with the need to invest large amounts of their carbon resources in the building of lignified, woody tissue in the form of trunks and branches to support leaves in the canopy. Populations of over 180 exotic vine and scrambler species have now been reported to Australian herbaria (Harris *et al.*, 2007). The majority of these species (61%) were introduced intentionally for the horticultural industry, chosen mainly for their showy ornamental flowers and rapid growth rates. Many climbers move around the landscape via wind-dispersed seeds which are launched from the top of the canopy; however, the ability to reproduce vegetatively through stem fragments also aids in their spread. Once established, climbing plants can severely degrade native vegetation by smothering host plants and disrupting the recruitment dynamics of native seedlings. For this reason they have become known as transformer weeds (see Box 6.4) which pose a serious threat, particularly in rainforest and riparian vegetation.

6.4.1.6   Aquatic weeds

It is not only Australia's terrestrial ecosystems which experience the detrimental effects of invasive introduced plants. Aquatic invaders are also a major challenge for natural resource managers, establishing in rivers, creeks, lakes, wetlands dams, estuaries and damp areas. Given the characteristically high availability of water and nutrients in these environments, aquatic weeds can rapidly become widespread and difficult to control using conventional herbicide treatments. The incorrect disposal of aquarium and pond plants, such as Green cabomba (*Cabomba caroliniana*) and *Sagittaria platyphylla*, is the primary source of aquatic invaders in Australia (Champion *et al.*, 2010). In the past 30 years, 400 species of freshwater aquatic plant have been freely traded in Australia and eight aquatic or semi-aquatic plants are now listed as WONS (Petroeschevsky & Champion, 2008; AWC, 2012). Large-scale infestations of aquatic plants reduce oxygen levels in water and light penetration into the water column, which in turn affect water temperatures and quality (Ruiz-Avila & Klemm, 1996; Perna & Burrows, 2005) and hence habitat quality for many native taxa.

The application of biocontrol agents has been highly effective in controlling infestations of some invasive aquatic plants throughout Australia (McFadyen, 1998). Most notable amongst the success stories is the use of the Salvinia weevil (*Cyrtobagous salviniae*) to control the spread of the imported pond weed Salvinia (Sullivan *et al.*, 2011). This weed has now been effectively controlled in the short-term in waterways of tropical northern Australia (Schooler *et al.*, 2006) and in temperate regions (Sullivan *et al.*, 2011), however the long-term efficacy of this biocontrol agent is still being investigated (Schooler *et al.*, 2011).

6.5 Managing weed threats across Australia

Managing Australia's weed problems requires an integrated approach, which combines prevention strategies and on-ground actions that are supported by effective legislation at the state and federal level. The most cost-effective option in all cases is to stop the introduction of potentially harmful species to the Australian landscape. This precautionary tactic is achieved through two primary mechanisms: weed risk assessment and quarantine inspection.

Weed risk assessment (WRA) is a procedure by which species are evaluated to determine the threat they pose as potential invaders, and ideally is performed prior to the granting of permission to import a new species (Pheloung, 1995). WRAs typically involve a series of questions about the ecology, invasive behaviour and biogeography (e.g. climatic suitability) of the organism being assessed. Responses to these questions are scored and tallied, and used to place the species into categories such as accept, reject or evaluate (further assessment following importation). Australian scientists have developed one of the world's most robust WRA protocols, which has been adopted in many other countries across the globe (Gordon *et al.*, 2008). However, the current system of border WRA was only introduced in Australia in 1997 and a large number of potentially problematic species were introduced before this time. Therefore, it is important to remain vigilant in efforts to detect emerging weed problems, particularly in the face of climate change and ongoing human alterations to the biosphere (see Box 6.5).

Quarantine inspection services also form a vital part of pre-border weed prevention strategies. The seizing of plant material (e.g. seeds, tubers, whole plants) at border entry points such as airports and shipyards blocks the importation of undesirable species and the pathogens and diseases they may carry. WRA and quarantine services work hand-in-hand under the auspices of the Commonwealth Government *Quarantine Act (1908)* to limit the introduction of new species. In recent decades, the upsurge in the use of the Internet to trade plant material has posed a novel threat to Australian biosecurity and the need for robust quarantine procedures has expanded.

Once weeds establish in the landscape there are three main methods for limiting their spread: chemical (e.g. herbicide application), physical/mechanical removal (e.g.

> ## Box 6.5 Weeds in a warmer world: climate change and Australia's introduced flora
>
> The Earth's climate is changing. Human-induced increases in the concentration of greenhouse gases in the atmosphere over the past 150 years are driving shifts in average temperature and precipitation patterns across the globe, as well as in the intensity and frequency of extreme climatic events. On average, global land surface temperatures rose by 0.74 °C (range 0.56–0.92 °C) between 1906 and 2005, and there have been both documented increases and declines in long-term average precipitation during this period dependent on geographic location (IPCC, 2007). In Australia, temperatures have risen by 0.9 °C since 1950 and there have been marked reductions in rainfall in the south-east portion of the continent (CSIRO, 2011). These changes in climate and atmospheric chemistry may have profound effects on the distribution, growth rates and spread of exotic plants in coming decades. In particular, invasive exotic plants are expected to be facilitated by climate change due to their general characteristics of wide tolerance of environmental conditions, capacity for rapid growth and ability to disperse over long distances.
>
> *Assessing weed risk using spatial models* Ecological niche models (ENMs) are popular tools for pre-emptively assessing the risk posed by naturalised and invasive plants under future climate regimes. These models combine data on long-term average environmental conditions (e.g. climate, soils, land-use) with data on species occurrences or physiological tolerances to describe the location of suitable habitat in the landscape. The modelled distributions are subsequently used to project changes in habitat suitability and location under alternative future scenarios. A key trend emerging from ENM studies of some of Australia's most problematic weed species is that few are yet to occupy the full complement of climatically suitable locations under current, average conditions across the continent. For example, models for *Anredera cordifolia* (Madeira vine), *Ipomoea* spp.(Morning glory) and *Parkinsonia aculeata* indicate that these exotic species could potentially spread to much larger regions of the Australian continent than are currently occupied (van Klinken *et al.*, 2009; Gallagher *et al.*, 2010). However, projections of how changing climates into the future may affect exotic plant distributions in Australia show mixed results. The extent of suitable habitat for some species is projected to contract (e.g. exotic perennial grasses, invasive vines), whilst others are projected to expand. These idiosyncratic results highlight the need to compare multiple species simultaneously in order to detect spatial patterns of invasion risk. One such study (O'Donnell *et al.*, 2012) aggregated projections of suitable habitat for 72 species declared or shortlisted as Weeds of National Significance in Australia. Hotspots of invasion risk under current and future climates (see Plate 5) were found to occur in the south-west and south-east corners of Australia. Studies of this kind can help invasion ecologists and land managers prioritise regions where multiple weed species can be targeted for prevention, eradication or control, simultaneously.
>
> *Elevated CO_2 and weed invasions* Carbon dioxide is a key ingredient for photosynthesis and increased concentrations in the atmosphere provide a stimulatory effect on plant growth rates and biomass allocation when other resources (e.g. nutrients, water, light) are not limiting (Curtis & Wang, 1998). This stimulatory effect

Box 6.5 (continued)

may be greatest in species which use the C3 photosynthetic pathway (Ainsworth & Long, 2005). There is some evidence that the productivity and competitive ability of key invasive species may increase under elevated CO_2 levels, from both glasshouse and open top chamber manipulative experiments (Sasek & Strain, 1988; Hattenschwiler & Korner, 2003; Weltzin *et al.*, 2003; Manea & Leishman, 2011) and from field-based studies (Smith *et al.*, 2000; Huxman & Smith, 2001; Belote *et al.*, 2004; Nagel *et al.*, 2004). However, results from a free-air carbon enrichment experiment conducted in grassland vegetation in Tasmania indicate that the combined effects of warming (+2 °C) and elevated CO_2 (550 ppm) lead to population declines in two exotic species, *Hypochaeris radicata* and *Leontodon taraxacoides* (Williams *et al.*, 2007). This result highlights the importance of considering elevated CO_2 concentrations in a climatic context; the effect of increasing temperature may have a more important effect on population persistence than the stimulatory effect of additional CO_2 on plant growth.

mowing, tilling, hand-removal, burning) and biological (biological control). Combinations of two or more of these tactics are often applied and are referred to collectively as integrated weed management. It is critical that any control measures are undertaken early in the invasion process in order to maximise their efficacy. This is because eradication is generally only feasible when weeds are at low density in localised populations. Therefore, the lag phase between establishment and naturalisation is the most effective time to eliminate new weed species. However, resources for weed control in natural systems are often limited and the detection and reporting of small, novel weed populations must compete with the management of larger, more damaging infestations. Herbariums play an important role in documenting the establishment of new weeds but populations are often not reported until they have become widespread and conspicuous in the landscape, by which time the window for effective eradication may have closed. New approaches for detecting isolated weed populations are currently being trialled, including remote sensing using satellite imagery and small, unmanned aircraft. Both these techniques have had some success in detecting new incursions of the rainforest weed *Miconia calvescens* in Australia's wet tropics bioregion.

Weeds which have become invasive in natural systems need active, on-ground management to limit their impact on native biodiversity. Some of the most valuable contributions to management are made by volunteer groups and non-governmental organisations involved in physical, mechanical or chemical control. Initiatives such as Bushcare and Landcare see thousands of volunteers actively participate in the removal of weed populations and revegetation of affected sites. In the longer term, a shift towards native plant species as garden plants and more responsible disposal of garden waste will help foster a culture of awareness of weed problems and their origins within Australia.

When effectively implemented, biological control (biocontrol) can also be a powerful tool for reducing the density of invasive plant infestations. This technique relies on the careful selection and introduction of pests and pathogens to the invaded region which use the target weed as a resource in the native range (Hoddle, 2004). Recent biocontrol successes in Australia include the introduction of three agents from South Africa to control Bridal creeper (*Asparagus asparagoides*): a leafhopper (*Zygina* sp.), a rust fungus (*Puccinia myrsiphylli*) and a leaf beetle (*Crioceris* sp.). The combined effects of these agents have helped to curb the spread of severe infestations of Bridal creeper by reducing photosynthesis, tuber production and inducing defoliation (Turner *et al.*, 2010). Well-executed biological control programmes of this kind offer many benefits over more conventional weed control methods. These include a reduction in herbicide use and the ability to control isolated or unreported infestations in cases where agents readily spread throughout the landscape. However, the potential for biocontrol agents to be ineffective, to host-switch to native species or to escape into the landscape, can have disastrous consequences for biodiversity.

6.5.1 The role of legislation and strategic planning in weed management

Various legislative instruments are used to prevent the introduction or to control the spread of weeds within Australia. Pre-border screening is currently mandated by the Commonwealth *Biological Control Act (1984)* and *Quarantine Act (1908)*, whereas the *Noxious Weed Act (1993)* is used to declare and order the control of invasive plants throughout Australia. Planning strategies to reduce the direct threat posed by weeds to native biodiversity are developed and implemented through the Commonwealth *Environment Protection and Biodiversity Conservation Act (1999)* and similar legislative arrangements in each state. In addition, a range of strategic initiatives aimed at prioritising which weeds should receive the most attention and funding have also been developed. Foremost amongst these are the Weeds of National Significance and national Alert List for Environmental Weeds programmes within the National Weeds Strategy, which identify and focus on a select suite of exotic plant species deemed to pose the greatest threat to Australia's natural and productive landscapes. At a regional scale, plans to minimise weed impact and maximise efficacy of control and prevention are produced by natural resource management authorities and weeds committees working in conjunction with local government. These plans have the advantage of being able to identify and target particularly problematic species at more local scales and provide a strategic approach to catchment or region-wide weed control.

6.6 Conclusions

Whilst exotic invasive plants are a global problem, Australia faces a range of distinct challenges in preventing and managing their impact on biodiversity. Biogeographic isolation over millions of years has endowed the Australian continent with a unique and varied endemic flora, which is threatened by the colonisation of taxonomically novel plant lineages from across the globe. It is clear that the use of plants by humans in horticulture and agriculture has played a critical role in the introduction and spread of exotic plants throughout Australia. Although mechanisms such as weed risk assessment and quarantine have helped to curb the flow of additional potentially problematic species, there remains a need to work cooperatively with the nursery and farming sectors to reduce the potential impact of those exotic plants already present for sale and being widely planted. Given that exotic plant species now outnumber native plant species in

Australia the ongoing need to devise and implement effective management practices is a key challenge for applied ecology.

Acknowledgements

We wish to thank Tanja Lenz for compiling data on invasive exotic species, Rod Randall for providing an updated version of Randall (2007), and Mark Hamilton, Natalie Rossiter-Rachor and Peter Cuneo for providing photographs of invasive plant species in the field.

REFERENCES

Adair, R. J. & Groves, R. H. (1998). *Impact of Environmental Weeds on Biodiversity: a Review and Development of a Methodology*. Canberra: Commonwealth of Australia.

Ainsworth, E. A. & Long, S. P. (2005). What have we learned from 15 years of free-air CO_2 enrichment (FACE)? A meta-analytic review of the responses of photosynthesis, canopy properties and plant production to rising CO_2, *New Phytologist*, **165**: 351–372.

Alvarez, M. E. & Cushman, J. H. (2002). Community-level consequences of a plant invasion: effects on three habitats in coastal California, *Ecological Applications*, **12**: 1434–1444.

AWC – Australian Weeds Committee (2012). *Weeds of National Significance 2012*. Canberra: Department of Agriculture, Fisheries and Forestry.

Bateman, I. J., Mace, G. M., Fezzi, C., Atkinson, G. & Turner, K. (2011). Economic analysis for ecosystem service assessments, *Environmental and Resource Economics*, **48**: 177–218.

Batianoff, G. N. and Butler, D. W. (2004). Impact assessment and analysis of sixty-six priority invasive weeds in south-east Queensland, *Plant Protection Quarterly*, **18**: 11–17.

Belote, R. T., Weltzin, J. F. & Norby, R. J. (2004). Response of an understory plant community to elevated $[CO_2]$ depends on differential responses of dominant invasive species and is mediated by soil water availability, *New Phytologist*, **161**: 827–835.

Birnbaum, C., Barrett, L. G., Thrall, P. H. & Leishman, M. R. (2012). Mutualisms are not constraining cross-continental invasion success of *Acacia* species within Australia, *Diversity and Distributions*, **18**: 962–976.

Brooks, K. J., Setterfield, S. A. & Douglas, M. M. (2010). Exotic grass invasions: applying a conceptual framework to the dynamics of degradation and restoration in Australia's tropical savannas, *Restoration Ecology*, **18**: 188–197.

Buckley, Y.M., Anderson, S., Catterall, C. P. *et al.* (2006). Management of plant invasions mediated by frugivore interactions, *Journal of Applied Ecology*, **43**: 848–857.

Caley, P., Groves, R. H. & Barker, R. (2008). Estimating the invasion success of introduced plants, *Diversity and Distributions*, **14**: 196–203.

Castro, S. A., Figueroa, J. A., Muñoz-Schick, M. & Jaksic, F. M. (2005). Minimum residence time, biogeographical origin, and life cycle as determinants of the geographical extent of naturalized plants in continental Chile, *Diversity and Distributions*, **11**: 183–191.

Champion, P. D., Clayton, J. S. & Hofstra, D. E. (2010). Nipping aquatic plant invasions in the bud: weed risk assessment and the trade, *Hydrobiologia*, **656**: 167–172.

CSIRO (Commonwealth Scientific and Industrial Research Organisation) (2011). *Climate Change: Science and Solutions for Australia*. Melbourne: CSIRO Publishing.

Cook, G. D. & Dias, L. (2006). Turner Review No. 12. It was no accident: deliberate plant introductions by Australian government agencies during the 20th century, *Australian Journal of Botany*, **54**: 601–625.

Coutts-Smith, A. J. & Downey, P. O. (2006) *The Impact of Weeds on Threatened Biodiversity in New South Wales*. Technical Series No. 11, Adelaide: Cooperative Research Centre for Australian Weed Management.

Csurshes, S. (2008). *Pest Plant Risk Assessment* – Nassella tenuissima *Mexican Feather Grass*. Brisbane: The State of Queensland, Department of Primary Industries and Fisheries.

Cuneo, P. & Leishman, M. R. (2012) Ecological impacts of invasive African olive (*Olea europaea* ssp. *cuspidata*) in Cumberland Plain Woodland, Sydney, Australia, *Austral Ecology*, **38**: 103–110

Cuneo, P., Offord, C. A. & Leishman, M. R. (2010). Seed ecology of the invasive woody plant African Olive (*Olea europaea* subsp. *cuspidata*): implications for management and restoration, *Australian Journal of Botany*, **58**: 342–348.

Curtis, P. S. & Wang, X. (1998). A meta-analysis of elevated CO_2 effects on woody plant mass, form, and physiology, *Oecologia*, **113**: 299–313.

D'Antonio, C. M. & Vitousek, P. M. (1992). Biological invasions by exotic grasses, the grass/fire cycle, and global change, *Annual Review of Ecology and Systematics*, **23**: 63–87.

Davis, M. A., Grime, J. P. & Thompson, K. (2001). Fluctuating resources in plant communities: a general theory of invisibility, *Journal of Ecology*, **88**: 528–534.

DECC – Department of Environment and Climate Change, NSW (2007). *Lord Howe Island Biodiversity Management Plan*. Sydney: Department of Environment and Climate Change (NSW).

Downey, P. & Turnbull, I. (2007). The biology of Australian weeds. 48. *Macfadyena unguis-cati* (L.) AH Gentry, *Plant Protection Quarterly*, **22**: 82–85.

Elton, C. S. (1958). *The Ecology of Invasions by Animals and Plants*. London: Methuen.

Firth, D. & Ensbey, R. (2009). *Primefact 733: Camphor Laurel*. Sydney: State of New South Wales through Department of Industry and Investment (Industry & Investment NSW).

Franks, A. J. (2002). The ecological consequences of Buffel Grass *Cenchrus ciliaris* establishment within remnant vegetation of Queensland, *Pacific Conservation Biology*, **8**: 99–107.

Gaertner, M., Breeyen, A. D., Hui, C. & Richardson, D. M. (2009). Impacts of alien plant invasions on species richness in mediterranean-type ecosystems: a meta-analysis, *Progress in Physical Geography*, **33**: 319–338.

Gallagher, R. V., Hughes L., Leishman, M. R. & Wilson, P. (2010). Predicted impact of exotic vines on an endangered ecological community under future climate change, *Biological Invasions*, **12**: 4049–4063.

Gallagher, R. V., Leishman, M. R., Miller, J. T. *et al.* (2011). Invasiveness in introduced Australian acacias: the role of species traits and genome size, *Diversity and Distributions*, **17**: 884–897.

Gallagher, R. V., Englert Duursma, D., O'Donnell, J. *et al.* (2012). The grass may not always be greener: projected reductions in climatic suitability for exotic grasses under future climates in Australia, *Biological Invasions*, **15**: 961–975.

Gordon, D. R., Onderdonk, D. A., Fox, A. M. & Stocker, R. K. (2008). Consistent accuracy of the Australian weed risk assessment system across varied geographies, *Diversity and Distributions*, **14**: 234–242.

Gosper, C. R. & Vivian-Smith, G. (2010). Fruit traits of vertebrate-dispersed alien plants: smaller seeds and more pulp sugar than indigenous species, *Biological Invasions*, **12**: 2153–2163.

Gosper, C. R., Stansbury, C. D. & Vivian-Smith, G. (2005). Seed dispersal of fleshy-fruited invasive plants by birds: contributing factors and management options, *Diversity and Distributions*, **11**: 549–558.

Griffin, A. R., Midgley, S. J., Bush, D., Cunningham, P. J. & Rinaudo, A. T. (2011). Global uses of Australian acacias – recent trends and future prospects, *Diversity and Distributions*, **17**: 837–847.

Groves, R. & Willis, A. (1999). Environmental weeds and loss of native plant biodiversity: some Australian examples, *Australian Journal of Environmental Management*, **6**: 164–171.

Groves, R. H. & Hosking, J. R. (1998). *Recent Incursions of Weeds to Australia 1971–1995*. Adelaide: Cooperative Research Centre for Australian Weed Management.

Groves, R. H., Lonsdale, M. & Boden, R. (2005). *Jumping the Garden Fence: Invasive Garden Plants in Australia and their Environmental and Agricultural Impacts*. Canberra: WWF-Australia.

Hamilton, M. A., Murray, B. R., Cadotte, M. W. *et al.* (2005). Life-history correlates of plant invasiveness at regional and continental scales, *Ecology Letters*, **8**: 1066–1074.

Hardesty, B. D., Metcalfe, S. S. & Westcott, D. A. (2011). Persistence and spread in a new landscape: Dispersal ecology and genetics of *Miconia* invasions in Australia, *Acta Oecologica*, **37**: 657–665.

Harris, C. J., Murray, B. R., Hose, G. C. & Hamilton, M. A. (2007). Introduction history and invasion success in exotic vines introduced to Australia, *Diversity and Distributions*, **13**: 467–475.

Harris, C. J., Dormontt, E. E., Le Roux, J. J., Lowe, A. & Leishman, M. R. (2012). No consistent association between changes in genetic diversity and adaptive responses of Australian acacias in novel ranges, *Evolutionary Ecology*, **26**: 1345–1360.

Hattenschwiler, S. & Korner, C. (2003). Does elevated CO_2 facilitate naturalization of the non-indigenous *Prunus laurocerasus* in Swiss temperate forests?, *Functional Ecology*, **17**: 778–785.

Hejda, M., Pyšek, P. & Jarošík, V. (2009). Impact of invasive plants on the species richness, diversity and composition of invaded communities, *Journal of Ecology*, **97**: 393–403.

Hoddle, M. S. (2004). Restoring balance: using exotic species to control invasive exotic species, *Conservation Biology*, **18**: 38–49.

Holmes, P. M. & Cowling, R. M. (1997). The effects of invasion by *Acacia saligna* on the guild structure and regeneration capabilities of South African fynbos shrublands, *Journal of Applied Ecology*, **34**: 317–332.

Hutley, L. B. & Setterfield, S. A. (2008). Savannas. In *Encyclopedia of Ecology*, Amsterdam: Elsevier, Ed. S. E. Jørgensen and B. Fath, pp. 3143–3154.

Hutton, I. (2003). *Management for Birds on Lord Howe Island*. Sydney: Department of Environment and Conservation.

Huxman, T. E. & Smith, S. D. (2001). Photosynthesis in an invasive grass and native forb at elevated CO_2 during an El Nino year in the Mojave Desert, *Oecologia*, **128**: 193–201.

IPCC (2007). *Climate Change 2007: The Physical Science Basis. Contribution of Working Group I to the Fourth Assessment Report of the Intergovernmental Panel on Climate Change*, Cambridge: Cambridge University Press.

Jackson, J. (2005). Is there a relationship between herbaceous species richness and buffel grass (*Cenchrus ciliaris*)?, *Austral Ecology*, **30**: 505–517.

Lake, J. & Leishman, M. R. (2004). Invasion success of exotic plants in natural ecosystems: the role of disturbance, plant attributes and freedom from herbivores, *Biological Conservation*, **117**: 215–226.

Leishman, M. (1990). Suburban development and resultant changes in the phosphorus status of soils in the area of Ku-ring-gai, Sydney, *Proceedings of the Linnean Society NSW*, **112**: 15–25.

Leishman, M. R. & Thomson, V. P. (2005). Experimental evidence for the effects of additional water, nutrients and physical disturbance on invasive plants in low fertility Hawkesbury Sandstone soils, Sydney, Australia, *Journal of Ecology*, **93**: 38–49.

Leishman, M. R., Haslehurst, T., Ares, A. & Baruch, Z. (2007). Leaf trait relationships of native and invasive plants: community- and global-scale comparisons, *New Phytologist*, **176**: 635–643.

Leishman, M. R., Hughes, M. T. & Gore, D. B. (2004). Soil phosphorus enhancement below stormwater outlets in urban bushland: spatial and temporal changes and the relationship with invasive plants, *Australian Journal of Soil Research*, **42**: 197–202.

Le Roux, J. J., Geerts, S., Ivey, P. *et al.* (2010). Molecular systematics and ecology of invasive Kangaroo Paws in South Africa: management implications for a horticulturally important genus, *Biological Invasions*, **12**: 3989–4002.

Levine, J. M. (2000). Species diversity and biological invasions: relating local process to community pattern, *Science*, **288**: 852–854.

Lonsdale, W. M. (1992). The biology of *Mimosa pigra*. In *Guide to the Management of* Mimosa pigra: *prepared for an international workshop held at Darwin, Australia*, Darwin, pp. 11–15.

Lonsdale, W. M. (1993). Rates of spread of an invading species – *Mimosa pigra* in Northern Australia, *Journal of Ecology*, **81**: 513–521.

Lonsdale, W. M. (1994). Inviting trouble: introduced pasture species in northern Australia, *Australian Journal of Ecology*, **19**: 345–354.

Lonsdale, W. M. (1999). Global patterns of plant invasions and the concept of invisibility, *Ecology*, **80**: 1522–1536.

McFadyen, R. E. C. (1998). Biological control of weeds, *Annual Review of Entomology*, **43**: 369–393.

McKinney, M. L. & Lockwood, J. L. (1999). Biotic homogenization: a few winners replacing many losers in the next mass extinction, *Trends in Ecology & Evolution*, **14**: 450–453.

Manea, A. & Leishman, M. R. (2011). Competitive interactions between native and invasive exotic plant species are altered under elevated carbon dioxide, *Oecologia*, **165**: 735–744.

Millennium Ecosystem Assessment (2005). *Ecosystems and Human Well-being: Biodiversity Synthesis*. Washington: World Resources Institute.

Murphy, H. T., Hardesty, B. D., Fletcher, C. S. *et al.* (2008). Predicting dispersal and recruitment of *Miconia calvescens* (Melastomataceae) in Australian tropical rainforests, *Biological Invasions*, **10**: 925–936.

Naeem, S., Knops, J. M. H., Tilman, D. *et al.* (2000). Plant diversity increases resistance to invasion in the absence of covarying extrinsic factors, *Oikos*, **91**: 97–108.

Nagel, J. M., Huxman, T. E., Griffin, K. L. & Smith, S. D. (2004). CO_2 enrichment reduces the energetic cost of biomass construction in an invasive desert grass, *Ecology*, **85**: 100–106.

Neilan, W., Catterall, C. P., Kanowski, J. & McKenna, S. (2006). Do frugivorous birds assist rainforest succession in weed dominated oldfield regrowth of subtropical Australia?, *Biological Conservation*, **129**: 393–407.

NRMMC – Natural Resource Management Ministerial Committee (2006). *Australian Weeds Strategy: a National Strategy for Weed Management in Australia*. Canberra: Commonwealth of Australia.

Oram, R. N. (1990) *Register of Australian Herbage Plant Cultivars*, 3rd edn. Melbourne: CSIRO Publishing.

O'Donnell, J., Gallagher, R. V., Wilson, P. D *et al.*, (2012). Invasion hotspots for non-native plants in Australia under current and future climates, *Global Change Biology*, **18**: 617–629.

Paul, M., Catterall, C. P., Pollard, P. C. & Kanowski, J. (2010a). Does soil variation between rainforest, pasture and different reforestation pathways affect the early growth of rainforest pioneer species?, *Forest Ecology and Management*, **260**: 370–377.

Paul, M., Catterall, C. P., Pollard, P. C. & Kanowski, J. (2010b). Recovery of soil properties and functions in different rainforest restoration pathways, *Forest Ecology and Management*, **259**: 2083–2092.

Paul, M., Catterall, C. P., Kanowski, J. & Pollard, P. C. (2012). Recovery of rain forest soil seed banks under different reforestation pathways in eastern Australia, *Ecological Management and Restoration*, **13**: 144–152.

Perna, C. & Burrows, D. (2005). Improved dissolved oxygen status following removal of exotic weed mats in important fish habitat lagoons of the tropical Burdekin River floodplain, Australia, *Marine Pollution Bulletin*, **51**: 138–148.

Petroeschevsky, A. & Champion, P. D. (2008). Preventing further introduction and spread of aquatic weeds through the ornamental plant trade. In *Sixteenth Australian Weed Conference, Cairns*, pp. 200–302.

Pheloung, P. (1995). *Determining the Weed Potential of New Plant Introductions to Australia*. Perth: WA Department of Agriculture.

Pysek, P. & Jarosík, V. (2005). Residence time determines the distribution of alien plants. In *Invasive Plants: Ecological and Agricultural Aspects*. Basel, ED: Inderjit, S., pp. 77–96.

Randall, R. P. (2007). *The Introduced Flora of Australia and its Weed Status*. Adelaide: Cooperative Research Centre for Australian Weed Management.

Richardson, D. M. & Pyšek, P. (2012). Naturalization of introduced plants: ecological drivers of biogeographical patterns, *New Phytologist*, **196**: 383–396.

Richardson, D. M., Pyšek, P., Rejmánek, M. *et al.* (2000). Naturalization and invasion of alien plants: Concepts and definitions, *Diversity and Distributions*, **6**: 93–107.

Richardson, D. M. & Rejmánek, M. (2011). Trees and shrubs as invasive alien species – a global review, *Diversity and Distributions*, **17**: 788–809.

Rodríguez-Echeverría, S., Crisóstomo, J. A., Nabais, C. & Freitas, H. (2009). Belowground mutualists and the invasive ability of *Acacia longifolia* in coastal dunes of Portugal, *Biological Invasions*, **11**: 651–661.

Rossiter, N. A., Setterfield, S. A., Douglas, M. M. & Hutley, L. B. (2003). Testing the grass-fire cycle: alien grass invasion in the tropical savannas of northern Australia, *Diversity and Distributions*, **9**: 169–176.

Ruiz-Avila, R. J. & Klemm, V. V. (1996). Management of *Hydrocotyle ranunculoides* L.f., an aquatic invasive weed of urban waterways in Western Australia, *Hydrobiologia*, **340**: 187–190.

Sasek, T. W. & Strain, B. R. (1988). Effects of carbon dioxide enrichment on the growth and morphology of kudzu (*Pueraria lobata*), *Weed Science*, **1**: 28–36.

Schooler, S., Julien, M. & Walsh, G. C. (2006). Predicting the response of *Cabomba caroliniana* populations to biological control agent damage, *Australian Journal of Entomology*, **45**: 327–330.

Schooler, S. S., Salau, B., Julien, M. H. & Ives, A. R. (2011). Alternative stable states explain unpredictable biological control of *Salvinia molesta* in Kakadu, *Nature*, **470**: 86–89.

Serbesoff-King, K. (2003). Melaleuca in Florida: a literature review on the taxonomy, distribution, biology, ecology, economic importance and control measures, *Journal of Aquatic Plant Management*, **41**: 98–111.

Setterfield, S. A., Rossiter-Rachor, N. A., Hutley, L. B., Douglas, M. M. & Williams, R. J. (2010). Turning up the heat: the impacts of *Andropogon gayanus* (gamba grass) invasion on fire behaviour in northern Australian savannas, *Diversity and Distributions*, **16**: 854–861.

Sinden, J., Jones, R., Hesterba, S. *et al.* (2004). *The Economic Impact of Weeds in Australia. A Report to the CRC for Australian Weed Management.* Adelaide: Cooperative Research Centre for Australian Weed Management.

Smith, S. D., Huxman, T. E., Zitzer, S. F. *et al.* (2000). Elevated CO_2 increases productivity and invasive species success in an arid ecosystem, *Nature*, **408**: 79–82.

Stephens, C. J., Facelli, J. M. & Austin, A. D. (2008). The impact of bridal creeper (*Asparagus asparagoides*) on native ground-cover plant diversity and habitat structure, *Plant Protection Quarterly*, **23**: 136–143.

Stohlgren, T. J., Binkley, D., Chong, G.W. *et al.* (1999). Exotic plant species invade hot spots of native plant diversity, *Ecological Monographs*, **69**: 25–46.

Stokes, K. E. & Cunningham, S. A. (2006). Predictors of recruitment for willows invading riparian environments in south-east Australia: implications for weed management, *Journal of Applied Ecology*, **43**: 909–921.

Sullivan, P. R., Postle, L. A. & Julien, M. (2011). Biological control of *Salvinia molesta* by *Cyrtobagous salviniae* in temperate Australia, *Biological Control*, **57**: 222–228.

Thomson, V. P. & Leishman, M. R. (2004). Survival of native plants of Hawkesbury Sandstone communities with additional nutrients: effect of plant age and habitat, *Australian Journal of Botany*, **52**: 141–147.

Tilman, D. (1999). The ecological consequences of changes in biodiversity: a search for general principles, *Ecology*, **80**: 1455–1474.

Turner, P. J., Scott, J. K. & Spafford, H. (2008). The ecological barriers to the recovery of bridal creeper (*Asparagus asparagoides* (L.) Druce) infested sites: impacts on vegetation and the potential increase in other exotic species, *Austral Ecology*, **33**: 713–722.

Turner, P. J., Morin, L., Williams, D. G. & Kriticos, D. J. (2010). Interactions between a leafhopper and rust fungus on the invasive plant *Asparagus asparagoides* in Australia: a case of two agents being better than one for biological control, *Biological Control*, **54**: 322–330.

van Kleunen, M., Weber, E., & Fischer, M. (2009). A meta-analysis of trait differences between invasive and non-invasive plant species, *Ecology Letters*, **13**: 235–245.

van Klinken, R. D., Lawson, B. E. & Zalucki, M. P. (2009). Predicting invasions in Australia by a neotropical shrub under climate change: the challenge of novel climates and parameter estimation, *Global Ecology and Biogeography*, **18**: 688–700.

Vilà, M., Espinar, J. L., Hejda, M. *et al.* (2011). Ecological impacts of invasive alien plants: a meta-analysis of their effects on species, communities and ecosystems, *Ecology Letters*, **14**: 702–708.

Weerheim, M. S., Klomp, N. I., Brunsting, A. M. & Komdeur, J. (2003). Population size, breeding habitat and nest site distribution of little penguins (*Eudyptula minor*) on Montague Island, New South Wales, *Wildlife Research*, **30**: 151–157.

Weltzin, J. F., Belote, R. T. & Sanders, N. J. (2003). Biological invaders in a greenhouse world: will elevated CO_2 fuel plant invasions?, *Frontiers in Ecology and the Environment*, **1**: 146–153.

White, E. & Vivian-Smith, G. (2011). Contagious dispersal of seeds of synchronously fruiting species beneath invasive and native fleshy-fruited trees, *Austral Ecology*, **36**: 195–202.

Williams, A. L., Wills, K. E., Janes, J. K. *et al.* (2007). Warming and free-air CO_2 enrichment alter demographics in four co-occurring grassland species, *New Phytologist*, **176**: 365–374.

Williamson, M. & Fitter, A. (1996). The varying success of invaders, *Ecology*, **77**: 1661–1666.

Willis, A. J., McKay, R., Vranjic, J. A., Kilby, M. J. & Groves, R. H. (2003). Comparative seed ecology of the endangered shrub, *Pimelea spicata* and a threatening weed, Bridal Creeper: smoke, heat and other fire-related germination cues, *Ecological Management and Restoration*, **4**: 55–65.

Winkler, M. A., Cherry, H. & Downey, P. O. (2008). *Bitou Bush Management Manual: Current Management and Control Options for Bitou Bush* (Chrysanthemoides monilifera *ssp.* rotundata) *in Australia*. Sydney: Department of Environment and Climate Change (NSW).

Yurkonis, K. A. & Meiners, S. J. (2004). Invasion impacts local species turnover in a successional system, *Ecology Letters*, **7**: 764–769.

CHAPTER 7

Environmental weeds in New Zealand: impacts and management

Margaret C. Stanley and Imogen E. Bassett

Summary

More than 25 000 plant species have been introduced to New Zealand, with 8.5% of those having naturalised so far, and an additional 20 species estimated to naturalise each year. Two-thirds of naturalised species were introduced as garden plants. The number of recognised environmental weeds in New Zealand has almost reached 400 species. Empirical impact data exists for less than 5% of our current environmental weeds, and these are almost all widespread or locally dominant species. The research that does exist has demonstrated a variety of negative impacts, ranging from reduced native species diversity, to altered nutrient regimes. However, management of weeds is more cost effective in the early stages of weed invasion, and there is rarely empirical impact research available at these stages of invasion. While it is not feasible to conduct empirical research on the impacts of all environmental weeds in New Zealand, the development of a comprehensive framework that evaluates known impacts would help inform timely weed management decisions.

7.1 Introduction

New Zealand currently has a native vascular flora of 2158 species (de Lange *et al.* 2006), although human-mediated additions have substantially increased New Zealand's total flora. Maori transported the first non-native plant species (estimated to be <10 spp.) to New Zealand during their immigration from the Pacific Islands around 1280 AD (Williams & Cameron 2006; Wilmshurst *et al.* 2008). European colonisation in the nineteenth century resulted in many more plant introductions, currently considered to be approximately 25 049 species (Diez *et al.* 2009). An estimated 2136 of these introduced species have naturalised (*sensu* Falk-Petersen *et al.* 2006; Diez *et al.* 2009), almost equalling the number of native vascular plant species (Howell 2008). Furthermore, at least 66% of these naturalised plant species were originally introduced

Austral Ark: The State of Wildlife in Australia and New Zealand, eds. A. Stow, N. Maclean and G. I. Holwell. Published by Cambridge University Press. © Cambridge University Press 2015.

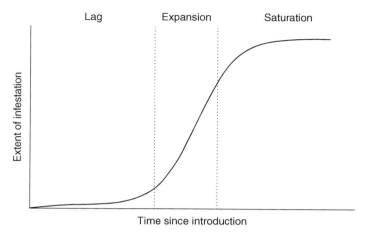

Figure 7.1 Infestation curve describing the relationship between the extent of a weed invasion and time since introduction, as described by a logistic population growth curve. Infestation phases are separated by dotted lines into: lag phase (relatively small area of infestation and slow population growth), expansion phase (area of infestation and population growth is increasing almost exponentially), and saturation phase (where almost all areas of suitable habitat are infested and growth is levelling off). Adapted from Williams (1997) and Groves (2006).

to New Zealand as garden plants (Howell 2008). Since 1950, around 20 additional plant species per year have naturalised from the existing pool of introduced species (Howell 2008) and some of these naturalised species then go on to become invasive, expanding their demographic and geographic range (Figure 7.1; Falk-Petersen *et al.* 2006). Of the naturalised species in New Zealand, 328 are considered to be environmental weeds by New Zealand's Department of Conservation (DOC) based on their negative impact in natural ecosystems, and are controlled where resources are available (Howell 2008). 'Environmental weed' is a fairly recent categorisation; agricultural weeds dominated New Zealand weed lists up until DOC produced lists of weeds in natural protected areas in the late 1980s (Howell 2008). Regional Councils also list specific environmental weeds, increasing the overall number of species nationally. Auckland Regional Council's Regional Pest Management Strategy 2007–2012 alone includes an additional 68 plant species not listed by the Department of Conservation. Of those plants that have naturalised, vines, trees and shrubs are more likely to be classified as environmental weeds than other plant forms, such as herbs (Howell 2008).

The pattern of invasion in New Zealand broadly follows the 'Tens Rule' proposed by Williamson (1993), with approximately 8.5% of introduced plants naturalised thus far. However, more than 20% of naturalised species have become invasive (recognised as weeds by a New Zealand government agency or primary industry) (Williams & Timmins 2002). This indicates that there is higher chance than usual that naturalised species will become invasive, however, we currently lack data on failure rates to assess this accurately (Diez *et al.* 2009).

It has been suggested that island systems are more vulnerable to invasion than continents (Sax & Gaines 2008). However, New Zealand appears no more vulnerable than Australia in terms of successful non-native plant naturalisations (Table 7.1). While a larger proportion of New Zealand's naturalised flora is non-native (due to substantially lower native plant diversity), the rate of non-native plant naturalisations is not significantly different from that in Australia (Diez *et al.* 2009; Hulme 2012). Ecosystem

Table 7.1   Comparative vulnerability of Australia and New Zealand to non-native plant naturalisation. Data sourced from Diez *et al.* (2009).

Country	No. of native plant spp.	No. of plant introductions	Proportion of (naturalised) flora that is non-native	Naturalisation success
Australia	15 822	28 866	15%	9.5% (2741)
New Zealand	1915	25 049	53%	8.5% (2136)

vulnerability to novel impacts, rather than naturalisations, may be driving the differences between islands and continents (Allen *et al.* 2006). However, as of 2005, the number of environmental weeds in Australia ($n = 1765$) was five times higher than that of New Zealand despite similar overall rates of plant introductions, suggesting that Australia is not any less vulnerable to weed impacts (Groves *et al.* 2005).

7.2 New Zealand's environmental weeds: the threat within

Unlike other pest organisms (e.g. invertebrates), the weed threat to New Zealand's biodiversity is primarily post-border, namely the more than 25 000 plant species that are already present in the country. The continuous change in the status of plants from 'introduced' to 'naturalised' to 'invasive', is akin to a 'conveyor belt' of weeds moving from residential and botanical gardens, and horticulture, into natural areas. At least 66% of environmental weeds were originally introduced as ornamental plants (Howell 2008). The advent of the Hazardous Substances and New Organisms (HSNO) Act in 1990, administered by the New Zealand Environmental Protection Agency (EPA), has essentially halted the introduction of new plant species to New Zealand. Since then, any person wanting to import a new species (or variety) of plant must apply to the EPA, and that species is put through a rigorous assessment, including passing the Weed Risk Assessment (WRA) as used by both New Zealand and Australia (Pheloung *et al.* 1999). Because of the seriousness of weed impacts in New Zealand, potential new imports are considered 'guilty' unless proven 'innocent'. Few applications ($n = 8$ applications; multiple species; www.epa.govt.nz) for new plant introductions have been approved under this legislation largely because of the significant costs incurred by importers applying for a new plant approval. While some have suggested this has led to an 'underground smuggling network', there is little evidence of this (Jack Craw, Auckland Council, personal communication). Plant breeders are focussing on creating new cultivars of species that already occur within New Zealand.

7.3 Vectors of spread

Humans are responsible not only for most plant introductions to New Zealand, but also for much of the dispersal and propagule pressure of non-native plants within New Zealand (leading to naturalisation and eventual invasion). Propagule pressure is a key determinant in establishment success of introduced species (Richardson & Rejmànek 1998; Duncan *et al.* 2006). In Britain, an analysis of ornamental plants for sale in British nursery catalogues over a hundred years showed that high market frequency and cheap seed prices had the greatest impact on the probability of a plant species becoming invasive more than 20 years after its first listing (Dehnen-Schmutz *et al.* 2007). In New Zealand, provinces with larger human populations have had more 'first naturalisations' (Williams & Cameron

2006). Although there may be some collector bias in this type of data, human population size is clearly far more important in driving naturalisation patterns than climate (Williams & Cameron 2006). Human-mediated dispersal of weeds into natural areas is becoming increasingly common as new subdivisions are created on their boundaries. Sullivan *et al.* (2005) found that the size and proximity of settlements are the dominating factors controlling the number of non-native plant species present in nearby native forests. Apart from the usual 'garden escapes' in terms of dispersal from these residential gardens into natural areas, garden dumpings (with seeds and vegetative shoots) were found in 25% of the areas surveyed. Furthermore, the New Zealand fondness for baches and cribs (holiday houses) in remote locations of natural beauty also provides opportunity for the transport of non-native plant species to natural environments. In effect, people are providing the critical long distance 'jump' dispersal for weed propagules, and facilitating the establishment of satellite populations (*sensu* Moody & Mack 1988).

Non-human dispersal vectors also contribute to the dispersal and spread of weeds in New Zealand. Birds probably play a major role in the dispersal of fleshy-fruited weed species, as has been found elsewhere in the world (Gosper *et al.* 2005). Blackbirds (*Turdus merula*) and silvereyes (*Zosterops lateralis*) may be particularly important, given that they are common throughout New Zealand, occur in many different ecosystems, and include a high proportion of exotic fruit in their diet (Williams 2006; Williams & Karl 1996). Starlings (*Sturnus vulgaris*) are also known to disperse large quantities of weed seeds in New Zealand (Ferguson & Drake 1999). Invasive mammals, such as possums (*Trichosurus vulpecula*) and feral pigs (*Sus scrofa*), are likely to disperse some weed seeds, although they often contribute to decreased plant reproduction and survival via herbivory or disturbance (Dungan *et al.* 2002; Williams 2003; Beavon 2007; Krull *et al.* 2013). While rodents consume weed seeds, most seeds are destroyed by the teeth and gut, apart from small hard seeds (Williams *et al.* 2000). Wind and water dispersal also contribute to weed spread in New Zealand. There are several weed species with wind-dispersed seeds that have spread widely from initial plantings, such as wilding pines (*Pinus radiata, P. contorta*) and grey willow (*Salix cinerea*) (Champion 1993; Hunter & Douglas 1984; Ledgard 2001). Additionally, neither crack willow (*Salix fragilis*) nor tradescantia (*Tradescantia fluminensis*) produce viable seed in New Zealand, but both have spread throughout the country via water dispersal of stem fragments (Timmins & Mackenzie 1995).

7.4 Negative impacts on New Zealand biodiversity

The plant species considered environmental weeds in New Zealand are classified as such because of their perceived impact in natural areas (Howell 2008). The financial cost of these impacts has been estimated at NZ$1.3 billion as a result of a 7% degradation of the conservation estate in New Zealand (Williams & Timmins 2002). However, quantifying the ecological impact of a weed on biodiversity is difficult and costly. Parker *et al.* (1999) proposed an equation to measure overall impact of invasive species in order to compare impact across taxa and geographic space: $I = R \times A \times E$, where I is the overall impact of a species, R is the range that the species occupies, A is the species' average abundance and E is the per capita effect. While this equation is useful conceptually, defining a species' range and abundance can be difficult and is seldom accurately measured. The per capita effect of an invader is the most difficult to measure, especially across different taxa and ecosystems and among different ecological levels (individuals, populations and ecosystems). Despite the often prohibitive cost, weed impact research is conducted, often because: (1) there is doubt as to the impact of the weed (no similar weeds or similar ecosystems have been studied for impact); (2) managers require further justification for spending management dollars and resources on weed control; (3) the Environmental Protection Agency (EPA) requires information about the benefits and costs of introducing a weed biocontrol agent, and this requires assessment of the target weed impact; or (4) there is conflict associated with controlling the weed because it has positive impacts (e.g. used for soil stability).

The impact of a weed can change temporally and spatially and effects quantified in one ecosystem cannot necessarily be transferred to other ecosystems. For example, subtropical palms, such as *Archontophoenix cunninghamiana*, that have naturalised in northern New Zealand, are unlikely to be a problem in the climatically unsuitable south (Sheppard 2013). Likewise, while heather (*Calluna vulgaris*) does not invade forest, it does have impacts in open subalpine ecosystems (Keesing 1995). Furthermore, the absence of an impact at the time of assessment does not mean that the weed will not have an impact in later years (e.g. with climate change) or as it increases in density and spreads to new localities. There is often a lag time in detecting weed impacts, usually because impacts are not perceived at low densities (Aikio *et al.* 2010). In most cases, it is only when a weed has become widespread, and is thus highly conspicuous, that it is deemed to have an impact.

Impacts on natural ecosystems can be classified into impacts on: (1) populations/ species; (2) communities; or (3) ecosystem processes. These are not mutually exclusive, and often the worst invaders have impacts in all three categories. Rigorous, published research data on the impact of weeds in New Zealand is available for approximately 22 environmental weed species (Table 7.2), comprising a mere 6.7% of the environmental weed species managed by the Department of Conservation. Most of these studies indicate significant negative impacts of weeds, although null results may be more difficult to publish. Demonstrated impacts almost invariably include reductions in native plant or invertebrate species diversity and/or changes in community composition. Almost half of the weed species were found to alter ecosystem processes in some way, such as reducing soil pH, or accelerating decomposition rates. Most of the species for which impact research is available are already widespread, or at least locally abundant, in New Zealand (Table 7.2; Figure 7.2). Thus the species we know most about have almost all passed the distribution threshold where eradication becomes technically and economically unfeasible. It would be more efficient and effective to know the potential impacts of newly naturalising species, so that management could be implemented at a stage when eradication is still feasible. However, it is unlikely that we will ever have complete impact information for the multitude of newly naturalising species in New Zealand, given the difficulty assessing weeds in the very early stages of spread when few ecosystems have been encountered. Furthermore, not all species have a history of naturalisation and impact elsewhere that could be used in predictive risk assessment models.

Impacts occur during the entire invasion process, and are usually only apparent once range expansion is well underway (Lockwood *et al.* 2007). At a basic level, impacts are assumed simply because the species is occupying an area of natural habitat that would otherwise be occupied by a native plant species (Williams & Newfield 2002). Declines in density of native plant species from an area is considered to have some impact for the remaining native species (albeit sometimes minor), in terms of resource availability and/or ecosystem function. While unrealistic to expect definitive impact research for all environmental weeds in New Zealand, it may be possible to use existing weed impact studies, both in New Zealand (Table 7.2) and globally (Pyšek *et al.* 2012), to

Table 7.2 Studies addressing the impact of weeds in New Zealand.

| Weed species | Distribution | | Type of impact | Reference |
	Extent of weed infestation	Recipient ecosystem		
Alligator weed (*Alternanthera philoxeroides*)	Localised dominance north of Auckland	Wetland Some open terrestrial sites (dispersal limited)	Increased abundance and species richness of fungivorous Coleoptera; altered trophic composition of Coleoptera	Bassett *et al.* (2011)
	Isolated early naturalisations elsewhere in upper-mid North Island		Altered invertebrate community composition	Bassett *et al.* (2012)
			Accelerated decomposition rates, altered patterns of biomass input	Bassett *et al.* (2010)
Buddleja (*Buddleja davidii*)	Localised dominance throughout most of North Island and upper/eastern South Island	Flood-plains Disturbed environments	Altered plant community composition	Smale (1990)
			Altered successional pathways	Bellingham *et al.* (2005) Smale (1990)
			Increased nutrient accumulation	Bellingham *et al.* (2005)
Climbing asparagus (*Asparagus scandens*)	Localised dominance upper North Island and less commonly coastal lower North Island and upper South Island	Disturbed environments Intact forest	Altered (increased or decreased) activity density of several invertebrate taxa	Bassett (2014)
Douglas fir (*Pseudotsuga menziesii*)	Abundant central North Island/Bay of Plenty and eastern parts of South Island	Grassland Forest	Reduced native seedling survival	Dehlin *et al.* (2008)
			Reduced nematode abundance and diversity, altered nematode community composition	Dehlin *et al.* (2008)
			Altered nutrient cycling. Reduced soil organic matter. Increased soil pH, NO_3 and NH_4	Dehlin *et al.* (2008)
Ginger (*Hedychium gardnerianum*)	Most extensive infestations north of Auckland. Also locally abundant in most of remaining North Island and upper South Island	Forest Disturbed environments	Altered invertebrate abundance (variable species and site interactions)	Yeates & Williams (2001) Bassett (2014)
			Reduced native seedling density and species richness, altered native seedling community composition	Williams *et al.* (2003)

Table 7.2 *(cont.)*

| Weed species | Distribution | | Type of impact | Reference |
	Extent of weed infestation	Recipient ecosystem		
Gorse (*Ulex europaeus*)	Widespread, abundant throughout entire country	Scrubland Grassland Disturbed environments	Altered invertebrate abundance (variable species and site interactions)	Yeates & Williams (2001)
			Increased invertebrate species richness. Altered invertebrate community composition and trophic structure	Harris *et al.* (2004)
			Increased soil nitrogen, higher rates of chemical cycling and dry matter accumulation compared to native beech forest.	Egunjobi (1969, 1971a, 1971b)
			Increased organic C and N	Sparling *et al.* (1994)
			Restricted native seedling regeneration following fire, but eventual facilitation of native regeneration	Lee *et al.* (1986)
			Altered native successional trajectories compared to native kanuka (*Kunzea ericoides*)	Sullivan *et al.* (2007)
Heather (*Calluna vulgaris*)	Locally abundant in central North Island.	Subalpine	Reduced native plant cover. Altered seasonal patterns of abundance for some invertebrate taxa	Keesing (1995)
Hieracium spp.	Abundant upper and eastern South Island. Also naturalised in middle parts of North Island.	High country Grasslands Forest	Reduced germination of rare endemic herb	Miller & Duncan (2004)
			Altered nutrient cycling. Increased soil organic C and N. Increased or decreased C:N ratio. Reduced soil pH	Knicker *et al.* (2000) Scott *et al.* (2001) McIntosh & Allen (1993) McIntosh *et al.* (1995) Saggar *et al.* (1999)
			Negligible impact on plant community	Meffin *et al.* (2010)

Species	Distribution	Habitat	Effects	Reference
Lodgepole pine (*Pinus contorta*)	Abundant in suitable habitat throughout much of South Island and central North Island	Grassland Alpine/subalpine	Reduced nematode abundance and diversity	Dehlin *et al.* (2008)
			Reduced species richness and nativeness of ectomycorrhizal fungal community	Dickie *et al.* (2010)
Marram grass (*Ammophila arenaria*)	Widespread throughout much of New Zealand	Coastal sand dunes Inland areas with low fertility	Native plant species displaced. Dune-forming processes affected	Hilton *et al.* (2005, 2006)
Mistflower (*Ageratina riparia*)	Most abundant north of Auckland. Also locally naturalised much of remaining mid-upper North Island, especially coastal provinces	Forest margins River banks Disturbed environments	Reduced native plant species richness and cover	Barton *et al.* (2007)
Old man's beard (*Clematis vitalba*)	Widespread throughout much of New Zealand	Coastal and lowland forest and shrubland	Contributes to a loss of forest structure Reduced native plant abundance and species richness	Ogle *et al.* (2000)
Radiata pine *Pinus radiata*	Locally naturalised throughout most of country. Most abundant central North Island/Bay of Plenty and upper and eastern parts of South Island	Grassland Disturbed environments	Altered invertebrate abundance (taxa-specific directionality) Altered invertebrate community composition Lower soil pH Lower soil organic C, total N, total P, exchangeable cations (indicates low organic matter content); lower soil microbial biomass	Yeates & Saggar (1998)
Spartina (*Spartina anglica*)	Locally abundant in scattered North and South Island locations	Intertidal areas with soft sediment	Increased sedimentation rates	Lee & Partridge (1983)
Tradescantia (*Tradescantia fluminensis*)	Abundant most of North Island. Some invasion of upper and western regions of South Island	Forest	Altered invertebrate abundance (variable species and site interactions); altered invertebrate community composition	Yeates & Williams (2001) Toft *et al.* (2001) Standish (2004) Bassett (2014)
			Accelerated decomposition; altered nutrient availability	Standish *et al.* (2004)

Table 7.2 (cont.)

Weed species	Distribution		Recipient ecosystem	Type of impact	Reference
	Extent of weed infestation				
				Reduced native seedling abundance and species richness	Standish et al. (2001) Kelly & Skipworth (1984)
Various aquatic weeds (Ceratophyllum demersum, Egeria densa, Elodea canadensis, Hydrilla verticillata, Hydrodictyon reticulatum, Lagarosiphon major)	Distribution varies with species		Submerged aquatic	Reduced abundance and species richness of native seeds in seed bank Reduced native plant species richness Localised macrophyte collapse and anoxic conditions	de Winton & Clayton (1996) Howard-Williams & Davies (1988) Wells et al. (1997) Wells & Clayton (2001)
Willow (Salix spp.)			Wetland	Increased invertebrate species richness and diversity; taxa-specific increases and decreases in invertebrate abundance Reduced total fish abundance, but increased trout abundance and biomass; site-specific effects on eel populations Reduced invertebrate abundance and biomass Increased beetle abundance, changes in beetle community trophic structure, increased proportion of introduced beetles	Glova & Sagar (1994) Lester et al. (1994) Watts et al. (2012)

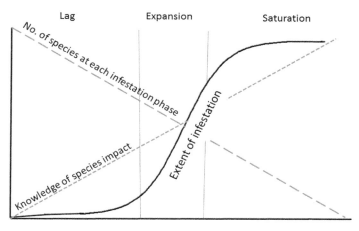

Figure 7.2 Relationship between the extent of weed infestation and the number of species at each phase (– – –) and the amount and certainty of knowledge available on the impact of each species (- - -). Derived from Williams (1997) and Groves (2006).

infer and predict impacts ('impact risk') based on taxonomy, growth form, preferred habitat, and other such characteristics. However, many reviews of weed impact studies conclude that impacts are highly variable and context-specific (e.g. Vilà *et al.* 2011; Pyšek *et al.* 2012; Hulme 2012). There are gaps and species biases in existing New Zealand weed impact studies (e.g. impact studies are primarily on entrenched weeds), and improvements are required in order to develop robust predictive tools (Hulme 2012).

In most cases, the only information we can use to assess potential impacts is the abundance and distribution of a species elsewhere in the world; although many of New Zealand's recent environmental weeds have no global history of invasiveness (Williams *et al.* 2001). In the absence of direct measures of impact, a wide range of assessment criteria can be used to prioritise species for management. Weed risk assessment systems have been developed in New Zealand, both for species not yet present (pre-border) and for species that are present (post-border) (Williams *et al.* 2002; Williams & Newfield 2002). Both include scoring systems based on expert opinion and current knowledge of weeds. The pre-border risk assessment system is based on the likelihood of a species establishing, what ecosystems it might establish in, and how easy it would be to control if it did establish (Williams *et al.* 2002). The post-border risk assessment system is based on weed history elsewhere; weediness of the species' relatives; potential interactions with native vegetation; history of the species in the area under consideration; and the technical considerations and social implications of attempting control (Williams & Newfield 2002). Carefully devised scoring systems such as these can be useful tools for predicting potential weed impacts.

Despite uncertainties about specific weed impacts, managers must make pragmatic decisions about controlling weeds. This process requires assessment of potential impacts and feasibility of control, in order to prioritise management according to the funds available. In practice, managers often make predictions about potential weed impacts on the basis of life form, assuming that similar growth forms will have broadly similar ecological impacts. For example, mistflower (*Ageratina riparia*) and tradescantia (*Tradescantia fluminensis*) are herbaceous ground-covering weeds, and both tend to reduce native plant species diversity and recruitment (Standish 2004; Barton *et al.* 2007; Table 7.2). Two other invasive

Table 7.3   Weeds used by native fauna in New Zealand

Weed	Perceived benefit to wildlife	Management outcome	Reference
Chinese windmill palm (*Trachycarpus fortunei*)	The endangered, endemic short-tailed bat (*Mystacina tuberculata*) visits these palm trees in the central North Island and consumes nectar (large amounts of pollen found in faeces)	Ongoing research to determine the extent of palm use and weed status; *M. tuberculata* are very opportunistic foragers and do not appear to rely on this weed	Molloy *et al.* (1995) Peterson *et al.* (2006)
Crack willow (*Salix fragilis*)	Willows in some regions used by birds such as the threatened Southern crested grebe (*Podiceps cristatus*). Also observations of use by the endemic New Zealand pigeon (*Hemiphaga novaeseelandiae*), the endemic rifleman (*Acanthis ittachloris*), little shags (*Phalacrocorax melanoleucos*) and by the endemic long-tailed bat (*Chalinolobus tuberculatus*)	Conventional control goes ahead – several specialised bird species in the same regions are threatened by willow invasion	O'Donnell (1982) Daniel & Williams (1984) Warren (1994) Clout *et al.* (1995) O'Donnell (2000)
Gorse (*Ulex europaeus*)	Used extensively by the threatened Mahoenui giant weta (Orthoptera: *Deinacrida mahoenui*) for food and shelter	Gorse is flammable and poses a high fire risk which would result in weta extinction; weta translocated to various sites in the region	Sherley & Hayes (1993) Sherley (1998a,b); C. Watts, personal communication)
Mistflower (*Ageratina riparia*)	Mistflower provides excellent cover for threatened endemic North Island brown kiwi (*Apteryx mantelli*); kiwis can move faster through this vegetation than dogs (predators) and mistflower also provides excellent cover from aerial predators (prey on juvenile kiwi)	Mistflower does not provide better cover than native vegetation such as parataniwha (*Elatostema rugosum*) and low shrubs (particularly from aerial predators)	H. Robertson, personal communication
Radiata pine (*Pinus radiata*)	An endemic parrot, kākā (*Nestor meridionalis septentrionalis*), utilises the high nutrient sap of pine trees by stripping the bark off trees to access the sap underneath. Pines may be targeted by kākā due to an increase in the sugar content of sap earlier than in native trees. Kākā also consume pine seeds	Competition with the introduced possum (*Trichosurus vulpecula*) for native resources may force kākā to utilise exotic trees; *Pinus radiata* is not targeted for control at sites where kākā utilise it	Beaven (1996)
	The threatened ground beetle *Holcaspis brevicula* is known only from three specimens, all of which have been found in a single pine plantation. Nearby native forest fragments have been searched for specimens	It is likely that this beetle species survived previous harvesting (every 27 years) or recolonised from adjacent stands. Restoration of kānuka shrubland suggested as an alternative	Brockerhoff *et al.* (2005) E. Brockerhoff, personal communication
Scotch broom (*Cytisus scoparius*)	Native New Zealand wood pigeons (Kererū; *Hemiphaga novaeseelandiae*) use scotch broom as a food source where little else is available due to forest clearance (particularly in the winter/spring months)	Control of scotch broom has occurred. Restoration of native food plants recommended	Clout *et al.* (1995) Fowler *et al.* (2000b)

Table 7.3   (cont.)

Weed	Perceived benefit to wildlife	Management outcome	Reference
Tree privet (*Ligustum lucidum*)	Native fantail (pīwakawaka; *Rhipidura fuliginosa*) use privet understorey to nest in the absence of native understorey	Where large privet control operations are undertaken, precautions are taken to ensure staggered removal and simultaneous replanting of native vegetation	D. Bardsley, personal communication Stone & Bettany (1999)
	Endemic puriri moth (*Aenetus virescens*) bores into privet trees and feeds on callus tissue	The puriri moth is a common moth species that has adapted well to European settlement. Puriri moths have a very broad host range and are unlikely to be affected by privet removal	Dugdale (1994)
	Kererū (New Zealand wood pigeon; *Hemiphaga novaeseelandiae*) feeds on privet fruit	Sections of the community in Te Aroha would like privet removed due to health concerns (asthma) while others want to retain privet because of the benefits to kererū – the community remains divided over whether to remove privet	H. Pene, personal communication

species (*Plectranthus ciliatus* and *Asparagus scandens*) similar in growth form to tradescantia and mistflower also reduce native plant species diversity and seedling recruitment (unpublished data, Kate McAlpine). Currently, advice given to community groups controlling weeds in natural areas is often based on assumed similar ecological impacts of weeds within life-form groups. For example, to first control the species that overtop intact ecosystems (e.g. vines), and then manage weeds that inhibit seedling recruitment subsequent to that (Williams 1997; Landcare Research 2004). Despite the practicality of this approach, the extent to which weed impacts can be predicted on the basis of life form remains uncertain. Validation of this approach through quantitative research could help to improve weed prioritisation procedures.

7.5   Potential positive consequences of weeds for New Zealand biodiversity

Weeds may have positive effects on components of native biodiversity and native ecosystems (Table 7.3). For example, weeds may provide food resources (Clout & Hay 1989) or shelter for native animals (Sherley & Hayes 1993), they may act as nurse plants for native plants (Porteous 1993), and facilitate forest succession (Williams 1997) (Table 7.3). Main (1992) has suggested some weeds may act as functional substitutes for extinct native species and have essential functions in ecosystems. Some threatened plants frequently coexist with weeds. For example, in New Zealand the mistletoe, *Tupeia antarctica*, uses introduced Scotch broom (*Cytisus scoparius*) as a host, although not exclusively, and broom also creates habitat for the orchid *Caladenia atradenia* (Reid 1998). The native mistletoe, *Ileostylus micranthus*, uses up to 87 non-native plant species as hosts, many of which are considered environmental weeds (Reid 1998).

Because of potential functional contributions such as those outlined above, rapid removal of weeds may sometimes disadvantage native flora and fauna by eliminating a suitable microclimate for native plants, or a resource, such as food or habitat, for native animals. Usually it is easier to observe a large, charismatic native bird (e.g. kererū, *Hemiphaga novaeseelandiae*) benefiting from a weed than it is to detect the negative impact of a weed on smaller, cryptic taxa. This can influence public perceptions of weed impacts (Fraser 2001). However, in many of these cases weeds are not essential resources, and use of weeds by native species is probably the result of availability or preference rather than dependence. Furthermore, weed exploitation by native species may have a number of cascade effects on ecosystems: (1) increased spread of weeds in natural areas through fruit consumption and seed dispersal (Williams & Karl 1996); (2) native species may no longer fulfil their function as pollinators and dispersers of native plant species (Pooley 1993; Scott-Shaw 1999); (3) extra resources provided by weed species could result in unnaturally high population numbers of native animals, which could in turn result in detrimental economic and ecological consequences (Bass 1995); or alternatively, (4) seasonal food shortages of native plant food sources for native fauna (Braithwaite *et al.* 1989). If, however, there is a risk that weed control could have a negative impact on dependent threatened or ecologically important species, management strategies should be put in place to ensure that alternative resources are available when the weed is controlled. In some situations, managers must decide whether to prioritise single species conservation or the retention of overall biodiversity.

Assessing the overall net effect of impacts of weeds on native species and ecosystems is becoming fundamental to weed control programmes. Although it is difficult and costly to assess the predicted positive and negative impacts of weeds and weed control on native species, it is often the only way to resolve conflicts and ensure that weed control does go ahead, even with caveats applied. Threat Reduction Assessment (TRA) is an approach that could be used by managers to assess the magnitude of the threat to species diversity posed by the presence/removal of the weed (Salafsky & Margoluis 1999). TRA is practical and cost-effective in terms of the data required for the assessment and it provides a means of assessing whether interventions, such as weed control, are working to reduce the threat to species diversity (Salafsky & Margoluis 1999). The conflicting positive and negative impacts of weeds often come to the fore when beneficial weeds are targeted for biocontrol programmes, but particularly if the weed also provides economic benefits (Stanley & Fowler 2004). In these cases, stakeholder involvement is critical, cost–benefit analyses are important tools, and occasionally more investigations are required into the level of specific benefits the weed provides (Stanley & Fowler 2004). In most cases, positive impacts of weeds are greatly outweighed by negative impacts (McFadyen, 1998) because weed abundance and spread usually displace the potentially dominant native biodiversity. Even if controlling the weed is likely to pose a risk to dependent native species, managers may decide that risks posed by *not* controlling the weed are greater. Efforts can then be made to mitigate risks to native species when conducting weed control.

7.6 Which weeds to manage?

7.6.1 Invasion as a demographic and geographic process

Weed invasion usually follows a logistic (sigmoidal) population growth curve, whereby the demographic and geographic expansion of weed populations increases over time.

However, extent of infestation (in hectares) is often used by managers *in lieu* of population growth, given that demographic data are not often available for most weed species. The relationship between infestation extent and time since introduction is often referred to as an infestation curve (Figure 7.1). There is usually a period of time where the population remains geographically small and localised (lag phase), after which it spreads at a near exponential rate (expansion phase) over many years until it asymptotes at a point where it has filled almost all available habitat in a particular region (saturation phase).

When the period of time between introduction and exponential growth is prolonged, while the weed remains in low densities, the population is in the lag phase (Groves 2006). Crooks & Soulé (1999) described three categories of lag phase: (1) Inherent lags caused by the nature of population growth and range expansion; (2) Environmental factors related to changes in ecological conditions that favour an exotic species; and (3) genetic factors related to the lack of fitness of the exotic species in a novel environment. All three processes appear to be operating in New Zealand. Bangalow palm (*Archontophoenix cunninghamiana*) took 94 years to naturalise from ornamental plantings, primarily due to long generation times (Cameron 2000). Williams *et al.* (2010) demonstrated that hawthorn (*Crataegus monogyna*) in Canterbury had increased its growth rate (r) 3.5 times following land-use changes that improved ecological conditions for the species. These changes included the cessation of burning, a decrease in the rabbit population (which reduced seedling browse), an increase in fertiliser use (which increased nutrient availability), and an increase in blackbird nesting site availability (which increased rates of seed dispersal). Genetic adaptation is also evident in weed invasion in New Zealand. The invasive cordgrass *Spartina anglica* arose from chromosome doubling in the sterile hybrid species *S. × townsendii* (Raybould *et al.* 2008). *S. anglica*, demonstrating considerable hybrid vigour, was widely planted to enhance sedimentation and 'reclamation' of estuarine tidal flats in New Zealand (Partridge 1987). *S. anglica* was also able to reproduce sexually and, for both of these reasons, is now far more widespread than other congeneric species present in New Zealand.

Lag phases appear common in New Zealand, with many plant species requiring several decades to invade. There is a strong relationship between the year an exotic plant species was detected as naturalised in New Zealand, and the number of regions it now occupies (Figure 7.3). However, in many cases, a perceived lag phase is merely due to a delay in detection of the spread, perhaps reflecting a lack of targeted surveillance, monitoring, or taxonomic awareness.

7.6.2 Targeting small weed infestations

The lag phase, whether real or perceived, is a critical issue for weed managers as early control or eradication is the most cost-effective approach (Figure 7.4; Grice 2009). This is when the most successful weed management programmes have occurred and the greatest benefits to costs have accrued (Grice 2009; Panetta 2009; Howell 2012). According to Rejmánek & Pitcairn (2002), the probability of eradication drops to 35% where the initial infestation size is greater than 1ha. In a review of 111 Department of Conservation weed management programmes in New Zealand, Howell (2012) found that eradication had only been achieved in four programmes, all involving infestations of less than 1 ha. Unfortunately, species in the early stages of naturalisation and range expansion tend to be inconspicuous and thus difficult to detect (Panetta & Timmins 2004). Additionally, it

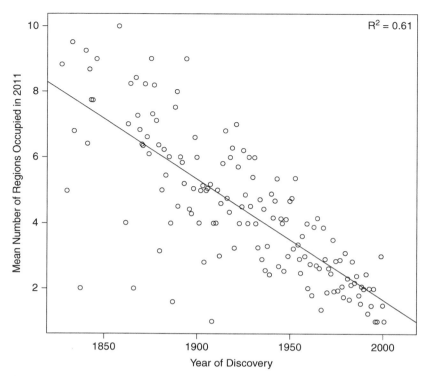

Figure 7.3 The average number of regions occupied for all fully naturalised species discovered each year between 1827 and 2011. N = 1789 species. Source: Gatehouse (2008).

can be difficult to gain public support for the management of these species, as their impact is perceived to be low and they may be primarily in gardens or parks.

Conversely, once weeds are abundant and highly conspicuous they are more difficult and expensive to control (Figure 7.4). Harris & Timmins (2009) demonstrated that if control is postponed until a species has become widespread, it is, on average, 40 times more expensive to remove and attempts are less likely to be successful. They concluded that it is worthwhile eradicating newly naturalising species in New Zealand even if the weed potential is unknown, as the cost and biodiversity benefits of removing bad weeds early more than compensates for eliminating non-weed species (Harris & Timmins 2009). The Department of Conservation recognises this principle in its strategic plan for managing invasive weeds, stating that 'early management of potential invasive weed species minimises both the future control costs and the possible degradation and loss of New Zealand's natural heritage' (Owen 1998). Thus, the goal of the Department of Conservation's 'weed-led' control programmes is to eradicate or contain emerging, potentially invasive weeds before they become a major problem. The Department of Conservation only controls widespread weeds if they fall within the ambit of 'site-led' control programmes which aim to protect the natural values of specific priority sites from invasive weeds (Owen 1998). In practice, the difference between weed-led and site-led control is less clear and widespread weeds are frequently controlled under the guise of weed-led programmes. An additional problem is that entrenched weeds are the most conspicuous weeds, and so are often the species the public wants controlled.

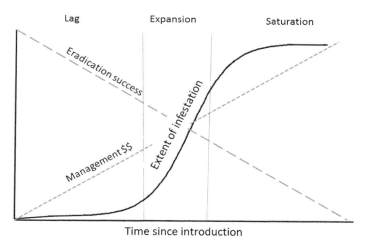

Figure 7.4 Relationship between the extent of weed infestation and the probability of eradication success (– – –) and the costs of weed management (- - -). Derived from Williams (1997); Grice (2009); and Howell (2012).

Several biological factors put weed eradication programmes at risk of failure including: (1) failure to find all original individuals from new or undiscovered infestations (detectability); (2) long time to maturity (surveillance stops before juveniles reach maturity); (3) propagule longevity (surveillance stops before buried seeds emerge and germinate); and (4) failure to understand dispersal mode/ability (new infestations not prevented or detected) (Panetta & Timmins 2004). In addition to these, there are a variety of socio-political issues (e.g. funding removed when weed no longer conspicuous) that pose a risk to eradication failure (Panetta & Timmins 2004; Parkes & Panetta 2009).

There are numerous examples of weed eradication programmes that have failed for one or more of these reasons (Panetta & Timmins 2004; Howell 2012). For example, a 3407 ha infestation of *Rhododendron ponticum* was targeted for eradication in Snowdonia National Park (Wales) in 1984 (Gritten 1995). Ten years later *R. ponticum* occupied a greater area of the park infested than in 1984, and £45m had been spent on the eradication programme. Reasons for failure were attributed to seedling re-invasion and a lack of coordinated control and follow-up (Gritten 1995). A similar situation occurred in New Zealand with a woolly nightshade (*Solanum mauritianum*) eradication programme in the Bay of Plenty. Enormous resources were put into the eradication without an initial estimate of the infestation size and without monitoring (Stanley 2003). Weed eradication programmes that have targeted large infestations (e.g. woolly nightshade in Bay of Plenty) are relegated to weed management programmes once they have been deemed to have failed (or have had funding removed) (Hulme 2006).

In many cases, the only potential option for the management of widespread weeds is biological control. This has occurred in New Zealand for a multitude of entrenched weeds, including broom (*Cytisus scoparius*), gorse (*Ulex europaeus*), old man's beard (*Clematis vitalba*) and woolly nightshade (*Solanum mauritianum*) (Fowler *et al.* 2000a). However, the development of a biological control programme is a slow and potentially long-term solution. It tends to be expensive (up-front research costs), fraught with potential difficulties, and has no guarantee of success. It frequently takes decades for agents to even establish, let alone cause significant damage to the target weed species, as has certainly been the case in New Zealand (Fowler *et al.* 2000a). However, biocontrol is often the only viable option, and has the potential in the long term to become a cost-effective, toxin-free alternative to chemical control (the primary means of control in New Zealand).

7.6.3   Deciding which environmental weeds to target

Numerous environmental weeds compete for finite weed management resources in New Zealand, and despite the fact that information about weed impacts in New Zealand is available for only a small proportion of environmental weeds, managers must make decisions about which weeds to control. As has been discussed above, priority is best given to weeds at the low end of the infestation curve where extent of infestation is smallest. Unfortunately, this group of species is the largest pool of species on the curve, particularly if species that have not yet naturalised are included (Figure 7.4). Therefore, there must be some assessment and prioritisation of the potential impact of those very low on the curve, including a cost–benefit analysis. The Department of Conservation recognises that weed-led programmes have the potential to quickly become both unachievable and expensive unless they are carefully chosen and planned. Initially, each weed species is assessed for eradication feasibility on the basis of seven criteria scored by expert opinion and current knowledge: (1) potential impacts on indigenous species and communities; (2) effectiveness and non-target impacts of available control methods; (3) current distribution; (4) likely benefits of control; (5) whether all legal requirements and necessary coordination can be gained; (6) detectability; and (7) risks involved (Owen 1998). Species competing for weed-led funds are then ranked on the basis of the species' 'weediness' (incorporating impacts on natural systems and likely rate of establishment and spread), and practicality of control. Weeds within sites are prioritised by: (1) the species' potential to establish, spread, and affect site values; (2) the practicality of controlling the species; and (3) the species' potential to create or exacerbate other weed problems (Owen 1998). The Department of Conservation is currently implementing a new approach to weed management, based on managing weed threats to native species or ecosystems within prioritised 'biodiversity management units' (BMUs). Weeds (and pest animals) will be managed within BMUs where they pose a threat to native species or ecosystems.

In New Zealand, each Regional Council or Unitary Authority must prepare a 'Regional Pest Management Plan' (RPMP – previously Regional Pest Management Strategies (RPMS)) under the Biosecurity Act (1993). The forthcoming National Policy Direction for Pest Management Plans and Programmes will set out changes to the structure of future RPMPs. A key change includes prescribed management programmes with consistent names and outcomes to be used in all pest management plans throughout the country. A new tool, Pathway Management Plans, will be available to manage vector pathways to complement the pest-led approach of RPMPs.

A draft plan must be prepared, listing proposed pests and corresponding management programmes. For example, the current Auckland RPMS includes 206 plant taxa, with management options ranging from banning of sale through to eradication at Council's expense. Identified benefits of managing the pest must outweigh the costs involved. Although the numbers used in the cost–benefits analyses (e.g., potential total area infested by the weed) are often projected estimates from experts and weed managers, conservative values are used for potential infestation extent and likelihood of impact. Nonetheless, even these conservative analyses produce very positive benefit to cost ratios for weeds with currently small infestations (Figure 7.4). The Auckland RPMS also includes 28 species classified as 'research organisms'. This category includes species requiring further research in order to determine whether they are capable of causing a 'serious adverse and unintented effect' (Section 72(1)(c) of the Act). It also provides the horticultural industry with prior notice that these species may be classified as pests in the future. New RPMPs will still be able to address research priorities, likely through their Regional Leadership Function. The draft plan goes through a period of public consultation, during which submissions can be made supporting or opposing the inclusion or treatment of particular species. The result of this process is that some species may be removed, or added, or their management altered if there is enough evidence that the benefits of managing the pest do not outweigh the costs of managing it. This can be a protracted process; for example, adoption of the Auckland RPMS (2007–2012) was delayed for seven months when it was challenged in the Environment Court by a palm nursery grower opposed to the proposed ban on phoenix palm (*Phoenix canariensis*).

Potential conflict between distributers and regulators of environmental weeds has led to the development of novel tools for managing weeds in New Zealand. The National Pest Plant Accord (NPPA) is a cooperative agreement between regional councils, government departments with biosecurity responsibilities and, since 2006, the Nursery and Garden Industry Association. Under the NPPA, all parties agree on a list of plant species that will be banned from sale, propagation and distribution on the basis of invasive potential or human health risk. Any person, including the general public, can propose that a species be added or removed, and the list is reviewed as required. Following the review in 2012, the list comprised 113 genera and at least 135 species (in some cases, the entire genus is banned, e.g. *Equisetum, Nassella*). The goal is to include only species that are in the early stages of naturalisation and spread, *and* are being (or are likely to be) actively sold or distributed. These criteria for inclusion have not, however, always been strictly applied, and there are a number of widespread species on the NPPA that may be removed in future.

7.7 Future challenges

Climate change will provide additional challenges for environmental weed management in New Zealand. For example, tropical and subtropical weeds are increasingly likely to establish and spread through the northern parts of New Zealand, bringing new challenges for weed managers (Sheppard *et al.* 2014). Specific areas of New Zealand, such as Hawkes Bay, are becoming more suitable for the establishment and spread of subtropical species, such as *Archontophoenix cunninghamiana*, *Psidium guajava* and *Schefflera actinophylla* (Sheppard 2013). Regional managers will be required to take into account potential distribution under climate change scenarios when undertaking weed risk assessments and cost–benefit analyses for their regions. The impetus to develop biofuels as a mitigation measure for climate change, may itself result in more invasive weed species. The ecological traits of ideal plant species for biofuel production (e.g. rapid growth, no known pests or diseases), are also some of the traits associated with invasive weeds (Raghu *et al.* 2006). Furthermore, propagule pressure as the result of mass plantings of biofuel crops is likely to increase the probability of naturalisation. Plant species with invasive properties being grown in New Zealand for biofuel production include *Jatropha curcas* (Pather 2012) and various *Salix* spp. (Snowdon *et al.* 2008).

Another challenge facing New Zealand weed managers is where native plant species are becoming invasive outside their natural range. For example, pōhutukawa (*Metrosideros excelsa*) has naturalised in the Wellington region, outside of its northern New Zealand range, while its congener, native to the Wellington region (northern rātā *Metrosideros robusta*), has become increasingly uncommon (Simpson 2005). The Department of Conservation and several Regional Councils recommend that people do not plant pōhutukawa outside its natural range. The Marlborough Regional Council extends this recommendation to additional native species that do not occur naturally in the area: pūriri (*Vitex lucens*), karo (*Pittosporum crassifolium*) and kauri (*Agathis australis*). While 'native weeds' is a more widely accepted concept in Australia (e.g. Costello *et al.* 2000; Gleadow & Ashton 1981; Price & Morgan 2008), it is likely that New Zealand managers will have to contend with this issue more in the coming decades. This issue has implications for public attitudes and buy-in to weed control generally. Currently, a common message being given to the public is 'remove weeds, plant natives'. The message can become confusing if it changes to 'remove weeds (including some natives), plant natives (but only some natives)'. This is particularly the case if the 'native weed' is an iconic native species such as the pōhutukawa. This issue is likely to be further complicated in the future, as climate change alters distribution limits for natives as well as existing weeds.

7.8 Conclusion

The number of recognised environmental weeds in New Zealand is upwards of 300 species and is increasing. Every year, more naturalised plant species spread from an initial localised population and become invasive. Empirical impact data exists for less than 7% of our current environmental weeds, and these are almost all widespread or locally dominant species. This is a likely consequence of the difficulties associated with studying newly naturalised or newly expanding species, as well as the elevated conspicuousness of widespread weeds. The research that has been completed on environmental weeds in New Zealand has demonstrated a variety of negative impacts, ranging from reduced native species diversity, to altered nutrient regimes. Although environmental weeds can have positive impacts, especially in terms of providing resources or habitat for native fauna, the negative impacts are usually considered to outweigh any perceived benefits. It is not feasible (in terms of monetary and time costs) to conduct empirical research on the impacts of all environmental weeds in New Zealand, but the development of a comprehensive framework that evaluates known impacts (both positive and negative) would help inform weed management decisions. While management of weeds is more cost effective in the early stages of weed invasion, there is rarely empirical impact research available at these stages of invasion. Hence, pragmatic management decisions are usually made based on expert opinion of potential distribution, and impact is predicted on the basis of a number of characteristics of the weed, the recipient ecological community, and degree of human assistance. While expert opinion on weed impact risk is not infallible, this is usually the only information available. Assessment of the risk posed by newly naturalised species (20% of which become environmental weeds) should be easier and more accurate than assessment of risk pre-border (a larger pool of species, most of which will not cross establishment and spread barriers and become environmental weeds). A useful research approach might be the development of general rules which link traits (e.g. growth forms) of weeds and characteristics (e.g. open vs. forest canopy) of ecosystems to specific impacts. Currently managers tend to link traits (e.g. vine vs. herb) to impact types, but no research exists that compares the impact types of multiple species (with a similar growth form) across different ecosystem types.

Acknowledgements

Many thanks to Kate McAlpine, Jack Craw, Hazel Gatehouse, Clayson Howell and Bill Lee for the information they provided and the helpful comments on various sections of the manuscript. We thank Carola Warner for editorial assistance. We would also like to acknowledge Peter Williams for inspiring and improving weed ecology research in New Zealand.

REFERENCES

Aikio, S., Duncan, R. P., Hulme, P. E. (2010). Lag-phases in alien plant invasions: separating the facts from the artefacts. *Oikos* **119**(2): 370–378.

Allen, R. B., Duncan, R. P., Lee, W. G. (2006). Updated perspectives on biological invasions. In: Allen, R. B., Lee, W. G. (eds.) *Biological Invasions in NZ*. Ecological Studies **186**. Springer-Verlag, Berlin; Heidelberg, pp. 437–451.

Barton, J., Fowler, S. V., Gianotti, A. F., *et al.* (2007). Successful biological control of mist flower (*Ageratina riparia*) in New Zealand: agent establishment, impact and benefits to the native flora. *Biological Control* **40**: 370–385.

Bass, D. A. (1995). Contribution of introduced fruits to the winter diet of pied currawongs in Armidale, New South Wales. *Corella* **19**: 127–132.

Bassett, I. E. (2014). Impacts on invertebrate fungivores: a predictable consequence of ground-cover weed invasion? *Biodiversity and Conservation* **23**: 791–810.

Bassett, I. E., Beggs, J. R., Paynter, Q. (2010). Decomposition dynamics of invasive alligator weed compared with native sedges in a Northland lake. *New Zealand Journal of Ecology* **34**(3): 324–331.

Bassett, I. E., Paynter, Q., Beggs, J. R. (2011). Invasive *Alternanthera philoxeroides* (alligator weed) associated with increased fungivore dominance in Coleoptera on decomposing leaf litter. *Biological Invasions* **13**: 1377–1385.

Bassett, I. E., Paynter, Q., Beggs, J. R. (2012). Invertebrate community composition differs between invasive herb alligator weed and native sedges. *Acta Oecologica* **41**: 65–73.

Beaven, B. M. (1996). Sap feeding behaviour of North Island kaka (*Nestor meridionalis septentrionalis*, Lorenz 1896) in plantation forests. Unpublished MSc Thesis. Waikato University, Hamilton.

Beavon, M. (2007). Pollination and dispersal of the noxious vine *Passiflora mollissima*. Unpublished MSc thesis, University of Canterbury, Christchurch, New Zealand.

Bellingham, P. J., Peltzer, D. A., Walker, L. R. (2005). Contrasting impacts of a native and an invasive exotic shrub on flood-plain succession. *Journal of Vegetation Science* **16**: 135–142.

Braithwaite, R. W., Lonsdale, W. M., Estbergs, J. A. (1989). Alien vegetation and native biota in tropical Australia: the impact of *Mimosa pigra*. *Biological Conservation* **48**, 189– 210.

Brockerhoff, V. E. G., Berndt, L. A., Jactel, H. (2005). Role of exotic pine forests in the conservation of the critically endangered ground beetle *Holcaspis brevicula* (Coleoptera: Carabidae). *New Zealand Journal of Ecology* **29**: 37–43.

Cameron, E. K. (2000). Bangalow palm (*Archontophoenix cunninghamiana*) begins to naturalise. *New Zealand Botanical Society Newsletter* **60**: 12–16.

Champion, P. D. (1993). Extent of willow invasion-the threats. In: **West, C. J.** (ed.) *Wild Willows in New Zealand: Proceedings of a Willow Control Workshop, 24–26 November, 1993.* Department of Conservation, Wellington, pp. 43–49.

Clout, M. N., Hay, J. R.(1989). The importance of birds as browsers, pollinators and seed dispersers in New Zealand forests. *New Zealand Journal of Ecology* **12** (suppl.): 27–33.

Clout, M. N., Karl, B. J., Pierce, R. J., Robertson, H. A. (1995). Breeding and survival of New Zealand pigeons *Hemiphaga novaeseelandiae. Ibis* **137**: 264–271.

Costello, D. A., Lunt, I. D., Williams, J. E. (2000). Effects of invasion by the indigenous shrub *Acacia sophorae* on plant composition of coastal grasslands in south-eastern Australia. *Biological Conservation* **96**: 113–121.

Crooks, J. A., Soulé, M. E. (1999). Lag times in population explosions of invasive species: Causes and implications. In: **Sandlund, O. T., Schei, P. J., Viken, Å.** (eds.) *Invasive Species and Biodiversity Management.* Kluwer Academic, Dordrecht, The Netherlands, pp. 103–125.

Daniel, M. J., Williams, G. R. (1984). A survey of the distribution, seasonal activity and roost sites of New Zealand bats. *New Zealand Journal of Ecology* **7**: 9–25.

Dehlin, H., Peltzer, D. A., Allison, V. J., *et al.* (2008). Tree seedling performance and below-ground properties in stands of invasive and native tree species. *New Zealand Journal of Ecology* **32**(1):67–79.

Dehnen-Schmutz, K., Touza, J., Perrings, C., Williamson, M. (2007). A century of the ornamental plant trade and its impact on invasion success. *Diversity and Distributions* **13**: 527–534.

de Lange, P. J., Sawyer, J. W. D., Rolfe, J. R. (2006). *New Zealand Indigenous Plant Checklist.* New Zealand Plant Conservation Network. Wellington. 94 pages.

de Winton, M. D., Clayton, J. S. (1996). The impact of an invasive submerged weed species on seed banks in lake sediments. *Aquatic Botany* **53**: 31–45.

Dickie, I. A., Bolstridge, N., Cooper, J. A., Peltzer, D. A. (2010). Co-invasion by *Pinus* and its mycorrhizal fungi. *New Phytologist* **187**: 475–484.

Diez, J. M., Williams, P. A., Randall, R. P., *et al.* (2009). Learning from failures: testing broad taxonomic hypotheses about plant naturalization. *Ecology Letters* **12**: 1174–1183.

Dugdale, J. S. (1994). Hepialidae (Insecta: Lepidoptera). *Fauna of New Zealand*, vol.**30**. Manaaki Whenua Press, Lincoln.

Duncan, R. P., Blackburn, T. M., Cassey, P. (2006). Factors affecting the release, establishment and spread of introduced birds in New Zealand. In: **Allen, R. B., Lee, W. G.** (eds.) *Biological Invasions in New Zealand.* Ecological Studies 186. Springer-Verlag, Berlin; Heidelberg, pp. 137–154.

Dungan, R. J., O'Cain, M. J., Lopez, M. L., Norton, D. A. (2002). Contribution by possums to seed rain and subsequent seed germination in successional vegetation, Canterbury, New Zealand. *New Zealand Journal of Ecology* **26**: 121–127.

Egunjobi, J. K. (1969). Dry matter and nitrogen accumulation in secondary successions involving gorse (*Ulex europaeus* L.) and associated shrubs and trees. *New Zealand Journal of Science* **12**: 175–193.

Egunjobi, J. K. (1971a). Ecosystem processes in a stand of *Ulex europaeus* L.: I. dry matter production, litter fall and efficiency of solar energy utilization. *The Journal of Ecology* **59**: 31–38.

Egunjobi, J. K. (1971b). Ecosystem processes in a stand of *Ulex europaeus* L.: II. The cycling of chemical elements in the ecosystem. *The Journal of Ecology* **59**: 669–678.

Falk-Petersen, J., Bohn, T., Sandlund, O. T. (2006). On the numerous concepts in invasion biology. *Biological Invasions* **8**: 1409–1424.

Ferguson, R. N., Drake, D. R. (1999). Influence of vegetation structure on spatial patterns of seed deposition by birds. *New Zealand Journal of Botany* **37**: 671–677.

Fowler, S. V., Syrett, P., Hill, R. L. (2000a). Success and safety in the biological control of environmental weeds in New Zealand. *Austral Ecology* **25**: 553–562.

Fowler, S. V., Syrett, P., Jarvis, P. (2000b). Will expected and unexpected non-target effects, and the new hazardous substances and new organisms act, cause biological control of broom to fail in New Zealand? In: **Spencer, N. R.** (ed.) *Proceedings of the X International Symposium on Biological Control of Weeds*. Montana State University, Bozeman, pp. 173–186.

Fraser, K. W. (2001). *Introduced Wildlife in New Zealand – a Survey of General Public Views*. Landcare Research Science Series No. 23. Landcare Research New Zealand, Christchurch.

Gatehouse, H. A. W. (2008). Ecology of the naturalisation and geographic distribution of the non-indigenous seed plant species of New Zealand. Unpubl. PhD thesis, Lincoln University, Lincoln. http://hdl.handle.net/10182/1009

Gleadow, R. M., Ashton, D. H. (1981). Invasion by *Pittosporum undulatum* of the forests of central Victoria. I. Invasion patterns and plant morphology. *Australian Journal of Botany* **2930**: 705–720.

Glova, G. J., Sagar, P. M. (1994). Comparison of fish and macroinvertebrate standing stocks in relation to riparian willows(*Salix* spp.) in three New Zealand streams. *New Zealand Journal of Marine and Freshwater Research* **28**: 255–266.

Gosper, C. R., Stansbury, C. D., Vivian-Smith, G. (2005). Seed dispersal of fleshy-fruited invasive plants by birds: contributing factors and management options. *Diversity and Distributions* **11**: 549.

Grice, T. (2009). Principles of containment and control of invasive species. In: **Clout, M. N., Williams, P. A.** (eds.) *Invasive Species Management. A Handbook of Principles and Techniques*, Oxford, UK, Oxford University Press, pp. 61–76.

Gritten, R. H. (1995). *Rhododendron ponticum* and some other invasive plants in the Snowdonia National Park. In: **Pyšek, P., Prach, K., Rejmánek, M., Wade, M.** (eds.) *Plant Invasions: General Aspects and Special Problems*. Workshop held at Kostelec nad Černými lesy, Czech Republic, 16–19 September 1993, pp. 213–219.

Groves, R. H. (2006). Are some weeds sleeping? Some concepts and reasons. *Euphytica* **148**: 111–120.

Groves, R. H., Boden, R., Lonsdale, W. M. (2005). *Jumping the Garden Fence: Invasive Garden Plants in Australia and their Environmental and Agricultural Impacts.* CSIRO report prepared for WWF-Australia. WWF-Australia. Sydney.

Harris, S., Timmins, S. M. (2009). *Estimating the Benefit of Early Control of all Newly Naturalised Plants.* Science for Conservation 292. Department of Conservation, Wellington. 25 pages.

Harris, R. J., Toft, R. J., Dugdale, J. S., Williams, P. A., Rees, J. S. (2004). Insect assemblages in a native (kanuka – *Kunzea ericoides*) and an invasive (gorse – *Ulex europaeus*) shrubland. *New Zealand Journal of Ecology* **28**(1): 35–47.

Hilton, M., Duncan, M., Jul, A. (2005). Processes of *Ammophila arenaria* (marram grass) invasion and indigenous species displacement, Stewart Island, New Zealand. *Journal of Coastal Research* **21**: 175–185.

Hilton, M., Harvey, N., Hart, A., James, K., Arbuckle, C. (2006). The impact of exotic dune grass species on foredune development in Australia and New Zealand: a case study of *Ammophila arenaria* and *Thinopyrum junceiforme*. *Australian Geographer* **37**: 313–334.

Howard-Williams, C., Davies, J. (1988). The invasion of Lake Taupo by the submerged water weed *Lagarosiphon major* and its impact on the native flora. *New Zealand Journal of Ecology* **11**: 13–19.

Howell, C. J. (2008). *Consolidated List of Environmental Weeds in New Zealand.* DOC Research & Development Series 292. Science & Technical Publishing, Department of Conservation, Wellington.

Howell, C. J. (2012). Progress toward environmental weed eradication in New Zealand. *Invasive Plant Science and Management* **5**(2): 249–258.

Hulme, P. E. (2006). Beyond control: wider implications for the management of biological invasions. *Journal of Applied Ecology* **43**: 835–847.

Hulme, P. E. (2012). Weed risk assessment: a way forward or a waste of time? *Journal of Applied Ecology* **49**: 10–19.

Hunter, G. G., Douglas, M. H. (1984). Spread of exotic conifers on South Island rangelands. *New Zealand Journal of Forestry* **29**: 78–96.

Keesing, V. F. (1995). Impacts of invasion on community structure: habitat and invertebrate assemblage responses to *Calluna vulgaris* (L.) Hull invasion in Tongariro National Park, New Zealand. Unpublished PhD thesis, Massey University, Palmerston North.

Kelly, D., Skipworth, J. P. (1984). *Tradescantia fluminensis* in a Manawatu (New Zealand) forest: I. Growth and effects on regeneration. *New Zealand Journal of Botany* **22**: 393–397.

Knicker, H., Saggar, S., Bäumler, R., McIntosh, P. D., Kögel-Knabner, I. (2000). Soil organic matter transformations induced by *Hieracium pilosella* L.in tussock grassland of New Zealand. *Biology and Fertility of Soils* **32**: 194–201.

Krull, C. R., Burns, B. R., Choquenot, D. & Stanley, M. C. (2013). Feral pigs in a temperate rainforest ecosystem: disturbance and ecological impacts. *Biological Invasions* **15**: 2193–2204.

Landcare Research. (2004) *Weed Management Guidelines.* Landcare Research Factsheet. http://www.landcareresearch.co.nz/__data/assets/pdf_file/0018/39042/weed_management_handout.pdf

Ledgard, N. (2001). The spread of lodgepole pine (*Pinus contorta*, Dougl.) in New Zealand. *Forest Ecology and Management* **141**: 43–57.

Lee, W. G., Partridge, T. R. (1983). Rates of spread of *Spartina anglica* and sediment accretion in the New River Estuary, Invercargill, New Zealand. *New Zealand Journal of Botany* **21**: 231–236.

Lee, W. G., Allen, R. B., Johnson, P. N. (1986). Succession and dynamics of gorse (*Ulex europaeus* L.) communities in the Dunedin Ecological District, South Island, New Zealand. *New Zealand Journal of Botany* **24**: 279–292.

Lester, P. J., Mitchell, S. F., Scott, D. (1994). Effects of riparian willow trees (*Salix fragilis*) on macroinvertebrate densities in two small central Otago, New Zealand, streams. *New Zealand Journal of Marine and Freshwater Research* **28**: 267–276.

Lockwood, J. L., Hoopes, M. F., Marchetti, M. P. (2007). *Invasion Ecology*. Blackwell Publishing, Oxford.

Main, A. R. (1992). The role of diversity in ecosystem function: an overview. In: Hobbs, R. J. (ed.). *Biodiversity in Mediterranean Ecosystems in Australia*. Surrey Beatty & Sons, Chipping Norton, pp. 77–93.

McFadyen, R. E. C. (1998). Biological control of weeds. *Annual Review of Entomology* **43**: 369–393.

McIntosh, P. D., Allen, R. B. (1993). Soil pH declines and organic carbon increases under hawkweed (*Hieracium pilosella*). *New Zealand Journal of Ecology* **17**: 59–60.

McIntosh, P. D., Loeseke, M., Bechler, K. (1995). Soil changes under mouse-ear hawkweed (*Hieracium pilosella*). *New Zealand Journal of Ecology* **19**: 29–34.

Meffin, R., Miller, A. L., Hulme, P. E., Duncan, R. P. (2010). Experimental introduction of the alien plant *Hieracium lepidulum* reveals no significant impact on montane plant communities in New Zealand. *Diversity and Distributions* **16**: 804–815.

Miller, A. L., Duncan, R. P. (2004). The impact of exotic weed competition on a rare New Zealand outcrop herb, *Pachycladon cheesemanii* (Brassicaceae). *New Zealand Journal of Ecology* **28**(1): 113–124.

Molloy, J. C., Daniel, M., O'Donnell, C., Lloyd, B., Roberts, A. (1995). *Bat (pekapeka) Recovery Plan (Mystacina, Chalinolobus)*. Threatened Species Recovery Plan No. 15. Department of Conservation, Wellington.

Moody, M. E., Mack, R. N. (1988). Controlling the spread of plant invasions: the importance of nascent foci. *Journal of Applied Ecology* **25**: 1009–1021.

O'Donnell, C. F. J. (1982). Food and feeding behaviour of the southern crested grebe on the Ashburton Lakes. *Notornis* **29**: 151–156.

O'Donnell, C. F. J. (2000). *The significance of river and open water habitats for indigenous birds in Canterbury, New Zealand*. Unpublished Report U00/37. Environment Canterbury, Christchurch.

Ogle, C. C., La Cock, G. D., Arnold, G., Mickleson, N. (2000). Impact of an exotic vine *Clematis vitalba* (F. Ranunculaceae) and of control measures on plant biodiversity in indigenous forest, Taihape, New Zealand. *Austral Ecology* **25**: 539–551.

Owen, S. J. (1998). *Department of Conservation Strategic Plan for Managing Invasive Weeds*. Department of Conservation, Wellington.

Panetta, F. D. (2009). Weed eradication – an economic perspective. *Invasive Plant Science and Management* **2**(4): 360–368.

Panetta, F. D., Timmins, S. M. (2004). Evaluating the feasibility of eradication for terrestrial weed incursions. *Plant Protection Quarterly* **19**(1): 5–11.

Parker, I. M., Simberloff, D., Lonsdale, W. M., *et al.* (1999). Impact: toward a framework for understanding the ecological effects of invaders. *Biological Invasions*, **1**(1), 3–19.

Parkes, J. P., Panetta, F. D. (2009). Eradication of invasive species: progress and emerging issues in the 21 century. In: **Clout, M. N., Williams, P. A.** (eds.) *Invasive Species Management. A Handbook of Principles and Techniques.* Oxford University Press, Oxford, UK, pp. 47–60.

Partridge, T. R. (1987). Spartina in New Zealand. *New Zealand Journal of Botany* **25**: 567– 575.

Pather, V. (2012). The ecology and future invasive potential of *Jatropha curcas* as a biofuel crop in New Zealand. *MSc Thesis*, School of Biological Sciences, University of Auckland.

Peterson, P. G., Robertson, A. W., Lloyd, B., McQueen, S. (2006).Non-native pollen found in short-tailed bat (*Mystacina tuberculata*) guano from the central North Island. *New Zealand Journal of Ecology* **30**(2): 267–272.

Pheloung, P. C., Williams, P. A., Halloy, S. R. (1999). A weed risk assessment model for use as a biosecurity tool evaluating plant introductions. *Journal of Environmental Management* **57**: 239–251.

Pooley, E. (1993). *The Complete Field Guide to Trees of Natal, Zululand and Transkei.* Natal Flora Publications Trust, Durban.

Porteous, T. (1993). *Native Forest Restoration: a Practical Guide for Landowners.* Queen Elizabeth the Second National Trust, Wellington.

Price, J. N., Morgan, J. W. (2008). Woody plant encroachment reduces species richness of herb-rich woodlands in southern Australia. *Austral Ecology* **33**: 278–289.

Pyšek, P., Jarošík, V., Hulme, P. E., *et al.* (2012). A global assessment of invasive plant impacts on resident species, communities and ecosystems: the interaction of impact measures, invading species' traits and environment. *Global Change Biology* **18**: 1725–1735.

Raghu, S., Anderson, R. C., Daehler, C. C., *et al.* (2006). Adding biofuels to the invasive species fire? *Science* **313**: 1742.

Raybould, A. F., Gray, A. J., Lawrence, M. J., Marshall, D. F. (2008). The evolution of *Spartina anglica* CE Hubbard (Gramineae): origin and genetic variability. *Biological Journal of the Linnean Society* **43**: 111–126.

Reid, V. A. (1998). *The Impact of Weeds on Threatened Plants.* Science and Research Internal Report No. 164. Department of Conservation, Wellington.

Rejmánek, M., Pitcairn, M. J. (2002). When is eradication of exotic pest plants a realistic goal? In: **Veitch, C. R. Clout, M. N.** (eds.) *Turning the Tide: The Eradication of Invasive Species.* Invasive Species Specialist Group of the World Conservation Union (IUCN). Auckland, New Zealand, pp 249–253.

Richardson, D. M., Rejmánek, M. (1998). *Metrosideros excelsa* takes off in the fynbos. *Veld & Flora* **85**: 14–16.

Saggar, S., McIntosh, P. D., Hedley, C. B., Knicker, H. (1999). Changes in soil microbial biomass, metabolic quotient, and organic matter turnover under *Hieracium* (*H. pilosella* L.). *Biology and Fertility of Soils* **30**: 232–238.

Salafsky, N., Margoluis, R. (1999). Threat reduction assessment: a practical and cost-effective approach to evaluating conservation and development projects. *Conservation Biology* **13**: 830–841.

Sax, D. F., Gaines, S. D. (2008). Species invasions and extinction: the future of native biodiversity on islands. *Proceedings of the Natural Academy of Sciences of the United States of America* **105**: 11490–11497.

Scott, N. A., Saggar, S., Mcintosh, P. D. (2001). Biogeochemical impact of *Hieracium* invasion in New Zealand's grazed tussock grasslands: sustainability implications. *Ecological Applications* **11**(5): 1311–1322.

Scott-Shaw, C. R. (1999). *Rare and Threatened Plants of KwaZulu-Natal and Neighbouring Regions*. KwaZulu-Natal Nature Conservation Service, Pietermaritzburg.

Sheppard, C. S., Burns, B. R., Stanley, M. C. (2014). Predicting plant invasions under climate change: are species distribution models validated by field trials? *Global Change Biology* **20**: 2800–2814.

Sherley, G. H. (1998a). Translocating a threatened New Zealand giant Orthopteran, *Deinacrida* sp. (Stenopelmatidae): some lessons. *Journal of Insect Conservation* **2**: 195–199.

Sherley, G. H. (1998b). *Threatened Weta Recovery Plan*. Threatened Species Recovery Plan No. 25. Department of Conservation, Wellington.

Sherley, G. H., Hayes, L. M. (1993). The conservation of a giant weta (*Deinacrida* n. sp. Orthoptera: Stenopelmatidae) at Mahoenui, King country: habitat use, and other aspects of its ecology. *New Zealand Entomologist* **16**: 55–68.

Simpson, P. (2005). *Pōhutukawa & Rātā: New Zealand's Iron-hearted Trees*. Te Papa Press, Wellington.

Smale, M. C. (1990). Ecological role of buddleia (*Buddleja davidii*) in streambeds in Te Urewera National Park. *New Zealand Journal of Ecology* **14**: 1–6.

Snowdon, K., McIvor, I., Nicholas, I. (2008). *Energy Farming with Willow in New Zealand*. Scion, Rotorua, New Zealand. 29 pp.

Sparling, G. P., Hart, P. B. S., August, J. A., Leslie, D. M. (1994). A comparison of soil and microbial carbon, nitrogen, and phosphorus contents, and macro-aggregate stability of a soil under native forest and after clearance for pastures and plantation forest. *Biology and Fertility of Soils* **17**: 91–100.

Standish, R. (2004). Impact of an invasive clonal herb on epigaeic invertebrates in forest remnants in New Zealand. *Biological Conservation* **116**: 49–58.

Standish, R. J., Robertson, A. W., Williams, P. A. (2001). The impact of an invasive weed *Tradescantia fluminensis* on native forest regeneration. *Journal of Applied Ecology* **38**: 1253–1263

Standish, R. J., Williams, P. A., Robertson, A. W., Scott, N. A., Hedderley, D. I. (2004). Invasion by a perennial herb increases decomposition rate and alters nutrient availability in warm temperate lowland forest remnants. *Biological Invasions* **6**(1): 71–81.

Stanley, M. (2003). *Review of the Woolly Nightshade (Solanum mauritianum) Management Programme in the Bay of Plenty*. Landcare Research Contract Report: LC0304/030

Stanley, M. C., Fowler, S. V. (2004). Conflicts of interest associated with the biological control of weeds. In: **Cullen, J. M., Briese, D. T., Kriticos, D. J.**, *et al.* (eds.) *Proceedings of the XI International Symposium on Biological Control of Weeds Canberra, Australia, 2003*, pp. 322–340.

Stone, G. S., Bettany, S. (1999). *Auckland Domain Weed Management Plan*. Te Ngahere Native Forest Management, Auckland.

Sullivan, J. J., Timmins, S. M., Williams, P. A. (2005). Movement of exotic plants into coastal native forests from gardens in northern New Zealand. *New Zealand Journal of Ecology* **29**(1): 1–10.

Sullivan, J. J., Williams, P. A., Timmins, S. M. (2007). Secondary forest succession differs through naturalised gorse and native kānuka near Wellington and Nelson. *New Zealand Journal of Ecology* **31**: 22–38.

Timmins, S. M., Mackenzie, I. W. (1995). *Weeds in New Zealand Protected Natural Areas Database*. Department of Conservation Technical Series No. 8. Department of Conservation, Wellington.

Toft, R. J., Harris, R. J., Williams, P. A. (2001). Impacts of the weed *Tradescantia fluminensis* on insect communities in fragmented forests in New Zealand. *Biological Conservation* **102**: 31–46.

Vilà, M., Espinar, J. L., Hejda, M., *et al.* (2011). Ecological impacts of invasive alien plants: a meta-analysis of their effects on species, communities and ecosystems. *Ecology Letters* **14**: 702–708.

Warren, A. (1994). Project river recovery. In: **West, C. J.** (ed.) *Wild Willows in New Zealand: Proceedings of a Willow Control Workshop*. Department of Conservation, Wellington, pp. 51–52.

Watts, C., Rohan, M., Thornburrow, D. (2012). Beetle community responses to grey willow (*Salix cinerea*) invasion within three New Zealand wetlands. *New Zealand Journal of Zoology* **39**(3): 209–227.

Wells, R. D. S., Clayton, J. S. (2001). Ecological impacts of water net (*Hydrodictyon reticulatum*) in Lake Aniwhenua, New Zealand. *New Zealand Journal of Ecology* **25**(2): 55–63.

Wells, R. D. S., de Winton, M. D., Clayton, J. S. (1997). Successive macrophyte invasions within the submerged flora of Lake Tarawera, central North Island, New Zealand. *New Zealand Journal of Marine and Freshwater Research* **31**: 449–459.

Williams, C. E. (1997). Potential valuable ecological functions of nonindigenous plants. In: **Luken, J. O., Thieret, J. W.** (eds.) *Assessment and Management of Plant Invasions*. Springer-Verlag, New York, pp. 26–34.

Williams, P. A. (1997). *Ecology and Management of Invasive Weeds*. Conservation Sciences Publication No. 7. Department of Conservation, Wellington.

Williams, P. A. (2003). Are possums important dispersers of large-seeded fruit? *New Zealand Journal of Ecology* **27**: 221–223.

Williams, P. A. (2006). The role of blackbirds (*Turdus merula*) in weed invasion in New Zealand. *New Zealand Journal of Ecology* **30**: 285–291.

Williams, P. A., Cameron, E. K. (2006). Creating gardens: the diversity and progression of European plant introductions. In: **Allen, R. B., Lee, W. G.** (eds.) *Biological Invasions in New Zealand.* Ecological Studies 186. Berlin & Heidelberg, Springer, pp. 33–48.

Williams, P. A., Karl, B. J. (1996). Fleshy fruits of indigenous and adventive plants in the diet of birds in forest remnants, Nelson, New Zealand. *New Zealand Journal of Ecology* **20**: 127–145.

Williams, P. A., Newfield, M. (2002). *A Weed Risk Assessment System for New Conservation Weeds in New Zealand.* Department of Conservation, Wellington.

Williams, P. A., Timmins, S. (2002). Economic impacts of weeds in New Zealand. In: **Pimentel, D.** (ed.). *Biological Invasions, Economic and Environmental Costs of Alien Plant, Animal, and Microbe Species.* CRC Press, London, pp. 175–184.

Williams, P. A., Karl, B. J., Bannister, P., Lee, W. G. (2000). Small mammals as potential seed dispersers in New Zealand. *Austral Ecology* **25**: 523–532.

Williams, P. A., Kean, J. M., Buxton, R. P. (2010). Multiple factors determine the rate of increase of an invading non-native tree in New Zealand. *Biological Invasions* **12**: 1377–1388.

Williams, P. A., Nicol, E., Newfield, M. (2001). Assessing the risk to indigenous New Zealand biota from new exotic plant taxa and genetic material. In: **Groves, R. H., Panetta, F. D., Virtue, J. G.** (eds.) *Weed Risk Assessment.* CSIRO Publishing, Collingwood, Australia, pp. 100–116.

Williams, P. A., Wilton, A., Spencer, N. (2002). *A Proposed Conservation Weed Risk Assessment System for the New Zealand Border.* Science for Conservation 209. Department of Conservation, Wellington. 47 pages.

Williams, P. A., Winks, C., Rijkse, W. (2003). Forest processes in the presence of wild ginger (*Hedychium gardnerianum*). *New Zealand Journal of Ecology* **27**(1):45–54.

Williamson, M. (1993). Invaders, weeds and the risk from genetically modified organisms. *Experientia* **49**: 219–224.

Wilmshurst, J. M., Anderson, A. J., Higham, T. F. G., Worthy, T. H. (2008). Dating the late prehistoric dispersal of humans to New Zealand using the commensal Pacific rat. *Proceedings of the National Academy of Sciences of the United States of America* **105**: 7676–7680.

Yeates, G. W., Saggar, S. (1998). Comparison of soil microbial properties and fauna under tussock-grassland and pine plantation. *Journal of the Royal Society of New Zealand* **28**(3): 523–535.

Yeates, G. W., Williams, P. A. (2001). Influence of three invasive weeds and site factors on soil microfauna in New Zealand. *Pedobiologia* **45**: 367–383.

CHAPTER 8

The insidious threat of invasive invertebrates

Darren F. Ward

Summary

Exotic invertebrates make up a sizeable, and growing, proportion of invertebrates in both Australia and New Zealand. However, there is a general lack of awareness of the impacts (realised or potential) of invasive invertebrates, and very few species have a high public profile. Border interception records show the sheer number and diversity of invertebrates being transported around the globe by human trade. In-depth studies on empty sea containers, ants and forestry insects, confirm that trade pathways are regularly contaminated with timber, agricultural and nuisance arthropod pests. A principal feature of the biosecurity systems in Australia and New Zealand is their holistic nature of managing invasive species through assessing threats pre-border, having high levels of surveillance at the border, and their rapid response to any incursions of new species. However, despite these systems many new species establish each year. Several issues are discussed that will be important over the next few decades: (i) the need for evidence of ecological impacts; (ii) climate change and trade liberalisation (which will affect which species will become invasive in the future); and (iii) improved technology and capability (needed to show there is the ability to manage invasive invertebrates).

8.1 Introduction

Intensification of human transportation and commerce around the world has led to the widespread movement of many species outside of their native range (Mack *et al*. 2000). As a result, biological invasions are now a global phenomenon, and are widely recognised as a significant component of global change, affecting agro-forestry industries, natural ecosystems and social activities (Mack *et al*. 2000; Pimentel 2002).

Despite forming a large part of the exotic fauna worldwide, invertebrates have received disproportionality less attention compared with the impacts of plants and vertebrates, especially for impacts associated with natural ecosystems (Kenis *et al*. 2009; Roy *et al*.

Austral Ark: The State of Wildlife in Australia and New Zealand, eds. A. Stow, N. Maclean and G. I. Holwell. Published by Cambridge University Press. © Cambridge University Press 2015.

2011). Although the *threat* posed by invasive invertebrates towards natural ecosystems is well recognised (Cook *et al.* 2002; Barlow & Goldson 2002), *evidence* is scare and limited to a few well-known examples (Kenis *et al.* 2009). Consequently, the ecological impacts of invasive invertebrates are poorly understood compared to vertebrate pests and weeds; and their impacts are often assumed rather than measured (Barlow & Goldson 2002).

However, the evidence for direct and indirect impacts of invasive invertebrates on natural ecosystems is growing (Kenis *et al.* 2009). In the past few years there has been several important syntheses (Snyder & Evans 2006; Kenis *et al.* 2009; Brockerhoff *et al.* 2010; Roy *et al.* 2011), the immense pan-European documentation of alien arthropods (Hulme & Roy 2010), and reviews for several major groups of invertebrate pests (Silverman & Brightwell 2008; Beggs *et al.* 2011; Rabitsch 2011). Together, these studies show that invasive invertebrates can negatively affect the population genetics and abundance of native species, the composition of natural communities and ecosystem processes (Kenis *et al.* 2009).

Exotic invertebrates make up a sizeable, and growing, proportion of invertebrates in both Australia and New Zealand. Cranston (2010) recently provided a review on the biodiversity and conservation of Australasian insects. Common threats were listed as habitat loss, introduced species and climate change, but he considered 'a more insidious threat comes from the ever-expanding distribution of invasive insects' (Cranston 2010, p. 65). In this chapter, I focus on the invasion of natural terrestrial ecosystems by invasive invertebrates in Australia and New Zealand. I summarise the ecological impacts of several well-known species, but also highlight a number of areas that are well developed for invasive invertebrates 'down under'; that is, the use of border interception records, a focus towards pre-border and border surveillance, eradication, and the development of innovative technologies. Key issues are discussed for invasive invertebrates that will be important over the next few decades. I follow Brockerhoff *et al.* (2010) and refer to species as invasive only if they are exotic *and* expanding their range, and use the term 'exotic' as a shorter synonym for non-indigenous species (i.e. outside its native range).

8.2 Invasion opportunity

The opportunity for dispersal is the fundamental stage of invasion upon which all other stages (establishment, spread, impact) are reliant. Only those species that successfully pass through the initial dispersal 'filter' proceed along the invasion pathway (Kolar & Lodge 2001; Puth & Post 2005). Global trade and transportation by humans has greatly extended the capacity of many species to become established in regions outside their natural range (Drake *et al.* 1989; Sandland *et al.* 1999; Floerl & Inglis 2004). Trade routes essentially represent pathways for invasion, with transport hubs (shipping ports, airports, mail centres) acting as important foci for the arrival and spread of exotic species.

Australia and New Zealand have a relatively similar history of invasive species introductions, mostly because of their similar history of European colonisation. Both countries were suddenly exposed to a raft of landscape changes with the arrival of European traders and settlers, especially the removal of native vegetation and its conversion to pasture. Without having a concept of 'biosecurity' as we know it today, and without any border inspection processes, many new species became quickly established. Most of

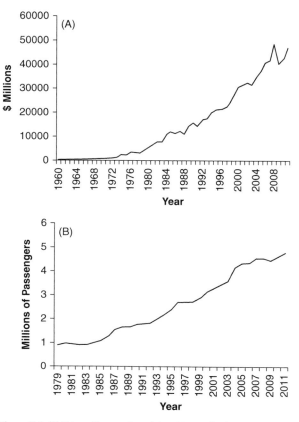

Figure 8.1 (A) Value of imported goods into New Zealand 1960–2011. Source: Statistics New Zealand (Table IMP054AA, downloaded 16/7/12). (B) Total annual passenger movements by all travel modes into New Zealand 1979–2011. Source: Statistics New Zealand (Table ITM236AA, downloaded 16/7/12).

these new species were primarily associated with the arrival of goods (e.g. machinery), stock and crops associated with agriculture systems.

However, introductions of new species into Australia and New Zealand steadily continued after the initial period of European settlement. A major factor in subsequent introductions was the increasing trend of mass transportation, of both people and goods, around the globe (Figure 8.1). Vessels became faster and larger, essentially negating the effect of isolation that Australia and New Zealand had previously relied on as a 'quarantining measure' to exclude exotic species. Coupled with increased efficiency, vessels were also increasingly connected to a much larger range of trading ports and stop-over points, a diversifying range of trading partners, and the removal of trade barriers. This combination of factors essentially filtered out time and distance as obstacles, but also provided the opportunity for invasion for an increasingly different set of species.

8.3 The insidious nature of invertebrate invasions

A number of studies have examined how initial arrival opportunity translates into establishment, but the majority of these studies have been based on species that have been *intentionally* introduced (e.g. insects released for biological control or vertebrates) (Duncan *et al.* 2003; Simberloff 2009). Historically, a large number of new animal and plant species were intentionally introduced into both Australia and New Zealand for farming, gardening, hunting, and subsequently became invasive (Clout 2002; Williams & Timmins 2002). Intentional introductions may not be representative of invasions, because (i) they bypass the first stage of invasion (opportunity), and (ii) they are deliberately selected for certain characteristics (Suarez *et al.* 2005). Furthermore, an even larger number of species arrived accidentally and largely unnoticed, that is, invertebrates, microorganisms and some plants (e.g. seeds). This is, in the vast majority of cases, a major difference between introductions of invertebrates and introductions of vertebrates and plants. Invertebrates were (and still are) typically accidental arrivals.

One way to measure invasion opportunity for accidental introductions is the use of border interceptions records (or port of entry records) accumulated over time. These records are well used in Australia and New Zealand for investigating the science of invasion biology, and for informing border authorities. Border interception records also serve to show the sheer number and diversity of invertebrates being transported around the globe by human trade. Frampton & Nalder (2009) examined >19 000 border interception records for New Zealand over four years (from 2003 to 2006); 96% were invertebrates. Stanaway *et al.* (2001) surveyed the floors of empty sea containers arriving in Brisbane over a six-month period. Thirty-nine per cent of containers were contaminated with arthropods. In all, more than 7400 live and dead insects were collected from 1174 containers, an average of ~6 per container. Stanaway *et al.* (2001) demonstrated that containers regularly contained timber, agricultural and nuisance arthropod pests.

Interception records of invasive ants have been particularly well studied for New Zealand; for factors governing establishment success (Lester 2005), the relative risk of different species (Harris *et al.* 2005) and invasion across trade pathways (Ward *et al.* 2006). Over the period 1955–2005 there were 4355 interception records, representing >115 species from 52 genera. However, there was no relationship between trade volumes and establishment, and the effectiveness of detecting exotic ant species was estimated at 48%–78% for different trade pathways. Analysis of border interception records for invasive ants in Australia (Department of the Environment and Heritage 2006; Figure 8.2) showed numbers are increasing, biased towards subtropical and tropical ports, and ants arrived on a large diversity of commodities and by many different pathways. As for New Zealand data, there was little relationship between trade volumes and the ports at which ants were arriving.

Border interceptions for forestry pests have also been well studied. Brockerhoff *et al.* (2006) examined interceptions of Scolytinae (Coleoptera) from 1950 to 2000 at the New Zealand border. Over 100 species were identified, including many pest species. Interceptions were primarily associated with wood (packaging, crates, and sawn timber), and originated from 59 countries around the globe. As for ants (Ward *et al.* 2006), interception frequency and establishment in New Zealand were not clearly related. Burnip *et al.* (2010) recently examined New Zealand border interception records for woodwasps (Siricidae), another major group of forestry pests. These records highlighted the value of long-term interception data for retrospective analyses of risk species and their probability of establishment. As for Scolytinae (Brockerhoff *et al.* 2006), virtually all siricid interceptions were associated with imported timber used as packaging (e.g.

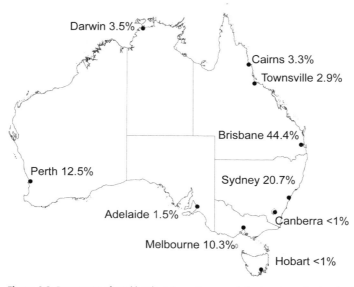

Figure 8.2 Percentage of total border interception records for invasive ants at major Australian cities expressed as proportion of total interception records (records from 1986–2003). Modified from Department of the Environment and Heritage (2006).

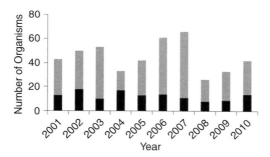

Figure 8.3 The number of new to New Zealand organisms from 2001 to 2010. Grey = microorganisms and viruses, black = invertebrates. Source: MPI PPIN database (accessed 16/7/12, only one plant and no vertebrates were recorded for this period).

wooden pallets, wooden crates and dunnage). These data are important for re-evaluating border inspection standards and processes.

The patterns and trends found in Australia and New Zealand have also been found in overseas studies (trade pathways in the USA; Work *et al.* (2005); invasive ants, Suarez *et al.* (2005)). In Europe, approximately 90% of terrestrial insects arrived unintentionally (75% associated with a commodity, 15% as stowaways) (Hulme & Roy 2010).

There is little doubt that each year numerous exotic species escape detection at the border and establish successfully. In New Zealand, Frampton & Nalder (2009) showed that 420 organisms were listed as new to New Zealand over an 18-year period (1990–2007); an average of 23 new exotic species every year! Subsequent data (2008–2010) shows another 101 organisms were listed as new to New Zealand (MPI PPIN database, unpub. data; see Figure 8.3). Interestingly, 65.7% were microorganisms and viruses, 33.8% were invertebrates, and only 0.5% were plants; no vertebrates were recorded

(Frampton & Nalder 2009). This overall percentage breakdown appears fairly consistent from year to year (Figure 8.3).

Given that there are already ~2000+ species of exotic invertebrates in New Zealand (Brockerhoff *et al.* 2010), and if on average ~10 new invertebrate species become established each year (see Figure 8.3), there could easily be in excess of 3000 exotic invertebrates in New Zealand by the year 2100.

8.4 Managing invasive species

The world's best biosecurity and pest management systems have been developed in Australia and New Zealand. This has come about from the desire to maintain the 'isolation effect' for species not present, and to minimise the damage caused by those species already established. The historical lessons of intentional introductions and impacts of existing pests are well engrained in the psyche of both countries.

A principal feature of the biosecurity systems in Australia and New Zealand is their holistic nature of managing invasive species through assessing threats pre-border, having high levels of surveillance at the border, and their rapid response to any incursions of new species. Kriticos *et al.* (2005) provided estimates of the economic costs of new exotic species predicted to establish in New Zealand for 2005–2017. Without further improvements to biosecurity systems, new exotic species would cost NZ$921 million in direct impacts and ongoing control costs (Kriticos *et al.* 2005). Improving border biosecurity is a major theme for government policy and also research agencies in both Australia and New Zealand. Further improvements to biosecurity systems may come in a number of ways.

8.4.1 Pre-border biosecurity

Pre-border security represents an opportunity to identify species with potential threats before they arrive. Threats are generally examined using a risk assessment, and although they are well used and an important first step in biosecurity, risk assessment methods suffer two main problems: (i) an inability to differentiate species with real threats from the enormous diversity of species with potential threats, and (ii) the difficulty in determining how invasive species will behave in a new region based on often limited information (their impacts could be less, or more, than is known from overseas).

Both New Zealand (Harris *et al.* 2005; Stanley *et al.* 2008) and Australia (Department of the Environment and Heritage 2006) have undertaken large and comprehensive risk assessments for invasive ants. The New Zealand example reviewed known information on 75 ant taxa and developed an 'assessment scorecard' to quantify and rank the threat of each. Subsequently, detailed risk assessments were prepared for the top eight species, which involved extremely detailed examination of trade pathways (commodities, volumes) from known border interceptions, climatic modelling, and potential impacts (Harris *et al.* 2005). The delivery of this pre-border risk assessment has subsequently translated into the 'National Invasive Ant Surveillance' (NIAS) programme specifically designed for the early detection and eradication of invasive ants (Burnip *et al.* 2007; Stringer *et al.* 2010). The approach in Australia was different: providing a broad overview of invasive ants and focussing on six well-known species (Department of the Environment and Heritage 2006). A number of wide-ranging objectives were set for invasive ants in order to prevent their entry and spread, prepare for rapid response to incursions, increase science-based knowledge, inform the Australian community, and provide better governmental coordination (Department of the Environment and Heritage 2006).

8.4.2 Moving risk offshore

An important part of biosecurity is to target specific risks and manage these offshore, that is, before they arrive at the border. While this is strongly tied into pre-border biosecurity, it differs because it first incorporates historical data (e.g. border interceptions, incursions) to determine risks, then uses these data to audit the pre-border processes to remove/minimise the threat.

For example, Ashcroft *et al.* (2008) summarise a demonstration project where invasive ants were monitored coming from several Pacific islands into New Zealand via shipping containers. Historically, a high percentage of empty containers from the Pacific islands are infested with invasive ants (Ward *et al.* 2006). In this trial project New Zealand authorities provided technical advice on ant management to port authorities at Port Moresby and Lae (Papua New Guinea) and Honiara (Solomon Islands). After treatment, the levels of invasive ants arriving at the New Zealand border decreased significantly, from ~17% of containers before treatment, to 0–1% afterwards (Ashcroft *et al.* 2008). Economic benefits accrued for the shipping company occur as a result of reduced inspection requirements on containers arriving into New Zealand. Economic benefits for the port companies occur as a result of less congestion and fewer container 'moves' at the port.

8.4.3 Border surveillance

Hellstrom (2008) indicated that border surveillance is probably one of the most profitable areas for reducing risks. Surveillance is an area where a considerable amount of activity, analysis and review take place (Froud *et al.* 2008a). This can include examination of very specific trade pathways, for example, the biosecurity associated with pineapple leaves and coconut husks (Popay *et al.* 2008), or the contaminants of upholstered furniture from China and Malaysia (Froud *et al.* 2008b); or broad analyses of, for example, tourist activities across New Zealand (Forer & McNeill 2008) or high-risk site surveillance (Stevens 2008). These examples show the complex and dynamic nature of biosecurity, and the need for a flexible system to keep pace with changes in trading partners, goods and risks.

One mechanism for enhancing biosecurity is the use of passive surveillance techniques. In New Zealand, passive surveillance via an emergency hotline phone number resulted in >26 000 calls in 2008–2009 (Froud & Bullians 2010). The general public was responsible for ~70% of these phone reports. Although responding to this many reports is time consuming, the public's use of this system is important, especially for invertebrates. Burnip & Froud (2008) showed that invertebrates contributed 81% of reports (insects 71%, spiders 10%), followed by fungi (7.3%). Several major pests were first found by members of the public in New Zealand (e.g. White-spotted tussock moth, Fall webworm moth, Red imported fire ant). However, the public's relationship with biosecurity is sometimes temperamental, particularly in association with the media, who are often very selective in their coverage of biosecurity, and focus the public's attention on selected examples. This 'attention-grabbing' is probably due in part to the often large financial costs associated with surveillance and managing incursions (Table 8.1).

Table 8.1. The big business of invertebrate pests. Examples of significant costs ($million) associated with biosecurity incursion and eradication responses for invertebrates[a]

Incursion and eradication – New Zealand	Year initiated	NZ$million
Southern Saltmarsh mosquito[b]	1998	$70
Painted apple moth[c]	1999	$65
White-spotted tussock moth[c]	1996	$12.4
Varroa bee mite[c]	2000	$10
Fall webworm moth[c]	2003	$7.8
Asian gypsy moth[c]	2003	$5.4
Varroa mite (Nelson only)[c]	2005	$4.3
Red imported fire ant (Auckland)[c]	2001	$2.6
Ant eradication – Australia (Federal)[d]	**Year Initiated**	**AU$million**
Red imported fire ant	2001	$118
Yellow Crazy ant, Christmas Is.	2011	$3.2
Electric ant (Cairns)	2006	$3.1
Yellow Crazy ant – Arnhem	2008	$0.25
Argentine ant (Norfolk)	2010	$0.15

[a] All costs are to the time of the publication of the source document (that is, additional costs may have occurred since).
[b] Biosecurity New Zealand Pests and Diseases [http://www.biosecurity.govt.nz/pests. Accessed 20/7/12].
[c] MAF Biosecurity New Zealand (2007).
[d] Australian Department of Sustainability, Environment Water, Population and Communities [Accessed 20/7/12. http://www.environment.gov.au/biodiversity/invasive/insects/index.html].

Unfortunately, the selected examples used by the media often portray biosecurity authorities in an overly negative manner, which does little to enhance the biosecurity system.

8.5   Impacts of invasive invertebrates

Kenis *et al.* (2009) provided the first global review of the ecological impacts of invasive invertebrates. Their literature search revealed >400 scientific publications on the topic. However, over half of these had been published recently (since 2000), indicating that our understanding of the ecological impacts of invasive invertebrates is still in its infancy. Nearly two-thirds of publications came from North America, but 8% of publications came from Australia/New Zealand, exceeding those from Europe and Asia.

The impacts of invasive ants contributed 41% of studies, but two species, the red imported fire ant and the Argentine ant, accounted for 18% and 14% of these studies, respectively (Kenis *et al.* 2009). Other impacts were divided between herbivores (24%), predators (19%), pollinators (10%), and parasitoids (6%). Only 72 species were represented in all of these studies, which is an extremely low proportion of the invasive invertebrates known from around the world. This limited impact is surprising because at least 2000 exotic invertebrates have become established in New Zealand (Brockerhoff *et al.* 2010), and large numbers are also present in other countries; Switzerland (311), USA (2000), and Hawaii (2500) (data from Kenis *et al.* 2009).

Generally, there is a lack of awareness of invasive invertebrates in Australia and New Zealand, and, as such, very few species have a high public profile. Apart from the relatively recent work on invasive ants, there is very little information on invasive invertebrates in Australian natural environments. New Zealand has a far better history of such research, which has been summarised in a recent review by Brockerhoff *et al.* (2010) who presented a long list of potential threats, but few with demonstrated impacts.

8.5.1 *Vespula* wasps

Without doubt the best-known invasive invertebrates in New Zealand are the Common wasp (*Vespula vulgaris*) and the German wasp (*Vespula germanica*) (Beggs 2001). Both species are also present in Australia, and although their densities are not as great as those in New Zealand, they are high enough densities to have previously warranted large-scale biological control attempts in Australia (Spradbery & Maywald 1992; Ward *et al.* 2002). The recent establishment of *Vespula germanica* in Argentina (Masciocchi *et al.* 2010) and *Vespa velutina* in France (Rasplus *et al.* 2010) continue to highlight the invasiveness of Vespidae (Beggs *et al.* 2011).

Vespula are extremely abundant (10 000/ha) in beech forests of the South Island (New Zealand) which are naturally infested with endemic scale insects (Beggs 2001). These scale insects provide an energy-rich food resource for *Vespula* wasps, fuelling their diverse array of ecological impacts, including predation of native invertebrates and nestling birds, competition with native species for honeydew and invertebrates for food, and flow-on effects to nutrient cycling and ecosystem carbon dynamics (Beggs & Wardle 2006; Elliott *et al.* 2010; Wardle *et al.* 2010). Elliott *et al.* (2010) recently showed that several common and widespread bird species have had significant declines in their abundance over the past 30 years; attributable to the impacts of a number of introduced species, but especially wasps. Wardle *et al.* (2010) found that wasps greatly influenced the storage of carbon, nitrogen and phosphorus in the soil humus, leading to increases in carbon sequestration.

8.5.2 Paper wasps

One group of invasive species in New Zealand not covered by Brockerhoff *et al.* (2010) was paper wasps. In New Zealand, two invasive paper wasps are present; the Asian paper wasp (*Polistes chinensis antennalis*) and the Australian paper wasp (*P. humilis*). Exotic paper wasps are also present in Australia (*Polistes chinensis antennalis* (ACT); *P. dominula* (WA)), but there is no information on their impacts (Beggs *et al.* 2011).

The Australian paper wasp is thought to have arrived in New Zealand accidentally in the mid to late 1800s. It is restricted to the North Island and is now much less abundant than the Asian paper wasp, possibly due to inter-specific competition (Clapperton *et al.* 1996). The Asian paper wasp was detected in 1979 near Auckland, and has spread rapidly across much of the North Island and several locations in the South Island. Clapperton *et al.* (1996) suggested the Asian paper wasp had not yet reached its maximum distribution, and this appears to be the case, as a recent 2012 survey has discovered Asian paper wasps present in a number of southern locations in the South Island (Ward & Morgan 2014).

In New Zealand, both species of paper wasp prefer 'open-canopy-habitats', such as flax and salt meadows, manuka/kanuka scrub and urban habitats. In these habitats

Table 8.2   List of taxa taken as prey by the Asian Paper Wasp, *Polistes chinensis antennalis*, in New Zealand

Order/family	Taxa
Lepidoptera	
Geometridae	*Chloroclytstis*[a], Apple Looper-*Phrissogonus laticostata*[b], *Scopula rubraria*[c]
Lycaenidae	Peablue-*Lampides boeticus*[c], Small Copper-*Lycaena*[c], Common Grass Blue-*Zizina labradus*[a]
Noctuidae[a, c]	Greasy Cutworm-*Agrotis ipsilon*[a], Green Garden Looper-*Chrysodeixis erisoma*[c], Scar bank gem-*Ctenoplusia limbirena*[b], *Dasypodia*[a], *Ectopatria*[c], *Graphania*[a], Tomato fruitworm-*Helicoverpa armigera*[a], Cosmopolitan Armyworm-*Mythimma separata*[a], Oriental Leafworm-*Spodoptera litura*[c], Soybean looper-*Thysanoplusia orichalcea*[c]
Nymphalidae	Monarch butterfly-*Danaus plexippus*[d], Yellow Admiral-*Vanessa itea*[c]
Pieridae	White butterfly-*Pieris rapae*[a]
Pterophoridae[c]	Plume moth-*Pterophorus*[a]
Tortricidae[b]	Light brown apple moth-*Epiphyas postvittana*[c]
Other Lepidoptera	Coleophorinae[c], Crambidae[c], Elachistidae[c]
Diptera	Syrphidae[a]: Hoverfly-*Platycheirus*[c]; Sarcophagidae[a]; Tipulidae[b]
Coleoptera	Chrysomelidae: Tortoise beetles-*Paropsis*[a]
Hemiptera[a]	Cicadelidae[a], Cicadidae[b]
Hymenoptera	Formicidae[a], 'Sphecidae'[a]
Mantodea[e]	*Miomantis caffra*[b]
Phasmatodea[e]	
Araneae[a]	Garden orbweb-*Eriophora pustulosa*[c], Pholcidae[a], *Tetragnatha* ?*pallescens*[c]

Source (first record of occurrence):
[a] Clapperton (1999);
[b] D. Ward, unpublished data (2012).
[c] Todd (2011);
[d] Ramsay (1964);
[e] Dymock (2000).

densities of paper wasps can reach up to 210 nests/ha, although densities of 20–40 nests/ha are more common. This translates into ~1000–2000 wasps/ha, who are responsible for many tens of thousands of prey captured per season. Clapperton (1999) found that 19 invertebrate taxa were prey for Asian paper wasps, but recent work using molecular identification techniques (Todd 2011; Ward & Ramon-Laca 2013) have shown a much wider range of prey, with over 40 taxa known (Table 8.2). A number of their prey are exotic species, and several are orchard pests, but this list is biased because the studies have largely been carried out in horticultural (Clapperton 1999; Todd 2011) and urban areas (Clapperton 1999; D. Ward, unpublished data) with less in natural environments. However, there are still concerns that paper wasps could have a significant impact on the native invertebrate fauna in native habitats (Clapperton 1999). Possibly by consuming a wide range of prey, the impacts on a few specific species may be less severe, but studies of population- and community-level impacts have not yet been undertaken.

8.5.3   Invasive ants

Globally, invasive ants have received considerable attention in the last decade, including a number of comprehensive reviews (Holway *et al.* 2002; Suarez *et al.* 2005; Silverman & Brightwell 2008; Lach *et al.* 2010; Sanders & Suarez 2011). Although, two species – the red imported fire ant (*Solenopsis invicta*) and the Argentine ant (*Linepithema humile*) (Kenis *et al.* 2009) dominate our knowledge of invasive ants, there are a number of other species that also have ecological impacts (Holway *et al.* 2002; Lach *et al.* 2010). As such, a number of invasive ant species are included in the top 100 'world's worst list' of invasive species (Lowe *et al.* 2000).

Research into the impacts of invasive ants in Australia and New Zealand is relatively new (Lach & Thomas 2008; Ward 2009; Ward *et al.* 2010). However, ecological impacts have been well documented for Australia (Lach & Thomas 2008), including that of competitive displacement with native ant species with potentially severe flow-on effects because of the diversity of ants in Australia and their roles in pollination and seed dispersal. In New Zealand, the focus has been on Argentine ants and, because of the very low diversity of native ants, research has been directed at impacts on non-ant invertebrates, decomposition and impacts on plants. Stanley & Ward (2012) found invasion by Argentine ants displaced landhoppers (an abundant and important decomposer in leaf litter), which resulted in less shredded litter for colonisation by microbes and other invertebrates, and consequently slower breakdown of leaf litter with less decomposition. Argentine ants have also been associated with increased seed production of a major weed, boneseed, *Chrysanthemoides monilifera monilifera* (Stanley *et al.* 2012). Also on boneseed, Argentine ants and introduced wasps (*Vespula*, *Polistes*) are a major cause of mortality of the boneseed leafroller (Lepidoptera: Tortricidae), a biological control agent introduced to attack boneseed (Paynter *et al.* in press). These last two studies demonstrate that ant invasions not only potentially affect biodiversity, but may also affect weed and pest management strategies.

8.5.4 Eradication and detection

An area that is well developed in Australia and New Zealand is emphasis on the eradication of invasive ants. A search of GERDA (the Global Eradication and Response Database) shows a total of 49 incursion or eradication responses for invasive ants, of which two-thirds occur either in Australia or in New Zealand (Keen *et al.* 2012).

Large-scale eradication programmes have been well developed in Australia (Vanderwoude *et al.* 2003; Hoffmann *et al.* 2006; Hoffmann 2010), although several projects attempting to eradicate Argentine ants are underway in New Zealand but have yet to be published. The eradication the African big headed ant, *Pheidole megacephala*, from northern Australia, is one of the few examples to document eradication but also the subsequent recovery of the native ant fauna (Hoffman 2010). In 2001 the red imported fire ant (*Solenopsis invicta*) was discovered in Brisbane over an area of 40 000 ha, and Australia's largest eradication programme for any invertebrate species was initiated (Vanderwoude *et al.* 2003). Ten years and almost AU$200 million later, the programme continues to finalise eradication; however, it is considered worthwhile because of the potential for enormous economic and ecological impacts from red imported fire ant in Australia (Moloney & Vanderwoude 2003).

The eradication programme for the red imported fire ant in Brisbane highlights a number of important points for managing and controlling invasive ants. First, there is little doubt we are still in the 'early stages' in our understanding of 'how to' with regards to the eradication of invasive ants. Similarities could be drawn to eradications of vertebrate pests from New Zealand in the 1970s and 1980s, where most eradications were considered too impractical but vertebrate pests are now being eradicated from islands of significant (>10 000 ha+) size (Clout & Veitch 2002). Furthermore, there are very few case studies of the eradication of invasive ants, and it appears that each ant species has certain characteristics that cannot always be transferred to eradication efforts for other species. A second issue relates to early detection and spread of new populations. Individual colonies or those in a relatively confined area (e.g. at the border) can be dealt with very effectively. The difficulty is where such colonies go unnoticed for a period of time and many hectares are consequently infested. This highlights the importance of early detection, and also the consequences of failing to detect a colony, and its subsequent spread.

Invasive ants are prime candidates for poor detection because of their small size, variable-foraging habits, cryptic nature (queens or incipient colonies), and strong association with human transportation (Stanley *et al.* 2008). Detection is important for surveillance at the border but also for finding small 'survivor' colonies that remain viable after a control operation has occurred. Detection of red imported fire ant colonies in Brisbane is an ongoing issue. A recent analysis by Schmidt *et al.* (2010) determined that more than twice as many colonies could have been found using Bayesian 'jump dispersal model' compared with the currently used 'proximity-based searching'. Recent work has shown that several of the well-used monitoring methods (baits, visual searching) for Argentine ants are having a much lower probability of detection than was expected (Ward & Stanley 2012). The inability to find all small colonies is currently a major obstacle preventing the successful eradication of Argentine ants at several locations in New Zealand.

8.5.5   Intentionally introduced invertebrates

The impacts of invertebrates intentionally introduced by humans (e.g. for biological control or pollination) has been the subject of debate for some time (Wajnberg *et al.* 2001; De Clercq *et al.* 2011). Although Kenis *et al.* (2009) and Roy *et al.* (2011) discuss many global examples of the non-target impacts of intentionally introduced invertebrates, there are relatively few examples in Australia and New Zealand. This is possibly the result of comparatively strict importation standards in both countries (Brockerhoff *et al.* 2010). While a small number of intentionally introduced species have been implicated in impacts on non-target native species (Wajnberg *et al.* 2001), the vast majority have not been found to have ecological impacts (Fowler *et al.* 2003; De Clercq *et al.* 2011).

One of the best examples of impacts on non-targets is the tachinid fly, *Trigonospila brevifacies*, which was introduced to control light brown apple moth (*Epiphyas postvittana*) in New Zealand. Munro & Henderson (2002) found this introduced species has become the dominant parasitoid in a tortricid food web of native Lepidoptera. Thus, not only does *T. brevifacies* attack non-target Lepidoptera, but it also competes for hosts with native parasitoids. Other intenionally introduced species have also been recorded attacking native species. For example, *Microctonus aethiopoides*, a braconid parasitoid, was released in New Zealand as a biological control agent against a weevil pasture pest, *Sitona discoideus*, despite laboratory host range testing showing it to be relatively polyphagous. *Microctonus aethiopoides* has now been found to parasitise a number of native weevil species and also several other exotic weevils to high levels (Barratt *et al.* 1997; Barratt 2004).

The intentional introduction of invertebrates as pollinators may also have unintended consequences (Kenis *et al.* 2009; Brockerhoff *et al.* 2010). The honeybee, *Apis mellifera*, can be very abundant and widespread in native ecosystems in both Australia and New Zealand, and also the bumblebee, *Bombus terrestris*, in New Zealand. Concerns have been raised over the competition of introduced pollinator species with native pollinators, potentially affecting pollination of native plant species. In New Zealand, the honeybees and bumblebees can represent a significant proportion of the visitors to native plants' flowers (Kelly *et al.* 2006). Hingston & McQuillan (1999) have shown that *Bombus terrestris*, accidentally introduced into Tasmania, affects the foraging of native bee species by deleting nectar resources, leading to the displacement of native bees.

Honeybees have also been implicated in the pollination and spread of several environmental weeds (Butz Huryn 1997). In Australia, Gross *et al.* (2010) showed that the honeybee was the primary visitor of *Phyla canescens*, a major weed, and that the abundance of both pollinator and weed was positively correlated. Honeybees are also the main pollinator of *Lantana camara*, a weed covering >40 000 km^2 in Australia (Goulson & Derwent 2004). In field and laboratory studies, honeybees were the only recorded flower visitors, and seed set was highest at sites where honeybees were present. Goulson and Derwent (2004) suggested that the management of honeybee hives is a sensible and cost-effective control option for *L. camara*. The contribution of honeybees to environmental weeds in New Zealand has also been assessed (Butz Huryn & Moller 1995). While for many weed species honeybees are seemingly unimportant, the success of several major weeds is facilitated by honeybee pollination (Butz Huryn & Moller 1995).

8.6 Future issues

In this section, five issues are summarised that will be important for invasive invertebrates over the next few decades. These issues are important for three reasons: (i) evidence of ecological impacts are needed to show the importance of invasive invertebrates; (ii) climate change and trade liberalisation are major themes that affect which species will become invasive in the future; and (iii) technology and capability are needed to show there is the ability to manage invasive invertebrates.

8.6.1 Evidence of impacts

The global review of ecological impacts of invasive invertebrates by Kenis *et al.* (2009) showed invasive invertebrates can affect native species and ecosystems through a number of mechanisms. However, evidence for ecological impacts exists for only a few, well-studied species. There is a clear need for more research in both Australia and New Zealand. Evidence of ecological impacts is vital to illustrate the negative roles invasive invertebrates can have.

Amongst those that are most well studied are generalist predators. These include the most well-known and notorious invaders in Australia and New Zealand, that is, social wasps (*Vespula, Polistes*) and ants. Such predators have been frequently mentioned in invasion ecology because they usually have a wide host range, feed at a range of trophic levels, can reach very high densities, and often have noticeable effects on prey (Snyder & Evans 2006). Other predatory invertebrates worthy of further investigation in New Zealand could include several solitary hymenoptera (e.g. *Ancistrocerus, Pison*), the South

African mantis (*Miomantis caffra*), and many species of spiders and beetles (carabid, staphylinids, coccinellids) (Brockerhoff *et al.* 2010). Kenis *et al.* (2009) also found that herbivorous insects that affect forest ecosystems are well studied. However, information is scarce for other invasive herbivores, although Brockerhoff *et al.* (2010) consider that good information is available for invertebrate herbivores in New Zealand. The fact that there are very few species with high-impact effects is used to support the idea that many New Zealand native plants are resistant to exotic invertebrate herbivores (Brockerhoff *et al.* 2010).

There is little doubt that ecological impacts receive far less attention than impacts that affect the economy or public health. Kenis *et al.* (2009) illustrated this through the high number of studies investigating the impact of exotic ants (nuisance value, economic), honey bees, plant pests (economic) or mosquitoes (public health), compared to studies solely on biodiversity. Studies on ecological impacts of invasive invertebrates appear to be difficult for a number of reasons: (i) biosecurity surveillance and response is chiefly aimed towards productive sectors; (ii) lack of public sympathy towards invertebrates; (iii) the huge diversity of invertebrates often means it is difficult to know whether a species is exotic or native; (iv) lack of diagnostic tools and people for identification of species; and (v) their small sizes and cryptic nature mean invertebrates can be very difficult to study in the field. However, it should be noted that Goldson *et al.* (2005) also stated that impacts of invertebrate pests in New Zealand pastures were also difficult to quantify, so the issue is not restricted just to natural ecosystems.

8.6.2 Climate change

Hellmann *et al.* (2008) identified five major consequences of climate change for invasive species: (i) altered transport and introduction pathways; (ii) establishment of new species; (iii) altered impact of existing species; (iv) altered distribution of existing species; and (v) altered effectiveness of control. These will have serious consequences for our ability to manage invasive invertebrates in the future. Increased abundance, spread and uncertainty is likely to mean increased 'out-breaks' – either of existing species turning up in new regions, or from new incursions into New Zealand – both further stretching resources for management.

The typical characteristics of invasive invertebrates, such as broad climatic tolerances, large geographic ranges, high fecundity and ecological plasticity, mean they are particularly well adapted to respond quickly to climate change (Hellmann *et al.* 2008). Cook *et al.* (2002) suggested that in New Zealand certain species of mosquitoes, ticks and ants would broaden their range to higher latitudes and higher altitudes, and breeding times would also be shorter. Mosquitoes, in particular, would benefit from increasing patterns of rainfall and rising sea levels, providing increased habitat for them to breed (Cook *et al.* 2002). Harris & Barker (2007) used climatic modelling to assess the ability of nine well-known invasive ant species to establish in New Zealand. Differences in climate, especially cooler temperatures, were a major barrier for these species to currently establish in New Zealand (Harris & Barker 2007). However, under the most likely climate change scenarios, this barrier would be reduced, or possibly even removed.

8.6.3 Trade liberalisation

With the advent of the World Trade Organization (WTO, and its predecessor, GATT, the General Agreement on Tariffs and Trade) there has been unprecedented growth in global trade, in part, through tariff reductions and trade liberalisation (Findlay & O'Rourke 2007). Despite the WTO having an objective of 'optimal use of the world's resources, sustainable development and environmental protection', trade liberalisation can often conflict with biosecurity and the management of invasive species.

It is widely recognised that the majority of exotic species arrive to new areas through human transportation and trade (Work *et al.* 2005; Suarez *et al.* 2005). With volumes of global trade predicted to continue to increase significantly, increases in the transportation of exotic species are also expected (Levine & D'Antonio 2003). A greater diversity of trading partners will also allow different species the opportunity to invade. Westphal *et al.* (2008) recently showed that the degree of international trade (merchandise imports) was the best factor for explaining the number of invasive species from a given area.

Both Australia and New Zealand are heavily involved in free trade agreements. Major trade agreements made recently include the Australia–United States Free Trade Agreement in 2004 and the New Zealand–China Free Trade Agreement in 2008. Both countries are also currently in free-trade negotiations with a range of countries, especially in Asia and South America. To minimise the risk of 'invasion opportunity' for exotic invertebrates from an increased array of trading partners is a difficult task. This requires a far better understanding of how exotic species are associated with trading networks (Levine & D'Antonio 2003; Westphal *et al.* 2008). Currently, a considerable hurdle to this task is the fact that border authorities do not collect data that is readily useable for robust scientific studies (Venette *et al.* 2002). Despite the growing number of studies that analyse 'biosecurity data' (e.g. border interception records), such data often has considerable shortcomings. Of primary concern is the uncertainty of how the data was collected (e.g. what was the sampling effort involved, what was *not* sampled, were all the data recorded, etc.), and how such factors may bias the data and thus interpretations. Overcoming this hurdle is key to a greater understanding of invasion opportunity and minimising the establishment of new exotic species.

8.6.4 Technology

The development and use of novel technologies has been an important part of the success of biosecurity and pest management systems in Australia and New Zealand (Goldson *et al.* 2005). Continued innovation will be important in the future.

The use of computer modelling will have increased importance for a range of applications, such as, predicting the risk of new species becoming established (Worner & Gevrey 2006; Harris & Barker 2007); predicting the spread of invading species particularly associated with humans (Forer & McNeill 2008; Roura-Pascal *et al.* 2011); and in detection of cryptic populations. Stringer *et al.* (2011) also examined detection techniques for red imported fire ants, as part of a review of the successful eradication programme in New Zealand. Their results showed that because small colonies forage in a very small area (no more than >3 m), far greater search effort and resources are needed to be confident of finding these very small colonies before they have an opportunity to grow and disperse. Such analyses have significant implications, not only for the efficiency of pest management operations, but also the long-term success of pest control, and in this case eradication.

New technologies are also important to improve the success and cost-effectiveness of surveillance. This may range from large projects which attempted to use sensor systems

'Sniffertech™' to detect biological threats in shipping containers at the border (Goldson *et al.* 2003), to 'sniffer-tech' options of training dogs for surveillance purposes. Sniffer dogs are being increasingly used for detection of a wide range of biosecurity contraband items like fruit and animal products, and also pests, such as snakes, gypsy moths, termites, bed bugs and fire ants (Lin *et al.* 2011). The world's first Argentine ant sniffer-dog has also just been trained in New Zealand (Romanos 2012). A great example of increased cost-effectiveness comes from Mattson (2008) who discussed the use of hand-held computer devices during the red imported fire eradication programme in Whirinaki, New Zealand. Improved data recording, processing and verification via handheld devices increased the placement of monitoring devices ten-fold, and eliminated the error prone and time-consuming data entry of paper-based systems.

Increased use of online, web-based tools, and smart-phone 'apps', is also inevitable. Bisby (2000) outlined the importance of the Internet to biodiversity; however, it appears that biosecurity has yet to appreciate and utilise the power of the Internet. It is likely that over the next decade *'biosecurity informatics'* will become increasingly important for delivering significant advances to a range of biosecurity systems. For biosecurity, the Internet has particular uses for the delivery of identifications (images, keys and fact-sheets); access to historical data; real-time information sharing; integration of data sources and avoiding duplication; and greater co-operations between institutions. One example is the Pest and Disease Image Library (PaDIL; http://www.padil.gov.au), an initiative of the Australian Government which began in 2005. PaDIL aims to create a virtual reference collection of pests and diseases, and by doing so, allows the rapid recognition of invasive species by a greater range of people.

8.6.5 Capability

Having the capacity and capability to assess and respond to biosecurity threats is a key part of the New Zealand biosecurity strategy. In both Australia and New Zealand, agricultural, horticultural and vertebrate pests have historically been of primary importance. Thus, it is not surprising that current capability in both countries is predominantly in these areas. Although the need to retain this capability is well recognised, the challenges for biosecurity and pest management change over time (Hellstrom *et al.* 2008). Now there is the need to increasingly improve the capability for managing new pest species.

There is little doubt that many new species will continue to arrive at the border and many of these will become established (see Figure 8.3). However, it is the taxonomic breakdown of these new species that will be different from historical introductions. Frampton & Nalder (2009) showed that two-thirds of new species were microorganisms and viruses, and one-third invertebrates. Over the ten-year period of their records, only one plant and no vertebrates were recorded. This clearly indicates that increased capability for the detection of microorganisms and invertebrates is greatly needed for the future.

However, several reviews in New Zealand have raised concerns about capability for biosecurity (Gerard *et al.* 1997; Green 2001; Hellstrom *et al.* 2008). One issue that has been raised is the increasing shortage of skills for virology, freshwater and marine biology, and entomology. This shortage of skills has been referred to as the 'taxonomic impediment', referring to the worlds of dwindling and geographically biased taxonomic expertise. The taxonomic impediment will affect biosecurity in Australia and New Zealand and the ability to minimise the establishment of new species, and manage those that do. Solutions to overcome this problem have begun, such as the use of remote diagnostics which, over desktop computers, display biological specimens for identification by experts that are based in other parts of the country, or the world (Kean *et al.* 2011). Such technologies could be used to build capability in biosecurity.

8.7 Conclusion

Exotic invertebrates make up a sizeable, and growing, proportion of invertebrates in both countries. Apart from a few well-studied species, we know very little about the ecological impacts of exotic invertebrates on the unique biota of either Australia or New Zealand. It is hoped that this chapter will stimulate further research on this topic.

Acknowledgements

Thanks to Greg Howell for inviting me to contribute to this book, to Therese Oliver (Ministry of Primary Industries) for information on the recent establishments of species in New Zealand. This research was supported by Core funding for Crown Research Institutes from the Ministry of Business, Innovation and Employment's Science and Innovation Group, through the 'Managing Invasive Weeds, Pests and Diseases' Portfolio.

REFERENCES

Ashcroft, T. T., Nendick, D., O'Connor, S. M., *et al.* 2008. Managing the risk of invasive exotic ants establishing in New Zealand, pp. 151–160. In: *Surveillance for Biosecurity: Preborder to Pest Management.* Froud, K. J., Popay, A. I., Zydenbos, S. M. (eds.). The New Zealand Plant Protection Society, Wickliffe, Christchurch.

Barlow, N. D., Goldson, S. L. 2002. Alien invertebrates in New Zealand, pp. 195–216. In: *Biological Invasions: Economic and Environmental Costs of Alien Plant, Animal and Microbe Species.* Pimentel D. (ed.). CRC Press, Florida.

Barratt, B. I. P. 2004. *Microctonus* parasitoids and New Zealand weevils: comparing laboratory estimates of host ranges to realized host ranges, pp. 103–120. In: Van Driesche, R. G., Reardon, R. (eds.). *Assessing Host Ranges for Parasitoids and Predators used for Classical Biological Control: A Guide to Best Practice.* Morgantown, West Virginia, USDA Forest Service.

Barratt, B. I. P., Evans, A. A., Ferguson, C. M., *et al.* 1997. Laboratory nontarget host range of the introduced parasitoids *Microctonus aethiopoides* and *Microctonus hyperodae* (Hymenoptera: Braconidae) compared with field parasitism in New Zealand. *Environmental Entomology* **26**: 694–702.

Beggs, J. 2001. The ecological consequences of social wasps (*Vespula* spp.) invading an ecosystem that has an abundant carbohydrate resource. *Biological Conservation* **99**(1): 17–28.

Beggs, J. R., Wardle, D. A. 2006. Keystone species: competition for honeydew among exotic and indigenous species, pp. 281–294. In: Allen, R. B., Lee, W. G. (eds.). *Biological Invasions in New Zealand.* Springer, Berlin.

Beggs, J. R., Brockerhoff, E. G., Corley, J. C., *et al.* 2011. Ecological effects and management of invasive alien Vespidae. *BioControl* **56**: 505–526.

Bisby, F. A. 2000. The quiet revolution: biodiversity informatics and the internet. *Science* **289**: 2309–2312.

Brockerhoff, E. G., Bain, J., Kimberley, M., Knízek, M. 2006. Interception frequency of exotic bark and ambrosia beetles (Coleoptera: Scolytinae) and relationship with establishment in New Zealand and worldwide. *Canadian Journal of Forestry Research* **36**: 289–298.

Brockerhoft, E. G., Barratt, B. I. P., *et al.* 2010. Impacts of exotic invertebrates on New Zealand's indigenous species and ecosystems. *New Zealand Journal of Ecology* **34**: 158–174.

Burnip, G. M., Froud, K. J. 2008. Characterisation of science inputs when investigating exotic organism incursions, pp. 143–149. *Surveillance for Biosecurity: Pre-border to Pest Management*. The New Zealand Plant Protection Society, Wickliffe, Christchurch.

Burnip, G. M., Sarty, M., Gunawardana, D., Green, O. 2007. A success story: early detection of exotic ants. *New Zealand Plant Protection Society* **60**: 313.

Burnip, G. M., Voice, D., Brockerhoff, E. G. 2010. Interceptions and incursions of exotic *Sirex* species and other siricids (Hymenoptera: Siricidae). *New Zealand Journal of Forestry Science* **40**: 133–140.

Butz Huryn, V. M. 1997. Ecological impacts of introduced honey bees. *The Quarterly Review of Biology* **72**: 275–297.

Butz Huryn, V. M., Moller, H. 1995. An assessment of the contribution of honey bees (*Apis mellifera*) to weed reproduction in New Zealand protected natural areas. *New Zealand Journal of Ecology* **19**: 111–122.

Clapperton, B. K. 1999. Abundance of wasps and prey consumption of paper wasps (Hymenoptera, Vespidae: Polistinae) in Northland, New Zealand. *New Zealand Journal of Ecology* **23**: 11–19.

Clapperton, B. K., Tilley, J. A. V., Pierce, R. J. 1996. Distribution and abundance of the Asian paper wasp *Polistes chinensis antennalis* Perez and the Australian paper wasp *P. humilis* (Fab) (Hymenoptera: Vespidae) in New Zealand. *New Zealand Journal of Zoology* **23**: 19–25.

Clout, M. N. 2002. Ecological and economic costs of alien vertebrates in New Zealand, pp. 185–194. In: *Biological Invasions: Economic and Environmental Costs of Alien Plant, Animal and Microbe Species*. Pimentel, D. (ed.). CRC Press, Florida.

Clout, M. N., Veitch, C. R. 2002. Turning the tide of biological invasion: the potential for eradicating invasive species, pp. 1–3. In: Veitch, C. R., Clout, M. N. (eds.). *Turning the Tide: the Eradication of Invasive Species*. IUCN SSC Invasive Species Specialist Group. IUCN, Gland, Switzerland and Cambridge, UK. 414 pp.

Cook, A., Weinstein, P., Woodward, A. 2002. The impacts of exotic insects in New Zealand, pp. 217–239. In: *Biological Invasions: Economic and Environmental Costs of Alien Plant, Animal and Microbe Species*. Pimentel, D. (ed.). CRC Press, Florida.

Cranston, P. S. 2010. Insect biodiversity and conservation in Australasia. *Annual Review of Entomology* **55**: 55–75.

De Clercq, P., Mason, P. G., Babendreier, D. 2011. Benefits and risks of exotic biological control agents. *BioControl* **56**: 681–698.

Department of the Environment and Heritage 2006. *Reduction in Impacts of Tramp Ants on Biodiversity in Australia and its Territories*. Also available at: http://www.environment. gov.au/biodiversity/threatened/publications/tap/trampants.html

Drake, J. A., Mooney, H. A., di Castri, F., *et al.* 1989. *Biological Invasions: A Global Perspective*. John Wiley & Sons, New York.

Duncan, R. P., Blackburn, T. M., Sol, D. 2003. The ecology of bird introductions. *Annual Review of Ecology and Systematics* **34**: 71–98.

Dymock, J. J. 2000. Risk assessment for establishment of polistine (*Polistes* spp.) and vespine (*Vespula* spp.) wasps on the Three Kings Islands in the Far North of New Zealand. *Science for Conservation* 156. Department of Conservation, Wellington. 18pp.

Elliott, G. P., Wilson, P. R., Taylor, R. H., Beggs, J. R. 2010. Declines in common, widespread native birds in a mature temperate forest. *Biological Conservation* **143**(9): 2119–2126.

Findlay, R., O'Rourke, K. H. 2007. *Power and Plenty: Trade, War, and the World Economy in the Second Millennium*. Princeton University Press, Princeton, New Jersey.

Floerl, O., Inglis, G. J. 2004. Starting the invasion pathway: the interaction between source populations and human transport vectors. *Biological Invasions* **7**: 589–606.

Forer, P. C., McNeill, M. R. 2008. Tourist flows, lagged arrivals and environmental consonance as factors in biological exchange risk in Aotearoa New Zealand, pp. 19–28. In: *Surveillance for Biosecurity: Pre-border to Pest Management*. Froud, K. J., Popay, A. I., Zydenbos, S. M. (eds.). The New Zealand Plant Protection Society, Wickliffe, Christchurch.

Fowler, S. V., Gourlay, A. H., Hill, R. L., Withers, T. 2003. Safety in New Zealand weed biocontrol: a retrospective analysis of host specificity testing and the predictability of impacts on non-target plants, pp. 265–270. In: Cullen, J., Briese, D. T., Kriticos, D. J., *et al.* (eds.). *Proceedings of the XI International Symposium on Biological Control of Weeds*. CSIRO Publishing, Melbourne.

Frampton, E. R., Nalder, K. 2009. A novel analysis of the risk of fresh produce imports. *New Zealand Plant Protection* **62**: 114–123.

Froud, K. J., Bullians, M. S. 2010. Investigation of biosecurity risk organisms for the plant and environmental domains in New Zealand for 2008 and 2009. *New Zealand Plant Protection* **63**: 262–269.

Froud, K. J., Popay, A. I., Zydenbos, S. M. 2008a. *Surveillance for Biosecurity: Pre-border to Pest Management*. The New Zealand Plant Protection Society, Wickliffe, Christchurch. 224 pp.

Froud, K. J., Pearson, H. G., McCarthy, B. J. T., Thompson, G. 2008b, pp. 63–76. In: *Surveillance for Biosecurity: Pre-border to Pest Management*. Froud, K. J., Popay, A. I., Zydenbos, S. M. (eds.). The New Zealand Plant Protection Society, Wickliffe, Christchurch.

Gerard, P., Popay, I., Rahman, A. 1997. *Plant Protection and Biosecurity: Science and Coordination Issues for New Zealand*. Report to the Ministry of Research, Science and Technology, Wellington. 68 pp.

Goldson, S. L., Frampton, E. R., Geddes, N. J., Braggins, T. J. 2003. The potential of sensor technologies to improve New Zealand's border security, pp. 63–71. In: *Defending the Green Oasis: New Zealand Biosecurity and Science*. Goldson, S. L., Suckling, D. M. (eds.). New Zealand Plant Protection Society.

Goldson, S. L., Rowarth, J. S., Caradus, J. R. 2005. The impact of invasive invertebrate pests in pastoral agriculture: a review. *New Zealand Journal of Agricultural Research* **48**: 401–415.

Goulson, D., Derwent, L. C. 2004. Synergistic interactions between an exotic honeybee and an exotic weed: pollination of *Lantana camara* in Australia. *Weed Research* **44**: 195–202.

Green, W. 2001. *Review of Current Biosecurity Research in New Zealand*. Report for the Biosecurity Strategy Development Team. Ecologic Conservation Consultants, Wellington. 116 pp.

Gross, C. L., Gorrell, L., MacDonald, M. J., Fatemi, M. 2010. Honeybees facilitate the invasion of *Phyla canescens* (Verbenaceae) in Australia – no bees, no seed! *Weed Research* **50**: 364–372.

Harris, R., Abbott, K., *et al.* 2005. Invasive ant pest risk assessment project for Biosecurity New Zealand. Series of unpublished Landcare Research contract reports to Biosecurity New Zealand. BAH/35/2004–1.

Harris, R. J., Barker, G. 2007. Relative risk of invasive ants (Hymenoptera: Formicidae) establishing in New Zealand. *New Zealand Journal of Zoology* **34**: 161–178.

Hellmann, J. J., Byers, J. E., Bierwagen, B. G., Dukes, J. S. 2008. Five potential consequences of climate change for invasive species. *Conservation Biology* **22**: 534–543.

Hellstrom, J. S. 2008. Biosecurity surveillance, pp. 1–10. In: *Surveillance for Biosecurity: Pre-border to Pest Management*. Froud, K. J., Popay, A. I., Zydenbos, S. M. (eds.). The New Zealand Plant Protection Society, Wickliffe, Christchurch.

Hellstrom, J., Moore, D., Melleny Black, M. 2008. *Think Piece on the Future of Pest Management in New Zealand*. Report for Ministry of Agriculture and Forestry, 76 pp.

Hingston, A. B., McQuillan, P. B. 1999. Displacement of Tasmanian native megachilid bees by the recently introduced bumblebee *Bombus terrestris* (Linnaeus, 1758) (Hymenoptera: Apidae). *Australian Journal of Zoology* **47**: 59–65.

Hoffmann, B. D. 2010. Ecological restoration following the local eradication of an invasive ant in northern Australia. *Biological Invasions* **12**: 959–969.

Hoffman, B. D., Abbott, K. L., Davies, P. 2006. Invasive ant management, pp. 287–304. In: *Ant Ecology*. Lach, L., Parr, K., Abbott, K. L. (eds.). Oxford University Press, Oxford.

Holway, D. A., Lach, L., Suarez, A., Tsutsui, N., Case, T. J. 2002. The causes and consequences of ant invasions. *Annual Review of Ecology and Systematics* **33**: 181–233.

Hulme, P. E., Roy, D. B. 2010. DAISIE and arthropod invasions in Europe. *BioRisk* **4**(1): 1–3.

Kean, J. M., Tobin, P. C., Lee, D. C., *et al.* 2012. Global eradication and response database. http://b3.net.nz/gerda (accessed 7 August 2012).

Kean, J. M., Vink, C. J., Till, C., *et al.* 2011. Real-time remote diagnostics for ecology: Wheeler *et al.*'s vision realized. *Frontiers in Ecology and the Environment* **10**: 99–104.

Kelly, D., Roberton, A. W., Ladley, J. J., Anderson, S. H., McKenzie, R. J. 2006. Relative (un)importance of introduced animals as pollinators and dispersers of native plants, pp. 227–245. In: *Biological Invasions in New Zealand*. Springer, Berlin, Heidelberg.

Kenis, M., Auger-Rozenberg, M. A., Roques, A., *et al.* 2009. Ecological effects of invasive alien insects. *Biological Invasions* **11**: 21–45.

Kolar, C. S., Lodge, D. M. 2001. Progress in invasion biology: predicting invaders. *Trends in Ecology and Evolution* **16**: 199–204.

Kriticos, D. J., Phillips, C. B., Suckling, D. M. 2005. Improving border biosecurity: potential economic benefits to New Zealand. *New Zealand Plant Protection* **58**: 1–6.

Lach, L., Thomas, M. L. 2008. Invasive ants in Australia: documented and potential ecological consequences. *Australian Journal of Entomology* **47**: 275–288.

Lach, L., Parr, K., Abbott, K. L. 2010. Invasive ants, pp. 231–304. In *Ant Ecology*. Lach, L., Parr, K., Abbott, K. L. (eds.). Oxford University Press, Oxford.

Lester, P. J. 2005. Determinants for the successful establishment of exotic ants in New Zealand. *Diversity and Distributions* **11**: 279–288.

Levine, J. M., D'Antonio, C. M. 2003. Forecasting biological invasions with increasing international trade. *Conservation Biology* **17**: 322–326.

Lin, H. M., Chi, W. L., Lin, C. C., *et al.* 2011. Fire ant-detecting canines: a complementary method in detecting red imported fire ants. *Journal of Economic Entomology* **104**: 225–231.

Lowe, S., Browne, M. Boudjelas, S., De Poorter, M. 2000. 100 of the world's worst invasive alien species. A selection from the Global Invasive Species Database. Invasive Species Specialist Group (ISSG) of the World Conservation Union (IUCN), 12 pp.

Mack, R. N., Simberloff, D., Lonsdale, W. M., *et al.* 2000. Biotic invasions: causes, epidemiology, global consequences, and control. *Ecological Applications* **10**: 689–710.

MAF Biosecurity. 2007. *Joint Decision-Making and Resourcing for Readiness and Incursion Responses*. MAF Biosecurity New Zealand, Surveillance and Incursion Response Working Group Discussion Paper No: 2007/02.

Masciocchi, M., Beggs, J. R., Carpenter, J. M., Corley, J. C. 2010. Primer registro de *Vespula vulgaris* (Hymenoptera: Vespidae) en la Argentina. *Revista de la Sociedad Entomológica Argentina* **69**: 267–270.

Mattson, L. T. W. 2008. Handheld computer technology – a vital link in New Zealand exotic ant programmes, pp. 161–166. In: *Surveillance for Biosecurity: Pre-border to Pest Management*. Froud, K. J., Popay, A. I., Zydenbos, S. M. (eds.). The New Zealand Plant Protection Society, Wickliffe, Christchurch.

Moloney, S. D., Vanderwoude, C. 2003. Potential ecological impacts of red imported fire ants in eastern Australia. *Journal of Agricultural and Urban Entomology* **20**: 131–142.

Munro, V. M. W., Henderson, I. M. 2002. Non target effect of entomophagous biocontrol: shared parasitism between native Lepidopteran parasitoids and the biocontrol agent *Trigonospila brevifascies* (Diptera: Tachnidae) in forest habitats. *Environmental Entomology* **31**: 388–396.

Paynter, Q., Forgie, S. A., Winks, C., *et al.* 2012. Biotic resistance limits boneseed leafroller *Tortrix* s.l. sp. *chyrsanthemoides* establishment in New Zealand. *Biocontrol* **63**: 188–194.

Pimentel, D. 2002. *Biological Invasions: Economic and Environmental Costs of Alien Plant, Animal and Microbe Species*. CRC Press, Florida.

Popay, A. I., James, T. K., Sarty, M., Dickson, M., Bullians, M. S. 2008. pp. 45–50. In: *Surveillance for Biosecurity: Pre-border to Pest Management*. Froud, K. J., Popay, A. I., Zydenbos, S. M. (eds.). The New Zealand Plant Protection Society, Wickliffe, Christchurch.

Puth, L. M., Post, D. M. 2005. Studying invasion: have we missed the boat? *Ecology Letters* **8**: 715–721.

Rabitsch, W. 2011. The hitchhiker's guide to alien ant invasions. *BioControl* **56**: 551–572.

Ramsay, G. W. 1964. Overwintering swarms of the Monarch butterfly (*Danaus plexippus* (L.)) in New Zealand. *New Zealand Entomologist* **3**: 10–16.

Rasplus, J. Y., Villemant, C., Paiva, M. R., Delvare, G., Roques, A. 2010. Hymenoptera. In: Roques, A., Kenis, M., Lees, D. (eds.). Arthropod invasions in Europe. *BioRisk* **4**: 669–776.

Romanos, M. 2012. Ant catcher is one of a kind. *New Zealand Dog World*. June: **8**.

Roura-Pascual, N., Hui, C., Ikeda, T., *et al.* 2011. The relative roles of climatic suitability and anthropogenic influence in determining the pattern of spread in a global invader. *Proceedings of the National Academy of Sciences of the United States of America* **108**: 220–225.

Roy, H. E., De Clercq, P., Handley, L. J., 2011. Alien arthropod predators and parasitoids: an ecological approach. *BioControl* **56**: 375–382.

Sanders, N. J., Suarez, A. V. 2011. Elton's insights into the ecology of ant invasions: lessons learned and lessons still to be learned, pp. 239–251. In: *Fifty Years of Invasion Ecology: the legacy of Charles Elton*. Richardson, D. M. (ed.). Blackwell, New York.

Sandland, O. T., Schei, P. J., Viken, A. 1999. *Invasive Species and Biodiversity Management*. Kluwer Academic Publishers, Dordrecht, the Netherlands.

Schmidt, D., Spring, D., Mac Nally, R., *et al.* 2010. Finding needles (or ants) in haystacks: predicting locations of invasive organisms to inform eradication and containment. *Ecological Applications* **20**: 1217–1227.

Silverman, J., Brightwell, R. J. 2008. The Argentine ant: challenges in managing an invasive unicolonial pest. *Annual Review of Entomology* **53**: 231–252.

Simberloff, D. 2009. We can eliminate invasions or live with them. Successful management projects. *Biological Invasions* **11**: 149–157.

Snyder, W. E., Evans, E. W. 2006. Ecological effects of invasive arthropod generalist predators. *Annual Review of Ecology and Evolutionary Systematics* **37**: 95–122.

Spradbery, J. P., Maywald, G. F. 1992. The distribution of the European or German wasp, *Vespula germanica* (F.) (Hymenoptera: Vespidae), in Australia: past, present and future. *Australian Journal of Zoology* **40**: 495–510.

Stanaway, M. A., Zalucki, M. P., Gillespie, P. S., Rodriguez, C. M., Maynard, G. V. 2001. Pest risk assessment of insects in sea cargo containers. *Australian Journal of Entomology* **40**: 180–192.

Stanley, M. C., Ward, D. F. 2012. Impacts of Argentine ants on invertebrate communities: below-ground consequences? *Biodiversity and Conservation* **21**: 2653–2669.

Stanley, M. C., Nathan, H., Phillips, L., *et al.* 2012. Invasive interactions: can Argentine ants indirectly increase the reproductive output of a weed? *Arthropod–Plant Interactions*. DOI: 10.1007/s11829-012-9215-2.

Stanley, M. C., Ward, D. F., Harris, R. J., *et al.* 2008. Optimising pitfall sampling for the detection of Argentine ants, *Linepithema humile* (Hymenoptera: Formicidae). *Sociobiology* **51**: 461–472.

Stevens, P. M. 2008. High risk site surveillance (HRSS) – and example of best practice plant pest surveillance, pp. 127–134. In: *Surveillance for Biosecurity: Pre-border to Pest Management*. Froud, K. J., Popay, A. I., Zydenbos, S. M. (eds.). The New Zealand Plant Protection Society, Wickliffe, Christchurch.

Stringer, L. D., Suckling, D. M., Baird, D., *et al.*, 2011. Sampling efficacy for the Red Imported Fire Ant *Solenopsis invicta* (Hymenoptera: Formicidae). *Environmental Entomology* **40**, 1276–1284.

Stringer, L. D., Suckling, D. M., Mattson, L. T. W., Peacock, L. R., 2010. Improving ant-surveillance trap design to reduce competitive exclusion. *New Zealand Plant Protection Society* **63**: 248–253.

Suarez, A. V., Holway, D. A., Ward, P. S. 2005. The role of opportunity in the unintentional introduction of non-native ants. *Proceedings of the National Academy of Sciences, USA* **102**: 17 032–17 035.

Todd, J. H. 2011. Selecting non-target invertebrates for risk assessment of biological control agents. Unpublished PhD thesis, University of Auckland.

Vanderwoude, C., Elson, H. M., Hargreaves, J., Harris, E., Plowman, K. P. 2003. An overview of the red imported fire ant (*Solenopsis invicta* Buren) eradication plan for Australia. *Records of the South Australian Museum Monograph Series* **7**: 11–16.

Venette, R. C., Moon, R. D., Hutchison, W. D. 2002. Strategies and statistics of sampling for rare individuals. *Annual Review of Entomology* **47**: 143–174.

Wajnberg, E., Scott, J. K., Quimby, P. C. 2001. *Evaluating Indirect Ecological Effects of Biological Control*. CAB International, Wallingford, 288 pp.

Ward, D. F. 2009. The potential distribution of the red imported fire ant, *Solenopsis invicta* Buren (Hymenoptera: Formicidae), in New Zealand. *New Zealand Entomologist* **32**: 67–75.

Ward, D. F., Morgan, F. 2014. Modelling the impacts of an invasive species across landscapes: a step-wise approach. *PeerJ* **2**:e435 http://dx.doi.org/10.7717/peerj.435.

Ward, D. F., Ramon-Laca, A. 2013. Molecular identification of the prey range of the invasive Asian paper wasp. *Ecology and Evolution*. DOI:10.1002/ece3.826.

Ward, D. F., Stanley, M. C. 2012. Site occupancy and detection probability of Argentine ant populations. *Journal of Applied Entomology* **137**: 197–203.

Ward, D. F., Beggs, J. R., Clout, M. N., Harris, R. J., O'Connor, S. 2006. The diversity and origin of exotic ants arriving to New Zealand via human-mediated dispersal. *Diversity and Distributions* **12**: 601–609.

Ward, D. F., Green, C., Harris, R. J., *et al.* 2010. Twenty years of Argentine ants in New Zealand: past research and future priorities for applied management. *New Zealand Entomologist* **33**: 67–78.

Ward, D. F., Honan, P., Lefoe, G. 2002. Colony structure and nest characteristics of European wasps *Vespula germanica* (F.) (Hymenoptera: Vespidae) in Victoria, Australia. *Australian Journal of Entomology* **31**: 306–309.

Wardle, D. A., Karl, B. J., Beggs, J. R., *et al.* 2010. Determining the impact of scale insect honeydew, and invasive wasps and rodents, on the decomposer subsystem in a New Zealand beech forest. *Biological Invasions* **12**: 2619–2638.

Westphal, M., Browne, M., MacKinnon, K., Noble, I. 2008. The link between international trade and the global distribution of invasive alien species. *Biological Invasions* **10**: 391–398.

Williams, P. A., Timmins, S. 2002. Ecological impacts of weeds in New Zealand, pp. 175–184. In: *Biological Invasions: Economic and Environmental Costs of Alien Plant, Animal and Microbe Species*. Pimentel, D. (ed.). CRC Press, Florida.

Work, T. T., McCullough, D. G., Cavey, J. F., Komsa, R. 2005. Arrival rate of nonindigenous insect species into the United States through foreign trade. *Biological Invasions* **7**: 323–332.

Worner, S. P., Gevrey, M. 2006. Modelling global insect species assemblages to determine risk of invasion. *Journal of Applied Ecology* **43**: 858–867.

CHAPTER 9

Pollution by antibiotics and resistance genes: dissemination into Australian wildlife

Michael Gillings

Summary

The effects of antibiotic use and antibiotic resistance are now spreading beyond hospitals and human-dominated landscapes to encompass the whole biosphere. Antibiotics disseminated by waste streams pollute aquatic systems worldwide. These waste streams are also contaminated by bacteria that carry genes for resistance to antibiotics on mobile DNA elements. This circumstance means that aquatic ecosystems are now an evolutionary reactor for DNA rearrangements where novel genes can be assembled into ever more complex DNA elements, and then transferred into a growing diversity of bacterial species. As a consequence, bacteria containing antibiotic resistance genes have now been identified in a range of marine and terrestrial organisms, including wild species and species harvested for human consumption. The dissemination of these resistance genes will have unpredictable consequences for both native organisms and human welfare.

9.1 Introduction

Water bodies connect soils, oceans and the atmosphere, and provide essential ecosystem services, but are mostly no longer naturally regulated, since the water cycle has been fundamentally altered by human activity. Further deterioration of water quality is caused by sedimentation, salinisation, eutrophication and contamination with chemical and microbial pollutants (Meybeck 2003). Management of water sources in Australia faces unique challenges as it is the driest inhabited continent on Earth, and is subject to unpredictable variations in rainfall. Extraction of water for human activities is having serious ecological consequences in Australia (Kingsford 2000). Furthermore, water

Austral Ark: The State of Wildlife in Australia and New Zealand, eds. A. Stow, N. Maclean and G. I. Holwell. Published by Cambridge University Press. © Cambridge University Press 2015.

bodies are subject to both diffuse and point sources of pollution from industry, sewage and urban run-off (Francey *et al.* 2010).

In this chapter, we examine the interface between humans, waste streams and water bodies. Specifically we will concentrate on the dissemination of antibiotics and antibiotic resistant bacteria into the general environment via aquatic routes. Using examples drawn from Australian environments and biota, we will show how pollution with bioactive molecules is having unintended evolutionary effects on non-target organisms, with consequences for the environment, human health and society.

9.2 Xenobiotic compounds

In the strict sense, xenobiotics are chemical compounds that do not occur naturally in the organisms in which they can now be found. In a broader sense, xenobiotics are chemical compounds that are foreign to biological systems, or are synthesised by humans. Humans make and distribute a wide variety of xenobiotics, with over 80 000 registered compounds readily available in the marketplace (Schwarzman & Wilson 2009). The sheer diversity of potential chemical pollutants makes it difficult to predict their effects, particularly when interactions between different compounds are considered (Rockstrom *et al.* 2009). Human activities generate enormous volumes of wastewater contaminated with xenobiotics. In some countries, these waters are released into natural environments untreated, but even with water treatment, such compounds are often difficult to remove (Schwarzenbach *et al.* 2010).

Many xenobiotic molecules can accumulate in food webs (Arnot & Gobas 2004) and have adverse effects (Schwarzenbach *et al.* 2010). In particular, we should be concerned about compounds that have direct biological effects at low concentrations. Endocrine disrupting chemicals are widely used in industry and agriculture, and exposure affects reproduction and development in a range of organisms (Diamanti-Kandarakis *et al.* 2009). Antibiotics are used both clinically and in agriculture, with significant proportions being excreted directly into natural watercourses, where their selective effects may alter bacterial assemblages and biodiversity (Lupo *et al.* 2012; Gillings 2013). Similarly, disinfectants have significant biotic effects and are common contaminants of wastewaters (Gaze *et al.* 2005; Gillings 2010).

9.3 Xenogenetic DNA elements

Human activities exert strong selective pressures on many organisms, to the extent that we may now be the greatest evolutionary force on the planet (Palumbi 2001). Possibly the best example of human-driven evolution in action is the rapid appearance and spread of antibiotic resistant bacteria. Use of antibiotics exerts strong selection pressure, so the only bacteria that survive are those which have accumulated specific mutations, or which have acquired resistance genes from another bacterial species (Davies & Davies 2010).

It is now clear that genes can be freely exchanged between different bacteria (lateral gene transfer, or LGT)(Siefert 2009), and that the enormous diversity of bacterial species in natural environments holds a correspondingly large pool of diverse genes, any of which can be sampled by LGT (Lapierre & Gogarten 2009). Amongst these genes are a huge number that can confer a phenotype of antibiotic resistance (Wright 2010; Forsberg *et al*. 2012).

Clinically important bacteria have frequently become resistant to multiple antibiotics and other selective agents by LGT of diverse resistance genes from the bacteria found in natural environments (Stokes & Gillings 2011; Wellington *et al*. 2013). Repeated acquisition of novel genes into a growing number of different mobile DNA elements has then resulted in the assembly of complex, mosaic DNA molecules composed of multiple genes, each with different phylogenetic origins (Garriss *et al*. 2009; Toleman & Walsh 2011). These complex DNA molecules have arisen as a direct consequence of human selection pressures mediated by antibiotics, disinfectants and heavy metals, and consequently can be thought of as being xenogenetic DNA elements, in the sense that they have arisen as a result of human activities (Gillings & Stokes 2012). See Box 9.1.

Box 9.1 Antibiotic resistance and integrons

Integrons are DNA elements often found embedded within larger xenogenetic DNAs. They are a gene acquisition and expression system that has a major role in the spread of antibiotic resistance in Gram-negative pathogens (Stokes & Gillings 2011; Stalder *et al*. 2012). Australian scientists made many of the key discoveries about integrons and about their evolution. The original terms 'integron' and the identity of the genes they capture as 'gene cassettes' were coined by Australian researchers (Stokes & Hall 1989; Hall *et al*. 1991).

In general, integrons encode an integrase protein (IntI) that catalyses recombination at an associated recombination site (*attI*) (Hall *et al*. 1991; Partridge *et al*. 2000). This reaction inserts a circular gene cassette adjacent to the integrase gene by recombination between *attI* and a similar recombination site, *attC*, found on the incoming gene cassette (Stokes *et al*. 1997). The newly inserted gene cassette is then expressed from a promoter, Pc, within the integron (Collis & Hall 1995). Repeated capture of new gene cassettes results in a cassette array, which in the case of clinical integrons, may encode resistance to multiple antibiotic classes (Plate 6).

It was not until more than a decade after the discovery of integrons in human pathogens that a range of genetically diverse integrons were discovered in natural environments, and that gene cassettes were first recovered from environmental DNA (Nield *et al*. 2001; Stokes *et al*. 2001). Gene cassettes in environmental samples rarely conferred antibiotic resistance, but encoded novel proteins that could be expressed (Nield *et al*. 2004), and crystallised to solve their structure (Robinson *et al*. 2005). As many as one in ten species of environmental bacteria had integrons on their chromosomes, and these integrons generally carried multiple gene cassettes of novel function (Holmes *et al*. 2003; Gillings *et al*. 2005; Boucher *et al*. 2006). Consequently, gene cassettes in environmental integrons provide a pool of diverse genes that can move between bacteria using LGT and integron activity. Human use of antibiotics then selected for integrons that had acquired gene cassettes encoding antibiotic resistance from this environmental pool (Gillings *et al*. 2008a,b). Thus bacteria have co-opted an ancient system of gene acquisition to deal with human selection pressure mediated by antibiotic use.

9.4 Pollution with antibiotics and resistance genes

Humans synthesise large quantities of antibiotics for use in medicine and animal production. Depending on the antibiotic class, some 30%–90% of the therapeutic dose is excreted essentially unchanged, to then be released into the general environment via waste streams (Sarmah *et al.* 2006). High concentrations of antibiotics are found in sewage, and these are often not removed by water treatment (Watkinson *et al.* 2007; Le-Minh *et al.* 2010). Antibiotics are also released from animal production facilities, from pharmaceutical plants and via spread manure (Chee-Sanford *et al.* 2009; Li *et al.* 2010; Heuer *et al.* 2011). The long-term consequences of antibiotic pollution in aquatic systems are of considerable concern (Taylor *et al.* 2011; Lupo *et al.* 2012).

Human waste streams also release large quantities of antibiotic resistant bacteria that carry diverse resistance genes. These genes and their associated mobile DNA elements should also be regarded as pollutants (Storteboom *et al.* 2010; Pruden *et al.* 2013), but with the critical distinction that xenogenetic pollutants can replicate and transfer to new hosts (Gillings & Stokes 2012). Antibiotic resistance genes can now be detected in places where antibiotics have never been used, such as remote jungle communities, polar regions, and in wild animals (Sjolund *et al.* 2008; Bartoloni *et al.* 2009; Stokes & Gillings 2011). Indeed, the concentration of resistance genes in soil has been increasing since the first use of antibiotics in the 1940s (Knapp *et al.* 2009).

Diverse bacteria containing antibiotic resistance genes are being released into the environment simultaneously with excreted antibiotics and other selective agents, such as disinfectants and heavy metals (Baker-Austin *et al.* 2006; Gaze *et al.* 2011). This creates a hotspot for complex interactions and selection events that affect the abundance and distribution of bacterial species, mobile DNA elements and the diverse resistance genes they carry (Taylor *et al.* 2011; Lupo *et al.* 2012; Stalder *et al.* 2012; Gillings 2013). In particular, wastewater and sewage treatment plants provide an active reaction vessel for extensive LGT, and for *de-novo* assembly of new gene combinations (Schlüter *et al.* 2008; Moura *et al.* 2010; Zhang *et al.* 2011).

The conclusion must be reached that there are second-order consequences arising from the antibiotic revolution, and that these include significant evolutionary effects on all aspects of the microbial biosphere, not just those microorganisms of clinical concern (Martinez 2012; Gillings 2013). Pollution with antimicrobials and heavy metals is affecting the population structure of microbial communities and is increasing the abundance of resistance genes and mobile DNA elements (Wright *et al.* 2008; Elsaied *et al.* 2011; Kristiansson *et al.* 2011). These selective agents are also affecting the general rates of evolution and the genetic variability of the microbial world (Gillings & Stokes 2012). Effects on LGT and the general rates of bacterial evolution are likely to have effects on natural ecosystems, wildlife, agriculture and human health.

9.5 Case studies

The issue of antibiotics and resistance genes in the general environment has been seriously investigated only in the past five years or so, and there is much that we do not know. We can shed light on some of the potential consequences by examining case studies, many of which were conducted in Australian environments.

9.5.1   Spread of resistance determinants in urban landscapes

Class 1 integrons characteristic of those found in clinical pathogens can be tracked in the general environment using molecular methods. Using such DNA analysis, the abundance of Class 1 integrons was found to be directly correlated with the degree of pollution and ecological condition in a peri-urban catchment (Cowan Creek, Sydney, NSW). The larger the human impact, the higher was the abundance of integrons (Hardwick *et al.* 2008). In the Greater Melbourne area, abundance of Class 1 integrons was also positively correlated with degree of ecosystem perturbation, and was found to be proportional to the discharge rate of sewage (Rosewarne *et al.* 2010). Similar results have now been reported from other continents (Nardelli *et al.* 2012).

One reason why these findings are of concern is that the antibiotic resistance genes now present in human pathogens originated from environmental bacteria, as did the clinical Class 1 integron (Gillings *et al.* 2008a). Environmental bacteria contain a vast and largely uncharacterised pool of genes that can confer resistance to diverse antimicrobial agents (Wright 2010; Forsberg *et al.* 2012), and these genes can be acquired by LGT to bring new resistance and virulence genes into the bacteria that pollute water bodies. We know that aquatic biofilms are sites where LGT and active rearrangement of resistance gene cassettes occur (Sorensen *et al.* 2005; Gillings *et al.* 2009a; Koenig *et al.* 2011). We also suspect that pollution with antibiotics and disinfectants provided the initial selective force to fix the Class 1 integron in clinical pathogens (Gaze *et al.* 2005; Gillings *et al.* 2008b). Environments that combine diverse selective agents, high bacterial abundance and complex DNA elements containing multiple resistance determinants are reactors for generating pathogens with a formidable genetic armoury.

9.5.2   Spread of resistance determinants into wild animals

Given the widespread dissemination of xenobiotic and xenogenetic pollutants in the environment, it is inevitable that resistance genes will find their way into wild animals. The growing number of such cases has recently been reviewed (Stokes & Gillings 2011). In general, the prevalence of antibiotic resistant bacteria in animals is directly correlated with their proximity to humans (Skurnik *et al.* 2006), demonstrating that such bacteria originate from human sources. It is of particular concern when endangered species acquire resistant bacteria, because this compromises disease management, and it demonstrates the potential for serious pathogens to infect populations.

In a recent investigation of the endangered brush-tailed rock wallaby (*Petrogale penicillata*), it was found that about half the individuals in captive breeding programmes carried resistance genes typical of those found in human pathogens. The diversity and distribution of these genes amongst individuals suggested at least five independent acquisitions of different resistance genes had occurred amongst a mere 14 animals (Power *et al.* 2013). This demonstrates the potential ease for such acquisitions when animals are raised in human-dominated ecosystems. In contrast, no animals from five isolated wild populations carried resistance genes. The captive animals have now been released to bolster the wild population, with the potential to transmit the antibiotic resistance determinants into wild animals.

Aquatic environments provide ideal locations for complex interactions between resistance genes, mobile DNA elements, bacterial species and their hosts (Taylor *et al.* 2011;

Lupo *et al.* 2012). The potential for such interactions to generate novel gene/host combinations is exacerbated by simultaneous pollution with bacteria containing resistance determinants and diverse selective agents, including antibiotics, disinfectants and heavy metals (Wright *et al.* 2008; Kristiansson *et al.* 2011; Stokes & Gillings 2011; Seiler & Berendonk 2012).

We have recently conducted a series of investigations examining the spread of resistance genes into natural environments. We reasoned that filter and particle feeders, such as molluscs and crustaceans, were likely endpoints for bacterial pollutants emanating from human activity. Bivalves do accumulate antibiotic resistant strains of *E. coli* from sewage discharge (M. Power, unpublished results), and antibiotic resistance genes can be readily detected in crustaceans (Gillings *et al.* 2009b; Sajjad *et al.* 2011). This latter observation can be used to illustrate the complexity of human–pollutant–environment interactions, as will be explained below.

To examine the fate of antibiotic resistance genes released into the environment from human sources, we tested a range of Australian seafood, including prawns, for the presence of Class 1 integrons typical of human pathogens. A number of prawn species were found to carry resistance elements within Class 1 integrons that were clearly identifiable as originating from human sources. However, these integrons had undergone LGT and modification during their passage from human to prawn. The most parsimonious explanation for our observations is as follows (Plate 7), although the order of events cannot necessarily be identified:

A typical clinical Class 1 integron, embedded in a Tn*402* type transposon and carrying a gene cassette that encoded streptomycin and spectinomycin resistance, was released into the environment. It is not known what organism originally carried the integron, but it is likely to have been *E. coli* or some other common human commensal. This integron then acquired a novel gene cassette that encoded two methionine sulphoxide reductases. These enzymes repair proteins after oxidative damage, and consequently help bacteria colonise and persist inside animals. The skeleton of the Tn*402* transposon was deleted, and replaced by a pair of miniature inverted repeat transposable elements (MITEs) that are capable of moving the integron between genomic locations (Gillings *et al.* 2009b). This appears to have happened, since the integron and flanking MITEs are now located on a much larger mobile DNA element, thought to be a genomic island (A. Sajjad, unpublished).

Why is this observation troubling? Over 75% of eastern king prawns tested from locations spanning the entire east coast of Australia now test positive for this integron/MITE/genomic island. The genomic island has made its way into at least three different species of *Acinetobacter* resident in the gut of prawns, and the genus *Acinetobacter* is an emerging nosocomial pathogen. Pathogenic members of this genus have been described that carry the same MITEs in clinical situations (Domingues *et al.* 2011). Consequently, an antibiotic resistance integron has become closely linked with a novel gene cassette that enhances colonisation of animal tissues; the integron is freely moving between at least three species of the genus *Acinetobacter* which is of serious concern for hospitals and for infection control; and it is present in the majority of individuals of a food species that is consumed with minimal cooking. There is a clear pathway for the integron originally released as a xenogenetic pollutant to return into human ecosystems as a part of a newly virulent and antibiotic resistant pathogen. Because integrons and other resistance determinants released via human waste streams can continue to accumulate novel genes, there are serious implications for natural ecosystems, marine organisms and human health.

9.6 Conclusion

Human use of antibiotics results in large quantities of these bioactive agents being released, essentially unchanged, into the environment. The same waste streams that disseminate antibiotics also disseminate bacteria and genes that confer resistance to antibiotics. This sets in train a complex series of ecological and evolutionary outcomes that may have consequences for human welfare. It is time to closely examine the environmental release of both genes and antibiotics, and to treat them as pollutants (Allen *et al.* 2010), especially as clinical resistance genes and xenogenetic elements are now being spread throughout the biosphere (Martinez 2009; Stokes & Gillings 2011). Pollution with antibiotics may have effects across the entire microbial biosphere, by changing the basal rates of bacterial evolution (Couce & Blázquez 2009; Gillings & Stokes 2012), and by fundamentally altering the abundance of mobile DNAs and resistance determinants (Gillings 2013).

REFERENCES

Allen, H. K., Donato, J., Wang, H. H., *et al.* (2010) Call of the wild: antibiotic resistance genes in natural environments. *Nature Reviews Microbiology* 8, 251–9.

Arnot, J. A. & Gobas, F. A. P. C. (2004) A food web bioaccumulation model for organic chemicals in aquatic ecosystems. *Environmental Toxicology and Chemistry* 23, 2343–55.

Baker-Austin, C., Wright, M. S., Stepanauskas, R. & McArthur, J. V. (2006) Co-selection of antibiotic and metal resistance. *Trends in Microbiology* 14, 176–82.

Bartoloni, A., Pallecchi, L., Rodríguez, H., *et al.* (2009) Antibiotic resistance in a very remote Amazonas community. *International Journal of Antimicrobial Agents* 33, 125–9.

Boucher, Y., Nesbø, C. L., Joss, M. J., *et al.* (2006) Recovery and evolutionary analysis of complete integron gene cassette arrays from Vibrio. *BMC Evolutionary Biology* 6, 3.

Chee-Sanford, J. C., Mackie, R. I., Koike, S., *et al.* (2009) Fate and transport of antibiotic residues and antibiotic resistance genes following land application of manure waste. *Journal of Environmental Qualities* 38, 1086–108.

Collis, C. M. & Hall, R. M. (1995) Expression of antibiotic resistance genes in the integrated cassettes of integrons. *Antimicrobial Agents and Chemotherapy* 39, 155–62.

Couce, A. & Blázquez, J. (2009) Side effects of antibiotics on genetic variability. *FEMS Microbiology Reviews* 33, 531–8.

Davies, J. & Davies, D. (2010) Origins and evolution of antibiotic resistance. *Microbiology and Molecular Biology Reviews* 74, 417–33.

Diamanti-Kandarakis, E., Bourguignon, J.-P., Giudice, L. C., *et al.* (2009) Endocrine-disrupting chemicals: an Endocrine Society scientific statement. *Endocrine Reviews* 30, 293–342.

Domingues, S., Nielsen, K. M. & da Silva, G. J. (2011) The blaIMP-5-carrying integron in a clinical *Acinetobacter baumannii* strain is flanked by miniature inverted-repeat transposable elements (MITEs). *Journal of Antimicrobial Chemotherapy* **66**, 2667–8.

Elsaied, H., Stokes, H. W., Kitamura, K., *et al.* (2011) Marine integrons containing novel integrase genes, attachment sites, attI, and associated gene cassettes in polluted sediments from Suez and Tokyo Bays. *The ISME Journal* **5**, 1162–77.

Forsberg, K. J., Reyes, A., Wang, B., *et al.* (2012) The shared antibiotic resistome of soil bacteria and human pathogens. *Science* **337**, 1107–11.

Francey, M., Fletcher, T. D., Deletic, A. & Duncan, H. (2010) New insights into the quality of urban storm water in south eastern Australia. *Journal of Environmental Engineering* **136**, 381–90.

Garriss, G., Waldor, M. K. & Burrus, V. (2009) Mobile antibiotic resistance encoding elements promote their own diversity. *PLoS Genetics* **5**, e1000775.

Gaze, W. H., Abdouslam, N., Hawkey, P. M. & Wellington, E. M. H. (2005) Incidence of class 1 integrons in a quaternary ammonium compound-polluted environment. *Antimicrobial Agents and Chemotherapy* **49**, 1802–7.

Gaze, W. H., Zhang, L., Abdouslam, N. A., *et al.* (2011) Impacts of anthropogenic activity on the ecology of class 1 integrons and integron-associated genes in the environment. *The ISME Journal* **5**, 1253–61.

Gillings, M. R. (2010) Biocide use, integrons and novel genetic elements. *Biocides in the Health Industry* **193**, 192.

Gillings, M. R. (2013) Evolutionary consequences of antibiotic use for the resistome, mobilome and microbial pangenome. *Frontiers in Microbiology* **4**.

Gillings, M. R. & Stokes, H. (2012) Are humans increasing bacterial evolvability? *Trends in Ecology & Evolution* **27**, 346–52.

Gillings, M., Boucher, Y., Labbate, M., *et al.* (2008a) The evolution of class 1 integrons and the rise of antibiotic resistance. *Journal of Bacteriology* **190**, 5095–100.

Gillings, M. R., Holley, M. P. & Stokes, H. W. (2009a) Evidence for dynamic exchange of qac gene cassettes between class 1 integrons and other integrons in freshwater biofilms. *FEMS Microbiology Letters* **296**, 282–8.

Gillings, M. R., Holley, M. P., Stokes, H. & Holmes, A. J. (2005) Integrons in Xanthomonas: a source of species genome diversity. *Proceedings of the National Academy of Sciences of the United States of America* **102**, 4419–24.

Gillings, M. R., Labbate, M., Sajjad, A., *et al.* (2009b) Mobilization of a Tn402-like class 1 integron with a novel cassette array via flanking miniature inverted-repeat transposable element-like structures. *Applied and Environmental Microbiology* **75**, 6002–4.

Gillings, M. R., Xuejun, D., Hardwick, S. A., Holley, M. P. & Stokes, H. W. (2008b) Gene cassettes encoding resistance to quaternary ammonium compounds: a role in the origin of clinical class 1 integrons? *The ISME Journal* **3**, 209–15.

Hall, R., Brookes, D. & Stokes, H. (1991) Site-specific insertion of genes into integrons: role of the 59-base element and determination of the recombination cross-over point. *Molecular Microbiology* **5**, 1941–59.

Hardwick, S. A., Stokes, H. W., Findlay, S., Taylor, M. & Gillings, M. R. (2008) Quantification of class 1 integron abundance in natural environments using real-time quantitative PCR. *FEMS Microbiology Letters* **278**, 207–12.

Heuer, H., Schmitt, H. & Smalla, K. (2011) Antibiotic resistance gene spread due to manure application on agricultural fields. *Current Opinion in Microbiology* **14**, 236–43.

Holmes, A. J., Holley, M. P., Mahon, A., *et al.* (2003) Recombination activity of a distinctive integron-gene cassette system associated with *Pseudomonas stutzeri* populations in soil. *Journal of Bacteriology* **185**, 918–28.

Kingsford, R. (2000) Ecological impacts of dams, water diversions and river management on floodplain wetlands in Australia. *Austral Ecology* **25**, 109–27.

Knapp, C. W., Dolfing, J., Ehlert, P. A. I. & Graham, D. W. (2009) Evidence of increasing antibiotic resistance gene abundances in archived soils since 1940. *Environmental Science & Technology* **44**, 580–7.

Koenig, J. E., Bourne, D. G., Curtis, B., *et al.* (2011) Coral-mucus-associated *Vibrio* integrons in the Great Barrier Reef: genomic hotspots for environmental adaptation. *The ISME journal* **5**, 962–72.

Kristiansson, E., Fick, J., Janzon, A., *et al.* (2011) Pyrosequencing of antibiotic-contaminated river sediments reveals high levels of resistance and gene transfer elements. *PLoS ONE* **6**, e17038.

Lapierre, P. & Gogarten, J. P. (2009) Estimating the size of the bacterial pan-genome. *Trends in Genetics* **25**, 107–10.

Le-Minh, N., Khan, S. J., Drewes, J. E. & Stuetz, R. M. (2010) Fate of antibiotics during municipal water recycling treatment processes. *Water Research* **44**, 4295–323.

Li, D., Yu, T., Zhang, Y., *et al.* (2010) Antibiotic resistance characteristics of environmental bacteria from an oxytetracycline production wastewater treatment plant and the receiving river. *Applied and Environmental Microbiology* **76**, 3444–51.

Lupo, A., Coyne, S. & Berendonk, T. U. (2012) Origin and evolution of antibiotic resistance: the common mechanisms of emergence and spread in water bodies. *Frontiers in Microbiology* **3**.

Martinez, J. L. (2009) Environmental pollution by antibiotics and by antibiotic resistance determinants. *Environmental Pollution* **157**, 2893–902.

Martinez, J. L. (2012) Natural antibiotic resistance and contamination by antibiotic resistance determinants: the two ages in the evolution of resistance to antimicrobials. *Frontiers in Microbiology* **3**.

Meybeck, M. (2003) Global analysis of river systems: from Earth system controls to Anthropocene syndromes. *Philosophical Transactions of the Royal Society of London. Series B: Biological Sciences* **358**, 1935–55.

Moura, A., Henriques, I., Smalla, K. & Correia, A. (2010) Wastewater bacterial communities bring together broad-host range plasmids, integrons and a wide diversity of uncharacterized gene cassettes. *Research in Microbiology* **161**, 58–66.

Nardelli, M., Scalzo, P. M., Ramírez, M. S., *et al.* (2012) Class 1 integrons in environments with different degrees of urbanization. *PLoS ONE* **7**, e39223.

Nield, B. S., Holmes, A. J., Gillings, M. R., *et al.* (2001) Recovery of new integron classes from environmental DNA. *FEMS Microbiology Letters* **195**, 59–65.

Nield, B. S., Willows, R. D., Torda, A. E., *et al.* (2004) New enzymes from environmental cassette arrays: functional attributes of a phosphotransferase and an RNA-methyltransferase. *Protein Science* **13**, 1651–9.

Palumbi, S. R. (2001) Humans as the world's greatest evolutionary force. *Science* **293**, 1786–90.

Partridge, S. R., Recchia, G. D., Scaramuzzi, C., *et al.* (2000) Definition of the attI1 site of class 1 integrons. *Microbiology* **146**, 2855–64.

Power, M., Emery, S. & Gillings, M. (2013) Into the wild: dissemination of antibiotic resistance dterminants via a species recovery program. *PLoS ONE* **8**, e63017.

Pruden, A., Larsson, D. J., Amézquita, A., *et al.* (2013) Management options for reducing the release of antibiotics and antibiotic resistance genes to the environment. *Environmental Health Perspectives*. doi:10.1289/eph.1206446.

Robinson, A., Wu, P. S.-C., Harrop, S. J., *et al.* (2005) Integron-associated mobile gene cassettes code for folded proteins: the structure of Bal32a, a new member of the adaptable [alpha]+[beta] barrel family. *Journal of Molecular Biology* **346**, 1229–41.

Rockstrom, J., Steffen, W., Noone, K., *et al.* (2009) Planetary boundaries: exploring the safe operating space for humanity. *Ecology and Society* **14**, 32.

Rosewarne, C. P., Pettigrove, V., Stokes, H. W. & Parsons, Y. M. (2010) Class 1 integrons in benthic bacterial communities: abundance, association with Tn402-like transposition modules and evidence for coselection with heavy-metal resistance. *FEMS Microbiology Ecology* **72**, 35–46.

Sajjad, A., Holley, M. P., Labbate, M., Stokes, H. & Gillings, M. R. (2011) Preclinical class 1 integron with a complete Tn402-like transposition module. *Applied and Environmental Microbiology* **77**, 335–7.

Sarmah, A. K., Meyer, M. T. & Boxall, A. B. A. (2006) A global perspective on the use, sales, exposure pathways, occurrence, fate and effects of veterinary antibiotics (VAs) in the environment. *Chemosphere* **65**, 725–59.

Schlüter, A., Krause, L., Szczepanowski, R., Goesmann, A. & Pühler, A. (2008) Genetic diversity and composition of a plasmid metagenome from a wastewater treatment plant. *Journal of Biotechnology* **136**, 65–76.

Schwarzenbach, R. P., Egli, T., Hofstetter, T. B., Von Gunten, U. & Wehrli, B. (2010) Global water pollution and human health. *Annual Review of Environment and Resources* **35**, 109–36.

Schwarzman, M. R. & Wilson, M. P. (2009) New science for chemicals policy. *Science* **326**, 1065–6.

Seiler, C. & Berendonk, T. U. (2012) Heavy metal driven co-selection of antibiotic resistance in soil and water bodies impacted by agriculture and aquaculture. *Frontiers in Microbiology* **3**.

Siefert, J. L. (2009) Defining the mobilome. In: *Horizontal Gene Transfer: Genomes in Flux* (eds. **Gogarten, M. B., Gogarten, J. P. & Olendzenski, L. C.**), pp. 13–27. Humana Press.

Sjolund, M., Bonnedahl, J., Hernandez, J., *et al.* (2008) Dissemination of multidrug-resistant bacteria into the Arctic. *Emerging Infectious Diseases* **14**, 70–2.

Skurnik, D., Ruimy, R., Andremont, A., *et al.* (2006) Effect of human vicinity on antimicrobial resistance and integrons in animal faecal *Escherichia coli*. *Journal of Antimicrobial Chemotherapy* **57**, 1215–19.

Sorensen, S. J., Bailey, M., Hansen, L. H., Kroer, N. & Wuertz, S. (2005) Studying plasmid horizontal transfer in situ: a critical review. *Nature Reviews Microbiology* **3**, 700–10.

Stalder, T., Barraud, O., Casellas, M., Dagot, C. & Ploy, M.-C. (2012) Integron involvement in environmental spread of antibiotic resistance. *Frontiers in Microbiology* **3**.

Stokes, H. W. & Gillings, M. R. (2011) Gene flow, mobile genetic elements and the recruitment of antibiotic resistance genes into Gram-negative pathogens. *FEMS Microbiology Reviews* **35**, 790–819.

Stokes, H. & Hall, R. (1989) A novel family of potentially mobile DNA elements encoding site-specific gene-integration functions: integrons. *Molecular Microbiology* **3**, 1669–83.

Stokes, H., Holmes, A. J., Nield, B. S., *et al.* (2001) Gene cassette PCR: sequence-independent recovery of entire genes from environmental DNA. *Applied and Environmental Microbiology* **67**, 5240–6.

Stokes, H., O'Gorman, D., Recchia, G. D., Parsekhian, M. & Hall, R. M. (1997) Structure and function of 59-base element recombination sites associated with mobile gene cassettes. *Molecular Microbiology* **26**, 731–45.

Storteboom, H., Arabi, M., Davis, J. G., Crimi, B. & Pruden, A. (2010) Identification of antibiotic-resistance-gene molecular signatures suitable as tracers of pristine river, urban, and agricultural sources. *Environmental Science and Technology* **44**, 1947–53.

Taylor, N. G. H., Verner-Jeffreys, D. W. & Baker-Austin, C. (2011) Aquatic systems: maintaining, mixing and mobilising antimicrobial resistance? *Trends in Ecology & Evolution* **26**, 278–84.

Toleman, M. A. & Walsh, T. R. (2011) Combinatorial events of insertion sequences and ICE in Gram-negative bacteria. *FEMS Microbiology Reviews* **35**, 912–35.

Watkinson, A. J., Murby, E. J. & Costanzo, S. D. (2007) Removal of antibiotics in conventional and advanced wastewater treatment: Implications for environmental discharge and wastewater recycling. *Water Research* **41**, 4164–76.

Wellington, E. M., Boxall, A., Cross, P., *et al.* (2013) The role of the natural environment in the emergence of antibiotic resistance in Gram-negative bacteria. *The Lancet Infectious Diseases* **13**, 155–65.

Wright, G. D. (2010) The antibiotic resistome. *Expert Opinion in Drug Discovery* **5**, 779–88.

Wright, M. S., Baker-Austin, C., Lindell, A. H., *et al.* (2008) Influence of industrial contamination on mobile genetic elements: class 1 integron abundance and gene cassette structure in aquatic bacterial communities. *The ISME Journal* **2**, 417–28.

Zhang, T., Zhang, X.-X. & Ye, L. (2011) Plasmid metagenome reveals high levels of antibiotic resistance genes and mobile genetic elements in activated sludge. *PLoS ONE* **6**, e26041.

CHAPTER 10

Invasive vertebrates in Australia and New Zealand

Cheryl R. Krull, Josie A. Galbraith, Al S. Glen and Helen W. Nathan

Summary

Invasive vertebrate species have had a dramatic impact on the unique native ecosystems of both Australia and New Zealand. Some of these species were accidentally introduced, though many were introduced deliberately for a number of reasons: as a food resource, for hunting and trade, as a mode of transportation, as a control tool for other pests, and by acclimatisation societies to remind colonists of home. Regardless of the method of introduction, these invasive species have had major negative impacts on the native flora and fauna, including herbivory, predation, competition, disease, hybridisation and habitat change, and have also affected human health and industry. In both countries the aim is now to prevent establishment of new invasive species and preserve key areas of high biodiversity value through the control or eradication of invasive species.

10.1 Introduction

Invasions of vertebrate species into habitats outside their natural range have had major impacts across the globe and particularly in Australia and New Zealand (Simberloff & Rejmánek, 2011). Preventing the arrival of these species is the best protection for native ecosystems, but once introduction and spread have taken place, effective and efficient management of entrenched species is the goal. Sound management decisions rely on detailed information on the invasive population, the type and degree of impacts, and the strategic, science-based application of control.

Recently, the management of invasive vertebrate species has shifted its emphasis from reducing numbers to mitigating the damage they cause (Hone, 2007). Increasingly, invasive species management is conducted to reduce impacts on native biodiversity. Invasive species may have many impacts on their new environment, affecting individual species or populations, through hybridisation, behavioural and morphological shifts, as well as changes in density and distribution resulting from competitive exclusion (Falk-Petersen *et al.*, 2006). Invasive species may also impact

Austral Ark: The State of Wildlife in Australia and New Zealand, eds. A. Stow, N. Maclean and G. I. Holwell. Published by Cambridge University Press. © Cambridge University Press 2015.

ecosystems by reducing species diversity, changing environmental conditions and altering ecological processes (Chapin III *et al.*, 2000). They are also a threat to island ecosystems including New Zealand, where species have evolved in isolation from certain taxa (e.g. mammalian predators and herbivores), leaving them defenceless against such invaders (Lee *et al.*, 2006).

Here we discuss the introduction history, ecological impacts and management of the predominant vertebrate invaders in Australia and New Zealand. Invasive species that have established in Australia and/or New Zealand and are currently considered a threat are listed in Table 10.1, along with their associated impacts. Note that there are numerous introduced species that are not considered to be invasive. For some this is because their impacts on biodiversity have not yet been determined, consequently the threat they pose is unknown. These species are not discussed here. Instead we focus on those species whose negative effects on native biodiversity and ecosystems have been documented, or whose impacts are predicted to be ecologically significant, particularly mammals (see also Box 10.1.) The impacts of invasive fish, amphibians and reptiles are touched on only briefly; however, a full chapter of this volume is dedicated to the cane toad problem in Australia.

10.2 Ungulates

The majority of feral ungulates arrived with the early settlers and were spread throughout Australia and New Zealand by unrestrained stock escapes (sheep *Ovis aries*, cattle *Bos taurus*, goats *Capra hircus*, and pigs *Sus scrofa*) (King, 2005; Choquenot *et al.*, 1996; Parkes *et al.*, 1996; Bomford & Hart, 2002; Page *et al.*, 2009). However, some species were introduced as modes of transport (camels *Camelus dromedarius*, donkeys *Equus asinus*, and horses *Equus caballus*), and as transportation methods improved, liberations occurred in large numbers (Saalfield & Edwards, 2010; Sharp & Saunders, 2012). Other species were deliberately introduced for recreational hunting purposes (e.g. deer Cervidae) (McKnight, 1976; King, 2005). Vegetation in New Zealand evolved in the absence of ungulates and is therefore highly vulnerable to their methods of browsing. Large herbivorous mammals were present in Australia until their extinction in the Pleistocene; however, feral ungulates reach population densities that exceed those in their native range, which may intensify the impacts on vegetation (Freeland, 1990).

Sheep and cattle have established feral populations in both Australia and New Zealand (King, 2005; Bomford & Hart, 2002). As grazers, feral sheep have little impact on mainland New Zealand; feral cattle, in contrast, severely modify vegetation by browsing and trampling (King, 2005). In Australia the main impacts occur when feral cattle encroach into agricultural landscapes (Bomford & Hart, 2002). Feral ungulates in Australia (including buffalo *Bubalus bubalis*, donkey, horse, and camel) foul watering holes, destroy fences, reduce the productivity of livestock by denuding pasture and increasing erosion, and impact native flora and fauna (Dobbie *et al.*, 1993; Skeat *et al.*, 1996; Bomford & Hart, 2002; Edwards *et al.*, 2010). However, the potential for these species to carry disease to livestock is a shared impact for both Australia and New Zealand (King, 2005; Bomford & Hart, 2002). Whilst Australia is free of bovine tuberculosis or TB (*Mycobacterium bovis*), feral ungulate populations in New Zealand act as a reservoir for this disease (Tweddle & Livingston, 1994).

Table 10.1 The primary invasive vertebrates of Australia and New Zealand, and their main impacts in each country.

Present in the wild				Impacts on native ecosystems							Impacts on modified ecosystems	
Species	Common name	Australia	New Zealand	Herbivory	Predation	Competition	Disease	Hybridisation	Habitat change	Toxicity	Human health	Industry
	Ungulates											
Capra hircus	Feral goat	✓	✓	BOTH		AUS			BOTH			BOTH
Sus scrofa	Feral pig	✓	✓	BOTH	BOTH	AUS	BOTH		BOTH		BOTH	BOTH
Equus caballus	Feral horse	✓	✓	BOTH		AUS			BOTH			BOTH
Equus asinus	Feral donkey	✓		AUS		AUS			AUS			AUS
Bubalus bubalis	Feral buffalo	✓		AUS		AUS			AUS			AUS
Camelus dromedarius	Feral camel	✓		AUS		AUS						AUS
Bos taurus	Feral cattle	✓	✓	BOTH								BOTH
Ovis aries	Feral sheep	✓	✓	BOTH								BOTH
Axis axis	Axis deer	✓		AUS		AUS			AUS			AUS
Cervus porcinus	Hog deer	✓		AUS		AUS			AUS			AUS
Cervus elaphus scoticus	Red deer	✓	✓	BOTH		AUS			BOTH			BOTH
Cervus unicolor	Sambar deer	✓	✓	BOTH		AUS			BOTH			BOTH
Cervus timorensis	Rusa deer	✓	✓	BOTH		AUS			BOTH			BOTH
Cervus nippon	Sika deer		✓	NZ					NZ			NZ
Dama dama	Fallow deer	✓	✓	BOTH		AUS			BOTH			BOTH
Cervus elaphus nelsoni	Wapiti deer	✓	✓	BOTH					BOTH			BOTH
Odocoileus virginianus	White-tailed deer		✓	NZ					NZ			NZ
Rupicapra rupicapra	Chamois		✓	NZ					NZ			NZ
Hemitragus jemlahicus	Himalayan tahr		✓	NZ					NZ			NZ

Table 10.1 (cont.)

Present in the wild				Impacts on native ecosystems							Impacts on modified ecosystems	
Species	Common name	Australia	New Zealand	Herbivory	Predation	Competition	Disease	Hybridisation	Habitat change	Toxicity	Human health	Industry
Rodents												
Rattus rattus	Ship rat/ black rat	✓	✓	BOTH	BOTH	BOTH					BOTH	BOTH
Rattus norvegicus	Norway rat/ brown rat	✓	✓	BOTH	BOTH	BOTH					BOTH	BOTH
Rattus exulans	Kiore/Polynesian rat		✓	NZ	NZ	NZ					NZ	NZ
Mus musculus	House mouse	✓	✓	BOTH	BOTH	BOTH					BOTH	BOTH
Lagomorphs												
Lepus europaeus	European brown hare	✓	✓	BOTH		AUS			BOTH			BOTH
Oryctolagus cuniculus	European rabbit	✓	✓	BOTH	BOTH	AUS			BOTH			BOTH
Erinaceomorphs												
Erinaceus europaeus occidentalis	European hedgehog		✓	NZ	NZ	NZ						
Marsupials												
Macropus eugenii*	Tammar (dama) wallaby	✓	✓	NZ					NZ			
Macropus rufogriseus rufogriseus*	Bennetts wallaby	✓	✓	NZ					NZ			
Trichosurus vulpecula*	Common brushtail possum	✓	✓	NZ	NZ		NZ		NZ		NZ	NZ
Carnivores												
Vulpes vulpes	European red fox	✓			AUS	AUS					AUS	AUS
Canis lupus ssp.	Wild dog	✓			AUS						AUS	AUS
Felis catus	Feral cat	✓	✓		BOTH	AUS	AUS				BOTH	BOTH
Mustelids												
Mustela erminea	Stoat		✓		NZ							NZ
Mustela nivalis vulgaris	Weasel		✓		NZ							NZ
Mustela furo	Ferret		✓		NZ							NZ

Birds

Scientific name	Common name						
Branta canadensis	Canada goose	✓					NZ
Anas platyrhynchos	Mallard	✓		NZ	NZ	BOTH	
Columba livia	Rock pigeon	✓		BOTH	BOTH		BOTH
Streptopelia chinensis	Spotted dove	✓		BOTH	BOTH		
Cacatua galerita*	Sulphur-crested cockatoo	✓		NZ	NZ		
Trichoglossus haematodus*	Rainbow lorikeet	✓		NZ	NZ		
Platycercus eximius*	Eastern rosella	✓		NZ	NZ		NZ
Gymnorhina tibicen**	Australian magpie	✓	NZ	NZ	NZ		NZ
Corvus frugilegus	Rook	✓					NZ
Pycnonotus jocosus	Red-whiskered bulbul	✓		AUS	AUS		AUS
Turdus merula	Blackbird	✓					BOTH
Sturnus vulgaris	Common starling	✓		BOTH	BOTH	BOTH	BOTH
Acridotheres tristis	Common/Indian myna	✓	BOTH	BOTH	BOTH		BOTH
Passer domesticus	House sparrow	✓		BOTH	BOTH		BOTH
Lonchura punctulata	Nutmeg manikin	✓		AUS	AUS		

Fish

Scientific name	Common name						
Tilapia mariae and *Oreochromis mossambicus*	Tilapia	✓	AUS	AUS	AUS	AUS	AUS
Misgurnus anguillicaudatus	Oriental weatherloach	✓		AUS	AUS	AUS	
Gambusia spp.	Gambusia	✓		BOTH	BOTH	BOTH	
Cyprinus carpio	Koi carp	✓	BOTH	BOTH	BOTH	BOTH	BOTH
Salmo trutta	Brown trout	✓		BOTH	BOTH		
Oncorhynchus mykiss	Rainbow trout	✓		BOTH	BOTH		

Table 10.1 (cont.)

Present in the wild			Impacts on native ecosystems						Impacts on modified ecosystems		
Species	Australia	New Zealand	Herbivory	Predation	Competition	Disease	Hybridisation	Habitat change	Toxicity	Human health	Industry
Perca fluviatilis	✓	✓		BOTH	BOTH						
Carassius auratus	✓	✓	BOTH			BOTH		BOTH			
Scardinius erythrophthalmus		✓	NZ		NZ					NZ	
Ameiurus nebulosus		✓		NZ	NZ						
Reptiles											
Trachemys scripta elegans	✓				AUS	AUS				AUS	
Hemidactylus frenatus	✓				AUS	AUS					
Lampropholis delicata		✓			NZ						
Amphibians											
Rhinella marina	✓			AUS					AUS	AUS	

* Native to Australia but invasive in New Zealand.

** Native to Australia and considered a pest, invasive in New Zealand.

Box 10.1 Eradication success – restoring Macquarie Island

Macquarie Island (128 km^2) is a biodiversity hotspot and the breeding ground for many species of marine mammals and ground-nesting seabirds. In the 1800s cats (*Felis catus*), black rats (*Rattus rattus*), rabbits (*Oryctolagus cuniculus*) and house mice (*Mus musculus*) were introduced to the island by sealers and explorers. These species established populations rapidly and seriously impacted the biodiversity of the island. In 1933 Macquarie Island was recognised as a wildlife sanctuary and in 1997 it was listed as a World Heritage Area.

   The rabbit population was estimated at 150 000 in 1978 and concerns about the impacts of their grazing prompted the release of the myxoma virus, which reduced the population by 90% in the following years. Cats were eradicated from the Island in June 2000. Subsequently, the rabbit population began to expand again and the pressure of rabbit grazing resulted in a reduction of vegetation cover, erosion and landslips on the steep coastal slopes. In 2007 the Tasmanian Parks and Wildlife Service (responsible for management of the island) and the Australian Government funded a $24.7 million (AUD) eradication programme for rabbits, rats and mice. The eradication was planned in two stages; aerial baiting followed by intensive hunting with trained dogs as a remnant population of rabbit survivors was expected; therefore, it was considered that the hunting effort to find and eradicate the last rabbits would be critical to the project's success. In the months before the eradication programme, rabbit haemorrhagic disease virus (RHDV) was released on the island to provide an initial knockdown of rabbit numbers and ensure sufficient bait was available for rats and mice. The first phase was undertaken in April 2011. In two bait drops, a total of 305 tonnes of Pestoff 20R bait containing brodifacoum was applied to the island. The second phase of the operation involved hunting teams and dogs continuing the eradication of the rabbits. A total of 13 rabbits were removed following the baiting. No sign of rats or mice was observed after the bait drop and this was confirmed in 2013 by the deployment of rodent-detecting dogs. The total eradication of rabbits, rats and mice was officially declared a success in April 2014.

   Seabird breeding success has now increased on the island and Macquarie's unique vegetation has been released from the grazing pressure of rabbits. However, the island ecosystem had suffered serious damage as a result of decades of intense grazing pressure and it may take 20 years for the island to recover. Macquarie Island is the largest multi-pest eradication undertaken to date.

Adapted from Wren (2012).

   In Australia, feral goats cause huge economic losses for agriculturalists through pasture loss and competition with grazing livestock. They also have many negative impacts on natural ecosystems including soil erosion, herbivory, damage to native vegetation (debarking trees) and also competition with native fauna (e.g. rock-wallabies *Petrogale* spp.) (Parkes *et al.*, 1996). However, their damage is most severe on islands. Feral goats on Kangaroo Island overgrazed ground plants and depleted the food source of the endangered glossy black cockatoo (*Calyptorhynchus lathami*). A successful eradication programme was conducted on Kangaroo Island using Judas goats (sterilised goats fitted with GPS collars) to locate the herds, which were then dispatched by ground and aerial shooting (IACRC, 2013). Feral goats have similar impacts in New Zealand where the negative effect of their browsing on native vegetation is especially evident on offshore islands. On Raoul Island goats removed all palatable seedlings, particulary those of the coastal canopy species *Metrosideros exelsa* (King, 2005). The eradication of goats on this island has allowed the recovery of endangered insular endemic plants (Parkes, 1990).

A number of deer species have established in New Zealand and Australia (see Table 10.1). Their browsing has caused a successional trend towards browse-tolerant species, and structural and compositional changes in forest vegetation (King, 2005). In New Zealand, deer also act as a reservoir for bovine tuberculosis which is a concern for the farming industry (Nugent *et al.*, 2001). Deer in New Zealand reached peak numbers in the 1920s and, since then, widespread control and recreational hunting has reduced most deer populations to below carrying capacity. It is also important to note that the negative impacts of deer may be increased in the presence of other invasive ungulates and common brushtail possums (*Trichosurus vulpecula*). Therefore single species deer control in New Zealand may not reverse deer impacts (Coomes *et al.*, 2003). Deer are an emerging issue in Australia and knowledge about their impacts is limited. Control operations are also minimal; consequently populations continue to increase in most areas (Dolman & Wäber, 2008).

The distribution of feral camels (*Camelus dromedarius*) covers much of the rangelands in Western Australia, South Australia, the Northern Territory and far Western Queensland (Saalfield & Edwards, 2010). Camels cause damage to pastoral infrastructure, including fences, yards and water equipment. Camels are thought to drive the local extinction of highly palatable plant species and in arid areas even low camel densities can cause severe vegetation damage. Camels frequent wetlands in arid areas, dominating the water supply, often depleting it. They may also become stuck and die in boggy areas, leading to pollution and eutrophication. This has severe impacts on native species that rely on these areas as refugia in the arid landscape. Aboriginal landholders have raised concerns about the impacts of camels on wetlands in regards to the declining abundance of bushmeat species, such as red kangaroos (*Macropus rufus*), emu (*Dromaius novaehollandiae*) and bustards (*Ardeotis australis*) via competition for grass and water. Many wetlands are culturally significant sites for desert Aboriginal people and therefore the negative impacts of camels on wetlands also leads to cultural and social impacts (Edwards *et al.*, 2010).

Feral pigs (*Sus scrofa*) are widely distributed in Australia and New Zealand and their direct impacts mainly involve herbivory and predation. Pig rooting and disturbance also have indirect effects on biodiversity and plant assemblages in Australia and New Zealand (Choquenot *et al.*, 1996; Krull *et al.*, 2013). There is growing evidence that feral pigs aid the dispersal of *Phytophthora*, responsible for dieback in native vegetation in Australia (*Phytophthora cinnamomi*) and New Zealand (*Phytophthora* 'taxon *Agathis*') (Li *et al.*, 2010; Krull *et al.*, 2012) (see Box 10.2). The requirement for protein, especially for breeding and growth, may be what drives predation of animals by feral pigs. In New Zealand feral pigs are known predators of birds, particularly terrestrial breeding colonies of seabirds, and of *Powelliphanta*, indigenous, giant land snails. In Australia feral pigs are a major agricultural problem due to their predation of lambs (Choquenot *et al.*, 1997). Preferred control methods for feral pigs are highly variable and dependent on habitat structure. Helicopter shooting is particularly effective in more open Australian rangelands, though where vegetation cover

Box 10.2 Feral pigs as vectors of soil-borne plant pathogens

Invasive vertebrate species (particularly ungulates) are a serious threat to forest eco-systems in Australia and New Zealand through herbivory. Many forests are also being attacked by microscopic soil-borne invaders. Assisted movement of soil-borne disease is required for rapid spread over large distances as natural spread is slower and localised (Litchman, 2010). Therefore, invasive vertebrates may cause further damage to forest ecosystems by acting as vectors for soil-borne disease. The likelihood of pathogen establishment can be increased by a vector in a number of ways (Figure 10.1). It has become increasingly apparent that feral pigs are a significant vector of soil-borne plant pathogens. Pigs may carry large amounts of soil on their bodies, which can contain thousands of microscopic pathogenic propagules. Pigs disturb the ground whilst digging for food and may then deposit these propagules in areas close to plant root zones; they may also physically damage these roots, assisting pathogen infection. Pigs may disturb a number of different root zones in a short space of time and can travel large distances quickly. Research in New Zealand found that feral pigs are vectors of 19 species of plant pathogen in soil adhered to their snouts and trotters (Krull *et al.*, 2012). Feral pigs were also found to spread *Phytophthora cinnamomi*, which is the causal agent of jarrah dieback in Australia and are a likely vector of a *Phytophthora* strain that causes kauri dieback in New Zealand (Krull *et al.*, 2012). Feral pigs may also carry soil-borne diseases in their faeces, and can pass viable spores of *Phytophthora cinnamomi* through their gut for up to 7 days after ingestion (Li *et al.*, 2010). The combined threat to forests from herbivory and soil-borne pathogens could lead to intensified negative impacts on forest ecosystems.

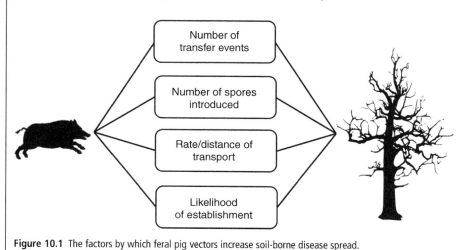

Figure 10.1 The factors by which feral pig vectors increase soil-borne disease spread.

becomes taller and denser this method's efficiency declines (Choquenot *et al.*, 1999), and other methods must be employed. Trapping has been used in Australia, but its effectiveness against some sections of a pig population can be limited as males are more difficult to trap (Choquenot *et al.*, 1993) and establishing traps in rugged terrain with little or no vehicle access is expensive. Ground-based hunting with dogs is the most common approach to pig control in New Zealand, as no toxin is currently registered for use against feral pigs.

10.3 Rodents

A similar suite of invasive rodents is present in Australia and New Zealand, consisting of ship rats (also known as black rats), Norway rats (*Rattus norvegicus*), and house mice. All were introduced as stowaways on European sealing, whaling and exploration ships in the late eighteenth and early nineteenth centuries (King, 2005). In New Zealand, a fourth species is present; the kiore or Pacific rat (*Rattus exulans*) was introduced with early Polynesian explorers and retains cultural importance with some Māori iwi today (King, 2005).

Rats are omnivorous predators impacting on native ecosystems through consumption of seeds, fruit, invertebrates, small reptiles, eggs, birds and small mammals. Impacts on biodiversity in New Zealand, where native biota evolved in the absence of natural mammalian predators, have generally been more obvious and widespread than those experienced in Australia and evidence linking ship rats to native species declines or extinctions on the Australian mainland is largely circumstantial (Banks & Hughes, 2012). However, Australian islands have experienced more evident impacts. Ship rats are thought to have caused the local extinction of five bird species on Lord Howe Island between 1919 and 1938 (Hindwood, 1940). The role of rats as competitors of native mammals is also of concern in Australia (Morris, 2002).

In New Zealand, rat impacts have been keenly felt, and devastating effects on native biota are common. Ship rat invasion of Taukihepa (Big South Cape) Island in 1962 directly caused the extinction of the New Zealand bush wren (*Xenicus longipes*) as well as local extinctions of five bird, one bat and one invertebrate species (Towns *et al.*, 2006). The greater short-tailed bat (*Mystacina robusta*), one of only three land mammals native to New Zealand, was probably also driven to extinction by rats during the Taukihepa invasion (Towns *et al.*, 2006).

The impacts of house mice, also omnivorous and opportunistic feeders, have often been overlooked or underestimated in conservation management and there is a general paucity of knowledge of mouse impacts compared to other invasive species (Simberloff, 2009). Mice are often present as part of a suite of introduced mammals in an ecosystem where their populations may be suppressed by predation or competition with other invasive species (Caut *et al.*, 2007). However, where mice are the only introduced mammal species present, impacts are more severe and more readily observed (Angel *et al.*, 2009). Invertebrates are particularly vulnerable to mouse predation. For instance, on Mana Island (New Zealand) abundance indices of Cook's Strait giant weta (*Deinacrida rugosa*) increased more than tenfold after mice were eradicated (Newman, 1994). Mice also indirectly impact New Zealand bird species by supporting high populations of predators, such as stoats (*Mustela erminea*). See also Box 10.3.

Mice also cause economic impacts as a major agricultural pest. In Australia, periodic 'mouse plagues' can result in densities of more than 800 mice per hectare. Farmers are affected through substantially reduced yields of grain crops as well as damage to equipment and property (Bomford & Hart, 2002).

While all rodent species are able to swim to some extent, Norway rats distinguish themselves in this regard. They are capable of swimming distances of 1–2 km, depending on conditions, so any offshore islands within this distance of a source population are at risk of invasion (Russell *et al.*, 2008). This highlights the importance of monitoring pest-free areas to ensure timely detection of new incursions, allowing contingency responses to take place before a new population is firmly established.

Box 10.3 Hyperpredation of mohua in New Zealand beech forest

Direct interactions between native and introduced species are the most easily under-stood and the best-studied impacts of invasive species. However, systems with only one introduced species are atypical and the combined impacts of a group of invaders may be greater than the summed direct impacts of the individual invasive species (Simberloff & Von Holle, 1999). Hyperpredation is the process by which an intro-duced prey species indirectly increases predation pressure on a native prey species by supporting increased numbers of an introduced predator, potentially leading to extinction of the native prey species (Smith & Quin, 1996).

One well-described example of hyperpredation occurs in New Zealand southern beech (*Fuscospora* spp. and *Lophozonia* spp.) forests. Here, periodic heavy beech seed production (known as 'masting') prompts increased reproduction of introduced house mice and an unusually rapid and steep increase in mouse population size (Ruscoe *et al.*, 2005). The abundance of stoats, an introduced predator, soon increases in response to the burgeoning mouse population (King *et al.*, 2003). The endangered endemic mohua (*Mohua ochrocephala*) is a hole-nesting bird, rendering adult females as well as chicks and eggs vulnerable to predation by stoats. Predation pressure on mohua is greatly increased when stoat populations grow following a mouse population irruption (O'Donnell & Phillipson, 1996). The number of breeding pairs of mohua was reduced by 75% at one site following stoat population irruptions (Elliott & O'Donnell, 1988) and mohua fledging success was halved (O'Donnell & Phillipson, 1996). Over time and space the effects of such intense predation pressure can be dramatic. Stoat predation is thought to be the primary cause for a 75% reduction in the geographical range of mohua (O'Donnell, 1996) and models predict that, without stoat control, any given population of mohua would have a 75% probability of extinction within 120 years of the introduction of stoats (Choquenot, 2006). Fortunately, effective control of stoats following population irruptions can drastically reduce impacts on mohua populations (O'Donnell & Phillipson, 1996). Furthermore, recently improved ability to predict masting events (Kelly *et al.*, 2013) may allow pre-emptive rodent control in mast years to prevent mouse outbreaks.

10.4 Lagomorphs

Two lagomorph species have been introduced to Australia and New Zealand: the European rabbit *(Oryctolagus cuniculus)* and the brown hare *(Lepus europaeus)*. Hares naturally occur at lower densities than rabbits, do not dig burrows, and tend to graze plants lightly without killing them (King, 2005). Overall, the impacts of hares on natural and productive environments are substantially less severe than those of rabbits (Flux *et al.*, 1990). For this reason, discussion in this section focusses on rabbits.

In New Zealand and Australia domesticated forms of the European rabbit were among the first animals imported by Europeans in the 1770s and wild rabbits were introduced to both countries from the mid 1800s for sport, the fur trade and by acclimatisation societies (Williams *et al.*, 1995; King, 2005). Rabbits are now a serious agricultural and ecological pest in both countries.

In the production sector, rabbits cause economic losses by grazing emerging crops, reducing yields, and contributing to soil degradation by scratching and burrowing (Williams *et al.*, 1995). They also compete with livestock for food by grazing pasture, which may reduce the number of livestock that can be supported on a property.

Impacts on native biodiversity have also been considerable. Rabbits directly impact native plant species through herbivory, inhibiting regeneration of palatable trees, shrubs and grasses (Edwards *et al.*, 2004). It is suspected that in some Australian grasslands this behaviour has led to permanent changes in vegetation composition, with only those species able to withstand intensive rabbit grazing surviving (Leigh *et al.*, 1989). Native animals are directly impacted through competition for food and shelter; most Australian mammals that have become extinct or limited in range have been small to medium-sized herbivores in the arid zone, where rabbits are the dominant competitor during periods of drought (Williams *et al.*, 1995). Indirect impacts on native prey species also occur, as high densities of rabbits support non-native predators, which prey upon native species in addition to rabbits (Williams *et al.*, 1995).

Biological control has featured heavily in attempts to control the rabbit populations in both Australia and New Zealand. Early attempts to introduce mustelids to New Zealand as natural predators both failed to control the rabbit population and produced catastrophic conservation outcomes (King, 2005). The introduction of rabbit pathogens began in Australia with the release of the myxoma virus (the causal agent of the disease myxomatosis) in 1950. Initially, the virus was highly successful, spreading at an average rate of 5 km per day and dramatically reducing Australia's rabbit population (Williams *et al.*, 1995). However, co-evolution between virus and rabbits has led to both reduced virulence of the virus and resistance in rabbit populations. Consequently, while the myxoma virus persists in Australia today, many rabbits survive infection (CSIRO, 2011). In New Zealand, attempts were also made to introduce the myxoma virus in the 1950s but these failed due to a lack of a suitable disease vector (King, 2005).

A second viral pathogen, rabbit haemorrhagic disease virus (RHDV, also known as rabbit calicivirus) escaped from a quarantine facility in Australia in 1995 and was introduced illegally to New Zealand in 1997. In both countries, the efficacy of the disease has been somewhat variable by region (Story *et al.*, 2004; Parkes *et al.*, 2002) but overall has been successful in substantially reducing rabbit populations. There are, however, signs of disease resistance developing in rabbit populations in both Australia and New Zealand, calling into question the long-term efficacy of RHDV as a control agent (Elsworth *et al.*, 2012; Parkes *et al.*, 2008).

10.5 Hedgehogs

European hedgehogs (*Erinaceus europaeus occidentalis*) were first imported to New Zealand in the 1870s by acclimatisation societies to remind British settlers of their homeland (King, 2005). Since then, the species has colonised much of the country through natural spread and deliberate introductions by humans. The hedgehog's reputation as a predator of garden pests has no doubt encouraged these deliberate liberations. Hedgehogs are absent from Australia.

Relative to other invasive mammals, the importance of hedgehog impacts on the New Zealand environment has only recently been recognised. Hedgehogs are predominantly insectivorous but also opportunistically depredate lizards and ground-nesting birds (Jones *et al.*, 2013; Sanders & Maloney, 2002). Hedgehogs have a tendency to focus foraging effort on locally abundant prey, as evidenced by large numbers of particular prey items in gut content analyses. This implies that hedgehog effects on small, localised prey populations could be severe (Jones *et al.*, 2013). Furthermore, concerns have been raised that this extensive local foraging on invertebrates may seriously impact native insectivores, such as kiwi, through competition for food. Hedgehogs are increasingly being targeted for control by trapping and/or poison as their impacts become better known. Lethal doses for poisoning are currently unknown and there is evidence that at least some individuals display high tolerance to the commonly used anti-coagulant brodifacoum (Berry, 1999).

10.6 Possums

Brushtail possums *(Trichosurus vulpecula)* are of major ecological concern in New Zealand, due to herbivory and predation. They were introduced from their native Australia in 1837 in the attempt to develop a fur trade (King, 2005).

Possums are versatile feeders and can survive on poor and irregular food supplies, enabling them to establish and spread to 90% of the country (Ritchie, 2000). In Australia most plants have evolved physical or chemical defences to limit the amount of browsing (McKnight, 1976). However, New Zealand species have few defences and possums can consume large amounts of palatable species (e.g. *Metrosideros* spp., *Fuchsia* spp., and mistletoe spp.) (Cowan & Wadington, 1990). Possum browse leads to a change in the structure and composition of native forests. Catastrophic dieback of most, and in extreme cases all, of the canopy can occur in areas dominated by a few highly palatable tree species (*Metrosideros* spp. and *Weinmannia racemosa* forests) (Batcheler & Cowan, 1988). In more diverse forest communities depletion of palatable species may occur over decades and result in a change in forest composition (Campbell, 1990). Regeneration of possum-preferred plant species is vastly decreased in the presence of ungulates (which target these palatable species as seedlings and saplings) and therefore the additive effect of possum and ungulate browse can be devastating to forest composition. Possums are also large consumers of fruit and flowers. There is evidence to suggest that possums function as both seed dispersers and seed predators (King, 2005). However, New Zealand plant species are adapted to bird dispersal and the longer gut passage in possum consumption may render most seeds non-viable.

Possums are also known to kill the eggs, chicks and adults of a number of native bird species and also prey upon invertebrates, specifically large-bodied, slow-moving, nocturnal species such as *Powelliphanta* snails (Cowan, 2001). Competition with native fauna may occur when possums deplete certain plant species and dominate food resources. Possum control is conducted widely in New Zealand. Over half of all control operations are conducted by the Animal Health Board (AHB) as possums consume agricultural crops and also carry and spread bovine tuberculosis (Ritchie, 2000). Successful possum control is carried out through the aerial application of poison baits (mainly 1080), which can achieve a rapid reduction in possum numbers (Cowan, 2001). However, public aversion to 1080 aerial drops is common and often leads to the need to seek other methods of control such as ground poisoning (brodifacoum, cynanide or 1080) and trapping. Possum-specific, resettable, multi-kill devices are currently being developed to increase trapping efficiency (Blackie *et al.*, 2011).

10.7 Red foxes

The European red fox (*Vulpes vulpes*) was introduced to Australia in the 1870s and spread rapidly, covering most of southern Australia within 50–60 years (Abbott *et al.*, 2014). Foxes were absent from Tasmania until around the turn of the twenty-first century but have since been detected across much of the island.

Foxes in Australia eat a wide variety of mammals, birds, reptiles and invertebrates, as well as plant material and human refuse. This generalist diet not only allows foxes to persist in most Australian environments, but also makes them a serious threat to livestock and native biodiversity. Foxes have been implicated in the declines and extinctions of many native animals, and fox control has led to the dramatic recovery of some species such as black-footed rock-wallabies (*Petrogale lateralis*), long-nosed potoroos (*Potorous tridactylus*) and southern brown bandicoots (*Isoodon obesulus*). In addition to predation, foxes can also affect native fauna through competition (see Box 10.4) and spread of disease, including round-worm, scabies and hydatid tapeworms (Saunders *et al.*, 2010).

Foxes are also predators of livestock such as lambs and poultry. However, their agricultural impacts are difficult to measure as an unknown proportion of the domestic animals eaten by foxes are already dead or dying of other causes. The most widespread and

---

Box 10.4 Impacts of foxes on other carnivores

The devastating effects of fox predation on native Australian prey are well documented. However, in recent years it has become increasingly clear that foxes also have impacts on other predators. These impacts occur through competition for resources as well as direct aggression by foxes towards smaller predators (Glen & Dickman, 2005) (Figure 10.2).

Predator populations are often strongly influenced by the availability of prey, which in turn can be affected by other predators competing for the same resources. Foxes have very similar diets to many other predators in Australia. For example, rabbits are a staple prey for foxes in many areas, as well as for feral cats and native predators such as quolls (*Dasyurus* spp.), goannas (*Varanus* spp.) and birds of prey (Glen, 2014). Such similarity in diet often leads to competition for limited numbers of prey. Competition could also occur for other resources such as den sites (Glen & Dickman, 2011).

Perhaps more important than competition for resources is direct aggression by foxes towards smaller predators. Aggressive competition is common among carnivores, and as a result, smaller predators often suffer serious harm or death (Palomares & Caro, 1999). In Australia, foxes are known or suspected to kill quolls, goannas and cats. This may not only cause increased mortality, but can also force smaller predators to alter their behaviour to avoid direct encounters with foxes. For example, lace monitors (*Varanus varius*) modify their foraging behaviour in the presence of foxes, causing them to have lower body condition compared to areas where fox numbers are suppressed (Anson *et al.*, 2013).

There is also evidence of negative relationships in abundance between foxes and smaller predators. For example, fox baiting at Shark Bay in Western Australia was followed by a three-fold increase in spotlight counts of cats (Risbey *et al.*, 2000), while fox control in southwest Western Australia has led to a dramatic recovery of western quoll populations (Morris *et al.*, 2003). In New South Wales, it has been observed that spotted-tailed quolls are most abundant where foxes are scarce or absent (Glen & Dickman, 2011). As illustrated by these examples, the introduction of foxes to Australia has had far-reaching consequences not only for prey species, but also for other predators.

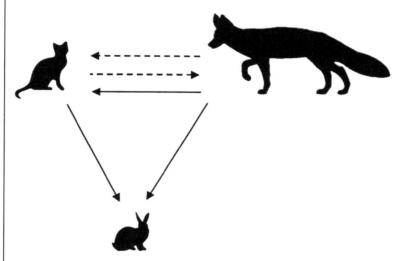

Figure 10.2 Diagram representing possible interactions between foxes (*Vulpes vulpes*), cats (*Felis catus*) and rabbits (*Oryctolagus cuniculus*). Solid arrows indicate that one species kills the other; dashed arrows show competition for shared resources.

effective method of controlling foxes in Australia is the use of toxic meat baits, however various other methods such as shooting, trapping and fencing are also used in some areas (Saunders *et al.*, 2010).

10.8 Wild dogs

In Australia the term 'wild dog' is used to refer to dingoes (*Canis lupus dingo*), feral dogs (*C. l. familiaris*) and hybrids of the two. Dingoes arrived in Australia 3500–5000 years ago and are thought to have been introduced by people from southeast Asia (Corbett, 1995). They may have contributed to the extinction from the Australian mainland of the thylacine (*Thylacinus cynocephalus*), Tasmanian devil (*Sarcophilus harrisii*) and Tasmanian native hen (*Gallinula mortierii*). Domestic dogs *(Canis lupus familiaris)* were brought to Australia by European settlers, and have subsequently hybridised with dingoes (Corbett, 1995).

Wild dogs are considered serious pests of the grazing industry, in particular because of their impacts on sheep, which are sometimes killed or mauled in large numbers. Wild dogs are widely controlled in pastoral areas through poisoning, fencing, shooting and trapping (Fleming *et al.*, 2001). Non-lethal methods to protect livestock from attack by wild dogs are also being tested. Guardian animals, including dogs bred specifically for livestock protection, show considerable promise as a means to prevent wild dog attacks (van Bommel & Johnson, 2012).

The impacts of wild dogs on biodiversity are the subject of debate. There is strong evidence that wild dogs can suppress numbers of foxes and feral cats, as well as large herbivores such as kangaroos and feral pigs. This may have indirect benefits for native plants and animals threatened by predation and overgrazing (Letnic *et al.*, 2012). However, wild dogs could also pose a threat to some native prey populations, especially those reduced to low numbers by human influences such as habitat alteration (Fleming *et al.*, 2012).

10.9 Cats

Domestic cats *(Felis catus)* are thought to have arrived in Australia with early European settlers and established wild populations throughout the country by the 1890s (Abbott *et al.*, 2014). Cats reached New Zealand with European explorers late in the eighteenth century, and had established wild populations across much of the country by the 1850s (King, 2005).

Cats are strictly carnivorous, consuming invertebrates, birds, reptiles and mammals weighing up to 2 kg, but mostly less than 200 g. Amphibians are also eaten occasionally, and human refuse is an important source of food for some cats. Cats have been implicated in widespread faunal declines on the Australian mainland. However, their clearest impacts have been on islands, where they have been a leading cause of mammal extinctions in Australia and avian extinctions in New Zealand (King, 2005; Van Dyck & Strahan, 2008).

Unlike dogs and foxes, cats have a low propensity to consume meat baits, generally preferring live prey. Widespread control of cats is therefore much more difficult, although baiting methods are being developed and tested for this purpose (e.g. Hetherington *et al.*, 2007). On a more localised scale, feral cats can be controlled through trapping and shooting, or with predator-proof fencing. Cats have been eradicated from a number of islands in Australia and New Zealand (Campbell *et al.*, 2011).

10.10 Mustelids

Three mustelid species, native to Europe, are considered to be conservation pests in New Zealand. These are stoats, ferrets *(Mustela furo)* and weasels *(M. nivalis)*. Stoats and weasels are absent from Australia; however, since 1998 it has been legal to keep ferrets as pets in all states except Queensland and Northern Territory. One isolated feral ferret population has been recorded in Tasmania (Wilson *et al.*, 1992) but, as yet, ferrets are not regarded as a major pest species in Australia.

Mustelids were introduced to New Zealand from the late 1870s as a biological control agent for rabbits. They proved to be ineffective at controlling rabbit populations but have had devastating impacts on native ecosystems; depredating native birds, inverte-brates, and reptiles. Stoats, in particular, have been instrumental in declines of native bird species. In one study, pre-adult mortality of kiwi (*Apteryx* spp.) was estimated at 94%. Predators, primarily stoats, were responsible for around half of these deaths (McLennan *et al.*, 1996). Stoat predation has also led to dramatically skewed sex ratios through disproportionate predation on nesting female kaka (*Nestor meridionalis*) (Wilson *et al.*, 1998), and mohua (Elliott, 1996), exacerbating population declines.

Ferrets and weasels have similar feeding habits to stoats; their diets being made up mostly of small mammals, birds, lizards, invertebrates and sometimes eggs (King, 2005). Weasels tend to eat smaller prey items than stoats and, being much less common, have a smaller impact on New Zealand ecosystems overall (King, 2005). However, they are still considered a significant threat to localised populations. Ferrets are an important predator of many native bird species and also pose an economic threat as in high population densities they may act as a maintenance host of bovine tuberculosis (Caley & Hone, 2005).

Island sanctuaries cleared of mustelids and other predators have proved to be an invaluable conservation resource in New Zealand. However, the risk that island sanctua-ries will be reinvaded is constant. In 2009, a multi-pest eradication was undertaken on Rangitoto and Motutapu islands in Auckland's Hauraki Gulf but in 2010, a single male stoat was caught in a trap. Genetic population assignment techniques showed that the stoat was not a survivor of the eradication but a re-invader, which probably swam at least 3 km from the mainland; a distance previously thought to be outside the swimming range of stoats (Veale *et al.*, 2012). This highlights the importance of monitoring pest-free sanctuary islands to detect new incursions.

10.11 Birds

Over the past 160 years a multitude of exotic bird species have been introduced to both Australia and New Zealand – at least 55 and 137 species respectively (Thomson, 1922; Veltman *et al.*, 1996; Duncan *et al.*, 2001). Approximately 36% of those introduced to Australia survived to establish wild populations (Duncan *et al.*, 2001) and 20% of those introduced to New Zealand (Veltman *et al.*, 1996). In general, the impacts of the intro-duced bird species considered invasive are poorly understood in Australia and New Zealand, with little empirical evidence of their effects on native species and systems. In contrast to the direct and fairly immediate effects of impacts such as predation, many invasive bird species affect natives through competitive interactions. These interactions are far more difficult to study and quantify, particularly when the effects may take decades, even centuries, to become apparent (Strubbe *et al.*, 2011; Grarock *et al.*, 2012). The evidence presented for invasiveness in birds is frequently anecdotal or speculative (Bauer & Woog, 2010; Strubbe *et al.*, 2011), and it is important to bear in mind that some of the potential impacts of the birds discussed below have derived from such anecdotal accounts. The lack of demonstrable effects on native populations is, in part, the reason that invasive birds are generally confined to the lower reaches of many conservation management lists, with invasive mammals taking priority.

There is one exception to this on both sides of the Tasman; the common (or Indian) myna (*Acridotheres tristis*). Now considered among the 100 worst invasive species in the world (Lowe *et al.*, 2000), the common myna was deliberately introduced on multiple occasions to Australia during the 1860s and New Zealand in the 1870s, in both cases to control agricultural insect pests (Cunningham, 1948; Hone, 1978). It subsequently spread to much of eastern and southeastern Australia, and throughout the North Island of New Zealand. Its range is still expanding in Australia, however it appears to be contracting in the lower North Island of New Zealand (Martin, 1996; Robertson *et al.*, 2007). There is growing evidence, both anecdotal and empirical, that suggests mynas are a serious threat to native biodiversity in their introduced range (Tindall *et al.*, 2007; Dhami & Nagle, 2009; Feare, 2010; Grarock *et al.*, 2012). Competition appears to be their primary mode of impact, with the species frequently referred to as aggressive and dominant. Of serious concern is their potential to monopolise nest-cavity resources (e.g. Pell & Tidemann, 1997), in addition to food and territories. Mynas can also impact through other mechanisms including predation and disease. For example, they have been observed depredating open nests of passerines and some seabirds, consuming eggs and young chicks (Feare, 2010). They can carry a number of pathogens and parasites that may be detrimental to native bird populations, including avian malarial parasites, and some which are transmissible to humans, e.g. *Salmonella* and avian influenza (Dhami & Nagle, 2009; Saavedra, 2010). In public perception mynas are arguably the most concerning invasive bird. In 2005 they were voted as the 'most significant pest' in Australia (ABC, 2005). As well as causing damage to crops and orchards (Tracey *et al.*, 2007), mynas are considered a public nuisance and aggravate people with their loud, raucous contact calls. Some attempts have been made to manage myna numbers, typically via trapping, shooting or poisoning (Dhami & Nagle, 2009; Saavedra, 2010); however, mynas are intelligent animals and notoriously difficult targets. These operations are small scale in New Zealand, with individuals undertaking myna control for specific purposes (e.g. Tindall *et al.*, 2007). However in Canberra, Australia, a coordinated community-driven operation is showing promising outcomes. A reduction in local myna numbers has been achieved through the use of specifically designed traps placed in backyards (Handke, 2009), though it remains unclear whether regional-scale reductions are attainable with this approach.

Another cavity-nesting Sturnid like the myna, the starling (*Sturnus vulgaris*) was repeatedly introduced to both countries beginning *c.* 1862 (Thomson, 1922; Woolnough *et al.*, 2005), again with the aim of controlling insect pests in agricultural systems. Starlings are now widespread and abundant in eastern Australia and throughout New Zealand (Woolnough *et al.*, 2005; Robertson *et al.*, 2007). They are considered a serious threat primarily due to their impacts on crops (Bomford & Sinclair, 2002; Kirkpatrick & Woolnough, 2007; Tracey *et al.*, 2007), which can result in substantial economic losses. However, they also compete with native species for food and cavity-nest sites (Pell & Tidemann, 1997), may spread weeds, transmit a plethora of diseases (Linz *et al.*, 2007), and can cause damage to infrastructure through fouling (Bomford & Sinclair, 2002; Linz *et al.*, 2007). One crucial behaviour of the starling is their tendency to amass into huge flocks, particularly at roost sites where tens of thousands of individuals can gather (Caccamise *et al.*, 1983). The product of these large flocks is a greatly exacerbated level of impact on agriculture and, very likely, biodiversity. In Western Australia starling abundance is still comparatively low. The Government of Western Australia is currently controlling starling populations to prevent further colonisation and population growth, with the ultimate intention of eradicating starlings from Western Australia (Kirkpatrick & Woolnough, 2007).

The house sparrow (*Passer domesticus*) is one of the most successful avian invaders globally. The first releases in Australia and New Zealand occurred in the early 1860s (Thomson, 1922; Long, 1981). It is now widespread in the eastern half of Australia (though recent declines have been observed), with few records from Western Australia (Woodall, 1996; Barrett *et al.*, 2003). In New Zealand it is one of the most abundant bird species in urban areas (Spurr, 2012), and found throughout the country (Robertson *et al.*, 2007). It potentially threatens native species via competition for food and cavity-nest resources (Massam, 2005; McGregor & McGregor, 2008). Competition for food resources is more concerning in Australia, where the foraging niche of house sparrows overlaps with that of native granivorous birds (native New Zealand avifauna lacks obligate granivorous species). Because of their diet and their tendency to form large flocks, house sparrows are considered one of the primary bird pests of arable crops, eating grain as well as fruit crops (MacMillan, 1981; Tracey *et al.*, 2007). They also pose a serious disease risk; with some pathogens they carry causing human illness (e.g. *Salmonella* Typhimurium DT160) (Chilvers *et al.*, 1998; Alley *et al.*, 2002). Sparrow management is typically limited to local-scale operations, carried out by individuals with specific aims; for instance a farmer wanting to reduce crop damage on their own farm.

The key hybridisation threat from an invasive bird falls to the mallard (*Anas platyrhynchos*). Mallard were introduced to Australia as early as 1862, and New Zealand five years later, with numerous releases into the 1950s. An estimated minimum of 30 000 individuals were released during this time in New Zealand alone (Dyer & Williams, 2010). Hybridisation was noted as early as 1913 (Dyer & Williams, 2010). The Pacific grey (or black) duck (*Anas superciliosa*), native to both Australia and New Zealand, is at risk of genetic dilution, and possibly genetic extinction, from interbreeding with the more prolific, dominant mallard (Guay & Tracey, 2009). Indeed, severe declines of Pacific grey duck populations via genetic introgression have occurred in New Zealand, and on Lord Howe Island (Australia) where it is thought that no genetically distinct grey ducks remain (Tracey *et al.*, 2008). It is argued that, in addition to the impact from hybridisation, the larger size, higher fecundity and longer lifespan of the mallard, and their utilisation of human-modified landscapes, have all contributed to their displacement of the native grey duck (Williams & Basse, 2006).

The pet trade is responsible for the spread of many bird species around the world, and many subsequent invasions, those from the charismatic Psittacine family in particular with nearly two-thirds of parrot species transported to novel environments (Cassey *et al.*, 2004). A number of Australian parrots have made their way to New Zealand, aided by humans, and established wild populations after both deliberate and accidental releases. Two larger species, the sulphur-crested cockatoo (*Cacatua galerita*) and galah (*Eolophus roseicallipa*) are currently fairly localised in parts of the North Island, and have formed small, scattered flocks (Heather & Robertson, 1996). There is considerable risk from these species with potential future expansion, through competition with native parrots including kaka, through the spread of avian diseases, and through damage to crops. One Australian parrot has been successfully eradicated from the wild in New Zealand; a small number of rainbow lorikeets (*Trichoglossus haematodus*) had established in the Auckland region in the 1990s, growing to a population of 150–200 individuals by 1999. A control programme was initiated by the Department of Conservation in 2000 to eradicate this population before it spread more widely, as it posed a serious threat to native honey-eaters such as the tui (*Prosthemadera novaeseelandiae*) and to native cavity-nesting birds. New incursions of rainbow lorikeet are dealt with promptly to prevent re-establishment.

The current primary threat to native biodiversity from exotic parrots in New Zealand is the eastern rosella (*Platycercus eximius*). Also native to Australia, the eastern rosella is the most widespread parrot species in the North Island, and continues to expand geographically (Robertson *et al.*, 2007). It was first introduced to Dunedin in 1910, though this population has remained relatively confined. Subsequent populations established in Auckland in 1920 and Wellington in the 1960s from both deliberate and accidental releases of cage birds (Heather & Robertson, 1996). The eastern rosella poses a risk to communities of native cavity-nesting birds by using and potentially competing for cavity resources (Galbraith, 2010), competing with native species for food resources (Fraser, 2008), and is implicated in the spread of the parrot-specific Beak and Feather Disease Virus (BFDV) which could threaten populations of native parrots (Massaro *et al.*, 2012). As with most other invasive bird species in New Zealand, there is no national- or regional-scale management of eastern rosella populations. Some culling has been done on small, local scales, usually by shooting.

In addition to their more visible, discernible impacts on agricultural systems, quantitative evidence is emerging of the long-term detrimental effects of invasive birds on communities of native species through competitive interactions and other mechanisms (e.g. Pell & Tidemann, 1997; Tindall *et al.*, 2007; Tracey *et al.*, 2008; Grarock *et al.*, 2012). It is important to consider that while the impacts of invasive birds, in many cases, may only be observable over extended time periods and difficult to disentangle from other environmental changes, the outcomes for native species may be just as severe. Although the empirical evidence demonstrating ecological impacts is still lacking for most species, it is prudent to assume that invasive birds have had, and are still having, considerable repercussions for native biota. Nevertheless, current management is focussed on preventing new incursions, interception at the border, and control at the early stages of establishment, rather than tackling well-established species (but see Box 10.5 for one exception). This ultimately reflects the financial and logistical realities of invasive species management.

10.12 Fish, amphibians and reptiles

Invasive fish, amphibians and reptiles cause significant ecological and economic impacts in both Australia and New Zealand. In New Zealand, there is growing evidence that introduced trout species (*Salmo trutta* and *Oncorhynchus mykiss*) seriously impact native species through predation and competition (McDowall, 2003; Lintermans & Raadik, 2001). Despite this, the high value placed on trout as a recreational resource means that little has been achieved to protect native freshwater ecosystems from these species (Clout, 2011). Introduced carp (*Cyprinus carpio*) in Australia degrade water quality through their feeding habits and graze and uproot aquatic plants. In Australia the cane toad (*Rhinella marina*), toxic to native predators, is rapidly expanding its range across Australia (Chapter 5).

Box 10.5 Drivers of bird control – rook case study

Many birds attain pest status through processes that affect people and property (Nelson, 1990; Bomford & Sinclair, 2002), including fouling, building damage, crop damage and behaviour that the public considers an annoyance (e.g. loud, raucous calls). The economic and social dimensions of pest impacts frequently drive management prioritisation, as public pressure mounts for government agencies to take action.

Damage to agricultural systems is one key driver of bird control. The ongoing control of rook (*Corvus frugilegus*) populations in New Zealand is an example of this, and one of the few exceptions to the current lack of management actions against well-established invasive birds. Rook populations were slow to reach pest proportions in New Zealand after they were first released in 1862 (Heather & Robertson, 1996). Rooks successfully established in the Hawke's Bay and Canterbury regions where they are most abundant today. Control operations (via shooting and poisoning) initiated in the 1960s and 1970s went some way to reducing rook abundance, and complaints of damage fell (Porter *et al.*, 2008). However, the disturbance caused by control operations potentially contributed to a subsequent range expansion, as disturbance to rookeries can cause birds to disperse (Porter *et al.*, 2008).

Like most introduced bird species, however, there have been few studies dedicated to identifying and understanding rook impacts. The effects of rooks on native birds, for instance, have not been examined. Instead, the primary impetus for this control has been damage to agricultural systems. Pasture is damaged when turf is ripped up by rooks foraging for invertebrates in the topsoil (McLennan & MacMillan, 1983). There is also reported damage to a wide range of crops including cereals and fruit (Heather & Robertson, 1996; Cowan *et al.*, 2012). Yet even for these observed agricultural impacts there have been no robust economic analyses on the cost of rook damage, limiting the capacity for cost–benefit-based decision-making. Regardless, control of rook populations has been maintained, primarily by way of poisoning rookeries (Porter *et al.*, 2008). Control is likely to continue into the future, with regional eradication a possibility (Cowan *et al.*, 2012).

10.13 Conclusion

Australia and New Zealand are unique ecosystems facing similar threats from a suite of invasive species that cause numerous impacts, as highlighted in this chapter. Whilst invasive species composition in both countries is somewhat different, the impacts these species have on native populations remain similar in regards to invasive ungulates, birds and lagomorphs. The devastating effects of invasive species predation of native prey in Australia and New Zealand are well documented. However, invasive predators in Australia have an added impact in terms of competition with native predators. In both countries the aim is now to prevent establishment of new invasive species and preserve key areas of high biodiversity value through the control or eradication of invasive species. The New Zealand and Australian governments spend millions annually in an attempt to mitigate the establishment of unwanted organisms, as eradicating and managing invasive species can be much more costly. However, preventing the arrival and establishment of new organisms is not always possible, and in many cases, controlling the spread and impacts of invasive species becomes the overall management goal. Recently there has been a growth in community participation in ongoing control programmes in New Zealand and Australia. Biodiversity in Australia and New Zealand will depend on their continued efforts.

LITERATURE CITED

Abbott, I., Peacock, D., & Short, J. 2014, 'The new guard: the arrival and impacts of cats and foxes' in A. S. Glen & C. R. Dickman (eds.), *Carnivores of Australia: Past, Present and Future*, Collingwood: CSIRO Publishing, pp. 69–104.

ABC. 2005, *Wildwatch Australia Survey*, viewed 13 May 2013, http://www.abc.net.au/tv/wildwatch.

Alley, M. R., Connolly, J. H., Fenwick, S. G. *et al.* 2002, 'An epidemic of salmonellosis caused by *Salmonella* Typhimurium DT160 in wild birds and humans in New Zealand', *New Zealand Veterinary Journal*, **50**, 170–176.

Angel, A., Wanless, R. M. & Cooper J. 2009, 'Review of impacts of the introduced house mouse on islands in the Southern Ocean: are mice equivalent to rats?', *Biological Invasions*, **11**, 1743–1754.

Anson, J. R., Dickman, C. R., Boonstra, R. & Jessop, T. S. 2013, 'Stress triangle: do introduced predators exert indirect costs on native predators and prey?' *PLoS ONE*, **8** (4), e60916.

Banks, P. B. & Hughes, N. K. 2012, 'A review of the evidence for potential impacts of black rats (*Rattus rattus*) on wildlife and humans in Australia', *Wildlife Research*, **39** (1), 78–88.

Barrett, G., Silcocks, A., Barry, S., Cunningham, R. & Poulter, R. 2003, *The New Atlas of Australian Birds*, Melbourne: Royal Australasian Ornithologists Union.

Batcheler, C. L. & Cowan, P. E. 1988, *Review of the Status of the Possum* (Trichosurus vulpecula) *in New Zealand*, Contract report for the Technical Advisory Committee (Animal Pests), Wellington: Ministry of Agriculture & Fisheries.

Bauer, H. G. & Woog, F. 2010, 'On the 'invasiveness' of non-native bird species', *Ibis*, **153** (1), 204–206.

Berry, C. 1999, *Potential Interactions of Hedgehogs with North Island Brown Kiwi at Boundary Stream Mainland Island*. Conservation Advisory Science, no. 268. Department of Conservation, Wellington.

Blackie, H. M., Woodhead, I., Diegel, H. *et al.* 2011, 'Integrating ecology and technology to create innovative pest control devices', *Proceedings of the 8th European Vertebrate Pest Management Conference*, Berlin, pp. 152–153.

Bomford, M. & Hart, Q. 2002, 'Non-indigenous vertebrates in Australia' in D. Pimental (ed.) *Biological Invasions: Economic and Environmental Costs of Alien Plant, Animal, and Microbe Species*, 1st edition, Boca Raton: CRC Press, pp. 25–41.

Bomford, M. & Sinclair, R. 2002, 'Australian research on bird pests: impact, management and future directions', *Emu*, **102** (1), 29–45.

Caccamise, D. F., Lyon, L. A. & Fischl, J. 1983, 'Seasonal patterns in roosting flocks of starlings and common grackles', *The Condor*, **85**, 474–481.

Caley, P. & Hone, J. 2005, 'Assessing the host disease status of wildlife and the implications for disease control: *Mycobacterium bovis* infection in feral ferrets', *Journal of Applied Ecology*, **42** (4), 708–719.

Campbell, D. J. 1990, 'Changes in the structure and composition of a New Zealand lowland forest inhabited by brushtail possums', *Pacific Science*, **44** (3), 277–296.

Campbell, K. J., Harper, G., Algar, D., *et al.* 2011, 'Review of feral cat eradications on islands' in C. R. Veitch, M. N. Clout & D. R. Towns (eds.), *Island Invasives: Eradication and Management. Proceedings of the International Conference on Island Invasives*. IUCN, Gland, Switzerland and Auckland, New Zealand. pp. 37–46.

Cassey, P., Blackburn, T. M., Russell, G. J., Jones, K. E. & Lockwood, J. L. 2004, 'Influences on the transport and establishment of exotic bird species: an analysis of the parrots (Psittaciformes) of the world', *Global Change Biology*, **10** (4), 417–426.

Caut, S., Casanovas, J. G., Virgos, E., *et al.* 2007, 'Rats dying for mice: modelling the competitor release effect', *Austral Ecology*, **32** (8), 858–868.

Chapin III, F. S., Zavaleta, E. S., Eviner, V. T. *et al.* 2000, 'Consequences of changing biodiversity', *Nature*, vol. **405**, no. 6783, pp. 234–242.

Chilvers, B. L., Cowan, P. E., Waddington, D. C., Kelly, P. J. & Brown, T. J. 1998, 'The prevalence of infection of *Giardia* spp. and *Cryptosporidium* spp. in wild animals on farmland, southeastern North Island, New Zealand', *International Journal of Environmental Health Research*, **8** (1), 59–64.

Choquenot, D. 2006, 'Bioeconomic modeling in conservation pest management: effect of stoat control on extinction risk of an indigenous New Zealand passerine, *Mohua ochrocephala*', *Conservation Biology*, **20** (2), 480–489.

Choquenot, D., Hone, J. & Saunders, G. 1999, 'Using aspects of predator-prey theory to evaluate helicopter shooting for feral pig control', *Wildlife Research*, **26** (3), 251–261.

Choquenot, D., Kilgour, R. J. & Lukins B. S. 1993, 'An evaluation of feral pig trapping', *Wildlife Research*, **20** (1), 15–22.

Choquenot, D., Lukins, B. & Curran, G. 1997, 'Assessing lamb predation by feral pigs in Australia's semi-arid rangelands' *Journal of Applied Ecology*, **34** (6), 1445–1454.

Choquenot, D., McIlroy, J. & Korn, T. 1996, *Managing Vertebrate Pests: Feral Pigs*, Canberra: Australian Government Publishing Service.

Clout, M. N. 2011, 'Ecological and economic costs of alien vertebrates in New Zealand', in D. Pimental (ed.), *Biological Invasions: Economic and Environmental Costs of Alien Plant, Animal, and Microbe Species*, Boca Raton: CRC Press, pp. 283–291.

Coomes, D. A., Allen, R. B., Forsyth, D. M. & Lee, W. G. 2003, 'Factors preventing the recovery of New Zealand forests following control of invasive deer', *Conservation Biology*, **17** (2), 450–459.

Corbett, L. K. 1995, *The Dingo in Australia and Asia*, UNSW Press, Sydney, Australia.

Cowan, P., Glen, A., Barron, M., Duckworth, J. & Booth, L. 2012, 'Recent developments in the control of rooks (*Corvus frugilegus*) in New Zealand', in T. J. Korn (ed.), *Abstracts of the Australasian Wildlife Management Society Conference*, Australasian Wildlife Management Society, Adelaide. p. 89.

Cowan, P. E. 2001, 'Advances in New Zealand mammalogy 1990–2000: brushtail possum', *Journal of the Royal Society of New Zealand*, **31** (1), 15–29.

Cowan, P. E. & Waddington, D. C. 1990, 'Suppression of fruit production of the endemic forest tree, *Elaeocarpus dentatus*, by introduced marsupial brushtail possums, *Trichosurus vulpecula*', *New Zealand Journal of Botany*, **28** (3), 217–224.

CSIRO. 2011, *Myxomatosis and Rabbits in Australia Today*, viewed 27 May 2013, http://www.csiro.au/en/Outcomes/Safeguarding-Australia/Myxomatosis.aspx>

Cunningham, J. M. 1948, 'Distribution of myna in N.Z', *New Zealand Bird Notes*, **3**, 57–64.

Dhami, M. K. & Nagle, B. 2009, 'Review of the biology and ecology of the Common Myna (*Acridotheres tristis*) and some implications for management of this invasive species' Pacific Invasives Initiative, The University of Auckland, Auckland.

Dobbie, W. R., Berman, D. M. & Braysher, M. L. 1993, *Managing Vertebrate Pests: Feral Horses*, Canberra: Australian Government Publishing Service.

Dolman, P. M. & Wäber, K. 2008. 'Ecosystem and competition impacts of introduced deer', *Wildlife Research*, **35** (3), 202–214.

Duncan, R. P., Bomford, M., Forsyth, D. M. & Conibear, L. 2001, 'High predictability in introduction outcomes and the geographical range size of introduced Australian birds: a role for climate', *Journal of Animal Ecology*, **70** (4), 621–632.

Dyer, J. & Williams, M. 2010, 'An introduction most determined: Mallard (*Anas platyrhynchos*) to New Zealand', *Notornis*, **57**, 178–195.

Edwards, G. P., Pople, A. R., Saalfeld, K. & Caley, P. 2004, 'Introduced mammals in Australian rangelands: future threats and the role of monitoring programmes in management strategies', *Austral Ecology*, **29** (1), 40–50.

Edwards, G. P., Zeng, B., Saalfeld, W. K. & Vaarzon-Morel, P. 2010, 'Evaluation of the impacts of feral camels', *The Rangeland Journal*, **32** (1), 43–54.

Elliott, G. & O'Donnell, C. 1988, *Recent Decline in Yellowhead Populations*, Science & Research internal report, no. 29, Wellington: Department of Conservation.

Elliott, G. P. 1996, 'Productivity and mortality of mohua (*Mohoua ochrocephala*)', *New Zealand Journal of Zoology*, **23** (3), 229–237.

Elsworth, P. G., Kovaliski, J. & Cooke, B. D. 2012, 'Rabbit haemorrhagic disease: are Australian rabbits (*Oryctolagus cuniculus*) evolving resistance to infection with Czech CAPM 351 RHDV?', *Epidemiology and Infection*, **11** (11), 1972–1981.

Falk-Petersen, J., Bohn, T. & Sandlund, O. T. 2006, 'On the numerous concepts in invasion biology', *Biological Invasions*, **8** (6), 1409–1424.

Feare, C. J. 2010, 'The use of Starlicide in preliminary trials to control invasive common myna *Acridotheres tristis* populations on St Helena and Ascension Islands, Atlantic Ocean' *Conservation Evidence*, **7**, 52–61.

Fleming, P., Corbett, L., Harden, R. & Thomson, P. 2001, *Managing the Impacts of Dingoes and Other Wild Dogs*. Canberra: Bureau of Rural Sciences.

Fleming, P. J. S., Allen, B. L. & Ballard, G. A. 2012, 'Seven considerations about dingoes as biodiversity engineers: the socioecological niches of dogs in Australia', *Australian Mammalogy*, **34** (1), 119–131.

Flux, J. E. C., Duthie, A. G., Robinson, T. J. & Chapman, J. A. 1990, 'Exotic populations' in J. A. Chapman & J. E. C. Flux (eds), *Rabbits, Hares and Pikas: Status Survey and Conservation Action Plan*. Gland: IUCN, pp. 147–158.

Fraser, E. A. 2008, The winter ecology of the eastern rosella (*Platycercus eximius*) in New Zealand. B.Sc. (Hons) Dissertation, University of Auckland.

Freeland, W. J. 1990, 'Large herbivorous mammals: exotic species in northern Australia', *Journal of Biogeography*, **17** (4/5), 445–449.

Galbraith, J. A. 2010, The ecology and impact of the introduced eastern rosella (*Platycercus eximius*) in New Zealand. M.Sc. Thesis, University of Auckland.

Glen, A. S. 2014, 'Fur, feathers and scales: the interactions between mammalian, reptilian and avian predators', in A. S. Glen & C. R. Dickman (eds.), *Carnivores of Australia: Past, Present and Future*. Collingwood: CSIRO Publishing, pp. 279–299.

Glen, A. S. & Dickman, C. R. 2005, 'Complex interactions among mammalian carnivores in Australia, and their implications for wildlife management', *Biological Reviews*, **80** (3), 387–401.

Glen, A. S. & Dickman, C. R. 2011, 'Why are there so many spotted-tailed quolls *Dasyurus maculatus* in parts of north-eastern New South Wales?', *Australian Zoologist*, **35** (3), 711–718.

Grarock, K., Tidemann, C. R., Wood, J. & Lindenmayer, D. B. 2012, 'Is it benign or is it a pariah? Empirical evidence for the impact of the common myna (*Acridotheres tristis*) on Australian birds', *PLoS ONE*, **7**, e40622.

Guay, P. J. & Tracey, J. P. 2009, 'Feral mallards: A risk for hybridisation with wild pacific black ducks in Australia?', *Victorian Naturalist*, **126** (3), 87–91.

Handke, B. 2009, The Canberra Indian Myna Action Group – A Study in community-action, viewed 3 May 2013, http://www.indianmynaaction.org.au/

Heather, B. & Robertson, H. 1996, *The Field Guide to the Birds of New Zealand*. Auckland: Viking.

Hetherington, C. A., Algar, D., Mills, H. & Bencini, R. 2007, 'Increasing the target-specificity of ERADICAT® for feral cat (*Felis catus*) control by encapsulating a toxicant' *Wildlife Research*, **34** (6), 467–471.

Hindwood, K. A. 1940, 'The birds of Lord Howe Island', *Emu*, **40** (1), 1–86.

Hone, J. 1978, 'Introduction and spread of the Common Myna in New South Wales', *Emu*, **78** (4), 227–230.

Hone, J. 2007, *Wildlife Damage Control*, Victoria: CSIRO Publishing.

IACRC. 2013, *Pestsmart Case Study: Feral Goat Eradication on Kangaroo Island*, Canberra: Invasive Animals Cooperative Research Centre.

Jones, C., Norbury, G. & Bell, T. 2013, 'Impacts of introduced European hedgehogs on endemic skinks and weta in tussock grassland', *Wildlife Research*, **40** (1), 36–44.

Kelly, D., Geldenhuis, A., James, A. *et al*. 2013, 'Of mast and mean: differential temperature cue makes mast seeding insensitive to climate change', *Ecology Letters*, **16** (1), 90–98.

King, C. M. 2005, *The Handbook of New Zealand Mammals*, 2nd edition, Melbourne: Oxford University Press.

King, C. M., Piran, C. L., White, D. C. & Lawrence, B. 2003, 'Matching productivity to resource availability in a small predator, the stoat (*Mustela erminea*)', *Canadian Journal of Zoology*, **81** (4), 662–669.

Kirkpatrick, W. & Woolnough, A. 2007, *Common Starling*, Pestnote, no. 253, Department of Agriculture and Food, Perth: Government of Western Australia.

Krull, C. R., Choquenot, D., Burns, B. R. & Stanley, M. C. 2013, 'Feral pigs in a temperate rainforest ecosystem: disturbance and ecological impacts', *Biological Invasions*, **15**, 2193–2204.

Krull, C. R., Waipara, N. W., Choquenot, D., *et al*. 2012, 'Absence of evidence is not evidence of absence: Feral pigs as vectors of soil-borne pathogens', *Austral Ecology*, **38** (5), 534–542.

Lee, W. G., Allen, R. B. & Tompkins, D. M. 2006, 'Paradise lost – the last major colonization', in R. B. Allen & W. G. Lee (eds.), *Biological Invasions in New Zealand*, Germany: Springer-Verlag, pp. 1–16.

Leigh, J. H., Wood, D. H., Holgate, M. D., Slee, A. & Stanger, M. G. 1989, 'Effects of rabbit and kangaroo grazing on two semi-arid grassland communities in central-western New South Wales', *Australian Journal of Botany*, **37** (5), 375–396.

Letnic, M., Ritchie, E. G. & Dickman, C. R. 2012, 'Top predators as biodiversity regulators: the dingo *Canis lupus dingo* as a case study' *Biological Reviews*, **87** (2), 390–413.

Li, A. Y., Williams, N., Adams, P. J., Fenwick, S. & Hardy, G. E. S. J. 2010, 'The spread of *Phytophthora cinnamomi* by feral pigs' *5th IUFRO Phytophthora Diseases in Forests and Natural Ecosystems*, Rotorua, New Zealand.

Lintermans, M. & Raadik, T. 2001, 'Local eradication of trout from streams using rotenone: the Australian experience' *Proceedings of the Managing Invasive Freshwater Fish in New Zealand Workshop*, Hamilton: Department of Conservation, pp. 95–111.

Linz, G. M., Homan, H. J., Gaulker, S. M., Penry, L. B. & Bleier, W. J. 2007, 'European starlings: a review of an invasive species with far-reaching impacts', *Managing Vertebrate Invasive Species: Proceedings of an International Symposium*, Paper 24. USDA/APHIS Wildlife Services, National Wildlife Research Center, Fort Collins, Colorado, USA.

Litchman, E. 2010, 'Invisible invaders: non-pathogenic invasive microbes in aquatic and terrestrial ecosystems', *Ecology Letters*, **13** (12), 1560–1572.

Long, J. L. 1981, *Introduced Birds of the World*. New York Universe Books, New York.

Lowe, S., Browne, M., Boudjelas, M. & De Poorter, M. (2000). *100 of the World's Worst Invasive Alien Species*. Invasive Species Specialist Group (International Union for the Conservation of Nature), Auckland. www.issg.org/booklet.pdf

MacMillan, B. W. H. 1981, 'Food of house sparrows and greenfinches in a mixed farming district, Hawke's Bay, New Zealand', *New Zealand Journal of Zoology*, **8** (1), 93–104.

Martin, W. K. 1996, 'The current and potential distribution of the Common Myna *Acridotheres tristis* in Australia', *Emu*, **96** (3), 166–173.

Massam, M. 2005, *Sparrows*, Farmnote, No. 117/99, Department of Agriculture, Perth: Government of Western Australia.

Massaro, M., Ortiz-Catedral, L., Julian, L., *et al.* 2012, 'Molecular characterisation of beak and feather disease virus (BFDV) in New Zealand and its implications for managing an infectious disease', *Archives of Virology*, **157** (9), 1651–1663.

McDowall, R. M. 2003, 'Impacts of introduced salmonids on native galaxiids in New Zealand upland streams: a new look at an old problem', *Transactions of the American Fisheries Society*, **132** (2), 229–238.

McGregor, D. B. & McGregor, B. A. 2008, 'Eliminating an avian pest (House Sparrow *Passer domesticus*) population: the role of trapping at a homestead scale', *Victorian Naturalist*, **125** (1), 4.

McKnight, T. L. 1976, *Friendly Vermin: a Survey of Feral Livestock in Australia* (Vol. **21**), California: University of California Press.

McLennan, J. A. & MacMillan, B. W. H. 1983, 'Predation by the rook, *Corvus frugilegus* L., on larvae of the grass grub, *Costelytra zealandica* (White), in Hawke's Bay, New Zealand', *New Zealand Journal of Agricultural Research*, **26** (1), 139–145.

McLennan, J. A., Potter, M. A., Robertson, H. A. *et al.* 1996, 'Role of predation in the decline of kiwi, *Apteryx* spp., in New Zealand', *New Zealand Journal of Ecology*, **20** (1), 27–35.

Morris, K. D. 2002, 'The eradication of the black rat (*Rattus rattus*) on Barrow and adjacent islands off the north-west coast of Western Australia', in C. R. Veitch & M. N. Clout (eds.), *Turning the Tide: the Eradication of Invasive Species*. Gland: IUCN SSC Invasive Species Specialist Group, pp. 219–225.

Morris, K., Johnson, B., Orell, P. *et al.* 2003, 'Recovery of the threatened chuditch (*Dasyurus geoffroii*): a case study', in M. Jones, C. Dickman & M. Archer (eds.), *Predators with Pouches: the Biology of Carnivorous Marsupials*. Collingwood: CSIRO Publishing, pp. 435–451.

Nelson, P. C. 1990, 'Bird problems in New Zealand – methods of control', *Proceedings of the Fourteenth Vertebrate Pest Conference 1990*, p. 63.

Newman, D. G. 1994, 'Effects of a mouse, *Mus musculus*, eradication programme and habitat change on lizard populations of Mana Island, New Zealand, with special reference to McGregor's skink, *Cyclodina macgregori*', *New Zealand Journal of Zoology*, **21** (4), 443–456.

Nugent, G., Fraser, K. W., Asher, G. W. & Tustin, K. G. 2001, 'Advances in New Zealand mammalogy 1990–2000: deer', *Journal of the Royal Society of New Zealand*, **31** (1), 263–298.

O'Donnell, C. F. J. 1996, 'Monitoring mohua (yellowhead) populations in the South Island, New Zealand, 1983–1993', *New Zealand Journal of Zoology*, **23** (3), 221–228.

O'Donnell, C. F. J. & Phillipson, S. M. 1996, 'Predicting the incidence of mohua predation from the seedfall, mouse, and predator fluctuations in beech forests', *New Zealand Journal of Zoology*, **23** (3), 287–293.

Page, A., Kirkpatrick, W. & Massam, M. 2009, *Risk Assessment for Australia – Domestic Sheep* (Ovis aries). Department of Agriculture and Food, Government of Western Australia.

Palomares, F. & Caro, T. M. 1999, 'Interspecific killing among mammalian carnivores', *American Naturalist*, **153** (5), 492–508.

Parkes, J. P. 1990, 'Eradication of feral goats on islands and habitat islands', *Journal of the Royal Society of New Zealand*, **20** (3), 297–304.

Parkes, J., Henzell, R. & Pickles, G. S. 1996, *Managing Vertebrate Pests: Feral Goats*, Canberra: Australian Government Publishing Service.

Parkes, J. P., Glentworth, B. & Sullivan, G. 2008, 'Changes in immunity to rabbit haemorrhagic disease virus, and in abundance and rates of increase of wild rabbits in Mackenzie Basin, New Zealand' *Wildlife Research*, **35** (8), 775–779.

Parkes, J. P., Norbury, G. L., Heyward, R. P. & Sullivan, G. 2002, 'Epidemiology of rabbit haemorrhagic disease (RHD) in the South Island, New Zealand, 1997–2001', *Wildlife Research*, **29** (6), 543–555.

Pell, A. S. & Tidemann, C. R. 1997, 'The impact of two exotic hollow-nesting birds on two native parrots in Savannah and Woodland in eastern Australia', *Biological Conservation*, **79** (2), 145–153.

Porter, R. E. R., Clapperton, B. K. & Coleman, J. D. 2008, 'Distribution, abundance and control of the rook (*Corvus frugilegus* L.) in Hawke's Bay, New Zealand, 1969–2006', *Journal of the Royal Society of New Zealand*, **38** (1), 25–36.

Risbey, D. A., Calver, M. C., Short, J., Bradley, J. S. & Wright, I. W. 2000, 'The impact of cats and foxes on the small vertebrate fauna of Heirisson Prong, Western Australia. II. A field experiment', *Wildlife Research*, **27** (3), 223–235.

Ritchie, J. 2000, *Possum: Everybody's Problem*, viewed 13 May 2013, http://www.doc. govt.nz/documents/science-and-technical/everybodyspossum.pdf

Robertson, C. J. R., Hyvonen, P., Fraser, M. J. & Prichard, C. R. 2007, *Atlas of Bird Distribution in New Zealand*, Wellington: Ornithological Society of New Zealand.

Ruscoe W. A., Elkinton J. S., Choquenot D. & Allen R. B. 2005, 'Predation of beech seed by mice: effects of numerical and functional responses', *Journal of Animal Ecology*, **74** (6), 1005–1019.

Russell, J. C., Beaven, B. M., MacKay, J. W. B., Towns, D. R. & Clout, M. N. 2008, 'Testing island biosecurity systems for invasive rats', *Wildlife Research*, **35** (3), 215–221.

Saalfield, W. K. & Edwards, G. P. 2010, 'Distribution and abundance of the feral camel (*Camelus dromedarius*) in Australia', *The Rangeland Journal*, **32** (1), 1–9.

Saavedra, S. 2010, 'Eradication of invasive Mynas from islands. Is it possible?', *Aliens: The Invasive Species Bulletin*, **29**, 40–47.

Sanders, M. D. & Maloney, R. F. 2002, 'Causes of mortality at nests of ground-nesting birds in the Upper Waitaki Basin, South Island, New Zealand: a 5-year video study', *Biological Conservation*, **106** (2), 225–236.

Saunders, G. R., Gentle, M. N. & Dickman, C. R. 2010, 'The impacts and management of foxes *Vulpes vulpes* in Australia', *Mammal Review*, **40** (3), 181–211.

Sharp, T. & Saunders, G. 2012, *Model Code of Practice for the Humane Control of Feral Donkeys*, Australia: Invasive Animals Cooperative Research Centre.

Simberloff, D. 2009, 'Rats are not the only introduced rodents producing ecosystem impacts on islands', *Biological Invasions*, **11** (7), 1735–1742.

Simberloff, D. & Rejmánek, M. 2011, *Encyclopedia of Biological Invasions*, Berkley: University of California Press.

Simberloff, D. & Von Holle, B. 1999, 'Positive interactions of nonindigenous species: invasional meltdown?', *Biological Invasions*, **1** (1), 21–32.

Skeat, A. J., East, T. J. & Corbett, L. K. 1996, 'Impact of feral water buffalo', *Landscape and Vegetation Ecology of the Kakadu Region, Northern Australia*, pp. 155–177.

Smith, A. P. & Quin, D. G. 1996, 'Patterns and causes of extinction and decline in Australian conilurine rodents', *Biological Conservation*, **77** (2), 243–267.

Spurr, E. B. 2012, 'New Zealand Garden Bird Survey – analysis of the first four years', *New Zealand Journal of Ecology*, **36** (3), 1–5A.

Story, G., Berman, D., Palmer, R. & Scanlan, J. 2004, 'The impact of rabbit haemorrhagic disease on wild rabbit (*Oryctolagus cuniculus*) populations in Queensland', *Wildlife Research*, **31** (2), 183–193.

Strubbe, D., Shwartz, A. & Chiron, F. 2011, 'Concerns regarding the scientific evidence informing impact risk assessment and management recommendations for invasive birds', *Biological Conservation*, **144** (8), 2112–2118.

Thomson, G. M. 1922, *'The Naturalization of Plants and Animals in New Zealand'*, Cambridge: Cambridge University Press.

Tindall, S., Ralph, C. & Clout, M. N. 2007, 'Changes in bird abundance following common myna control on a New Zealand island' *Pacific Conservation Biology*, **13** (3), 202–212.

Towns, D. R., Atkinson, I. A. E. & Daugherty, C. H. 2006, 'Have the harmful effects of introduced rats on islands been exaggerated?', *Biological Invasions*, **8** (4), 863–891.

Tracey, J., Bomford, M., Hart, Q., Saunders, G. & Sinclair, R. 2007, *Managing Bird Damage to Fruit and Other Horticultural Crops*, Canberra: Bureau of Rural Sciences.

Tracey, J. P., Lukins, B. S. & Haselden, C. 2008, 'Hybridisation between mallard (*Anas platyrhynchos*) and grey duck (*A. superciliosa*) on Lord Howe Island and management options', *Notornis*, **55**, 1–7.

Tweddle, N. E. & Livingstone, P. 1994, 'Bovine tuberculosis control and eradication programs in Australia and New Zealand', *Veterinary Microbiology*, **40** (1), 23–39.

van Bommel, L. & Johnson, C. N. 2012, 'Good dog! Using livestock guardian dogs to protect livestock from predators in Australia's extensive grazing systems', *Wildlife Research*, **39** (3), 220–229.

Van Dyck, S. & Strahan, R. 2008, *The Mammals of Australia*, Chatswood: Reed New Holland.

Veale, A. J., Clout, M. N. & Gleeson, D. M. 2012, 'Genetic population assignment reveals a long-distance incursion to an island by a stoat (*Mustela erminea*)', *Biological Invasions*, **14** (3), 735–742.

Veltman, C. J., Nee, S. & Crawley, M. J. 1996, 'Correlates of introduction success in exotic New Zealand birds', *The American Naturalist*, **147**, 542–557.

Williams, K., Parer, I., Coman, B., Burley, J. & Braysher, M. 1995, *Managing Vertebrate Pests: Rabbits*, Bureau of Resource Sciences/CSIRO Division of Wildlife and Ecology, Canberra: Australian Government Publishing Service.

Williams, M. & Basse, B. 2006, 'Indigenous gray ducks, *Anas superciliosa*, and introduced mallards, *A. platyrhynchos*, in New Zealand: processes and outcome of a deliberate encounter', *Acta Zoologica Sinica*, **52**, 579–582.

Wilson, G., Dexter, N., O'Brien, P. & Bomford, M. 1992, *Pest Animals in Australia: A Survey of Introduced Wild Mammals*, Kenthurst: Bureau of Rural Resources and Kangaroo Press.

Wilson, P. R., Karl, B. J., Toft, R. J., Beggs, J. R. & Taylor, R. H. 1998, 'The role of introduced predators and competitors in the decline of kaka (*Nestor meridionalis*) populations in New Zealand', *Biological Conservation*, **83** (2), 175–185.

Woodall, P. F. 1996, 'Limits to the distribution of the House Sparrow *Passer domesticus* in suburban Brisbane, Australia', *Ibis*, **138** (2), 337–340.

Woolnough, A. P., Massam, M. C., Payne, R. L. & Pickles, G. S. 2005, 'Out on the border: keeping starlings out of Western Australia', *13th Australasian Vertebrate Pest Conference*, pp. 183–189.

Wren, L. 2012, 'Macquarie Island on the road to recovery following eradication' *Australasian Wildlife Management Society Newsletter*, **26** (3), 4–5.

CHAPTER 11

Freshwaters in New Zealand

Mike J. Joy

Summary

Freshwater ecosystems in New Zealand have been under considerable stress since European colonisation. In the past 150 years the draining of 90% of wetlands and the removal of a similar amount of indigenous vegetation has placed much strain on the health of freshwaters through the loss of the crucial hydrologic and biological functions performed by intact wetland and forest ecosystems. This loss has been exacerbated by the more recent intensification of farming with the concomitant addition of excess nutrients and sediment to water as well as the effects of urbanisation and introductions of exotic fish species. The cumulative impacts of all these changes can be seen with declining water chemistry measures and the biological status of freshwater ecosystems. The most obvious impacts are revealed by biological indicators with 68% of the native freshwater fish species listed as threatened, and 90% of lowland waterways failing bathing standards. Lowland lakes are under immense pressure; 44% of monitored lakes are eutrophic or worse and they are mostly the lowland lakes. The legislative response from central and local government to the obvious declines has failed to halt or even reduce the deterioration. In contrast, government initiatives to increase farming intensification mean there is little or no chance of improvement in the future for New Zealand freshwaters.

11.1 Introduction

Over the past century, New Zealand's freshwater ecosystems have undergone significant and obvious deterioration, both chemically and ecologically. The decline is revealed in a multitude of ways, including severe reductions in biodiversity and aesthetic values corroborated by declining physicochemical measures taken at most lowland waterways (Larned *et al.* 2004). One of the starkest indications of the extent of the deterioration is the fact that New Zealand now has proportionally more threatened freshwater fish species than almost any country globally (Allibone *et al.* 2010; IUCN 2010). While in global terms

Austral Ark: The State of Wildlife in Australia and New Zealand, eds. A. Stow, N. Maclean and G. I. Holwell. Published by Cambridge University Press. © Cambridge University Press 2015.

the freshwater declines in New Zealand are relatively recent, they mirror declines world-wide where the symptoms and drivers of deterioration are similar but have occurred over much longer time periods. The primary driver of decline has been the unrestrained agricultural intensification (Williams 2004) with attendant increases in nutrient and sediment entering lakes, rivers and groundwater and to a lesser extent from the numerous impacts of urbanisation. Apart from a few exceptions to protect iconic waters, there is little indication of any government initiative to halt this erosion in the health of freshwater ecosystems. In fact, the reverse is happening, with considerable effort by central and local government going into increasing farming intensification through irrigation schemes and weakening of the legislation designed to protect freshwaters.

11.2   The condition of New Zealand rivers and streams

Flowing waters in New Zealand have been regularly monitored since 1990 at a set of 77 river sites known as the National River Water Quality Network (NRWQN) (note 35 of these sites are control sites, that is they are above developed areas). Analysis of these data reveals consistent declines in measured water quality parameters in most lowland rivers, particularly those measures related to diffuse nutrient and faecal pollution and sediment (Ballantine and Davies-Colley 2010) although the real extent of declines has been masked by the reporting agencies averaging results over control and impact sites. Another dataset of more than 300 lowland waterways collected by local government revealed that 96% of the sites in lowland pastoral catchments, and all sites in urban catchments, failed the pathogen standard considered safe for swimming and more than 80% exceeded nutrient guideline levels (Larned *et al*. 2004). The human impact of these high levels of pathogen contamination is revealed from estimates by the Ministry of Health that 18 000–34 000 people annually contract waterborne diseases (Ball 2006). While damning, these human health impacts occur despite the fact that many lowland waterways and estuaries have health warning signs, and these signs are now a common sight around much of lowland New Zealand.

11.3   Lakes and groundwater

Lakes and groundwaters show parallel declines to flowing waters, in that they suffer from excess nutrient inputs from agricultural intensification as well as urban waste and storm water inputs. The difference for lakes and groundwater from flowing waters, however, is that impacts are longer lasting but more easily measured. The high level of nutrients are demonstrated by 44% of monitored lakes in New Zealand now classed as polluted (that is they are now eutrophic or worse) and almost all of these polluted lakes are in lowland areas and in agricultural or urban catchments (Verburg *et al*. 2010). In groundwater, nitrate levels are rising at 39% of monitored sites and groundwater pathogen levels exceeded human drinking standards at 21% of monitored sites (Daughney and Wall 2007). See also Box 11.1.

11.4   Freshwater biodiversity

Any changes in freshwater ecosystem health are ultimately and most comprehensively revealed by changes in freshwater biodiversity. Nationally, native freshwater fish

Box 11.1 'Water-quality' or measuring the wrong things the wrong way?

The term 'water quality' suggests a comprehensive measurement of freshwater condition that would encompass some aspect of habitat and freshwater health and integrity but, in reality, it is a managerial rather than an ecological assessment. The parameters used to assess water quality are more closely related to ease of sampling than any genuine representation of waterway condition. 'Water quality' assessment is prescribed by the Ministry for the Environment as consisting of a suite of five physicochemical measures and suggests some minimal biological assessment. The physicochemical factors are: suspended sediment, nitrogen, phosphorus, temperature and dissolved oxygen, and the biological assessment is: faecal coliforms and, occasionally, macro-invertebrate metrics and visual assessment of periphyton (Ballantine and Davies-Colley 2010). Remarkably, this 'water quality' assessment examines neither function, nor habitat quality and biodiversity. What is worse though, is that this limited set of measures are collected as one-off 'snap-shot' samples when it has long been known that the parameters become progressively more variable as impacts accumulate in freshwater systems. For example, oxygen levels are known to fluctuate through diurnal cycles due to algal photosynthesis and that the fluctuations become more extreme as nutrient levels increase with eutrophication.

Crucially, the impacts on freshwater biology are often not directly related to the parameters that are measured. The biological effects are often secondary; for example, when nutrients in rivers increase, fish at first are not affected directly (although at very high levels these nutrients are toxic), but first the ensuing increase in algal growth can lead to extreme fluctuations in oxygen availability. As a regional example, oxygen saturation fluctuates enormously in the Manawatu River. At one point in the river (Homelands Road, below an intensively farmed catchment) oxygen saturation levels in summer vary from less than forty percent in the early morning to more than one hundred and forty percent in the late afternoon of the same day (Clapcott and Young 2009). These extremes (both low and high) are potentially lethal for all stream life, or at least harmful, but because guidelines and measurements are based on 'snapshot' sampling, all this diurnal variability is overlooked, and thus the detrimental consequences are generally not apparent to resource managers. The other 'water-quality' parameters; nutrient levels, pH, suspended sediments and temperature, also vary in degraded systems; however, unlike oxygen the changes are not always diurnal but also relate to flow and biological instream processes. For example, the bulk of the phosphorus entering flowing systems is during flood events and both phosphorous and nitrogen levels can vary as these nutrients are taken up and released by instream plant life. Obviously, assessing such variability using one-off snap-shot sampling is not scientifically robust.

Crucially other key indicators of ecological decline are not measured at a national scale, including physical alteration of habitat by deposited sediment, which infills interstitial spaces in the substrate that are known to be crucial to fish and invertebrate life (McEwan and Joy 2011, 2014). As well as the physical instream engineering of rivers for flood control using heavy machinery and the associated confining of rivers within stop-banks, there is the loss of habitat to migrating fish and the blockage of downstream passages to complete life-cycles caused by dams for hydroelectricity and irrigation.

abundance and diversity have been declining for at least the past century but the rate has accelerated over the past 40 years. While only one species (the grayling *Prototroctes oxyrhynchus*) has become extinct, the range and abundance of almost all species has diminished. The declines are revealed by the increase in the number of species listed as threatened over the past 20 years, with the proviso that the criteria for threat rankings change over time and data for the listings inevitably lag behind actual declines. In 1992 the New Zealand Department of Conservation (DOC) recorded ten species as threatened; by 2002 that number had risen to 16 species (Hitchmough 2002). Three years later, in 2005, 24 species were listed as threatened (Hitchmough and Cromarty 2007). In 2007 a new threat classification scheme was established (Townsend *et al.* 2008) using a reduced set of categories but retaining the key threat descriptors from previous classifications. Under this new system 68% of all extant native taxa and 76% of all non-diadromous taxa are considered threatened or at risk (Allibone *et al.* 2010). This proportion of threatened fish species is one of the highest globally and gives a strong indication of the true extent of freshwater ecosystem decline in New Zealand.

These reductions in freshwater fish diversity have been paralleled by declines in invertebrate diversity and distribution. The number of invertebrate taxa that might be considered at risk to some degree increased from 69 in 2002, to 139 in 2005, to 295 in 2010 and includes New Zealand's only freshwater crayfish and mussel species. Although, some of this increase in invertebrates listed as declining reflects increasing knowledge of taxonomy and distribution, the number of nationally critical taxa has increased over the same time from four in 2002, to 11 in 2005, to 58 in 2010 (Joy and Death 2013). However, even within this biodiversity assessment there are some clear anomalies, with the crayfish (*Paranephrops planifrons*) listed, but its commensal platyhelminth flatworm (*Temnohaswellia novaezelandiae*) not listed. See also Box 11.2.

11.5 Biodiversity impacts

One of the dominant natural patterns of native fish distributions is that species richness and abundance are greatest near the coast (in unimpacted waterways) and both decrease inland (Joy *et al.* 2000; Joy and Death 2001). This arises from the movement of diadromous species between ocean and freshwater biomes, and has major implications for fish distribution and biodiversity. The problem for the native fish is that freshwater ecosystem health generally deteriorates downstream, so the lower reaches are generally more degraded, but this is where biodiversity potential is highest (McDowall 1990; Joy *et al.* 2000); conversely, the healthiest habitat tends to be in the upper reaches of rivers where diversity is naturally lowest. In New Zealand diadromous fish comprise a large part of freshwater fish biodiversity, therefore changes in land-use, chemical barriers, or physical barriers like dams will affect these fish in particular. Thus, pristine upper reaches of waterways are not available to the majority of New Zealand fish. While, in the geological past, having part of the population out at sea at any one time was a good strategy (for example when major volcanic disturbance occurs), the impacts recent land-use changes have wrought on rivers in New Zealand mean this is no longer the case.

Another major impact is accelerated sediment deposition caused by forest clearance and poor management of agriculture in hill country land (Quinn and Stroud 2002). Suspended sediment receives some attention (see Box 11.1) but a major, and probably more important, issue for native fish is deposited sediment. Most New Zealand fish

> **Box 11.2** Freshwater fish biodiversity trends
>
> To evaluate trends in freshwater fish biodiversity in New Zealand, an internationally used measure of the integrity of fish communities, the Index of Biotic Integrity (IBI) (Joy and Death 2004) was applied to a huge database of freshwater fish distribution records collected throughout New Zealand over the past 40 years (Joy 2009). A trend analysis of these IBI scores clearly showed a statistically significant decline in fish communities for all catchment land-use types in New Zealand over the past 40 years, especially in the past decade. The strongest declines were at sites with catchments in agricultural and urban land-cover (Joy 2009).
>
> Biodiversity trends over four decades were assessed by using fish assemblages with the IBI, but an alternative is to look at trends in the distribution of the individual fish species. To do this for the past four decades, more than 22 000 distribution records from 26 species found in flowing waters were gathered from the New Zealand freshwater fish database (Richardson 1989). Changes in the proportions of sites containing each species over this time were assessed using the Mann–Kendall trend test. Twenty-six (77%) species had negative coefficients, meaning the proportion of sites at which these species had been found had decreased over the 40 years. After correcting for false discovery rates (FDR) (Benjamini and Hochberg 1995), nine (35%) of the 26 species had significant trends and they were all declines. Of the nine species, eight were native; six were endemic, and one was non-native (brown trout). All nine species were migratory: five amphidromous (black flounder, torrentfish, common bully, bluegill bully, and koaro), two anadromous (brown trout and common smelt), and one catadromous (longfin eel). Trends for each of the species were also assessed in the two major land-cover classes; namely, native vegetation (indigenous forest and scrub) and pasture. To visualise the relationships between species and land-cover the coefficients for the trend tests were plotted for these two land-use types (Figure 11.1). The plot shows the individual species trends in relation to land-cover.

species are benthic and some spend a considerable proportion of their time in the substrate below the stream bed (McEwan and Joy 2013); this makes them susceptible to sediment build-up due to deposited sediment filling the interstitial spaces in which they live, severely reducing the amount of available habitat. Many New Zealand streams are now impacted by the extent of deposition of fine sediment, reducing the number of individuals that can occupy any reach of a waterway.

Other impacts on freshwater fish biodiversity include competition from, and predation by, exotic fish. The New Zealand freshwater fish fauna evolved without large pelagic species like salmonids, and this has increased the likelihood of negative interactions with these introduced species (McDowall 2000). On the other hand, the economic and sport values of trout mean that without them freshwaters would potentially have less protection and be in a worse state (Joy and Atkinson 2012).

11.6 Drivers of freshwater declines

The decline of freshwater biodiversity in New Zealand resembles global declines in biodiversity. This is not surprising given the drivers of decline in New Zealand and

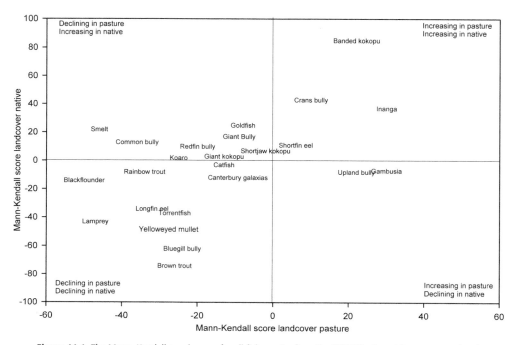

Figure 11.1 The Mann–Kendall trend scores for all fish species from the NZFFDB, plotted for pasture and native vegetation catchment classes from the River Environment Classification (Snelder *et al.* 2004). For each of the fish species a negative score indicates that the species is declining, i.e. the proportion of sites occupied by the species has decreased, a positive score means an increase in proportional occupancy.

their impacts on freshwater biodiversity are similar to those occurring globally. These pressures include eutrophication, habitat loss and population isolation caused by the damming of rivers, habitat destruction, species invasions, overharvesting and climate change (Allan and Flecker 1993). While this list of pressures is not comprehensive, it does include the major impacts; however, ascertaining how they interact, particularly the question of whether they are additive or multiplicative, is difficult to assess (Ormerod *et al.* 2010).

In New Zealand the declines in the health of freshwaters are for the most part related to agricultural impacts; excess sediment, phosphorus and nitrogen, as well as faecal pathogens (Ballantine and Davies-Colley 2010). Thus, the major contemporary driver of the deterioration in the health of New Zealand's lakes, groundwater, rivers and streams is associated with increases in nutrients, mainly nitrogen from the virtually uncontrolled intensification of dairy production. This escalation in intensity is driven by a farming system based on a strategy of low-cost production which, in the absence of any central government leadership, has inevitably led to many unsustainable practices (Baskaran *et al.* 2009). The main issue for freshwaters from this intensification is diffuse-source nutrient and pathogen pollution of waterways from the pasture-based livestock farming model. This diffuse pollution is the run-off or seepage through soils of nutrient-laden water and urine due to high stocking rates. The remarkable stocking rates now found in New Zealand have been achieved by increasing use of 'off-farm' feed supplements like palm kernel, and fossil fuel-derived nitrogenous fertiliser and imported fossil phosphate.

As an example of the magnitude of intensification of dairy farming in New Zealand, between 1990 and 2010 the number of dairy cows in the South Island increased seven-fold, with an obvious massive impact on the quality of lowland streams. During the same period the number of cows in the Waikato River catchment increased by 37% and over that period nitrogen levels in that river increased by 40% and phosphorus by 25% (NIWA 2010). Dairy cow numbers reached 6.5 million in 2012 (Statistics New Zealand) and, given that each cow excretes more waste than 15 humans (Waikato Regional Council 2008), the human-equivalent population of New Zealand would be more than 90 million. The actual human population of New Zealand is less than 4.5 million, thus, these statistics put the relative volume of human versus animal wastes into perspective.

11.7 Environmental protection failures

11.7.1 Legislative approaches

At the same time that the Resource Management Act (RMA, 1991) legislation was passed into law, New Zealand committed internationally to halt environmental declines at the United Nations Rio Earth Summit (UN Conference 1992). But in the ensuing two decades there has been a comprehensive failure to achieve any of those commitments. The list of failures begins with Principle 16 which declared that *'authorities should endeavour to promote the internalization of environmental costs and the use of economic instruments'* and further that *'the polluter should bear the cost of pollution'*. To date there has not been a fee applied, or any attempt to internalise the costs of the pollution of freshwaters in New Zealand. The only cost for 'out of pipe' (point source) polluters is a one-off consent fee which is essentially an administration charge required by local government. The problem for freshwater health is that the biggest pollution source in New Zealand is not at point source; rather it is diffuse and this form of freshwater contamination is not controlled. Diffuse pollution is the nutrient, urine and faecal contamination that make their way into lakes and rivers through and over the soil, mainly via cow urine patches and seeping of faeces. The resulting additions of nutrients and microbial contaminants to lakes, rivers and streams has led to many ecological and human health impacts (outlined above) but these are not paid for by the polluters. To date the Lake Taupo catchment is the only one in New Zealand where an attempt has been made to reduce diffuse pollution and protect this iconic lake from nutrient pollution, through regulation using a nitrogen cap and trade system.

Apart from the Lake Taupo example, local authorities have failed to use the capacity they have had under the RMA to control the obvious impacts of farming intensification on freshwaters. Instead they have chosen only to control the much less significant impact of dairy-shed wastewater. The main reason for local government (councils') failure to address the main impact on freshwater quality in New Zealand lies to some extent with the failure of central government to implement a National Policy Statement (NPS) for freshwater management since the RMA (1991). This was despite a legislative requirement to do so soon after the act was passed. This would undoubtedly have given guidance to Regional Councils and confidence that they wouldn't be picked off individually as protective legislation was developed. The NPS was finally put in place in 2011 but it has been criticised as being 'too little and too late' and 'unlikely to produce any improvement in water quality' (Sinner 2011). The only other response from central government around freshwater protection was to set up a stakeholder group known as the Land and Water Forum (LAWF). The LAWF was proposed and set up by central government as a collaborative approach to managing freshwaters into the future. In reality the forum membership was heavily weighted toward very well-resourced stakeholders, with minimal representation from freshwater protectors and conservationists. The LAWF worked through many issues over four years and produced three reports and made many recommendations, but none thus far that has any chance of halting freshwater declines has made it into legislation.

In terms of biodiversity declines, ironically, none of the threatened native fish species have any legal protection; indeed, at least five threatened fish species are harvested commercially and recreationally. Absurdly, the Freshwater Fisheries Act (1983) formally protects the extinct grayling last seen in the 1930s, some introduced fish, mainly trout and salmon, but not native fish. The native fish are only protected if they are not used for 'human consumption or scientific purposes', thus, in reality they have no protection. In addition, four of the five species that make up the whitebait catch (juveniles of the migratory galaxiids; a popular recreational and commercial seasonal harvest in New Zealand) are listed as threatened. To summarise, 50 years after its extinction a law was passed to protect the endemic grayling, and other native fish species have no legal protection apart from harvesting rules.

11.7.2 Voluntary approaches

Other than the National Policy Statement on freshwater management the only significant response to date from central government to the many freshwater issues was the negotiation of a voluntary accord with the largest dairy company in New Zealand, Fonterra, signed originally in 2003 (and regularly updated since). This agreement, originally called the Clean Streams Accord, was between Fonterra, Regional Councils and the Ministry for the Environment, and required that farmers undertake a number of measures to reduce their impacts on freshwater. The agreement at first appeared impressive but closer investigation revealed many failings. These failings include that the accord lacks any ability to enforce requirements, and the stream fencing requirements ignore the smaller streams where actions could be most effective. A further crucial flaw is that all the monitoring requirements are for assessing whether the accord requirements are being implemented rather than any assessment of whether they are in fact improving water quality.

The result has been that while the accord progressed stream fencing, it did not include riparian buffer zones and mostly only occurred on larger waterways. What it did do, however, was to focus publicity on the continuing problems of dairy effluent management; and it resulted in the uptake of farm nutrient budgeting. The down side, however, was that it gave Regional Councils a pretext to continue to defer introducing rules to address the impacts of farm intensification and diffuse pollution. So the result was that while the accord was a great public relations tool for the industry to suppress criticism there is no evidence that it has done anything to halt the decline of water quality. The updates to the accord that have occurred since its inception have not addressed any of the issues raised above and in 2013 the phrase 'clean streams' was removed and it is now called the Water Accord.

11.8 Solutions

The precedent and example for reducing farming intensity, and thus impacts on freshwater, have been set with a cap and trade limits on nutrient loads to Lake Taupo. While this is just one of a range of options, it does give a model for the sort of approach that could be adopted nationally. The other potentially positive change is the agreements made to protect and improve the health of the Waikato River resulting from a claim over the river taken to the Waitangi Tribunal by Tainui-Raupatu. Under this co-management agreement (Waikato-Tainui 2010), Tainui-Raupatu iwi have had their vision and strategy for cleaning up the Waikato River legislated. The vision statement is aspirational and is summarised in this statement from the report: 'Our vision is for a future where a healthy Waikato River sustains abundant life and prosperous communities who, in turn, are all responsible for restoring and protecting the health and wellbeing of the Waikato River, and all it embraces, for generations to come'. As an example of just how far reaching this vision is, one of the objectives is: (objective k) 'the restoration of water quality within the Waikato River so that it is safe for people to swim in and take food from over its entire length'. To achieve these aspirational objectives however, substantial changes would be required on land-use in the Waikato River catchment. However, at the time of writing, three years after the enactment of this legislation, there is little sign of any changes to rules necessary to achieve the vision and strategy. Nevertheless this vision and strategy may still lead the way for the changes in land-use required to ultimately improve water quality in New Zealand.

On farms, the only solution for protecting the natural capital of New Zealand, while at the same time producing large quantities of low-value milk powder or any other agricultural product is simply to 'close the loop' and ensure that nutrients and soils stay on farms and are cycled within the system, and that fossil fuel use for fertiliser or energy is reduced (Nelson 1996). Examples of this truly sustainable style of farming are occurring in North America and Europe and these farms can be used to show the way in New Zealand.

11.9 The future

The future prospects for freshwater health in New Zealand are bleak, there is no prospect of any 'polluter-pays' legislation and the lag times for nutrients entering waterways are often decades. Thus even if moves to cap and reduce nutrient loads were immediate and applied nationally, water quality will continue to decline for some time. If the polluter-pays principle had been applied in New Zealand some decades ago as promised at the Rio Summit (1992) then the massive intensification of dairy farming would likely not have occurred and increases in profit would instead have come through adding value. Thus, the hands-off approach over the past two decades has led to a huge overshoot of the carrying capacity of soils and water and the withdrawal from this situation will be difficult and expensive. Even worse than the failures to limit intensification, councils are now involved in funding irrigation schemes that will inevitably increase nutrient loadings on freshwater systems. The Regional Councils have a crucial legislative frontline role in protecting the environment but several have entered into a conflict of interest when they have become investors in schemes that will result in further freshwater degradation. This development has seen moves by these councils to massively increase nutrients limits in waterways. The justification being used is that by controlling one of the two nutrients (nitrogen and phosphorus) required for algal proliferation they can allow excessive levels of another without impacts occurring. This limiting nutrient scenario only happens in rare extreme cases and allowing build-up of one nutrient will set up the potential for a major algal bloom, as it is simply not possible to stop nutrient loss from the landscape.

11.10 Conclusions

The deterioration in freshwater health in New Zealand developed through the twentieth century but accelerated over time, especially in the past few decades. At the same time the United States and many European countries have implemented regulatory changes and halted declines, resulting in improvements in water quality in many cases. New Zealand has ignored this and continued with unconstrained intensification of farming in association with exponential increases in fertiliser use and importation of stock feed. The relationship between land-cover (a surrogate for land-use) and fish communities reveals the obvious causes of declines (Joy 2009). In general, deterioration in the health of freshwaters is related to agricultural impacts: excess sediment, phosphorus and nitrogen, as well as faecal pathogens (NIWA 2010). The major driver of this deterioration is the expansion and intensification of agriculture, particularly dairy farming (Wright 2007). The decline in fish biodiversity is also related to the loss of habitat, a result of barriers to migration such as hydroelectric dams and weirs and the draining of more than 90% of wetlands, mainly for agriculture (Joy 2012).

In 1991 the Resource Management Act (RMA) was passed into legislation. It encompassed the lofty ideals of a generation of New Zealanders committed to a healthy and environmentally sustainable future. Sadly the work of the authors of the RMA proved futile because over the following two decades the RMA was systematically diluted by a lack of enforcement and then later weakened through the Resource Management Simplifying and Streamlining Act (2009), with further proposals now being considered to further weaken the protection intended by the RMA. This legislation put emphasis on speeding up the consent process and thus, less emphasis on the quality of decisions. This weakening of the law, combined with a failure to address the most pervasive impact on water quality – the intensification and industrialisation of dairy farming – has in part resulted in New Zealand's slide to the lowest levels of environmental performance globally (Bradshaw *et al.* 2010). This study, based on a suite of measures including fertiliser use, biodiversity loss, marine captures, water quality and more, ranked New Zealand around 130 of 180 countries.

The only indication of a future move to improve water quality in New Zealand is the involvement of Māori in freshwater management (the Waikato co-management example) and the economic value of tourism, leading to moves to protect Lake Taupo by reducing dairy farming intensity. However, while this co-management has been mandated there is little evidence of changes in regional plans to meet the aspirations.

The conflicting needs of agricultural intensification, biodiversity conservation, sport fisheries management, and urban spread, have created many pressures on water resources. These show no sign of abating – in fact, all are increasing. Despite the many unequivocal measured impacts on fresh water from intensification of farming, the government is backing further intensification, mainly of dairy farming, through irrigation in drier areas. Consequently, impacts on freshwater biodiversity will inevitably accelerate. Irrigation has already increased; for example, from 1999 to 2006 water allocation grew by 50%, mostly for irrigation, and this is likely to increase substantially. Undoubtedly the combination of climate change, agricultural intensification and further urban spread will have very serious consequences for freshwater biodiversity in New Zealand (Ling 2010).

REFERENCES

Allan, J. D. and A. S. Flecker. 1993. Biodiversity conservation in running waters. *BioScience* **43**:32–43.

Allibone, R., B. David, R. Hitchmough, *et al.* 2010. Conservation status of New Zealand freshwater fish, 2009. *New Zealand Journal of Marine and Freshwater Research* **44**:271–287.

Ball, A. 2006. *Estimation of the Burden of Water-borne Disease in New Zealand: Preliminary Report.* Prepared as part of a Ministry of Health Contract for scientific services by ESR, Wellington.

Ballantine, D. J. and R. J. Davies-Colley. 2010. *Water Quality Trends at NRWQN Sites for the Period 1989–2007.* National Institute of Water & Atmospheric Research Ltd.

Baskaran, R., R. Cullen, and R. Colombo. 2009. Estimating values of environmental impacts of dairy farming in New Zealand. *New Zealand Journal of Agricultural Research* **52**:377–389.

Benjamini, Y. and Y. Hochberg. 1995. Controlling the false discovery rate – a practical and powerful approach to multiple testing. *Journal of the Royal Statistical Society Series B – Methodological* **57**:289–300.

Bradshaw, C. J. A., X. Giam, and N. S. Sodhi. 2010. Evaluating the relative environmental impact of countries. *PLoS ONE* **5**:1–16.

Clapcott, J. and R. G. Young. 2009. *Temporal Variability in Ecosystem Metabolism of Rivers in the Manawatu–Wanganui Region.* Cawthron Institute.

Daughney, C. J. and M. Wall. 2007. *Groundwater Quality in New Zealand: State and Trends 1995–2006.* GNS Science Consultancy Report 2007/23.

Hitchmough, R. 2002. *New Zealand Threat Classification System Lists.* Department of Conservation, Wellington.

Hitchmough, R. and P. Cromarty. 2007. *New Zealand Threat Classification System lists.*

IUCN. 2010. Redlist. http://www.iucnredlist.org/.

Joy, M. K. 2009. *Temporal and Land-cover Trends in Freshwater Fish Communities in New Zealand's Rivers: An Analysis of Data from the New Zealand Freshwater Database – 1970–2007.* A report to the Ministry for the Ministry for the Environment. Massey University.

Joy, M. K. 2012. Water quality. Chapter 2 in World Wildlife Fund, *Beyond Rio: New Zealand's Environmental Record Since the Original Earth Summit.* World Wildlife Fund Wellington.

Joy, M. K. and N. K. Atkinson. 2012. *Salmonids and Native Fish in New Zealand; Are Trout to Blame for the Decline in Native Fish?* A report prepared for Fish and Game New Zealand; Wairesearch Ltd., Wellington.

Joy, M. K. and R. G. Death. 2001. Control of freshwater fish and crayfish community structure in Taranaki, New Zealand: dams, diodromy or habitat structure? *Freshwater Biology*, **46**(3),417–429.

Joy, M. K. and R. G. Death. 2004. Application of the index of biotic integrity methodology to New Zealand freshwater fish communities. *Environmental Management* **34**:415–428.

Joy, M. K. and R. G. Death. 2013. Freshwater biodiversity. In **J. Dymond**, editor. *Ecosystem Services in New Zealand*. Landcare Research, Palmerston North.

Joy, M. K., I. M. Henderson, and R. G. Death. 2000. Diadromy and longitudinal patterns of upstream penetration of freshwater fish in Taranaki, New Zealand. *New Zealand Journal of Marine and Freshwater Research* **34**:531–543.

Larned, S. T., M. R. Scarsbrook, T. H. Snelder, N. J. Norton, and B. J. F. Biggs. 2004. Water quality in low-elevation streams and rivers of New Zealand: recent state and trends in contrasting land-cover classes. *New Zealand Journal of Marine and Freshwater Research* **38**:347–366.

Ling, N. 2010. Socio-economic drivers of freshwater fish declines in a changing climate: a New Zealand perspective. *Journal of Fish Biology* **77**:1983–1992.

McDowall, R. M. 1990. *New Zealand Freshwater Fishes: A Natural History and Guide.* Heinemann Reed, Auckland.

McDowall, R. M. 2000. *The Reed Field Guide to New Zealand Freshwater Fishes.* Reed

McEwan, A. J. and M. K. Joy. 2011. Monitoring a New Zealand freshwater fish community using passive integrated transponder (PIT) technology; lessons learned and recommendations for future use. *New Zealand Journal of Marine and Freshwater Research* **45**:121–133.

McEwan, A. J. and M. K. Joy. 2014. Habitat use of redfin bullies *(Gobiomorphus huttoni)* in a small upland stream in Manawatu, New Zealand. *Environmental Biology of Fishes* **97**:121–132.

Nelson, T. 1996. Closing the nutrient loop. *World Watch*; Volume 9: Issue 3.

National Institute of Water and Atmospheric Research. 2010. How clean are our rivers? *Water and Atmosphere.* Issue 1, July.

Ormerod, S. J., M. Dobson, A. G. Hildrew, and C. R. Townsend. 2010. Multiple stressors in freshwater ecosystems. *Freshwater Biology*, **55**(s1) 1–4.

Quinn, J. M. and M. J. Stroud. 2002. Water quality and sediment and nutrient export from New Zealand hill-land catchments of contrasting land use. *New Zealand Journal of Marine and Freshwater Research* **36**:409–429.

Richardson, J. 1989. The all-new freshwater fish database. *Freshwater Catch* **41**:20–21.

RMA (Resource Management Act) 1991. http://www.legislation.govt.nz/act/public/1991/0069/latest/DLM230265.html

Sinner, J. 2011. *Implications of the National Policy Statement on Freshwater Management,* Prepared for Fish & Game New Zealand., Cawthron Report No. 1965, Nelson.

Snelder, T. H., F. Cattaneo, A. M. Suren, and B. J. E. Biggs. 2004. Is the River Environment Classification an improved landscape-scale classification of rivers? *Journal of the North American Benthological Society* **23**:580–598.

Townsend, A. J., P. J. de Lange, C. A. J. Duffy, *et al.* 2008. *New Zealand Threat Classification System Manual.* Report 978-0-478-14363-8, Department of Conservation, Wellington.

UN Conference 1992. *Report of the United Nations Conference on Environment and Development*, Rio de Janeiro, Brazil, 3–14 June 1992.

Verburg, P., K. Hamill, M. Unwin, and J. Abell. 2010. *Lake Water Quality in New Zealand 2010: Status and Trends.* NIWA.

Waikato-Tainui. 2010. Waikato-Tainui Raupatu Claims (Waikato River) Settlement Act 2010 No 24.

Waikato Regional Council. 2008. *The Condition of Rural Water and Soil in the Waikato Region; Risks and Opportunities.* Waikato Regional Council, Hamilton.

Williams, M. 2004. *Growing for Good, Intensive Farming, Sustainability and New Zealand's Environment.* Parliamentary Commissioner for the Environment.

Wright, J. 2007. *Dairy Farming Impact on Water Quantity and Quality;* Briefing to the Primary Production Committee. Wellington

A garden at the edge of the world; the diversity and conservation status of the New Zealand flora

Carlos A. Lehnebach

Summary

After drifting away from Gondawana, the land that we presently know as New Zealand went through several geological events such as the Oligocene drowning, the upheaval of the Southern Alps and several glacial–interglacial periods. These events have dramatically shaped New Zealand's flora and fauna, causing the extinction of some lineages but also promoting speciation and ecological diversification in others. Nowadays, the New Zealand vascular flora includes over 2230 species; most of them endemic (c. 80%). Despite being one of the last places on the world to be reached by people, damage to its native flora and fauna has been significant. Currently, over 38% of the native vascular flora is of conservation concern and six species became extinct. The main threats to New Zealand plants are habitat destruction, herbivory, competition from weeds and the disruption of ecological interactions. Birds are important pollinators and seed dispersers in New Zealand so a decline in their number, caused by the introduction of mammalian predators, has severely damaged plant–bird interactions and the effects of reproduction and dispersal failure are already being noticed. The spread of introduced plant pathogens is also a threat to native plants and their effect on native trees, in particular, is a matter of current concern. Preservation of New Zealand's flora is a matter of national and international relevance and many initiatives are working towards this goal. Examples include: habitat protection and restoration, conservation by cultivation, and establishment of seed banks. Future research, however, should focus on the autoecology of threatened plants, particularly in those species that depend on mutualistic interactions. This research will

Austral Ark: The State of Wildlife in Australia and New Zealand, eds. A. Stow, N. Maclean and G. I. Holwell. Published by Cambridge University Press. © Cambridge University Press 2015.

help to maximise resources allocated to their conservation and secure their survival and functioning in the long term.

12.1 Origin of New Zealand flora

The origin of the New Zealand flora is critical to understanding the country's plant biodiversity, patterns of distribution and the peculiarity of some species. The origin of New Zealand biota has been a topic of much interest and debate, and over time several hypotheses have been put forward to explain the affinity of New Zealand plants with those found in other landmasses of the Southern Hemisphere (Gibbs, 2006). Traditionally, opposing views such as vicariance and dispersal have been evoked to explain such biogeographic patterns. However, with the advent of molecular techniques, new geological evidence and the increasing number of phylogenetic studies of New Zealand species, interesting details about the origin, colonisation dynamics and evolution of the New Zealand flora have been uncovered.

New Zealand has its origin in the supercontinent of Gondwana. It was situated on the eastern side of Gondwana along with Australia, Antarctica and part of southern South America. These four regions started to drift apart from the rest of Gondwana about 130 million years ago but remained in contact with each other until *c.* 80 million years ago. At this point, the Tasman Sea started to develop between Australia and New Zealand, leaving New Zealand completely isolated from other landmasses (Gibbs, 2006).

It was commonly believed that many of the ancestors of plants and animals currently occurring in the country originated in Gondwana and much of the current biota has evolved in isolation. The presence of members of southern hemisphere groups such as *Agathis*, *Metrosideros* and *Nothofagus* was commonly cited as evidence to support this theory. Recent geological evidence, however, indicates most of New Zealand, if not all, was submerged during the Oligocene–early Miocene, about 25 million years ago. The drowning of New Zealand would have caused the extinction of most of the resident terrestrial fauna and flora at the time. Therefore, it is conceivable that much of the current biota evolved largely from fortuitous arrivals during the past 22 million years coinciding with tectonic upheavals pushing the current landscape above sea level.

Interestingly, molecular studies of emblematic Gondwanan genera have provided contrasting evidence. For instance, molecular studies on *Agathis australis* (kauri) by Stöckler *et al.* (2002) indicate this is the oldest living species in the genus, that it survived the Oligocene drowning and that the ancestor of extant *Agathis* species in Australia, Melanesia, Vanuatu and Fiji originated in New Zealand. Furthermore, survival of *Agathis australis*, along with the presence of pollen of several modern plants in the fossil record, implies that sufficient land may have remained above sea level during this time for these plants to survive. On the other hand, molecular studies of southern beech (*Nothofagus*) species by Knapp *et al.* (2005), suggest their presence in New Zealand has been intermittent with several arrivals, establishments, and multiple extinction events since New Zealand separated from Gondwana.

During the past ten years phylogenetic studies of a number of lowland and alpine plant species have indicated long distance dispersal (LDD) from other land masses to New Zealand, which have played an important role in shaping the current flora. Dispersal into New Zealand has occurred as a direct event or 'via stepping stones', using other land masses, or mountain tops for alpine lineages, to reach New Zealand. Direct LDD events have been considered improbable (Winkworth *et al.*, 2002b); however, recent phylogenetic studies of plants with disjunct distribution support their occurrence. For example, direct LDD from Eurasia into New Zealand may explain the bipolar distribution of *Ceratocepahala* (Ranunculaceae), a genus with only three species; two native to Eurasia and one endemic to New Zealand (Lehnebach, 2008). Direct LDD events between North America and New Zealand–Australia, during the late Tertiary and Quaternary, have also been suggested for the origin of other native plant genera such as *Lepidium* (Brassicaceae) and *Microseris* (Asteraceae) (Mummenhoff and Franke, 2007).

Dispersal into New Zealand from the Northern Hemisphere, using southern Asia, New Guinea and Australia as 'stepping stones' was hypothesised over 30 years ago to explain the presence of northern hemisphere elements in the New Zealand flora (Raven, 1973). Nowadays, molecular phylogenetic studies have also confirmed such dispersal routes for the ancestors of the genus *Corynocarpus* (Wagstaff and Dawson, 2000), *Gunnera* (Wanntorp and Wanntorp, 2003), a group of Australian *Ranunculus* (Armstrong, 2003), and members of the Australian Chenopodiaceae (Kadereit *et al.*, 2005). During the late Pliocene and Pleistocene (2–7 Mya) the collision of the Australian plate with the Asian plate resulted in the uplift of mountains in Southeast Asia, New Guinea and Australia. Raven (1973) suggested the uplift of these mountain ranges would have created suitable habitats for the migration of sub-alpine and alpine plants between Asia and Australia.

It is important to point out that New Zealand has also served as a source for dispersal into other land masses in the Pacific region. Dispersal from New Zealand to Australia, New Caledonia, and South America has been inferred from several phylogenetic studies (see revision by Winkworth *et al.*, 2002b; Meudt and Bayley, 2008). Two Australian buttercups (*Ranunculus anemoneous* and *R. gunnianus*), for example, have their closest relatives in the Southern Alps of New Zealand (Lockhart *et al.* 2001). Both species belong to different lineages within the New Zealand alpine *Ranunculus* radiation, suggesting dispersal across the Tasman from New Zealand has occurred at least twice (Lockhart *et al.*, 2001). A similar situation has been reported more recently in *Muehlenbeckia* (Schuster *et al.*, 2013).

Local diversification has also been important in the evolution of the New Zealand flora, particularly so in the alpine habitats. Alpine areas in the North and South Island host over 700 plant species and *c.* 90% of them are endemic (Mark, 2012). Phylogenetic studies using sequences of nuclear and chloroplast markers have revealed that a number of these species belong to groups that originated after the arrival of a single ancestor (Breitwieser *et al.*, 1999; Lockhart *et al.*, 2001; Wagstaff *et al.*, 2002; Winkworth *et al.*, 2002a). Some of the best-studied examples are *Myosotis* and *Ranunculus* (Box 12.1), where molecular dating studies suggest diversification of these groups coincided with the uplift of the Southern Alps. The relatively recent origin of the alpine habitats, 4–2 million years ago, and the great extent of morphological and ecological diversification observed in these two genera has suggested diversification of these groups occurred very rapidly (Lockhart *et al.*, 2001; Gibbs 2006; Linder, 2008). Of interest is also the small genetic change accompanying the diversification of these groups. In fact, many of the chloroplast and nuclear markers commonly employed in phylogenetic studies overseas are not informative when used on New Zealand alpine plants, and generally a poorly resolved and uninformative phylogeny is recovered (e.g. Lockhart *et al.*, 2001; Winkworth *et al.*, 2002a; Ford *et al.*, 2007; Prebble *et al.* 2012).

Box 12.1

There are about 40 species of buttercups (*Ranunculus*) in New Zealand. They are found from coastal ephemeral pools, wetlands, and the forest floor to alpine areas up to 2800 m.a.s.l. The most conspicuous species are those restricted to the sub-alpine and alpine areas of the North and South Island. This group includes about 20 taxa that have colonised different habitats such as mobile scree, snowbanks, herbfields, tarns and pockets of fine clay soils (Figure 12.1, Plate 30). Phylogenetic studies using chloroplast and nuclear markers have confirmed there are four lineages within this group, all descend from a single common ancestor (Lockhart *et al.*, 2001). The closest relative to this alpine radiation is found outside New Zealand; this is *R. pseudotrullifolius*, a small buttercup found in a few sub-Antarctic Islands and southern South America (Lehnebach, 2008). This alpine *Ranunculus* group has been regarded as a typical example of an adaptive radiation and it has been equated to plant radiations such as the silver swords alliance or lobeliads from Hawai'i. Similar to these iconic radiations, New Zealand alpine buttercups have diversified very rapidly (within the past 5 million years), they all share a common ancestor and they exhibit aremarkable morphological and ecological diversification, despite the low genetic

Figure 12.1 (Plate 30) Four species of buttercups (*Ranunculus*) endemic to the sub-alpine and alpine areas of the New Zealand Southern Alps (South Island). Clockwise, top left corner: *Ranunculus lyallii, R. haasti* var. *haasti, R. gracilipes, R. crithmifolius* var. *crithmifolius*. A black and white version of this figure will appear in some formats. For the colour version, please refer to the plate section.

> **Box 12.1 (continued)**
>
> divergence detected among the species in this group, suggesting this radiation has been adaptive. Lockhart *et al.* (2001) suggested radiation of this group coincided with the time the main mountain ranges began to develop, which provided a range of novel and unoccupied habitats available for speciation to occur.

12.2 Plant biodiversity in New Zealand

The New Zealand Botanical Region includes the North and South Island, Stewart Island and a number of outlying islands: Kermadec, Chatham and sub-Antarctic Islands. Overall, it comprises an area of *c.* 265 000km^2 with extremely diverse landscapes and habitats. The most up to date list of native plants indicates about 2230 vascular plant species occur in New Zealand (Breitwieser *et al.*, 2012) and over 80% of them are endemic. Ironically, the number of introduced plants is higher and 2471 taxa have been listed (Breitwieser *et al.*, 2012). The most numerous plant group in the flora is the dicotyledonous (1513 species), followed by the monocotyledonous (617 species), and ferns and fern allies (201 species). Gymnosperms are represented by 21 species. The largest family of flowering plants in the country is the Asteraceae, with 292 indigenous species, followed by the Poaceae (188 species) and the Cyperaceae (178) (Breitwieser *et al.*, 2012). Interestingly, most endemics occur at the species level and only 50 of 1249 genera are endemic. There are no endemic plant families. The high number of endemic species in New Zealand has generally been linked to its isolation from other land-masses, suggesting most of them are the result of local speciation events. There are three endemic-rich centres in New Zealand; the upper North Island (*c.* 125 species) and the upper and lower South Island (189 and 90 species, respectively). These areas are believed to be older and geologically more stable (Wardle, 1991; McGlone *et al.*, 2001).

Making an inventory of New Zealand's plant biodiversity is still far from complete and it is believed that over 200 taxonomically indeterminate plants still await formal description or recognition as distinct taxa. Some of these plants can be easily recognised in the field and they have been included in plant surveys and national checklists over many years (e.g. see list in de Lange *et al.*, 2009) but the shortage of professionals in plant biosystematics in New Zealand and limited funding for systematic and taxonomic studies (Black, 2008) has hindered further investigation into their taxonomic status (Box 12.2). Unfortunately, some of these plants are already of conservation concern and their conservation status has been listed as Nationally Critical. For instance, the taxonomic status of the sun orchid *Thelymitra* 'Ahipara' has remained unresolved since its discovery in 1987. Shortly after its discovery, the wetland supporting the orchid was subject to clearing and draining and converted into pasture. Almost 300 plants were rescued from the site and transferred to wetlands administered by the Department of Conservation (de Lange *et al.*, 1991). Twenty-six years after its discovery, *Thelymitra* 'Ahipara' still remains nameless.

12.3 Conservation of New Zealand plants

New Zealand was one of the last places on the world to be reached by people; the first records for human occupation date from *c.* 700–800 years ago. Despite this rather recent arrival, the impact of human settlement on the native flora and fauna has been

Box 12.2

Myosotis (forget-me-not) is one of New Zealand's plant genera with the highest number of taxa of conservation concern. Currently this genus comprises about 40 species and over 75% of them have been ranked either as Threatened or At Risk (de Lange *et al.*, 2010). Furthermore, one species and one variety are already considered Extinct. Unfortunately, *Myosotis* also requires considerable taxonomic revision and boundaries between numerous species pairs are uncertain. More than ten taxonomically indeterminate entities have been detected in the wild and in historical herbarium collections. Besides the great diversity of habitat requirements, plant habit, flower colour and breeding system exhibited by New Zealand forget-me-nots, the highly restricted distribution of some species and the small size of each population at a single site are the most striking features of this genus. In fact, several of the currently accepted species are known only from a few specimens collected at the type locality. *Myosotis mooreana* is among the most extreme examples. This species is known from only a single population where six individuals were counted at the time of its discovery (Lehnebach, 2012). Techniques such as DNA fingerprinting and next generation sequencing, along with traditional morphological studies, are currently being used by botanists to detect species boundaries between closely related species and measure genetic diversity within populations of selected threatened species.

considerable. The most recent assessment of the conservation status of New Zealand's vascular plants listed 901 described taxa of conservation concern, which represents *c.* 38% of the total native vascular flora (de Lange *et al.*, 2010). This figure includes six species classified by the New Zealand Threat Classification System (NZTCS, Box 12.3) as Extinct, 184 species as Threatened, and *c.* 650 species in the At Risk category. A comprehensive assessment of New Zealand threatened and uncommon plants, their life form, habitat and distribution has been prepared by Peter de Lange and colleagues (de Lange *et al.*, 2010). Conservation status of the species included in this list is reassessed every three years and the list updated accordingly. In the latest assessment, the Asteraceae was the family with the greatest number of threatened taxa, followed by the Boraginaceae, Plantaginaceae and Poaceae (de Lange *et al.*, 2010). Most of New Zealand's threatened plants occur within lowland, montane habitats and coastal habitats and the least are found in sub-alpine and alpine settings. Overall, most threatened species were recorded in the South Island (Southland and Canterbury) and the far north region of the North Island (Northland).

12.4   Threats to the survival of New Zealand flora

There are many reasons why New Zealand native plants are threatened with extinction. Some of these threats are: habitat destruction; herbivory; plant pathogens; competition; and disruption of ecological interactions either with their pollinators or seed dispersers. Generally these threats do not operate independently and two or more could be affecting a species simultaneously.

Box 12.3

New Zealand Threat Classification System (NZTCS) sorts all native vascular plants into Extinct, Threatened (with the subcategories Nationally Critical, Nationally Endangered, Nationally Vulnerable), At Risk (with the subcategories Declining, Recovering, Relict, Naturally Uncommon) and Not Threatened (de Lange *et al.*, 2010). This system is comparable with the IUCN Red List conservation status listing and ranking system, but a few modifications are introduced to account for the unique geographical features of New Zealand and the numerous taxa with naturally restricted distributions and small populations. The NZTCS also uses a number of qualifiers to provide additional information on the plant's management requirements and status. Some of the qualifiers used are 'Conservation Dependent', 'Data Poor' or 'Extinct in the Wild'. In contrast to other rare plant lists overseas, species included in this list do not have legal protection. Rather, the list is used as guide to help the Department of Conservation to manage these threatened species, allocate resources and prioritise conservation efforts. In fact, there is no legal protection for threatened plants in New Zealand; except where found on Conservation land. The only legislation on native plant protection dates from 1934 (The Native Plants Protection Act) and it is considered simplistic and unlikely to provide any real protection (de Lange *et al.*, 2010).

12.4.1 Habitat destruction

Habitat destruction is one of the most serious threats to New Zealand flora and fauna. New Zealand's landscape has been subject to considerable change since human settlement and this has occurred very rapidly. Transformation of the landscape began with the arrival of Māori to New Zealand about 700–800 years ago and it has been estimated about 40% of native forests were cleared after their arrival (see McWethy *et al.*, 2010). Recent studies suggest this was a period of intense local fire activity, and it has been referred to as the 'Initial Burning Period' by McWethy *et al.* (2010). During this time, forests in the eastern side of the South Island and part of the North Island were burned and converted into grasslands or fernlands, and maintained this way by repeated burning (Given, 1981; McWethy *et al.*, 2010; Perry *et al.*, 2012). The rate at which habitat modification occurred was greatly increased by the permanent influx of European settlers that occurred from 1830 onwards. During this period, most of the lowland and accessible parts of the country were cleared and replaced by crops and pasture of introduced grasses, timber was extracted from native forests, and native tussock grasslands were subjected to sheep and cattle grazing (Given, 1981). Furthermore, some areas were completely cleared for mining, road construction and urban development. Overall, since human settlement, the indigenous forest cover of New Zealand has been reduced from an estimated 70–80% (McWethy *et al.*, 2010) to only 30% (Ministry for the Environment, 2010). Nowadays grasslands cover half of the country's land area, most of them are devoted to agricultural grazing (Ministry for the Environment, 2010). Damage to wetlands has also been substantial and over 90% of the wetlands occurring in New Zealand pre-human settlement has been destroyed (Hunt, 2007). Habitat destruction is still ongoing (Statistics New Zealand, 2009) and in the past ten years about 50 000 hectares of native forest have been converted to grassland and forest plantations, and about 100 hectares of wetlands lost

---

Box 12.4

The umbrella fern, *Sticherus tener*, is a Nationally Critical species. This fern was initially known in New Zealand from a single record made in 1980s in Fiordland, in the south-western portion of the South Island. Because of this sole record and its occurrence also in Australia, *S. tener* was considered a Vagrant species in New Zealand, and its conservation status was not prioritised (de Lange *et al.*, 2009). A recent revision of New Zealand umbrella ferns has reported the discovery of several other sites for *S. tener* in New Zealand, all of them in the South Island (Brownsey *et al.*, 2013). Despite the increase in the number of known sites for this fern, abundance at each site is low and the authors of this study believe there are fewer than 250 individuals of *S. tener* in the country, which is the upper threshold of the Nationally Critical category. Currently, the conservation of this species is a matter of great concern since some of the newly discovered populations, those representing the northern limit of its distribution; occur in an area where a *c.* 200 hectare opencast coal mine has been proposed (Nichol and Overmars, 2008). Interestingly, the presence of this umbrella fern had been reported before from this area but it was misidentified as *S. flabellatus*, a Non Threatened species. Cultivation of *S. tener* has not been attempted in New Zealand yet but reports from Australia claim it can be easily grown from spores.

(Ministry for the Environment, 2010). Habitat transformation has also extended to coastal areas where housing development has modified dunes systems or destroyed habitats, such as ephemeral pools and coastal wetlands.

Habitat loss is considered a threat to the survival of over 50% of the plants currently listed under the Nationally Critical category (de Lange *et al.*, 2010). Habitat destruction has also been linked to the extinction of herbaceous species such *Lepidium obtusatum* and *Stellaria elatinoides*, which were once found in coastal habitats and wetlands/riverbanks, respectively. Securing the habitat of rare or threatened species outside protected areas is difficult to achieve (Box 12.4). Fortunately, many private landowners in New Zealand have an interest and commitment towards conservation of native species. Initiatives such as the Queen Elizabeth II Trust (www.openspace.org.nz), which helps landowners secure long-term protection of natural and cultural features on their land through open space covenants; or Nga Whenua Rahui, a fund available to support protection of native forest and other indigenous ecosystems on Maori land, are available. These initiatives are further supported by the Statements of National Priorities set by the Ministry for the Environment and Department of Conservation to provide the regional and district councils with a framework for decision-making about biodiversity on private land. These Statements focus on the protection of habitats supporting threatened native species, native vegetation associated with ecosystems types that have become uncommon due to human activity (wetlands, sand dunes), or those ecosystems originally rare (e.g. geothermal systems).

12.4.2   Herbivory

After habitat destruction, browsing by animals is perhaps the second most serious threat to native plants in New Zealand. Numerous exotic animals have been introduced into the country. For instance, animals such as deer, Himalayan tahr and chamois were released in the country to promoted game hunting (King, 2005) while other animals such as goat and pigs were introduced as early as 1773 during Captain James Cook's first voyage to New Zealand to provide meat and fibre for castaways. The damage these animals have caused on the vegetation is variable and ranges from altering successional patterns and species composition (e.g. Krull *et al.*, 2013) to reducing reproductive success and recruitment of palatable species. Species with subterranean organs, such as orchids, have also been affected by the introduction of feral pigs which commonly uproot entire populations. Other animals, such as rats, mostly inadvertently introduced to the country, have limited the reproduction and establishment of native plants by feeding on their flowers, fruits or seeds.

Of all introduced animals in the country, possums (*Trichosurus vulpecula*) have caused the most damage to native forest species, including plants, trees, birds and invertebrates. Possums were introduced from Australia in the 1850s to establish a fur industry. Almost 100 years later, in 1947, possums were officially recognised as a pest in New Zealand (Clout, 2006) and currently, it is estimated 30 million possums occur in the country (Warburton *et al.*, 2009). Possums are selective browsers and focus on a single tree, and over time they can defoliate it completely and cause its death. In fact, shortly after establishing at a site, a possum population can cause the entire forest canopy to die back. Besides destroying localised tree stands, possums can change the overall structure and composition of native forests and other ecosystems. Possums can also consume buds, flowers, fruits, seedling bark and seedlings as well as ferns and fungi (Atkinson *et al.*, 1995). Although they feed on a wide range of plants and trees (up to 70 species), only three or four species will represent over 50% of their diet (Atkinson *et al.*, 1995). Some of the species commonly consumed by possums are the trees kamahi (*Weinnmania racemosa*), and northern and southern rata (*Metrosideros robusta* and *M. umbellata*, respectively), but the most preferred species are commonly soft-leaved seral species such as the tree fuchsia (*Fuchsia excorticata*) and wine-berry (*Aristotelia serrata*), and mistletoes (Clout, 2006). In fact, browsing by possums has been identified as the main cause for the decline of several of the surviving species of native mistletoes and an important contributor to the extinction of Adam's mistletoe (*Trilepidea adamsii*), which was last seen in the wild in 1954 (de Lange *et al.*, 2010).

12.4.3 Competition from weeds

Over 2470 exotic plants are found in New Zealand (Breitwieser *et al.*, 2012). Many of these plants were intentionally introduced as garden plants and many have now naturalised (*n* = 1753) (Breitwieser *et al.*, 2012); that means they have established in the wild and are successfully reproducing and spreading. Gorse (*Ulex europaeus*), for instance, was originally introduced as a hedging plant and ornamental, and it is now widespread throughout New Zealand (Roy *et al.*, 2004). Weeds compete with native plants at different levels; they compete for resources, space or services from other organisms such as pollinators, mycorrhiza or seed dispersers. Some weeds may also have a bigger impact in ecosystems as they may change the structure, function and composition of the vegetation, animal communities and the overall landscape. Weeds able to trigger such changes have been classified as *environmental weeds* and there are about 300 species of these listed for New Zealand (Howell, 2008). These can be categorised into five types: climbers which smother native plants (e.g. old man's beard *Clematis vitalba*); shade-tolerant herbs that form extensive clones and suppress native regeneration (e.g. ginger *Hedychium* spp.); species with long-lived seed banks (e.g. Himalayan honeysuckle *Lycesteria formosa*); woody seral species (*Acacia* spp); and bird-dispersed trees (e.g. lilly pilly *Syzygium smithii*).

Currently, more than half of the native species ranked as Nationally Critical (94 species) are threatened by weeds. For many of these plants, recruitment has been dramatically affected as weed encroachment prevents seeds germinating or seedlings establishing. Weeds have also extended to distinctive habitats such as limestone rocks or ultramafic soils where several Nationally Critical plants occur. For instance, the habitat of the Castle Hill's forget-me-not (*Myosotis colensoi*) (Figure 12.2, Plate 31), has been invaded by weeds such as hawkweed (*Pilosella officinarum*), Chewing's fescue (*Festuca rubra* subsp. *communata*), browntop (*Agrostis capillaris*) and cocksfoot (*Dactylis glomerata*) (de Lange *et al.*, 2010).

Controlling the spread of these exotic plants is a difficult task and until now only seven weed species have been successfully eradicated from the country. These are the creeping knapweed (*Acroptilon repens*), skeleton weed (*Chondrilla juncea*), bogbean (*Menyanthes trifoliate*), fringed water lily (*Nymphoides peltata*), water lettuce (*Pistia stratiotes*), clasped pondweed (*Potamogeton perfoliatus*) and annual wild rice (*Zizania palustris*) (Clayson Howell, Department of Conservation, personal communication). Weed control can be mechanical (using weed-eaters), manual (hand-weeding),

Figure 12.2 (Plate 31) Competition from weeds is one of the main threats to the Castle Hill's forget-me-not (*Myosotis colensoi*); one of the many Threatened forget-me-nots found in New Zealand. A black and white version of this figure will appear in some formats. For the colour version, please refer to the plate section.

biological (using insects or fungi) or chemical (using herbicide). Unfortunately, some of these methods can damage non-target species and spray drift from herbicides, for instance, has been linked to the decline of Nationally Critical species such as pink broom (*Carmichaelia carmichaeliae*) and tainui (*Pomaderris apetala* subsp. *maritima*) after aerial spraying for gorse (*Ulex europaeus*) and broom (*Cytisus scoparius*) control has occurred (de Lange *et al.*, 2010).

12.4.4 Plant collectors

Collection of plants from the wild has greatly contributed to the decline of species around the world, especially of those with horticultural potential or of interest for plant collectors such as carnivorous plants and orchids. Collection of wild plants has already been listed as an important threat to the native orchids' survival (de Lange *et al.*, 2007). Illegal collection of native orchids was exposed in 2004 when two Czech citizens were intercepted while trying to smuggle several terrestrial and epiphytic orchids out of the country. Collection of native orchids, however, is not limited to foreign plant collectors, and it is believed populations of several epiphytic orchids in the forests of the lower North Island have declined because of local orchid collectors (de Lange *et al.*, 2007). Earlier botanists also over-collected some native species and a classic example is the demise of almost all extant plants of the spiral sun orchid, *Thelymitra matthewsii*, before it was even formally described. New populations were rediscovered in the 1980s and this orchid is currently ranked as Nationally Critical.

12.4.5 Milling

Harvesting of native species for timber has also taken its toll on many native trees such as kauri (*Agathis australis*), kahikatea (*Dacrycarpus dacrydioides*) and rimu (*Dacrydium cupressinum*). Although none of these trees are currently threatened, their removal means a reduction in habitat for other species growing in the understorey or as epiphytes. Forests dominated by these species have been heavily logged in the past 200 years and the area they once occupied has been dramatically reduced (Given, 1981). For instance, kauri forest once covered over a million hectares of the far north of New Zealand. Demand for its high-quality timber (used for construction) and its gum (collected to produce varnish and linoleum), devastated many stands. Currently only 7500 hectares of mature kauri forest remain, scattered in isolated patches in the country, mainly in conservation reserves (Steward and Beveridge, 2010).

12.4.6 Plant pathogens

Diseases caused by plant pathogens can be as devastating as any human-induced threat. The effect pathogens may have on plants and particularly trees has recently become of great concern after a fungus-like disease killed numerous kauri (*Agathis australis*) trees; an iconic and culturally important tree species, in northern New Zealand (Steward and Beveridge, 2010). The disease was formally identified in 2008 and DNA studies indicate it is caused by an undescribed species of water mould (oomycete), currently known as *Phytophthora* taxon *Agathis*, or PTA (Dawson and Lucas, 2011). The damage PTA causes on the trees is extensive, ranging from yellowing and loss of leaves, canopy thinning, dead branches and lesions at the base of the trunk that bleed resin. PTA affects trees of all ages and every infected tree dies. Research to understand the origin of PTA, detection, dispersal routes, and control, is currently underway (e.g. Than *et al.*, 2013) and methods to propagate kauri for *ex-situ* conservation have been developed (Gough *et al.*, 2012). Also, a strong public awareness campaign has been launched to prevent the spread of PTA between forest remnants and to involve local communities in the protection of this treasured native tree (e.g. www.kauridieback.co.nz).

There are many other pathogenic fungi that can have a similarly detrimental effect on New Zealand flora. The potential arrival of the fungus *Puccinia psidii* from Australia, for instance, and the consequences that an outbreak of this disease could have in New Zealand have already been investigated by the Ministry of Primary Industries (Clark, 2011). The organism, known as myrtle rust, is easily recognised by the production of masses of powdery bright yellow or orange-yellow spores on infected plant parts. Although some species may withstand the infection, it is believed highly susceptible species may easily die. The impact of this disease may be considerable as it could potentially affect any of the 22 members of the myrtle family occurring in New Zealand, ranging from common iconic species such as *Metrosideros excelsa* to Critically Threatened species such as *M. barletii*. Species of economic importance such as feijoa (*Acca sellowiana*; export fruit), eucalyptus plantations and manuka (*Leptospermum scoparium*; honey industry) could also be affected. It is believed eradication of the myrtle rust from New Zealand would not be feasible.

12.4.7   Disruption of ecological interactions

Unlike other threats, the disruption of ecological interactions may have effects on plant species, or entire plant communities, which could remain unnoticed for a long time. Disruption of these interactions may have complex cascading effects and their restoration can be extremely challenging. For instance, disruption of plant–pollinator interactions may lead to reproductive failure (low or no fruit-set), boost self-pollination rates and cause inbreeding and low fitness offspring, and ultimately halt regeneration. Failure of seed dispersal by birds may prevent seed germination, lessen local recruitment, limit a species' ability to colonise other areas, and reduced gene flow between populations. Disruption of less evident interactions such as plant–mycorrhizal associations, which enhance nutrient up-take, plant growth and may even facilitate establishment, will limit survival of those species that are fully dependent on their fungal partner (mycoheterotrophic species) or those with a species-specific interaction, such as orchids (Box 12.5). In the absence of the correct fungal partner, orchid seeds will not germinate, plants will not establish and any restoration or translocation attempt will be futile.

Pollination and dispersal failure is already evident in New Zealand and it has been detected in several shrub and tree species (de Lange and Jones, 2000; Robertson *et al.*, 2008; Anderson *et al.*, 2011). The main cause for the failure of these interactions is the extinction, or functional extinction, of native birds which are important pollinators and seed dispersers in New Zealand. The number of birds has declined considerably since human arrival and it is now believed about 40% of endemic birds have disappeared since human settlement (Innes *et al.*, 2010; Kelly *et al.*, 2010). Introduction of mammalian predators such as stoats, rats, cats and possums, in addition

Box 12.5

The lady's tresses orchid, *Spiranthes novae-zelandiae* is a Nationally Vulnerable species. This terrestrial orchid is endemic to New Zealand and although it is found in several sites throughout the North and South Island and also on Chatham Island, it is never abundant at each site. It is usually confined to wetlands, stream banks or lake margins and occasionally seepages within tussock grassland. When flowering, plants form a spike 20–150 cm long, with numerous small, pink, self-pollinating flowers. Habitat destruction is the main threat to this orchid. Currently, about 90% of the wetlands existing in New Zealand before human settlement have disappeared; many of them drained for urban or rural development (Hunt 2007). Unlike many other threatened plants in the country, cultivation and propagation of orchids for conservation purposes has never been attempted. This is mainly because the skills to execute these techniques do not exist in New Zealand and the knowledge of the orchid–fungal partner interaction of native orchids is lacking. Since orchids are one of the top five plant groups with conservation issues in New Zealand (de Lange *et al.*, 2009) implementing such techniques is fundamental for their conservation. A study to identify the fungal partner of *S. novae-zelandiae* using molecular markers, and to germinate orchid's seeds *in-situ* and *ex-situ*, is currently underway at Victoria University in collaboration with the Museum of New Zealand Te Papa Tongarewa and Otari Wilton's Bush (Frericks and Lehnebach, 2013). This study is a part of greater effort to understand New Zealand orchids' interaction with their fungal partners and factors limiting orchid distribution and establishment in the wild.

to habitat destruction, are the main causes for the decline of birds in New Zealand (Craig *et al.*, 2000).

Currently, only one of the three endemic bird pollinators, tui (*Prosthemadera novae-zeelandiae*), is found in the upper North Island (Anderson *et al.*, 2011). Other endemic pollinators, such as stitchbirds (*Notiomystis cincta*) and bellbirds (*Anthornis melanura*), have been exterminated from most of the upper North Island and they are now only confined to small predator-free islands or fenced predator-free protected areas. The poor status of the pollination service in the North Island has been recently highlighted by a study comparing the reproductive success of the bird-pollinated shrub *Rhabdothamnus solandri* in the mainland (North Island) with those populations on islands about 20 km away (Anderson *et al.*, 2011). On the mainland, this plant has experienced considerable reduction of seed output and also low recruitment; 84% and 55% less than on the islands, respectively. It is interesting to note that pollination failure in this shrub has occurred despite one of its three native pollinators, the tui, still being present on the mainland. Anderson *et al.* (2011) suggest changes in the tui's feeding behaviour, e.g. feeding higher in the canopy to avoid predators or on introduced nectar-richer species, may have contributed to the low fruit-set observed in mainland populations of *R. solandri*.

The demise of frugivorous birds is also of concern, and especially so for those large-seeded tree species such as taraire (*Beilshmiedia tarairi*) or karaka (*Corynocarpus laevigatus*). These trees form fruits of >14 mm width and the only bird capable of dispersing such a large seed in the country is the New Zealand pigeon (*Hemiphaga novaeseelandiae*). The New Zealand pigeon, or kereru, is an endemic bird currently classified as 'near threatened' by the IUCN (IUCN, 2013). Habitat loss, introduced mammalian predators and illegal hunting have been listed as some of the causes behind their gradual decline (cited in Wotton and Kelly, 2011). Hunting has also caused the dramatic decline of another frugivorous bird in New Zealand, the kea (*Nestor notabilis*), which is the only species of alpine parrot in the world. Owing to kea attacks on sheep, a bounty was placed on kea beaks by the government in the late 1800s that lasted until 1971. During this time, more than 150 000 kea were killed (Cunningham, 1948). From 1986 the kea has been given full protection, but currently fewer than 5000 individuals remain in the wild (Anderson, 1986). A recent study shows kea play an unexpectedly important role as seed dispersers for fruiting plants in alpine ecosystems, and particularly over long distances, since kea are the only bird that make frequent long-distance flights between, and within, mountain ranges (Young *et al.*, 2012). This study also noted that kea could play an important role in the dispersal and reintroduction of sub-alpine plant species from more intact areas into those now destined for restoration.

12.5 Conservation initiatives and future research

Most of the plant species found in New Zealand occur nowhere else in the world; some are of cultural importance, and others have become iconic species. Furthermore, New Zealand is a biodiversity hotspot and, along with the other 24 regions of the world, it provides habitat for 44% of the world's plant species (Myers *et al.*, 2000). Therefore, preserving New Zealand flora is a matter of national and international significance. Fortunately, a number of initiatives have been implemented towards this end. Habitat protection and ecological restoration, monitoring of small populations, conservation by cultivation (Box 12.6), and the establishment of *ex-situ* collections and seed banks, to secure species under immediate threat, are just some of the many activities community groups, governmental and non-governmental organisations and regional councils are involved with in the country.

Public awareness and education are also important components of plant conservation and the New Zealand Plant Conservation Network (NZPCN) (http://www.nzpcn.org.nz/) has played a critical role in this. In the past ten years the NZPCN has been actively involved in the dissemination of information about New Zealand's native flora, promoting activities to protect threatened plants and coordinating *ex-situ* conservation activities. The network has also prepared and delivered a number of programmes to provide plant conservation training to the community.

Research into threatened plants' autoecology, however, is urgently needed. Information on habitat preferences, breeding systems, and ecological interactions with pollinators (insects or birds), seed dispersers and mycorrhizal partners is lacking for the great majority of New Zealand threatened plants, and this information is critical to stop their decline and, ultimately, secure their survival both *in-situ* or *ex-situ*. Studies aiming to resolve taxonomic uncertainty within several plant groups are also urgently needed as they will provide the basic information that underpins any biological research and effective conservation actions. Finally, conservation genetics should be further

Box 12.6

The shrubby tororaro or wiggywig (*Muehlenbeckia astonii*) is an endemic divaricating shrub found in the North and South Islands in coastal to lowland settings. This species is currently listed as Nationally Endangered and about ten years ago only 2600 individuals were estimated to exist in the wild (de Lange and Jones, 2000). Little is known about the autoecology of this species and much of the information available is anecdotal. This species has plants with hermaphrodite and male flowers and others with female flowers only. Flowers are insect pollinated and after pollination takes place, fruits are formed within 15 days. Not much is known about dispersal of its seeds but it is likely birds or geckos are involved. Several threats to the conservation of this species have been identified but habitat fragmentation and destruction, predation, and recruitment failure are considered the most significant. Lack of regeneration has been noted throughout its range, and failure to set fruits and inbreeding depression caused by changes in the ratio of female to male/hermaphrodite plants within populations are likely to be the main reasons. In fact, in the five North Island populations and some of those in the South Island no out-crossing has been reported (de Lange and Silbery, 1993). It is believed habitat fragmentation and destruction has increased spatial isolation between plants of the opposite gender preventing pollen reaching plants bearing female flowers. Fortunately, *M. astonii* can be easily propagated from cuttings and seeds, and a number of *ex-situ* collections from some of the existing population have been established in botanical gardens. Furthermore, it is being promoted as a garden plant and is now commonly seen in many urban and private gardens (Figure 12.3, Plate 58).

Figure 12.3 (Plate 58)   The shrubby tororaro (*Muehlenbeckia astonii*), a species Nationally Endangered, has become a popular garden plant and is now commonly seen in traffic islands. A black and white version of this figure will appear in some formats. For the colour version, please refer to the plate section.

developed in New Zealand and used to inform practical conservation management of uncommon species with highly restricted distributions. Survival of many plant species may depend on these research actions.

Acknowledgements

Thanks to Rebecca Stanley, Jeremy Rolfe, Clayson Howell and Leon Perrie for sharing their knowledge on New Zealand plants and to Vivienne McGlynn, Andreas Zeller and Margaret Stanley for comments and constructive criticism on an earlier version of this chapter.

REFERENCES

Anderson, R. 1986. Keas for keeps. *Forest and Bird* **17**: 2–5.

Anderson, S. H.; Kelly, D.; Ladley, J. J.; Mollow, S. and J. Terry. 2011. Cascading effects of bird functional extinction reduce pollination and plant density. *Science* **331**: 1068–1070.

Armstrong, T. J. B. 2003. Hybridisation and adaptive radiation in Australian alpine *Ranunculus*. PhD thesis. Australian National University. Australia.

Atkinson, I. A. E.; Campbell, D. J.; Fitzgerald, B. M.; Flux, J. E. C. and M. J. Meads. 1995. *Possums and Possum Control; Effects on Lowland Forest Ecosystems: A Literature Review with Specific Reference to the Use of 1080*. Department of Conservation, Wellington. New Zealand.

Black, J. 2008. The creature crisis. *The Listener* **4**: 24–28.

Breitwieser, I.; Glenny, D. S.; Thorne, A. and S. J. Wagstaff. 1999. Phylogenetic relationships in Australasian Gnaphalieae (Compositae) inferred from ITS sequences. *New Zealand Journal of Botany* **37**: 399–412.

Breitwieser, I.; Brownsey, P. J; Garnock-Jones, P. J.; Perrie, L. R. and A. D. Wilton. 2012. Phylum Tracheophyta: vascular plants. In: *New Zealand Inventory of Biodiversity*. Gordon, D. P. (Ed.). Canterbury University Press. Christchurch.

Brownsey, P. J.; Ewans, R.; Rance, B.; Walls, S. and L. R. Perrie. 2013. A review of the fern genus *Sticherus* (Gleicheniaceae) in New Zealand with confirmation of two new species records. *New Zealand Journal of Botany* **51**: 104–115.

Clark, S. 2011. *Risk Analysis of the* Puccinia psidii/*Guava Rust Fungal Complex (Including* Uredo rangelii/*Myrtle Rust) on Nursery Stock*. Biosecurity Risk Analyses Group; Ministry of Agriculture and Forestry. Wellington. New Zealand.

Clout, M. N. 2006. Keystone Aliens? The multiple impacts of brushtail possums. In: Allen, R. B. and W. G. Lee (Eds.) Biological Invasions in New Zealand. *Ecological Studies* **186**: 265–279.

Craig, J.; Anderson, S.; Clout, M.; *et al.* 2000. Conservation issues in New Zealand. *Annual Review of Ecology and Systematics* **31**: 61–78.

Cunningham, J. M. 1948. Number of keas. *New Zealand Bird Notes* **2**: 154.

Dawson, J. and R. Lucas. 2011. *New Zealand's Native Trees*. Craig Potton Publishing, Nelson. New Zealand.

de Lange, P. and C. Jones. 2000. Shrubby tororaro (*Muehlenbeckia astonii* Petrie) recovery plan 2000–2010. Threatened Species Recovery Plan 31. Biodiversity Recovery Unit, Department of Conservation. Wellington.

de Lange, P. and T. Silbery. 1993. Saving the shrubby tororaro (*Muehlenbeckia astonii* Petrie) – an urban approach to threatened plant conservation. In: **Oates, M.** (Ed.) *People, Plants and Conservations: Botanic Gardens into the 21st Century*, Christchurch, Royal New Zealand Institute of Horticulture.

de Lange, P.; Crowcroft, G. M. and L. J. Forester. 1991. Thelymitra *"Ahipara" and Endangered Orchid Transferred, Notes on its Taxonomic Status, Distribution and Ecology*. Science and Research Internal Report No. 113. Department of Conservation. Wellington. New Zealand.

de Lange, P.; Rolfe, J.; St. George, I. and J. Sawyer. 2007. *Wild Orchids of the Lower North Island*. Department of Conservation, Wellington. New Zealand.

de Lange, P.; Norton, D. A.; Courtney, S. P.; *et al.* 2009. Threatened and uncommon plants of New Zealand. *New Zealand Journal of Botany* **47**: 61–96.

de Lange, P.; Heenan, P.; Norton, D.; Rolfe, J. and J. Sawyer. 2010. *Threatened Plants of New Zealand*. Christchurch: Canterbury Press.

Ford, K. A.; Ward, J. M.; Smissen, R. D.; Wagstaff, S. J. and I. Breitwieser. 2007. Phylogeny and biogeography of *Craspedia* (Asteraceae: Gnaphalieae) based on ITS, ETS and *psb*A-*trn*H sequence data. *Taxon* **56**: 783–794.

Frericks, J. and C. A. Lehnebach. 2013. Growing native terrestrial orchids from seed; the first steps towards orchid conservation in New Zealand. *Trilepidea* **117**: 3–5.

Gibbs, G. 2006. *Ghosts of Gondwana. The History of the Life in New Zealand*. Craig Potton Publishing, Nelson. New Zealand.

Given, D. R. 1981. *Rare and Endangered Plants of New Zealand*. Reed, Wellington.

Gough, K.; Hargreaves, C.; Steward, G.; *et al.* 2012. Micropropagation of kauri (*Agathis australis* (D.Don.(Lindl.): *in vitro* stimulation of shoot and root development and the effect of rooting hormone application method. *New Zealand Journal of Forestry Sciences* **42**: 107–116.

Howell, C. J. 2008. *Consolidated List of Environmental Weeds in New Zealand*. Science & Technical Publishing, Department of Conservation. Wellington.

Hunt, J. 2007. *Wetlands of New Zealand: A Bitter-sweet Story*. Random House, Auckland.

Innes, J.; Kelly, D.; Overton, J. M. and C. Gillies. 2010. Predation and other factors currently limiting New Zealand forest birds. *New Zealand Journal of Ecology* **34**: 86–114.

IUCN. 2013. IUCN Red list of Threatened Species. Version 2013.1 http://www.iucnredlist.org/ (accessed 10 July 2013)

Kadereit, G.; Gotzek, D.; Jacobs, S. and H. Freitag. 2005. Origin and age of Australian Chenopodiaceae. *Organisms, Diversity and Evolution* **5**: 59–80.

Kelly, D., Ladley, J. J., Robertson, A. W.; *et al.* 2010. Mutualisms with the wreckage of an avifauna: the status of bird pollination and fruit dispersal in New Zealand. *New Zealand Journal of Ecology* **34**: 65–85.

King, C. M. 2005. *The Handbook of New Zealand Mammals*. Oxford University Press, Auckland. New Zealand.

Knapp, M.; Stöckler, K.; Havell, D.; Delsuc, F.; Sebastiani, F. and P. J. Lockhart. 2005. Relaxed molecular clock provides evidence for long-distance dispersal of *Nothofagus* (Southern Beech). *PLoS Biology* **3**: e14. doi:10.1371/journal.pbio.0030014.

Krull, C. R.; Burns, B. R.; Choquenot, D. and M. C. Stanley. 2013. Feral pigs in a temperate rainforest ecosystem: disturbance and ecological impacts. *Biological Invasions*. DOI: 10.1007/s10530-013-0444-9.

Lehnebach, C. 2008. Phylogenetic affinities, species delimitation and adaptive radiation of New Zealand *Ranunculus*: PhD thesis. Massey University, Palmerston North, New Zealand.

Lehnebach, C. A. 2012. Two new species of forget-me-nots (*Myosotis*, Boraginaceae) from New Zealand. *PhytoKeys* **16**: 53–64.

Linder, H. P. 2008. Plant species radiations: where, when, why? *Philosophical Transactions of the Royal Society B: Biological Sciences* **363**: 3097–3105.

Lockhart, P.; McLechnanan, P. A.; Havell, D.; *et al.* 2001. Phylogeny, dispersal and radiation of New Zealand alpine buttercups: molecular evidence under split decomposition. *Annals of the Missouri Botanical Garden* **88**: 458–477.

Mark, A. F. 2012. *Above the Treeline: A Nature Guide to Alpine New Zealand*. Craig Potton Publishing. Nelson, New Zealand.

McGlone, M. S.; R. P. Duncan and P. B. Heenan. 2001. Endemism, species selection and the origin and distribution of the vascular plant flora of New Zealand. *Journal of Biogeography* **28**: 199–216.

McWethya, D. B.; Whitlock, C.; Wilmshurst, J. M.; *et al.* 2010. Rapid landscape transformation in South Island, New Zealand, following initial Polynesian settlement. *PNAS* **107**: 21 343–21 348.

Meudt, H. M. and M. J. Bayley. 2008. Phylogeographic patterns in the Australasian genus *Chionohebe* (*Veronica* s.l., Plantaginaceae) based on AFLP and chloroplast DNA sequences. *Molecular Phylogenetics and Evolution* **47**: 319–338.

Ministry for the Environment. 2010. *Land Use Environmental Snapshot*. INFO 472. Ministry for the Environment. Wellington (http://www.mfe.govt.nz/publications/land/).

Mummenhoff, K. and A. Franzke. 2007. Gone with the bird: late tertiary and quaternary intercontinental long-distance dispersal and allopolyploidization in plants. *Systematics and Biodiversity* **5**(3): 255–260.

Myers, N.; Mittermeier, R. A.; Mittermeier, C. G.; da Fonseca, G. A. B. and J. Kent. 2000. Biodiversity hotspots for conservation priorities. *Nature* **403**: 853–858.

Nicol, R. and Overmars, F. B. 2008. Vegetation and Flora Baseline Survey – Whareatea Mine Access Road, L and M Coal Ltd Escarpment Mine Project, Denniston Plateau. Prepared for Resource and Environmental Management Ltd.

Perry, G. L. W.; Wilmshurst, J. M.; McGlone, M. S. and A. Napier. 2012. Reconstructing spatial vulnerability to forest loss by fire in pre-historic New Zealand. *Global Ecology and Biogeography* **21**: 1029–1041.

Prebble, J. M., Meudt, H. M. and P. J. Garnock-Jones. 2012. An expanded molecular phylogeny of the southern bluebells (*Wahlenbergia*, Campanulaceae) from Australia and New Zealand. *Australian Systematic Botany* **25**:11–30.

Raven, P. H. 1973. Evolution of the subalpine and alpine plant groups in New Zealand. *New Zealand Journal of Botany* **11**: 177–200.

Robertson, A. W.; Ladley, J. J; Kelly, D.; *et al.* 2008. Pollination and fruit-dispersal limitation in *Fuchsia excorticata* (Onagraceae) on the New Zealand mainland. *New Zealand Journal of Botany* **46**: 299–314.

Roy, B.; Popay, I.; Champion, P.; James, T. and A. Rahman. 2004. *An Illustrated Guide to Common Weeds of New Zealand.* 2nd edition. New Zealand Plant Protection Society (Inc).

Schuster, T. M.; Setaro, S. D. and K. A. Kron. 2013. Age estimates for the buckwheat family Polygonaceae based on sequence data calibrated by fossils and with a focus on the amphi-Pacific *Muehlenbeckia. PLoS ONE* **8**: e61261.

Statistics New Zealand. 2009. *Measuring New Zealand's Progress Using Sustainable Development Approach: 2008.* Statistics New Zealand. Wellington.

Steward, G. A. and E. Beveridge. 2010. A review of New Zealand kauri (*Agathis australis* (D.Don (Lindl.): its ecology, history, growth and potential for management for timber. *New Zealand Journal of Forestry Science* **40**: 33–59.

Stöckler, K.; Daniel, I. L. and P. J. Lockhart. 2002. New Zealand Kauri (*Agathis australis* (D.Don) Lindl., Araucariaceae) survives Oligocene drowning. *Systematic Biology* **51**: 827–832.

Than, D. J.; Hughes, K. J. D.; Boonhan, N.; *et al.* 2013. A TaqMan real-time PCR assay for the detection of *Phytophthora* 'taxon *Agathis*' in soil, pathogen of kauri in New Zealand. *Forest Pathology* **43**: 324–330.

Wagstaff, S. J. and M. A. Dawson. 2000. Classification, origin, and patterns of diversification of *Corynocarpus* (Corynocarpaceae) inferred from DNA sequences. *Systematic Botany* **25**: 134–149.

Wagstaff, S. J.; Bayly, M. J.; Garnock-Jones, P. J. and D. C. Albach. 2002. Classification, origin and diversification of the New Zealand hebes (Scrophulariaceae). *Annals of the Missouri Botanical Garden* **89**:38–63.

Wanntorp, L. and H. Wanntorp. 2003. The biogeography of *Gunnera* L.: vicariance and dispersal. *Journal of Biogeography* **30**: 979–987.

Warburton, B.; Cowan, P. and J. Shepherd. 2009. *How Many Possums are now in New Zealand Following Control and How Many Would it be Without?* Landcare Research Contract Report LC0910/060. Landcare Research New Zealand Ltd.

Wardle, P. 1991. *Vegetation of New Zealand.* Cambridge University Press. Cambridge.

Winkworth, R. C.; Grau, J.; Robertson, A. W. and P. J. Lockhart. 2002a. The origins and evolution of the genus *Myosotis* L. (Boraginaceae). *Molecular Phylogenetics and Evolution* **24**: 180–193.

Winkworth, R. C; Wagstaff, S. J.; Glenny, D. and P. J. Lockhart. 2002b. Plant dispersal N.E.W.S. from New Zealand. *Trends in Ecology and Evolution* **17**: 514–520.

Wotton, D. M. and D. Kelly. 2011. Frugivore loss limits recruitment of large-seeded trees. *Proceedings of the Royal Society B* **278**: 3345–3354.

Young, L. M.; Kelly, D. and X. J. Nelson. 2012. Alpine flora may depend on declining frugivorous parrot for seed dispersal. *Biological Conservation* **147**: 133–142.

CHAPTER 13

The evolutionary history of the Australian flora and its relevance to biodiversity conservation

Maurizio Rossetto

Summary

Australia is a repository of a unique flora once widespread throughout Gondwana, the southern supercontinent. As the connectivity to other landmasses was lost, an indigenous Australian flora evolved in response to changing environmental conditions from prevalently wet to prevalently dry. Although the Tertiary fossil record suggests the continent-wide dominance of rainforest communities, it also documents an adaptive trend towards cooler, drier and more seasonal climates. More recently, the Quaternary saw a succession of glacial cycles that produced transitional plant communities that were considerably different from current ones. On the whole, major natural disturbances and long-term landscape modifications have steadily contributed to Australia's floristic diversity.

However, the arrival of people has introduced increasing and more rapid pressure on natural habitats through intensive land usage and degradation, as well as the introduction of pests and weeds resulting in a new set of threats to biodiversity. Protecting and conserving species as well as the condition of natural ecosystems is becoming an increasingly pressing concern that requires an integrated approach. Describing the extent and distribution of diversity and endemism is important, but should represent only a preliminary step towards the development of conservation criteria. Biodiversity management is increasingly reliant on our understanding of the relative vulnerability of species and communities to current, past and future threats. As we better understand how species and assemblages respond to temporal changes in environmental disturbance, it will become increasingly possible to explore proactive conservation and restoration strategies.

13.1 Origins and distribution of Australia's floristic diversity

Australia is a unique repository of plant diversity representative of a flora once widespread throughout Gondwana, the southern supercontinent. An indigenous Australian flora evolved as the connectivity to other landmasses was lost and the continental plate moved from polar to equatorial latitudes. A continent-wide change from prevalently wet to prevalently dry conditions stimulated floristic diversification by driving the rapid radiation of dry-adapted vegetation types. However, conditions have not been consistently uniform across the continent. Environmental and biogeographic barriers have impacted on connectivity and diversification, particularly during the recent climatic cycles of the Quaternary.

13.1.1 A very brief account of the major events that shaped the early Australian flora

The earliest recognised vascular flora appeared around the Early Silurian (over 400 Mya) and by the end of the Devonian lycophytes (an ancient plant group including club-mosses and quillworts) started to develop latitudinal and regional differentiation across emerged landmasses, attaining astonishing size and emulating the structure of trees within the Tropics (McLoughlin, 2001). These early vascular plants dominated the flora through to the later part of the Carboniferous (to 300 Mya) when a shift of the southern landmasses toward polar latitudes cooled the climate and triggered the ascendancy of seed ferns. During the early Permian, the retreat of glacial ice-sheets brought about further global transformations that contributed to the southern radiation of the distinctive Glossopterids (an extinct order of seed ferns; Hill *et al.*, 1999).

Through the Permian and into the Mesozoic (between 300 and 200 Mya), emerged landforms across the globe were connected into a large, continuous landmass, Pangaea. The considerable latitudinal extent of this supercontinent became one of the primary drivers of floristic differentiation, resulting in distinct southern and northern plant assemblages. At that time, landscape-level transformations instigated major floristic restructures in the south, and the radiation of ferns, seed ferns, cycads and conifers resulted in a net increase in biodiversity. This floristic transition represented the culmination of a long period of geological and environmental transformation.

The physiological innovations that provided embryos with greater protection from the elements were critical to the evolution of our modern flora. The Gymnosperms (seed-producing plants) were dominant and most diverse throughout the Mesozoic. Yet, although the fossil record contains an abundance of conifer lineages affiliated to extant ones (such as Aracauriaceae and Podocarpaceae), many of the Gymnosperms that dominated the Jurassic went extinct towards the end of the Cretaceous. A set of unique environmental conditions combining warmer climates with extreme photoperiods (Australia and Antarctica were connected at very high latitudes) were conducive to further ecological and evolutionary innovations and the increasing dominance of the flowering plants (Angiosperms), as exemplified by the diversification of the ancient genus *Nothofagus* (Hill *et al.*, 1999; Figure 13.1, Plate 32).

13.1.2 The rise of flowering plants

During the Tertiary the Southern-Gondwanan flora was dominated by broad-leaved vegetation. A phase of global warming during the Eocene (between 55 and 35 Mya)

Figure 13.1 (Plate 32) *Nothofagus moorei* in the Border Ranges (NSW). This is the most northerly Australian member of a quintessential Gondwanan genus extensively represented within the fossil record. The genus is still present from South America, through to New Zealand and Papua New Guinea. (Photo by M. Rossetto.) A black and white version of this figure will appear in some formats. For the colour version, please refer to the plate section.

resulted in the continent-wide dominance of rainforest communities. Some of the lineages present in these extensive southern forests can still be found within rainforest remnants scattered along the eastern coast of Australia (Weston & Kooyman, 2002). This surprising taxonomic conservatism has sometimes inspired incorrect narratives of Australian rainforests being static replicates of those ancient communities.

The morphological characteristics of macrofossils from the Australian flora of the Tertiary also document a trend towards the cooler, drier and more seasonal climates that still shape our flora (Greenwood & Christophel, 2005). By the end of the Eocene, the separation of Australia from Antarctica resulted in the establishment of circumpolar currents and in the formation of polar ice caps. As a result, the post-Eocene flora was increasingly variable across the landscape with dry-adapted (scleromorphic) and at a later stage arid-adapted (xeromorphic) vegetation types becoming more common (Hill, 1998).

Although typical southern families such as the Proteaceae (currently including banksias, grevilleas and waratahs) appear in the fossil record from around 90 Mya, it was only by the Early Oligocene (around 30 Mya) that floristic assemblages resembling

---

Box 13.1 Recent insights on the origins of the Australian flora

Gondwanan origins   Recent discoveries have expanded and sometimes modified our understanding of evolutionary patterns within the Gondwanan flora. The fossil record that represents the Mesozoic and Cenozoic vegetation of what is now Patagonia (South America) is uncovering taxa that are unexpectedly similar to those now almost exclusively restricted to Australian mesic forests (Wilf *et al.*, 2005). These findings support the increasingly popular notion that Australia's unique flora is representative of ancient vegetation types once widely distributed across the southern supercontinent. Interestingly, the fossil deposits from southern Argentina have also uncovered the oldest known *Eucalyptus* macrofossils (Gandolfo *et al.*, 2011). Such findings stress the complex evolutionary history of Australia's characteristic dry-adapted flora, and its links to ancient marginal Gondwanan habitats.

Ancient families, recent radiations   The available fossil record shows that ancient plant groups, such as the cycads, have preserved consistent morphological patterns through time. This has lead to the general acknowledgement that maximum cycad diversity in Australia evolved during the Jurassic and has since decreased under competitive pressure from the Angiosperms. Surprisingly, recent molecular studies revealed that most of the existing cycad diversity in Australia originated as recently as 12 Mya (Nagalingum *et al.*, 2011). Similarly, population-level studies in *Macrozamia* found that current distributional patterns in Central Australia are the result of even more recent dispersal events (Ingham *et al.*, 2013). These studies demonstrate that ancient clades can still actively respond to new evolutionary opportunities.

Major diversification episodes   As the Antarctic Circumpolar Current was established after the separation of Australia and Antarctica in the Oligocene, closed forests were gradually replaced by sclerophyllous open vegetation continent-wide. Molecular dating studies have shown that this period of increased continental drying corresponded to the timing of major evolutionary radiations among some of the most iconic dry-adapted lineages (*Allocasuarina, Banksia, Eucalyptus, Tetratheca* for example) and to the timing of extinction of many rainforest lineages (Crisp *et al.*, 2004; Crayn *et al.*, 2006). Molecular dating also identified coincident patterns between East–West speciation and diversification, and the aridification and elevation of the Nullarbor Plain (13–14 Mya), an important barrier to East–West migration (Crisp & Cook, 2007).

---

extant communities started to appear, such as heaths and peat-swamps. As the Australian continent (Sahul) came into contact with the Philippines sea-plate around 25 Mya, and progressively closed-in on the Sundaland continent to the north, opportunities for floristic exchange emerged (Sniderman & Jordan, 2011). In the Late Oligocene–Early Miocene, Australia crossed into subtropical latitudes and the marked shift in dominance from *Nothofagus* to Myrtaceae and Casuarinaceae symbolizes the extent of relentless environmental transformation. The aridification of Central Australia started around 15 Mya and intensified during the past 4 My. By then, forests had changed from closed to open canopy across the continent, and warm/wet plant communities were mostly relegated to coastal areas. See also Box 13.1.

13.1.3   The cyclical events of the Quaternary

During the past 2 My, a succession of glacial cycles has brought about rapid worldwide shifts in habitat suitability and availability. During the past 700 kya climatic fluctuations were particularly intense (Hays *et al.*, 1976). Globally cooler, drier climates became more prevalent and tropical deserts expanded. In the Northern Hemisphere the ice-ages were characterised by the expansion of extensive ice sheets that reduced the habitat available to whole communities, a process that left distinct signatures on the fossil record and in the genetics of populations (Petit *et al.*, 2005).

The ecological and physiological responses of single species to these extreme cycles have played an important role in defining current floristic patterns. Some species adapted to changing conditions or dispersed to more suitable locations, while others only persisted in marginal portions of their former range or became extinct (Hewitt, 2000). Consequently, much can be learned about species- and community-level resilience to ongoing change from the exploration of historical dynamics.

In Australia glacial maxima did not bring about extensive ice sheets; however, an overall decline in moisture shaped the distribution of vegetation by increasing aridity, fire frequency and fire intensity. Well-dated fossil pollen records from northern Queensland suggest that increased climatic instability in the past 200 ky brought about the replacement of araucarian forests by true sclerophyllous vegetation dominated by eucalypts (Kershaw *et al.*, 2007). The last glacial maximum was particularly intense, as evidenced by the appearance of large dune fields throughout the continent, and by the severity of habitat fragmentation, exemplified by *Eucalyptus* intrusions in tropical rainforest as recently as 8 kya (Hopkins *et al.*, 1993).

The brief interglacial periods (such as the current one) brought about warmer, moister and more predictable climatic conditions. Populations constrained to localised refugia by the previous glacial peak could take advantage of these new circumstances to expand into newly available habitat. These cyclical expansion/contraction events produced transitional plant communities that were considerably different from current ones. For instance, the combination of genetic and environmental data suggests that some species were able to survive in habitats otherwise considered as unsuitable (Worth *et al.*, 2009). Interestingly, an increasing number of studies investigating temporal changes in distributional patterns suggest that the loss of a species' 'core' habitat does not necessarily imply its localised extinction (although it certainly indicates a greater level of vulnerability; Stewart & Lister, 2001).

13.1.4   Patterns of diversity: why are species distributed and assembled the way they are?

The absence of direct land-bridges between Australia and other continents for over 20 My has led to the endurance of a distinctive local flora (over 90% of the 30 000+ flowering plants are endemic; Chapman, 2005). The prevalence of low-nutrient soils, the scarcity of altitudinal relief, and the continuing impact of large-scale disturbance events (such as fires, floods, droughts and cyclones) contributed to diversification, but also resulted in an uneven spread of diversity (Box 13.2). At present most of the arid centre remains diversity-poor, while coastal regions such as the Wet Tropics and the Southwest are particularly species-rich (Hopper & Gioia, 2004).

Box 13.2 Learning from the past to predict the future: what regulates rainforest diversity?

The distribution and diversity of the Australian rainforests have been transformed since the break-up of the Gondwanan supercontinent and through the extreme climatic cycles of the Quaternary. Broad-leaved vegetation has contracted continent-wide and is now restricted to a 'continental archipelago' of remnants along the east coast where the altitudinal relief provided by the Great Dividing Range preserves suitable habitat. Yet, despite occupying only 1% of the continental landmass, rainforests are still characterised by high floristic diversity and endemism.

While assemblage-level richness decreases significantly to the South and to the West, many rainforest species have distributions that transcend those of areas traditionally identified as rainforest refugia (Plate 10). Such patterns show how the selective filtering brought about by large-scale environmental change has impacted differently on species and assemblages through time.

Palaeoecological evidence suggests that rainforest vegetation contracted during the cooler, drier conditions of the glacial peaks (Kershaw *et al.*, 2007). The impact of these recurrent events varied according to landscape-level stability and to the availability of micro-refugia (Graham *et al.*, 2006; Mellick *et al.*, 2012). It is possible to differentiate between stable refugia and re-colonisation areas by interpreting regional measures of phylogenetic diversity in the light of recurrent disturbance events (Kooyman *et al.*, 2011). Fine-scale population genetic studies can then ascribe unambiguously conflicting distributional patterns in closely related species to differences in dispersal potential (Rossetto *et al.*, 2009).

Functional groups are likely to respond to large-scale environmental disturbances in different manners (Rossetto & Kooyman, 2005). At a species level, the survival of rainforest tree populations is dependent on their capacity to persist locally (by resprouting after disturbance for example; Rossetto *et al.*, 2004a) or to move to newly available habitat (Rossetto *et al.*, 2004b). Interestingly, functional fitness varies with local conditions. For instance, the absence of large frugivorous vertebrate (such as the cassowary) in northern NSW limits the distribution of taxa that are otherwise more widely dispersed in the Wet Tropics (Rossetto *et al.*, 2008).

Continental-scale functional biogeographic patterns can also reveal local disturbance history. A comparative study between regional floristic pools discovered a prevalence of easily dispersed rainforest lineages within areas that have endured significant levels of long-term disturbance (Kooyman *et al.*, 2011). These associative patterns highlight how re-colonisation processes can vary according to local environmental and ecological histories.

Overall, the Australian rainforest flora demonstrates a considerable degree of resilience despite having been significantly impacted by climatic and geological change. As direct and indirect anthropogenic pressures increase, understanding how temporal transformations in selective pressures impact on species and functional groups can help us assess their longer-term vulnerability.

By investigating the distribution and assemblage of species across the landscape, we can identify the major biotic and abiotic factors filtering the distribution of lineages. The study of biodiversity turn-over across environmental gradients is becoming an increasingly popular research topic for evolutionary and conservation ecologists. Three main measures of biodiversity are commonly referred to: gene diversity, lineage diversity and assemblage diversity. Each of these measures has unique evolutionary significance and informs conservation in a distinct manner.

Quantifying biodiversity at the gene level can identify the distribution of important adaptive variation (Hoffmann & Willi, 2008). Lineage diversity can be used to circumscribe the geographic boundaries of evolutionary differentiated provenances (Evolutionary Significant Units, ESUs; Moritz, 1994) and guide translocation and restoration processes. At a broader geographic level, quantifying species richness or measuring phylogenetic diversity and endemism can help identify regions of high conservation priority (Forest *et al.*, 2007). Finally, the definition of floristic and functional assemblages can be used to explore repeated landscape-level patterns and identify vulnerable communities.

It must be remembered, though, that describing the distribution of diversity is only an initial step. In order to interpret the distribution of biodiversity and develop adequate management strategies, the relative impact of current (adaptive) vs. historical (biogeographic) processes needs to be investigated.

13.2 Conservation challenges

The Gondwanan heritage is still a conspicuous element of Australia's flora, with major natural disturbances and long-term landscape modifications steadily contributing to current floristic diversity. The arrival of humans between 65 kya and 40 kya brought further and arguably more rapid and extreme change. The landing of early hunter-gatherer communities has impacted directly (via locally altered fire regimes) and indirectly (via the extinction of the megafauna; Rule *et al.*, 2012) on native vegetation. The latest wave of colonisation over the past 200 years has introduced increasing pressure on natural habitats through intensive land usage and degradation, as well as the introduction of pests and weeds (see also Chapter 6) resulting in a new set of threats to biodiversity.

13.2.1 Major threats to the Australian flora

The development of effective biodiversity conservation strategies relies on the identification of the biotic and abiotic processes that are a threat to the abundance, distribution and survival of taxa (Table 13.1). Key threatening processes are defined as those directly causing the decline of natural plant populations (Falk, 1990). These include the destruction of habitat, competition by invasive species, the loss of associated fauna (pollinators, dispersal vectors and other symbionts), the loss of genetic diversity and adaptive potential, and direct extirpation (via clearing, foraging or disease).

In order to develop adequate flora management strategies, it is important to appreciate the broader temporal context of threats, as the impact of external stresses can change through time, particularly as a result of ever increasing anthropogenic pressures.

Table 13.1   Listed threatening processes: an example from New South Wales (Australia). The list of 26 key threatening processes currently (August 2013) listed as impacting on native flora in NSW under the *Threatened Species Conservation Act 1995* (shown in alphabetical order) www.environment.nsw.gov.au/threatenedspecies

Threatening process impacting on native flora	
Alteration of habitat following subsidence due to longwall mining	Introduction and establishment of exotic rust fungi of the order Pucciniales pathogenic on plants of the family Myrtaceae
Alteration to the natural flow regimes of rivers and streams and their floodplains and wetlands	Introduction of the large earth bumblebee (*Bombus terrestris*)
Anthropogenic climate change	Invasion and establishment of exotic vines and scramblers
Bushrock removal	Invasion and establishment of Scotch broom (*Cytisus scoparius*)
Clearing of native vegetation	Invasion and establishment of the cane toad (*Bufo marinus*)
Competition and grazing by the feral European rabbit (*Oryctolagus cuniculus*)	Invasion of native plant communities by African Olive *Olea europaea* L. subsp. *cuspidata*
Competition and habitat degradation by feral goats (*Capra hircus*)	Invasion, establishment and spread of *Lantana camara*
Competition from feral honey bees (*Apis mellifera*)	Invasion of native plant communities by *Chrysanthemoides monilifera* (bitou bush and boneseed)
Forest Eucalypt dieback associated with over-abundant psyllids and bell miners	Invasion of native plant communities by exotic perennial grasses
High-frequency fire resulting in the disruption of life cycle processes in plants and animals and loss of vegetation structure and composition	Loss and degradation of native plant and animal habitat by invasion of escaped garden plants, including aquatic plants
Herbivory and environmental degradation caused by feral deer	Predation by the ship rat (*Rattus rattus*) on Lord Howe Island
Importation of red imported fire ants (*Solenopsis invicta*)	Predation, habitat degradation, competition and disease transmission by feral pigs (*Sus scrofa*)
Infection of native plants by *Phytophthora cinnamomi*	Removal of dead wood and dead trees

The combined effect of threatening processes needs also to be considered at the community-level as well as at the species-level. For instance a number of Australian ecosystems, such as tropical savannas and temperate eucalypt forests, have been identified as particularly vulnerable to multiple combined threats including changes in hydrology, extreme weather events, salinisation, sea-level rise, pollution and overexploitation (Laurance *et al.*, 2011).

Protecting and conserving species as well as the condition of natural ecosystems requires an integrated approach. This involves support from pertinent research, the development of long-term monitoring strategies, the upgrade of existing protected areas (including their location and status), the use of predictive conservation management approaches (that incorporate climate change scenarios), and the continuing development of adequate regulatory and educational tools. Increasing consideration has been given to human-induced climate change, and Lindenmayer *et al.* (2010) suggested a range of landscape-level biodiversity conservation strategies specifically targeted to this issue. These include the significant reduction of greenhouse gas emissions, improved tackling of pre-existing stressors, preparing for the effects of major natural disturbances, significant improvements to off-reserve conservation, and making the existing reserve system more comprehensive, representative and forward-looking.

13.2.2   Conservation of threatened plants

Within a setting of limited time and resources, an important issue arising from conservation planning is the prioritisation of what needs to be conserved. Such priority has traditionally been given to species and communities that are rare and potentially already in trouble from an evolutionary perspective.

A central management tool employed by environment protection agencies is the development of recovery plans listing and prioritising conservation actions for individual threatened species. In principle, such plans provide a comprehensive and strategic inventory of actions such as threat abatement, surveys, monitoring and research that should ultimately lead to the removal of the species from the threatened species list. The listing of recovery actions relies on a basic understanding of the ecological, genetic and physiological factors likely to influence long-term survival. Unfortunately, a compromise needs often to be struck between the urgency for action, the availability of resources and the need for critical research-based information.

Recovery activities for threatened species usually involve one or more of the following steps: habitat protection (including grazing, fire and weed control), disease containment, *ex-situ* storage and propagation (including seed banking; Offord & Meagher, 2009), habitat restoration, and finally various forms of translocation (Box 13.3; See also Figure 13.2, Plate 49). These activities are normally prioritised according to the degree of impact they will have on the recovery of this and other species, and according to resource requirements and feasibility. Generally though, the recovery of a species is never a simple short-term project, and successful down-listing can rarely be achieved without the early establishment of monitoring strategies and performance measures (in NSW, for example, no rare plants have been down-listed due to recovery actions alone).

Because of the limited amount of resources available for the recovery of such a considerable number of threatened species (612 plant species and 103 ecological communities are currently listed in NSW alone), single-species recovery plans and actions are not always the best option. In fact, recent studies have shown that the development of a recovery plan does not necessarily influence the status of a threatened species, mostly because of the paucity of resources dedicated to its implementation, and the failure of planning for long-term maintenance (Bottrill *et al.*, 2011). Consequently, investing in actions that benefit more than one species is increasingly being recommended as an important means to improve biodiversity conservation outcomes.

However, multi-species recovery plans should not bring together a random group of threatened taxa. The approach through which species are combined within a single recovery plan needs to be considered carefully, because even closely related taxa can be ecologically different. Consequently the development of truly shared recovery actions can be difficult. The recognition that not all species contribute equally to ecosystem function has shifted the focus of ecological and conservation research from taxonomic to functional grouping (Westoby & Wright, 2006).

Box 13.3 Major considerations for translocation as a conservation tool for rare plants (from Vallee *et al.*, 2004)

Translocations can be defined as the transfer of plants from an *ex-situ* collection or a natural population to a location in the wild. There are three main types of translocation actions for threatened species.

- Enhancement: where an existing population is increased in size and in genetic diversity via the planting of additional individuals.

- Re-introduction: where a new population is established in a site where it formerly occurred but it is now extinct.

- Conservation introduction: where a new population is established in an environmentally and ecologically suitable site outside its known range.

Many translocation attempts have been unsuccessful in the past, consequently the need to translocate and the likelihood of success need to be carefully considered before starting (Godefroid *et al.*, 2011). The main objective of a translocation programme for a threatened plant should be to directly support the conservation of the target species through the establishment and maintenance of self-sustaining populations. Translocation can be deemed as successful by the following criteria.

- Management and control of the threats responsible for the rarity of the species (including site management and protection).

- Long-term establishment of translocated individuals.

- Replication across multiple sites in order to avoid risks associated with stochastic events.

- Successful recruitment within translocated populations.

Once it has been decided that translocation is justified and likely to succeed, a number of preliminary steps need to be followed. These include organising a relevant recovery team with the necessary skills, and gathering relevant information including collating research findings where necessary. The type of biological information needed includes the following.

- Understanding the species' taxonomic status and evolutionary potential. This includes the species' population genetics measuring diversity, inbreeding, as well as current and historical dynamics.

- Understanding the species' reproductive biology. This includes the species' primary mode of regeneration, breeding system, pollination and dispersal mechanisms, seed biology, and demography.

- Understanding the species' ecology. This includes the species' distribution range, its edaphic and climatic requirements, associated flora and fauna, relevant functional and life-history characteristics, response to fire, and its susceptibility to local threats.

- Understanding optimal propagation and collection strategies. This includes being aware of *ex-situ* propagation and seed storage techniques, and *in situ* planting methodologies.

Figure 13.2 (Plate 49) Successful reintroduction of the Corrigin grevillea (*Grevillea scapigera*) in the Western Australian wheatbelt. This site was weeded and partially restored. Reintroduced plants are successfully growing and reproducing. (Photo by Robert Dixon. Reproduced with permission.) A black and white version of this figure will appear in some formats. For the colour version, please refer to the plate section.

Once these issues have been considered, recipient sites have been selected and a suitable experimental design has been developed, a translocation proposal can be prepared and the necessary resources can be gathered. Source material can then be collected based on the available understanding of the species' biology, sites can be prepared by removing threats and restoring native vegetation (if necessary), and plants can be planted being mindful of timing and experimental design. Finally, a translocation project can only be successful if appropriate monitoring, evaluation of success and ongoing management are planned from the start (Monks, 2008).

As an additional note, it is important to remember that climate change brings to the fore further considerations when planning rare species' translocations. Prioritisation criteria should consider the ability of the species to track predicted change based on its functional capacities and the characteristics of its ideal habitat. 'Stranded' species will raise a range of issues relating to planting outside the species' natural range. The risks involved with the potential of hybridisation and localised invasions needs to be balanced with extinction risk. These issues further highlight the importance of including evolutionary research and long-term monitoring in threatened species conservation plans.

Research based on ecological and functional traits indicates that if the recovery objective is to reinstate functioning ecosystems, community diversity (species richness) is not as relevant as functional diversity. Within this context, functional grouping of threatened species can be used to identify shared threatening processes and to coordinate recovery efforts across entire regions, while simultaneously responding to the requirement for cost-effectiveness (Kooyman & Rossetto, 2008). Such progressive multi-species approaches are likely to provide more resource-efficient information gathering and lead to more focussed implementation strategies.

13.2.3 Biodiversity and climate change

Human-induced climate change is likely to further impact on a biosphere already heavily perturbed by human-derived threatening processes such as decreased availability of water, clearing and fragmentation, invasive species and pollution (Steffen *et al.*, 2009). Rapid climate change can also act directly on the physiology and life-cycle of plants. One of the critical issues of increasing concern for conservation biologists is to understand how plant species and floristic assemblages will respond to rising CO_2 levels. For instance, simulated future scenarios project strong floristic shifts towards more woody communities across Africa (Higgins & Scheiter, 2012). These predictive models support the expectation that atmospheric CO_2 has been and will be a major factor shaping future vegetation patterns (Manea & Leishman, 2011).

Changing temperature gradients can directly impact on the reproductive phenology of species and establish new barriers to gene flow among geographically separated populations (Elzinga *et al.*, 2007). This, in turn, can cause genetic isolation and localised reproductive failures. As a consequence, the longer-term evolutionary context needs to be carefully considered when outlining new protected areas in preparation for climate change, so that selected sites can improve connectivity and anticipate possible modifications in key environmental and climatic criteria (Bellard *et al.*, 2012).

Plants have a range of options to counteract climate change. They can take advantage of existing genetic plasticity, or they can rely on gene flow from other localities to complement their adaptive response. Either of these two options relies on the presence of sufficient genetic variation accessible within short time frames. Alternatively, individuals will need to be able to disperse to more suitable habitats, provided that these are available and accessible (Corlett & Wescott, 2013). Such range shifts have already been described (Parmesan, 2006) but these are only possible for functional groups that are suited to dispersal (such as species with wind- or vertebrate-dispersed fruits for example). Furthermore, landscape-level connectivity can be particularly difficult in areas where habitat fragmentation and land development are increasingly impacting on natural processes.

Overall, it can be expected that complex case-specific interactions between the functional capacities of species and local environmental conditions will result in significant changes in the composition and distribution of plant assemblages. As local environmental conditions change, species with a sufficiently broad evolutionary spectrum will be more likely to persist locally either through plasticity or through adaptive changes. If and when species run out of options, direct management actions (including assisted colonisation, which is the focus of growing conservation debate) will need to be considered before anthropogenic needs exclusively dictate landscape conservation criteria.

13.2.4 Restoring communities: achieving a balance between integrity and evolutionary potential

Strictly speaking, ecological restoration projects endeavor to return a community to a perceived historical state. Unfortunately, good planning and long-term monitoring are not always considered in restoration projects, as these often have numerical or aesthetic targets rather than ecological ones. A more realistic (and arguably more evolutionarily fitting) goal may be to move a degraded community towards a less disturbed state that will enable the restored ecosystem to recover compositional structure and function, as well as dynamics (Falk *et al.*, 2006). Consequently, restoration projects should aim to re-establish the dominant species of key functional groups first (such as nitrogen fixers or early successional woody species), rather than focus on number of species alone. Targeting key functional elements becomes the first step towards the recovery of the ecosystem services that are critical to broader-level biodiversity conservation (seed dispersal, pollination, pest control, for example).

Restoration practice and research have made significant advances in fields such as ecophysiology and functional ecology, food webs and the distribution of resources, the impact of invasive species and biodiversity dynamics (see relevant chapters within Falk *et al.*, 2006). There has also been increasing consciousness about the importance of including an evolutionary context within restoration ecology. Although the objective of restoration projects is to maximise the self-sustainability and adaptive potential of re-established populations, this does not necessarily translate to trying to maximise local genetic diversity. Inbreeding is not always a negative factor and heterozygosity is not always a positive one. The introduction of novel diversity can potentially reduce the fitness of locally adapted populations (a process known as outbreeding depression; Templeton, 1994) and, accordingly, the sourcing of re-vegetation material should be considered in light of explicit evolutionary criteria. However, because of the limited resources devoted to obtaining relevant data (as well as the inevitable practical limitations), the selection of provenance material is often based on basic distributional patterns.

Much has been written about provenance sourcing and it is now evident that the definition of 'suitable' provenance varies from species to species (Broadhurst *et al.*, 2008). The issue is not only with identifying provenance boundaries but also with defining the relative evolutionary impact of those boundaries. The perceived benefits of retaining the genetic 'integrity' of a site are based on the assumption that locally sourced plants are better adapted to local conditions and will therefore survive longer, grow faster and have reproductive advantages over non-local plants. However, plants that are locally adapted now may not be so in the future.

Furthermore, the risks associated with the mixing of material from geographically distant populations are generally overestimated, particularly within recently fragmented landscapes (Frankham *et al.*, 2011). Excessively conservative sourcing criteria could in fact increase the risk of inbreeding and further erode an already low genetic pool. As a consequence, the availability of information on genetically diverse and climatically suitable non-local provenance material will become ever more important in the future (Hancock & Hughes, 2012).

Rapid climate change can accentuate existing disequilibria between a species' geographic spread and the availability of suitable habitat, with or without direct human intervention. Under the current circumstances of high anthropogenic disturbance, biodiversity conservation should increasingly focus on managing change as restoring past assemblages will become gradually more difficult. Thomas (2011) argued that a philosophy based on restoring the composition of biological communities as they were is out of sync with the reality of current environmental and biological change. An alternative option for restoring evolutionary fitting ecosystems and conserve biodiversity is the translocation of species outside their normal distributional range. Although assisted colonisation is generally considered only if traditional conservation strategies are insufficient, the risks involved might be contained if planting takes place within the same broad geographic region and the destination sites lack related endemics.

Populations stranded in increasingly unsuitable refugia, particularly those surrounded by areas mainly dedicated to human use, might be suitable targets for active management. Moving individuals from warm-adapted populations to cooler locations may increase the probability of more widespread resilience under a warming environment (Webber *et al.*, 2011). Such approaches can potentially be applied to species naturally distributed across wide environmental gradients or to species that have been shown to display genetically based clines under different climatic conditions.

These are still relatively untested concepts and the current level of scientific debate on the benefits and pitfalls of artificial modifications in a species' distribution is understandable and justified (Ricciardi & Simberloff, 2009; Hannah, 2010). Interestingly though, reshuffling events have been relatively common in natural populations and communities, particularly during the current glacial cycles (Willis & Niklas, 2004). An increasing body of scientific work is showing that current distributional ranges sometimes describe past dynamic process rather than represent a state of equilibrium with current habitat availability (Box 13.4).

Box 13.4 Reshuffling of genes, individuals and assemblages: evidence from the Proteaceae

A novel set of analytical tools can tease apart associations between the distribution of genetic diversity and a variety of relevant environmental variables. Combining molecular data with environmental niche models can track how temporal variations in habitat suitability are associated with changes in connectivity and gene flow.

A study combining molecular, morphological and bio-climatic analyses across the entire distribution of *Telopea speciosissima* (the NSW waratah, Proteaceae) discovered strong associations between population-level differentiation and localised climatic conditions (Rossetto *et al.*, 2011). Fine-scale population genetics showed that as postglacial warming broke down the phenological barriers that separated upland and lowland populations, gene flow was established between these morphologically and genetically differentiated populations.

A follow-up investigation extending to the other three continental waratahs demonstrated that within- and between-species divergences were caused by the instability of their respective climatic envelopes during glacial cycles (Rossetto *et al.*, 2012; Plates 8 and 9). These studies recognised the allopatric basis of speciation in *Telopea*, as well as the transitional nature of species-level differentiation. Within such systems, the adaptive morphological and genetic segregation between lineages is strengthened during periods of environmentally enforced isolation, while lineage-level plasticity is enhanced during times of admixture.

During the Quaternary, recurring adjustments in the availability of suitable habitat impacted on the distribution and assemblage of many other species. These reshufflings brought together closely related species and led to landscape-level hybridisation events, such as those described for a number of eucalypts from the subgenus *Symphomyrtus* in Tasmania (McKinnon *et al.*, 2001).

More recent phylogeographic research has uncovered evidence of hybridisation among species that are highly morphologically and ecologically distinct. By mapping the distribution of chloroplast haplotypes shared among continental *Lomatia* species (Proteaceae), it was possible to identify refugial areas that protected diversity during unfavourable climatic conditions, distributional overlap zones where hybridisation was common, and biogeographic breaks that consistently prevented genetic connectivity (Milner *et al.*, 2012). Such information greatly advances our understanding of the evolutionary history of this genus, as well as being critical to the detection of geographic areas of high conservation value.

Further fine-scale morphological and genetic studies on selected species of *Lomatia* are showing that genetic exchange at hybrid fronts continues for multiple generations. However, despite continuing gene exchange the strength of the selective drivers that influence differentiation among species is such that the adaptive architecture of the original species is eventually restored (McIntosh *et al.*, 2014).

These and other studies highlight the transitional nature of species-level differentiation in plants. Understanding the causational processes responsible for the establishment of distinct evolutionary units is a critical step towards the development of predictive biodiversity conservation strategies.

13.3 Conclusion

Historically geological and climatic instability have been important factors influencing the local reshuffle of genes, species and communities. Yet, despite the fact that the Australian flora has evolved considerable resilience to change, the speed and intensity of human-induced disturbance present significant threats to the preservation of biodiversity.

As anthropogenic pressure is accelerating and intensifying, there is increasing awareness about the need to move towards more 'evolutionary aware' biodiversity management strategies. Describing the extent and distribution of diversity and endemism is important, but should only represent a preliminary step towards the development of conservation criteria. Biodiversity management is increasingly reliant on our understanding of the relative vulnerability of species and communities to current, past and future threats.

As changes in the distribution and assemblage of species take place in response to local selective pressures, it is important to differentiate between long-term dynamic processes and a state of equilibrium with currently available niche space. As management actions and landscape conservation criteria are increasingly impacted by urgent anthropogenic needs, proactive restoration strategies will need to be explored more regularly.

REFERENCES

Bellard, C., C. Bertelsmeier, P. Leadley, W. Thuiller, F. Courchamp (2012). Impacts of climate change on the future of biodiversity. *Ecology Letters*, **15**, 365–377.

Bottrill, M. C., J. C. Walsh, J. E. M. Watson, *et al.* (2011). Does recovery planning improve the status of threatened species? *Biological Conservation*, **144**, 1595–1601.

Broadhurst, L. M., A. Lowe, D. J. Coates, *et al.* (2008). Seed supply for broadscale restoration: maximising evolutionary potential. *Evolutionary Applications*, **1**, 587–597.

Chapman, A. D. (2005). *Number of Living Species in Australia and the World*. Report to the Department of the Environment and Heritage, Canberra.

Corlett, R. T., D. A. Westcott (2013). Will plant movements keep up with climate change? *Trends in Ecology and Evolution*, **28**, 482–488.

Crayn, D. M., M. Rossetto, D. J. Maynard (2006). Molecular phylogeny and dating reveals an Oligo-Miocene radiation of dry-adapted shrubs (former Tremandraceae) from rainforest tree progenitors (Elaeocarpaceae) in Australia. *American Journal of Botany*, **93**, 1328–1342.

Crisp, M., L. Cook, D. Steane (2004). Radiation of the Australian flora: what can comparisons of molecular phylogenies across multiple taxa tell us about the evolution of diversity in present-day communities? *Philosophical Transactions of the Royal Society, London B*, **359**, 1551–1571.

Crisp, M. D., L. G. Cook (2007). A congruent molecular signature of vicariance across multiple plant lineages. *Molecular Phylogenetics and Evolution*, **43**, 1106–1117.

Elzinga, J. A., A. Atlan, A. Biere *et al.* (2007). Time after time: flowering phenology and biotic interactions. *Trends in Ecology and Evolution*, **22**, 432–439.

Falk, D. A. (1990). Endangered forest resources in the US: integrated strategies for conservation of rare species and genetic diversity. *Forest Ecology Management*, **35**, 91–107.

Falk, D. A., M. A. Palmer, J. B. Zedler (2006). *Foundations of Restoration Ecology*. Island Press, Washington USA.

Forest, F., G. Grenyer, M. Rouget, *et al.* (2007). Preserving the evolutionary potential of floras in biodiversity hotspots. *Nature*, **445**, 757–760.

Frankham, R., J. D. Ballou, M. D. B. Eldridge, *et al.* (2011). Predicting the probability of outbreeding depression. *Conservation Biology*, **25**, 465–475.

Gandolfo, M. A., E. J. Hermesen, M. C. Zamaloa, *et al.* (2011). Oldest known *Eucalyptus* macrofossils are from South America. *PLoS ONE*, **6**, e21084.

Godefroid, S., C. Piazza, G. Rossi, *et al.* (2011). How successful are plant species reintroductions? *Biological Conservation*, **144**, 672–682.

Graham, C. H., C. Moritz, S. E. Williams (2006). Habitat history improves prediction of biodiversity in rainforest fauna. *PNAS*, **103**, 632–636.

Greenwood, D. R., D. C. Christophel (2005). The origins and Tertiary history of Australian Tropical rainforests. In *Tropical Rainforests: Past, Present and Future*. E. Bermingham, C. W. Dick. C. Moritz (eds.). The University of Chicago Press, USA, pp. 336–373.

Hancock, N., L. Hughes (2012). How far is it to your local? A survey on local provenance use in New South Wales. *Ecological Management & Restoration*, **13**, 259–266.

Hannah, L. (2010). A global conservation system for climate-change adaptation. *Conservation Biology*, **24**, 70–77.

Hays, J. D., J. Imbrie, N. J. Shackelton (1976). Variations in the Earth's orbit: pacemaker of the ice ages. *Science*, **194**, 1121–1132.

Hewitt, G. M. (2000). The genetic legacy of ice ages. *Nature*, **405**, 907–913.

Higgins, S. I., S. Scheiter (2012). Atmospheric CO_2 forces abrupt vegetation shifts locally, but not globally. *Nature*, **488**, 209–213.

Hill, R. S. (1998). Fossil evidence for the onset of xeromorphy and scleromorphy in Australian Proteaceae. *Australian Systematic Botany*, **11**, 391–400.

Hill, R. S., E. M. Truswell, S. McLoughlin, M. E. Dettmann (1999). The evolution of the Australian flora: fossil evidence. In *Flora of Australia*, 2nd edition A. E. Orchard (ed). CSIRO Publishing, Melbourne, pp. 251–320.

Hoffmann, A. A., Y. Willi (2008). Detecting genetic responses to environmental change. *Nature Reviews Genetics*, **9**, 421–432.

Hopkins, M. S., J. Ash, A. W. Graham, J. Head, R. K. Hewett (1993). Charcoal evidence of the spatial extent of *Eucalyptus* woodland expansions and rainforest contractions in North Queensland during the Late Pleistocene. *Journal of Biogeography*, **20**, 357–372.

Hopper, S. D., P. Gioia (2004). The Southwest Australian Floristic Region: evolution and conservation of a global hotspot of biodiversity. *Annual Review of Ecology and Systematics*, **35**, 623–650.

Ingham, J. A., P. I. Forster, M. D. Crisp, L. G. Cook (2013). Ancient relicts or recent dispersal: how long have cycads been in central Australia? *Diversity and Distributions*, **19**, 307–316.

Kershaw, A. P., S. C. Bretherton, S. van der Kaars (2007). A complete pollen record of the last 230Ka from Lynch's crater, northern Australia. *Palaeogeography, Palaeoclimatology, Palaeoecology*, **251**, 23–45.

Kooyman, R., M. Rossetto (2008). Definition of plant functional groups for informing implementation scenarios in resource-limited multi-species recovery planning. *Biodiversity and Conservation*, **17**, 2917–2937.

Kooyman, R., M. Rossetto, W. Cornwell, M. Westoby (2011). Phylogenetic tests of community assembly across regional to continental scales in tropical and subtropical rain forests. *Global Ecology and Biogeography*, **20**, 707–716.

Laurance, W. F., B. Dell, S. M. Turton, *et al.* (2011). The 10 Australian ecosystems most vulnerable to tipping points. *Biological Conservation*, **144**, 1472–1480.

Lindenmayer, D. B., W. Steffen, A. A. Burbidge, *et al.* (2010). Conservation strategies in response to rapid climate change: Australia as a case study. *Biological Conservation*, **143**, 1587–1593.

Manea, A., M. R. Leishman (2011). Competitive interactions between native and invasive exotic plant species are altered under elevated carbon dioxide. *Oecologia*, **3**, 1–10.

McIntosh, E. J., M. Rossetto, P. H. Weston, G. M. Wardle (2014). Maintenance of strong morphological differentiation despite ongoing natural hybridization between sympatric species of *Lomatia* (Proteaceae). *Annals of Botany*, doi:10.1093/aob/mct314.

McKinnon, G. E., R. E. Vaillancourt, H. D. Jackson, B. M. Potts (2001). Chloroplast sharing in the Tasmanian eucalypts. *Evolution*, **55**, 703–711.

McLoughlin, S. (2001). The breakup history of Gondwana and its impact on pre-Cenozoic floristic provincialism. *Australian Journal of Botany*, **49**, 271–300.

Mellick, R., A. Lowe, C. Allen, R. S. Hill, M. Rossetto (2012). Palaeodistribution modelling and genetic evidence highlight differential post-glacial range shifts of a rain forest conifer distributed across a latitudinal gradient. *Journal of Biogeography*, **39**, 2292–2302.

Milner, M. L., M. Rossetto, M. D. Crisp, P. H. Weston (2012). The impact of multiple biogeographic barriers and hybridization on species-level differentiation. *American Journal of Botany*, **99**, 2045–2057.

Monks, L. (2008). Experimental approaches in threatened plant translocations: how failures can still lead to success. *Australasian Plant Conservation*, **17**, 8–10.

Moritz, C. (1994). Defining 'Evolutionary Significant Units' for conservation. *Trends in Ecology and Evolution*, **9**, 373–375.

Nagalingum, N. S., C. R. Marshall, T. B. Quental, *et al.* (2011). Recent synchronous radiation of a living fossil. *Science*, **334**, 796–799.

Offord, C. A., P. F. Meagher (2009). *Plant Germplasm Conservation in Australia*. Australian Network for Plant Conservation, Canberra.

Parmesan, C. (2006). Ecological and evolutionary responses to recent climate change. *Annuual Review of Ecology, Evolution, and Systematics*, **37**, 637–669.

Petit, R. J., I. Aguinagalde, J. L. de Beaulieu, *et al.* (2005). Glacial refugia: hotspots but not melting pots of genetic diversity. *Science*, **300**, 1563–1565.

Ricciardi, A., D. Simberloff (2009) Assisted migration is not a viable conservation strategy. *Trends in Ecology and Evolution*, **24**, 248–253.

Rossetto, M., R. Kooyman (2005). The tension between dispersal and persistence regulates the current distribution of rare palaeo-endemic rain forest flora: a case study. *Journal of Ecology*, **93**, 906–917.

Rossetto, M., C. L. Gross, R. Jones, J. Hunter (2004a). The impact of clonality on an endangered tree (*Elaeocarpus williamsianus*) in a fragmented rainforest. *Biological Conservation*, **117**, 33–39.

Rossetto, M., R. Jones, J. Hunter (2004b). Genetic effects of rainforest fragmentation in an early successional tree (*Elaeocarpus grandis*). *Heredity*, **93**, 610–618.

Rossetto, M., R. Kooyman, W. Sherwin, R. Jones (2008). Dispersal limitations rather than bottlenecks or habitat specificity can restrict the distribution of rare and endemic rainforest trees. *American Journal of Botany*, **95**, 321–329.

Rossetto, M., D. Crayn, A. Ford, R. Mellick, K. Sommerville (2009). The influence of environmental and life-history traits on the distribution of genes and individuals: a comparative study on 11 rainforest trees. *Molecular Ecology*, **18**, 1422–1438.

Rossetto, M., K. A. G. Thurlby, C. A. Offord, C. B. Allen, P. H. Weston (2011). The impact of distance and a shifting temperature gradient on genetic connectivity across a heterogeneous landscape. *BMC Evolutionary Biology*, **11**, 126.

Rossetto, M., C. B. Allen, K. A. G. Thurlby, P. H. Weston, M. L. Milner (2012). Genetic structure and bio-climatic modeling support allopatric over parapatric speciation along a latitudinal gradient. *BMC Evolutionary Biology*, **12**,149.

Rule, S., B. W. Brook, S. G. Haberle, *et al.* (2012). The aftermath of megafaunal extinction: ecosystem transformation in Pleistocene Australia. *Science*, **335**, 1483–1486.

Sniderman, J. M. K., G. J. Jordan (2011). Extent and timing of floristic exchange between Australia and Asian rain forests. *Journal of Biogeography*, **38**, 1445–1455.

Steffen, W., A. A. Burbidge, L. Hughes, *et al.* (2009). *Australia's Biodiversity and Climate Change*. CSIRO Publishing, Melbourne.

Stewart, J. R., A. M. Lister (2001). Cryptic northern refugia and the origins of the modern biota. *Trends in Ecology and Evolution*, **16**, 608–613.

Templeton, A. R. (1994). Coadaptation, local adaptation, and outbreeding depression. In *Principles of Conservation Biology*, G. K. Meffe, C. R. Carroll (eds.). Sinauer Associates, Sunderland USA, pp. 152–153.

Thomas, C. D. (2011). Translocation of species, climate change, and the end of trying to recreate past ecological communities. *Trends in Ecology and Evolution*, **26**, 216–221.

Vallee, L., T. Hogbin, L. Monks, *et al.* (2004). *Guidelines for the Translocation of Threatened Plants in Australia*. Australian Network for Plant Conservation, Canberra.

Webber, B. L., J. K. Scott, R. K. Didham (2011). Translocation or bust! A new acclimatization agenda for the 21st century? *Trends in Ecology and Evolution*, **26**, 495–496.

Westoby, M., I. J. Wright (2006). Land-plant ecology on the basis of functional traits. *Trends in Ecology and Evolution*, **21**, 261–268.

Weston, P. H., R. M. Kooyman (2002). Systematics of *Eidothea* (Proteaceae), with the description of a new species, *E. hardeniana*, from the Nightcap Range, north-eastern New South Wales. *Telopea*, **9**, 821–832.

Wilf, P., K. R. Johnson, N. R. Cuneo, *et al.* (2005). Eocene plant diversity at Laguna del Hunco and Rio Pichileufu, Argentina. *American Naturalist*, **165**, 634–650.

Willis, K. J., K. J. Niklas (2004). The role of Quaternary environmental change in plant macroevolution: the exception or the rule? *Philosophical Transactions of the Royal Society, London B*, **359**, 159–172.

Worth, J. P., G. J. Jordan, G. E. McKinnon, R. E. Vaillancourt (2009). The major Australian cool rainforest tree *Nothofagus cunninghamii* withstood Pleistocene glacial aridity within multiple regions: evidence from the chloroplast. *New Phytologist*, **182**, 519–523.

CHAPTER 14

Protecting the small majority: insect conservation in Australia and New Zealand

Gregory I. Holwell and Nigel R. Andrew

Summary

Insects represent the largest component of Australasia's animal diversity. While the uniqueness and conservation needs of Australia and New Zealand's vertebrates are generally understood, the importance of our insects and the threats they face are less appreciated. Some groups, including locally endemic butterflies and flightless giants, such as giant weta, are important for raising public awareness of insect conservation. However, our understanding of how broad processes influence insect populations and communities is in its infancy. Part of the issue is due to a complete lack of knowledge of the biology of the vast majority of insect species, as most insects in Australasia remain undescribed. In this chapter we discuss insect biodiversity in Australia and New Zealand and discuss both insect species and diversity conservation, contrasting patterns in Australia and New Zealand. We then discuss some of the major threats facing insect species and diversity, specifically focussing on the impacts of habitat loss and fragmentation, predation by invasive rodents and climate change. Lastly, we discuss interactions between insects and humans including the provision of ecosystem services by insects in an agricultural context, human consumption of insects (entomophagy) and concerns surrounding the lack of taxonomic expertise for insects in Australasia.

14.1 Insect biodiversity in Australia and New Zealand

The uniqueness of the Australian fauna has been known for centuries, and since the first European explorers returned from voyages to the Antipodes, naturalists have remarked on the diversity of Australasian life, and its peculiarity. While most people are familiar with the stories of European incredulity when faced with a stuffed platypus or kiwi, many may not appreciate that the insect fauna of Australia and New Zealand is equally unique, and far more diverse. The first insect formally identified in Australia was the charismatic Botany Bay weevil (*Chrysolopus spectabilis*) by Joseph Banks who

Austral Ark: The State of Wildlife in Australia and New Zealand, eds. A. Stow, N. Maclean and G. I. Holwell. Published by Cambridge University Press. © Cambridge University Press 2015.

accompanied James Cook in 1770, but since then over 60 000 species have been described from Australia and New Zealand. Estimates for the species-richness of Australia's terrestrial insects range between 84 000 species (CSIRO, 1991) and 205 000 (Yeates *et al.*, 2003), of which 75% are yet to be described, and given a name. New Zealand has lower diversity owing to its smaller landmass and its more temperate latitudinal range, but still holds an estimated 20 000 species of insects with 10 000 still requiring description (Cranston, 2010). The diversity of the Australasian invertebrate fauna can partly be explained by the sheer diversity of habitats present in the region, ranging from tropical rainforests and monsoonal grasslands, to deserts and dry woodlands, to tall temperate forests and alpine ranges (particularly in New Zealand). Because of this environmental diversity, it is perhaps no surprise that the invertebrate fauna has radiated to fill so many habitats and niches. The diversity of the Australasian invertebrate fauna can also be considered in light of the diversity of its origins. Australasian insects can be divided into ancient globally distributed groups, Gondwanan lineages being shared with other southern hemisphere landmasses, Austral taxa shared between Australia and New Zealand, and species that are widely dispersed throughout the globe either through natural or human-mediated dispersal (Austin *et al.*, 2004).

Species- and genus-level endemicity is very high with many groups specific to either Australia, New Zealand, or spanning both (and sometimes including fauna in New Caledonia). More noteworthy perhaps are the number of invertebrate groups that are endemic at the family level, reflecting an ancient 'uniqueness' to the region. Endemic Australian families of insects include the Hemipteran families Henicocoridae, Hyocephalidae, Aphylidae, Lestoniidae, and Tettigarctidae; the five beetle families Rhinorhipidae, Acanthocnemidae, Lamingtoniidae, Tasmosalpingidae and Myraboliidae; the fly families Ironomyiidae and Valeseguyidae; the moth families Lophocoronidae, Palaeosetidae, Anomosetidae, Hypertrophidae, Cyclotornidae, Carthaeidae, Anthelidae, Oenosandridae; and the wasp family Austrocynipidae (Austin *et al.*, 2004). In New Zealand, the ectoparasitic bat-fly is the sole member of the family Mystacinobiidae, being entirely restricted to the bodies, roosts and guano of one of New Zealand's two native bats, the short-tailed bat *Mystacina tuberculata*. Another fly family, the Huttoninidae is endemic to New Zealand along with the moth family Mnesarchaeidae and the mayfly family Siphlaenigmatidae.

There are a number of notable absences among the insect families present in Australia and New Zealand, and some families within each of the major orders are poorly represented in Australasia, despite being otherwise common and cosmopolitan (Cranston, 2009). The orders Zoraptera, Mantophasmatodea and Grylloblattodea are absent from the region and the Embioptera are absent from New Zealand. Also in New Zealand, a number of insect groups are poorly represented, with few species and no major radiations. This is particularly obvious for groups such as ants, termites, butterflies and dragonflies, who are all dominant components of the insect fauna elsewhere, including Australia.

14.2 Conservation of insect diversity and species

Throughout the world it is becoming increasingly apparent that effective conservation of insects will rely more heavily on efforts to conserve insect diversity, rather than focussed single-species approaches (Samways, 2005). The sheer diversity of the insect fauna in most of the world's biomes makes effective assessment and monitoring of the conservation needs of individual insect species prohibitive in the vast number of cases. But, as we discuss later, a number of insect species have received legislative protection, and recovery plans are in place in Australasia as in the rest of the world (New, 2009).

Australian and New Zealand conservation legislation attempts to protect insects in a variety of different ways. Species-specific listings occur in both nations, and most states of Australia, but whole communities may be listed (e.g. Mt Piper threatened butterfly community). Plant communities such as the Cumberland Plains Woodland in NSW, which was listed as an Endangered Ecological Community under the Threatened Species Conservation Act of 1995, may act as an umbrella for the conservation of a range of insect species. In the Australian Capital Territory, legislation follows the Nature Conservation Act 1980, which closely follows the IUCN declaration of a threatened species, ecological community or threatening process. This allows communities to be listed such that, although these endangered community listings are based primarily on plant assemblages, many insect species will be conserved.

One of the most difficult issues regarding the preservation of insect diversity and insect communities is how to identify them, and how to assess an assemblage (a group of organisms found together, but not necessarily interacting) versus a community (a group of organisms interacting with each other, but not necessarily reliant on each other), and how robust the interactions are among species. There have been many assessments of insect assemblages and how these groups associate together, but if a species within the group becomes locally extinct, or shifts its range outside of a particular area, what impact does that have on the workings of that assemblage. A key aspect of preserving communities is to understand the ecology of species interactions. This is particularly pertinent for assessing the impacts of habitat removal, invasive species and climate change on species interactions.

For example, one of the major impacts of climate change is the potential movement of species either polewards or altitudinally to stay within their thermal range. This may mean changing interactions among species, and also changing interactions with different populations of the same species. Such change in community structure is extremely difficult to manage and to predict. Current conservation practices do not take into account changes in community structure, or what implications such changes may have on the conservation status of an endangered ecological community.

A recent method to assess insect conservation priorities for diversity is based on the use of plant surrogates (Moir & Leng, 2013). They found that 70 insect species were under immediate conservation threat, based on the threatened status of their host plant. Sessile feeders, such as mealybugs, and brachypterous (shortened or reduced wings) taxa, such as cicadellids and weevils, may be highly vulnerable, even on widespread host plants, as their dispersal is restricted; but mobile taxa such as psyllids may also be vulnerable. Such methods have merits in assessing the potential threatened status of species; however, it can lead to mislabelling of species as threatened if they have not been sampled adequately. A good example is the psyllid *Acizzia keithi*, which was thought to be found only on a threatened plant species, *Pultenaea glabra*, at Cataract Falls NSW, with models identifying an 80%–100% probability of *A. keithi* being monophagous (Moir *et al.*, 2010). However, further sampling along the coast and westwards found its distribution both geographically and across host plants to be extended (Powell *et al.*, 2011).

This highlights the need to sample from widespread congenerics when assessing the host plant relationships of threatened plant species.

Active conservation of individual insect species is perhaps more rare than for vertebrates in Australia and New Zealand, but a variety of species particularly at risk are recognised by legislative protection, have specific recovery plans and are the focus of more active management through the eradication of pests and weeds and the creation of specific conservation reserves (Eltham copper butterfly – see Box 14.2; Mahoenui giant weta), captive breeding programmes (Lord Howe Island stick insect; giant weta – see Box 14.1) and the restoration of key habitats through planting of host plants (Richmond birdwing). A comparison of the number of species receiving conservation status in Australia (via national or state conservation organisations) and New Zealand reveals some points of interest (Table 14.1; Figures 14.1 and 14.2). First, the number of species officially recognised as threatened is somewhat similar between both countries, but only if you

Box 14.1 Protecting the giants

Both Australia and New Zealand have had recent concerted efforts to protect and conserve highly threatened iconic giant insects. New Zealand's largest insects, giant weta (*Deinacrida* sp.), rival the heaviest insects in the world, and have been the focus of major translocation efforts, and the dedication of specific reserves for their conservation. In Australia, or more specifically, Lord Howe Island, the recent rediscovery of the extraordinary Lord Howe Island stick insect has received great publicity, and the ongoing efforts to ensure its survival have attracted much needed public enthusiasm backed by high-profile supporters. Giant insects likely act as important flagships for insect conservation, because their size makes them more charismatic and captures the imaginations of even otherwise entomologically uninterested citizens.

Giant weta (*Deinacrida* spp.) have declined significantly in all parts of their former range in New Zealand where invasive rodents have spread. Predation by rats, combined with habitat loss, has led to the current conservation concerns over this extraordinary and endemic group of insects (Watts *et al.*, 2008; Watts & Thornburrow, 2009). The Mahoenui giant weta (*Deinacrida mahoenui*) must be one of only a few insects worldwide to have a dedicated conservation reserve at the Mahoenui Giant Weta Scientific Reserve, in the central north island, New Zealand. This giant was discovered here in a patch of gorse (*Urex europaeus*: a dense and thorny invasive weed) on farmland. The gorse, which has grown into a dense thicket encouraged by browsing from goats, appears to have provided protection from predators for the weta. This is an unusual example of such a heavily modified environment fortuitously providing the ideal habitat for a threatened species, and perhaps the only protected patch of gorse in the world! Since this area was protected as a reserve, a number of translocations of *D. mahoenui* have been attempted, with mixed success (Watts *et al.* 2008; Watts & Thornburrow, 2009). Two translocated populations are now thriving on Mahurangi Island and at Warrenheip, which are both rat-free, but attempts to establish populations at other locations appear to have failed, apparently due to a lack of effective rodent control. Populations of translocated *D. mahoenui* have been monitored intermittently, and to varying degrees, however another species of giant weta *D. rugosa* has been successfully translocated and followed by an intensive monitoring regime (Watts *et al.* 2008; Watts & Thornburrow, 2009).

Box 14.1 (continued)

One hundred *D. rugosa*, the Cook Strait giant weta, were translocated to Karori Wildlife Sanctuary, in suburban Wellington, a mammal-controlled reserve surrounded by a predator-proof fence. Twenty individuals were released with radio-transmitters and tracked over two months, with all individuals being located each day. Insect translocations, themselves rare occurrences, are even more infrequently followed up by such intensive monitoring, mostly due to the logistics of tracking small animals. Developments in tracking technology, and the large size of these insects allowed the focussed, post-translocation monitoring, that is usually only possible for birds and mammals. The monitoring yielded important information about the ecology of giant weta. It showed that the weta moved much further than expected and, of those that died during the tracking period, none was killed by predators. Translocation is likely to continue to be an important component of ongoing conservation efforts for these extraordinary giants.

The Lord Howe Island stick insect, or 'land lobster', *Dryococelus australis* has become an iconic flagship species for conservation on Lord Howe Island, a small volcanic island in the Tasman Sea (Cranston, 2010). The stick insect, a large flightless herbivore whose nearest relative is from New Guinea, was common on the island in the nineteenth century. Black rats (*Rattus rattus*) arrived on Lord Howe Island, presumably when the trading vessel Makambo was damaged offshore on the 14th of June 1918, and cargo was brought ashore and, after their establishment, a number of vertebrate and invertebrate species quickly declined, many to the point of extinction, the land lobster among them (Priddel *et al.*, 2003). No land lobsters were seen for decades and were officially described as extinct in the IUCN Red List, until in 1964 a rock climber successfully scaled Ball's Pyramid: an extraordinary 550 m high volcanic outcrop of rock, approximately 25 km south-east of Lord Howe, and photographed a recently dead adult female stick insect. A further two dead specimens were subsequently collected and positively identified as *Dryococelus australis*. Despite this, it was only in 2001 that an official expedition was undertaken to determine if a population of the stick insects was persisting on Ball's Pyramid. Living land lobsters were found in 2001 and 2002, and another expedition in 2003 allowed the collection of two pairs of adults to return to mainland Australia for captive breeding. Melbourne Zoo has led the captive breeding and management of the land lobsters and, despite the challenges of breeding a species whose biology is completely unknown, this has been a successful venture, and the long-term aim of translocating captive-bred land lobsters back to Lord Howe Island, or some of the other offshore islets surrounding it, appear to be tenable (Honan, 2008). One of the key aspects of the land lobster story is that the public support for the conservation of an insect species has been the main driver of what will likely be a major rodent eradication programme on the main Lord Howe Island. Despite the extinction of many species and the persistent threat of rats to remaining fauna on the island, rat control was not given sufficient attention until public support for the re-release of stick insects gave it momentum. Many species will likely benefit from such an eradication, and as such the Lord Howe Island stick insect represents a wonderful example of a flagship for conservation.

Box 14.2 Butterfly conservation: the importance of community participation

Butterflies are perhaps the most common flagship group for insect conservation in Australia and internationally. Public enthusiasm for butterflies is often easy to muster and community involvement is often the driving force in the conservation of threatened butterflies (New, 2010). This is in part because many threatened butterfly species are local endemics, limited to very narrow distributional ranges, which fosters strong community pride. As with threatened butterfly species throughout the world, conservation of the insects is strongly linked to understanding their host plants, and often the ants that tend their larvae. The Richmond birdwing (*Ornithoptera richmondia*) has received a huge benefit from the involvement of schools. The 'adopt a caterpillar' programme has contributed greatly to the conservation of *O. richmondia*, along with major plantings of their host vines (*Pararistolochia praevenosa*) and removal of the invasive vine, dutchman's pipe (*Aristolochia elegans*), which is an attractive site for oviposition by adult females, but toxic to developing caterpillars (Sands, 2008).

The Eltham copper (*Paralucia pyrodiscus lucida*, a brightly coloured sub-species of the more widely distributed dull copper) is another excellent example of how community participation can drive practical insect conservation (New, 1990). *P. p. lucida* was commonly collected in the north-eastern suburbs of Melbourne in the middle of the twentieth century but declined in the 1960s to the point where it was believed to be locally extinct. In 1987, a large population was discovered on land scheduled for housing development. The government and the developers were approached, and a large community-led fund-raising effort led to the purchase of the land exclusively for the purpose of butterfly conservation. This is an excellent example of community participation in insect conservation. Major financial contributions were made by state and local government, but the importance of community involvement was crucial. Awareness was raised by passionate community members who formed the 'Friends of the Eltham copper butterfly' conservation group. Conservation management plans involved promoting the larval food plant (*Bursaria spinosa*), and the ants that attend the caterpillars of *P. p. lucida*, but the initial response was largely around protecting the site from a range of factors associated with urban encroachment: weed invasion, disturbance by people and domestic animals, rubbish and pollutants. The seriousness with which government conservation agencies have taken Eltham copper conservation provides an excellent example of how it is possible to instigate species-level conservation efforts for invertebrates, and strong support and involvement from the community was likely a major contributing factor.

A very similar story surrounds the conservation efforts for another Lycaenid butterfly *Paralucia spinifera*, known locally as the Bathurst copper, or the Lithgow copper, depending on which town your allegiances lie with. Both communities have led community conservation efforts, and claim the butterfly as 'their own', highlighting the importance of a sense of ownership in community conservation, but also highlighting how local endemism can at first contribute to the threats posed to narrowly distributed insects, but can eventually be their saviour as local communities invest more emotional attachment to their special insect. Other threatened butterflies receive the benefits of community conservation groups, not

Box 14.2 (continued)

focussed on the butterflies themselves, but on their habitats and the threatened communities of which they are a part. The Altona skipper (*Hesperilla flavescens flavescens*), restricted to swamps that possess their larval food plant, the sedge *Gahnia filum*, benefit from the activities of habitat-focussed community groups such as the Friends of Westona Wetlands, and similarly the swordgrass brown butterfly hybrid complex *Tisiphone abeona joanna*, occurring in swamps surrounding Port Macquarie, NSW, benefits from the activities of local environmental action groups such as Knox Environment Group (Sands & New, 2002). Conservation of habitats and communities of species is likely to be the most useful conservation measure for insects in Australasia, as with the rest of the world. By protecting diverse areas of habitat, benefits can be achieved simultaneously for multiple species, and stakeholders. Victoria is again somewhat unique in affording protected status to a 'threatened community' of butterflies, at Mt Piper, north of Melbourne (Britton & New, 1995). This isolated volcanic peak provides an important hilltopping site for a variety of breeding butterflies, the fungi and lichens their larvae feed on, and the ants that attend many of them. The action statement for this threatened community specifically calls for involvement by the community, and the raising of environmental awareness surrounding the site and its unique butterfly community.

include those recognised at the Australian state level. The number which have received national threatened status is considerably fewer for Australia than New Zealand. Second, if you consider the number of species considered threatened as a proportion of the number described, and the estimated total number of species, New Zealand has a far greater proportion of threatened insect species than Australia. It may be that a greater proportion of New Zealand insect species are under threat, but it seems more likely that this simply reflects a more comprehensive knowledge of the insect fauna and their conservation requirements in New Zealand. In Australia, only an estimated quarter of insect species are described, and few of these have been studied in any detail, such that the conservation requirements of most species are unknown. This is not to say that the conservation requirements of New Zealand's insect fauna are particularly well known either, but more than 1% of described New Zealand insect species have official threatened species status, and a further 6% are classified as 'at risk' due to declining numbers, restricted geographic range or naturally small populations.

14.3 Threats to insects in Australasia

The threats to insect species and insect diversity in Australia and New Zealand are diverse. Some factors, such as habitat loss, influence insect diversity as a whole, driving the local extinction of many species and leaving behind only those that are able to tolerate the heavily modified environment. Others, such as the invasion of mammalian predators, may only influence those insect species large enough to be targeted as prey. In the following sections of this chapter, we discuss three key threats to insects in Australasia: habitat loss and fragmentation, invasive mammalian predators and climate change.

Table 14.1　Total number of species of each major group of insects and other terrestrial arthropods, with threatened species status for Australia and each Australian State, and for New Zealand (including those designated as at risk)

	Order	Common name	Aust.	NSW	Vic	Tas	Qld	SA	WA	NT	ACT	NZ – threatened	NZ – at risk
Insects	Coleoptera	Beetles	15	14	35	39	7	0	6	2	2	125	596
	Lepidoptera	Butterflies	4	4	1	13	7					45	267
		Moths	5	4	18	5				1		49	69
	Hymenoptera	Bees	2	2	3	3			2	1	1	1	2
		Ants			1								
		Wasps	2		2							1	18
	Diptera	Flies			1							1	145
	Hemiptera	Bugs							3			9	49
	Phasmatodea	Stick insects	1	1									
	Plecoptera	Stoneflies	1	1	6								
	Trichoptera	Caddisflies			2	13							
	Odonata	Dragonflies & Damselflies		2	1								
	Blattodea	Cockroaches		1									
	Orthoptera	Crickets, Weta, Grasshoppers		1	1	5					1	6	40
	Mantodea	Praying Mantises											1
	Phthiraptera	Lice										12	5
Arachnids	Araneae	Spiders	1			7			22			12	156
	Pseudoscorpiones	Pseudoscorpions	1			4			11			3	155
	Opiliones	Harvestmen				1			1			1	
	Acari	Mites				2						8	1
	Schizomida	Schizomida							10				
Others	Onychophora	Peripatus	1			2			22			3	
	Diplopoda	Millipedes	1			2			22				
	Chilopoda	Centipedes										2	
	Diplura	Diplura										1	

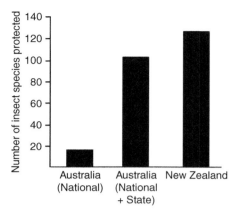

Figure 14.1 Comparison of the number of insect species classified as threatened and therefore receiving conservation protection in Australia and New Zealand. Many species are classified as threatened at a state level in Australia.

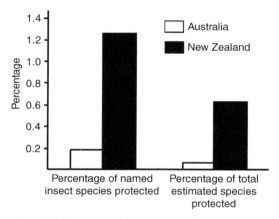

Figure 14.2 Comparison of the percentage of insect species receiving protection via threatened species status in Australia and New Zealand. Percentages are shown for the total number of named species and the estimated number of species for each country.

14.3.1   Habitat loss and fragmentation

Habitat loss is likely to be the largest threat to biodiversity worldwide, and the removal and conversion of indigenous habitats into agricultural or urban landscapes has, and will have, profound effects on biodiversity (Samways, 2005). Invertebrates will be likely to suffer from landscape modification as all the resources they depend upon are severely modified when vegetation is cleared. Deforestation has historically led to the highly modified landscapes now found in both Australia and New Zealand, and continues to be a major threat to biodiversity, including insects, in some parts of Australia. Along with the removal of forest, conversion of native grasslands into the monocultures of agro-ecosystems has led to them being among the most highly threatened ecosystems in Australia. One consequence of habitat clearing is the increased fragmentation of suitable remnants of native vegetation on which animals rely, and the impacts of fragmentation have become a major focus for conservation biologists in recent decades (See Lakeman

Fraser and Ewers, Chapter 3 of this book). Habitat corridors also play a role in connectivity with some species utilising roadside strips of vegetation to move between remnants.

In Australia, a number of studies have focussed on invertebrate responses to habitat fragmentation. For example, major differences in diversity were not apparent for wolf spider communities between small and larger remnants and roadside strips in woodland remnants within the wheatbelt of NSW (Major *et al.*, 2006). Similarly, arboreal beetle and bug communities on *Callitris* pines were not influenced by remnant area or configuration (Major *et al.*, 2003), and beetle and ant species richness and abundance did not differ between remnants and roadside strips (Major *et al.*, 1999). One way that fragmentation has been frequently shown to influence invertebrate communities are through edge effects. Edge communities tend to reflect a mix of remnant-specialists, matrix-specialists and generalists that move between the two. Deeper in the remnants, particular species and assemblages are more likely, but they tend to represent lower species diversity.

One of the most exciting datasets on the effects of habitat fragmentation comes from a long-term experimental study at Mt WogWog in the rainforest of southern NSW. Here, Margules and colleagues investigated the influence of habitat fragmentation on various invertebrates, after the experimental fragmentation of the forest into both small and larger remnants amidst a matrix of cleared and planted pine plantation (Davies & Margules, 2000). The results were intriguing, as some species were negatively affected by the fragmentation process, but others remained unaffected, or in fact increased in abundance. Overall, after seven years the total relative abundance per trap, and the diversity of beetles among the fragments did not differ from the continuous neighbouring forest. Variation in beetle communities instead were more affected by 'within patch' processes such as edge effects. Additionally, a synergistic effect was found between habitat specialisation and rarity. Species that were both rare and specialised seemed most at risk of extinction due to fragmentation, and this effect was above and beyond the effect of either rarity or specialisation acting alone.

14.3.2   Impacts of mammalian predators

One of the major drivers behind the evolution of fauna on oceanic islands, such as New Zealand, is the lack of native mammalian predators (Gibbs, 2009). In response to the relaxed predation risk that comes with a lack of mammals, many island insects have lost their ability to fly, as is the case for many island birds. Additionally, island insects are often large, showing a gigantism that is also a likely response to the lack of mammalian predators. The major threat to these charismatically large, but defenceless, insects in Australasian islands, as in other island ecosystems, is the relatively recent (over the past few centuries) arrival of mammalian insect-feeding pests such as rats, mice, stoats, hedgehogs and cats.

New Zealand has a particularly notable giant insect fauna, most exemplified by giant weta (*Deinacrida* spp.) but also including smaller, but still impressive, tree weta, cave weta and tusked weta, and a diverse range of large and flightless beetles, including ground beetles (Carabidae), longhorn beetles (Cerambycidae), stag beetles (Lucanidae), click beetles (Elateridae) and weevils (Curculionidae). Stratigraphic studies looking at the presence of large beetles in sites representing fauna prior to, and after, arrival of kiore, in New Zealand have demonstrated that a number of large species were present in good numbers, but notably absent after the arrival of kiore (*Rattus exulans*) (Gibbs, 2009). In New Zealand there are a number of comparative studies demonstrating major differences between the insect fauna of otherwise similar rodent-free and infested islands. Similarly, a number of natural experiments demonstrate how insects have been affected by the successful removal of mammals from offshore islands and fenced 'mainland islands' (St Clair, 2011).

For example, the Mercury Island tusked weta (*Motuweta isolata*) was present, until recently, only on Middle Island, a small 13 hectare island in the Mercury group off the Coromandel coast of the north island of New Zealand, which remained rodent free, in contrast to the other islands of the group which had populations of kiore (Watts *et al.*, 2008). After successful eradication of rodents for other islands in the group, *M. isolata* were successfully translocated to, and became established on, both Red Mercury and Double islands. This has turned out to be very fortunate for tusked weta, as the Middle Island population has now potentially become extinct with no individuals seen there since 2001. Other translocation programmes for weta have similarly been successful (see Box 14.1) to both offshore islands, and pest-controlled mainland islands. On one such mainland island, Maungatautiri Ecological Island, the installation of a predator-proof fence and subsequent eradication of rats, stoats and possums has allowed for the translocation of a variety of fauna but has also greatly benefited fauna already existing in the reserve. Both weta and ground-dwelling beetle numbers increased dramatically following mammal eradication, with an astonishing 12-fold increase in tree weta abundance within only two years of mammal eradication (Watts *et al.*, 2011).

Comparison of rat-infested and rat-free islands dominated by seabird colonies exhibited significant differences in invertebrate abundances. Eight orders of macro-invertebrates and even some soil micro-invertebates (that would be unlikely to be preyed upon by rats) were significantly more abundant on islands that had not been invaded by rats than on rodent-free islands. This indicated that rats were having indirect effects on invertebrate abundance as well as direct predation effects, likely brought about by the reduction in seabirds, and subsequent changes in island ecology (Towns *et al.*, 2009). Mouse (*Mus musculus*) eradications from islands have similarly led to increased abundances of invertebrates. After mouse eradication from Allports Island, New Zealand, two spiders (Udolidae), two moths (*Meringa* sp.) and eight beetle species all increased in abundance (St Clair, 2011).

14.3.3 Climate change

Climate change is one of the most pressing issues facing humanity. Across Australia, each decade since the 1950s has been warmer than the previous. Average daily maximum temperatures have increased by 0.75 °C since 1910, and overnight minimum temperatures have increased by more than 1.1 °C in the same period (CSIRO-ABM, 2012). Future predictions are for a generally warmer and drier continent by 2030 (CSIRO, 2007), but with the likely impacts of climate change being complex and highly variable across Australia and worldwide (Dunlop *et al.*, 2012). This is likely to lead to more extreme maximum temperature events, and fewer yet more intense, minimum temperature extremes (WMO, 2011). For insects, this means that seasons become less predictable, and fluctuations from the mean may also become more variable.

Temperature and resource availability are of fundamental importance to animal physiology, behaviour and ecology. This is particularly true for insects (Addo-Bediako et al., 2001). Insects can respond to mean temperature variation, predictability, and extremes in the thermal environment. This can be quantified due to the strong mechanistic links between metabolism, development and performance and the thermal environment. The body temperature of insects influences how they adapt to their environment, their capability to survive, grow, reproduce and disperse. In order to determine the influence of climate change on insects, an understanding of differences in key performance traits (physiological, behavioural and ecological traits) is required at different life stages of populations, within species put under different climatic pressures, and among species facing similar climatic pressures (Andrew and Terblanche, 2013). Resource availability (including food, shelter and egg-laying sites) can also play a critical role in interactions among species including competition, cooperation and exposure to predators and pathogens.

Ectotherms, particularly insects in the tropics (i.e. from warmer climates and lower latitudes) have been argued to be more susceptible to climate warming, due to their relatively low warming tolerance (WT) than more temperate or high-latitude species (Deutsch et al., 2008). This pattern is largely a result of the fact that upper temperature limits for survival and activity are relatively constant across the planet (e.g. Deutsch et al., 2008; Overgaard et al., 2011), and tropical species live in environments that are warmer and are therefore closer to their upper thermal limits and optimal performance temperatures (T_{opt}). In addition, plasticity of tolerance may be intrinsically coupled to basal tolerance levels, further complicating prediction and forecasting efforts (e.g. Stillman, 2003).

Furthermore, there has been a lack of information incorporating physiological tolerances with microclimatic and behavioural optimisation of insects in the field when assessing their vulnerability to climate change, and the data for tropical and temperate zone species are, in some cases, notably sparse (e.g. Clusella-Trullas et al., 2011; Kearney et al., 2009). This is further exacerbated by the lack of information of physiological tolerances at different life stages (egg, larvae, pupae, adult) and when individuals are under competition stress, both inter- and intra-specifically.

There will be winners and losers with a changing climate but the predictions of who will benefit or not is difficult to predict for individual regions, functional groups, taxonomic groups, species, or populations. This can stem from individuals being more susceptible to climatic extremes at different parts of their life-cycle. Much of the work on climate change in Australia has focussed on Diptera (mainly Drosophila) and Lepidoptera species (Andrew et al., 2013). Much less information has been carried out on other native species, including common species which are ecosystem drivers (e.g. ants, dung beetles and pollinators), and in particular little information has been gathered assessing physiological tolerances or behavioural changes of insects to adapt to a rapidly changing climate (Andrew, 2013; Andrew et al., 2013; Andrew & Terblanche, 2013).

14.4 Insects in agricultural landscapes

Land-use intensification is one of the greatest threats to biodiversity outside of reserves (Cunningham *et al.*, 2013). Remnant vegetation on farms is a key resource for ecosystem service providers such as pollinators, predators and parasitoids. Restoration of managed land is also used to ameliorate soil salinity and erosion, but also has a key role in increasing diversity in these landscapes (Gibb & Cunningham, 2010; Vesk *et al.*, 2008). On farms, for those insects which are important ecosystem service providers, there are key issues that threaten their efficiency, the biggest of which is on-farm chemical usage. A pertinent example of this relates to insect pollinators.

Honeybees (*Apis mellifera*), pollinate human food crops worth around $4 billion annually in Australia. The principal honey-producing areas of Australia are temperate lands between south-eastern Queensland and central Victoria, with New South Wales producing 41% of the total annual honey yield. While pollination of commercial crops by managed honeybee colonies is common, much of the pollination of crops is done by wild bees found in native areas, which are providing a free ecosystem service (Garibaldi *et al.*, 2013). This includes crops like broad beans, cotton and canola. Therefore, any negative impacts on the honeybee industry, either through honey production or as pollinators of Australian crops and pastures, has huge implications for Australia's food security.

Neonicotinoids are some of the most widely used insecticides in Australia and worldwide (Jeschke & Nauen, 2008). They are systemic insecticides which protect all parts of the plant and have been used widely as seed dressings, as well as for protection against wood borers and root feeders (Goulson, 2013). Their usage has increased dramatically as they are more toxic to insects than they are to vertebrates (although, see Kimura-Kuroda *et al.*, 2012), and they can be effective in minute amounts. However, due to their widespread use, there has been a strong reliance on neonicotinoids for insect pest management over the past 30 years (Goulson, 2013). The use of neonicotinoids as the key driver of insect pest control has led to a range of concerns across agricultural landscapes (Stokstad, 2013), the most dominant of which is, currently, pollinator declines and impacts on other non-target taxa (Goulson, 2013). Neonicotinoids have also been found in flowers and pollen, and so are taken up by pollinators when foraging. Most of the current research has been carried out in the USA and Europe where there have been impacts on colony reproduction in bumblebees (Whitehorn *et al.*, 2012) and honeybee foraging (Henry *et al.*, 2012). In Australia, much of our understanding of neonicotinoid impacts on bees comes from the Northern Hemisphere, which makes predictions difficult, as currently, Australian bees are not under the same pressures from other immediate pollinator threats such as Varroa mites or Colony Collapse Disorder.

For honeybees, another potential future problem is hives being infested with Varroa mites. Australia is considered Varroa mite-free but recently the Asian honeybee was introduced into Australia and carries Varroa mite. The bees were introduced via a ship into Queensland, became established in Townsville and are now moving south. There have been calls for more use of native pollinators, a preparedness for the introduction of *Varroa destructor*, and the preservation of native remnant vegetation to protect the Australian industry (Cunningham *et al.*, 2002; Cunningham *et al.*, 2013).

However Australia's pollination and honey industry is still vulnerable. Australia is also the only major honeybee-producing nation on Earth that does not have a dedicated research institute or centre devoted to researching these threats and developing control measures, and assisting to secure a sustainable future for the industry.

One of the general problems is that we are so reliant on honeybees under managed hive conditions. Most of our agricultural practices derive from Europe, and the European honeybee is the primary insect pollinator of these crops (Cunningham *et al.*, 2002). There are attempts to use native pollinator species: such as the Blue-banded bee (*Amegilla cingulata*) to pollinate glasshouse tomatoes (Hogendoorn *et al.*, 2007). However, rich native bee faunas can be found in intensively cropped agricultural areas if untilled native vegetation is found close by (Lentini *et al.*, 2012) to maintain a variety of food sources. The role that native pollinators play in cropping systems is most likely underutilized and definitely undervalued.

Dung beetles (Coleoptera: Scarabeinae) are also important ecosystem service providers (Nichols *et al.*, 2008). In America, dung beetles are worth $0.38 billion for dung burial (Losey & Vaughan, 2006), and would have a similar, if not higher, value in Australia due to our less seasonal climate. They are frequently studied as model species (Simmons & Ridsdill-Smith, 2011) and they are providers of significant ecosystem services. They are critical to the maintenance of agricultural ecosystems in Australia. As ecosystem service providers, dung beetles provide an opportunity to offset a substantial proportion of the greenhouse gas (GHG) emissions of cattle by burying the manure from these animals. Dung beetles also increase soil carbon and soil health, and reduce fly numbers. Even though Australian dung beetles are found in agricultural systems, they have evolved to deal with small dry and hard dung pellets from native mammals. Overseas species were introduced as only a few native species are able to break down the large, soft and wet dung pats. How natives and exotics co-exist and compete for resources is currently unclear. Dung beetles may respond to climate change directly via changes in temperature and rainfall (Terblanche *et al.*, 2010), but also indirectly via changes in dung quality and chemistry. Dung beetles also compete for nesting sites, mates and food (both adults and larvae) within a dung pat. A changing climate may influence the quality of dung as a food and nesting resource for the beetles, changing the way individuals interact intra- and inter-specifically. In addition, stress levels from interactions with other individuals may also be inflated, but is currently poorly understood. For example, up to 100 dung beetles may be attracted to just one litre of cattle dung (Ridsdill-Smith *et al.*, 1982) thus making competition for food and breeding sites high.

Farmlands occupy 61% of Australia's landmass and they occupy a range of remnant and managed vegetation types. Traditionally, agricultural production and conservation of species and ecosystems have been seen as polarised viewpoints, especially when production intensifies (Macfadyen *et al.*, 2012). Conservation of native species and remnant vegetation will intersect with agricultural practices if there is a tangible benefit and yields some value to the farm, in both the short and long term. This value may come in the form of paid resources, such as ecosystem services provided, carbon sequestration credits, or tourism; or from a national protected listing/ legislation; or public esteem from giving the area protection.

14.5   Entomophagy – should we be eating insects to save them?

Entomophagy, or the use of insects for human and livestock consumption, has a long and important history (van Huis *et al.*, 2013). However, since modern agricultural practices have developed, western nations have broadly seen insects more as pests, than as a resource; apart from domesticated species such as the honeybee, *Apis mellifera*, and silkworm, *Bombyx mori*. With the human population increasing at a rapid rate, and land availability to produce current staple crops and livestock in decline, insects are being reassessed as a viable alternative food source (Defoliart, 2005).

Over 1900 species of insects have been used as a food source for humans worldwide (van Huis *et al.*, 2013) with species from at least nine orders and 22 families and 49 species known in Australia (Defoliart, 2005; Yen, 2005). In Australia, Aboriginal communities have used insects as an important food source, relying on a variety of species across the continent (Yen, 2009). Indigenous Australians include honey ants, bogong moths, wichetty grubs, and sugar-bag bees in their diet, and also other beetle larvae, crickets, grasshoppers and termites (Yen, 2005). Lerps, the sugary covering made by psyllids (Hemiptera) by the excretion of excess sugar, is also eaten (Yen, 2005).

Edible insects generally inhabit a naturally large area, and are easily found as a food resource, even if they are ephemeral (Boulidam, 2010). However, in highly populated areas, such as south east Asia, popular insects have become locally extinct and as their habitat is degraded they are becoming more difficult to find in their natural environments, so forest conservation is crucial to the long-term management of these food supplies (Boulidam, 2010; van Huis *et al.*, 2013; Yen, 2009).

From a nutritional perspective, insects provide superior energy output per unit input compared to livestock. Insects can be reared on waste products and in small restricted areas, making their usage viable in a range of areas (Khusro *et al.*, 2012). However, many of these insects are harvested from the wild, and have not been harvested in a sustainable manner for large-scale consumption. Khusro *et al.* (2012) assessed the viability of using insects to feed livestock, such as poultry, where viable industries could use local insects (e.g. crickets and grasshoppers) to manage and feed to free-range animals as long as the total cost of rearing and feeding insects is no higher than current protein sources, such as soybean meal and grains. Entomophagy is in its infancy as a legitimate method of providing humans and livestock with nutrition in Australia; however, it is starting to be assessed as a viable food production industry for humans, livestock and fisheries (Khusro *et al.*, 2012; Yen, 2005).

14.6   Insect taxonomists: a threatened species

In the Mt Wog Wog study described earlier, an important factor is highlighted that relates significantly to all aspects of insect conservation: the importance of taxonomic expertise. Without expert taxonomic input into this large collaborative project, the identities of the beetles collected, and the information on their life histories, habitats and degrees of specialisation would not have been known, and much of the work would not have been possible. Therefore, perhaps one of the most insidious threats to insect biodiversity in Australasia, as with the rest of the world, is the declining number of entomologists with expertise to accurately identify the growing number of insect species requiring conservation. While the number of species receiving active management for conservation represents a fraction of the number of species actually under threat, what is perhaps more frightening is the fact that the vast majority of species requiring conservation attention, could not be accurately identified by anybody, and do not even have a name. This relates to the 'taxonomic impediment' (Taylor, 1983), and is seen as a worldwide phenomenon. The sheer number of insect species requiring description, naming and revision is so overwhelmingly large, and the number of practising taxonomists is so few, that it is possible we will lose the majority of these species, before we even know they exist (Samways, 1993). The decline in taxonomic expertise (Cheesman & Key, 2007) and the lack of young scientists choosing a career in taxonomy, is of international concern, and one that will require concerted efforts from governments to provide funding for baseline taxonomic work over sustained periods of time.

14.7 Conclusions

While the threats to insect species and diversity are many, there is cause for optimism when considering the future of insect conservation in Australasia. Efforts to reduce the impacts of climate change, habitat loss and agriculture on biodiversity in general, are likely to benefit insects, but some threats such as predation by invasive rodents and the introduction of unpalatable weeds that outcompete crucial larval host plants, have major and species-specific effects requiring targeted conservation management. In such cases, particularly when threatened insect species are large or charismatic, Australia and New Zealand have instituted successful recovery programmes, and community awareness and participation in both countries has been important. Perhaps the greatest concern is that the vast majority of our insect species remain undescribed, and all critical aspects of their biology are entirely unknown. Unless Australia and New Zealand invest in taxonomic and conservation efforts, thousands of insect species are likely to become unnamed casualties, without us ever knowing their intrinsic value or their ecological importance.

REFERENCES

Addo-Bediako, A., Chown, S. L., Gaston, K. J. (2001) Revisiting water loss in insects: a large scale view. *Journal of Insect Physiology*, **47**, 1377–1388.

Andrew, N. R. (2013). Population dynamics of insects: impacts of a changing climate. In *The Balance of Nature and Human Impact*. Rohde, K. (Ed.), Cambridge University Press, pp. 311–324.

Andrew, N. R. & Terblanche, J. S. (2013). The response of insects to climate change. In: *Climate of Change: Living in a Warmer World*. Salinger, J. (Ed.), David Bateman Ltd Auckland, pp. 311–323.

Andrew, N. R., Hill, S. J., Binns, M. *et al.* (2013) Assessing insect responses to climate change: What are we testing for? Where should we be heading? *PeerJ*, **1**, e11.

Austin, A. D., Yeates, D. K., Cassis, G. *et al.* (2004). Insects 'down under'–diversity, endemism and evolution of the Australian insect fauna: examples from select orders. *Australian Journal of Entomology*, **43**(3), 216–234.

Boulidam, S. (2010). Edible insects in a Lao market economy. In: *Forest Insects as Food: Humans Bite Back*, Proceedings of a workshop on Asia-Pacific resources and their potential for development, Bangkok, Thailand, FAO Regional Office for Asia and the Pacific, Durst, P. B., Johnson, D. V., Leslie, R. L. & Shono, K. (Eds.), pp. 131–140.

Britton, D. R. & New, T. R. (1995). Rare Lepidoptera at Mount Piper, Victoria – the role of a threatened butterfly community in advancing understanding of insect conservation. *Journal of the Lepidopterists' Society*, **49**, 97–113.

Cheesman, O. D. & Key, R. S. (2007). The extinction of experience: a threat to insect conservation? In *Insect Conservation Biology: Proceeding of the Royal Entomological Society's 23nd Symposium* (No. 232, p. 322). CABI.

Clusella-Trullas, S., Blackburn, T. M. & Chown, S. L. (2011) Climatic predictors of temperature performance curve parameters in ectotherms imply complex responses to climate change. *The American Naturalist*, **177**, 738–751.

Cranston, P. S. (2009). Biodiversity of Australasian insects. In *Insect Biodiversity: Science and Society*, **Foottit, R. G. & Adler, P. H.** (Eds.). John Wiley & Sons, pp. 83–105.

Cranston, P. S. (2010). Insect biodiversity and conservation in Australasia. *Annual Review of Entomology*, **55**, 55–75.

CSIRO (1991). *Insects of Australia.*

CSIRO (2007). *Climate Change in Australia: Observed Changes and Projections.* Available at: www.climatechangeinaustralia.gov.au.

CSIRO-ABM (2012). *State of the Climate 2012.* CSIRO and the Australian Bureau of Meteorology, Canberra. Available at http://www.csiro.au/Outcomes/Climate/Understanding/State-of-the-Climate-2012.aspx.

Cunningham, S. A., Fitzgibbon, F. & Heard, T. A. (2002). The future of pollinators for Australian agriculture. *Australian Journal of Agricultural Research*, **53**, 893–900.

Cunningham, S. A., Attwood, S. J., Bawa, K. S. *et al.* (2013). To close the yield-gap while saving biodiversity will require multiple locally relevant strategies. *Agriculture, Ecosystems and Environment*, **173**, 20–27.

Davies, K. F. & Margules, C. R. (2000). The beetles at Wog Wog: a contribution of Coleoptera systematics to an ecological field experiment. *Invertebrate Systematics*, **14**(6), 953–956.

Defoliart, G. R. (2005). Overview of role of edible insects in preserving biodiversity. In *Ecological Implications of Minilivestock: Potential of Insects, Rodents, Frogs and Snails.* Paoletti, M. G. (Ed.). Science Publishers, Inc., Enfield, NH, pp. 123–139.

Deutsch, C. A., Tewksbury, J. J., Huey, R. B. *et al.* (2008). Impacts of climate warming on terrestrial ectotherms across latitude. *Proceedings of the National Academy of Sciences of the United States of America*, **105**, 6668–6672.

Dunlop, M. *et al.* (2012). *The Implications of Climate Change for Biodiversity Conservation and the National Reserve System: Final Synthesis.* A report prepared for the Department of Sustainability, Environment, Water, Population and Communities, and the Department of Climate Change and Energy Efficiency. CSIRO Climate Adaptation Flagship, Canberra.

Garibaldi, L. A., Steffan-Dewenter, I., Winfree, R. *et al.* (2013). Wild pollinators enhance fruit set of crops regardless of honey bee abundance. *Science*, **339**, 1608–1611.

Gibb, H. & Cunningham, S. A. (2010) Revegetation of farmland restores function and composition of epigaeic beetle assemblages? *Biological Conservation*, **143**, 677–687.

Gibbs, G. W. (2009). The end of an 80-million year experiment: a review of evidence describing the impact of introduced rodents on New Zealand's 'mammal-free' invertebrate fauna. *Biological Invasions*, **11**, 1587–1593.

Goulson, D. (2013). An overview of the environmental risks posed by neonicotinoid insecticides? *Journal of Applied Ecology*, **50**, 977–987.

Henry, M., Béguin, M., Requier, F. *et al.* (2012) A common pesticide decreases foraging success and survival in honey bees. *Science*, **336**, 348–350.

Hogendoorn, K., Coventry, S. & Keller, M. (2007). Foraging behaviour of a blue banded bee, *Amegilla chlorocyanea* in greenhouses: implications for use as tomato pollinators. *Apidologie*, **38**, 86–92.

Honan, P. (2008). Notes on the biology, captive management and conservation status of the Lord Howe Island Stick Insect (*Dryococelus australis*)(Phasmatodea). In *Insect Conservation and Islands*. Springer Netherlands, pp. 205–219.

Jeschke, P. & Nauen, R. (2008). Neonicotinoids – from zero to hero in insecticide chemistry. *Pest Management Science*, **64**, 1084–1098.

Kearney, M., Shine, R. & Porter, W. P. (2009). The potential for behavioral thermoregulation to buffer "cold-blooded" animals against climate warming. *Proceedings of the National Academy of Sciences of the United States of America*, **106**, 3835–3840.

Khusro, M., Andrew, N. R. & Nicholas, A. (2012). Insects as poultry feed: a scoping study for poultry production systems in Australia. *World's Poultry Science Journal*, **68**, 435–446.

Kimura-Kuroda, J., Komuta, Y., Kuroda, Y., Hayashi, M. & Kawano, H. (2012). Nicotine-like effects of the neonicotinoid insecticides Acetamiprid and Imidacloprid on cerebellar neurons from neonatal rats. *PLoS ONE*, **7**, e32432.

Lentini, P. E., Martin, T. G., Gibbons, P., Fischer, J. & Cunningham, S. A. (2012). Supporting wild pollinators in a temperate agricultural landscape: Maintaining mosaics of natural features and production. *Biological Conservation*, **149**, 84–92.

Losey, J. E. & Vaughan, M. (2006) The economic value of ecological services provided by insects. *BioScience*, **56**, 311–323.

Macfadyen, S., Cunningham, S. A., Costamagna, A. C. & Schellhorn, N. A. (2012). Managing ecosystem services and biodiversity conservation in agricultural landscapes: are the solutions the same? *Journal of Applied Ecology*, **49**, 690–694.

Major, R. E., Smith, D., Cassis, G., Gray, M. & Colgan, D. J. (1999). Are roadside strips important reservoirs of invertebrate diversity? A comparison of the ant and beetle faunas of roadside strips and large remnant woodlands. *Australian Journal of Zoology*, **47**(6), 611–624.

Major, R. E., Christie, F. J., Gowing, G., Cassis, G. & Reid, C. A. (2003). The effect of habitat configuration on arboreal insect in fragmented woodlands of south-eastern Australian. *Biological Conservation*, **133**(1), 35–48.

Major, R. E., Gowing, G., Christie, F. J., Gray, M. & Colgan, D.(2006). Variation in wolf spider (Araneae: Lycosidae) distribution and abundance in response to the size and shape of woodland fragment. *Biology Conservation*, **132**(1), 98–108.

Moir, M. L. & Leng, M. C. (2013). *Developing Management Strategies to Combat Increased Coextinction Rates of Plant-dwelling Insects Through Global Climate Change*. National Climate Change Adaptation Research Facility, Gold Coast, p. 111.

Moir, M. L., Vesk, P. A., Brennan, K. E. *et al*. (2011). Identifying and managing threatened invertebrates through assessment of coexinction risk. *Conservation Biology*, **25**(4), 787–796.

New, T. R. (1990). Conservation of butterflies in Australia. *Journal of Research of the Lepidoptera*, **29**, 237–255.

New, T. R. (2009). *Insect Species Conservation*. Cambridge University Press.

New, T. R. (2010). Butterfly conservation in Australia: the importance of community participation? *Journal of Insect Conservation*, **14**, 305–311.

Nichols, E., Spector, S., Louzada, J., *et al*. (2008). Ecological functions and ecosystem services provided by Scarabaeinae dung beetles. *Biological Conservation*, **141**, 1461–1474.

Overgaard, J., Kristensen, T. N., Mitchell, K. A. & Hoffmann, A. A. (2011). Thermal tolerance in widespread and tropical *Drosophila* species: does phenotypic plasticity increase with latitude. *The American Naturalist*, **178**, S80–96.

Powell, F. A., Hochuli, D. F. & Cassis, G. (2011). A new host and additional localities for the rare psyllid *Acizzia keithi* Taylor and Moir (Hemiptera: Psyllidae). *Australian Journal of Entomology*, **50**, 441–444.

Priddel, D., Carlile, N., Humphrey, M., Fellenberg, S. & Hiscox, D. (2003). Rediscovery of the 'extinct'Lord Howe Island stick-insect (*Dryococelus australis* (Montrouzier)) (Phasmatodea) and recommendations for its conservation. *Biodiversity & Conservation*, **12**, 1391–1403.

Ridsdill-Smith, T. J., Hall, G. P. & Craig, G. F. (1982). Effect of population density on reproduction and dung dispersal by the dung beetle *Onthophagus binodis* in the laboratory. *Entomologia Experimentalis et Applicata*, **32**, 80–85.

Samways, M. J. (1993). Insects in biodiversity conservation: some perspectives and directives. *Biodiversity & Conservation*, **2**(3), 258–282.

Samways, M. J. (2005). *Insect Diversity Conservation*. Cambridge University Press.

Sands, D. (2008). Conserving the Richmond Birdwing Butterfly over two decades: where to next? *Ecological Management & Restoration*, **9**, 4–16.

Sands, D. P. A. & New, T. R. (2002). *The Action Plan for Australian Butterflies*. Canberra: Environment Australia.

Simmons, L. W. & Ridsdill-Smith, T. J. (2011). Reproductive competition and its impact on the evolution and ecology of dung beetles. In *Ecology and Evolution of Dung Beetles*. Simmons, L. W. & Ridsdill-Smith, T. J. (Eds.). Wiley-Blackwell, Oxford, pp. 1–20.

St Clair, J. J. (2011). The impacts of invasive rodents on island invertebrates. *Biological Conservation*, **144**(1), 68–81.

Stillman, J. H. (2003). Acclimation capacity underlies susceptibility to climate change. *Science* **301**, 65.

Stokstad, E. (2013). Pesticides under fire for risks to pollinators. *Science*, **340**, 674–676.

Taylor, R. W. (1983). Descriptive taxonomy: past, present, and future. *Australian Systematic Entomology: A Bicentenary Berspective*, **93**, 134.

Terblanche, J. S., Clusella-Trullas, S. & Chown, S. L. (2010). Phenotypic plasticity of gas exchange pattern and water loss in *Scarabaeus spretus* (Coleóptera: Scarabaeidae): deconstructing the basis for metabolic rate variation. *Journal of Experimental Biology*, **213**, 2940–2949.

Towns, D. R., Wardle, D. A., Mulder, C. P. H., *et al.* (2009). Predation of seabirds by invasive rats: multiple indirect consequences for invertebrate communities. *Oikos*, **118**, 420–430.

van Huis, A., van Itterbaeeck, J., Klunder, H. *et al.* (2013). *Edible Insects: Future Prospects for Food and Food Security*. Food and Agriculture Organisation of the United Nations, Rome, p. 171.

Vesk, P. A., Nolan, R., Thomson, J. R., Dorrough, J. W. & Nally, R. M. (2008). Time lags in provision of habitat resources through revegetation. *Biological Conservation*, **141**, 174–186.

Watts, C. & Thornburrow, D. (2009). Where have all the weta gone? Results after two decades of transferring a threatened New Zealand giant weta, *Deinacrida mahoenui*. *Journal of Insect Conservation*, **13**, 287–295.

Watts, C., Stringer, I., Sherley, G., Gibbs, G., & Green, C. (2008). History of weta (Orthoptera: Anostostomatidae) translocation in New Zealand: lessons learned, islands as sanctuaries and the future. *Journal of Insect Conservation*, **12**, 359–370.

Watts, C. H., Armstrong, D. P., Innes, J. & Thornburrow, D. (2011). Dramatic increases in weta (Orthoptera) following mammal eradication on Maungatautari – evidence from pitfalls and tracking tunnels. *New Zealand Journal of Ecology*, **35**(3), 261.

Whitehorn, P. R., O'Connor, S., Wackers, F. L. & Goulson, D. (2012). Neonicotinoid pesticide reduces bumble bee colony growth and queen production. *Science*, **336**, 351–352.

WMO (2011). *Weather Extremes in a Changing Climate*. World Meteorological Organization, Switzerland. Available at http://www.wmo.int/pages/mediacentre/news/documents/1075_en.pdf.

Yeates, D. K., Harvey, M. S. & Austin, A. D. (2003). New estimates for terrestrial arthropod species-richness in Australia. *Records of the South Australian Museum Monograph Series*, **7**, 231–241.

Yen, A. (2009). Entomophagy and insect conservation: some thoughts for digestion. *Journal of Insect Conservation*, **13**, 667–670.

Yen, A. L. (2005). Insect and other invertebrate foods of the Australian aborigines. In *Ecological Implications of Minilivestock: Potential of Insects, Rodents, Frogs and Snails*. Paoletti, M. G. (Ed.). Science Publishers, Inc., Enfield, NH, pp. 367–387.

CHAPTER 15

Terrestrial mammal diversity, conservation and management in Australia

Mark D. B. Eldridge and Catherine A. Herbert

Summary

As a result of its long-term isolation as an island continent, Australia's mammal fauna is exceptional both for its evolutionary diversity and high endemism. It is the only place where representatives of the three surviving major mammal lineages coexist and is the only continent dominated by marsupials. Endemism is also high amongst the Australian rodent and microbat radiations. Over recent millennia and especially in the past 200 years, the trajectory of this unique mammal fauna has been one of decline and extinction, leaving the ecosystems of the continent profoundly altered. While much unique diversity been lost, increased scientific knowledge and growing management expertise has prevented many further extinctions. In such a dynamic and altered landscape, managing Australia's unique mammals is a formidable challenge that includes encouraging the persistence of threatened species, as well as suppressing introduced mammalian competitors and predators and some endemic species that are now over-abundant. While many threats to Australia's mammals are ongoing and novel threats continue to arise, it is hoped this unique fauna will persist and continue to fascinate.

15.1 Introduction: why Australian mammals are different

Ever since Australia was first visited by Europeans in the seventeenth and eighteenth centuries, the continent's unusual mammals have fascinated and perplexed western science (Deakin *et al.*, 2012; Olsen, 2010). While such bizarre species as the platypus, koala and kangaroo are now iconic to Australians and are internationally recognised symbols of the island continent, many unique features of the Australian mammal fauna remain unappreciated.

Australia and New Guinea are the only places on Earth where representatives of the three surviving major mammal lineages (monotremes, marsupials and eutherians) coexist.

Austral Ark: The State of Wildlife in Australia and New Zealand, eds. A. Stow, N. Maclean and G. I. Holwell. Published by Cambridge University Press. © Cambridge University Press 2015.

Table 15.1   Number of terrestrial mammal species from Australia, showing the proportion (%) endemic, extinct and threatened

	Number of species	Endemic Australia (%)	Endemic Australasia[a] (%)	Extinct[b] (%)	Threatened[c] (%)
Monotremes	3	67	100	33	–
Marsupials	160	93	100	6.3	24
Carnivorous[d]	61	97	100	1.6	30
Bandicoots[e]	11	81	100	27	36
Diprotodontids[f]	88	90	100	6.8	19
Eutherians	149	75	95	10	17
Megabats	11	18	90	9	18
Microbats[g]	71	73	92	1.4	7
Murid rodents	66	91	100	18	26
Soricid Shrews	1	100	100	0	100
Introduced[h]	22	0	0	–	–

Data from (Baker *et al.*, 2012; Churchill, 2008; Helgen *et al.*, 2012; Menkhorst & Knight, 2011; Van Dyck & Strahan, 2008).
[a] Australia, New Guinea and adjacent islands.
[b] Species that have become extinct since European settlement of Australia in 1788.
[c] Listed under the Australian *Environmental Protection and Biodiversity Conservation Act* 1999. http://www.environment.
gov.au/cgi-bin/sprat/public/publicthreatenedlist.pl
[d] Families Dasyuridae, Myrmecobiidae, Thylacinidae, Notoryctidae.
[e] Families Peramelidae, Thylacomyidae, Chaeropodidae.
[f] Families Phascolarctidae, Vombatidae, Phalangeridae, Hypsiprymnodontidae, Potoroidae, Macropodidae, Burramyidae, Pseudocheiridae, Petauridae, Tarsipedidae, Acrobatidae.
[g] Families Pteropodidae, Megadermatidae, Rhinolophidae, Hipposideridae, Emballonuridae, Molossidae, Miniopteridae, Vespertilionidae.
[h] Families Muridae, Sciuridae, Canidae, Felidae, Leporidae, Equidae, Suidae, Camelidae, Bovidae, Cervidae.

While early Paleocene monotreme fossils are known from South America (Pascual *et al.*, 1992) and fossil marsupials from most continents (Luo *et al.*, 1992), only in Australasia did both these ancient mammalian lineages survive to the present day (Holocene).

Australia is also unique because its mammal fauna is dominated by marsupials, both in terms of the number of species (Table 15.1) and physical size since they represent the largest animals in most ecosystems (Van Dyck & Strahan, 2008). As a consequence, marsupials fill many of the mammalian niches (e.g., browser, grazer, folivore, insectivore, carnivore) occupied by eutherians in other continents. Absent from Australia's native fauna are the familiar ungulates, canids, felids, mustelids, lagomorphs and insectivores of Europe, Asia, Africa and the Americas, but present instead are marsupial equivalents of each, resulting in many stunning examples of convergent evolution (Tyndale-Biscoe, 2005). South America is the only other continent that currently hosts a marsupial fauna, but it is less diverse than Australia's both in number of extant species and major lineages; with three families and ~90 species of marsupials known from South America, compared to 18 families and ~160 species in Australia (Wilson & Reeder, 2005).

Australia's mammal fauna is also unique amongst the continents because for most of the past 40 million years it has evolved in isolation (Long *et al.*, 2002). The fragmentation of the great southern supercontinent Gondwana commenced ~170 million years ago (MYA). When Australia broke free of Antarctica ~40 MYA it began a slow drift northwards carrying with it a Gondwanan fauna and flora, which then continued to evolve with very little influence from the rest of the continents. Present on this great continental ark were the ancestors of the modern monotremes and marsupials which would come to characterise the unique Australian mammal fauna (Long *et al.*, 2002). Whether representatives of other terrestrial mammal lineages (e.g. eutherians) were also originally present on the continent, but subsequently died out, remains debated (Godthelp *et al.*, 1992; Long *et al.*, 2002; Rich *et al.*, 1997), but appears likely since eutherians were radiating within Gondwana during the Cretaceous (Murphy *et al.*, 2001). What is clear, however, is that the marsupials radiated spectacularly (Kirsch *et al.*, 1997; Meredith *et al.*, 2008) and came to dominate and mould the Australian landscape.

Australia's isolation was interrupted ~5 MYA when it drifted close enough to Asia to be colonised by dispersing murid rodents (Geffen *et al.*, 2011). Murids appear to have successfully established in Australia at least twice, in the Pliocene (Rowe *et al.*, 2008) and more recently in the Pleistocene (Rowe *et al.*, 2011). Each colonisation produced a spectacular burst of rapid diversification and speciation resulting in a unique and diverse native Australian rodent fauna (~66 species: Table 15.1), which remains largely underappreciated by the Australian public and in the international scientific literature.

The origins of Australia's diverse bat fauna, which comprises eight families, is less well understood. Some microbats were present in Australia before the continent separated from Antarctica (Hand *et al.*, 1994). Others appear to have arrived at varying times as Australia drifted further north, most likely dispersing over water from Asia, though some lineages may also have entered Australia from the south (Long *et al.*, 2002). Nevertheless, most lineages have diversified within Australia resulting in a substantial endemic fauna (~82 species: Table 15.1).

As a consequence of Australia's long history of isolation, most mammal species are unique, with exceptionally high rates of endemism (Ceballos & Brown, 1995), especially for marsupials and rodents (>90%: Table 15.1). These reach 100% when the adjacent island of New Guinea is included, since periodic faunal exchanges have occurred between these landmasses via a land-bridge exposed during the glacial cycles of the Pleistocene (Van Dyck & Strahan, 2008). Rates of endemism are lower in bats (Table 15.1), due to their greater dispersal abilities, but also because of limited understanding of species boundaries in many Southeast Asian and Australasian bat genera. Nevertheless >90% of known bat species are endemic to Australia/New Guinea. The high rates of species endemism found within Australian mammals also extend to the genus and family level (Ceballos & Brown, 1995).

Although well advanced, our understanding of extant Australian mammal diversity remains incomplete, with new species continuing to be discovered regularly (e.g. Baker *et al.*, 2012; Helgen *et al.*, 2012; Parnaby, 2009). Indeed unlike birds, the discovery curve for Australian mammal species is yet to plateau (Woinarski *et al.*, 2014). Taxonomic uncertainty is highest amongst microbats, but is widespread, including some groups of rodents, dasyurids, possums and macropods (Woinarski *et al.*, 2014). Clearly the true diversity of Australian mammals is higher than currently understood.

While the extant Australian mammal fauna is famously unique, for a large continental landmass it is comparatively depauperate (Ceballos & Ehrlich, 2006; Schipper *et al.*, 2008). For example, Australia has ~312 terrestrial mammal species (Table 15.1) compared with ~264 in New Guinea, an island only one-eighth the size (Lavery *et al.*, 2013).

This difference is partly a consequence of Australia's low topography, infertile soils and dry variable climate (Geffen *et al.*, 2011). Australia's increasing aridity, as it drifted north into higher latitudes has been profoundly influential, leading to a major contraction of highly biodiverse mesic habitats and the spread of the arid zone which now dominates the continent (Byrne *et al.*, 2008, 2011). Reduction of wet forests and increased aridity led to major changes in mammal diversity with diversification of some arid-adapted lineages but declines in many mesic-adapted groups (Archer *et al.*, 1991; Hocknull *et al.*, 2007; Long *et al.*, 2002). More recently, however, the diversity of the Australian mammal fauna has been seriously impacted by a late Pleistocene megafaunal extinction, which resulted in the loss of at least 45 mammal species, most >50 kg, including many large (>100 kg) marsupial herbivores, the ecological equivalents of rhinoceros, tapirs, hippopotamus and large antelopes (Johnson, 2006; Long *et al.*, 2002). Australia's megafaunal extinction coincides with the arrival of modern humans to the continent ~50 000 years ago, and although human involvement seems likely (reviewed in Johnson, 2006), direct evidence remains elusive (Flannery, 2012).

Another major change to Australia's mammal fauna associated with the arrival of humans occurred much later during the mid-Holocene, with the introduction of the dingo (*Canis lupus dingo*) ~4000 years ago (Johnson, 2006). The spread of dingoes throughout Australia coincided with the extinction from the mainland of the two largest surviving marsupial carnivores, the thylacine (*Thylacinus cynocephalus*) and Tasmanian devil (*Sarcophilus harrisii*), which were previously found across the continent (Johnson, 2006). However, both species survived in dingo-free Tasmania until modern times (Van Dyck & Strahan, 2008).

15.2 Recent impacts on Australian mammals

Australia's relative isolation came to an abrupt end with the European colonisation of the continent in 1788, unleashing a host of environmental changes which continue to the present day. The widespread loss, degradation and fragmentation of native vegetation, introduction of European agricultural practices, alteration of fire regimes and the introduction of numerous exotic animal and plant species (including 22 mammals; Table 15.1), had and continues to have a profound impact on the native mammal fauna (Kennedy, 1992; Maxwell *et al.*, 1996; Woinarski *et al.*, 2014).

Within the past 200 years at least 25 species of mammals are known to have become extinct in Australia (Table 15.2) and many more have declined significantly (e.g., Table 15.3). This rate of recent mammal extinction is the highest in the world (Short & Smith, 1994); a tragic and unenviable record that represents a significant loss of unique biodiversity. These extinctions have been unevenly distributed across the Australian mammal fauna (Tables 15.1 and 15.2) and the landscape, with extinctions concentrated on arid/semi-arid zone rodents, bandicoots and small macropods (Burbidge & McKenzie, 1989; McKenzie *et al.*, 2007; Short & Smith, 1994). However, the magnitude of mammal decline in the mesic southeast may have been underestimated (Bilney *et al.*, 2010). Across Australia, the most impacted have been terrestrial (i.e., non-arboreal), small–medium-sized (35–5500 g) rodents and marsupials which have been termed Critical Weight Range (CWR) species (Burbidge & McKenzie, 1989; Johnson & Isaac, 2009) and comprise the bulk of recently extinct (87%; Table 15.2) and threatened mammal (80%) species in

Table 15.2   Species of mammals that have become extinct in Australia since European settlement in 1788

Taxon	Last record	CWR
Monotremes		
Western long-beaked echidna *Zaglossus bruijnii*	1901	N
Marsupials		
Thylacine *Thylacinus cynocephalus*	1930	N
Lesser bilby *Macrotis leucura*	1960s	Y
Pig-footed bandicoot *Chaeropus ecaudatus*	1950s	Y
Desert bandicoot *Perameles eremiana*	1960s	Y
Broad-faced potoroo *Potorous platyops*	1875	Y
Desert rat-kangaroo *Caloprymnus campestris*	1950s	Y
Central hare-wallaby *Lagorchestes asomatus*	1960	Y
Eastern hare-wallaby *Lagorchestes leporides*	1889	Y
Crescent nailtail wallaby *Onychogalea lunata*	1956	Y
Toolache wallaby *Macropus greyi*	1939	N
Rodents		
White-footed tree-rat *Conilurus albipes*	1845	Y
Lesser stick-nest rat *Leporilus apicalis*	1933	Y
Short-tailed hopping mouse *Notomys amplus*	1896	Y
Long-tailed hopping mouse *Notomys longicaudata*	1901	Y
Big-eared hopping mouse *Notomys macrotis*	1843	Y
Darling Downs hopping mouse *Notomys mordax*	1840s	Y?
Broad-cheeked hopping mouse *Notomys robustus*	1850s?	Y
Long-eared mouse *Pseudomys auritus*	1850s	Y
Blue-grey mouse *Pseudomys glaucus*	1956	?
Gould's mouse *Pseudomys gouldi*	1857	Y
Maclear's rat *Rattus macleari*	1902	Y
Bulldog rat *Rattus nativitatis*	1902	Y
Bats		
Percy Island flying-fox *Pteropus brunneus*	1874	Y
Lord Howe long-eared bat *Nyctophilus howensis*	1920s?	N?

CWR = Critical Weight Range, see text for details.
Data from Helgen *et al.*, 2012; Johnson, 2006; Van Dyck & Strahan, 2008.

Australia. The disproportionate vulnerability of CWR species appears a direct consequence of their size and therefore susceptibility to predation by the introduced European red fox (*Vulpes vulpes*) and feral cat (*Felis catus*) (Burbidge & McKenzie, 1989; Johnson, 2006). Both these invasive species are now widespread in Australia (Van Dyck & Strahan, 2008), existing in most environments (except for foxes in the tropics), often at high density. They have effectively eliminated CWR species from vast swathes of southern and central Australia (e.g. Dickman *et al.*, 1993). It appears that the impact of the fox and cat on CWR species was exacerbated by persecution and elimination of the dingo (i.e., mesopredator release: Ritchie & Johnson, 2009) throughout much of southern and central Australia

Table 15.3   Species of Australian mammals that recently became extinct on the mainland but persisted on off-shore islands

Taxon	Last mainland record	Islands where persisted	CWR
Marsupials			
Tasmanian devil *Sarcophilus harrisii*	430ybp	Tasmania	N
Eastern quoll *Dasyurus viverrinus*	1966	Tasmania	Y
Western barred bandicoot *Perameles bougainville*	1930s	Bernier, Dorre	Y
*Eastern barred bandicoot *Perameles gunnii*	2002	Tasmania	Y
Tasmanian bettong *Bettongia gaimardi*	1919	Tasmania	Y
Burrowing bettong *Bettongia lesueur*	1960s	Barrow, Bernier, Dorre	Y
Banded hare-wallaby *Lagostrophus fasciatus*	1906	Bernier, Dorre	Y
Tasmanian pademelon *Thylogale billardierii*	1930s	Tasmania	Y
*Rufous hare-wallaby *Lagorchestes hirsutus*	1991	Bernier, Dorre	Y
Rodents			
Greater stick-nest rat *Leporillus conditor*	1938	Franklin	Y
Shark bay mouse *Pseudomys fieldi*	1895	Bernier	Y

CWR = Critical Weight Range, see text for details; *Mainland population survives in enclosures.
Data from DSE, 2009; Johnson, 2006; Van Dyck & Strahan, 2008.

(Johnson *et al.*, 2007; Letnic *et al.*, 2009). 'Predation by the European red fox' and 'Predation by the feral cat' are listed as Key Threatening Processes (KTP) under Australian legislation (*Environmental Protection and Biodiversity Conservation (EPBC) Act* 1999).

The recent (~1999) deliberate introduction of the European red fox to Tasmania (Sarre *et al.*, 2012), in an apparent act of eco-terrorism, is therefore cause for considerable concern, given the impact of this species on the biodiversity of mainland Australia. Tasmania currently supports relatively healthy populations of 15 species of CWR mammals (Van Dyck & Strahan, 2008), four of which have already become extinct on the Australian mainland (Table 15.3). Should the fox become established in Tasmania, many of these species are likely to decline, some to extinction.

While introduced species are the major threat to Australia's mammal fauna, this is relatively unusual in a global context (Woinarski *et al.*, 2011) where exotic species are recognised as a relatively minor threat to mammalian diversity compared to other processes (Schipper *et al.*, 2008). This disparity is likely a consequence of Australia's long history of isolation which appears to have made its fauna especially vulnerable to the negative impact of introduced predators and competitors.

The major threat to mammal diversity (and biodiversity) worldwide is habitat loss and degradation (Schipper *et al.*, 2008). This has also significantly impacted Australian mammals with the majority of species having lost significant areas of habitat in the past 200 years (Kennedy, 1992; Short & Smith, 1994; Woinarski *et al.*, 2014). Only a few of the larger macropod species appear to have benefited from European settlement and increased their range (Calaby & Grigg, 1989; Pople *et al.*, 2010). The clearing and fragmentation of native vegetation for agriculture and urbanisation has been most severe in the more densely settled east, southeast and southwest mesic areas of Australia. Habitat loss and degradation is especially a threat to species with naturally restricted distributions and was likely a factor in the extinction of the toolache wallaby (*Macropus greyi*) in southeastern South Australia (SA), as well as precipitating a major decline in the now 'endangered' mahogany glider (*Petaurus gracilis*) and northern hairy nosed wombat (*Lasiorhinus krefftii*) in Queensland (Short & Smith, 1994; Van Dyck & Strahan, 2008). 'Land clearing' is listed as a KTP under Australian legislation (*EPBC Act* 1999).

Habitat degradation has also been widespread in Australia with the introduction of exotic herbivores, both domestic (e.g. sheep, cattle) and wild (e.g. rabbits, goats, deer, buffalo, horses, donkeys, camels, pigs) and alterations of fire regimes significantly impacting native vegetation across most of the continent (SoEC, 2011). The European rabbit (*Oryctolagus cuniculus*), introduced to Australia in 1858, spread rapidly across the southern two-thirds of Australia and has profoundly impacted native vegetation communities leading to serious land degradation, especially in the arid and semi-arid zone (Van Dyck & Strahan, 2008). 'Competition and land degradation by rabbits' is listed as a KTP under Australian legislation (*EPBC Act* 1999). In many areas of Australia populations of the fox and cat are sustained at high density by rabbits, which leads to additional predation pressure on surviving native mammals and other fauna (Johnson, 2006; Van Dyck & Strahan, 2008). The rabbit and other introduced herbivores have also reduced and simplified vegetative ground cover in many areas, not only competing with native herbivores for food but leaving them more exposed to predation (Johnson, 2006). For example, the simplification of grassland habitat by introduced herbivores resulted in increased predation and the near extinction of the eastern barred bandicoot (*Perameles gunnii*) in western Victoria (Vic) (Seebeck *et al.*, 1990). Conversely, the removal of introduced herbivores from a central Kimberley cattle station, in northwest Western Australia (WA), led to a rapid increase in small mammal density and diversity (Legge *et al.*, 2011).

Altered fire regimes can also have a similar impact by fundamentally changing vegetation structure, removing resources from native species and increasing their vulnerability to predation. The change in the past ~150 years from many small fires to few large hot fires has been implicated in the decline of arid zone (Burbidge *et al.*, 1988; Johnson *et al.*, 1989; Letnic & Dickman, 2006) and northern Australian (Legge *et al.*, 2011; Woinarski *et al.*, 2011) mammals. Altered fire regimes in forested areas can also degrade habitat for mammals by removing old hollow-bearing trees (Parnaby *et al.*, 2010) which are an important and limited resource for many species (Gibbons & Lindenmayer, 2002). Unsympathetic forestry practices which remove old hollow-bearing trees and logging intervals that prevent new hollow formation are also significant threats to hollow-dependent arboreal marsupial and bat species (Lindenmayer *et al.*, 2012), such as Leadbeater's possum (*Gymnobelidius leadbeateri*) (Lindenmayer & Possingham, 1996) and south-eastern long-eared bat (*Nyctophilus corbeni*) (Van Dyck & Strahan, 2008).

Other suggested causes for the widespread decline and extinction of Australian mammals include hunting and disease (Kennedy, 1992; Maxwell *et al.*, 1996). In Australia the hunting of native mammals for fur and as supposed agricultural pests was widespread in the first 150 years of European settlement, resulting in the loss of millions of individuals and the suppression or extinction of many local populations (Hrdina & Gordon, 2004; Lunney *et al.*, 1997; Short & Smith, 1994). However, hunting appears to have been a major factor in the global extinction of relatively few Australian mammal species with only the thylacine and toolache wallaby being potential examples (Short & Smith, 1994;

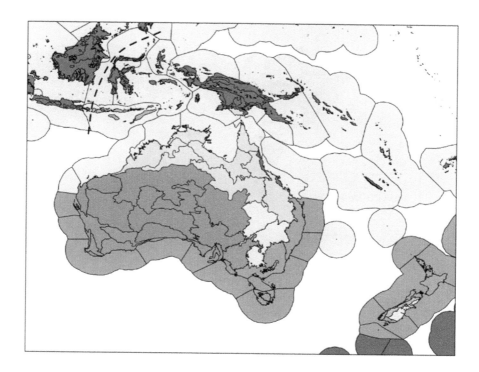

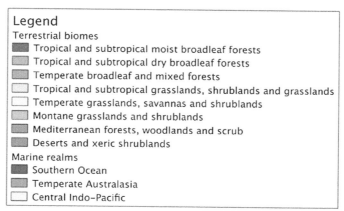

Legend

Terrestrial biomes
- Tropical and subtropical moist broadleaf forests
- Tropical and subtropical dry broadleaf forests
- Temperate broadleaf and mixed forests
- Tropical and subtropical grasslands, shrublands and grasslands
- Temperate grasslands, savannas and shrublands
- Montane grasslands and shrublands
- Mediterranean forests, woodlands and scrub
- Deserts and xeric shrublands

Marine realms
- Southern Ocean
- Temperate Australasia
- Central Indo–Pacific

Chapter 1, **Plate 1** Biogeographic units in the vicinity of Australia. Dashed line indicates Wallace's Line – the boundary between the Oriental and Australian biogeographic realms of Wallace (1876). Fine solid lines delineate terrestrial and marine ecoregions. Ecoregions are coloured according to terrestrial biome or marine realm. Terrestrial ecoregions and biomes are as defined by Olson *et al.* (2001). Marine ecoregions and realms are as defined by Spalding *et al.* (2007).

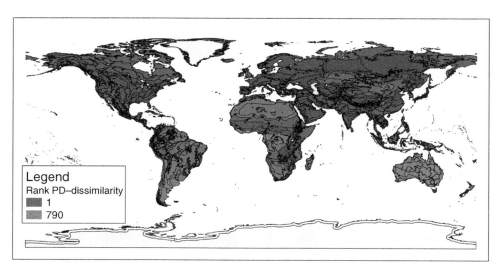

Chapter 1, **Plate 2** Map of evolutionary distinctiveness of mammal faunas of terrestrial ecoregions. Ecoregions are coloured according to rank order of mean dissimilarity – red indicates high dissimilarity in phylogenetic diversity with other ecoregions while blue indicates low dissimilarity. Grey areas have no data. Distribution data were sourced from the WildFinder database (http://worldwildlife.org/pages/wildfinder). Phylogenetic data was sourced from the mammal supertree of Fritz *et al.* (2009).

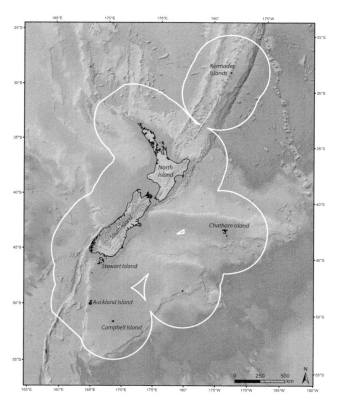

Chapter 2, **Plate 3** The New Zealand biogeographic region based on the Exclusive Economic Zone, modified from Gordon (2009) showing terrestrial environments and the extent of the submerged continental shelf. Data sourced from NIWA: New Zealand 250 Bathymetry Rainbow (2008).

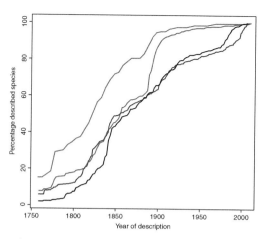

Chapter 1, **Plate 4** Rates of species description for mammals in Australia (black) compared to other countries of comparable area – Canada (red), United States (blue), and Brazil (green). Lines show the cumulative proportion of currently known species described by a particular year. All curves begin with the publication of the 10th edition of *Systema Naturae* (Linnaeus, 1758). Data sourced from the IUCN Red List (www.iucnredlist.org).

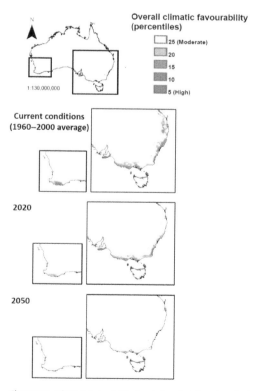

Chapter 6, **Plate 5** Potential hotspots for plant invasions in Australia under current conditions and future climate change scenarios for 2020 and 2050. Each map shows invasion hotspots in Australia's south-east and south-west based on the top 5th, 10th, 15th, 20th and 25th percentiles of combined climatic suitability values for the 72 Australian Weeds of National Significance. Areas with equivalent climate suitability to the combined current climatic suitability percentile bands are identified for the composite maps for projected climates for 2020 and 2050. Reproduced from O'Donnell *et al.* (2012).

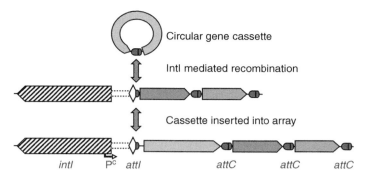

Chapter 9, **Plate 6** Generalised structure of an integron: integrons contain a gene (*intI*) encoding a site-specific tyrosine recombinase, called an integron-integrase (*IntI*). The integrase protein catalyses the recombination of gene cassettes into the *attI* site (white diamond). Gene cassettes are composed of an open reading frame (solid arrows) and a secondary recombination site (*attC*). This recombination activity results in a tandem array of gene cassettes, that can in some circumstances contain hundreds of different genes, represented here by different colors. Expression of the inserted gene cassettes is driven by the integron encoded promoter, Pc.

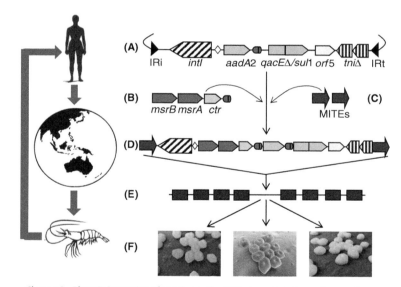

Chapter 9, **Plate 7** Generation of novel, complex DNA elements in the environment as a consequence of pollution with xenogenetic DNAs. (A) Structure of a typical Class 1 integron found in human pathogenic and commensal bacteria. It consists of: terminal inverted DNA repeats IRi and IRt; the integron integrase gene *intI1*; a gene cassette encoding *aadA2*, which confers streptomycin resistance; and a 3′ conserved segment consisting of fused genes for disinfectant and sulfonamide resistance (*qacEΔ/sul*1), *orf*5 and the remnants of genes encoding transposition functions (tniΔ). A DNA element of this type was released into an aquatic environment where it (B) acquired a novel gene cassette encoding two methionine sulphoxide reductases (*msrB* and *msrA*), and where (C) the terminal inverted repeats were replaced by miniature inverted-repeat transposable elements (MITEs). This event gave the compound MITE/integron element (D) added mobility, allowing it to (E) insert into a genomic island. The genomic island with inserted integron has now moved into at least three different species of the genus *Acinetobacter* (F). Consequently, resistance determinants released from human waste streams may interact with gene cassettes and mobile DNA elements in aquatic ecosystems to generate new combinations of potential virulence genes in environmental bacteria. The presence of these bacteria in food items provides a readily accessible route for contamination of the food chain and the emergence of novel, virulent pathogens.

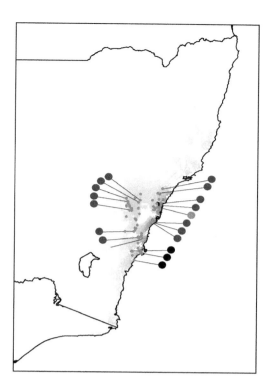

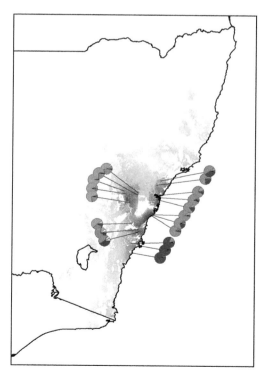

Chapter 13, **Plates 8 and 9** Changes in habitat availability to the waratah from the last glacial maximum (left) to the current period (right). Chloroplast (Plate 8, top) and nuclear (Plate 9, bottom) DNA diversity across the distribution of the species are also displayed to show how genetic structure corresponds to historical variation in distributional patterns. (Modified from Rossetto *et al.* 2012).

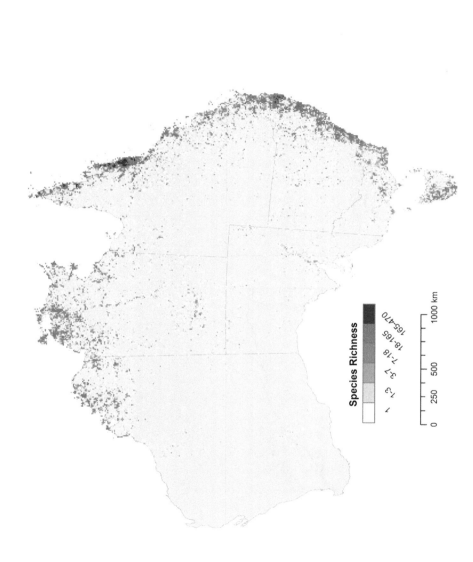

Species Richness

| 1 | 1–3 | 3–7 | 7–18 | 18–165 | 165–470 |

0 250 500 1000 km

Chapter 13, **Plate 10** The distribution of Australian rainforest trees based on herbarium records (corrected data obtained from the Australian Virtual Herbarium). Green represents high species richness and identifies the regions of major diversity such as the Wet Tropics and northern NSW. Red identifies the areas of lowest species richness (often representing a single species per geographic unit) and shows how some species can have wide distributions including in marginal habitat across the continent.

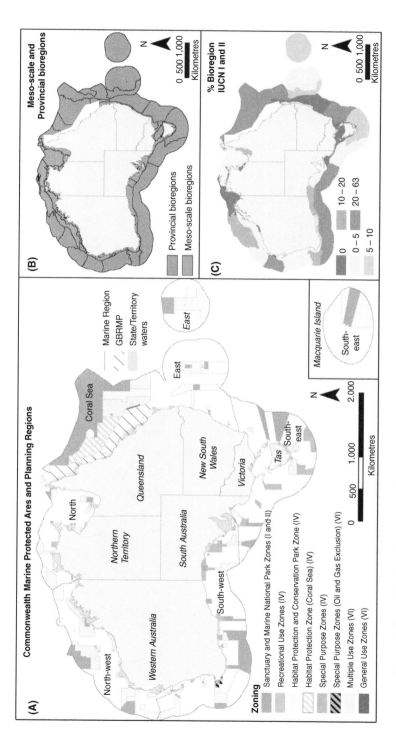

Chapter 27, **Plate 11** The Australian marine jurisdiction and marine protected areas (MPAs). Note that Australian marine waters surrounding Heard, McDonald, Cocos (Keeling) and Christmas Islands are not shown here because they were not included in the Commonwealth's recent bioregional planning exercise to establish new MPAs. (A) Zoning map of Commonwealth MPAs designated in 2007 and 2012, and their IUCN categories (see Table 27.1). (B) The boundaries of meso-scale and provincial bioregions ('bioregions') that informed the design of the Commonwealth MPAs. (C) Percentages of meso-scale and provincial bioregions protected within 'no-take' marine reserves (IUCN categories I and II; see Table 27.1). IUCN = International Union for the Conservation of Nature; Tas = Tasmania; GBRMP = Great Barrier Reef Marine Park. The white area in (B) and (C) is the Great Barrier Reef Marine Park, excluded from the bioregional planning processes that were completed in 2007 and 2012.

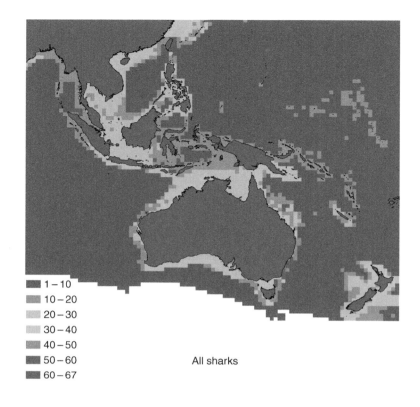

1 – 10
10 – 20
20 – 30
30 – 40
40 – 50
50 – 60
60 – 67

All sharks

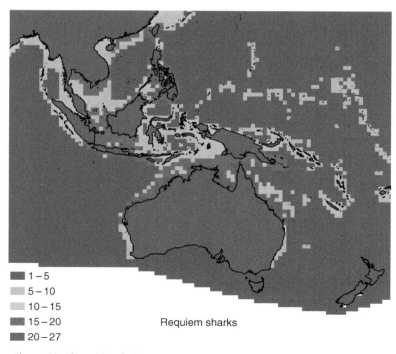

1 – 5
5 – 10
10 – 15
15 – 20
20 – 27

Requiem sharks

Chapter 22, **Plates 12 and 13** Heat map representing shark species richness in the Indo-Australasian region. All shark species (Plate 12, top) and whaler sharks (Plate 13, bottom). Data on species distribution were obtained from the Global Shark Distribution Database (http://www.globalshark.ca/gs_distribution_db/data.php), and were originally published by Lucifora *et al.* (2011).

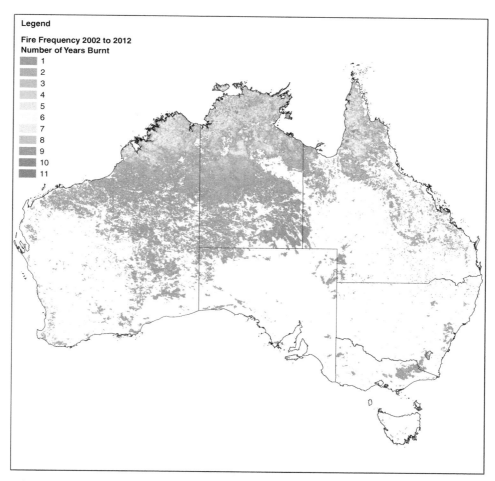

Chapter 25, **Plate 14** Map of fire frequency in Australia, 2002–2012, illustrating the major regional differences in fire prevalence across the continent, with very high frequency in the seasonal tropics of northern Australia. (Reproduced from Landgate – Western Australian Land Information Authority – firewatch.landgate.wa.gov.au, with thanks to Carolyn McMillan.)

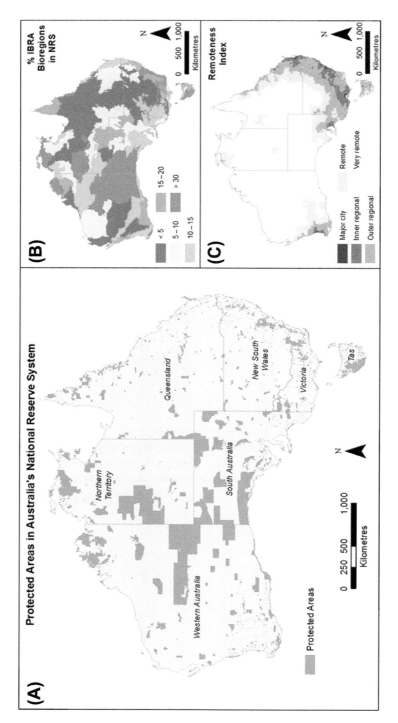

Chapter 26, **Plate 15** The distribution of protected areas in Australia. (A) Map of all protected areas. (B) Map showing the proportion of each Australian bioregion (from IBRA, or the Interim Biogeographic Regionalisation for Australia) covered by protected areas. (C) Map showing the remoteness index, a measure of distance to major human settlements and infrastructure. Data sources: CAPAD 2010 (all States and Territories except Victoria), CAPAD 2008 (Victoria), Indigenous Protected Areas August 2013, Interim Biogeographic Regionalisation for Australia (Commonwealth of Australia); and, Australian Bureau of Statistics. Note: Protected areas that were classified in CAPAD as not being in the NRS were removed.

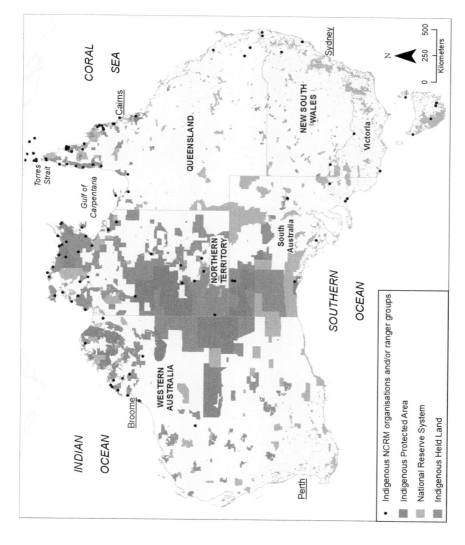

Chapter 26, **Plate 16** The distribution of Indigenous natural and cultural resource management (NCRM) organisations and ranger groups, Indigenous Protected Areas (IPAs), the National Reserve System additional to IPAs, and Indigenous-held lands in Australia. IPAs (red) occur on Indigenous-held land or sea where Indigenous communities have entered into an agreement with the Australian government to manage their Country as a protected area. Indigenous-held lands (brown) are owned or controlled by Indigenous people, and include areas that are communally owned, as required by native title and some State land rights acts.

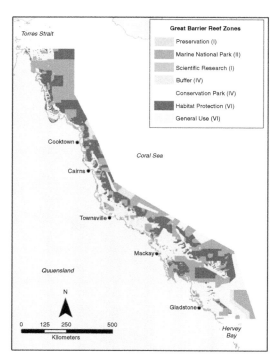

Chapter 27, **Plate 17** The Australian Government's Great Barrier Reef Marine Park Zoning Plan and Queensland's contiguous Great Barrier Reef Coast Marine Park Zoning Plan.

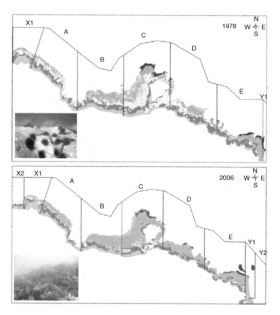

Chapter 28, **Plate 18** Large-scale habitat changes in a marine reserve. Large increases in the biomass of snapper and lobster inside the Cape Rodney to Okakari Point (or Leigh) Marine Reserve eventually triggered a cascade of indirect responses to marine reserve protection. This resulted in large-scale changes in reef habitats within the reserve (Leleu *et al.* 2012). The maps above show the dominant reef habitat types when the reserve was established (top) and after ~30 years of marine reserve protection (below). The marine reserve includes areas A–E; the land is shown as white stipple and sandy habitats in yellow; the major reef habitat types are shallow mixed brown macroalgae (red), urchin barrens (pale blue, inset photo at top) and kelp forest (green, inset photo below). When the reserve was established approximately 30% of the reef was classified as urchin barrens. This habitat has now all but gone from within the reserve (<1%) and it has been replaced predominantly by kelp forest, due to the suppression of grazing sea urchins by snapper and lobster.

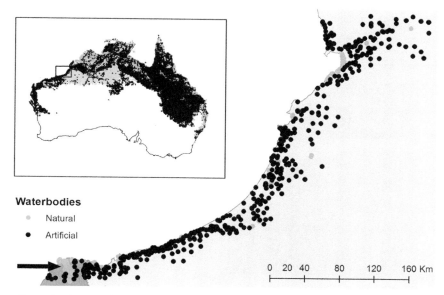

Waterbodies

- Natural
- Artificial

0   20 40   80   120   160 Km

Chapter 5, **Plate 19 (Figure 5.4)** Artificial waterbodies from which toads can be excluded to prevent their spread into Australia's Pilbara region. For full caption text see page [95].

Invasive vines can 'transform' the ecosystems they invade

(1) Vines establish in edges or gaps

(3) leading to host tree breakage and gap creation

(2) they rapidly smother host vegetation

Chapter 6, **Plate 20 (Figure 6.6)** The positive feedback loop which facilitates exotic vine invasions.

Chapter 6, **Plate 21 (Figure 6.7)** Gamba grass invading ecosystems in Australia's north. For full caption text see page [120].

Chapter 6, **Plate 22 (Figure 6.4)** Invasive Australian acacias. For full caption text see page [114].

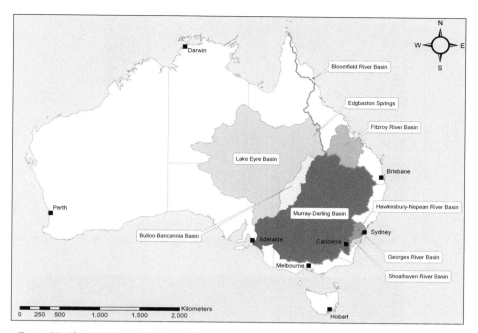

Chapter 23, **Plate 23 (Figure 23.2)** Map of Australia showing the major drainage basins as well as geographical features and locations mentioned throughout the chapter. The great dividing range is indicated by the red line.

Chapter 3, **Plate 24 (Figure 3.1)** The removal of native vegetation by people can leave remnant patches, often located along rivers or roads or difficult to develop areas such as steep hill sides. For full caption text see page [47].

Chapter 29, **Plate 25 (Figure 29.1)** Individual Grey Nurse shark have a unique arrangement of dots that can be used for identification and monitoring this critically endangered species. Image kindly provided by Rob Harcourt.

Chapter 5, **Plate 26 (Figure 5.1)** The cane toad (*Rhinella marina*) was brought from South America to Australia and released in northeastern Queensland in 1935. For full caption text see page [84].

Chapter 5, **Plate 27 (Figure 5.2)** A radio-tracked death adder (*Acanthophis praelongus*) found dead in the field after it was fatally poisoned by a very large cane toad (*Rhinella marina*). Photograph by Ben Phillips.

Chapter 6, **Plate 28 (Figure 6.1)** Australia's Weeds of National Significance (WONS). Thirty-two weeds, or suites of weeds, have been declared as WONS some of which are shown above. For full caption text see page [107].

Chapter 6, **Plate 29 (Figure 6.5)** African Olive invasion and control at the Mount Annan Botanic Gardens. For full caption text see page [118].

Chapter 12, **Plate 30 (Figure 12.1)** Four species of buttercups (*Ranunculus*) endemic to the sub-alpine and alpine areas of the New Zealand Southern Alps (South Island). For full caption text see page [243].

Chapter 12, **Plate 31 (Figure 12.2)** Competition from weeds is one of the main threats to the Castle Hill's forget-me-not (*Myosotis colensoi*); one of the many Threatened forget-me-nots found in New Zealand.

Chapter 13, **Plate 32 (Figure 13.1)** *Nothofagus moorei* in the Border Ranges (NSW). For full caption text see page [261].

Chapter 16, **Plate 33 (Figure 16.1)** New Zealand fur seal male resting on the rocks. For full caption text see page [336].

Chapter 17, **Plate 34 (Figure 17.1)** The broad-headed snake is highly prized by snake enthusiasts due to its rarity and beautiful colouration. Photograph by Jonathan Webb.

Chapter 17, **Plate 35 (Figure 17.3)** This captive Christmas Island forest skink (*Emoia nativatatis*) may be the last individual. For full caption text see page [368].

Chapter 17, **Plate 36 (Figure 17.4)** The green sea turtle (*Chelonia mydas*) is the most widespread of the sea turtles nesting in Australia, and is thus exposed to a wide range of climatic conditions and local threats. For full caption text see page [371].

Chapter 18, **Plate 37 (Figure 18.1)** Lizard pollinators: *Woodworthia maculate* carrying out an 'ecosystem service'.

Chapter 20, **Plate 38 (Figure 20.1)** An alarming number of Australian bird species have been seriously challenged by the sudden shift in fire regime during the 200 years of European settlement. For full caption text see page [427].

Chapter 21, **Plate 39 (Figure 21.1)** Fijian Ground Frog leaves its eggs on land, young froglets hatch directly from large, fertilised eggs.

Chapter 21, **Plate 40 (Figure 21.2)** Most Cophixalus have highly localised distributions with over half restricted to mountaintops.

Chapter 21, **Plate 41 (Figure 21.3a)** *Litoria lorica* a once widespread species was not detected for 16 years.

Chapter 21, **Plate 42 (Figure 21.3b)** The habitat in which *L. lorica* was rediscovered.

Chapter 5, **Plate 43 (Figure 5.5)** One promising new approach to toad control involves curtailing recruitment, by trapping toad tadpoles. For full caption text see page [95].

Chapter 22, **Plate 44 (Figure 22.4)** The grey reef shark, the quintessential coral reef associated shark. Photo by Robert Harcourt. Reproduced with permission.

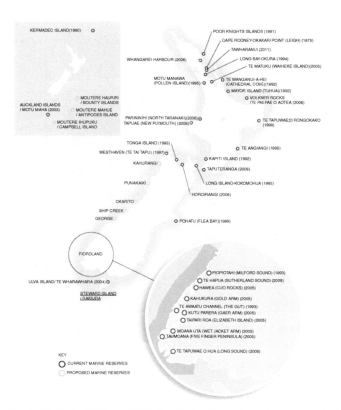

Chapter 28, **Plate 45 (Figure 28.1)** New Zealand's no-take marine reserves. For full caption text see page [602].

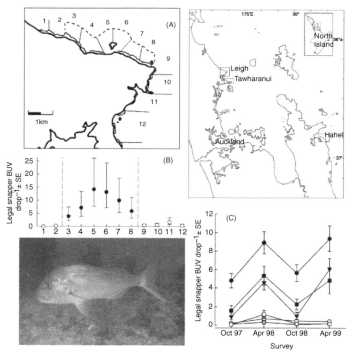

Chapter 28, **Plate 46 (Figure 28.3)** Direct response of snapper *Pagrus auratus* to marine reserve protection in northern New Zealand marine reserves. For full caption text see page [605].

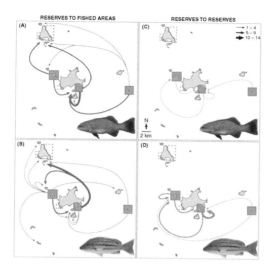

Chapter 28, **Plate 47 (Figure 28.5)** Export of larvae from marine reserves to fished populations. For full caption text see page [612].

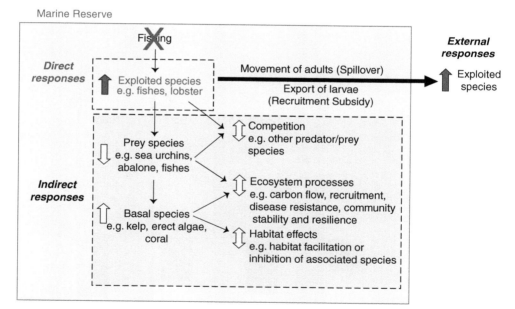

Chapter 28, **Plate 48 (Figure 28.2)** The ecological responses of species and ecosystems to marine reserve protection. For full caption text see page [603].

Chapter 13, **Plate 49 (Figure 13.2)** Successful reintroduction of the Corrigin grevillea (*Grevillea scapigera*) in the Western Australian wheatbelt. For full caption text see page [269].

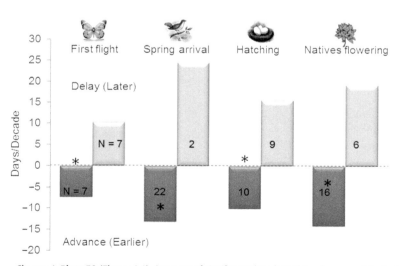

Chapter 4, **Plate 50 (Figure 4.1)** Summary of significant phenological trends amongst Australian species (modified from data in Chambers *et al.*, 2014). For full caption text see page [70].

Chapter 4, **Plate 51 (Figure 4.3)** A tuatara (*Sphenodon punctatus*).

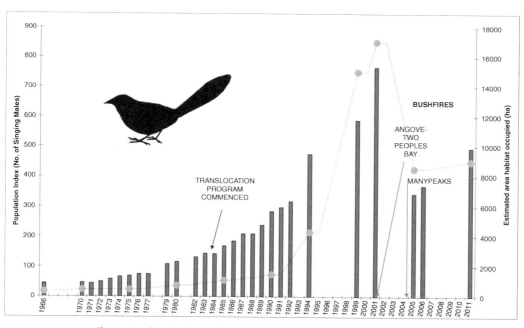

Chapter 25, **Plate 52 (Figure 25.2)** Recovery of the Noisy Scrub-bird population 1966–2011. For full caption text see page [549].

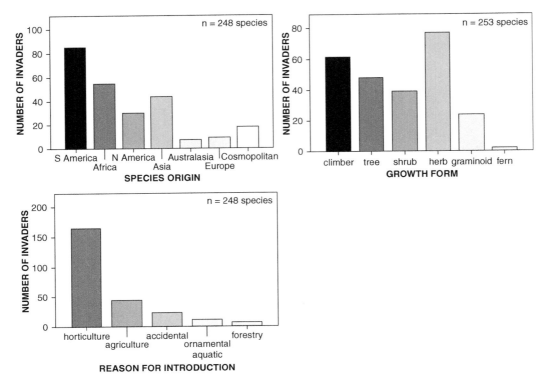

Chapter 6, **Plate 53 (Figure 6.2)** Australia's invasive exotic flora: origins, reasons for introduction and growth forms. For full caption text see page [108].

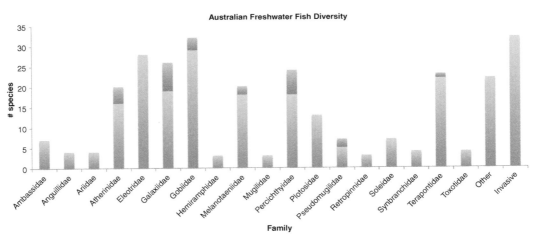

Chapter 23, **Plate 54 (Figure 23.1)** Diversity of Australian freshwater fishes by family. For full caption text see page [493].

Chapter 17, **Plate 55 (Figure 17.2)** In some cases it may be necessary to restrict access to endangered reptile populations. For full caption text see page [361].

Chapter 24, **Plate 56 (Figure 24.1)** Upper, *Crenisopus* sp. from Koolan Island, WA; middle, Phreatoicid isopod from WA; lower, *Brevisomabathynella uramurdahensis* (Parabathynellidae) from calcrete aquifer in Western Australia. For full caption text see page [518].

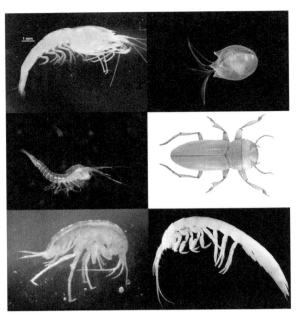

Chapter 24, **Plate 57 (Figure 24.2)** Clockwise from upper left: *Stygiocaris* sp. (Atyidae); *Danielopolina baltanasi*; *Paroster byroensis*; *Pygolabis* sp. Tainisopodidae); Paramelitid amphipod; *Halosbaena tulki* (Thermosbaenacea). For full caption text see page [521].

Chapter 12, **Plate 58 (Figure 12.3)** The shrubby tororaro (*Muehlenbeckia astonii*), a species Nationally Endangered, has become a popular garden plant and is now commonly seen in traffic islands.

Van Dyck & Strahan, 2008). Disease has also been suggested as another major agent of decline for Australian mammals (Abbott, 2006), which, given the continent's history of isolation, is highly plausible. However, conclusive evidence has proved difficult to obtain, especially for past epidemics, although some clear examples include trypanosome infection in the now-extinct Christmas Island rats (*Rattus macleari, Rattus nativitatis*: Pickering & Norris, 1996; Wyatt *et al.*, 2008) and the deadly devil facial tumour disease (DFTD) currently threatening Tasmanian Devil populations (Hawkins *et al.*, 2006).

The initial decline and extinction of Australian mammal species post European settlement was largely confined to the southern two-thirds of Australia (Short & Smith, 1994), leaving the mammal communities of northern Australia (where the fox and rabbit are absent) relatively intact. However, in the past few decades it has become apparent that the small–medium-sized mammal communities of Australia's northern monsoon tropics are now in precipitous decline and urgent action is required to avert another wave of extinctions engulfing Australia's unique mammal fauna (Fitzsimons *et al.*, 2010; Woinarski *et al.*, 2011). The reasons for this decline are currently unclear but complex interactions between factors seem likely (Woinarski *et al.*, 2011). Altered fire regimes, widespread cattle grazing and feral cat predation are probable candidates, although the region is also being impacted by cane toad (*Bufo marinus*) and weed invasion, as well as a range of feral species including pigs, horses and donkeys (Fitzsimons *et al.*, 2010; Legge *et al.*, 2011; Woinarski *et al.*, 2011).

15.3 Managing threatened Australian mammals

Currently 20% of Australian mammal species are listed as threatened (Table 15.1). The highest proportion of threatened species occurs in the marsupials (24%) and rodents (26%) with microbats appearing the least threatened (7%: Table 15.1). However, given the current poor state of knowledge of Australian microbat fauna this is most certainly an underestimate. This rate of threat is typical of mammals globally (~22%: Hoffmann *et al.*, 2011), though fails to take into account the considerable losses (25 species extinct) Australia has suffered in the past 200 years (Table 15.2).

Many of these threatened species are now the subject of Recovery Plans and active management (Maxwell *et al.*, 1996; Woinarski *et al.*, 2014). For some, considerable success has been achieved in addressing threatening processes through the preservation and restoration of habitat, as well as the control of exotic predators and competitors (Maxwell *et al.*, 1996; Woinarski *et al.*, 2014). However, species loss is still occurring, for example the Christmas Island pipistrelle (*Pipistrellus murrayi*: Lumsden, 2009; Martin *et al.*, 2012), which was last recorded in 2009.

The influential pioneering work of Jack Kinnear and colleagues in southwest WA (Kinnear *et al.*, 1988, 1998) demonstrated that effective fox control can lead to the dramatic recovery of CWR mammal populations, including the black-footed rock-wallaby (*Petrogale lateralis lateralis*), brush-tailed bettong (*Bettongia penicillata*), numbat (*Myrmecobius fasciatus*) and western quoll (*Dasyurus geoffroii*) (Friend & Thomas, 2003; Kinnear *et al.*, 2002; Morris *et al.*, 2003). This discovery resulted in the widespread implementation of broad-scale fox control on conservation lands across Australia (e.g., Western Shield in WA, Bounceback in SA, Southern Ark in Vic, Fox Threat Abatement Program in New South Wales) and increases in the abundance of many species (e.g. Arkell, 1999; Dexter *et al.*, 2007; Dexter & Murray, 2009; Kinnear *et al.*, 2002). In WA, the recovery of tammar wallaby (*Macropus eugenii*), brush-tailed bettong, numbat, western quoll and southern brown bandicoot (*Isoodon obesulus*) populations was sufficient to have their threatened status downgraded (Maxwell *et al.*, 1996). However, in areas of eastern Australia some populations of CWR species showed little response to fox control (e.g., brush-tailed rock-wallaby *Petrogale penicillata*: DECC, 2008), suggesting that either the level of fox control was insufficient or that factors in addition to fox predation were suppressing populations.

In southwest WA, populations of some species that initially showed strong recovery under fox control have begun declining again, for example the brush-tailed bettong (now listed as Critically Endangered), black-footed rock-wallaby, and the numbat (IUCN, 2012; Pearson, 2012; Wayne *et al.*, 2013). The apparent causes of these new declines are diverse and multifaceted (e.g. increased cat predation, habitat degradation due to drought and overgrazing, disease: Botero *et al.*, 2013; IUCN, 2012; Kinnear *et al.*, 2010; Pearson, 2012; Yeatman & Groom, 2012), illustrating the difficulty in sustaining species recovery in such highly disturbed and fragmented landscapes as typifies much of southern Australia. It also demonstrates the absolute necessity of regularly monitoring the populations of potentially vulnerable species so that dangerous declines do not go undetected (Groom, 2010).

The growing recognition of the impact of feral cat predation on the Australian mammal fauna (Burbidge & Manly, 2002; Christensen & Burrows, 1994; Dickman, 1996; Gibson *et al.*, 1994; Glen *et al.*, 2010; Johnson, 2006; Priddel & Wheeler, 2004; Woinarski *et al.*, 2011; Yeatman & Groom, 2012) presents a major challenge, since currently no cost-effective broad-scale cat control method is available (Dickman, 1996), despite significant recent progress (Algar & Burrows, 2004; Veitch *et al.*, 2011). In addition, there is evidence of increased cat abundance as a consequence of effective fox control (e.g. Priddel & Wheeler, 2004; Risbey *et al.*, 2000) through mesopredator release (Ritchie & Johnson, 2009). The phenomenon of mesopredator release also suggests a solution, since the restoration of the dingo should then suppress both cats and foxes (Dickman *et al.*, 2009; Ritchie & Johnson, 2009). However, restoring dingoes to those Australian ecosystems from which they have been eliminated or suppressed for over 150 years would be difficult and controversial (Dickman *et al.*, 2009; Johnson & Ritchie, 2013).

As well as managing the threats to remaining populations in the field, more intensive interventions have also been widely employed, including captive breeding, translocations, reintroductions and supplementations (Burbidge, 1999; Friend & Thomas, 2003; Johnson, 1999; Maxwell *et al.*, 1996; Morris *et al.*, 2003; Short & Smith, 1994; Woinarski *et al.*, 2014). Translocations/reintroductions of Australian mammals have had mixed results, with most failures relating to an inability to successfully control introduced predators (foxes and cats) at the release site (Finlayson *et al.*, 2010; Fischer & Lindenmayer, 2000; Moseby *et al.*, 2011; Short, 2009; Short *et al.*, 1992). These difficulties have encouraged the establishment or use of isolated areas (fenced enclosures, fenced peninsulas, islands) by both government agencies and private entities, from which introduced species can be more effectively controlled or eliminated (Johnson, 1999, 2006). This approach has resulted in the successful establishment of additional populations of a variety of threatened mammal species (e.g. Groom, 2010; Moseby *et al.*, 2011; Pizzuto *et al.*, 2007; Richards & Short, 2003; Short & Turner, 2000; Vieira *et al.*, 2007). However, issues of exotic predator incursion, long-term population viability, maintenance of gene flow, overabundance and the absence of most native terrestrial predators remain areas of concern.

15.3.1 Genetic management of threatened Australian mammal populations

A major consequence of the widespread decline of mammal species in Australia is the resulting fragmentation of surviving populations. The continued survival of isolated populations often now confined to habitat patches is uncertain, as small population size and isolation will increase their probability of extinction through demographic, environmental and genetic stochasticity (Frankham *et al.*, 2010). The loss of genetic diversity, the raw material of evolution, reduces a population's ability to evolve and adapt to changing environmental conditions (Frankham *et al.*, 2010). Inbreeding (mating between relatives) increases homozygosity throughout the genome and so elevates the expression of deleterious recessive alleles. This typically results in the production of offspring with reduced fitness (i.e., inbreeding depression) and so increases an inbred population's vulnerability to extinction (Frankham *et al.*, 2010). These genetic factors can then act to increase the likelihood of population extinction (Briskie & Mackingtosh, 2004; Frankham, 2005; Saccheri *et al.*, 1998). Therefore, if gene flow is not re-established, many small isolated mammal populations will unnecessarily go extinct, since the negative impact of inbreeding and drift is often underestimated, while concerns about outbreeding depression, which may hinder supplementation, appear to be overstated in most cases involving recently isolated populations (Frankham *et al.*, 2011).

It is important then that genetic factors are taken into account when managing threatened mammal populations, and to avoid management strategies (e.g. limited founders, serial bottlenecking and lack of ongoing supplementation) that will further deplete variation. Genetic management can also be used to increase the resilience of threatened mammal populations and to prepare them for the environmental changes associated with ongoing climate change (Weeks *et al.*, 2011).

Historically many management actions did result in the loss of genetic diversity, with a recent review of the population genetics of Australian marsupials showing that captive and reintroduced populations typically had significantly reduced diversity compared to conspecific wild populations (Eldridge, 2010). Koala (*Phascolarctos cinereus*) populations in southern Australia provide a good example of the consequences of poor management, with many populations established historically via small numbers of founders and serial bottlenecking now showing significantly reduced diversity, high levels of inbreeding and reduced fitness (Cristescu *et al.*, 2009; Houlden *et al.*, 1996; Seymour *et al.*, 2001). For example, the Eyre Peninsula (SA) koala population has an effective inbreeding coefficient of 0.75, and 24% of males have abnormal testes (Seymour *et al.*, 2001). Koalas (*n* = 6) were introduced to Eyre Peninsula from Kangaroo Island (SA) in the 1960s. The Kangaroo Island population had itself been introduced in the 1920s using 18 individuals sourced from French Island (Vic), where a population had originally been established in the 1880s using two or three mainland animals as founders (Seymour *et al.*, 2001). By contrast, populations of northern quoll recently introduced to two islands off Arnhem Land, Northern Territory and the captive western barred bandicoot (*Perameles bougainville*) population established at Dryandra (WA), were established using founders from multiple wild populations and have resulted in new relatively diverse populations (Cardoso *et al.*, 2009; Smith & Hughes, 2008).

The reduced genetic diversity observed in many threatened species of Australian mammals (Eldridge, 2010) is most likely a direct consequence of their persistence at small effective population size for multiple generations (Spielman *et al.*, 2004). This is also likely to make population recovery more difficult (Evans & Sheldon, 2008) and increase their vulnerability to extinction in the future (Frankham *et al.*, 2010). This suggests that the genetic quality of individuals may be more important in ensuring long-term survival than simply the quantity of individuals, and that large but genetically uniform populations are likely to be inherently vulnerable to extinction.

15.3.2 The role of islands in Australian mammal conservation

Australia's already high recent mammal extinction rate would have been considerably higher had not populations of ten species (eight marsupial, two rodent) that became extinct on the mainland survived on off-shore islands (Table 15.3). While island populations are known to be highly susceptible to extinction in the long term (Frankham, 1998; Loehle & Eschenbach, 2012), in this case the survival of mammal populations was a consequence of the island environment acting to isolate populations from the threatening processes at work on the mainland (Burbidge & Manly, 2002). In cases where these threatening processes (typically introduced predators and/or competitors) also reached islands, then their impact was similarly devastating (Burbidge & Manly, 2002; Burbidge *et al.*, 1997). These island populations have therefore been highly important in Australian mammal conservation (Burbidge, 1999) and increasingly are being used to source animals for conservation initiatives including captive breeding, reintroductions and translocations (Christensen & Burrows, 1994; Finlayson & Moseby, 2004; Moro, 2002; Richards & Short, 2003; Smith & Hughes, 2008; Stanley *et al.*, 2010; Woinarski *et al.*, 2014).

Australia also has a long history of using offshore islands, free from exotic predators and competitors, to 'maroon' (Williams, 1977) populations of threatened mammal species (Burbidge, 1999). This management technique has been used inappropriately in the past (Copley, 1994) resulting in additional wildlife management problems, for example the overabundant introduced koala population on Kangaroo Island, SA (Masters *et al.*, 2009). Nevertheless, threatened mammal species in Australia continue to be introduced to islands as a management strategy to establish 'back-up' populations; for example the Proserpine rock-wallaby (*Petrogale persephone*) (Johnson *et al.*, 2003), mala (*Lagorchestes hirsutus*) (Langford & Burbidge, 2001), Gilbert's potoroo (*Potorous gilbertii*) (Finlayson *et al.*, 2010), northern quoll (*Dasyurus hallucatus*) (Cardoso *et al.*, 2009), dibbler (*Parantechinus apicalis*) (Moro, 2002) and Tasmanian devil (DPIPWE, 2012).

While islands have acted, and can be used, as short-term refuges for some Australian mammal species (Burbidge, 1999), the very nature of the island environment also brings with it long-term challenges. The unique island environment (e.g. isolation, limited size, reduced ecological complexity) can also negatively affect the long-term evolutionary potential of resident populations, through the erosion of genetic diversity (as a consequence of finite size and isolation; Frankham *et al.*, 2002), inbreeding (Eldridge *et al.*, 2004; Frankham, 1998), reduced fitness (Eldridge *et al.*, 1999) and adaptations to the island environment (e.g. reduced dispersal and fecundity, predator naivety; Grant, 1998). These factors make most island populations inherently vulnerable to extinction in the long term, as history has powerfully demonstrated

(Frankham, 1998), and caution against an overreliance on islands and island populations for conservation (Eldridge, 1998).

The predicted pattern of reduced genetic diversity in island populations (Frankham, 1997) is strongly supported by data from Australian mammals, with reductions in diversity of up to 90% compared to conspecific mainland populations being observed (Eldridge, 2010). For most island populations, erosion of diversity, increased inbreeding and an increased risk of extinction is inevitable and populations recently marooned on islands as a conservation strategy will, in the absence of gene flow, follow a similar trajectory. The same is true for any small isolated population, whether inhabiting an isolated habitat patch, a fenced enclosure or a captive colony, and makes the regular exchange of individuals or the re-establishment of gene flow a high priority and a major challenge for wildlife managers and conservation biologists.

For those mammal species naturally found on multiple islands, the potential exists to recreate more genetically diverse and robust populations for use in conservation initiatives. For although individual islands retain little diversity, each island is likely to preserve a different subset of mainland alleles (e.g. Cardoso *et al.*, 2009; Eldridge *et al.*, 1999; How *et al.*, 2009), due to the random nature of genetic drift. As a consequence, a population established with founders from multiple island populations will have significantly increased diversity (e.g. Smith & Hughes, 2008), while the addition of unrelated individuals is also likely to increase the fitness of those populations suffering inbreeding depression (Eldridge *et al.*, 1999), as has been reported in a wide variety of species (Madsen *et al.*, 1999; Pimm *et al.*, 2006; Saccheri *et al.*, 1996; Vila *et al.*, 2003; Westemeier *et al.*, 1998).

Most island populations of Australian mammals were only relatively recently isolated by rising sea levels after the last glacial maximum ~15 000 years ago (Eldridge *et al.*, 1999). Nevertheless, the consequences of genetic drift and adaptation to the island environment has frequently led to morphological changes (differences in size, pelage) in island populations. Historically many of these morphologically distinct populations have been described as separate subspecies (or sometime species; Walton, 1988) and therefore became of conservation concern. Increasingly, however, it is being recognised that this relatively rapid morphological response to an island environment may not represent a major evolutionary or genetic change and that many morphologically distinct island populations, while biologically interesting, should not be a high priority for conservation as they do not represent a significant component of intra-specific genetic diversity (Keogh *et al.*, 2005; Miller *et al.*, 2011; Neaves *et al.*, 2012; Zenger *et al.*, 2003).

15.4 Managing overabundant Australian mammals

Not all mammalian species have been disadvantaged by human-induced changes in the past 200 years, with increases in population density for some species stimulating debate over how (or if) they should be managed. These can be divided into species that have increased throughout their range (especially the large kangaroo species), and those that demonstrate local or regional overabundance in some areas of their range, but with variable conservation status in other locales (e.g., koalas and common brushtail possums, *Trichosurus vulpecula*).

The large kangaroo species are generally believed to have increased in abundance, and possibly range, since European settlement, through the provision of improved pasture, artificial watering points and protection from the dingo (Maxwell *et al.*, 1996; Pople & Grigg, 1999; Pople *et al.*, 2010). Population management generally takes the form of culling at a local or regional scale, and is regulated by the relevant state or territory government wildlife management agencies, which may or may not include a commercial harvesting component (reviewed by Herbert *et al.*, 2010). The iconic status of kangaroos, both nationally and internationally, means that management operations are highly vexatious and often cause significant conflict between stakeholders (e.g. Cooney *et al.*, 2012).

Other marsupial species show increases in abundance within some areas of their range, but widespread declines in other areas. Two prime examples are the common brushtail possum and the koala. Common brushtail possums are one of the few marsupial species to have adapted well to the process of urbanisation, and they are generally believed to have increased within urban centres along the east coast due to their ability to exploit human-modified environments to their benefit (e.g. additional food resources and denning sites; Eymann *et al.* 2006). However, this adaptability has not benefited them throughout their entire range, and they have declined in most non-urban areas, especially in the arid and semi-arid zones, tropical woodlands and south-western Australia (How & Hillcox, 2000; Kerle, 2001). The reduction in tree density in many of these areas forces possums to spend more time on the ground where they are more susceptible to predation (How & Hillcox, 2000). These dichotomous situations highlight the need for context-specific approaches to wildlife conservation and management. In urban areas where they are common, possums are frequently viewed as a 'nuisance', and there is much debate about the need for and approach to management. Current government-legislated guidelines for possum management place significant restrictions on the primary management tool, i.e. translocation, and the success (or otherwise) of this management approach is questionable (Eymann *et al.*, 2006).

Perhaps the most dichotomous of all marsupial species is the koala. The conservation status of the koala varies on a national scale, with populations within the south of the range being considered overabundant and in need of urgent management to reduce population densities (reviewed by Herbert, 2007), while populations in the north of the range are listed as 'Vulnerable' (SEWPC, 2013). Overabundance of southern koala populations has been linked to habitat fragmentation, removal of hunting and predation pressure, and changes to vegetation structure from altered fire regimes. Fragmentation is probably one of the most significant issues for ongoing koala management. Most overabundant southern populations are located on oceanic islands, or 'virtual' islands by way of habitat fragmentation (Herbert, 2007), and this isolation has limited the ability of burgeoning populations to disperse, causing widespread habitat degradation within vegetation isolates. In northern populations, habitat fragmentation has been a source of direct mortality, and is causing ongoing declines post-clearance due to the reduced long-term viability of small, isolated populations (SEWPC, 2013). When this is coupled with the impacts of urbanisation, in particular 'encounter mortality' with dogs and cars, the prognosis for northern koala populations is a cause for concern. The variable status of the koala has made management operations for southern populations particularly vexatious, so much so that culling has been universally rejected as a potential management option (Herbert, 2007). This has led to significant investment of time and resources to intensive translocation and fertility control programmes in the states of Vic and SA (Duka & Masters, 2005; Menkhorst, 2008). The success of these non-lethal management operations is to a large extent unknown. Translocation, while effective as a short-term fix, has probably exacerbated the problem, and wildlife managers will need to intensively manage this species in perpetuity.

15.5 Conclusion

Australia's mammal fauna is exceptional both for its evolutionary diversity, unique history and high endemism. Yet over the past ~50 000 years its trajectory has primarily been one of decline and extinction, leaving the ecosystems of the continent profoundly altered and disturbed. While much has been lost, our increased biological understanding and management expertise has, over the past 50 years, prevented an even larger tragedy from unfolding. In such a dynamic and altered landscape, managing Australia's unique mammals is a formidable challenge, especially with limited (and currently declining) resources. While many threats to Australia's mammals are ongoing and novel threats continue to arise, it is hoped that this unique fauna will continue to fascinate and amaze for centuries to come.

REFERENCES

Abbott, I. (2006) Mammalian faunal collapse in Western Australia, 1875–1925: the hypothesised role of epizootic disease and a conceptual model of its origin, introduction, transmission, and spread. *Australian Zoologist* 33, 530–561.

Algar, D., Burrows, N. D. (2004) Feral cat control research: Western Shield review – February 2003. *Conservation Science Western Australia* 5, 131–163.

Archer, M., Hand, S. J., Godthelp, H. (1991) *Riversleigh. The Story of Animals in Ancient Rainforests of Inland Australia*. Reed Books, Sydney.

Arkell, B. (1999) *Flinders Ranges Bounce Back*. Department of Environment, Heritage and Aboriginal Affairs, Adelaide.

Baker, A. M., Mutton, T., Van Dyck, S. (2012) A new dasyurid marsupial from eastern Queensland: the buff-footed antechinus *Antechinus mysticus* sp. nov. (Marsupialia: Dasyuridae). *Zootaxa* 3515, 1–37.

Bilney, R. J., Cooke, R., White, J. G. (2010) Underestimated and severe: small mammal decline from the forests of south-eastern Australia since European settlement, as revealed by a top-order predator. *Biological Conservation* 143, 52–59.

Botero, A., Thompson, C. K., Peacock, C. S., *et al.* (2013) Trypanosomes genetic diversity, polyparasitism and the population decline of the critically endangered Australian marsupial, the brush tailed bettong or woylie (*Bettongia penicillata*). *International Journal for Parasitology: Parasites and Wildlife* 2, 77–89.

Briskie, J. V., Mackingtosh, M. (2004) Hatching failure increases with severity of population bottlenecks in birds. *Proceedings of the National Academy of Sciences of the United States of America* 101, 558–561.

Burbidge, A. A. (1999) Conservation values and management of Australian islands for non-volant mammal conservation. *Australian Mammalogy* 21, 67–74.

Burbidge, A. A., Manly, B. F. J. (2002) Mammal extinctions on Australian islands: causes and conservation implications. *Journal of Biogeography* **29**, 465–473.

Burbidge, A. A., McKenzie, L. M. (1989) Patterns in the modern decline of Western Australia's vertebrate fauna: causes and conservation implications. *Biological Conservation* **50**, 143–198.

Burbidge, A. A., Johnson, K. A., Fuller, P. J., Southgate, R. I. (1988) Aboriginal knowledge of the mammals of the central deserts of Australia. *Australian Wildlife Research* **15**, 9–39.

Burbidge, A. A., Williams, M. R., Abbott, I. (1997) Mammals of Australian islands: factors influencing species richness. *Journal of Biogeography* **24**, 703–715.

Byrne, M., Steane, D. A., Joseph, L., *et al.* (2011) Decline of a biome: evolution, contraction, fragmentation, extinction and invasion of the Australian mesic zone biota. *Journal of Biogeography* **38**, 1635–1656.

Byrne, M., Yeates, D. K., Joseph, L., *et al.* (2008) Birth of a biome: insights into the assembly and maintenance of the Australian arid zone biota. *Molecular Ecology* **17**, 4398–4417.

Calaby, J. H., Grigg, G. (1989) Changes in macropod communities and populations in the last 200 years, and the future. In: *Kangaroos, Wallabies and Rat-kangaroos* (eds. Grigg, G., Jarman, P. J., Hume, I.), pp. 813–820. Surrey Beatty and Sons, Sydney.

Cardoso, M., Eldridge, M. D. B., Oakwood, M., *et al.* (2009) Effects of founder events on the genetic variation of translocated island populations: implications for conservation management of the northern quoll. *Conservation Genetics* **10**, 1719–1733.

Ceballos, G., Brown, J. H. (1995) Global patterns of mammalian diversity, endemism, and endangerment. *Conservation Biology* **9**, 559–568.

Ceballos, G., Ehrlich, P. R. (2006) Global mammal distributions, biodiversity hotspots, and conservation. *Proceedings of the National Academy of Sciences, USA* **103**, 19 374–19 379.

Christensen, P., Burrows, N. (1994) Project Desert Dreaming: an experimental reintroduction of mammals to the Gibson Desert, Western Australia. In: *Reintroduction Biology of Australian and New Zealand Fauna* (ed. Serena, M.), pp. 199–207. Surrey Beatty and Sons, Sydney.

Churchill, S. (2008) *Australian Bats*. Second edition. Allen and Unwin, Crows Nest.

Cooney, R., Archer, M., Baumber, A., *et al.* (2012) THINKK again: getting the facts straight on kangaroo harvesting and conservation. In: *Science Under Siege: Zoology Under Threat* (eds. Banks, P., Lunney, D., Dickman, C.), pp. 150–160. Royal Zoological Society of New South Wales, Mosman.

Copley, P. B. (1994) Translocation of native vertebrates in South Australia: A review. In: *Reintroduction Biology of Australian and New Zealand Fauna* (ed. Serena, M.), pp. 35–42. Surrey Beatty and Sons, Sydney.

Cristescu, R., Cahill, V., Sherwin, W. B., *et al.* (2009) Inbreeding and testicular abnormalities in a bottlenecked population of koalas (*Phascolarctos cinereus*). *Wildlife Research* **36**, 299–308.

Deakin, J. E., Graves, J. A. M., Rens, W. (2012) The evolution of marsupial and monotreme chromosomes. *Cytogenetic and Genome Research* **137**, 113–129.

DECC (2008) *Recovery plan for the brush-tailed rock-wallaby* (Petrogale penicillata). Department of Environment and Climate Change, Sydney, New South Wales, Sydney.

Dexter, N., Murray, A. (2009) The impact of fox control on the relative abundance of forest mammals in East Gippsland, Victoria. *Wildlife Research* **36**, 252–261.

Dexter, N., Meek, P. D., Moore, S., Hudson, M., Richardson, H. (2007) Population responses of small and medium sized mammals to fox control at Jervis Bay, southeastern Australia. *Pacific Conservation Biology* **13**, 283–292.

Dickman, C. R. (1996) *Overview of the Impacts of Feral Cats on Australian Native Fauna.* Nature Conservation Agency, Canberra.

Dickman, C. R., Glen, A. S., Letnic, M. (2009) Reintroducing the dingo: can Australia's conservation wastelands be restored? In: *Reintroduction of Top-order Predators* (eds. Hayward, M. W., Somers, M. J.), pp. 238–269. Wiley-Blackwell, Oxford.

Dickman, C. R., Pressey, R. L., Lim, L., Parnaby, H. E. (1993) Mammals of particular conservation concern in the western division of New South Wales. *Biological Conservation* **65**, 219–248.

DPIPWE (2012) *Devil Refugees Make Maria Island Home.* Department of Primary Industries, Parks, Water and Environment, Hobart.

DSE (2009) Eastern barred bandicoot *Perameles gunnii*. Action Statement #4. Department of Sustainability and Environment, Victoria, Melbourne.

Duka, T., Masters, P. (2005) Confronting a tough issue: fertility control and translocation for over-abundant koalas on Kangaroo Island. South Australia. *Ecological Management and Restoration* **6**, 172–181.

Eldridge, M. D. B. (1998) Trouble in paradise? *Nature Australia* **26**, 24–31.

Eldridge, M. D. B. (2010) Population and conservation genetics of marsupials. In: *Marsupial Genetics and Genomics* (eds. Deakin, J. E., Waters, P. D., Graves, J. A. M.), pp. 461–497. Springer, Dordrecht.

Eldridge, M. D. B., King, J. M., Loupis, A. K., *et al.* (1999) Unprecedented low levels of genetic variation and inbreeding depression in an island population of the black-footed rock-wallaby. *Conservation Biology* **13**, 531–541.

Eldridge, M. D. B., Kinnear, J. E., Zenger, K. R., McKenzie, L. M., Spencer, P. B. S. (2004) Genetic diversity in remnant mainland and 'pristine' island populations of three endemic Australian macropodids (Marsupialia): *Macropus eugenii, Lagorchestes hirsutus* and *Petrogale lateralis. Conservation Genetics* **5**, 325–338.

Evans, S. R., Sheldon, B. C. (2008) Interspecific patterns of genetic diversity in birds: correlations with extinction risk. *Conservation Biology* **22**, 1016–1025.

Eymann, J., Herbert, C. A., Cooper, D. W. (2006) Management issues of urban brushtail possums (*Trichosurus vulpecula*): a loved or hated neighbour. *Australian Mammalogy* **28**, 153–171.

Finlayson, G. R., Moseby, K. E. (2004) Managing confined populations: the influence of density on the home range and habitat use of burrowing bettongs (*Bettongia lesueur*). *Wildlife Research* **31**, 457–463.

Finlayson, G. R., Finlayson, S. T., Dickman, C. R. (2010) Returning the rat-kangaroos: translocation attempts in the family Potoroidae (superfamily Macropodoidea) and recommendations for conservation. In: *Macropods: the Biology of Kangaroos, Wallabies and Rat-kangaroos* (eds. Coulson, G. M., Eldridge, M. D. B.), pp. 245–262. CSIRO Publishing, Melbourne.

Fischer, J., Lindenmayer, D. B. (2000) An assessment of the published results of animal relocations. *Biological Conservation* **96**, 1–11.

Fitzsimons, J., Legge, S., Trail, B., Woinarski, J. (2010) *Into Oblivion? The Disappearing Native Mammals of Northern Australia.* The Nature Conservancy, Melbourne.

Flannery, T. F. (2012) After the future. Australia's new extinction crisis. *Quarterly Essay* **48**, 1–80.

Frankham, R. (1997) Do island populations have less genetic variation than mainland populations? *Heredity* **78**, 311–327.

Frankham, R. (1998) Inbreeding and extinction: Island populations. *Conservation Biology* **12**, 665–675.

Frankham, R. (2005) Genetics and extinction. *Biological Conservation* **126**, 131–140.

Frankham, R., Ballou, J. D., Briscoe, D. A. (2002) *Introduction to Conservation Genetics* Cambridge University Press, Cambridge, UK.

Frankham, R., Ballou, J. D., Briscoe, D. A. (2010) *Introduction to Conservation Genetics.* Second edition. Cambridge University Press, Cambridge, UK.

Frankham, R., Ballou, J. D., Eldridge, M. D. B., *et al.* (2011) Predicting the probability of outbreeding depression. *Conservation Biology* **25**, 465–475.

Friend, J. A., Thomas, N. D. (2003) Conservation of the numbat (*Myrmecobius fasciatus*). In: *Predators with Pouches: The Biology of Carnivorous Marsupials* (eds. Jones, M., Dickman, C., Archer, M.), pp. 542–463. CSIRO Publishing, Melbourne.

Geffen, E., Rowe, K. C., Yom-Tov, Y. (2011) Reproductive rates in Australian rodents are related to phylogeny. *PLoS ONE* **6**, e19199.

Gibbons, P., Lindenmayer, D. B. (2002) *Tree Hollows and Wildlife Conservation in Australia.* CSIRO Publishing, Melbourne.

Gibson, D. F., Johnson, K. A., Langford, D. C., *et al.* (1994) The rufous hare-wallaby *Lagorchestes hirsutus:* A history of experimental reintroduction in the Tanami Desert, Northern Territory. In: *Reintroduction Biology of Australian and New Zealand Fauna* (ed. Serena, M.), pp. 171–176. Surrey Beatty and Sons, Sydney.

Glen, A. S., Berry, O., Sutherland, D. R., *et al.* (2010) Forensic DNA confirms intraguild killing of a chuditch (*Dasyurus geoffroii*) by a feral cat (*Felis catus*). *Conservation Genetics* **11**, 1099–1101.

Godthelp, H., Archer, M., Cifelli, R., Hand, S. J., Gilkeson, C. F. (1992) Earliest known Australian Tertiary mammal fauna. *Nature* **356**, 514–516.

Grant, P. R. (1998) Patterns on islands and microevolution. In: *Evolution on Islands* (ed. Grant, P. R.), pp. 1–17. Oxford University Press, Oxford, UK.

Groom, C. (2010) Justification for continued conservation efforts following the delisting of a threatened species: a case study of the woylie, *Bettongia penicillata ogilbyi* (Marsupialia : Potoroidae). *Wildlife Research* **37**, 183–193.

Hand, S. J., Novacek, M. J., Godthelp, H., Archer, M. (1994) First Eocene bat from Australia. *Journal of Vertebrate Paleontology* **14**, 375–381.

Hawkins, C. E., Baars, C., Hesterman, H., *et al.* (2006) Emerging disease and population decline of an island endemic, the Tasmanian devil *Sarcophilus harrisii*. *Biological Conservation* **131**, 307–324.

Helgen, K. M., Miguez, R. P., Kohen, J. L., Helgen, L. E. (2012) Twentieth century occurrence of the long-beaked echidna *Zaglossus bruijnii* in the Kimberley region of Australia. *ZooKeys* **255**, 103–132.

Herbert, C. A. (2007) From the urban fringe to the Abrolhos Islands: Management challenges of burgeoning marsupial populations. In: *Pest or Guest: the Zoology of Overabundance* (eds. Lunney, D., Eby, P., Hutchings, P., Burgin, S.), pp. 129–141. Royal Zoological Society of New South Wales, Mosman.

Herbert, C. A., Renfree, M. B., Coulson, G., *et al.* (2010) Advances in the development of fertility control technologies for managing of overabundant macropodid populations. In: *Macropods: the Biology of Kangaroos, Wallabies and Rat-kangaroos* (eds. Coulson, G. M., Eldridge, M. D. B.), pp. 313–324. CSIRO Publishing, Melbourne.

Hocknull, S. A., Zhao, J., Feng, Y., Webb, G. E. (2007) Responses of Quaternary rainforest vertebrates to climate change in Australia. *Earth and Planetary Science Letters* **264**, 317–331.

Hoffmann, M., Hilton-Taylor, C., Angulo, A., *et al.* (2011) The impact of conservation on the status of the world's vertebrates. *Science* **330**, 1503–1509.

Houlden, B. A., England, P. R., Taylor, A. C., Greville, W. D., Sherwin, W. B. (1996) Low genetic variability of the koala *Phascolarctos cinereus* in south-eastern Australia following a severe population bottleneck. *Molecular Ecology* **5**, 269–281.

How, R. A., Hillcox, S. J. (2000) Brushtail possum, *Trichosurus vulpecula*, populations in south-western Australia: demography, diet and conservation status. *Wildlife Research* **27**, 81–89.

How, R. A., Spencer, P. B. S., Schmitt, L. H. (2009) Island populations have high conservation value for northern Australia's top marsupial predator ahead of a threatening process. *Journal of Zoology, London* **278**, 206–217.

Hrdina, F., Gordon, G. (2004) The koala and possum trade in Queensland. 1906–1936. *Australian Zoologist* **32**, 543–585.

IUCN (2012) *Red List of Threatened Species*. Version 2012.2. IUCN, Gland, Switzerland.

Johnson, C. N. (2006) *Australia's Mammal Extinctions: a 50 000 Year History*. Cambridge University Press, Melbourne.

Johnson, C. N., Isaac, J. L. (2009) Body mass and extinction risk in Australian marsupials: the 'Critical Weight Range' revisited. *Austral Ecology* **34**, 35–40.

Johnson, C. N., Ritchie, E. G. (2013) The dingo and biodiversity conservation: response to Fleming *et al.* (2012). *Australian Mammalogy* **35**, 8–14.

Johnson, C. N., Isaac, J. L., Fisher, D. O. (2007) Rarity of a top predator triggers continent-wide collapse of mammal prey: dingoes and marsupials in Australia. *Proceedings of the Royal Society of London. Series B: Biological Sciences* **274**, 341–346.

Johnson, K. A. (1999) Recovery and discovery: where we have been and where we might go with species recovery. *Australian Mammalogy* **21**, 75–86.

Johnson, K. A., Burbidge, A. A., McKenzie, N. L. (1989) Australian Macropodoidea: Status, causes of decline and future research and management. In: *Kangaroos, Wallabies and Rat-kangaroos* (eds. Grigg, G., Jarman, P. J., Hume, I.), pp. 641–657. Surrey Beatty and Sons, Sydney.

Johnson, P. M., Nolan, B. J., Schaper, D. N. (2003) Introduction of the Proserpine rock-wallaby *Petrogale persephone* from mainland Queensland to nearby Hayman Island. *Australian Mammalogy* **25**, 61–71.

Kennedy, M. (1992) *Australasian Marsupials and Monotremes: an Action Plan for Their Conservation*. IUCN, Gland, Switzerland.

Keogh, J. S., Scott, I. A. W., Hayes, C. (2005) Rapid and repeated origin of insular gigantism and dwarfism in Australian tiger snakes. *Evolution* **59**, 226–233.

Kerle, J. A. (2001) *Possums – the Brushtails, Ringtails and Greater Glider*. University of NSW Press, Sydney.

Kinnear, J. E., Krebs, C. J., Pentland, C., *et al.* (2010) Predator-baiting experiments for the conservation of rock-wallabies in Western Australia; a 25-year review with recent advances. *Wildlife Research* **37**, 57–67.

Kinnear, J. E., Onus, M. L., Bromilow, R. N. (1988) Fox control and rock-wallaby population dynamics. *Australian Wildlife Research* **15**, 435–450.

Kinnear, J. E., Onus, M. L., Sumner, N. R. (1998) Fox control and rock-wallaby population dynamics. II. An update. *Wildlife Research* **25**, 81–88.

Kinnear, J. E., Sumner, N. R., Onus, M. L. (2002) The red fox in Australia – an exotic predator turned biocontrol agent. *Biological Conservation* **108**, 335–359.

Kirsch, J. A. W., Lapointe, F. J., Springer, M. S. (1997) DNA-hybridisation studies of marsupials and their implications for metatherian classification. *Australian Journal of Zoology* **45**, 211–280.

Langford, D. C., Burbidge, A. A. (2001) Translocation of mala (*Lagorchestes hirsutus*) from the Tanami Desert, Northern Territory to Trimouille Island, Western Australia. *Australian Mammalogy* **23**, 37–46.

Lavery, T. H., Fisher, D. O., Flannery, T. F., Leung, L. K.-P. (2013) Higher extinction rates of dasyurids on Australo-Papuan continental shelf islands and the zoogeography of New Guinea mammals. *Journal of Biogeography* **40**, 747–758.

Legge, S., Kennedy, M. S., Lloyd, R., Murphy, S. A., Fisher, A. (2011) Rapid recovery of mammal fauna in the central Kimberley, northern Australia, following the removal of introduced herbivores. *Austral Ecology* **36**, 791–799.

Letnic, M., Dickman, C. R. (2006) Boom means bust: interactions between the El Nino/Southern Oscillation (ENSO), rainfall and the processes threatening mammal species in arid Australia. *Biodiversity and Conservation* **15**, 3847–3880.

Letnic, M., Koch, A., Gordon, C., Crowther, M. S., Dickman, C. R. (2009) Keystone effects of an alien top-predator stem extinctions of native mammals. *Proceedings of the Royal Society of London. Series B: Biological Sciences* **276**, 3249–3256.

Lindenmayer, D. B., Possingham, H. P. (1996) Ranking conservation and timber management options for Leadbeater's possum in southeastern Australia using population viability analysis. *Conservation Biology* **10**, 235–251.

Lindenmayer, D. B., Blanchard, W., McBurney, L., *et al.* (2012) Interacting factors driving a major loss of large trees with cavities in a forest ecosystem. *PLoS ONE* **7**, e41864.

Loehle, C., Eschenbach, W. (2012) Historical bird and terrestrial mammal extinction rates and causes. *Diversity and Distributions* **18**, 84–91.

Long, J. L., Archer, M., Flannery, T. F., Hand, S. J. (2002) *Prehistoric Mammals of Australia and New Guinea: One Hundred Million Years of Evolution.* University of NSW Press, Sydney.

Lumsden, L. (2009) The extinction of the Christmas Island pipistrelle. *Australian Bat Society Newsletter* **33**, 21–25.

Lunney, D., Law, B., Rummery, C. (1997) An ecological interpretation of the historical decline of the Brush-tailed Rock-wallaby *Petrogale penicillata* in New South Wales. *Australian Mammalogy* **19**, 281–296.

Luo, Z.-X., Ji, Q., Wible, J. R., Yuan, C.-X. (1992) An early Cretaceous triosphenic mammal and metatherian evolution. *Science* **302**, 1934–1940.

Madsen, T., Shine, R., Olsson, M., Wittzell, H. (1999) Restoration of an inbred adder population. *Nature* **402**, 34–35.

Martin, T. G., Nally, S., Burbidge, A. A., *et al.* (2012) Acting fast helps avoid extinction. *Conservation Letters* **5**, 274–280.

Masters, P., Duka, T., Berris, S., Moss, G. (2009) Koalas on Kangaroo Island: from introduction to pest status in less than a century. *Wildlife Research* **34**, 267–272.

Maxwell, S., Burbidge, A. A., Morris, K. (1996) *The 1996 Action Plan for Australian Marsupials and Monotremes.* Wildlife Australia, Canberra.

McKenzie, N. L., Burbidge, A. A., Baynes, A., *et al.* (2007) Analysis of factors implicated in the recent decline of Australia's mammal fauna. *Journal of Biogeography* **34**, 597–611.

Menkhorst, P. (2008) Hunted, marooned, re-introduced, contracepted: a history of koala management in Victoria. In: *Too Close for Comfort: Contentious Issues in Human–Wildlife Encounters* (eds. Lunney, D., Munn, A., Meikle, W.), pp. 73–92. Royal Zoological Society of New South Wales, Mosman.

Menkhorst, P. W., Knight, F. (2011) *A Field Guide to the Mammals of Australia.* Third edition. Oxford University Press, Melbourne.

Meredith, R. W., Westerman, M., Case, J. A., Springer, M. S. (2008) A phylogeny and timescale for marsupial evolution based on sequences for five nuclear genes. *Journal of Mammalian Evolution* **15**, 1–36.

Miller, E. J., Eldridge, M. D. B., Morris, K. D., Zenger, K. R., Herbert, C. A. (2011) Genetic consequences of isolation: island tammar wallaby (*Macropus eugenii*) populations and the conservation of threatened species. *Conservation Genetics* **12**, 1619–1631.

Moro, D. (2002) Translocation of captive-bred dibblers *Parantechinus apicalis* (Marsupialia: Dasyuridae) to Escape Island, Western Australia. *Biological Conservation* **111**, 305–315.

Morris, K. D., Johnston, B., Orell, P., *et al.* (2003) Recovery of the threatened chuditch (*Dasyurus geoffroii*): a case study. In: *Predators with Pouches: The Biology of Carnivorous Marsupials* (eds. Jones, M., Dickman, C., Archer, M.), pp. 435–451 CSIRO Publishing, Melbourne.

Moseby, K. E., Read, J. L., Paton, D. C., *et al.* (2011) Predation determines the outcome of 10 reintroduction attempts in arid South Australia. *Biological Conservation* **144**, 2863–2872.

Murphy, W. J., Eizirik, E., O'Brien, S. J., *et al.* (2001) Resolution of the early placental mammal radiation using Bayesian phylogenetics. *Science* **294**, 2348–2351

Neaves, L. E., Zenger, K. R., Prince, R. I. T., Eldridge, M. D. B. (2012) Impact of Pleistocene aridity oscillations on the population history of a widespread, vagile Australian mammal, *Macropus fuliginosus*. *Journal of Biogeography* **39**, 1545–1563.

Olsen, P. (2010) *Upside Down World: Early European Impressions of Australia's Curious Animals.* National Library of Australia, Canberra.

Parnaby, H. E. (2009) A taxonomic review of Australian greater long-eared bats previously known as *Nyctophilus timoriensis* (Chiroptera: Vespertilionidae) and some associated taxa. *Australian Zoologist* **35**, 39–81.

Parnaby, H. E., Lunney, D., Shannon, I., Fleming, M. (2010) Collapse rates of hollow-bearing trees following low intensity prescription burns in the Pilliga forests, New South Wales. *Pacific Conservation Biology* **16**, 209–220.

Pascual, R., Archer, M., Jaureguizar, E. O., *et al.* (1992) The first non-Australian monotreme: and early Paleocene South American platypus (Monotremata, Ornithorhynchidae). In: *Platypus and Echidnas* (ed. Augee, M. L.), pp. 2–15. Royal Zoological Society of NSW, Sydney.

Pearson, D. J. (2012) Recovery Plan for five species of rock-wallabies: black-flanked rock-wallaby (*Petrogale lateralis*), Rothschild's rock-wallaby (*Petrogale rothschildi*), short-ear rock wallaby (*Petrogale brachyotis*), monjon (*Petrogale burbidgei*) and nabarlek (*Petrogale concinna*). Department of Environment and Conservation, Perth, WA.

Pickering, J., Norris, C. A. (1996) New evidence concerning the extinction of the endemic murid *Rattus macleari* from Christmas Island, Indian Ocean. *Australian Mammalogy* **19**, 19–25.

Pimm, S. L., Dollar, L., Bass Jr, O. L. (2006) The genetic rescue of the Florida panther. *Animal Conservation* **9**, 115–122.

Pizzuto, T. A., Finlayson, G. R., Crowther, M. S., Dickman, C. R. (2007) Microhabitat use by the brush-tailed bettong (*Bettongia penicillata*) and burrowing bettong (*B. lesueur*) in semiarid New South Wales: implications for reintroduction programs. *Wildlife Research* **34**, 271–279.

Pople, A. R., Grigg, G. C. (1999) *Commercial Harvesting of Kangaroos in Australia.* Environment Australia, Canberra.

Pople, A. R., Grigg, G. C., Phinn, S. R., *et al.* (2010) Reassessing the spatial and temporal dynamics of kangaroo populations. In: *Macropods: the Biology of Kangaroos, Wallabies and Rat-kangaroos* (eds. Coulson, G. M., Eldridge, M. D. B.), pp. 197–210. CSIRO Publishing, Melbourne.

Priddel, D., Wheeler, R. (2004) An experimental translocation of brush-tailed bettongs (*Bettongia penicillata*) to western New South Wales. *Wildlife Research* **31**, 421–432.

Rich, T. H., Vickers-Rich, P., Constantine, A., *et al.* (1997) A triospenic mammal from the Mesozoic of Australia. *Science* **278**, 1438–1442.

Richards, J. D., Short, J. (2003) Reintroduction and establishment of the western barred bandicoot *Perameles bougainville* (Marsupialia: Peramelidae) at Shark Bay, Western Australia. *Biological Conservation* **109**, 181–195.

Risbey, D. A., Calver, M. C., Short, J., Bradley, J. S., Wright, I. W. (2000) The impact of cats and foxes on the small vertebrate fauna of Heirisson Prong, Western Australia. II. A field experiment. *Wildlife Research* **27**, 223–235.

Ritchie, E. G., Johnson, C. N. (2009) Predator interactions, mesopredator release and biodiversity conservation. *Ecology Letters* **12**, 982–998.

Rowe, K. C., Aplin, K. P., Baverstock, P. R., Moritz, C. (2011) Recent and rapid speciation with limited morphological disparity in the genus *Rattus*. *Systematic Biology* **60**, 188–203.

Rowe, K. C., Reno, M. L., Richmond, D. M., Adkins, R. M., Steppan, S. J. (2008) Pliocene colonization and adaptive radiations in Australia and New Guinea (Sahul): Multilocus systematics of the old endemic rodents (Muroidea: Murinae). *Molecular Phylogenetics and Evolution* **47**, 84–101.

Saccheri, I., Brakefield, P. M., Nichols, R. A. (1996) Severe inbreeding depression and rapid fitness rebound in the butterfly *Bicyclus anynana* (Satyridae). *Evolution* **50**, 2000–2013.

Saccheri, I., Kuussaari, M., Kankare, M., *et al.* (1998) Inbreeding and extinction in a butterfly metapopulation. *Nature* **392**, 491–494.

Sarre, S. D., MacDonald, A. J., Barclay, C., Saunders, G. R., Ramsey, D. S. L. (2012) Foxes are now widespread in Tasmania: DNA detection defines the distribution of this rare but invasive carnivore. *Journal of Applied Ecology* **50**, 459–468.

Schipper, J., Chanson, J. S., Chiozza, F., *et al.* (2008) The status of the world's land and marine mammals: diversity, threat, and knowledge. *Science* **322**, 225–230.

Seebeck, J. H., Bennett, A. F., Dufty, A. C. (1990) Status distribution and biogeography of the eastern barred bandicoot, *Perameles gunnii* in Victoria. In: *Management and Conservation of Small Populations* (eds. Clark, T. W., Seebeck, J. H.), pp. 21–32. Chicago Zoological Society, Brookfield, Illinois.

SEWPC (2013) *Phascolarctos cinereus* (combined populations of Qld, NSW and the ACT). In: *Species Profile and Threats Database*. Department of Sustainability, Environment, Water, Population and Communities, Canberra.

Seymour, A. M., Montgomery, M. E., Costello, B. H., *et al.* (2001) High effective inbreeding coefficients correlate with morphological abnormalities in populations of South Australian koalas (*Phascolarctos cinereus*). *Animal Conservation* **4**, 211–219.

Short, J. (2009) *The Characteristics and Success of Vertebrate Translocations Within Australia*. Department of Agriculture, Fisheries and Forestry, Canberra.

Short, J., Smith, A. (1994) Mammal decline and recovery in Australia. *Journal of Mammalogy* **75**, 288–297.

Short, J., Turner, B. (2000) Reintroduction of the burrowing bettong *Bettongia lesueur* (Marsupialia: Potoroidae) to mainland Australia. *Biological Conservation* **96**, 185–196.

Short, J., Bradshaw, S. D., Giles, J., Prince, R. I. T., Wilson, G. R. (1992) Reintroduction of macropods (Marsupialia: Macropodoidea) in Australia – a review. *Biological Conservation* **62**, 189–204.

Smith, S., Hughes, J. (2008) Microsatellite and mitochondrial DNA variation defines island genetic reservoirs for reintroductions of an endangered Australian marsupial, *Perameles bougainville*. *Conservation Genetics* **9**, 547–557.

SoEC (2011) *Australia: State of the Environment 2011*. Department of Sustainability, Environment, Water, Population and Communities, Canberra.

Spielman, D., Brook, B. W., Frankham, R. (2004) Most species are not driven to extinction before genetic factors impact them. *Proceedings of the National Academy of Sciences of the United States of America* **101**, 15 261–15 264.

Stanley, F., Morris, K., Holmes, T., Moore, J. (2010) Giant steps: industry and conservation make history through Gorgon. *Landscape* **25**, 10–16.

Tyndale-Biscoe, C. H. (2005) *Life of Marsupials*. CSIRO Publishing, Melbourne.

Van Dyck, S., Strahan, R. (2008) *The Mammals of Australia*. Third edition. Reed New Holland, Sydney.

Veitch, C. R., Clout, M. N., Towns, D. R. (2011) *Island Invasives: Eradication and Management. Proceedings of the International Conference on Island Invasives*. IUCN, Gland, Switzerland.

Vieira, E. M., Finlayson, G. R., Dickman, C. R. (2007) Habitat use and density of numbats (*Myrmecobius fasciatus*) reintroduced in an area of mallee vegetation, New South Wales. *Australian Mammalogy* **29**, 17–24.

Vila, C., Sundqvist, A.-K., Flagstad, O., *et al.* (2003) Rescue of a severely bottlenecked wolf (*Canis lupus*) population by a single immigrant. *Proceedings of the Royal Society of London. Series B: Biological Sciences* **270**, 91–97.

Walton, D. W. (1988) *Zoological Catalogue of Australia. 5. Mammalia*. Australian Government Publishing Service, Canberra.

Wayne, A. F., Maxwell, M. A., Ward, C. G., *et al.* (2013) Importance of getting the numbers right: quantifying the rapid and substantial decline in an abundant marsupial, *Bettongia penicillata*. *Wildlife Research* **40**, 169–183.

Weeks, A. R., Sgro, C. M., Young, A. G., *et al.* (2011) Assessing the benefits and risks of translocations in changing environments: a genetic perspective. *Evolutionary Applications* **4**, 709–725.

Westemeier, R. L., Brawn, J. D., Simpson, S. A., *et al.* (1998) Tracking the long-term decline and recovery of an isolated population. *Science* **282**, 1695–1698.

Williams, G. R. (1977) Marooning, a technique for saving threatened species from extinction. In: *International Zoo Yearbook* (ed. Olney, P. J. S.), pp. 102–106. Zoological Society of London, London.

Wilson, D. E., Reeder, D. M. (2005) *Mammal Species of the World: a Taxonomic and Geographic Reference*. Third edition. Johns Hopkins University Press, Baltimore.

Woinarski, J. C. Z., *et al.* (2014). *The 2013 Action Plan for Australian Mammals*. CSIRO Publishing, Melbourne.

Woinarski, J. C. Z., Legge, S., Fitzsimons, J. A., *et al.* (2011) The disappearing mammal fauna of northern Australia: context, cause, and response. *Conservation Letters* **4**, 192–201.

Wyatt, K. B., Campos, P. F., Gilbert, T. P., *et al.* (2008) Historical mammal extinction on Christmas Island (Indian Ocean) correlates with introduced infectious disease. *PLoS ONE* **3**, e3602.

Yeatman, G. J., Groom, C. J. (2012) *National Recovery Plan for the Woylie (*Bettongia penicillata*).* Department of Environment and Conservation, Perth.

Zenger, K. R., Eldridge, M. D. B., Cooper, D. W. (2003) Intraspecific variation, sex-biased dispersal and phylogeography of the eastern grey kangaroo (*Macropus giganteus*). *Heredity* **92**, 153–162.

CHAPTER 16

Marine mammals, back from the brink? Contemporary conservation issues

Robert Harcourt, Helene Marsh, David Slip, Louise Chilvers, Mike Noad
and Rebecca Dunlop

Summary

The extensive territorial waters of Australia and New Zealands (NZ) (over 8 million km^2 for Australia and a further 4 million km^2 for NZ) are home to approximately 49 species of whales and dolphins, 11 species of seals and the dugong. Within Australia, at least eight species are listed as *threatened*, though there is insufficient information on a further 25 to determine their conservation status, while in NZ eight species are listed as *threatened*. The relationship between humans and Australasia's marine mammals is culturally diverse and has changed significantly in recent years. Dugongs and stranded whales have been important both spiritually and as a source of nutrition to some Aborigines and Torres Strait Islanders for thousands of years; seals and whales had a similar role for Maori in NZ. In recent history, exploitation of baleen whales, elephant seals and fur seals was an important driver for much of the earliest European settlement of Australasia. The success of the whaling and sealing industries came at the expense of marine mammal populations, leading to the near extirpation of many species by the mid twentieth century. In more recent decades there has been a fundamental shift in public attitudes towards marine mammals, in particular the great whales and dolphins. All marine mammals are protected within Australia and NZ waters. Traditional hunting of dugongs is legal in Australia for Native Title holders.

Marine mammal protection is an important platform of government foreign policy with strong bipartisan support. Both Australia and NZ play key roles in the International Whaling Commission and the Commission for the Conservation of Aquatic Marine Living Resources. Despite the strong government and public focus on marine mammal conservation, species remain vulnerable to a number of threats including fisheries interactions, vessel disturbance, coastal and offshore development and climate change. Managing these threats can be particularly difficult for marine mammals as many species are migratory and so only inhabit areas managed by Australia and NZ for part of their

Austral Ark: The State of Wildlife in Australia and New Zealand, eds. A. Stow, N. Maclean and G. I. Holwell. Published by Cambridge University Press. © Cambridge University Press 2015.

life-cycle. Interactions between threats are often poorly understood, and even individual threats can have severe consequences if not well managed. Despite these challenges, marine mammals represent some of the most successful examples of effective conservation in contemporary Australia and NZ, with rapid recovery of many of the great whales and fur seals. Yet some of our most iconic species face an uncertain future. A few are in immediate jeopardy, including the Maui's dolphin and the New Zealand sea lion. Others face a very uncertain future due as much to lack of knowledge of cumulative effects of threatening processes as to any specific threat. To increase the resilience of marine mammals, manageable threats need to be investigated, understood and carefully managed or, where possible, ameliorated.

16.1 Introduction: status of marine mammals in Australasia

Marine mammals occupy all of the world's oceans, and include both top predators and important low trophic level consumers and grazers. Australasia is recognised as a hotspot for marine mammal species richness (Pompa *et al.* 2011) and many species occur in the waters of both countries. There are currently 59 recognised marine mammals that occur around Australia's coast, including its sub-Antarctic islands, for at least some part of the year. Of these, 48 are cetaceans (whales, dolphins and porpoises), ten are pinnipeds (seals and sea lions), and one is a sirenian, the dugong *Dugong dugong*. New Zealand is also exceptionally species rich with marine mammals (56 species) including two cetaceans and one pinniped not found in Australia.

Australian waters encompass the entire range for four endemic species, the Australian snubfin dolphin (*Orcaella heinsohni*) (Beasley *et al.* 2006); the Burrunan dolphin (*Tursiops australis*) (Charlton-Robb *et al.* 2011), Australian humpback dolphin (*Sousa sahulensis*) (Jefferson & Rosenbaum, 2014) and the Australian sea lion (*Neophoca cinerea*) (Gales *et al.* 1994). Furthermore, multiple lines of genetic evidence suggest that Australian humpback dolphins are reproductively isolated and genetically distinct from those in south-east Asia (Frère *et al.* 2011). For other species they represent key breeding and/or feeding habitat, e.g. southern right whales (*Eubaelana australis*), blue whales (*Balaenoptera musculus*), dugongs (see Box 16.2), while others are only occasional visitors, e.g. leopard seals (*Hydrurga leptonyx*). For most Australian marine mammals however, there is little information on the nature and extent of their use of Australian waters.

New Zealand is similar. Hector's dolphin (*Cephalorhynchus hectori)* and the subspecies, Maui's dolphin (*C. hectori maui*), are endemic to New Zealand as are New Zealand sea lions (*Phocarctos hookeri*), except for vagrants on Macquarie Island. New Zealand is an important habitat for coastal dolphins such as common bottlenose dolphins (*Tursiops truncatus*), short-beaked common dolphins (*Delphinus delphis*) dusky dolphins (*Lagenorhynchus obscurus*), killer whales (*Orcinus orca*), and whales including sperm whales (*Physeter macrocephalus*) and Bryde's whales (*Balaenoptera edeni*), while the sub-Antarctic Auckland Islands are important habitat for southern right whales. New Zealand is also frequently visited by offshore pelagics such as southern right whale dolphins (*Lissodelphis peronei*), short finned pilot whales (*Globicephala macrorhyncus*), and vagrants such as leopard and southern elephant seals (*Mirounga leonina*).

According to the IUCN Red List (IUCN 2011), and under the *Environment Protection and Biodiversity Conservation Act 1999* (EPBC Act) at least eight species of marine mammals are considered threatened in Australia, although there is some difference between the two classifications because they apply at different spatial scales and use slightly different categories and criteria. IUCN lists blue, sei (*Balaenoptera borealis*), and fin whales (*B. physalus*), and Australian sea lions as *Endangered*, sperm whales and dugongs as *Vulnerable*, and Australian humpback (*Sousa sahulensis*) and Australian snubfin dolphins as *Near Threatened*. By contrast, under the EPBC Act, blue and southern right whales are listed as *Endangered* while humpback (*Megaptera novaeangliae*) sei and fin whales, sub-Antarctic fur seals (*Arctophoca tropicalis*), southern elephant seals and Australian sea lions are all listed as *Vulnerable*. Furthermore, owing to insufficient data the IUCN conservation status of 25 cetacean species is unknown (a *Data Deficient* category is not available under the EPBC Act). In NZ the conservation status of species is determined by the NZ Threat Classification System rather than through specific legislation, and this system is designed to complement the IUCN Red List (Baker *et al.* 2010). Of the 56 taxa recorded in NZ, eight are *Threatened* (five *Nationally Critical* and three *Nationally Endangered*); while 13 are *Data Deficient*. All three endemic NZ marine mammals (the NZ sea lion, Hector's dolphin and Maui's dolphin) are now considered threatened with extinction (Baker *et al.* 2010).

Marine mammals in Australasia are protected by extensive legislation at multiple levels. In Australia all marine mammals within Australia's Exclusive Economic Zone (8 505 348 km^2) are provided with clear protection under the EPBC Act. This legislation extends to claimed waters off the Antarctic continent, a further 2 million km^2; however, the latter is only recognised by the United Kingdom, New Zealand, France and Norway, and as a whole this area forms the Australian Whale Sanctuary. Around the continent of Australia there is additional state legislative protection within state boundaries (from the coast to three nautical miles) that varies slightly from state to state. In New Zealand all marine mammals within its EEZ (4 083 744 km^2) are protected under the Marine Mammal Protection Act 1978.

Legislative protection extends beyond simply a ban on commercial harvesting, since it is also a legal offence to injure, take, trade, keep, move or interfere with any marine mammal (EPBC Act).[1] Within Environmental Impact Assessments the EPBC Act enacts strict stipulations in order to mitigate impacts on marine mammals from seismic surveys, coastal development, boating activities and fishing. Further to this, Australia and NZ play key roles in the International Whaling Commission and the Commission for the Conservation of Aquatic Marine Living Resources (CCAMLR). CCAMLR specifically manages all fisheries in the southern ocean with the objective of maintaining ecological relationships and preventing irreversible changes to the ecosystem by taking the requirements of other predators including marine mammals into account (Agnew 1997).

Marine mammals in Australasian waters have a varied history and are likely to have a varied future. Smaller coastal marine mammals including dugongs and dolphins in Australia and seals in New Zealand, along with stranded whales, had great spiritual importance as well as being an important source of protein to Aborigines, Torres Strait Islanders (Marsh 1996), and the Maori (Smith 1989), respectively. Dugongs have been hunted across northern Australia for thousands of years and remain an important component of the culture of many groups of Aborigines and Torres Strait Islanders (see Box 16.2). Dolphins and whales feature prominently in the culture of the Maori and coastal Indigenous Australians. Whales were not thought to be hunted at sea by Maori but were driven ashore to be killed and thereby provided an important source of food and other resources such as

---

[1] Apart from legal Indigenous hunting of dugongs.

bone for carving and fashioning weapons. Whales are a feature of tribal traditions in particular as guardians for the ancestors' canoe journeys to Aotearoa (New Zealand). Oral histories recall interactions between people and whales, and whale riding is also a common theme in Maori oral storytelling. At the same time, fur seals (kekeno) and the less numerous sea lions were hunted extensively around both North and South Islands until restricted to remote areas of the far South West and the sub-Antarctic Islands by the 1700s (Smith 1989).

Exploitation of baleen whales, elephant seals and fur seals was an important driver for much of the earliest European settlement of Australasia. Humpback, southern right, blue and sperm whales were all important whale fisheries. Whaling ceased in NZ in 1964 when it became uneconomic, and in Australia the last whaling station closed in Albany, Western Australia in 1978 (Suter 1981). Since cessation of whaling, dramatic recoveries have been seen in the humpback whale and southern right whales in Australasian waters (see Box 16.1 and Carroll *et al.* 2011). Less confidence

---

Box 16.1 Australian humpback whales: truly back from the brink

In the Southern Hemisphere, humpback whales (*Megaptera novaeangliae*) form many populations, but all feed in the Southern Ocean in the Austral summer. In autumn and winter they migrate to separate tropical breeding grounds where they mate and calve, returning to the feeding grounds in spring and early summer (Chittleborough 1965; Dawbin 1966). Australia is associated with two of these populations. One migrates along the western coast of Australia, wintering in the Kimberly region of north-western Australia while the other migrates along the east coast of Australia, wintering in the lagoon of the Great Barrier Reef. Both populations are thought to have some level of connection with each other as well as, especially in the case of the east coast population, with other neighbouring populations.

Historically, the Southern Hemisphere populations were divided up as 'Groups' by the International Whaling Commission (IWC). Around the Southern Hemisphere, Groups I–VI were assumed to feed in corresponding 'Areas' of the Southern Ocean. The western Australian whales were known as Group IV while the east Australian whales were seen as part of a larger population, Group V, which also included whales that wintered in the Chesterfield Reef system, New Caledonia, Fiji and Tonga. More recently the IWC realised that these Groups did not necessarily reflect the breeding populations of humpbacks, and they were redesignated as breeding stocks A–G, with the western Australian whales stock D and the eastern Australian whales E1, with New Caledonia E2 and Tonga E3 (IWC 2006). This separation, particularly of the old Group V into three sub-populations was based on genetic studies, mark recapture studies and demographic differences (IWC 2006).

In 2008 the species was downgraded to 'least concern' by the IUCN Red List of threatened species (Reilly *et al.* 2008) but is still listed in Australia as *Vulnerable* under the *Environment Protection and Biodiversity Conservation Act 1999*.

History of whaling

The humpback whales that migrate along the coastlines of Australia were hunted to near-extinction in the 1950s and early 1960s (Chittleborough 1965; Bannister 1964; Bannister *et al.* 1991; Bannister & Hedley, 2001). Prior to the 1950s, there was little

Box 16.1 (continued)

exploitation of the Australian humpback whale populations. In 1952, industrial shore-based whaling commenced, and together with massive illegal pelagic whaling in the Southern Ocean (Yablokov 1994; Mikhalev 2000), took whales in such abundance that the populations collapsed by 1962 (more than 22 000 humpback whales were taken from areas south of eastern Australia and Oceania between 1959 and 1963 alone; Clapham *et al.* 2009). Taking into account the large illegal Soviet catches, Jackson *et al.* (2008) estimated the pre-exploitation east Australian population to have been 22 000 and 25 700. Estimates of the Group V population size (East Australia plus the western South Pacific populations) in the early to mid 1960s include 104 (Bannister & Hedley, 2001) and 400–500 (Chittleborough 1965), while Paterson *et al.* (1994) estimated there were 34–137 whales remaining in the east Australian population alone. Although the distribution of surviving whales was not known, the rapid recovery of east Australian whales and apparent lack of recovery of whales migrating past New Zealand or of the Oceania populations suggests that most of these were from the east Australian population.

Since the end of the whaling era, there has been an apparent rapid increase in the population on the both west and east coasts of Australia.

Current status

The migratory corridor used by humpback whales is especially narrow along the southern coastline of Queensland. Off Pt Lookout of North Stradbroke Island, and Cape Moreton of Moreton Island, most whales pass within 10 km of land (Noad *et al.* 2011) making the whales available for land-based counts. Post-whaling surveys of the east Australian population were initiated at Pt Lookout in 1978, and occurred most years until around 2001 (Bryden surveys, Paterson surveys). In 2004 they were taken over by a team led by Michael Noad from the University of Queensland. These surveys show that the population has grown rapidly and consistently with an average rate of 10.9% per annum (95% CI 10.5%–11.3%) with a population estimate for 2010 of 14 522 whales (95% CI 12 777–16 504) (Noad *et al.* 2011). These growth rates are among the highest recorded for any humpback whale population and are close to the theoretical reproductive limit of the species (Best 1993; Brandao *et al.* 2000; Bannister & Hedley 2001). The set of surveys now comprise one of the best records of absolute and relative population size for any group of whales in the world.

On the Western Australian (WA) coast, the migratory pathway of humpback whales is further offshore than on the east coast of Australia making population surveys more difficult. Following increasing reports of humpback whale sightings in the early to mid 1970s, aerial surveys of humpback whales during their northward migration were undertaken from Carnarvon, close to an old whaling station. Early aerial surveys (up to 1988; Bannister *et al.* 1991) showed evidence of an increase in population from an all time low in 1963 (estimated to be between 500 and 600 animals). Further surveys, in 1991 and 1994 provided enough data to calculate a rate of increase of 10.15 ± 4.6% and a population size of 4000–5000 animals (Bannister 1995).

Aerial surveys remain the predominant method used for estimating population abundance on the west coast of Australia. In 2008 the abundance of northward migrating whales was estimated by Hedley *et al.* (2011) at 21 750 (95% CI: (17 550–43 000)). A separate study (also aerial-based) estimated the population to be 26 100 animals (95% CI: (20 152–33 272)) in 2008 (Salgado Kent *et al.* 2012). Growth rates are similar to those of the east coast albeit with less confidence: 12.5% (Hedley *et al.* 2011); 13% (95% CI 5.6%–18.1%) (Salgado Kent *et al.* 2012).

Prognosis

Despite being hunted to near-extirpation, the east and west coast Australian humpback whale populations are booming and prognosis for continued increase is good. However, with such high and increasing numbers of whales migrating close to the Australian coast, combined with an increase in major coastal developments (especially off the west coast), the potential for conflicts between humpbacks and humans may also increase. In 2011, the Department of Environment and Conservation reported a notable increase in the number of humpback stranding events on the west coast of Australia. Whether this increase was a direct result of increasing conflict with humans (boat strikes, entanglements in shark and fishing nets, etc.), or a natural and unavoidable result of a booming population, remains to be seen.

Box 16.2 The status of the dugong in Australia: a spatially variable situation

Dugongs occur across Australia's northern coastline from Shark Bay in Western Australia (25° S) to Moreton Bay in Queensland (27° S) (Marsh *et al.* 2002, 2011). There are occasional sightings south to 32–33.5° S on the east coast in summer and stranded carcasses are sometimes found south to ~36.5° S. The winter range, which extends to 25–27° S on the east coast encompasses about a quarter of the dugong's global 'Extent of Occurrence' (EOO, IUCN Red List definition) based on coastline (Marsh *et al.* 2011).

Population size and trends

Estimates of population size are based on standardised aerial surveys that attempt to correct for various visibility biases (Marsh & Sinclair 1989; Pollock *et al.* 2006), but are known to be underestimates (Marsh *et al.* 2011; Hagihara *et al.* in press). Nonetheless, these surveys indicate that the dugong is the most abundant marine mammal in the coastal waters of northern Australia; estimates from the more than 120 000 km^2 area surveyed since 2005 total almost 70 000 dugongs. Population estimates are unavailable or outdated for large regions including the Western Australian coast north of Exmouth Gulf, most of the Northern Territory coast outside of the Gulf of Carpentaria, and Australia's tropical offshore territories (Marsh *et al.* 2011).

Catch per unit effort data collected by the Queensland Shark Control Program suggest that the dugong population on the urban coast of Queensland (Cairns to NSW border) declined precipitously between the 1960s and early 1980s (Marsh *et al.* 2005). In contrast, the standardised aerial surveys conducted since the mid 1980s suggest that populations are stable in Shark Bay, the Exmouth/Ningaloo Reef region

Box 16.2 (continued)

of Western Australia, the Gulf of Carpentaria, and the northern Great Barrier Reef (Marsh *et al.* 2007, 2011; Hodgson *et al.* 2008). Nonetheless, the power of these surveys to detect low levels of decline is generally weak. The surveys suggest also that dugong numbers in Moreton Bay, Hervey Bay and Torres Strait fluctuate over time (Marsh *et al.* 2007, 2011; Sobtzick *et al.* 2012). The fluctuations in estimates for Hervey Bay and Moreton Bay are attributable to dugongs moving between the two bays or from shallow to deeper water within bays, after seagrass dieback associated with extreme weather events (Marsh *et al.* 2011; Sobtzick *et al.* 2012). Movements of dugongs from shallow to deeper water within the survey region is a plausible explanation for the population fluctuations observed in Torres Strait (Marsh *et al.* 2004, 2007, 2011), although the contribution of possible overharvest cannot be eliminated (Heinsohn *et al.* 2004; Marsh *et al.* 2004).

Threats

Significant Indigenous harvest of dugongs pre-dates European contact. For example, archaeological evidence indicates that dugong hunting in Torres Strait dates from at least 4000 years BP (Crouch *et al.* 2007). At one site on Mabuiag Island, the remains of an estimated 10 000–11 000 dugongs caught between *c.* 1600 AD and *c.* 1900 AD have been documented (McNiven & Beddingfield 2008).

Throughout remote northern Australia, the greatest contemporary source of dugong mortality remains legal Indigenous hunting (Marsh *et al.* 2011), which is localised around communities. Dugongs are also killed by illegal poaching and incidental capture in nets. However, the sale of dugong meat is illegal and the imperative to sell incidental catch is much less than in countries where food insecurity is a problem (Marsh *et al.* 2011).

Commercial fishing for dugongs by non-indigenous operators occurred intermittently at various places along the coast of Queensland from 1847 to 1969 (Daley *et al.* 2008). This cottage industry was principally for dugong oil, which was sought for medicine, cooking and cosmetics. Dugong hides, tusks, bones and meat were also sold; meat was sometimes given away as a by-product (Daley *et al.* 2008). Some Aboriginal missions in Queensland also earned income from dugong oil that was sold for distribution to Aboriginal communities for medicinal use until the mid 1970s (Daley *et al.* 2008).

The following threats to the dugongs' seagrass habitats along the urban coast of the Great Barrier Reef region were identified by Grech *et al.* (2011) using expert opinion: agricultural, urban and industrial run-off; urban and port infrastructure development; dredging; shipping accidents; trawling; recreational and commercial boat damage; and commercial fishing other than trawling. The remote tropical waters of most of the remainder of the dugong's range in Australia are subject to very low levels of human impact (Halpern *et al.* 2008) and threats to most dugong habitats in this region are low (Marsh *et al.* 2011). The increase in the number of commodity ports and associated vessel traffic is cause for concern.

Conservation actions

Australia has implemented significant measures to protect dugongs at national, state/territory and local levels as detailed in Marsh *et al.* (2011), even though the species is not listed as threatened under the *Environment Protection and Biodiversity Conservation Act 1999*. The most effective measures are fisheries closures in dugong hotpots (Marsh *et al.* 2011). Nonetheless, the status of the dugong is very variable. The situation is most serious on the urban coast of the Great Barrier Reef region (Marsh *et al.* 2011). This situation was exacerbated by the severe weather events in the summer of 2010/11, which followed several wet summers, further impacting the seagrass that was already in poor condition (McKenzie *et al.* 2012). An aerial survey in 2011 (Sobtzick *et al.* 2012) indicated that the estimated size of the dugong population in this region in November 2011 was the lowest since surveys began in 1986. No calves were seen, suggesting that the dugongs may have stopped breeding because of the seagrass dieback (Marsh *et al.* 2011).

Prognosis

Despite the public and scientific concern about the sustainability of hunting in the northern Great Barrier Reef and Torres Strait (e.g. Marsh *et al.* 1997, 2004; Heinsohn *et al.* 2004), declines have not been detected after 25 years of aerial surveys, probably because most of the high-quality dugong habitat in these regions is not exposed to hunting or most other human impacts (Marsh *et al.* 2011). The status of the dugong in the vast region from the northern tip of Cape York west to the Northern Territory border is unknown due to data deficiency. The population in Shark Bay in Western Australia is likely to be the most secure dugong population in the world because of low levels of human impact and high levels of protection (Marsh *et al.* 2011).

abounds the recovery of blue and sperm whales with little evidence of recovery of the heavily harvested sperm whale population off Albany (Carroll *et al.* 2014). New Zealand enacted its Marine Mammal Protection Act in 1978 and Australia committed to whale protection in 1979 following the recommendations of the Frost inquiry into whales and whaling. Elephant seals were subject to a commercial harvest on Macquarie Island from their discovery in 1810 until 1919 (Hindell & Burton 1988). Fur seals and sea lions were heavily exploited throughout Australia and New Zealands' southern shores and offshore islands from 1792 until full protection in 1948/49 (Harcourt 2001; Ling 2002). Fur seals of both species are now recovering at a rapid rate, but sea lions in Australia and in New Zealand are in a much more precarious state (Kirkwood *et al.* 2005; Shaughnessy *et al.* 2006; See also Boxes 16.3 and 16.4).

While all marine mammals in Australia and New Zealand are now, and have for some time, been subject to legal protection, and the dramatic recovery of a few species is cause for celebration, this should not lead to complacency. Not all populations of marine mammals are stable, and while there have been outstanding success stories (see Box 16.1), there are a number of populations and indeed species that remain in a perilous position. In the following section we outline the major threats to marine mammals in the region, and their potential impacts.

Box 16.3 The southern elephant seal not quite back from a regional brink: recent population stability might indicate a system shift in the Southern Ocean

The issue

The southern elephant seal (*Mirounga leonina*) is a top order predator of the Southern Ocean (Hindell *et al.* 2003). It has a circumpolar distribution, spending most of its time at sea sometimes travelling over large distances to forage and diving deeply, but returns to sub-Antarctic islands or the South American mainland to breed (Biuw *et al.* 2007). There are four genetically distinct groups: (1) the South Georgia stock in the South Atlantic, 98% of which breed on South Georgia with smaller populations on the Falklands, South Shetlands, and South Orkneys; (2) the Kerguelen stock in the southern Indian Ocean where 70% breed on Îles Kerguelen, 25% breed on Heard Island, and smaller populations occur at Marion Island, Prince Edward Island and Îles Crozet; (3) the Macquarie stock in the South Pacific Ocean 99% of which breeds at Macquarie Island with small populations on Campbell Island and Antipodes Island; and (4) the continental South American stock which breed at Peninsula Valdés (Laws 1994; Hoelzel *et al.* 2001).

During the 1950s, southern elephant seal populations declined at many locations (Laws 1994). Most populations within the Kerguelen and Macquarie stocks declined by about 50% or more between 1949 and 1986 (McMahon *et al.* 2005a). At the same time, the larger South Georgia population stayed relatively stable (Boyd *et al.* 1996), while the Peninsula Valdés population increased.

Current status

The population decline to the order of 50% at both Macquarie Island and Heard Island was of sufficient concern that in 2001 the southern elephant seal was listed as 'vulnerable' under the *Environment Protection and Biodiversity Conservation Act 1999* because it was considered likely to be regionally extinct in 100 years (Threatened Species Scientific Committee 2001). At Heard Island the population had declined by 47% between 1949 and 1986 (Slip & Burton 1999), and other islands of the Indian Ocean stock had shown declines of similar or greater magnitude over the same time frame. At Îles Kerguelen the population declined by 48% between 1970 and 1987 (Guinet *et al.* 1999), while the small populations of Marion Island and Îles Crozet declined by 83% and 70% respectively over similar time frames (Laws 1994; Guinet *et al.* 1999). At Macquarie Island the population declined by about 50% between 1949 and 1985 (Hindell & Burton 1987), and the small Campbell Island population declined by about 96% (Laws 1994).

There is now evidence that the Indian Ocean populations at Kerguelen, Crozet, Heard and Marion Islands all stabilised in the 1990s and are now either stable or may be increasing (Slip & Burton 1999; Guinet *et al.* 1999; McMahon *et al.* 2009; Authier *et al.* 2011). The Macquarie Island population declined from 1952 to 1998 at a rate of 1.95% per annum but has been either stable or slightly increasing from 1999 to 2004.

The estimate for the world population is 720 000 animals, with the South Georgia stock contributing 55%, the Kerguelen stock 30%, the Macquarie stock 9% and Peninsula Valdés 6% (McMahon *et al.* 2005a). Globally, the species is listed in the 'least concern' category on the IUCN Red List, largely due to the population at South

Georgia remaining stable from the 1950s and due to the increase in the population at Peninsula Valdés (Campagna 2008).

Knowledge and vulnerability

Concern at the extent of the decline at several breeding sites resulted in the instigation of a range of detailed research targeted at understanding the role of this species in the Southern Ocean ecosystem, with a view to explaining the reasons behind the decline. While a number of hypotheses have been proposed to explain the decline of two of the four discrete sub-populations, the two most plausible are interspecific competition for food resources and environmental change influencing absolute food supply (McMahon *et al.* 2005a). However, understanding the impact of changing climatic conditions on marine predators is difficult because of the lag between the change in the climate variable and the response of the predator, as well as the complex interaction between intrinsic and extrinsic factors as they influence fluctuations in population size (de Little *et al.* 2007).

Southern elephant seals are highly sexually dimorphic, polygamous, colonial breeders and with an annual cycle with two well-defined pelagic phases, with transitions being marked by moult and reproduction (Slip & Burton 1999). Their foraging areas cover both mesopelagic and benthic ecosystems from the northern reaches of the Southern Ocean to high-latitude Antarctic waters pack-ice and sea-ice zones (Biuw *et al.* 2007, 2010). They are opportunistic feeders with a broad foraging niche and their main prey are squid and fish (Bradshaw *et al.* 2003), including some species that are caught commercially (Burton & van den Hoff 2002). Seals that feed pelagically and those with a more northerly range consume predominantly squid and mesopelagic myctophid fish, while fish dominates the diet when seals are over the Antarctic continental shelf (Bradshaw *et al.* 2003). Myctophid fish are the primary component of the diet of juveniles (Newland *et al.* 2011).

The frontal zones, which are major contributors to the production of the Southern Ocean, and the continental shelf and slope areas, the edges of cyclonic eddies, and temperature/salinity gradients under winter pack ice are important foraging habitats of elephant seals (Biuw *et al.* 2007). Seals tend to return to the same general foraging regions year after year even though there is high variation in annual foraging success but they show fidelity to regions that have high long-term productivity (Bradshaw *et al.* 2004).

Southern elephant seals are deep divers and while dive depth and duration varies by season and between the sexes, they typically dive from 300 to 600 m deep and from 20 to 30 minutes in duration, and individuals can dive up to 1700 m and remain submerged for more than one hour (Slip *et al.* 1994).

Survival of first-year pups is influenced by climatic fluctuations and mass at weaning. For example, cooler El Niño years have more sea-ice and hence more prey available which enables females to acquire and store resources for the pup more easily, resulting in pups weaning at a greater mass and hence, are better equipped for longer foraging trips and deeper dives (McMahon & Burton 2005). Small changes in survival and fecundity had dramatic effects on population growth rates. At Macquarie Island for example, a small change (*c.* 5%) in survival and fecundity rates resulted in the population reverting from a decreasing one to a population that increased. The life-history characteristics that had the greatest impact on

Box 16.3 (continued)

population trajectory were, in order of importance: (1) juvenile survival, (2) adult survival, (3) adult fecundity, and (4) juvenile fecundity (McMahon *et al.* 2005b).

Threats

While the recently arrested declines in some populations were of concern for some years, the impacts of fisheries or climatic and oceanographic change on prey availability have been identified as impending or secondary threats (Kovacs *et al.* 2012).

There is no evidence that the declines for the Macquarie and Kerguelen stocks are related to fisheries in the Southern Ocean (SCAR 1991). However, future development or expansion of fisheries at high latitudes have a significant impact on elephant seal populations and must be managed to avoid this (SCAR 1991). While elephant seal entanglements in fishing gear are reported rarely, there have been interactions with trawl fishery, longline fishery and squid fishery that have either resulted in mortality or would have without intervention (Burton & van den Hoff 2002; Campagna *et al.* 2007).

Management

The breeding habitat of southern elephant seals has been well protected. Heard Island and Macquarie Island are protected areas that were inscribed on the World Heritage List in 1997. In additional both islands are surrounded by substantial marine reserves, most of the waters out to 200 nautical miles to the east of Macquarie Island and a 65 000 km marine reserve surrounding Heard Island.

Commercial fishing in the sub-Antarctic is regulated by the international Commission for the Conservation of Antarctic Living Marine Resources (CCAMLR) under the Convention of the same name (CAMLR Convention) which has as its objectives conserving including rational use of marine living resources south of the Antarctic Polar Front through precautionary and ecosystem management (Miller 2007). Elephant seals from all populations spend some time foraging in CCAMLR-managed areas (Biuw *et al.* 2007), and while recent commercial fishery catch within these zones is minimal relative to the prey consumed by elephant seals, increases in fishing activity in these zones may result in competition for marine resources (van den Hoff *et al.* 2002; Hindell *et al.* 2003).

Prognosis

Elephant seals have behavioural flexibility that may allow them to adapt to the changing climate and oceanographic conditions. They have a broad diet and can adapt feeding strategies to take advantage of benthic ecosystems or the diel vertical migrations of the mesopelagic system (Biuw *et al.* 2010).

The two impacts of climate change on southern elephant seals are food availability, through the link between ocean temperature/sea-ice and productivity, and access to breeding areas. One seemingly anomalous result of climate change is the increase in sea-ice in some regions. A probable mechanism is through warmer temperatures causing the ice sheets to melt and this additional cooler melt water reduces sea surface and facilitates an increase in sea-ice (Bintanja *et al.* 2013). This may increase prey availability in the areas where sea-ice is increasing such as the Indian and Pacific sectors of the Southern Ocean. The opposite might occur in the Antarctic Peninsula and the Bellingshausen Sea where increasing sea temperatures are reducing sea-ice extent. But the loss of sea-ice and the retreat of ice fields and glaciers in the Antarctic Peninsula has created additional beach areas for haul outs, moulting and breeding and may be a positive influence on the elephant seal population there.

While the most recent analyses suggest that the Macquarie and Heard/Kerguelen populations are no longer declining, changes in oceanography, climate cycles, sea-ice conditions, and the potential future increase in demand for fisheries products means that this species is surviving in a dynamic environment. As population trends can be detected with certainty only when long-term data are available, and given that the regional decline in population size of around 50% occurred in less than three decades, continued monitoring of populations is justified.

Box 16.4 A precarious future for the New Zealand sea lion?

The issue

The New Zealand sea lion is New Zealand's only endemic seal species. The population at their main breeding area has declined rapidly in the past 15 years. Disease epizootics and fisheries impacts have been identified as the most plausible influences in decline. Management measures to date have been ineffective in reversing the decline and the prognosis for New Zealand sea lions is not positive.

Status

The New Zealand (NZ) sea lion (*Phocarctos hookeri* – formerly known as Hooker's sea lion), is listed as *Nationally Critical* under the NZ threat classification system (Baker *et al.* 2010; equivalent to IUCN's *Critically Endangered*). Historically, NZ sea lions bred along the entire NZ coastline and in the NZ sub-Antarctic Islands. Now, the species breeds predominantly on the sub-Antarctic Auckland and Campbell Islands. In the past 15 years, pup production at the Auckland Islands (>70% of the species) has declined by 44%, from 3021 (1998) to 1683 pups (2012). Adult numbers have also shown a significant decline (Chilvers 2012a). Conversely, Campbell Island pup production appears to be increasing, but survey timing and methodology have varied, hindering a true trend comparison. Auxiliary information, including the extremely high early morality rates of pups (36%–63% at one month), suggest it is unlikely that this population will maintain its present size (Maloney *et al.* 2012).

Knowledge and vulnerability

New Zealand sea lions are sexually dimorphic, polygamous, colonial breeders and have a highly synchronized breeding season. Females give birth to one pup in mid to late December, then divide their time over the next 10 months between foraging at sea (~2 days) and spending time ashore feeding their dependent pup (1.5 days) (Chilvers *et al.* 2005). Males congregate at the sub-Antarctic breeding areas during breeding and then disperse from Macquarie Island to the South Island of NZ (Robertson *et al.* 2006). Compared with other sea lion species, NZ sea lions have

Box 16.4 (continued)

low reproductive rates (Chilvers *et al.* 2010), unusually low mean milk lipid content during early lactation and low pup growth rates (Chilvers *et al.* 2007).

NZ sea lions predominantly forage at the edges of and over the continental shelves near their breeding areas (Chilvers 2009). Females have the deepest and longest dive durations recorded, plus the longest distances travelled during foraging trips, of any otariid (Chilvers *et al.* 2005, 2006). They dive almost continuously when at sea, and their diving behaviour is at or close to their physiological limits (Chilvers *et al.* 2006). Individuals show either benthic (foraging on the sea floor) or mesopelagic (foraging at various, deep, depths in the water column) diving behaviours and show strong fidelity to foraging areas and strategies (i.e. either benthic or mesopelagic) within and between breeding seasons and throughout the year (Chilvers & Wilkinson 2009).

The NZ sea lion has naturally occurring, epizootic mass mortality events (Baker 1999,) and have been affected by two epizootics in the past 15 years (one epizootic lasted two years). Epizootics reduced pup production with 53%, 32% and 21% of pups dying in their first month in 1998, 2002 and 2003, respectively. The 1998 epizootic killed at least 74 adult females. An age-structured VORTEX population viability analysis model indicated that, although naturally occurring epizootics reduce the population growth rate, disease alone could not account for the observed sea lion decline (Chilvers 2012b).

Threats

Robertson & Chilvers (2011) assessed nine possible reasons for the NZ sea lion's decline, including disease epizootics, predation, migration, environmental change, population 'overshoot', genetic effects, effects of contaminants, indirect effects of fisheries (i.e. resource competition) and direct effects of fisheries (i.e. by-catch deaths). Of these, direct and indirect fisheries impacts were identified as the most plausible.

Auckland Island trawl fisheries spatially and temporally overlap completely with female NZ sea lions foraging locations and depths during the breeding and lactation periods (Chilvers 2009). Competition for prey with fisheries and low energy content of available prey are considered factors that could be influencing the reproduction rates, milk fat contents and pup growth rates within NZ sea lion populations (Robertson & Chilvers 2011). Fisheries by-catch is the main anthropogenic cause of mortality known for NZ sea lions, and the majority of mortalities in recent years have been females (Robertson & Chilvers 2011). This increased fisheries mortality is thought to be manifest in the lower survival of adult female sea lions than males (Chilvers & Mackenzie 2010).

Management

Marine mammals in NZ waters are protected under the NZ Marine Mammal Protection Act (1978). Current management specific to NZ sea lions includes a NZ sea lion Species Management Plan (DOC 2009), a 12 nautical mile Marine Mammal Sanctuary and Marine Reserve surrounding the Auckland Islands, by-catch limit controls enforced under the Fisheries Act (1996), and the use of sea lion exclusion devices (SLED). A SLED is a metal grid fixed inside the trawl net that theoretically directs large objects, such as sea lions, to an escape hatch opening in the net (thereby reducing sea lion by-catch) (Chilvers 2009).

Suspected misreporting of sea lion by-catch and the use of SLEDs necessitates modelling of catch rates (estimated as number of sea lions incidentally killed per 100 tows). An annual fishing-related mortality limit (FRML) dictates the fishing effort in the Auckland Island squid trawl fishery (Chilvers 2009). When the FRML is exceeded the fishery is closed under legislation.

Prognosis

Management measures to date have been ineffective with pup production at the Auckland Islands continuing to decline. The marine sanctuary and reserve, while well intended, does not cover the foraging area of any satellite-tracked adult female (Chilvers 2009). It is unknown if SLEDs allow sea lions to escape nets unharmed, or whether the devices passively release sea lions already dead, thereby inadvertently masking sea lion by-catch and compounding the problem (Robertson & Chilvers 2011).

Sea lion management remains controversial. For example, while direct and indirect fisheries impacts are deemed plausible influences on the species (Robertson & Chilvers 2011), the NZ Ministry of Primary Industry concluded that fishing is not impacting the species (www.mpi.govt.nz). Based on this decision, squid fishing effort increased by 140% to 6415 tows in the 2012/13 season amounting to three times the effort seen in any season since 1998 (Robertson & Chilvers 2011). Furthermore, population ecology research and species monitoring has been severely reduced with the result that as yet unknown factors contributing to the decline in pup production of the NZ sea lion will remain undetected. Sadly, the prognosis for the NZ sea lions is not positive.

16.2 Threats to marine mammals in Australasia
16.2.1 Fisheries interactions

Marine mammals risk entanglement or being by-caught in fisheries operations either by being attracted to the nets as a source of food (Shaughnessy & Davenport 1996; Chilvers & Corkeron 2001; see Box 16.4), or by incidentally becoming entangled in fishing gear or debris (Shaughnessy *et al.* 2003; Kemper *et al.* 2005; Figure 16.1, Plate 33). While the great whales (except sperm whales) are usually not attracted to vessels to feed, humpback and southern right whales are at particular risk of accidental entanglement in set nets and lobster lines as a result of their habit of migrating close to shore each year (Kemper *et al.* 2008; Groom & Coughran 2012; Harcourt *et al.* 2014). Humpback whales, dugongs and dolphins are incidentally caught, and sometimes killed, in shark meshing programmes on the east coast of Australia in most years (Gribble *et al.* 1998). Dolphins and seals are even more vulnerable to fishing activities as they readily learn that active fishing can provide an easy meal and are thus attracted to fishing operations. Australian fur seals enter trawl nets in Bass Strait (Shaughnessy & Davenport 1996) and off Tasmania where they are incidentally caught (Hamer & Goldsworthy 2006). Australian sea lions are particularly susceptible to gill net fisheries (Hamer *et al.* 2011). Both Australian sea lions and New Zealand fur seals suffer high entanglement rates in marine debris, to the extent that this appears to be preventing the recovery of Australian sea lions (Goldsworthy & Page 2007).

Figure 16.1 (Plate 33) New Zealand fur seal male resting on the rocks. New Zealand fur seals are subject to significant bycatch in both New Zealand and Australian waters, particularly off the West Coast of Southland, New Zealand and the south west of Tasmania. A black and white version of this figure will appear in some formats. For the colour version, please refer to the plate section.

New Zealand fur seals in New Zealand are frequently by-caught in the Hoki fishery (Harcourt 2001; Dickie & Dawson 2003) and New Zealand sea lions have frequent interactions with the Auckland Island arrow squid fishery (see Box 16.4). Mortalities of common dolphins in interactions with the South Australian fishing industry (Kemper *et al.* 2005; Hamer *et al.* 2008) have been flagged as being of major concern due to levels of population sub-division (Bilgmann *et al.* 2008), and with regional fisheries closures implemented by management agencies in recent years (AFMA 2011, 2012). In New Zealand, Hector's dolphins are subject to mortality in gill nets (Slooten *et al.* 2006) and as a threatening process this may lead to extinction of the *Critically Endangered* Maui's dolphin (Slooten & Dawson 2009; Slooten & Davies 2012). Dugongs are also extremely vulnerable to entanglement in nets such as gill, shark and fishing nets (Marsh *et al.* 1996, 2001, 2005, 2011).

16.2.2 Prey depletion, competition and harvest

The Southern Ocean ecosystem has been profoundly affected by human harvesting of marine mammals, almost certainly producing substantial changes in ecosystem structure (Croxall *et al.* 1992; Trathan *et al.* 2007). This makes it particularly difficult to disentangle population responses to ongoing effects within ecosystems such as krill fishing or climate change from the effects of past exploitation in the region (Croxall *et al.* 1992; Trathan *et al.* 2007). A small harvest of minke (*B. bonaerensis*) and fin whales continues within the Japanese scientific whaling programme. While only a small harvest compared to the early days of whaling, the lack of accepted abundance estimates makes it difficult to determine the long-term effects of this harvest (Leaper & Miller 2011). Modelling suggested that current levels of indigenous harvesting of Torres Strait and Cape York dugongs was unsustainable (Heinsohn *et al.* 2004; Marsh *et al.* 2004) but new data suggest that this assessment may not be correct (see Box 16.2).

16.2.3   Vessel disturbance and collisions

Vessel disturbance can occur in the form of collisions or by disturbance of the behaviour of animals. Southern right whales appear to be the primary species involved in vessel collisions in the Southern Hemisphere, although there are low numbers of recorded strikes in Australasian waters (Van Waerebeek *et al.* 2007). Vessel collisions fall into four categories – indeterminate collisions with the bow or hull of a vessel where the animal suffers blunt trauma; bow bulb draping where animals become wedged on the front of ships; propeller strike; and collisions where animals bump into vessels. There were two fatal vessel collisions and three non-fatal collisions with southern right whales recorded in Australian waters in the period 1950–2006 (Kemper *et al.* 2008). Ship strike is most likely to be fatal in baleen whales, when ships move faster than 14 knots (Laist *et al.* 2001; Leaper & Miller 2011). Ship strike has been identified as an important threat to the local population of Bryde's whales in the Hauraki Gulf, New Zealand (Baker *et al.* 2010). They may also disturb other marine mammals (Lawler *et al.* 2007). Dugongs may cease feeding and move away in response to boats passing within 50 m (Hodgson & Marsh 2007).

16.2.4   Tourism

Tourist activities may change the behaviour of marine mammals. Migrating humpback whales respond to boat approaches from distances as far as 1000 m (Gulessarian *et al.* 2011). As whale watching continues to grow, compliance with regulations worsens potentially compounding effects (Kessler & Harcourt 2013). Repeated exposure of bottlenose dolphins to boats in Port Stephens, south-eastern Australia, may prevent dolphins from maintaining large groups (Möller *et al.* 2002; Steckenreuter *et al.* 2011, 2012). In Shark Bay, Western Australia, an increase in vessel-based dolphin watching tours resulted in a significant decline in numbers of bottlenose dolphins, presumably due to more sensitive animals moving away (Bejder *et al.* 2006). Likewise, travelling Indo-Pacific bottlenose dolphins (*T. aduncus*) in Jervis Bay, New South Wales, alter their behaviour to avoid boats at distances of 100 m (Lemon *et al.* 2006). In New Zealand in recent years, tour boats have started to use auditory stimuli to attract Hector's dolphins to boats with unknown effects (Martinez *et al.* 2012). Boat disturbance also influences resting and suckling behaviour in fur seals, and may cause groups to flee into the water (Boren *et al.* 2002). However, fur seals may eventually become habituated to boat-based ecotourism (Black 2010) and dugongs appear to be relatively unaffected when feeding (Hodgson & Marsh 2007).

16.2.5   Coastal and offshore development

If poorly managed, coastal development can pose a significant threat to inshore dolphins and dugongs. Development in sensitive areas such as prime feeding grounds or calving areas need particular attention (Smith *et al.* 2012). Dredging near ports and harbour entrances alters coastal habitats, and this may result in degradation or complete removal of seagrass beds and mangrove systems (Bannister *et al.* 1996). Likewise, land clearing for urban, agricultural and industrial developments can remove or disturb coastal wetlands and mangrove communities and when they are lost, urban discharge and agricultural runoff may increase, with a resultant degradation of seagrass habitats (Hale 1997). Seagrass is essential forage for dugongs, and seagrass beds and mangrove areas act as nursery grounds for prey for inshore dolphins. Cumulative, permanent loss of seagrass can therefore have bottom-up trophic cascade effects or complete removal of essential habitat.

Contaminants from industrial, human and agricultural waste also produce increased inshore nutrient loads. These pose a particular threat to inshore marine mammal species that occupy coastal regions near major contaminant sources, such as coastal industrial centres, through the bioaccumulation of contaminants within the trophic web (Bannister *et al.* 1996). Moderate to high levels of heavy metals and/or organochlorines have been recorded in several cetacean species, dugongs and Australian fur seals in Australia (Kemper *et al.* 1994), and in Hector's and common dolphins in New Zealand (Stockin *et al.* 2007). It is not known how marine mammals are affected by these pollutants but there is some evidence that they increase susceptibility to diseases (see below and Kemper *et al.* 1994). In addition, entanglement in, and ingestion of, plastic debris is a significant problem, causing injury and death to marine mammals throughout the world (Harcourt *et al.* 1994; Goldsworthy & Page 2007).

16.2.6 Noise interference

Anthropogenic noise in the marine environment is an increasing threat to marine mammals (Hatch & Wright 2007). Loud noises or long exposure can lead to call masking and interruption of communication between individuals. Whales, dolphins and seals are known to respond to high noise conditions by changing the frequency and volume of calls, call duration, and call rate (Nachtigall *et al.* 2003, Kastak *et al.* 2005, Parks *et al.* 2011). Noise may make preferred habitats and migration routes less suitable and, in extreme situations, can cause physical damage (Southall *et al.* 2007). Noise interference is of particular concern within, or near to, aggregation areas where young calves/pups are present or where marine mammals are resident for long periods of time. Noise may also deter recovering populations from establishing aggregations in otherwise suitable but currently unused areas. Potential forms of noise interference in Australasian waters include seismic operations, mining, dredging, infrastructure construction and operation, defence activities, boats, and low flying aircraft. Some forms of noise are temporary (e.g. during construction of offshore facilities) but some forms may effectively be permanent (e.g. ongoing mining operations, established ports, ferry routes)

16.2.7 Petroleum and gas exploration activities

Petroleum and gas exploration including seismic activities, and defence training, are one of the main noise interference risks (Acoustic Ecology Institute 2009). The activity has the potential to affect many individuals at one time. Animals may be exposed at vulnerable periods, such as during calving, pupping and breeding, or may be displaced from major feeding areas. As migratory travel routes to and from the calving/breeding grounds remain unknown for many migratory species, seismic activity may also potentially disrupt migratory routes.

16.2.8 Industrial noise

An increasing potential threat is interference from construction noise from offshore or near-shore renewable energy technologies such as wind farms and tidal turbines, and from seafloor mining and subsea oil and gas processing units (Bailey *et al.* 2010). Noise also comes from activities such as pile driving and dredging (Richardson & Wursig 1997; Wright *et al.* 2007). Some noise is temporary but once established, some operations will continue to produce large amounts of noise in the marine environment. For example, mining units that operate on the ocean floor or in the water column have the potential to produce large amounts of ongoing noise. Attenuation of noise and therefore the scale of any impact will vary depending on the volume and frequency of the sound and the dispersal characteristics of the local environment (Thompson *et al.* 2010).

16.2.9   Shipping noise

Chronic noise exposure is primarily due to increased shipping activity, including the use of tender vessels and port development. The total distance travelled by trading vessels at sea in the Australian EEZ is estimated as 58 million ship-km per year with a projected annual growth estimated at 3.6% (AMSA 2011). Although substantially lower in frequency than the Northern Hemisphere, growth in shipping movements in Australian waters are highest in areas where we have least information on species assemblages and composition (e.g. northwest Australia and northeast Australia).

16.2.10   Aircraft noise

Low flying aircraft, such as those used for scenic tours or naval exercises, propagate large amounts of sound along the ocean surface and into the water column. The volume and extent of propagation varies depending on the type of aircraft and the length of time the aircraft is in the area. Aircraft sounds may displace feeding whales (Croll *et al.* 2001) and cause fur seals to stampede into the water, sometimes with catastrophic mortalities of young animals (Richardson & Thomson 1998). Aircraft could have an impact in areas of high aggregation or where there is repeated exposure, especially when whales are spending significant time at the surface, such as mother and calf humpback or right whales in calving and resting areas, or blue whales in feeding areas such as the Bonney Upwelling in Victoria (Gill *et al.*2012).

16.2.11   Climate change

Projected warmer air and ocean temperatures are likely to have profound effects on marine mammals. The ambient environment can have significant influences on breeding behaviour in pinnipeds (Carey 1991) with the potential for major changes in reproductive skew as climate change alters patterns of sunshine and precipitation (Twiss *et al.* 2007). Warming seas are also likely to affect distribution through influences on thermal tolerance. The ranges of species currently associated with tropical to temperate waters are likely to expand southwards (Lawler *et al.* 2007; Trathan *et al.* 2007; Nicol *et al.* 2008; MacLeod 2009), while species with ranges limited by warm and cold water boundaries such as the Antarctic Polar Front are likely to shift southwards (MacLeod 2009). This may result in a loss of diversity in tropical cetacean communities, and an increase in diversity at higher latitudes (Whitehead *et al.* 2008). For coastal species inhabiting the continental shelves of Australia and New Zealand, there are constraints on how far south their ranges may shift. Given that biological productivity is also likely to be affected by the predicted warming scenarios, significant declines may occur in southern areas bounded by deep waters such as Tasmania and South Island, New Zealand (MacLeod *et al.* 2005). Even for pelagic species, such as southern right whales, it is possible that their feeding grounds may shift southward, increasing their migratory path and therefore the sum energetic cost of their reproductive cycle.

In the Southern Ocean ambient temperatures are likely to have a profound influence on marine mammals due to the predicted effects on sea-ice extent and the important link between sea-ice and productivity. Krill stocks in the South Atlantic have decreased in accordance with loss of sea-ice resulting from the profound warming in that sector (Atkinson *et al.* 2004). Many baleen whales feed near, or within, the Antarctic pack ice (Nicol *et al.* 2008) as do the ice seals including southern elephant seals (Biuw *et al.* 2010). Any decrease in the extent of sea-ice with increased air temperatures is likely to have a detrimental impact on these species (Nicol *et al.* 2008). Rorquals such as blue and fin whales may be particularly vulnerable to changes in prey abundance since the high costs of their lunge-feeding behaviour confines them to areas of dense aggregations of krill (Acevedo-Gutiérrez *et al.* 2002). These costs may well flow through to fitness, and recent modelling of reductions in krill abundance in response to sea-ice extent suggests that birth rates in blue whales may decline as a result of reduced krill availability (Wiedenmann *et al.* 2011). Warmer temperatures may have direct effects on fecundity and recruitment regardless of ice association. Studies on Macquarie Island have shown increased survival in southern elephant seal pups during El Niño events probably because the more abundant sea-ice increased food availability (McMahon & Burton 2005). Fur seals and sea lions in general are vulnerable to extreme climatic events such as El Niño, with decreased recruitment and increased mortality (Trillmich *et al.* 1991). Climate events have already been linked to the frequency of cetacean strandings in southern Australia (Evans *et al.* 2005), and dugongs are vulnerable to stranding during cyclones (Marsh 1989; Marsh *et al.* 2011). Moreover, as anthropogenic climate change projections indicate that the intensity of storms and cyclones will increase, the associated rainfall and flooding may enhance transportation of pathogens and pollutants into coastal waters potentially affecting or even displacing marine mammals (Lawler *et al.* 2007; Fury & Harrison 2011). Declines in the abundance and extent of seagrasses can also be expected under projected climatic changes, particularly in shallow waters, due to storms (Connelly 2009). This is likely to adversely affect marine mammal species which are associated with seagrass habitats, such as dugongs and snubfin dolphins (Marsh & Kwan 2008; Parra 2006; Marsh *et al.* 2011).

16.2.12 Disease

Many marine mammal species are susceptible to mass mortalities due to disease outbreaks, as they are colonial during vulnerable periods such as breeding, facilitating the spread of infectious disease (Harwood & Hall 1990). Moreover, marine mammals have long lifespans, feed at a high trophic level, and have extensive fat stores that can serve as depots for anthropogenic toxins, leading potentially to increased vulnerability to disease. Newly described or re-emerging disease agents or diseases affecting marine mammals include various papillomaviruses, dolphin poxvirus, and other viral infections; lobomycosis; various neoplastic diseases; algal bloom biointoxications; manatee cold stress syndrome; and the idiopathic cardio-myopathy of pygmy and dwarf sperm whales (Bossart 2007). In the USA, recent pathologic studies of dolphins and manatees indicate that these emerging or resurging infectious and neoplastic diseases may reflect environmental distress (Bossart 2006).

Within the Australasian region, marine mammals also appear vulnerable. Mass die-offs of New Zealand sea lions have occurred twice in the last decade, further increasing their vulnerability to extirpation (see Box 16.4). Morbilliviruses have been implicated in mass dolphin dieoffs in several Australian states with the first confirmed case reported in 2011 (Stone *et al.* 2011). As climate change induces thermal stress on marine mammals, particularly those at the edge of their range, the physiological stress may compromise host resistance and increase the frequency of opportunistic diseases (Harvell *et al.* 1999). Given that new diseases have typically emerged through host and/or range shifts of known pathogens, the synergy of increased exposure to human activities and climate-mediated stress, means that disease events are likely to be of increased importance.

16.3 Conclusions

Marine mammals in Australia and New Zealand have gone through dramatic changes in their interactions with humans. Some groups, particularly the great whales and the pinnipeds, were severely exploited and extirpated from much of their ranges. In the last few decades these groups have responded to a high level of protection and some have shown great recovery. These recoveries show that with removal of significant threats, at least some marine mammal species exhibit substantial resilience. However, we cannot afford to be complacent. Not all marine mammal populations have responded to protection and new and arising threats have pushed some protected populations to the brink. Moreover the extent to which we are modifying our marine environment shows no sign of abatement, especially as human population growth heightens the ever-increasing demand for resources from the natural environment.

Critically, threatening processes do not act in isolation, and to understand their synergies is a critical first step to mitigation. While our understanding of the scale of interactions between multiple threats is poor, strategic amelioration of manageable threats will almost certainly add resilience to the most vulnerable species and populations. We suggest that despite best efforts, refinement of policies designed to protect and assist marine mammals to adapt to changing environmental conditions is still hampered by a lack of knowledge. Most of the marine mammals in Australasian waters are data deficient, including animals found in areas of major development. Enhanced information on trends in abundance and habitat requirements in relation to anthropogenic use of the marine environment is urgently required. But with long-lived animals, many of which inhabit extensive ranges at low densities, collecting precise data on demography and abundance trends is necessarily a long-term and difficult process. We may not have the luxury of waiting that long. Given the paucity of current knowledge and the rate of environmental change, new approaches are required. Using available data in new and innovative ways may enhance our ability to avert further species loss. Torres *et al.* (2013) show the utility of incorporating historical exploitation data into species distribution models to inform management and help combat biodiversity loss. Enhanced information combined with adoption of the precautionary principle where likely or known threats overlap marine mammal ranges are our best options for continued existence of all of our marine mammal species.

REFERENCES

Acevedo-Gutiérrez, A., Croll, D. A. & Tershy, B. R. (2002). High feeding costs limit dive time in the largest whales. *Journal of Experimental Biology*, **205**, 1747–1753.

Acoustic Ecology Institute (2009). *Ocean Noise 2009 Science, Policy, Legal Developments*. Acoustic Ecology Institute, Santa Fe, 45 pages.

AFMA (2011). *Protecting Marine Wildlife in the Southern and Eastern Scalefish and Shark Fishery* (27 April 2011). Australian Fisheries Management Authority (AFMA), Australian Government, Canberra. http://www.afma.gov.au/wp-content/uploads/2011/04/media_release_27apri.pdf

AFMA (2012). *Australian Sea Lion Bycatch Triggers – Changes to Fisheries Management Arrangements to Further Protect Australian Sea Lion Sub-populations in the Gillnet, Hook and Trap Fishery*. Australian Fisheries Management Authority (AFMA), Australian Government, Canberra. http://www.afma.gov.au/wp-content/uploads/2012/01/Revised-ASL-Bycatch-Triggers-and-Zones.pdf (accessed 26 April 2012).

Agnew, D. J. (1997). Review – The CCAMLR ecosystem monitoring programme. *Antarctic Science*, **9**(03), 235–242.

Atkinson, A., Siegel, V., Pakhomov, E. & Rothery, P. (2004). Long-term decline in krill stock and increase in salps within the Southern Ocean. *Nature*, **432**(7013), 100–103.

Australian Maritime Safety Authority (AMSA) (2011). Shipping and Offshore activity data for Australian ports and waters 14 Dec 2011.

Authier, M., Delord, K. & Guinet, C. (2011). Population trends of female Elephant Seals breeding on the Courbet Peninsula, îles Kerguelen. *Polar Biology*, **34**(3), 319–328.

Bailey, H., Senior, B., Simmons, D., *et al.*(2010). Assessing underwater noise levels during pile-driving at an offshore windfarm and its potential effects on marine mammals. *Marine Pollution Bulletin*, **60**, 888–897.

Baker, A. (1999). *Unusual Mortality of the New Zealand Sea Lion*, Phocarctos hookeri, *Auckland Islands, January – February 1998*. Department of Conservation, Wellington.

Baker, C. S., Chilvers, B. L., Constantine, R., *et al.*(2010). Conservation status of New Zealand marine mammals (suborders Cetacea and Pinnipedia), 2009. *New Zealand Journal of Marine and Freshwater Research*, **44**(2), 101–115.

Bannister, J. L. (1964). Australian whaling 1963, catch results and research. *CSIRO Division of Fisheries and Oceanography Reports*, **38**, 13 pages.

Bannister, J. L. (1995). Report on aerial survey and photoidentification of humpback whales off Western Australia, 1994. Report to the *Australian Nature Conservation Agency (unpublished)*, 17 pages. [Available from Dept of Environment and Heritage, PO Box 787, Canberra, Australia.]

Bannister, J. L. & Hedley, S. L. (2001). Southern Hemisphere Group IV humpback whales: their status from recent aerial surveys. *Memoirs of the Queensland Museum*, **47**, 587–598.

Bannister, J. L., Kirkwood, G. P. & Wayte, S. E. (1991). Increase in humpback whales off Western Australia. *Reports of the International Whaling Commission*, **41**, 461–465.

Bannister, J. L., Kemper, C. M. & Warneke, R. M. (1996). *The Action Plan for Australian Cetaceans*. Australian Nature Conservation Agency, Canberra.

Beasley, I., Robertson, K. M. & Arnold, P. (2006). Description of a new dolphin, the Australian snubfin dolphin *Orcaella heinsohni* sp. n.(Cetacea, Delphinidae). *Marine Mammal Science*, **21**(3), 365–400.

Bejder, L., Samuels, A., Whitehead, H., *et al.* (2006). Decline in relative abundance of bottlenose dolphins exposed to long-term disturbance. *Conservation Biology*, **20**, 1791–1798.

Best, P. B. (1993). Increase rates in severely depleted stocks of baleen whales. *ICES Journal of Marine Science*, **50**, 169–186

Bilgmann, K., Möller, L. M., Harcourt, R. G., Gales, R. & Beheregaray, L. B. (2008). Common dolphins subject to fisheries impacts in Southern Australia are genetically differentiated: implications for conservation. *Animal Conservation*, **11**(6), 518–528.

Bintanja, R., van Oldenborgh, G. J., Drijfhout, S. S., Wouters, B. & Katsman, C. A. (2013). Important role for ocean warming and increased ice-shelf melt in Antarctic sea-ice expansion. *Nature Geoscience*, **6**, 376–379.

Biuw, M., Boehme, L., Guinet, C., *et al.* (2007). Variations in behaviour and condition in a Southern Ocean top predator in relation to *in situ* oceanographic conditions. *Proceedings of the National Academy of Sciences*, **104**, 13705–13710.

Biuw, M., Nøst, O. A., Stien, A., *et al.* (2010). Effects of hydrographic variability on the spatial, seasonal and diel diving patterns of Southern Elephant Seals in the Eastern Weddell Sea. *PLoS ONE*, **5**, e13816.

Black, J. J. (2010), Behavioural responses of Australian fur seals to boat-based ecotourism. Msc thesis, School of Life and Environmental Sciences, Deakin University.

Boren, L. J., Gemmell, N. J. & Barton, K. J. (2002). Tourist disturbance on New Zealand fur seals *Arctocephalus forsteri*. *Australian Mammalogy*, **24**, 85–95.

Bossart, G. D. (2006). Marine mammals as sentinel species for oceans and human health. *Oceanography*, **19**(2), 134–137.

Bossart, G. D. (2007). Emerging diseases in marine mammals: from dolphins to manatees. *Microbe-American Society for Microbiology*, **2**(11), 544–549.

Boyd, I. L., Walker, T. R. & Poncet, J. (1996). Status of southern elephant seals at South Georgia. *Antarctic Science*, **8**, 237–244

Bradshaw, C. J., Hindell, M. A., Best, N. J., *et al.* (2003). You are what you eat: describing the foraging ecology of southern elephant seals (*Mirounga leonina*) using blubber fatty acids. *Proceedings of the Royal Society of London. Series B: Biological Sciences*, **270**(1521), 1283–1292.

Bradshaw, C. J., Hindell, M. A., Sumner, M. D. & Michael, K. J. (2004). Loyalty pays: potential life history consequences of fidelity to marine foraging regions by southern elephant seals. *Animal Behaviour*, **68**(6), 1349–1360.

Brandao, A., Butterworth, D. S. & Brown, M. R. (2000). Maximum possible humpback whale increase rates as a function of biological parameter values. *Journal of Cetacean Research and Management, (Suppl.)*, **2**, 192–193.

Burton, H. & van den Hoff, J. (2002). Humans and the southern elephant seal *Mirounga leonina*. *Australian Mammalogy*, **24**(1), 127–139.

Campagna, C. (IUCN SSC Pinniped Specialist Group) (2008). *Mirounga leonina*. In: IUCN 2012. IUCN Red List of Threatened Species. Version 2012.2. www.iucnredlist.org. Downloaded on 10 April 2013.

Campagna, C., Falabella, V. & Lewis, M. (2007). Entanglement of southern elephant seals in squid fishing gear. *Marine Mammal Science*, **23**, 414–418.

Carey, P. W. (1991). Resource-defense polygyny and male territory quality in the New Zealand fur seal. *Ethology*, **88**, 63–79.

Carroll, E., Patenaude, N., Alexander, A., *et al.* (2011). Population structure and individual movement of southern right whales around New Zealand and Australia. *Marine Ecology Progress Series*, **432**, 257–268.

Carroll, G., Hedley, S., Bannister, J., Ensor, P. & Harcourt, R. (2014). No evidence for recovery in the population of sperm whale bulls off Western Australia, 30 years since the cessation of whaling. *Endangered Species Research*, **24**, 33–43.

Charlton-Robb, K., Gershwin, L. A., Thompson, R., *et al.* (2011). A new dolphin species, the Burrunan dolphin *Tursiops australis* sp. endemic to Southern Australian coastal waters. *PloS ONE*, **6**(9), e24047

Chilvers, B. L. (2009). Foraging locations of a decreasing colony of New Zealand sea lions (*Phocarctos hookeri*). *New Zealand Journal of Ecology*, **33**, 106–113.

Chilvers, B. L. (2012a). Life-history traits of New Zealand sea lions, Auckland Islands, during a period of significant pup production decline. *Journal of Zoology, London*, **287**, 240–249.

Chilvers, B. L. (2012b). Population viability analysis of New Zealand sea lions, Auckland Islands, New Zealand's sub-Antarctic's: assessing relative impacts and uncertainty. *Polar Biology*, **35**, 1607–1615.

Chilvers, B. L. & Corkeron, P. J. (2001). Trawling and bottlenose dolphins' social structure. *Proceedings of the Royal Society of London. Series B: Biological Sciences*, **268**(1479), 1901–1905.

Chilvers, B. L. & Mackenzie, D. (2010). Age and sex specific survival estimates incorporating tag loss for New Zealand sea lions, *Phocarctos hookeri*. *Journal of Mammology*, **91**, 758–767.

Chilvers, B. L. & Wilkinson, I. S. (2009). Divers foraging strategies in lactating New Zealand sea lions. *Marine Ecology Progress Series*, **378**, 299–308.

Chilvers, B. L., Wilkinson, I. S. Duignan, P. J. & Gemmell, N. J. (2005). Identifying the distribution of summer foraging areas for lactating New Zealand sea lions, *Phocarctos hookeri*. *Marine Ecology Progress Series*, **304**, 235–247.

Chilvers, B. L., Wilkinson, I. S. Duignan, P. J. & Gemmell, N. (2006). Diving to extremes: are New Zealand sea lions (*Phocarctos hookeri*) pushing their limits in a marginal habitat? *Journal of Zoology, London*, **269**, 233–241.

Chilvers, B. L., Robertson, B. C., Wilkinson, I. S. & Duignan, P. (2007). Growth and survival of New Zealand sea lions, *Phocarctos hookeri*: birth to 3 months. *Polar Biology*, **30**, 459–469.

Chilvers, B. L., Wilkinson, I. S. & McKenzie, D. (2010). Predicting life-history traits for female New Zealand sea lions, *Phocarctos hookeri*: integrating short-term mark-recapture data and population modelling. *Journal of Agricultural, Biological and Ecological Statistics*, **15**, 259–264.

Chittleborough, R. G. (1965). Dynamics of two populations of the humpback whale, *Megaptera novaeangliae* (Borowski). *Australian Journal of Marine and Freshwater Research*, **16**, 33–128.

Clapham, P., Mikhalev, Yu, Franklin, W., *et al.* (2009). Catches of humpback whales by the Soviet Union and other nations in the Southern Ocean, 1947–1973. *Marine Fisheries Review*, **71**, 39–43.

Connelly, R. (2009). Seagrass. In *A Marine Climate Change Impacts and Adaptation Report Card for Australia 2009* (Eds. E. S. Poloczanska, A. J. Hobday & A. J. Richardson). NCCARF Publication 05/09, ISBN 978-921609-03-9.

Croll, D. A., Clark, C. W., Calambokidis, J., Ellison, W. T. & Tershy, B. R. (2001). Effect of anthropogenic low-frequency noise on the foraging ecology of Balaenoptera whales. *Animal Conservation*, **4**, 13–27.

Crouch, J., McNiven, I. J., David, B., Rowe, C. & Weisler, M. I. (2007). Berberass: Marine resource specialisation and environmental change in Torres Strait during the past 4000 years. *Archaeology in Oceania*, **42**, 49–64.

Croxall, J. P., Callaghan, T., Cervellati, R. & Walton, D. W. H. (1992). Southern Ocean environmental changes: effects on seabird, seal and whale populations [and discussion]. *Philosophical Transactions of the Royal Society of London. Series B: Biological Sciences*, **338**, 319–328.

Daley, B., Griggs, P. & Marsh, H. (2008). Exploiting marine wildlife in Queensland: The commercial dugong and marine turtle fisheries, 1847–1969. *Australian Economic History Review*, **48**, 227–265.

Dawbin, W. H. (1966). The seasonal migratory cycle of Humpback whales.. In: *Whales, Dolphins and Porpoises*, (Ed. K. S. Norris), pp. 145–170. University of California Press: Berkeley and Los Angeles.

de Little, S. C., Bradshaw, C. J., McMahon, C. R. & Hindell, M. A. (2007). Complex interplay between intrinsic and extrinsic drivers of long-term survival trends in southern elephant seals. *BMC Ecology*, **7**, 3.

Dickie, G. S. & Dawson, S. M. (2003). Age, growth and reproduction of New Zealand fur seals. *Marine Mammal Science*, **19**, 173–185

DOC (2009). *New Zealand Sea Lion Species Management Plan: 2009–2014*. Department of Conservation, Science & Technical Report. http://www.doc.govt.nz/documents/science-and-technical/sap251entire.pdf

Evans, K., Thresher, R., Warneke, R. M., *et al.* (2005). Periodic variabiltiy in cetacean strandings: links to large-scale climate events. *Biology Letters*, **1**, 147–150.

Frère, C., Seddon, J., Palmer, C., Porter, L. J. & Parra, G. J. (2011). Multiple lines of evidence for an Australasian geographic boundary in the Indo-Pacific humpback dolphin (*Sousa chinensis*): population or species divergence? *Conservation Genetics*, **1**, 1–6.

Fury, C. A. & Harrison, P. L. (2011). Impact of flood events on dolphin occupancy patterns. *Marine Mammal Science*, **27**, E185–E205.

Gales, N. J., Shaughnessy, P. D. & Dennis, T. E. (1994). Distribution, abundance and breeding cycle of the Australian sea lion *Neophoca cinerea* (Mammalia: Pinnipedia). *Journal of Zoology, London*, **234**, 353–370.

Gill, P. C., Morrice, M. G., Page, B., *et al.* (2012). Blue Whale habitat selection and within-season distribution in a regional upwelling system off southern Australia. *Marine Ecology Progress Series*, **421**, 243–263.

Goldsworthy, S. D. & Page, B. (2007). A risk-assessment approach to evaluating the significance of seal bycatch in two Australian fisheries. *Biological Conservation*, **139**, 269–285.

Grech, A., Coles, R. & Marsh, H. (2011). A broad-scale assessment of the risk to coastal seagrasses from cumulative threats. *Marine Policy*, **35**, 560–567.

Gribble, N. A., McPherson, G. & Lane, B. (1998). Effect of the Queensland Shark Control Program on non-target species: whale, dugong, turtle and dolphin: a review. *Marine and Freshwater Research*, **49**, 645–651.

Groom, C. & Coughran, D. (2012). Entanglements of baleen whales off the coast of Western Australia between 1982 and 2010: patterns of occurrence, outcomes and management responses. *Pacific Conservation Biology*, **18**(3), 203.

Guinet, C., Jouventin, P. & Weimerskirch, H. (1999). Recent population change of the southern elephant seal at Îles Crozet and Îles Kerguelen: the end of the decrease? *Antarctic Science*, 11, 193–197.

Gulesserian, M., Slip, D., Heller, G. & Harcourt, R. (2011). Modelling the behaviour state of humpback whales *Megaptera novaeangliae* in response to vessel presence off Sydney, Australia. *Endangered Species Research*, **15**(3), 255–264.

Hagihara, R., Jones, R., Sheppard, J., *et al.* (*in press*). Improving population estimates by quantifying diving and surfacing patterns: a dugong example. *Marine Mammal Science*.

Hale, P. (1997). Conservation of inshore dolphins in Australia. *Asian Marine Biology*, **14**, 83–91.

Halpern, B. S., Walbridge, S., Selkoe, K. A. *et al.* (2008). A global map of human impact on marine ecosystems. *Science*, **319**, 948–952.

Hamer, D. J. & Goldsworthy, S. D. (2006). Seal-fishery operational interactions: identifying the environmental and operational aspects of a trawl fishery that contribute to by-catch and mortality of Australian fur seals (*Arctocephalus pusillus doriferus*). *Biological Conservation*, **130**, 517–529.

Hamer, D. J., Ward, T. M. & McGarvey, R. (2008). Measurement, management and mitigation of operational interactions between the South Australian Sardine Fishery and short-beaked common dolphins (*Delphinus delphis*). *Biological Conservation*, **141**, 2865–2878.

Hamer, D. J., Ward, T. M., Shaughnessy, P. D. & Clark, S. R. (2011). Assessing the effectiveness of the Great Australian Bight Marine Park in protecting the endangered Australian sea lion *Neophoca cinerea* from bycatch mortality in shark gillnets. *Endangered Species Research*, **14**, 203–216.

Harcourt, R., Aurioles, D. & Sanchez, J. (1994). Entanglement in man-made debris of California sea lions at Los Islotes, Bay of La Paz, Mexico. *Marine Mammal Science*, 10, 122–125.

Harcourt, R. G. (2001). Advances in New Zealand mammalogy 1990–2000: pinnipeds. *Journal of the Royal Society of New Zealand*, **31**, 135–160.

Harcourt, R. G., Pirotta, V., Heller, G., Peddemors, V. & Slip, D. (2014). Whale alarms fail to deter migrating humpback whales: an empirical test. *Endangered Species Research*. doi: 10.33541esr00614.

Harvell, C. D., Kim, K., Burkholder, J. M., *et al.* (1999). Emerging marine diseases – climate links and anthropogenic factors. *Science*, **285**, 1505–1510.

Harwood, J. & Hall, A. (1990). Mass mortality in marine mammals: its implications for population dynamics and genetics. *Trends in Ecology & Evolution*, **5**, 254–257.

Hatch, L. & Wright, A. (2007). A brief review of anthropogenic sound in the oceans. *International Journal of Comparative Psychology*, **20**, 121–133.

Hedley, S. L., Bannister, J. L. & Dunlop, R. A. (2011). Abundance estimates of Breeding Stock 'D' Humpback Whales from aerial and land-based surveys off Shark Bay, Western Australia, 2008. *Journal of Cetacean Research and Management, Humpback Whale Special Issue*, **3**, 209–221.

Heinsohn, R., Lacy, R. C., Lindenmayer, D. B., *et al.* (2004). Unsustainable harvest of dugongs in Torres Strait and Cape York (Australia) waters: two case studies using population viability analysis. *Animal Conservation*, **7**, 417–425.

Hindell, M. A. & Burton, H. R. (1987). Past and present status of the southern elephant seal (*Mirounga leonina*) at Macquarie Island. *Journal of Zoology*, **213**(2), 365–380.

Hindell, M. A. & Burton, H. R. (1988). The history of the elephant seal industry at Macquarie Island and an estimate of the pre-sealing numbers. *Papers and Proceedings of the Royal Society of Tasmania*, **122**, 159–176.

Hindell, M. A., Bradshaw, C. J. A., Sumner, M. D., Michael, K. J. & Burton, H. R. (2003). Dispersal of female southern elephant seals and their prey consumption during the Austral summer: relevance to management and oceanographic zones. *Journal of Applied Ecology*, **40**, 703–715.

Hoelzel, A. R., Campagna, C. & Arnbom, T. (2001). Genetic and morphometric differentiation between island and mainland southern elephant seal populations. *Proceedings of the Royal Society of London. Series B: Biological Sciences*, **268**, 325–332.

Hodgson, A. J. & Marsh, H. (2007). Response of dugongs to boat traffic: the risk of disturbance and displacement. *Journal of Experimental Marine Biology and Ecology*, **340**, 50–61.

Hodgson, A. J., Marsh, H., Gales, N., Holley, D. K. & Lawler, I. (2008). *Dugong Population Trends Across Two Decades in Shark Bay, Ningaloo Reef and Exmouth Gulf*. Denham, Western Australia: WA Department of Environment and Conservation. 38 pages.

International Whaling Commission (IWC) (2006). *Report of the Workshop on the Comprehensive Assessment of Southern Humpback Whales*. Paper SC/58/Rep 5 presented to the IWC Scientific Committee, June 2006 (unpublished), 77 pages.

IUCN (2011). 'IUCN Red List of Threatened Species (Version 2011.2).' Available at http://www.iucnredlist.org/ [accessed 10 March 2012].

Jackson, J. A., Patenaude, N. J., Carroll, E. L. & Baker, C. S. (2008). How few whales were there after whaling? Inference from contemporary mtDNA diversity. *Molecular Ecology*, **17**, 236–251.

Jefferson, T. A. & Rosenbaum, H. C. (2014). Taxonomic revision of the humpback dolphins (*sousa* spp.), and description of a new species from Australia. *Marine Mammal Science*. doi:10.1111/MMS-12152.

Kastak, D., Southall, B. L., Schusterman, R. J., & Kastak, C. R. (2005). Underwater temporary threshold shift in pinnipeds: Effects of noise level and duration. *The Journal of the Acoustical Society of America*, **118**, 3154.

Kemper, C., Gibbs, P., Obendorf, D., Marvanek, S. & Lenghaus, C. (1994). A review of heavy metal and organochlorine levels in marine mammals in Australia. *The Science of the Total Environment*, **154**, 129–139.

Kemper, C. M., Flaherty, A., Gibbs, S. E., *et al.* (2005). Cetacean captures, strandings and mortalities in South Australia 1881–2000, with special reference to human interactions. *Australian Mammalogy*, **27**, 37–47.

Kemper, C., Coughran, D., Warneke, R., *et al.* (2008). Southern right whale (*Eubalaena australis*) mortalities and human interactions in Australia, 1950–2006. *Journal of Cetacean Research and Management*, **10**(1), 1–8.

Kessler, M. & Harcourt, R. G. (2013). Whale watching regulation compliance trends and the implications for management off Sydney, Australia. *Marine Policy*, **42**, 14–19.

Kirkwood, R., Gales, R., Terauds, A., *et al.* (2005). Pup production and population trends of the Australian fur seal (*Arctocephalus pusillus doriferus*). *Marine Mammal Science*, **21**, 260–282.

Kovacs, K. M., Aguilar, A., Aurioles, D., *et al.* (2012). Global threats to pinnipeds. *Marine Mammal Science*, **28**(2), 414–436.

Laist, D. W., Knowlton, A. R., Mead, J. G., Collet, A. S. & Podesta, M. (2001). Collisions between ships and whales. *Marine Mammal Science*, **17**, 35–75.

Lawler, I. R., Parra, G. & Noad, M. (2007). Vulnerability of marine mammals in the Great Barrier Reef to climate change. In *Climate Change and the Great Barrier Reef: a Vulnerability Assessment*. (Eds. J. E. Johnson, & A. Marshall), pp. 497–513. Great Barrier Reef Marine Park Authority: Townsville, Queensland.

Laws, R. M. (1994). History and present status of southern elephant seal populations. In: Le Boeuf, B. J. & Laws, R. M. (eds.) *Elephant Seals: Population Ecology, Behavior, and Physiology*. University of California Press, Berkeley, pp. 49–65.

Leaper, R. & Miller, C. (2011). Management of Antarctic baleen whales amid past exploitation, current threats and complex marine ecosystmes. *Antarctic Science*, **23**, 503–529.

Lemon, M., Lynch, T. P., Cato, D. H. & Harcourt, R. G. (2006). Response of travelling bottlenose dolphins (*Tursiops aduncus*) to experimental approaches by a powerboat in Jervis Bay, New South Wales, Australia. *Biological Conservation*, **127**, 363–372.

Ling, J. K. (2002). Impact of colonial sealing on seal stocks around Australia, New Zealand and subantarctic islands between 150 and 170 degrees East. *Australian Mammalogy*, **24**, 117–126.

MacLeod, C. D. (2009). Global climate change, range changes and potential implications for the conservation of marine cetaceans: a review and synthesis. *Endangered Species Research*, **7**, 125–136.

MacLeod, C. D., Bannon, S. M., Pierce, G. J., *et al.* (2005). Climate change and the cetacean community of north-west Scotland. *Biological Conservation*, **124**, 477–483.

Maloney, A., Chilvers, B. L., Muller, C. G. & Haley, M. (2012). Increasing pup production of New Zealand sea lions at Campbell Island/Motu Ihupuku: can it continue? *New Zealand Journal of Zoology*, **39**, 19–29.

Marsh, H. (1989). Mass stranding of dugongs by a tropical cyclone in Northern Australia. *Marine Mammal Science*, **5**, 78–84.

Marsh, H. (1996). Progress towards the sustainable use of dugongs by indigenous peoples in Queensland. In: *The Sustainable Use of Wildlife by Aboriginal and Torres Strait Islander People*, pp. 139–151.

Marsh, H. & Kwan, D. (2008). Temporal variability in the life history and reproductive biology of female dugongs in Torres Strait: the likely role of sea grass dieback. *Continental Shelf Research*, **28**, 2152–2159.

Marsh, H. & Sinclair, D. F. (1989). Correcting for visibility bias in strip transect aerial surveys of aquatic fauna. *Journal of Wildlife Management*, **53**(4), 1017–1024.

Marsh, H., Corkeron, P., Lawler, I. R., Lanyon, J. M. & Preen, A. R. (1996). *The Status of the Dugong in the Southern Great Barrier Reef Marine Park*. GBRMPA Research Publication No. 41, Great Barrier Reef Marine Park Authority, Townsville, Queensland.

Marsh, H., Harris, A. N. M. & Lawler, I. R. (1997). The sustainability of the indigenous dugong fishery in Torres Strait, Australia/Papua New Guinea. *Conservation Biology*, **11**, 1375–1386.

Marsh, H., De'ath, G., Gribble, N., Lane, B. & Lawler, I. R. (2001). *Shark Control Records Hindcast Serious Decline in Dugong Numbers off the Urban Coast of Queensland/Dugong Distribution and Abundance in the Southern Great Barrier Reef Marine Park and Hervey Bay: Results of an Aerial Survey in October-December 1999*. Great Barrier Reef Marine Park Authority.

Marsh, H., Penrose, H., Eros, C. & Hugues, J. (2002). *The Dugong* (Dugong dugon) *Status Reports and Action Plans for Countries and Territories in its Range. Early Warning and Assessment Reports*. Nairobi: United Nations Environment Programme. 162 pages.

Marsh, H., Lawler, I. R., Kwan, D., *et al.* (2004). Aerial surveys and the potential biological removal technique indicate that the Torres Strait dugong fishery is unsustainable. *Animal Conservation*, **7**, 435–443.

Marsh, H., De'ath, G., Gribble, N. & Lane, B. (2005). Historical marine population estimates: triggers or targets for conservation? The dugong case study. *Ecological Applications*, **15**, 481–492.

Marsh, H., Hodgson, A., Lawler, I., Grech, A. & Delean, S. (2007). Condition, status and trends and projected futures of the dugong in the Northern Great Barrier Reef and Torres Strait; including identification and evaluation of the key threats and evaluation of available management option to improve its status. *Final Report to Marine and Tropical Science Research Facility*, 73 pp. Downloaded from the internet 10 October 2010. http://www.rrrc.org.au/publications/downloads/141-JCU-Marsh-2007-NGBR–Torres-Strait-Final-Report.pdf.

Marsh, H., O'Shea, T. J. & Reynolds, J. R. (2011). *The Ecology and Conservation of Sirenia; Dugongs and Manatees*. Cambridge University Press, 521 pages.

Martinez, E., Orams, M. B., Pawley, M. D. M. & Stockin, K. A. (2012). The use of auditory stimulants during swim encounters with hector's dolphins in Akaroa harbour, New Zealand. *Marine Mammal Science*, **28**, E295–315.

McKenzie, L. J., Collier, C. & Waycott, M. (2012). *Reef Rescue Marine Monitoring Program – Inshore Seagrass*, Annual Report for the sampling period 1st July 2010–31st May 2011. Fisheries Queensland, Cairns. 230 pages.

McMahon, C. R. & Burton, H. R. (2005). Climate change and seal surival: evidence for environmentally mediated changes in elephant seal, *Mirounga leonina*, pup survival. *Proceedings of the Royal Society B*, **272**, 923–928.

McMahon, C. R., Bester, M. N., Burton, H. R., Hindell, M. A. & Bradshaw, C. J. (2005a). Population status, trends and a re-examination of the hypotheses explaining the recent declines of the southern elephant seal *Mirounga leonina*. *Mammal Review*, **35**, 82–100.

McMahon, C. R., Hindell, M. A., Burton, H. R. & Bester, M. N. (2005b). Comparison of southern elephant seal populations, and observations of a population on a demographic knife-edge. *Marine Ecology Progress Series*, **288**, 273–283.

McMahon, C. R., Bester, M. N., Hindell, M. A., Brook, B. W. & Bradshaw, C. J. A. (2009). Shifting trends: detecting environmentally mediated regulation in long-lived marine vertebrates using time-series data. *Oecologia*, **159**, 69–82

McNiven, I. J. & Bedingfield, A. C. (2008). Past and present marine mammal hunting rates and abundances: dugong (*Dugong dugon*) evidence from Dabangai Bone Mound, Torres Strait. *Journal of Archaeological Science*, **35**, 505–515.

Mikhalev, Y. A. (2000). Biological characteristics of humpbacks taken in Antarctic Area V by the whaling fleets *Slava* and *Sovietskaya Ukraina*. *Paper SC/52/IA12 presented to the IWC Scientific Committee*, May 2003 (unpublished), 18 pages.

Miller, D. G. M. (2007). Managing fishing in the sub-Antarctic. *Papers and Proceedings of the Royal Society of Tasmania*, **141**, 121–140.

Möller, L. M., Allen, S. J. & Harcourt, R. G. (2002). Group characteristics, site fidelity and seasonal abundance of bottlenose dolphins *Tursiops aduncus* in Jervis Bay and Port Stephens, south-eastern Australia. *Australian Mammalogy*, **24**, 11–21.

Nachtigall, P. E., Pawloski, J. L. & Au, W. W. (2003). Temporary threshold shifts and recovery following noise exposure in the Atlantic bottlenosed dolphin (*Tursiops truncatus*). *The Journal of the Acoustical Society of America*, **113**, 3425.

Newland, C. B., Field, I. C., Cherel, Y., *et al.* (2011). Diet of juvenile southern elephant seals reappraised by stable isotopes in whiskers. *Marine Ecology-Progress Series*, **424**, 247–258.

Noad, M., Dunlop, R., Paton, D. & Cato, D. (2011). Absolute and relative abundance estimates of Australian east coast humpback whales (*Megaptera novaeangliae*). *Journal of Cetacean Research and Management; Humpback Whale Special Issue*, **3**, 243–252.

Nicol, S., Worby, A. & Leaper, R. (2008). Changes in the Antarctic sea ice ecosystem: potential effects on krill and baleen whales. *Marine and Freshwater Research*, **59**, 361–382.

Parks, S., Johnson, M., Nowacek, D. & Tyack, P. (2011). Individual right whales call louder in increased environmental noise *Biology Letters*, **7**, 33–35.

Parra, G. J. (2006). Resource partitioning in sympatric delphinids: space use and habitat preferences of Australian snubfin and Indo-Pacific humpback dolphins. *Journal of Animal Ecology*, **75**, 862–874.

Paterson, R., Paterson, P. & Cato, D. H. (1994). The status of humpback whales *Megaptera novaeangliae* in east Australia thirty years after whaling. *Biological Conservation*, **70**, 135–142.

Pollock, K., Marsh, H., Lawler, I. & Alldredge, M. (2006). Modelling availability and perception processes for strip and line transects: an application to dugong aerial surveys. *Journal of Wildlife Management*, **70**, 255–262.

Pompa, S., Ehrlich, P. R. & Ceballos, G. (2011). Global distribution and conservation of marine mammals. *Proceedings of the National Academy of Sciences*, **108**, 13 600–13 605.

Reilly, S. B., Bannister, J. L., Best, P. B., *et al.* (2008). *Megaptera novaeangliae*. In: IUCN 2012. *IUCN Red List of Threatened Species*. Version 2012.2. www.iucnredlist.org. Downloaded on 11 March 2013.

Richardson, W. J. & Würsig, B. (1997). Influences of man-made noise and other human actions on cetacean behaviour. *Marine & Freshwater Behaviour & Physiology*, **29**(1–4), 183–209.

Richardson, W. J. & Thomson, D. H. (1998). *Marine Mammals and Noise*. San Diego; Toronto: Academic Press.

Robertson, B. C. & Chilvers, B. L. (2011). New Zealand sea lions *Phocarctos hookeri* possible causes of population decline. *Mammal Review*, **41**, 253–275.

Robertson, B. C., Chilvers, B. L., Duignan, P. J., Wilkinson, I. S. & Gemmell, N. J. (2006). Dispersal of breeding, adult male *Phocarctos hookeri*: Implications for disease transmission, population management and species recovery. *Biological Conservation*, **127**, 227–236.

Salgado Kent, C., Jenner, C., Jenner, M., Bouchet, P. & Rexstad, E. (2012). Southern Hemisphere Breeding Stock D humpback whale population estimates from North West Cape, Western Australia. *Journal of Cetacean Research and Management*, **12**, 29–38.

SCAR (1991). Report of the workshop or southern elephant seals, Monterey, California, USA (unpublished).

Shaughnessy, P. D. & Davenport, S. R. (1996). Underwater videographic observations and incidental mortality of fur seals around fishing equipment in south-eastern Australia. *Marine and Freshwater Research*, **47**, 553–556.

Shaughnessy, P., Kirkwood, R., Cawthorn, M., Kemper, C. & Pemberton, D. (2003). 7 Pinnipeds, cetaceans and fisheries in Australia: a review of operational interactions. *Books Online*, **2006**, 136–152.

Shaughnessy, P. D., McIntosh, R. R., Goldsworthy, S. D., Dennis, T. E. & Berris, M. (2006). Trends in abundance of Australian sea lions, *Neophoca cinerea*, at Seal Bay, Kangaroo Island, south Australia. In *Sea Lions of the World*. (Eds. A. W. Trites, S. K. Atkinson, D. P. DeMaster, *et al.*,) pp. 325–351. A Laska Sea Grant Collage Program, University of Alaska Fairbanks.

Slip, D. J. & Burton, H. R. (1999). Population status and seasonal haulout patterns of the Southern elephant seals (*Mirounga leonina*) at Heard Island. *Antarctic Science*, **11**, 38–47.

Slip, D. J., Hindell, M. A. & Burton, H. R. (1994). Diving behavior of southern elephant seals from Macquarie Island: an overview. In: *Elephant Seals: Population Ecology, Behavior and Physiology*, pp. 253–270.

Slooten, E. & Davies, N. (2012). Hector's dolphin risk assessments: old and new analyses show consistent results. *Journal of the Royal Society of New Zealand*, **42**, 49–60.

Slooten, E. & Dawson, S. M. (2009). Assessing the effectiveness of conservation management decisions: likely effects of new protection measures for Hector's dolphin

(*Cephalorhynchus hectori*). *Aquatic Conservation: Marine and Freshwater Ecosystems*, **20**(3), 334–347.

Slooten, E., Rayment, W. & Dawson, S. (2006). Offshore distribution of Hector's dolphins at Banks Peninsula, New Zealand: is the Banks Peninsula Marine Mammal sanctuary large enough? *New Zealand Journal of Marine and Freshwater Research*, **40**, 333–343.

Smith, I. W. G. (1989). Maori impact on the marine megafauna: pre-European distributions of New Zealand sea mammals. In *Saying So Doesn't Make It So. Papers in Honour of B. Foss Leach*, (Ed. D. G. Sutton), pp. 76–108. (New Zealand Archaeological Association Monograph 17.)

Smith, J. N., Grantham, H. S., Gales, N., *et al.* (2012). Identification of humpback whale breeding and calving habitat in the Great Barrier Reef. *Marine Ecology Progress Series*, **447**, 259–272.

Sobtzick, S., Hagihara, R., Grech, A. & Marsh, H. (2012). Aerial survey of the urban coast of Queensland to evaluate the response of the dugong population to the widespread effects of the extreme weather events of the summer of 2010–11. Final Report to the Australian Marine Mammal Censnvironment Research Program June 1 2012.

Southall, B. L., Bowles, A. E., Ellison, W. T., *et al.* (2007). Marine mammal noise special issue. Exposure criteria: initial scientific recommendations. *Aquatic Mammals Special Edition*, **33**, 411–522.

Steckenreuter, A., Harcourt, R. & Moller, L. (2011). Distances does matter: close approaches by boats impede feeding and resting behaviour of Indo-Pacific bottlenose dolphins. *Wildlife Research*, **38**, 455–463.

Steckenreuter, A. Möller, L. & Harcourt, R. (2012). How does Australia's largest dolphin-watching industry affect the behaviour of a small and resident population of Indo-Pacific bottlenose dolphins? *Journal of Environmental Management*, **97**, 14–21.

Stockin, K. A., Law, R. J., Duignan, P. J., *et al.* (2007). Trace elements, PCBs and organochlorine pesticides in New Zealand common dolphins (*Delphinus* sp.). *The Science of the Total Environment*, **387**, 333–345.

Stone, B. M., Blyde, D. J., Saliki, J. T., *et al.* (2011). Fatal cetacean morbillivirus infection in an Australian offshore bottlenose dolphin (*Tursiops truncatus*). *Australian Veterinary Journal*, **89**, 452–457.

Suter, K. D. (1981). The international politics of saving the whale. *Australian Journal of International Affairs*, **35**, 283–294.

Thompson, P. M., Lusseau, D., Barton, T., *et al.* (2010). Assessing the responses of coastal cetaceans to the construction of offshore wind turbines. *Marine Pollution Bulletin*, **60**, 1200–1208.

Torres, L. G., Smith, T. D., Sutton, P., *et al.* (2013). From exploitation to conservation: habitat models using whaling data predict distribution patterns and threat exposure of an endangered whale. *Diversity and Distributions*, **19**, 1138–1152.

Trathan, P. N., Forcada, J. & Murphy, E. J. (2007). Environmental forcing and Southern Ocean marine predator populations: effects of climate change and variability. *Philosophical Transactions of the Royal Society B: Biological Sciences*, **362**, 2351–2365.

Trillmich, F., Ono, K. A., Costa, D. P., *et al.* (Eds.) (1991). *The Effects of El Nino on Pinniped Populations in the Eastern Pacific*, pp. 247–270. Berlin Heidelberg, Springer.

Twiss, S. D., Thomas, C., Poland, V., Graves, J. A. & Pomeroy, P. (2007). The impact of climatic variation on the opportunity for sexual selection. *Biology Letters*, **3**, 12–15.

van den Hoff, J., Burton, H. R., Hindell, M. A., Sumner, M. D. & McMahon, C. R. (2002). Migrations and foraging of juvenile southern elephant seals from Macquarie Island within CCAMLR managed areas. *Antarctic Science*, **14**, 134–145.

Van Waerebeek, K., Baker, A. N., Felix, F., *et al.* (2007). Vessel collisions with small cetaceans worldwide and with large whales in the Southern Hemisphere, an initial assessment. *Latin American Journal of Aquatic Mammals*, **6**, 43–69.

Whitehead, H., McGill, B. & Worm, B. (2008). Diversity of deep-water cetaceans in relation to temperature: implications for ocean warming. *Ecology Letters*, **11**, 1198–1207.

Wiedenmann, J., Cresswell, K. A., Goldbogen, J., Potvin, J. & Mangel, M. (2011). Exploring the effects of reductions in krill biomass in the Southern Ocean on blue whales using a state-dependent foraging model. *Ecological Modelling*, **222**, 3366–3379.

Wright, A. J., Soto, N. A., Baldwin, A. L., *et al.* (2007). Do marine mammals experience stress related to anthropogenic noise?. *International Journal of Comparative Psychology*, **20**, 274–316.

Yablokov, A. V. (1994). Validity of whaling data. *Nature*, **367**, 108.

CHAPTER 17

Australian reptiles and their conservation

Jonathan K. Webb, Peter S. Harlow and David A. Pike

Summary

Australia has a spectacular and diverse reptile fauna approaching 1000 species, 93% of which are endemic to the continent. Despite this, there is a paucity of information on the biology of Australian reptiles compared with mammals and birds. The single greatest threat to Australian reptiles is the removal of native vegetation, most of which has occurred in the state of Queensland during the past few decades. Since European settlement in Australia, land clearing for stock grazing and other agricultural activities has reduced the extent of native vegetation, and resulted in extensive habitat fragmentation. Ultimately, habitat fragmentation leads to species loss and local extinctions. Other threats to Australian reptiles include livestock grazing, which occurs on 55% of the continent, coupled with changing fire regimes and predation by exotic predators, especially foxes and feral cats. Currently, we know little about the long-term impacts of pastoralism, fire and introduced predators on reptile communities. The conservation of Australian reptiles requires urgent changes in government policy to reduce rates of vegetation clearing. A critical challenge is the conservation of reptiles in the vast arid and semi-arid regions, where reptile diversity is remarkably high. This will require coordinated management of threatening processes across multiple land tenures, including pastoral leases, crown lands, Aboriginal lands and conservation reserves. In southern Australia, the conservation of reptiles in fragmented landscapes will require strategic tree planting to increase the sizes of habitat remnants and their connectivity, in addition to retaining important structural habitat features such as rock outcrops, old growth trees and fallen timber. In addition to *in situ* conservation practices, breeding programmes are being employed to prevent the extinction of imperilled species.

17.1 Introducing Australia's reptiles

Australia has close to 1000 species of reptiles and at least 189 described subspecies, representing 18 families and 163 genera, which equates to almost 10% of the world's reptile fauna. Numerically, Australia has the most endemic species of any country, with 93% of its

Austral Ark: The State of Wildlife in Australia and New Zealand, eds. A. Stow, N. Maclean and G. I. Holwell. Published by Cambridge University Press. © Cambridge University Press 2015.

reptile species unique to the continent. The arid zone, which covers two-thirds of the continent, harbours one of the most diverse lizard assemblages on the planet. In the Great Victoria Desert, you can find 47 species of lizards living together at the same sand ridge site. No other deserts come close to matching this diversity; North American deserts harbour just 12 lizard species, while only 20 lizard species occur in the Kalahari Desert in southern Africa (Pianka, 1986). Roughly 6% of Australia's reptile species are threatened (Table 17.1), comprising seven species listed as critically endangered, 17 species listed as endangered, and 34 species listed as vulnerable (EPBC, 2013). Taxa with disproportionate numbers of threatened species include marine turtles (100%: six of six species), freshwater turtles (22%: five of 23 species), and pygopodid lizards (18%: seven of approximately 40 species). Why so many legless lizards are threatened is perplexing. This small family, endemic to Australia and New Guinea, displays an extraordinarily high diversity of diets and foraging modes.

Table 17.1   Threatened Australian reptiles (EPBC Act List of Threatened fauna 2014)

Group	Scientific Name	Classification
Sea Turtles		
Loggerhead Turtle	*Caretta caretta*	Endangered
Leatherback Turtle	*Dermochelys coriacea*	Endangered
Olive Ridley Turtle	*Lepidochelys olivacea*	Endangered
Hawksbill Turtle	*Eretmochelys imbricata*	Vulnerable
Green Turtle	*Chelonia mydas*	Vulnerable
Flatback Turtle	*Natator depressus*	Vulnerable
Freshwater Turtles		
Western Swamp Turtle	*Pseudemydura umbrina*	Critically Endangered
Gulf Snapping Turtle	*Elseya lavarackorum*	Endangered
Mary River Turtle	*Elusor macrurus*	Endangered
Bell's Turtle (Namoi River)	*Myuchelys bellii = Wollumbinia belli*	Vulnerable
Fitzroy River Turtle	*Rheodytes leukops*	Vulnerable
Snakes		
Short-nosed Seasnake	*Aipysurus apraefrontalis*	Critically Endangered
Leaf-scaled Seasnake	*Aipysurus foliosquama*	Critically Endangered
Plains Death Adder	*Acanthophis hawkei*	Vulnerable
Ornamental Snake	*Denisonia maculata*	Vulnerable
Dunmall's Snake	*Furina dunmalli*	Vulnerable
Broad-headed Snake	*Hoplocephalus bungaroides*	Vulnerable
Olive Python (Pilbara subspecies)	*Liasis olivaceus barroni*	Vulnerable
Krefft's Tiger Snake (Flinders Ranges)	*Notechis scutatus ater*	Vulnerable
Christmas Island Blind Snake	*Ramphotyphlops exocoeti*	Vulnerable
Lizards		
Nangur Spiny Skink	*Nangura spinosa = Concinnia spinosa*	Critically Endangered
Christmas Island Blue-tailed Skink	*Cryptoblepharus egeria*	Critically Endangered
Christmas Island Forest Skink	*Emoia nativitatis*	Critically Endangered
Lister's Gecko (Christmas Island)	*Lepidodactylus listeri*	Critically Endangered

Table 17.1 (*cont.*)

Group	Scientific Name	Classification
Christmas Island Giant Gecko	*Cyrtodactylus sadleiri*	Endangered
Arnhem Land Egernia	*Bellatorias obiri*	Endangered
Alpine She-oak Skink	*Cyclodomorphus praealtus*	Endangered
Baudin Island Spiny-tailed Skink	*Egernia stokesii badia*	Endangered
Blue Mountains Water Skink	*Eulamprus leuraensis*	Endangered
Corangamite Water Skink	*Eulamprus tympanum marnieae*	Endangered
Allan's Lerista	*Lerista allanae*	Endangered
Guthega Skink	*Liopholis guthega*	Endangered
Slater's Skink	*Liopholis slateri slateri*	Endangered
Yellow-snouted Gecko	*Lucasium occultum*	Endangered
Adelaide Bluetongue Lizard	*Tiliqua adelaidensis*	Endangered
Grassland Earless Dragon	*Tympanocryptis pinguicolla*	Endangered
Five-clawed Worm-skink	*Anomalopus mackayi*	Vulnerable
Pink-tailed Legless Lizard	*Aprasia parapulchella*	Vulnerable
Flinders Ranges Worm-lizard	*Aprasia pseudopulchella*	Vulnerable
Hermite Island Worm-lizard	*Aprasia rostrata rostrata*	Vulnerable
Lord Howe Island Gecko	*Christinus guentheri*	Vulnerable
Three-toed Snake-tooth Skink	*Coeranoscincus reticulatus*	Vulnerable
Yinnietharra Rock-Dragon	*Ctenophorus yinnietharra*	Vulnerable
Airlie Island Ctenotus	*Ctenotus angusticeps*	Vulnerable
Lancelin Island Skink	*Ctenotus lancelini*	Vulnerable
Hamelin Ctenotus	*Ctenotus zastictus*	Vulnerable
Striped Legless Lizard	*Delma impar*	Vulnerable
Atherton Delma	*Delma mitella*	Vulnerable
Collared Delma	*Delma torquata*	Vulnerable
Yakka Skink	*Egernia rugosa*	Vulnerable
Houtman Abrolhos Spiny-tailed Skink	*Egernia stokesii aethiops = Egernia stokesii badia*	Vulnerable
Mount Cooper Striped Lerista	*Lerista vittata*	Vulnerable
Great Desert Skink	*Liopholis kintorei*	Vulnerable
Jurien Bay Skink	*Liopholis pulchra longicauda*	Vulnerable
Pedra Branca Skink	*Niveoscincus palfreymani*	Vulnerable
Lord Howe Island Skink	*Oligosoma lichenigera*	Vulnerable
Bronzeback Snake-lizard	*Ophidiocephalus taeniatus*	Vulnerable
Granite Belt Thick-tailed Gecko	*Uvidicolus sphyrurus*	Vulnerable

Some small worm-like fossorial species gorge themselves on ant pupae and larvae, much like blind snakes; some diurnally active foragers feed mostly on spiders; while others feed on insects. One species, Burton's snake lizard, is a snake analogue with specialised hinged teeth that allows it to subdue scincid lizards, which it ambushes from leaf litter (Greer, 1997). If space permitted, we could go on describing the fascinating and remarkable reptiles in Australia, but this is not our aim. Instead, we highlight the major threatening processes which endanger Australian reptiles, and offer some potential solutions.

17.2 Threatening processes and mitigating actions
17.2.1 Removal of native vegetation

Clearing of native vegetation poses the single greatest threat to Australian reptiles (Cogger *et al.*, 1993, 2003). Despite the well documented problems associated with broad-scale vegetation removal, such as soil erosion, hydrological changes, and dry land salinity (Taylor & Hoxley, 2003), Australia has one of the highest rates of native vegetation clearing in the world. Incredibly, more vegetation has been removed in recent decades than at any other time in Australia's history (Bradshaw, 2012). Since 1988, most vegetation clearing has occurred in the state of Queensland, where >6 million ha of native vegetation have been felled for livestock grazing and other agricultural activities (Bradshaw, 2012, DERM, 2010). Rates of land clearing in QLD remained high until 2008; approximately 700 000 ha were cleared annually between 1988 and 1990, while over 300 000 ha were cleared annually between 1990 and 2006 (DERM, 2010). This fell to 99 000 ha per year in the period 2008–2009, but the recent relaxation of vegetation clearing laws by the Queensland government may lead to a resurgence of vegetation removal. These losses are particularly troubling when one considers that Queensland supports over half of Australia's terrestrial endemic reptile species, and is a recognised hotspot for diverse reptile groups including geckos, skinks and snakes (Cogger *et al.*, 1993).

Vegetation removal affects reptiles via direct and indirect pathways, which operate over short and long timescales. In the short term, many reptiles are killed or receive life-threatening injuries from clearing activities (Cogger *et al.*, 2003). Following the removal of vegetation, many surviving reptiles will likely be killed by aerial predators (raptors, corvids, kookaburras, owls) or feral cats, foxes and dingoes. Although some reptiles may find some temporary shelter in woodpiles, most of these animals will die when the piles are subsequently burnt. Reptile deaths caused by vegetation clearing are staggering; during the period 1997–1999 it was estimated that 89 million individuals were killed in Queensland each year (Cogger *et al.*, 2003). Such losses cannot be replaced once the habitat is destroyed; indeed, you only have to walk through a treeless paddock to see that very few reptiles persist in these desolate landscapes (Driscoll, 2004).

Over longer timescales, local extinctions will continue to occur in any remaining habitat remnants (Tilman *et al.*, 1994). Such extinctions will occur due to multiple factors including edge effects, stochastic events (such as wildfires), mortality from motor vehicles, habitat degradation from livestock grazing, predation, and the inability of some species to use or recolonise the remnant patches (Driscoll, 2004). Ultimately, the effects of land clearing in Queensland will not be fully realised for several decades, and unless we take actions to curtail current rates of clearing, many endemic reptile species will likely be extinct in the next 30 years.

17.2.2 Habitat fragmentation

Habitat fragmentation, the end result of vegetation clearing, poses a serious threat to Australian reptiles (Cogger *et al.*, 1993, 2003). Indeed, habitat fragmentation was believed to have caused the extinction of the pygmy bluetongue lizard until it was rediscovered in a small habitat remnant in 1992 (Box 17.1).

Reptiles are particularly sensitive to habitat fragmentation due to their poor dispersal abilities (Williams *et al.*, 2012). Some reptiles rarely disperse across cleared areas, which in

Box 17.1 Rediscovery and conservation of the pygmy bluetongue lizard *Tiliqua adelaidensis*

The pygmy bluetongue lizard, *Tiliqua adelaidensis*, is the smallest (to 20 cm long) member of the skink genus *Tiliqua*, and was once widely distributed in South Australia. Habitat loss and fragmentation decimated populations, and the lizard was thought to be extinct until it was rediscovered near Burra in South Australia in 1992. Incredibly, two South Australian herpetologists found a pygmy bluetongue inside the stomach of a dead road-killed brown snake! Searches in the area eventually led to the discovery of live specimens (Armstrong *et al.*, 1993). The species currently occurs on just 31 disjunct sites in a small farming region of South Australia (Duffy *et al.*, 2012).

Pygmy bluetongue lizards inhabit modified grasslands (dominated by exotic grasses), native grasslands, and grassy woodlands (Souter *et al.*, 2007). The lizards use empty burrows of wolf (lycosid) and trapdoor (mygalomorph) spiders as shelters, basking sites and ambush foraging sites (Milne & Bull, 2000, Hutchinson *et al.*, 1994). Population density ranges from 15 to 200 individuals per hectare, and is highly variable across sites (Duffy *et al.*, 2012). Female pygmy bluetongues are viviparous, and give birth to one to four young between January and March. Juveniles disperse to unoccupied burrows after birth, but fewer than 10% of juveniles survive to maturity (Milne, 1999). Pygmy bluetongues mature early (around one to two years) and once mature, adults seldom move further than 20 m from their burrows (Milne, 1999). These life-history traits, coupled with habitat fragmentation, may explain why gene flow is restricted both within and between populations (Smith *et al.*, 2009).

Threats to pygmy bluetongue lizards

Only 0.3% of the original native grasslands within the pygmy bluetongue's historical range remain; the remainder has been cleared and fragmented (Hyde, 1995). Extant populations of pygmy bluetongue lizards occur on private land and are threatened by inappropriate grazing and agricultural activities that disturb the soil. Ploughing and ripping of the soil can kill or injure lizards, and destroys the spider burrows that are crucial habitat requirements (Duffy *et al.*, 2012). Grazing helps to maintain basking sites around burrows and may prevent weed invasions, but overstocking sensitive grassland habitats could reduce prey availability and damage spider burrows (Souter *et al.*, 2007). Planting trees in grasslands could also provide roosting and nesting sites for birds, thereby increasing avian predation rates on lizards (Duffy *et al.*, 2012). Survival is a balancing act at many levels for the pygmy bluetongue lizard.

Because pygmy bluetongue lizards have specific habitat requirements, and extant populations are small and isolated, this species is particularly vulnerable to climate change. Low rates of gene flow in extant populations suggests that the species would have difficulty dispersing to new habitats should the current area of occupancy become unsuitable (Fordham *et al.*, 2012). Changes to fire regimes could also threaten extant lizard populations. For example, intense grass fires during the juvenile dispersal phase could increase juvenile mortality rates, which could reduce local populations or reduce dispersal rates across fragmented landscapes (Fenner & Bull, 2007).

Provision of burrows and mitigation of current threats

The future of the pygmy bluetongue lizard will depend on how well we can protect and maintain existing habitats and populations. Ultimately, recovery of this species will require working closely with landholders to actively manage grazing lands (Duffy *et al.*, 2012). Strategic rotational grazing may be necessary to maintain habitat quality (Clarke, 2000), and activities that disturb the soil will need to avoid areas occupied by lizards. Artificial burrows are a promising avenue for enhancing existing populations and establishing suitable reintroduction sites to increase the number of extant populations (Souter *et al.*, 2004). Increased burrow densities could reduce juvenile mortality rates and help increase lizard densities at established sites (Souter *et al.*, 2004). Although there are several challenges to conserving pygmy bluetongue lizards, with the cooperation of landholders, researchers and conservation groups, their future looks secure.

turn can alter the genetic structure of populations within remnants (Stow *et al.*, 2001). Over time, isolated populations may suffer from inbreeding or loss of genetic diversity, further increasing their vulnerability to extinction (Frankham, 2005). Species with specialised habitat requirements and/or small geographic ranges, such as the endangered broad-headed snake (Box 17.2) and endangered Nangur spiny skink (Box 17.3) are particularly sensitive to habitat fragmentation. However, fragmentation also affects habitat generalists. Many reptiles require leaf litter, fallen timber and rocks for shelter, thermoregulation and/or foraging, and the removal of fallen timber or paddock trees for firewood, coupled with livestock grazing, and the invasion of weeds, can degrade the quality of habitat patches over time (Cunningham *et al.*, 2007; Dorrough *et al.*, 2012). Hence, through chance events, small isolated reptile populations in an agricultural matrix may be on a path to extinction even if no further habitat loss occurs (Tilman *et al.*, 1994).

Studies in extensively cleared agricultural areas of southern Australia, where >90% of the original vegetation has been cleared, paint a particularly bleak future for reptiles. In gimlet *Eucalyptus salubris* woodland in the Western Australian wheatbelt, smaller remnants contained fewer reptile species than larger remnants (Kitchener & How, 1982; Smith *et al.*, 1996). In general, woodland remnants had a depauperate lizard fauna that was dominated by generalist species (Smith *et al.*, 1996). Driscoll (2004) found that the painted dragon *Ctenophorus pictus* and the hooded scalyfoot *Pygopus nigriceps* were locally extinct in habitat remnants in south-western NSW. An even more depressing situation was recorded by Brown *et al.* (2008). These authors sampled reptile assemblages in habitat remnants in the Victorian Riverina district, and found no reptiles at 22% of sites! Moreover, over half of the reptiles they observed were two common, widespread, generalist skinks (Brown *et al.*, 2008). Collectively, these results suggest that regional reptile extinctions have already occurred in fragmented agricultural landscapes.

Box 17.2 Restoring habitats for the broad-headed snake *Hoplocephalus bungaroides*

The broad–headed snake *Hoplocephalus bungaroides* is a small (to 90 cm snout–vent length), spectacularly coloured nocturnal elapid snake (Figure 17.1, Plate 34). The species is restricted to sandstone rock formations within a 200 km radius of Sydney, Australia's largest city (Cogger *et al.*, 1993). During the cooler months, broad-headed snakes thermoregulate underneath sun-exposed sandstone rocks or inside crevices, and during summer they shelter in tree hollows (Webb & Shine, 1998). In 1850, broad-headed snakes were common throughout the Sydney region, but by 1869 the species was becoming scarce due to the removal of 'bush rock' by builders and gardeners (Krefft, 1869). Today, the species is locally extinct in the Sydney metropolitan area, and is confined to a handful of disjunct populations south, west and north of Sydney.

Broad-headed snakes grow slowly, mature late, are long lived, and females reproduce infrequently. The snakes' slow life history, low juvenile dispersal and habitat specificity make it particularly vulnerable to extinction (Webb *et al.*, 2002b). Current threats include the removal of snakes for the illegal pet trade, the removal of sandstone rocks for supply to nurseries (below), the destruction of habitat associated with illegal reptile collecting activities, and overgrowth of rock outcrops by emergent vegetation (Webb *et al.*, 2002a; Pringle *et al.*, 2009; Pike *et al.*, 2010). To mitigate some of these threats, National Parks and Wildlife staff implemented a series of management actions including the erection of locked gates to exclude vehicular access to some populations (Figure 17.2, Plate 55), signage to inform the public that bush rock collection is illegal, and the installation of hidden remotely triggered cameras to record the number plates of vehicles used by snake collectors or bush rock collectors. Nonetheless, collectors continue to damage gates to gain access to such sites (Figure 17.2, Plate 55), suggesting that broad-headed snakes are prized by collectors.

Figure 17.1 (Plate 34) The broad-headed snake is highly prized by snake enthusiasts due to its rarity and beautiful colouration. Photograph by Jonathan Webb. A black and white version of this figure will appear in some formats. For the colour version, please refer to the plate section.

Restoration of rock outcrops degraded by bush rock collectors

One of the major threats to broad-headed snakes is the removal of 'bush rocks' for landscaping urban gardens. Bush rock removal is listed as a key threatening process under the NSW Threatened Species Conservation Act 1995 yet, incredibly, there is no legislation outlawing the collection or sale of bush rocks! Many rock outcrops in the Sydney region were stripped of their surface rocks during the 1960s and 1970s, and consequently, these degraded sites have few suitable shelters for broad-headed snakes or other rock-dwelling reptiles.

To restore degraded rock outcrops, researchers developed fibre-reinforced cement rocks with thermal attributes that mimic those of sandstone rocks favoured by broad-headed snakes (Croak *et al.*, 2010). These artificial rocks were used to restore degraded rock outcrops at several locations in the Sydney region. Encouragingly, velvet geckos, a major prey of juvenile broad-headed snakes, colonised the artificial rocks within months of deployment (Croak *et al.*, 2010). Just one year later, broad-headed snakes began using the rocks (Croak *et al.*, 2012). Future restoration of degraded rock outcrops, coupled with the translocation of juvenile broad-headed snakes to restored sites, could help to prevent the extinction of this iconic elapid snake.

Figure 17.2 (Plate 55) In some cases it may be necessary to restrict access to endangered reptile populations. Gates were erected to prevent vehicular access to populations of broad-headed snakes (*Hoplocephalus bungaroides*) and Nangur spiny skinks (*Nangura spinosa*). Although gates can receive frequent vandalism, if maintained they can deter human access. In the left-hand photo, the gate has been pulled out of the ground, despite being held in place with concrete. In the right-hand photo the gate has been pulled off the hinges, and is hanging open by the locked chain. These types of activities make conserving endangered reptiles more challenging. Photographs by David Pike. A black and white version of this figure will appear in some formats. For the colour version, please refer to the plate section.

17.2.2.1   Revegetation of fragmented agricultural landscapes

Australia has 42 highly fragmented subregions containing less than 30% of the original vegetation. These subregions occur in south-western Western Australia, south-eastern South Australia, central and western Victoria, the New England Tablelands of New South Wales and the southern and central parts of eastern Queensland. The vegetation types most affected by vegetation removal are eucalypt woodlands, eucalypt open forests, and mallee woodlands and shrublands (Bradshaw, 2012). To conserve reptiles in fragmented agricultural landscapes, we urgently need to revegetate and protect

> **Box 17.3** An uncertain future for the critically endangered Nangur spiny skink *Nangura spinosa*
>
> The Nangur spiny skink, *Nangura spinosa*, occurs only in Nangur National Park, west of Gympie in south eastern Queensland (Borsboom *et al.*, 2010). The two known populations support roughly 45 and 140 individuals in areas <8 ha and 868 ha in size, respectively (Borsboom *et al.*, 2010). These small disjunct populations are extremely vulnerable to extinction. Conserving such a narrowly distributed endemic species should in theory be straightforward; simply protect the known habitat from threats that could reduce population size or hinder population growth. In the case of the Nangur spiny skink, however, this problem is more complex; this species is vulnerable to many threats facing small populations *and* some of those that threaten widely distributed species. Local threats include poaching by collectors (Borsboom, 2012) and genetic bottlenecks due to inbreeding (Borsboom *et al.*, 2010). Widespread threats include habitat fragmentation and degradation (e.g. selective logging) and negative effects of invasive species (both animals and plants (Borsboom *et al.*, 2010, Borsboom, 2012)). Whether these threats combined will push Nangur spiny skinks to extinction is unknown, mainly due to a lack of the general ecological information necessary to adequately guide conservation and management efforts.

existing remnants (Driscoll, 2004). Thankfully, recent community-sponsored and government-funded initiatives have begun to protect vegetation remnants on some private lands, and develop replanting schemes and corridor plans for many areas of southern Australia.

Ideally, replanting projects should aim to enlarge and join existing forest remnants to maintain reptile species diversity (Driscoll, 2004). However, many reptile species require complex structural habitat features, such as logs, dead trees, large hollow-bearing trees, leaf litter, or rocks, which are often absent in newly revegetated habitat patches (Munro *et al.*, 2007). To conserve reptiles in agricultural landscapes, we need to retain these structural habitat components (Michael *et al.*, 2011). In areas where structural features are absent, the addition of coarse woody debris can enhance habitat suitability for reptiles, and may reduce the long time lag for the natural formation of such habitat features (Manning *et al.*, 2013). These conservation actions will require active participation by private landholders; in many states, private lands contain a significant proportion of remnant vegetation (Brown *et al.*, 2011b). Ultimately, we need to educate landholders about the benefits of retaining structural habitat features for native fauna if we are to conserve reptiles in fragmented agricultural landscapes.

17.2.3   The spread of the cane toad

The highly toxic cane toad *Rhinella marina* was introduced to Queensland in 1935 and has since spread across northern Australia. Many Australian reptiles lack physiological mechanisms to detoxify toad toxins, and die after mouthing or ingesting cane toads. Since cane toads invaded the Northern Territory, there have been massive declines in populations of varanid lizards (Doody *et al.*, 2009), freshwater crocodiles (Letnic *et al.*,

2008), and bluetongue lizards (Brown *et al.*, 2011a). As pointed out by Shine & Phillips (Chapter 5, this volume), some species may actually benefit from the arrival of cane toads because they no longer experience predation from varanid lizards. For example, in the Daly River region, green tree snakes increased in abundance after cane toads decimated populations of three species of varanid lizards (Doody *et al.*, 2013). Indirect effects of cane toads on reptiles could be positive or negative; for example, the removal of large varanid lizards might increase feral cat abundance, which, given the current declines of small mammals in northern Australia (Woinarski *et al.*, 2011) might increase predation on lizards and snakes. At present, we know little about the magnitude of these indirect effects of cane toads on reptiles (Shine & Phillips, Chapter 5, this volume).

Cane toads recently invaded the Kimberley region of Western Australia, which is recognised for its high reptile diversity and endemism (Cogger, 2000). The spread of cane toads will likely cause serious population declines of varanid lizards, some snakes and bluetongue lizards. However, given that we know virtually nothing about the interplay of cane toads with other threatening processes (predation, fire, and grazing), we hesitate to make any predictions about how cane toads may affect ecosystems in Western Australia.

17.2.3.1 Mitigating cane toad impacts

Despite much research on cane toads, we doubt that a method for eradicating cane toads will be developed in the foreseeable future. Traps baited with cane toad toxins will be useful for removing toad tadpoles from farm dams (Shine & Phillips, Chapter 5, this volume), but are impractical for reducing toad densities at a landscape scale. Encouragingly, replacing earthen farm dams with plastic water tanks could prevent cane toads from colonising semi-arid regions of the continent (Florance *et al.*, 2011). In fact, landscape-scale modelling demonstrated that strategic replacement of just 100 earthen dams with water tanks could prevent cane toads from reaching the Pilbara region of Western Australia (Tingley *et al.*, 2013). It would probably cost \$400 000 to keep cane toads out of the Pilbara (assuming poly tanks cost \$4000 each); this amount is trivial compared to the >\$1 million that has been squandered by community groups trying to eradicate cane toads via hand collection.

17.2.4 Changing fire regimes in tropical savannas and its impact on reptiles

Northern Australia is dominated by highly flammable tropical savannas which cover 1.9 million km^2. Temperatures are high year round, but most of the annual rainfall (400–1200 mm) falls in the four-month wet season (December–March), which is followed by an extended dry season. Prior to European settlement, savannas were populated by Aboriginal peoples who used fire for signalling, hunting, clearing country, and for promoting the growth of bush foods and vegetation that would attract macropods and other important prey species (Bowman, 1998). Most fires were lit early in the dry season (April–May), which created a mosaic of burnt and unburnt areas which prevented the spread of large, destructive late dry season fires. The loss of Aboriginal burning from this landscape in the 1960s resulted in a temporal shift to mid-to-late dry season fires that often burnt large tracts of savanna (Russell-Smith *et al.*, 2003). These regular, extensive fires have led to declines in flora and fauna across much of northern Australia (Vigilante & Bowman, 2004; Woinarski *et al.*, 2010).

Late dry season fires can cause direct mortality in frill-necked lizards, *Chlamydosaurus kingii* (Griffiths & Christian, 1996), while the removal of cover may increase the vulnerability of diurnal lizards to predation (Legge *et al.*, 2008). The major conservation challenge in savanna landscapes is to implement early dry season fires in a patchy manner across land tenures. Two exemplary projects in northern Australia – the Ecofire project (see Woinarski *et al.*, Chapter 25, this volume) and West Arnhem Land Fire Abatement (WALFA) project – demonstrate that this approach is not only possible, but can also yield significant biodiversity benefits. Importantly, such projects can empower Aboriginal people living in remote regions with few employment opportunities (Whitehead *et al.*, 2008). The WALFA project involves Aboriginal landowners and scientists working together to plan and manage fire on Aboriginal lands, using both on ground and aerial incendiaries to create fire breaks across the landscape. A recent study using Landsat imagery showed that the WALFA project has been highly successful; since the project was implemented, there has been a significant reduction in late dry season fires (from 29% to 12.5%) accompanied by a reduction in the mean annual proportion of country burnt from 38% to 30% (Price *et al.*, 2012).

17.2.5 Changing fire regimes in Australian deserts

The deserts of central Australia are dominated by fire-prone spinifex (*Triodia* spp.) landscapes, which contain a remarkably high diversity of reptiles (Pianka, 1986). The Aboriginal inhabitants of these landscapes used fire throughout the year for clearing country, signalling, hunting lizards, and for promoting the growth of bush foods. The movement of Aboriginal peoples from traditional lands to towns in Australian deserts during the 1960s resulted in a rapid shift from a patchwork mosaic of vegetation of different ages, to large patches of either long unburnt or recently burnt country (Burrows *et al.*, 2006). Current fire regimes are characterised by pulses of large, rainfall-driven wildfires which homogenise vast tracts of country (Edwards *et al.*, 2008). The loss of the fine-scale habitat mosaics created by traditional burning likely contributed to the extinction of small mammals in the region (Burrows *et al.*, 2006). Many desert reptiles are habitat specialists, so large-scale wildfires may benefit some species whilst negatively affecting others (Pianka, 1986). Woinarski *et al.* (Chapter 25, this volume) give some examples of reptiles that are affected by inappropriate fire regimes. More research is needed to understand the interactions among fires, rainfall, grazing and predators in Australian deserts (Pianka & Goodyear, 2012). Managing fire regimes in isolated, uninhabited regions of arid Australia is logistically difficult, and will require communication and collaboration between pastoralists, national parks and Aboriginal landholders (Woinarski *et al.*, Chapter 25, this volume).

17.2.6 Changes to fire regimes in temperate regions

In temperate regions of Australia, there is little consensus about whether Aboriginal peoples used fire as a tool to manage vegetation, and whether fire regimes and vegetation have changed substantially since European colonisation (Pringle *et al.*, 2009). Nonetheless, analysis of charcoal deposits and the records of early settlers suggest that

Aboriginal peoples used fire frequently in some parts of southern Australia (Black *et al.*, 2008; McLoughlin, 1998). By contrast, Europeans adopted a policy of fire suppression to protect property and grazing lands; such policies resulted in changes in the severity and extent of wildfires (Shea *et al.*, 1981). Whether these changes have affected reptiles remains unknown, although some species could be disadvantaged by fire-mediated vegetation changes (Pringle *et al.*, 2009, 2012). Recent studies suggest that, unlike mammals, reptiles do not show predictable responses to fire (Lindenmayer *et al.*, 2008); hence, maintaining a mosaic of habitats with different fire histories may be the best strategy for conserving reptiles (Driscoll & Henderson, 2008). Implementing effective fire management in temperate Australia remains a major challenge for reptile conservation.

17.2.7   Livestock grazing

Livestock grazing on natural vegetation occurs on 4.2 million square kilometres, or 55% of Australia. Grazing is the dominant land use on semi-arid and arid regions of the country, which are significant hotspots for reptile diversity (Pianka, 1969; Morton & James, 1988). Livestock grazing has resulted in loss of native vegetation, soil erosion, the degradation of riparian areas, and has contributed to declines of small mammals and birds (Martin & McIntye, 2007; Legge *et al.*, 2011). Although the impacts of grazing on reptiles are less clear, heavy grazing can cause reductions in reptile abundance at small spatial scales (James, 2003). For example, the heavily grazed and trampled bare ground (the piosphere) which surrounds bore-fed watering points is unsuitable for reptiles that rely on shrub layers or litter for cover, and may increase the risk of predation by aerial predators (James *et al.*, 1999). The provision of artificial watering points for cattle in arid Australia has also facilitated the spread of invasive cats, foxes (James *et al.*, 1999) and cane toads (Florance *et al.*, 2011), which can negatively affect reptile populations. Despite the ecological problems associated with livestock grazing, rangelands nonetheless contribute substantially to reptile conservation at regional scales (Woinarski *et al.*, 2013).

17.2.8   Predation by feral cats

Cats were deliberately released in Australia to control mice and rabbits in the nineteenth century, and they have since spread across the entire continent. Predation by feral cats is listed as a key threatening process under the *Environment Protection and Biodiversity Conservation Act 1999* (EPBC Act). Feral cats are a potential threat to vulnerable or threatened vertebrates, including at least 35 species of birds, 36 mammals, seven reptiles and three amphibians. Threatened reptiles at risk from cat predation include three Christmas Island lizards (the blue-tailed skink *Cryptoblepharus egeriae*, the forest skink *Emoia nativitatis* and Lister's gecko *Lepidodactylus listeria*), one legless lizard (striped legless lizard *Delma impar*), four skinks (great desert skink *Egernia kintorei*, Arnhem Land skink *Egernia obiri*, the Blue Mountains water skink *Eulamprus leuraensis*, and the Corangamite water skink *Eulamprus tympanum marnieae*) and the broad-headed snake *Hoplocephalus bungaroides* (Smith *et al.*, 2012, DEWHA, 2008a). Although cats have broad diets, and preferentially consume small mammals and birds, they can be significant predators of reptiles (Paltridge *et al.*, 1997). The recent decline of small mammals from savanna landscapes in northern Australia may result in cats including more reptiles in their diets. Potentially, cat predation could cause local extinctions of species which have suffered population declines due to cane toad poisoning, changes in fire regimes (Woinarski *et al.*, 2011), or other threatening processes. Species likely to be particularly vulnerable to cat predation in this respect are bluetongue lizards, which have suffered precipitous declines across northern Australia (Price-Rees *et al.*, 2010).

Controlling feral cats on the mainland is extraordinarily difficult due to their reluctance to consume toxic baits, and their low population densities in many landscapes. By contrast, feral cats have been eradicated on sub-Antarctic Macquarie Island and on the Montebello Islands off Western Australia (Nogales *et al.*, 2004). On the mainland, the best way to minimise the impacts of feral cats on vulnerable reptiles is to maintain appropriate fire regimes and habitats (i.e. cover) in the landscape; the possibility that dingoes might suppress cat (and fox) densities also warrants further investigation (Dickman, 1996).

17.2.9   Predation by European red foxes

The red fox was deliberately introduced to Australia in 1855, and has since spread across much of Australia (Dickman, 1996). Foxes prey on a diversity of animals, but are a major predator of small- and medium-sized mammals, ground-nesting birds and chelid turtles (Dickman, 1996). Predation by the European red fox is listed as a key threatening process under the EPBC Act. Foxes pose a threat to at least 12 species of threatened reptiles, including four species of marine turtle (loggerhead turtle *Caretta caretta*, green turtle *Chelonia mydas* (Figure 17.4, Plate 36), leatherback turtle *Dermochelys coriacea*, flatback turtle *Natator depressus*), four species of freshwater turtle (western swamp turtle *Pseudemydura umbrina*, Fitzroy River turtle *Rheodytes leukops*, Mary River turtle *Elusor macrurus* and Bellinger River turtle *Emydura signata*), two skinks (corangamite water skink *Eulamprus tympanum marnieae* and the great desert skink *Egernia kintorei*), one legless lizard (striped legless lizard *Delma impar*) and the broad-headed snake *Hoplocephalus bungaroides* (DEWHA, 2008b).

Reducing the impacts of foxes on native wildlife can be achieved with the use of extensive fencing or broad-scale 1080 poison baiting (Short & Turner, 2000). Baiting has been very successful in Western Australia, but less successful in eastern Australia (Saunders *et al.*, 2010). Fox control is expensive, and must be monitored and continued indefinitely to be successful. Increasing landholder participation and use of more efficient baiting techniques, such as broad-scale aerial baiting, are necessary to improve fox control in eastern Australia (Saunders *et al.*, 2010).

17.2.10   Invasive fire ants and yellow crazy ants

Two species of highly invasive ants, the crazy ant *Anoplolepis gracilipes* and the fire ant *Solenopsis invicta* pose a potential threat to Australian reptiles. Crazy ants form large nests and super-colonies that can cover large areas (750 ha) and they vigorously attack animals that disturb nests (Abbott, 2006). Although they lack stings, crazy ants kill invertebrates and vertebrates by biting and spraying formic acid. The ants prey on arthropods, earthworms, molluscs, land crabs, birds, mammals and reptiles (O'Dowd *et al.*, 2003). In Australia, crazy ants inhabit Christmas Island and a 2500 km^2 region of Arnhem Land (Young *et al.*, 2001). The ants thrive in human-disturbed areas, and can inhabit tropical and subtropical habitats, grasslands, savanna woodlands, woodlands and rainforests (O'Dowd *et al.*, 2003). Climatic modelling suggests that this species could inhabit most of northern Australia, eastern Queensland and parts of northern New South Wales (Chen, 2008).

The impacts of crazy ants have been particularly severe on Christmas Island. The ants were introduced in the 1930s but remained in low numbers until the 1990s. However, by 2002, the ants had formed super-colonies, with densities of 2000 ants per m^2 covering 2500 ha of the islands' forests (Abbott, 2006). The forest floor was literally crawling with carnivorous crazy ants, and was not the sort of place you would want to picnic in. The prolific crazy ants soon decimated the red crabs (*Gecarcoidea natalis*), the dominant consumers of the forest floor, which led to massive changes within rainforest habitats (O'Dowd *et al.*, 2003). Crazy ants threaten endemic Christmas Island reptiles, which have undergone massive population declines in recent decades (Box 17.4). To control crazy ants, Parks Australia carried out aerial baiting of super-colonies in 2002, 2009 and 2012. The baiting programme in 2009 was highly successful, with ant densities reduced by 99% at super-colony sites (Boland *et al.*, 2011). Nonetheless, continued monitoring and baiting will be necessary to control the ants on Christmas Island.

Box 17.4 The trouble with islands

Although no Australian reptile species has 'officially' gone extinct since European settlement, many are perilously close and several are almost definitely 'unofficially' extinct. The Australian reptile fauna has seemingly done better than other continents, with some authors suggesting that 19% of all reptile species worldwide are now vulnerable to extinction (Bohm *et al.*, 2013). In Australia, 58 reptile species are currently listed as Threatened (see Table 17.1), or about 6% of the approximately 970 current species (EPBC, 2013).

Island species are particularly vulnerable to extinction, with all 22 confirmed reptile extinctions worldwide occurring on islands (IUCN, 2013). The first Australian reptile extinctions will probably also be on islands. The Australian Indian Ocean territory of Christmas Island has seen a rapid and catastrophic decline in five of its six native reptile species (Smith *et al.*, 2012). Over 60% of the island (85 km^2) is a National Park, but the ecological changes wrought by super-colonies of yellow crazy ants (*Anoplolepis gracilipes*) together with introduced plant and animal species are immense (O'Dowd *et al.*, 2003). Hypotheses suggested for the rapid decline of the three skink, one gecko and a blind snake species from Christmas Island include habitat change, an introduced disease or pathogen, climate change, competition with introduced reptile species and exotic predators (Smith *et al.*, 2012; Maple *et al.*, 2012). Disease now seems unlikely (Hall *et al.*, 2011), but introduced predators like the giant centipede (*Scolopendra subspinipes*) and the specialist lizard-eating Asian wolf snake (*Lycodon capucinus*, introduced around 1987) are prime contenders. Four of the declining reptile species contracted to the south-western tip of the island, which is the most distant point from the single port facility and airport where the wolf snake presumably first arrived (Smith *et al.*, 2012). One needs only to consider the accidental introduction of the brown tree snake (*Boiga irregularis*) to Guam, which caused the extirpation of 13 of Guam's 22 native bird species, to realise how much we underestimate the ability of snakes to locate and devour naïve prey (Rodda & Savidge, 2007).

On Christmas Island the two *Emoia* species and the blue-tailed skink have only recently gone: the coastal skink (*E. atrocostata*) was last seen in 2004, the forest skink (*E. nativitatis*, Figure 17.3, Plate 35) and the blue-tailed skink (*Cryptoblepharus egeriae*) in mid 2010 (Smith *et al.*, 2012). Despite intensive surveying, tiny Lister's

Box 17.4 (continued)

Figure 17.3 (Plate 35) This captive Christmas Island forest skink (*Emoia nativatatis*) may be the last individual, as despite extensive field surveys this species has not been seen in the wild since mid 2010. Photograph by Peter Harlow. A black and white version of this figure will appear in some formats. For the colour version, please refer to the plate section.

gecko (*Lepidodactylus listeri*) has not been located in the wild since late 2012. Christmas Island National Park staff were helplessly watching and monitoring these declines, and in 2009–10 they captured 64 blue-tailed skinks and 43 Lister's geckos to begin captive breeding colonies. By late 2014 over 400 blue-tailed skinks and 200 Lister's gecko were in captivity on Christmas Island and at Taronga Zoo, Sydney. But what should be done with these lizards, as the threatening processes have not been identified, and thus no safe habitat remains to release them on Christmas Island? One solution may be 'assisted colonisation', to release these captive populations on another small tropical island.

The remarkable Pedra Branca skink (*Niveoscincus palfreymani*) survives in about 0.14 ha of available habitat on a 2.5 ha rock 26 km off the south-east coast of Tasmania. Three population estimates over 14 years show the population varies from about 290 to 560 individuals (Brothers *et al.*, 2003), and most evidence suggests that this island has been separated from Tasmania for at least 19 000 years (Banks, 1993). This species is a contender for natural extinction, with no human assistance, in the next few millennia as each stochastic population decline and genetic bottleneck increases the likelihood of inbreeding depression. One study that investigated the frequency and severity of catastrophic die-offs in 88 species of vertebrates suggested that the probability of a 50% or greater population decrease in any one year is approximately 14% per generation, or about one in every seven generations (Reed *et al.*, 2003). Biologists today are faced with increasing political and philosophical decisions; should we translocate Christmas Islands' captive lizards, or perhaps Pedra Branca skinks, to new islands or let extinction occur?

The invasive fire ant *Solenopsis invicta* builds earthen mounds that harbour between 200 000 and 400 000 workers, and can attain densities of up to 2600 mounds per hectare. The ants possess a powerful venom, and when attacking en masse, they can kill vertebrates, stock, domestic pets and humans (Moloney & Vanderwoude, 2003). In Australia, fire ants cover approximately 50 000 ha of the south-western suburbs of Brisbane and the eastern suburbs of Ipswich in south-eastern Queensland (Schmidt *et al.*, 2010). Already, the ants have reduced the abundance of invertebrates and reptiles, and are poised to invade the coastal belt and the more mesic inland areas of Australia. The spread of fire ants poses a significant risk to reptiles, particularly hatchling sea turtles and ground-dwelling lizards (Moloney & Vanderwoude, 2003).

To reduce the spread of fire ants, the Queensland Government has implemented movement controls to individuals and commercial operators in areas containing fire ants. The Australian Government has funded aerial detection of nests and deployment of baits in an attempt to eradicate fire ants from the Brisbane region. In Yarwun, central Queensland, fire ants were successfully eradicated. Nonetheless, continued surveillance and eradication programmes are necessary to prevent the spread of fire ants (Schmidt *et al.*, 2010).

17.2.11 Climate change and sea turtles

Climate change poses a major threat to Australian reptiles, particularly species which exhibit temperature-dependent sex-determination or which depend on rainfall for survival. For example, changing rainfall patterns in Western Australia threaten the survival of the critically endangered western swamp turtle, *Pseudemydura umbrina* (Box 17.5). Sea turtles in Australia could be especially vulnerable to climate change because many populations are depleted, or are harvested by traditional hunters (Box 17.6). Further population perturbations could push some populations to extinction, and thus understanding how and why sea turtle populations are vulnerable to increasing temperatures associated with climate change is an urgent conservation problem (Hamann *et al.*, 2013). Like all ectotherms, temperature influences every facet of the life history and ecology of sea turtles. This includes embryonic survival, hatchling sex, hatchling body size and performance, determining the rates of physiological processes, and influencing foraging distributions, food availability, nesting distributions, and nest-site availability (Hamann *et al.*, 2013). Although sea turtles spend the vast majority of their lives in the ocean, the terrestrial

---

Box 17.5 The critically endangered western swamp turtle *Pseudemydura umbrina*: promising initial recovery following decades of slow decline

Western swamp turtles are restricted to two ephemeral swamps of marginal quality on the fringe of Australia's fastest growing city, Perth, Western Australia. Swamps usually fill and remain wet during winter, when the carnivorous turtles are aquatic and forage for prey, but begin to dry from late winter through summer. As swamps dry, turtles migrate to nearby terrestrial aestivation sites, usually comprised of natural tunnels underground or beneath surface debris (Burbidge & Kuchling, 2004; Burbidge, 1981). Swamp filling and drying cycles are strongly tied to seasonal rainfall, which has

Box 17.5 (continued)

declined over the past three decades (Burbidge & Kuchling, 2004; Mitchell *et al.*, 2012a). In many recent years the ponds have dried before females are able to accumulate sufficient energy stores to produce eggs, resulting in the absence of population-level reproduction in those years (Mitchell *et al.*, 2012a,b).

Western swamp turtles have the slowest life history of any Australian turtle, which combined with a current population size of <50 adults in the wild, renders it vulnerable to extinction (Burbidge, 1981; Mitchell *et al.*, 2012a). Females are smaller than males, and reach maturity at 11–15 years of age and can live in excess of 60 years. During reproductive years, females lay a single clutch of only three to five eggs, but reproduction is strongly linked to environmental conditions and is thus less than annual (Burbidge & Kuchling, 2004; Burbidge, 1981). Variable and unpredictable seasonal rainfall contributes to slow growth rates, leading to delayed maturity and irregular and stochastic reproduction (Mitchell *et al.*, 2012b; Burbidge & Kuchling, 2004; Burbidge, 1981).

Western swamp turtles recently were on a trajectory towards extinction. At the larger of the two known populations (Twin Swamps Nature Reserve), the number of adult turtles known to be alive decreased from 38 in 1963 to only seven by 1984, an average loss of just over one adult per year (Burbidge & Kuchling, 2004). Clearly, this population decline and the current low number of adult animals exemplify that the survival and reproduction of every individual turtle is crucial to maintaining the entire species. There is hope for this species in the wild, however, because captive assurance colonies are now supplementing wild populations. Although the Twin Swamps population stayed below 10 individuals through 2001, this population has increased rapidly because of conservation efforts (Burbidge & Kuchling, 2004).

The rapid recovery of this species is encouraging, but increasing aridity could hinder population growth rates by constraining reproduction and foraging opportunities. Annually, turtles spend six or more months aestivating in terrestrial environments. The migration to and from wetlands is a period of high predation (Burbidge & Kuchling, 2004), and terrestrial aestivation substantially increases vulnerability to desiccation, energy depletion, and hyperthermia (Burbidge, 1981; King *et al.*, 1998). Two novel approaches are being used to: (1) predict how increasing temperatures and shorter, more variable hydroperiods could impact individual turtle growth rates (which influences age at maturity); and (2) identify wetland sites that will maintain favourable hydroperiods under climate change, and potentially translocate turtles to these sites. Increased water temperatures could increase growth rates of the hatchling and juvenile life stages, potentially allowing individuals to reach maturity at earlier ages (Mitchell *et al.*, 2012b). Assisted colonisation to high-quality ephemeral wetlands could help establish long-term, viable populations that are robust to the impacts of climate change. Selecting appropriate release sites can be difficult, but several candidate wetlands seem promising for establishing new populations (Mitchell *et al.*, 2012a). Anticipating the effects of climate change, and preparing for them will help ensure that western swamp turtle populations remain in the wild far into the future.

Box 17.6 Conserving endangered sea turtles and cultural values: the complexities of contemporary harvest

One challenging – and very real – goal is to balance science-based protection efforts with traditional use of wildlife by indigenous peoples (Kwan *et al.*, 2001; Wilson *et al.*, 2010; Nursey-Bray, 2009; Butler *et al.*, 2012). In Australia, the *Native Title Act* defines the rights of Indigenous Australians to continue traditional practices, even when these practices may be prohibited by contemporary law. Concerns over declining sea turtle populations led to a closure of turtle and egg harvest in Queensland in July 1968 through enactment of the Queensland Fisheries Act (Miller & Limpus, 2012). The *Native Title Act*, passed in 1993, reinstated the rights of Indigenous Australians to use native animals (including sea turtles and eggs, dugong and other endangered species) legally for communal, non-commercial purposes (Butler *et al.*, 2012; Kwan *et al.*, 2001). Traditional hunting provides an important, and under-utilised, opportunity for ecologists to learn from communities with extensive knowledge of ecology and animal behaviour.

Sea turtles and their eggs are culturally important foods for Torres Straight Islanders (Butler *et al.*, 2012). Green turtles *Chelonia mydas* are preferred (Figure 17.4, Plate 36), but other species and their eggs are also consumed (e.g. hawksbill *Eretmochelys imbricata*, flatback *Natator depressus*; olive ridley, *Lepidochelys olivacea*; (Butler *et al.*, 2012)). Female sea turtles are targeted because of their high fat content (Kwan *et al.*, 2001); thus, hunters selectively remove individuals that could have otherwise continued to lay eggs for decades. Information on the level of harvest is lacking, and we do not yet understand the potential impacts of traditional hunting on severely

Figure 17.4 (Plate 36) The green sea turtle (*Chelonia mydas*) is the most widespread of the sea turtles nesting in Australia, and is thus exposed to a wide range of climatic conditions and local threats, including hunting by indigenous groups. Although the impacts of climate change and traditional hunting on sea turtles is generally unknown, continuing efforts to protect this species will require an integrative approach that can minimise multiple threats. Photograph by Ian Bell. A black and white version of this figure will appear in some formats. For the colour version, please refer to the plate section.

Box 17.6 (continued)

reduced populations (Miller & Limpus, 2012; Kwan *et al.*, 2001). One concern raised by opponents of indigenous harvest is the manner by which turtles are located, captured, and despatched; today new technologies (motorboats, metal harpoons, knives for butchering) substantially increase harvest success. In some instances, however, animals are not despatched immediately, which has led to strong debate in the media about the ethical nature of traditional harvest. In 2012, Queensland closed loopholes to ensure that traditional harvest complies with animal cruelty laws. Although this debate is far from over, the larger conservation issue is whether sea turtle populations can withstand the pressures of traditional harvest when combined with climate change and other threats (Hamann *et al.*, 2013).

What are current, feasible solutions to ensure that sea turtle populations can still recover, while maintaining indigenous harvest? The latest approaches to responsible and sustainable management focus on blending scientific and traditional indigenous knowledge (Wilson *et al.*, 2010). This can enable local communities to develop their own harvest management plans using scientific input (e.g. self-imposed seasonal closures, restricted areas, catch limits, restrictions on harvest methods (Wilson *et al.*, 2010; Kwan *et al.*, 2001)), combined with community monitoring of the impacts to populations. Assisting traditional groups to develop their own adaptive management plans fully – which includes self-monitoring and applying traditional penalties for breaches – with the aid of scientific input provides enormous opportunities for conservation of species, ecosystems and culture. These benefits could not be achieved any other way, and provide a promising glimpse into a future where contemporary legislation and policy incorporate indigenous knowledge and values to maintain and conserve biodiversity.

environment is crucial for reproduction. Sea turtles bury their eggs on sandy beaches in tropical to temperate regions, and the incubation temperature of the nest influences hatching success and sex. Climate change is predicted to increase ambient temperatures above that of current decades by 1–5 °C by 2070 in Australia (Cabrelli *et al.* Chapter 4, this volume), which has the potential to influence sea turtle nest temperatures and population-level primary sex ratios. A global sea-level rise of up to 79 cm by 2100 could threaten several important nesting populations by reducing the availability of nesting habitat on low-lying islands or in areas limited by human development. Increased precipitation or changes in the severity or intensity of tropical cyclones could also impact nesting beaches (Hamann *et al.*, 2013). Changes in water temperatures could also influence the distribution and availability of food, migratory paths, inter-nesting intervals, and individual growth rates (Hamann *et al.*, 2013). Although some work has made progress in understanding these threats overseas (Witt *et al.*, 2010), Australian studies have yet to tackle climate change impacts at foraging grounds or on the potential changes in migratory routes. Sea turtles provide a wide range of ecosystem services, and protecting these charismatic megafauna under climate change will allow the continuation of important ecological, social, cultural and economic services, not only in Australia, but worldwide.

17.3 Conclusions

17.3.1 The challenges of conserving reptiles in a vast continent

Australia harbours a rich and diverse reptile fauna, which presents substantial challenges to conservation. Australia covers an area of 7 688 503 square km, and the dominant land use is livestock grazing on natural vegetation, which occurs on 55% of the continent. By contrast, only 7% of the continent is devoted to conservation reserves, while other protected areas, including indigenous uses, cover 13% of Australia. The current reserve system does not adequately protect threatened reptiles, nor is it likely to do so substantially in the future (Watson *et al.*, 2011). Hence, to conserve Australian reptiles, we need a coordinated approach which transcends land tenure and State/Territory boundaries (Woinarski & Fisher, 2003). This complex task requires setting clear long-term goals, installing appropriate monitoring programmes, and engaging in adaptive experimental management. Managers will need to respond to dynamic changes within systems, and account for future changes that are likely to occur under climate change (Lindenmayer & Hunter, 2010). Ultimately, reptile conservation will require goodwill and effective communication between a diversity of stakeholders; the WALFA and Ecofire projects demonstrate that this is possible.

Reptile conservation in Australia will increasingly be in the hands of the private sector, non-government organisations and concerned citizen groups. Many State and Territory wildlife departments have been endlessly 'restructured', and funding which could have been directed to staff salaries or conservation efforts has been funnelled to needless name changes on websites, logos and stationery. Indeed, many parks and wildlife departments have been so starved of funding that they can no longer adequately manage their own National Parks. The increasing activity of fringe animal rights groups has seen many State and Territory Wildlife Departments begin inane and resource-wasting bureaucracies that prescribe and legally enforce cage size regulations for pet lizard keepers! One has only to consider the thousands of reptiles that die each year on our roads, or from vegetation removal, to see the idiocy of such regulations.

The paucity of information concerning the effects of fire, grazing and introduced predators on reptiles, coupled with the absence of natural history data for many species, creates additional problems for conserving reptiles (Cogger *et al.*, 1993). Many threatening processes vary across broad biogeographic regions, so conservation actions must often be tailored to specific localities, and to particular species (see Boxes 17.1, 17.2 and 17.3). For example, Parks Australia's captive breeding programmes in partnership with Taronga Zoo have been implemented to prevent the extinction of imperilled Christmas Island reptiles (see Box 17.4). Captive breeding may also be a necessary step to prevent the extinction of the other species, such as the grassland earless dragon (*Tympanocryptis pinguicolla*), which has declined precipitously in recent years (Dimond *et al.*, 2012). More research is necessary to determine the causes of these recent declines. Despite the recent claim that conservation biologists do not need to collect any more data (Possingham, 2012), we clearly need more detailed natural history studies on Australian reptiles. Without basic information on the habitat requirements, diets, life history, and patterns of dispersal of threatened reptile species, it is difficult to diagnose, let alone reverse, population declines (Caughley & Gunn, 1996). Finally, we need to engender an awareness and appreciation of Australia's unique reptile fauna among all young Australians, who will ultimately be responsible for conserving our future.

REFERENCES

Abbott, K. L. (2006). Spatial dynamics of supercolonies of the invasive yellow crazy ant, *Anoplolepis gracilipes*, on Christmas Island, Indian Ocean. *Diversity and Distributions*, **12**: 101–110.

Armstrong, G., Reid, J. R. W. & Hutchinson, M. N. (1993). Discovery of a population of the rare scincid lizard *Tiliqua adelaidensis* (Peters). *Records of the South Australian Museum*, **36**: 153–155.

Banks, M. R. (1993). *Reconnaissance Geology and Geomorphology of the Major Islands South of Tasmania*. Tasmania: Report to Department of Parks, Wildlife and Heritage.

Black, M. P., Mooney, S. D. & Attenbrow, V. (2008). Implications of a 14 200 year contiguous fire record for understanding human-climate relationships at Goochs Swamp, New South Wales, Australia. *Holocene*, **18**: 437–447.

Bohm, M., Collen, B., Baillie, J. E. M., *et al.* (2013). The conservation status of the world's reptiles. *Biological Conservation*, **157**: 372–385.

Boland, C. R. J., Smith, M. J., Maple, D., *et al.* 2011. Heli-baiting using low concentration fipronil to control invasive yellow crazy ant supercolonies on Christmas Island, Indian Ocean. In: Veitch, C. R., Clout, M. N. & Towns, D. R. (eds.) *Island Invasives: Eradication and Management*. Gland, Switzerland: IUCN.

Borsboom, A. C. (2012). Nangur spiny skink. In: Curtis, L. K., Dennis, A. J., McDonald, K. R., Kyne, P. M. & Debus, S. J. S. (eds.) *Queensland's Threatened Animals*. Collingwood: CSIRO Publishing.

Borsboom, A. C., Couper, P. J., Amey, A. & Hoskin, C. J. (2010). Distribution and population genetic structure of the critically endangered skink *Nangura spinosa*, and the implications for management. *Australian Journal of Zoology*, **58**: 369–375.

Bowman, D. (1998). Tansley Review No. 101 – The impact of Aboriginal landscape burning on the Australian biota. *New Phytologist*, **140**: 385–410.

Bradshaw, C. J. A. (2012). Little left to lose: deforestation and forest degradation in Australia since European colonization. *Journal of Plant Ecology*, **5**: 109–120.

Brothers, N., Wiltshire, A., Pemberton, D., Mooney, N. & Green, B. (2003). The feeding ecology and field energetics of the Pedra Branca skink (*Niveoscincus palfreymani*). *Wildlife Research*, **30**: 81–87.

Brown, G. P., Phillips, B. L. & Shine, R. (2011a). The ecological impact of invasive cane toads on tropical snakes: field data do not support laboratory-based predictions. *Ecology*, **92**: 422–431.

Brown, G. W., Bennett, A. F. & Potts, J. M. (2008). Regional faunal decline – reptile occurrence in fragmented rural landscapes of south-eastern Australia. *Wildlife Research*, **35**: 8–18.

Brown, G. W., Dorrough, J. W. & Ramsey, D. S. L. (2011b). Landscape and local influences on patterns of reptile occurrence in grazed temperate woodlands of southern Australia. *Landscape and Urban Planning*, **103**: 277–288.

Burbidge, A. A. (1981). The ecology of the western swamp tortoise *Pseudemydura umbrina* (Testudines: Chelidae). *Australian Wildlife Research*, **8**: 203–223.

Burbidge, A. A. & Kuchling, G. (2004). *Western Swamp Tortoise* (Pseudemydura umbrina) *Recovery Plan*. Wanneroo, Western Australia: Western Australian Threatened Species and Communities Unit.

Burrows, N. D., Burbidge, A. A., Fuller, P. J. & Behn, G. (2006). Evidence of altered fire regimes in the Western Desert region of Australia. *Conservation Science Western Australia*, 5: 272–284.

Butler, J. R. A., Tawake, A., Skewes, T., Tawake, L. & Mcgrath, V. (2012). Integrating traditional ecological knowledge and fisheries management in the Torres Strait, Australia: the catalytic role of turtles and dugong as cultural keystone species. *Ecology and Society*, 17: 34.

Caughley, G. & Gunn, A. (1996). *Conservation Biology in Theory and Practice*. Cambridge, Massachusetts, Blackwell Science.

Chen, Y. H. (2008). Global potential distribution of an invasive species, the yellow crazy ant (*Anoplolepis gracilipes*) under climate change. *Integrative Zoology*, 3: 166–175.

Clarke, S. (2000). *Management of the Pygmy Bluetongue Lizard* (Tiliqua adelaidensis) *on Private Grazing Properties*, Mid-North SA. Canberra: Environment Australia.

Cogger, H., Cameron, E., Sadlier, R. & Eggler, P. (1993). *The Action Plan for Australian Reptiles*. The Australian Museum.

Cogger, H. G. (2000). *Reptiles and Amphibians of Australia*, Sydney, Reed New Holland.

Cogger, H. G., Ford, H., Johnson, C., Holman, J. & Butler, D. (2003). *Impacts of Land Clearing on Australian Wildlife in Queensland*. WWF Australia report.

Croak, B. M., Pike, D. A., Webb, J. K. & Shine, R. (2010). Using artificial rocks to restore nonrenewable shelter sites in human-degraded systems: colonization by fauna. *Restoration Ecology*, 18: 428–438.

Croak, B. M., Pike, D. A., Webb, J. K. & Shine, R. (2012). Habitat selection in a rocky landscape: experimentally decoupling the influence of retreat site attributes from that of landscape features. *PLoS ONE*, 7.

Cunningham, R. B., Lindenmayer, D. B., Crane, M., Michael, D. & Macgregor, C. (2007). Reptile and arboreal marsupial response to replanted vegetation in agricultural landscapes. *Ecological Applications*, 17: 609–619.

DERM (2010). Analysis of woody vegetation clearing rates in Queensland. Supplementary report to land cover change in Queensland 2008–09. In: *Vegetation Management*, D. O. E. A. R. M., QLD (ed.). Department of Environment and Resource Management.

DEWHA (2008a). *Threat Abatement Plan for Predation by Feral Cats*. Department of the Environment, Water, Heritage and the Arts, Canberrra.

DEWHA (2008b). *Threat Abatement Plan for Predation by the European Red Fox*. Canberra: Department of the Environment, Water, Heritage and the Arts.

Dickman, C. R. (1996). Impact of exotic generalist predators on the native fauna of Australia. *Wildlife Biology*, 2: 185–195.

Dimond, W. J., Osborne, W. S., Evans, M. C., Gruber, B. & Sarre, S. D. (2012). Back to the brink: population decline of the endangered grassland earless dragon (*Tympanocryptis pinguicolla*) following its rediscovery. *Herpetological Conservation and Biology*, 7: 132–149.

Doody, J. S., Green, B., Rhind, D., *et al.* (2009). Population-level declines in Australian predators caused by an invasive species. *Animal Conservation*, **12**: 46–53.

Doody, J. S., Castellano, C. M., Rhind, D. & Green, B. (2013). Indirect facilitation of a native mesopredator by an invasive species: are cane toads re-shaping tropical riparian communities? *Biological Invasions*, **15**: 559–568.

Dorrough, J., Mcintyre, S., Brown, G., *et al.* (2012). Differential responses of plants, reptiles and birds to grazing management, fertilizer and tree clearing. *Austral Ecology*, **37**: 569–582.

Driscoll, D. A. (2004). Extinction and outbreaks accompany fragmentation of a reptile community. *Ecological Applications*, **14**: 220–240.

Driscoll, D. A. & Henderson, M. K. (2008). How many common reptile species are fire specialists? A replicated natural experiment highlights the predictive weakness of a fire succession model. *Biological Conservation*, **141**: 460–471.

Duffy, A., Pound, L. & How, T. (2012). *Recovery Plan for the Pygmy Bluetongue Lizard* Tiliqua adelaidensis. South Australia: Department of Environment and Natural Resources.

Edwards, G. P., Allan, G. E., Brock, C., *et al.* (2008). Fire and its management in central Australia. *Rangeland Journal*, **30**: 109–121.

EPBC. (2014). *EPBC Act List of Threatened Fauna* [Online]. Australian Government. Available: http://www.environment.gov.au/cgi-bin/sprat/public/publicthreatenedlist. pl [Accessed 1 August 2014].

Fenner, A. & Bull, C. M. (2007). Short-term impact of grassland fire on the endangered pygmy bluetongue lizard. *Journal of Zoology*, **272**: 444–450.

Florance, D., Webb, J. K., Dempster, T., *et al.* (2011). Excluding access to invasion hubs can contain the spread of an invasive vertebrate. *Proceedings of the Royal Society B – Biological Sciences*, **278**: 2900–2908.

Fordham, D. A., Watts, M. J., Delean, S., *et al.* (2012). Managed relocation as an adaptation strategy for mitigating climate change threats to the persistence of an endangered lizard. *Global Change Biology*, **18**: 2743–2755.

Frankham, R. (2005). Genetics and extinction. *Biological Conservation*, **126**: 131–140.

Greer, A. E. (1997). *The Biology and Evolution of Australian Snakes*. Sydney, Surrey Beatty and Sons.

Griffiths, A. D. & Christian, K. A. (1996). The effects of fire on the frillneck lizard (*Chlamydosaurus kingii*) in northern Australia. *Australian Journal of Ecology*, **21**: 386–398.

Hall, J., Rose, K., Spratt, D., *et al.* 2011. *Assessment of Reptile and Mammal Disease Prevalence on Christmas Island*. Taronga Conservation Society Australia.

Hamann, M., Fuentes, M. M. P. B., Ban, N. C. & Mocellin, V. J. L. 2013. Climate change and marine turtles. In: Wyneken, J., Lohmann, K. J. & Musick, J. A. (eds.) *The Biology of Sea Turtles Volume III*. Boca Raton: CRC Press, 353–378.

Hutchinson, M. N., Milne, T. & Croft, T. (1994). Redescription and ecological notes on the Pygmy Bluetongue, *Tiliqua adelaidensis* (Squamata: Scincidae). *Transactions of the Royal Society of South Australia*, **118**: 217–226.

Hyde, M. (1995). *The Temperate Grasslands of South Australia their Composition and Conservation Status*. Sydney: World Wide Fund for Nature Australia.

IUCN (2013). *The IUCN Red List of Threatened Species*. Version 2013.1.

James, C. (2003). Response of vertebrates to fenceline contrasts in grazing intensity in semi-arid woodlands of eastern Australia. *Austral Ecology*, **28**: 137–151.

James, C. D., Landsberg, J. & Morton, Sr. (1999). Provision of watering points in the Australian arid zone: a review of effects on biota. *Journal of Arid Environments*, **41**: 87–121.

King, J. M., Kuchling, G. & Bradshaw, S. D. (1998). Thermal environment, behavior, and body condition of wild *Pseudemydura umbrina* (Testudines: Chelidae) during late winter and early spring. *Herpetologica*, **54**: 103–112.

Kitchener, D. J. & How, R. A. (1982). Lizard species in small mainland habitat isolates and islands off south-western Western Australia. *Australian Wildlife Research*, 9: 357–363.

Krefft, G. (1869). *The Snakes of Australia: an Illustrated and Descriptive Catalogue of all the Known Species*. Sydney, Thomas Richards, Government Printer.

Kwan, D., Dews, G., Bishop, M. & Garnier, H. (2001). Towards community based management of natural marine resources in Torres Strait. In: Baker, R., Davies, J. & Young, E. (eds.) *Working On Country: Indigenous Environmental Management in Australia*. Melbourne: Oxford University Press.

Legge, S., Murphy, S., Heathcote, J., *et al.* (2008). The short-term effects of an extensive and high-intensity fire on vertebrates in the tropical savannas of the central Kimberley, northern Australia. *Wildlife Research*, **35**: 33–43.

Legge, S., Kennedy, M. S., Lloyd, R., Murphy, S. A. & Fisher, A. (2011). Rapid recovery of mammal fauna in the central Kimberley, northern Australia, following the removal of introduced herbivores. *Austral Ecology*, **36**: 791–799.

Letnic, M., Webb, J. K. & Shine, R. (2008). Invasive cane toads (*Bufo marinus*) cause mass mortality of freshwater crocodiles (*Crocodylus johnstoni*) in tropical Australia. *Biological Conservation*, **141**: 1773–1782.

Lindenmayer, D. & Hunter, M. (2010). Some guiding concepts for conservation biology. *Conservation Biology*, **24**: 1459–1468.

Lindenmayer, D. B., Wood, J. T., Macgregor, C., *et al.* (2008). How predictable are reptile responses to wildfire? *Oikos*, **117**: 1086–1097.

Manning, A. D., Cunningham, R. B. & Lindenmayer, D. B. (2013). Bringing forward the benefits of coarse woody debris in ecosystem recovery under different levels of grazing and vegetation density. *Biological Conservation*, **157**: 204–214.

Maple, D. J., Barr, R. & Smith, M. J. (2012). A new record of the Christmas Island blind snake, *Ramphotyphlops exocoeti* (Reptilia: Squamata: Typhlopidae). *Records of the Western Australian Museum*, **27**: 156–160.

Martin, T. G. & McIntye, S. (2007). Impacts of livestock grazing and tree clearing on birds of woodland and riparian habitats. *Conservation Biology*, **21**: 504–514.

Mcloughlin, L. C. (1998). Season of burning in the Sydney region: the historical records compared with recent prescribed burning. *Australian Journal of Ecology*, **23**: 393–404.

Michael, D. R., Cunningham, R. B. & Lindenmayer, D. B. (2011). Regrowth and revegetation in temperate Australia presents a conservation challenge for reptile fauna in agricultural landscapes. *Biological Conservation*, **144**: 407–415.

Miller, J. & Limpus, C. (2012). Green turtle. In: Curtis, L. K., Dennis, A. J., McDonald, K. R., Kyne, P. M. & Debus, S. J. S. (eds.) *Queensland's Threatened Animals.* CSIRO Publishing.

Milne, T. (1999). Conservation and the ecology of the endangered Pygmy Bluetongue lizard *Tiliqua adelaidensis*. PhD, The Flinders University of South Australia.

Milne, T. & Bull, C. M. (2000). Burrow choice by individuals of different sizes in the endangered Pygmy Blue Tongue Lizard *Tiliqua adelaidensis*. *Biological Conservation*, **95**: 295–301.

Mitchell, N., Hipsey, M., Arnall, S., et al. (2012a). Linking eco-energetics and eco-hydrology to select sites for the assisted colonization of Australia's rarest reptile. *Biology*, **2**: 1–25.

Mitchell, N. J., Jones, T. V. & Kuchling, G. (2012b). Simulated climate change increases juvenile growth in a critically endangered tortoise. *Endangered Species Research*, **17**: 73–82.

Moloney, S. D. & Vanderwoude, C. (2003). Potential ecological impacts of red imported fire ants in eastern Australia. *Journal of Agricultural and Urban Entomology*, **20**: 131–142.

Morton, S. R. & James, C. D. (1988). The diversity and abundance of lizards in arid Australia: a new hypothesis. *American Naturalist*, **132**: 237–256.

Munro, N. T., Lindenmayer, D. B. & Fischer, J. (2007). Faunal response to revegetation in agricultural areas of Australia: a review. *Ecological Management & Restoration*, **8**: 199–207.

Nogales, M., Martin, A., Tershy, B. R., et al. (2004). A review of feral cat eradication on islands. *Conservation Biology*, **18**: 310–319.

Nursey-Bray, M. (2009). A Guugu Yimmithir Bam Wii: Ngawiya and Girrbithi: hunting, planning and management along the Great Barrier Reef, Australia. *Geoforum*, **40**: 442–453.

O'Dowd, D. J., Green, P. T. & Lake, P. S. (2003). Invasional 'meltdown' on an oceanic island. *Ecology Letters*, **6**: 812–817.

Paltridge, R., Gibson, D. & Edwards, G. (1997). Diet of the feral cat (*Felis catus*) in central Australia. *Wildlife Research*, **24**: 67–76.

Pianka, E. R. (1969). Habitat specificity, speciation, and species density in Australian desert lizards. *Ecology* **50**: 498–502.

Pianka, E. R. (1986). *Ecology and Natural History of Desert Lizards. Analyses of the Ecology, Niche and Community Structure*. Princeton University Press.

Pianka, E. R. & Goodyear, S. E. (2012). Lizard responses to wildfire in arid interior Australia: long-term experimental data and commonalities with other studies. *Austral Ecology*, **37**: 1–11.

Pike, D. A., Croak, B. M., Webb, J. K. & Shine, R. (2010). Subtle – but easily reversible – anthropogenic disturbance seriously degrades habitat quality for rock-dwelling reptiles. *Animal Conservation*, **13**: 411–418.

Pike, D. A., Webb, J. K. & Shine, R. (2012). Reply to comment on 'chainsawing for conservation: ecologically informed tree removal for habitat management'. *Ecological Management & Restoration*, **13**: e12–13.

Possingham, H. (2012). How can we sell evaluating, analyzing and synthesizing to young scientists? *Animal Conservation*, **15**: 229–230.

Price-Rees, S. J., Brown, G. P. & Shine, R. (2010). Predation on toxic cane toads (*Bufo marinus*) may imperil bluetongue lizards (*Tiliqua scincoides intermedia*, Scincidae) in tropical Australia. *Wildlife Research*, **37**: 166–173.

Price, O. F., Russell-Smith, J. & Watt, F. (2012). The influence of prescribed fire on the extent of wildfire in savanna landscapes of western Arnhem Land, Australia. *International Journal of Wildland Fire*, **21**: 297–305.

Pringle, R. M., Syfert, M., Webb, J. K. & Shine, R. (2009). Quantifying historical changes in habitat availability for endangered species: use of pixel- and object-based remote sensing. *Journal of Applied Ecology*, **46**: 544–553.

Reed, D. H., O'Grady, J. J., Ballou, J. D. & Frankham, R. (2003). The frequency and severity of catastrophic die-offs in vertebrates. *Animal Conservation*, **6**: 109–114.

Rodda, G. H. & Savidge, J. A. (2007). Biology and impacts of pacific island invasive species. 2. *Boiga irregularis*, the Brown Tree Snake (Reptilia : Colubiridae). *Pacific Science*, **61**: 307–324.

Russell-Smith, J., Yates, C., Edwards, A., *et al.* (2003). Contemporary fire regimes of northern Australia, 1997–2001: change since Aboriginal occupancy, challenges for sustainable management. *International Journal of Wildland Fire*, **12**: 283–297.

Saunders, G. R., Gentle, M. N. & Dickman, C. R. (2010). The impacts and management of foxes *Vulpes vulpes* in Australia. *Mammal Review*, **40**: 181–211.

Schmidt, D., Spring, D., Mac Nally, R., *et al.* (2010). Finding needles (or ants) in haystacks: predicting locations of invasive organisms to inform eradication and containment. *Ecological Applications*, **20**: 1217–1227.

Shea, S. R., Peet, G. B. & Cheney, N. P. (1981). The role of fire in forest management. In: Gill, A. M., Groves, R. H. & Noble, I. R. (eds.) *Fire and the Australian Biota*. Canberra: Australian Academy of Science.

Short, J. & Turner, B. (2000). Reintroduction of the burrowing bettong *Bettongia lesueur* (Marsupialia : Potoroidae) to mainland Australia. *Biological Conservation*, **96**: 185–196.

Smith, A. L., Gardner, M. G., Fenner, A. L. & Bull, C. M. (2009). Restricted gene flow in the endangered pygmy bluetongue lizard (*Tiliqua adelaidensis*) in a fragmented agricultural landscape. *Wildlife Research*, **36**: 466–478.

Smith, G. T., Arnold, G. W., Sarre, S., Abenspergtraun, M. & Steven, D. E. (1996). The effects of habitat fragmentation and livestock-grazing on animal communities in remnants of gimlet *Eucalyptus salubris* woodland in the Western Australian wheatbelt. 2. Lizards. *Journal of Applied Ecology*, **33**: 1302–1310.

Smith, M. J., Cogger, H., Tiernan, B., *et al.* (2012). An oceanic island reptile community under threat: the decline of reptiles on Christmas Island, Indian Ocean. *Herpetological Conservation and Biology*, **7**: 206–218.

Souter, N. J., Bull, C. M. & Hutchinson, M. N. (2004). Adding burrows to enhance a population of the endangered pygmy blue tongue lizard, *Tiliqua adelaidensis*. *Biological Conservation*, **116**: 403–408.

Souter, N. J., Bull, C. M., Lethbridge, M. R. & Hutchinson, M. N. (2007). Habitat requirements of the endangered pygmy bluetongue lizard, *Tiliqua adelaidensis*. *Biological Conservation*, **135**: 33–45.

Stow, A. J., Sunnucks, P., Briscoe, D. A. & Gardner, M. G. (2001). The impact of habitat fragmentation on dispersal of Cunningham's skink (*Egernia cunninghami*): evidence from allelic and genotypic analyses of microsatellites. *Molecular Ecology*, **10**: 867–878.

Taylor, R. J. & Hoxley, G. (2003). Dryland salinity in Western Australia: managing a changing water cycle. *Water Science and Technology*, **47**: 201–207.

Tilman, D., May, R. M., Lehman, C. L. & Nowak, M. A. (1994). Habitat destruction and the extinction debt. *Nature*, **371**: 65–66.

Tingley, R., Phillips, B. L., Letnic, M., *et al.* (2013). Identifying optimal barriers to halt the invasion of cane toads *Rhinella marina* in arid Australia. *Journal of Applied Ecology*, **50**: 129–137.

Vigilante, T. & Bowman, D. (2004). Effects of individual fire events on the flower production of fruit-bearing tree species, with reference to Aboriginal people's management and use, at Kalumburu, North Kimberley, Australia. *Australian Journal of Botany*, **52**: 405–415.

Watson, J. E. M., Evans, M. C., Carwardine, J., *et al.* (2011). The capacity of Australia's protected-area system to represent threatened species. *Conservation Biology*, **25**: 324–332.

Webb, J. K. & Shine, R. (1998). Using thermal ecology to predict retreat-site selection by an endangered snake species. *Biological Conservation*, **86**: 233–242.

Webb, J. K., Brook, B. W. & Shine, R. (2002a). Collectors endanger Australia's most threatened snake, the broad-headed snake *Hoplocephalus bungaroides*. *Oryx*, **36**: 170–181.

Webb, J. K., Brook, B. W. & Shine, R. (2002b). What makes a species vulnerable to extinction? Comparative life-history traits of two sympatric snakes. *Ecological Research*, **17**: 59–67.

Whitehead, P., Purdon, P., Russell-Smith, J., Cooke, P. & Sutton, S. (2008). The management of climate change through prescribed savanna burning: emerging contributions of indigenous people in northern Australia. *Public Administration and Development*, **28**: 374–385.

Williams, J. R., Driscoll, D. A. & Bull, C. M. (2012). Roadside connectivity does not increase reptile abundance or richness in a fragmented mallee landscape. *Austral Ecology*, **37**: 383–391.

Wilson, G. R., Edwards, M. J. & Smits, J. K. (2010). Support for Indigenous wildlife management in Australia to enable sustainable use. *Wildlife Research*, **37**: 255–263.

Witt, M. J., Hawkes, L. A., Godfrey, M. H., Godley, B. J. & Broderick, A. C. (2010). Predicting the impacts of climate change on a globally distributed sepcies: the case of the loggerhead turtle. *Journal of Experimenal Biology*, **213**: 901–911.

Woinarski, J. C. Z. & Fisher, A. (2003). Conservation and the maintenance of biodiversity in the rangelands. *Rangeland Journal*, **25**: 157–171.

Woinarski, J. C. Z., Armstrong, M., Brennan, K., *et al.* (2010). Monitoring indicates rapid and severe decline of native small mammals in Kakadu National Park, northern Australia. *Wildlife Research*, **37**: 116–126.

Woinarski, J. C. Z., Legge, S., Fitzsimons, J. A., *et al.* (2011). The disappearing mammal fauna of northern Australia: context, cause, and response. *Conservation Letters*, **4**: 192–201.

Woinarski, J. C. Z., Green, J., Fisher, A., Ensbey, M. & Mackey, B. (2013). The effectiveness of conservation reserves: land tenure impacts upon biodiversity across extensive natural landscapes in the tropical savannahs of the Northern Territory, Australia. *Land*, **2**: 20–36.

Young, G. R., Bellis, G. A., Brown, G. R. & Smith, E. S. C. (2001). The crazy ant *Anoplolepis gracilipes* (Smith) (Hymenoptera: Formicidae) in east Arnhem Land, Australia. *Australian Entomologist*, **28**: 97–104.

CHAPTER 18

New Zealand reptiles and their conservation

Nicola J. Nelson, Rod Hitchmough and Jo M. Monks

Summary

The reptile fauna of New Zealand consists of one endemic order, tuatara; at least 100 species of lizards (geckos – Diplodactylidae; skinks – Scincidae); invasive rainbow skinks; two migrant and three vagrant species of sea turtles; and one resident and three vagrant species of sea snake. The primary threat to the persistence of populations of New Zealand reptiles is predation by introduced mammals. The largest species of each lizard family are extinct, and surviving large-bodied and nocturnal lizards and tuatara are restricted to offshore islands free of introduced mammals or managed in mammal-proof enclosures on the mainland. Habitat loss, particularly through development initiatives, also threatens reptiles, as do poaching, lack of knowledge on potential diseases, climate change and potentially introduced reptiles as predators, competitors and hosts of new diseases. Research effort and targeted field surveys have resulted in taxonomic revisions, discoveries of new species, increased data availability on some populations, and changes in conservation management practices resulting in an evolving picture for species numbers and the threat status of New Zealand reptiles.

18.1 Introducing New Zealand reptiles

The reptile fauna of New Zealand consists of the endemic terrestrial tuatara and lizards, an invasive lizard, migrant sea turtles, a resident sea snake, and other occasional vagrant sea turtles and sea snakes (Gill & Whitaker, 1996; Jewell, 2008; Chapple *et al.*, 2009a; Hitchmough *et al.*, 2010). A review of the conservation efforts for New Zealand reptiles was published in 2001 (Towns *et al.*, 2001). Since that time considerable research effort and targeted field surveys have resulted in taxonomic revisions, discoveries of new species, increased data availability on some populations, and changes in conservation management practices (Chapple *et al.*, 2009b; Hitchmough *et al.*, 2010; Nielsen *et al.*, 2011). We provide an update on the conservation status of New Zealand reptiles and

Austral Ark: The State of Wildlife in Australia and New Zealand, eds. A. Stow, N. Maclean and G. I. Holwell. Published by Cambridge University Press. © Cambridge University Press 2015.

their conservation management and future priorities and present case studies that demonstrate conservation successes and challenges.

New Zealand reptiles include an endemic order, the Rhynchocephalia, of which tuatara are the sole living representatives (Hay *et al.*, 2010; Gaze, 2001). The native lizard fauna of at least 100 species is made up of two endemic radiations; geckos in the family Diplodactylidae (Nielsen *et al.*, 2011) and skinks in the family Scincidae (Chapple *et al.*, 2009b). The marine fauna comprises two migrant and three vagrant species of sea turtles and one resident and three vagrant species of sea snake (Hitchmough *et al.*, 2013). The only established introduced reptile is the rainbow skink (*Lampropholis delicata*), also known as the delicate or plague skink, which likely arrived with cargo from Australia in the 1960s (Robb, 1986; Gill & Whitaker, 1996; Chapple & Thompson, 2009).

The distribution of New Zealand reptiles is recorded in an electronic database on the Department of Conservation website (DOC, 2013a). All native New Zealand reptiles (terrestrial and marine) are protected by legislation under the Wildlife Act 1953 and its amendments. Introduced rainbow skinks (established, Gill & Whitaker, 1996) and red-eared slider turtles (potential to establish, Kikillus, 2010) are listed on Schedule 5 of the Act, specifying they have no protection status and are able to be actively controlled (Wildlife Act, 1953).

The increase in the number of lizard species since Gill & Whitaker's (1996) guidebook and Towns *et al.*'s review (2001) is due to new species discoveries and identification of cryptic species using molecular identification methods (Hay *et al.*, 2010; Nielsen *et al.*, 2011; Chapple *et al.*, 2009b). Changes in the number of taxa included in lists generated during conservation status reviews using the New Zealand Threat Classification System (Molloy *et al.*, 2002) show the pace of discovery, delimitation and formal description of new species: Hitchmough (2002) listed 80 extant native lizard taxa (45 described and 35 undescribed); Hitchmough *et al.* (2007) listed 88 (47 described and 41 undescribed); Hitchmough *et al.* (2010) listed 99 (54 described and 45 undescribed); and Hitchmough *et al.* (2013) again listed 99 taxa (58 described and 41 undescribed).

18.1.1 Tuatara

Tuatara are found only in New Zealand (Cree & Butler, 1993). They are now recognised as one species with distinctive geographic variants based on microsatellite and mitochondrial DNA evidence, rather than two species as previously reported (Hay *et al.*, 2010; Gaze, 2001; Cree & Butler, 1993). At their lowest ebb, they comprised 32 off-shore island populations, with the last extinction of a population in the early 1980s (Cree & Butler, 1993). Tuatara were functionally extinct on the mainland (defined as North, South and Stewart Islands) of New Zealand by about 1840 (Towns *et al.*, 2001). On islands with areas ranging from 0.4 to over 3000 hectares, populations varied in size from one to approximately 50 000 individuals (Cree & Butler, 1993). Historic records of subfossil remains indicate that tuatara were widely distributed throughout the North and South Islands of New Zealand in a wide variety of habitats, including areas that would have experienced snow in winter for periods of up to three months (Cree & Butler, 1993). Genetic diversity and population size of these historic populations are not known.

18.1.2 Geckos

The New Zealand geckos (Diplodactylidae) are a single endemic radiation from an ancestral colonist which arrived from Australia 40 million or more years ago (Nielsen

Figure 18.1 (Plate 37) Lizard pollinators: *Woodworthia maculate* carrying out an 'ecosystem service'. A black and white version of this figure will appear in some formats. For the colour version, please refer to the plate section.

et al., 2011). Previously it had been suggested that the closest relatives of New Zealand's geckos were in New Caledonia (Chambers *et al.*, 2001), but the work of Nielsen *et al.* (2011) showed that the New Caledonian and New Zealand gecko faunas were derived independently from Australia.

The New Zealand Diplodactylidae currently include 19 extant and one extinct species in seven genera, plus 22–23 identified taxa that have yet to be named, and further discoveries remain likely. From 1990 until 2011 only two genera were recognised – *Naultinus* for the very brightly-coloured diurnal green geckos and *Hoplodactylus* for the rest, which were mostly brown in colour and nocturnal or crepuscular. Nielsen *et al.* (2011) showed that these two genera are not sister groups. To retain *Naultinus* (eight recognised and one undescribed species) as a genus while avoiding having one genus contained within another it was necessary to recognise five new or resurrected genera – *Dactylocnemis* (one described and five undescribed species) for the Pacific gecko and its undescribed relatives, *Mokopirirakau* (four described and six or seven undescribed species) for the forest and black-eyed geckos and their relatives, *Toropuku* (one described and one undescribed species) for the striped geckos, *Tukutuku* (one described species) for the harlequin gecko, and *Woodworthia* (three described and nine undescribed species) for the 'common' and gold-stripe geckos and their relatives (Figure 18.1, Plate 37). *Hoplodactylus* now contains only Duvaucel's gecko (*H. duvaucelii*) and the giant extinct kawekaweau (*H. delcourti*).

Sympatric populations of geckos commonly comprise only two to (rarely) five species, with lower diversity at mainland sites reflecting the impacts of partial or total extinctions following changes associated with human settlement. Within each genus, species mainly or entirely replace each other geographically in a mosaic of abutting species distributions across the country, with no more than two species of one genus usually being found together (Nielsen *et al.*, 2011). A regional exception is southern Marlborough, where six *Woodworthia* species occur in the same region, but differences in habitat use mean that no more than two have been found at any single site. *Toropuku* (known from two islands in Cook Strait and a few localities on the Coromandel Peninsula) and *Tukutuku* (endemic to Stewart Island) have restricted distributions. *Naultinus* species have a mosaic of abutting species distributions across the entire country, with no known occurrences of more than one species at any site. *Hoplodactylus duvaucelii* was found over most of the country in pre-human times, but it is now confined to offshore islands in Cook Strait and northern New Zealand. Nothing concrete is known of the former distribution of *H. delcourti*, but the absence of fossil material for this extinct giant species (~370mm snout–vent length) suggests its distribution was restricted (Bauer & Russell, 1986). All three of the remaining genera can be found together in some places, but the highest species diversity and numerical dominance of *Dactylocnemis* is centred in the north of the North Island and it is not found south of Cook Strait. *Mokopirirakau* and *Woodworthia* are both found throughout the country except for the northern part of Northland, but *Woodworthia* predominates in the east of the South Island and south-east of the North Island, and *Mokopirirakau* in the higher rainfall areas in the west.

18.1.3 Skinks

From 1977 until 2009, two genera of New Zealand skinks were recognised (Hardy, 1977), but phylogenetic analysis by Chapple *et al.* (2009b) indicated that these genera were not clearly separated and that any subdivision of the New Zealand skink radiation into smaller groups was very difficult, so they recognised only the genus *Oligosoma* for all the New Zealand skinks. The New Zealand skinks, like the geckos, are the descendants of a single colonisation event, but they probably arrived here in the early Miocene (16–22.6 million years ago), more recently than the geckos and via a different route (Chapple *et al.*, 2009b). Skink bones are present in the Manuherikia fossil beds, believed to be 16–19 million years old, so if they were not present in New Zealand during the Oligocene, they must have arrived immediately after it ended (Lee *et al.*, 2009). The sister group of the New Zealand radiation is *Oligosoma lichenigera*, found on Lord Howe and Norfolk Islands, with the next most closely related groups in New Caledonia, indicating that skinks reached New Zealand by 'island hopping' south (Chapple *et al.*, 2009b). The recent meta-analysis of lizard and snake molecular systematics by Pyron *et al.* (2013) suggests that *Oligosoma* is the sister group of the New Caledonian radiation rather than derived from within it, as suggested by Chapple *et al.* (2009b).

The genus includes 39 named and about 17 known un-named New Zealand species. The species are ecologically quite diverse, including nocturnal and diurnal, and litter-dwelling, saxicolous and arboreal species, and ranging in adult size from 2 grams (*O. levidensum*) to extant species of 60 grams (*O. alani*). The extinct species *O. northlandi* was larger having femora 50% longer than those of extant adult *O. alani* (Worthy, 1991). The robustly built, primarily nocturnal, litter- and burrow-dwelling species formerly placed in the genus *Cyclodina* (Hardy, 1977) are confined to the North Island, whereas diurnal, sun-basking species are found throughout the country. Sympatric diversity of skinks is often higher than that of geckos. Up to six species can be found together on the mainland, and there are eight species on Little Barrier Island (DOC, 2013a), although they are generally separated by habitat and activity periods.

The abundance and diversity of lizards (both geckos and skinks) in the fossil record, on offshore islands free of introduced mammals, and in sites where habitat structure (such as extensive rock fields and/or dense divaricating vegetation) provides security from predators, strongly indicate that lizards have greatly reduced abundance in most mainland areas (Worthy, 1987, 1991; Towns, 1992a; Towns *et al.*, 2001). However, some native birds such as weka (*Gallirallus australis*) are also efficient lizard predators (Lettink *et al.*, 2013), so densities on the mainland in some habitats may never have been as high as they are on some offshore islands where these birds are absent. In general, the largest species have become extinct or nearly so on the mainland, and smaller species have declined greatly except at sites which provide security from mammalian predators (Towns *et al.*, 2001). A few species of small skinks remain reasonably common even in highly modified sites such as suburban gardens, but even these continue to decline in abundance and over much of the mainland it is very hard to find any lizards at all.

18.1.4 Marine fauna

Five marine turtles and four sea snakes (Hitchmough *et al.*, 2013) are natural visitors to New Zealand, arriving here without any human assistance (Gill & Whitaker, 1996; Gill *et al.*, 2001; Hitchmough *et al.*, 2010; Hitchmough *et al.*, 2013). Green turtles (*Chelonia mydas*) are resident at the Kermadec Islands for long periods but do not breed there, and also visit Northland waters as annual migrants. Leatherback turtles (*Dermochelys coriacea*) also visit regularly enough to be best regarded as migrants, allowing for the low detection probability at sea. The yellow-bellied sea snake (*Pelamis platurus*) is an entirely pelagic species associated with rafts of algae and debris in which it is usually concealed, so it is likely to be much more abundant than coastal records indicate. Modelling of its habitat and sea temperature requirements (Graham *et al.*, 1971) suggests that it is highly likely to be a permanent resident in the northernmost parts of the New Zealand Exclusive Economic Zone and, as it gives birth in its normal habitat at sea rather than returning to land to do so (Vallarino & Weldon, 1996), is also highly likely to breed there. Three other sea snakes of the genus *Laticauda* and three other turtles (hawksbill (*Eretmochelys imbricata*), loggerhead (*Caretta caretta*), and olive ridley (*Lepidochelys olivacea*)) visit less regularly and are regarded as vagrants (Gill & Whitaker, 1996; Gill *et al.*, 2001; Hitchmough *et al.*, 2010; Hitchmough *et al.*, 2013).

18.2 Threats

18.2.1 Introduced mammals

The primary threat to the persistence of populations of New Zealand reptiles is predation by introduced mammals (Towns *et al.*, 2001). Large-bodied and nocturnal species are the most vulnerable, as evidenced by extinctions of the largest gecko (kawekaweau) and the largest skink (*O. northlandi*) following human settlement and the introduction of mammalian predators (Worthy, 1987). Other large-bodied, nocturnal lizards and tuatara are restricted to offshore islands free of introduced mammals (Towns *et al.*, 2001); populations recently translocated into mammal-proof enclosures on the mainland are exceptions (see Box 18.1).

In the national threat listing process each recognised species is assigned a threat status based on criteria pertaining to number and size of populations, geographic extent and known threats (Townsend *et al.*, 2008). At the latest listing of reptiles, 32 species were considered 'Threatened' with extinction; three of which were assigned the highest threat

> ## Box 18.1 Recovery of tuatara
>
> Conservation efforts for tuatara have included actions driven by a Recovery Group and resulting Recovery Plans (Cree & Butler, 1993; Gaze, 2001). Four island populations have been augmented by artificial incubation of eggs, head starting and reintroduction of hatchlings. Reintroductions have occurred to islands following rodent eradications, where populations had been extirpated by rodents (Sherley *et al.*, 2010). Tuatara have recently been translocated to five mainland sites where introduced mammal control is in place (Nelson, unpublished data). Tuatara are in captivity at licensed facilities throughout New Zealand and in several overseas zoos, and are often used as advocates for conservation at educational events. They are recognised for their importance to extant reptilian diversity and representation of otherwise extinct linkages, and as treasures for Māori, and therefore enjoy a high public profile nationally and internationally. As such, they have attracted extensive research effort resulting in a substantial body of knowledge of their biology on which to base conservation action (Cree & Butler, 1993; Gaze, 2001).
>
> Many lizard populations on New Zealand's mainland, which hosts a suite of 11 introduced mammalian predators (Towns *et al.*, 2001), remain at threat of extinction due to predation pressure by introduced mammals. The reality of this continued pressure is exemplified by: (1) national threat status of many lizard species restricted to the mainland being considered higher than those with multiple secure offshore island populations (Hitchmough *et al.*, 2013); (2) mainland populations of vulnerable species continuing to decline at sites where predator management is at low intensity (e.g. Hoare *et al.*, 2007a); and (3) recovery of populations that previously showed drastic declines, as a result of intensive predator management (Reardon *et al.*, 2012; see Box 18.2).

category, 'Nationally Critical', ten were considered 'Nationally Endangered' and 19 were considered 'Nationally Vulnerable' (Hitchmough *et al.*, 2013). Of these 32 threatened species, 25 do not have secure, mammal-free populations and the other seven species have only one or a few populations free of mammalian predation pressure (Hitchmough *et al.*, 2013). A further 26 species are considered 'At Risk' of extinction and categorised as 'Declining' (Hitchmough *et al.*, 2013). For the majority of these 26 species (all but one species), the majority of known populations occur on the mainland and there is evidence for declines in several of them (Hitchmough *et al.*, 2013). In contrast, the 15 species listed as 'Recovering' or 'Relict' now occur entirely or primarily on offshore islands free of mammals and populations are either healthy (if always mammal-free) or recovering following the eradication of mammals (Hitchmough *et al.*, 2013).

At mainland sites where predator management is at low intensity or non-existent, the more vulnerable lizard populations (i.e. larger-bodied species) continue to decline. The best-documented case involves a mainland population of Whitaker's skink, *Oligosoma whitakeri*, a large-bodied (up to 101 mm snout–vent length), nocturnal skink vulnerable to predation by introduced mammals. The species is considered 'Nationally Endangered' under national threat listing criteria (Hitchmough *et al.*, 2013) and 'Vulnerable' under

Box 18.2 Protecting grand and Otago skinks from introduced mammalian predators

Grand and Otago skinks (*Oligosoma grande* and *O. otagense*, respectively) are two of New Zealand's largest and most critically endangered species (Tocher, 2009; Hitchmough *et al.*, 2010) for which no secure offshore island population exists. Both species are associated with the deeply fractured schist rock outcrop landscapes of Central Otago in the South Island. They have shown an apparent decline to about 8% of their presumed former ranges (Whitaker & Loh, 1995). Relict populations of both species now occur in native tussock grasslands at the extreme east and west of their former range (Reardon *et al.*, 2012), with localised extinctions of populations being recorded within their current range since the mid-1970s (Whitaker & Loh, 1995).

Molecular evidence for grand skinks supports the hypothesis that the recent and dramatic declines witnessed are of anthropogenic origin (Berry & Gleeson, 2005). Specifically, declines are attributed to a history of hyperabundant introduced rabbits, *Oryctolagus cuniculus*, and the suite of introduced mammals that prey on them and switch to eating native species (including skinks) when rabbit populations crash (Tocher, 2006). Initial conservation management aiming to selectively remove top predators (feral cats, *Felis catus*, and ferrets, *Mustela furo*) through trapping failed to produce population-level recovery (Tocher, 2006). Consequently, a recovery programme including *in situ* management was instigated to evaluate whether remnant populations of grand and Otago skinks could be protected by either eradication of five mammalian predator species (cat, ferret, stoat (*M. erminea*), weasel (*M. nivalis*) and hedgehog (*Erinaceus europaeus*)) within mammal-proof fences or landscape-scale suppression of the same suite of predators (Reardon *et al.*, 2012).

Skink populations inside mammal-proof fences and those protected by trapping of predators within a 2100 ha area were compared with populations where predators were unmanaged (Reardon *et al.*, 2012). Monitoring of skink populations was by non-invasive photo-resight methods. At one of the unmanaged sites, the grand skink population underwent a catastrophic decline that was attributed to episodic predation by a stoat.[1] Comparison of managed and unmanaged populations confirmed that predation by introduced mammals is a key driver in the decline of these skinks and that the use of either mammal-proof fences or intensive predator control over a large enough area allowed skink populations to recover (Reardon *et al.*, 2012). Reardon *et al.*'s (2012) study provides an important precedent as it documents the first successful demonstration that introduced predators can be controlled at a mainland site to a level allowing recovery of vulnerable lizard populations. This result has strong implications for the feasibility and design of predator management to protect similarly vulnerable lizard species elsewhere on the New Zealand mainland.

World Conservation Union criteria (IUCN, 2012). Three natural, disjunct populations of Whitaker's skinks remained (Miller *et al.*, 2009) until recently when translocations to three further offshore islands were conducted to improve the security of the

---

[1] A stoat was seen actively searching for prey at the site during the monitoring period and both pre- and post-stoat event survival estimates and evidence of tail autotomy among survivors support the hypothesis that the skinks underwent a significant predation event (Reardon *et al.*, 2012).

species.[2] Long-term monitoring of the only mainland population of Whitaker's skink demonstrated ongoing decline to the point that they are now barely detectable (Towns, 1992a; Towns & Elliott, 1996; Miskelly, 1997; Hoare *et al.*, 2007a).

A similar trend is evident in the large-bodied speckled skinks (*O. infrapunctatum*) that were once relatively abundant at mainland sites near Lake Rotoiti in the northern South Island. Despite recent predator control in a small area in which they occur, the speckled skink population has continued to decline (C. T. Dumont, University of Canterbury, personal communication).

The magnitude of the threat to New Zealand's reptiles of predation by introduced mammals is highlighted by results of management of grand and Otago skinks (*Oligosoma grande* and *O. otagense*, respectively) at Macraes Flat in Central Otago, south-eastern South Island. Both species are characterised by slow maturity and high longevity (3–5 and 4–6 years to produce the first offspring and longevity of up to 18 and 13.5 years in the wild for grand and Otago skinks, respectively (Tocher, 2009)), and were declining precipitously prior to management intervention (Whitaker & Loh, 1995). Intensive management of the range of mammalian predators present at Macraes Flat, either by predator exclusion or large-scale predator control, improved survival of both skink species (Reardon *et al.*, 2012; see Box 18.2).

18.2.2   Habitat loss

While predation by introduced mammals is considered the primary threat to New Zealand's reptiles on a national scale, habitat loss is extremely important at local and regional scales in areas under development pressure (Böhm *et al.*, 2013). Developments threatening reptile populations in New Zealand include those that stem from urbanisation, land conversion for forestry or farming, wind farms, hydro-electric dams and mining (Towns *et al.*, 2001).

All native reptile species are legally protected (Wildlife Act, 1953) and reptile habitats are protected by the Resource Management Act 1991. However, these statutory obligations are not consistently understood or enforced by the regional authorities responsible for evaluating development proposals (Chapman *et al.*, 2010; Anderson *et al.*, 2012). Further, even when effects of development on lizards are considered by the Environment Court and salvage of lizards or other mitigation is decided on, the outcomes of these actions are often poorly monitored and we have little knowledge of their success at protecting lizards. Unfortunately, this issue is poorly documented in the scientific literature and the debate is mainly occurring in Environment Court cases with expert witnesses' evidence and reports (Corbett, 1997).

Urbanisation is a reality in the northern North Island, particularly in parts of Auckland and Northland, and is having a marked impact on biodiversity values, including lizard populations (Frances & Warren, 1999). Elsewhere in New Zealand, poor guidelines for protecting biodiversity values while encouraging both sustainable energy developments (wind farms and hydro-electric power stations) and more controversial mineral

---

[2] Whitaker's skinks have established on Korapuki Island (Towns, 1999), but the outcome of translocations to the other two locations has not been confirmed, despite reasonable search effort (R. Chappell, Department of Conservation, personal communication).

exploitation mean that high-value sites from a biodiversity perspective are being proposed as development sites. Recent examples from the South Island include a wind farm proposed for Mt Cass, the most intact and best remaining limestone ecosystem in the eastern South Island (N. Head, Department of Conservation, personal communication) and a coal mine approved on the Denniston Plateau, a unique ecosystem with high biodiversity values on the northern West Coast (Lettink, 2012). Both places host nationally important populations of threatened species. Mt Cass has a large population of Canterbury geckos, *Woodworthia* 'Canterbury' (considered 'At Risk – Declining' under New Zealand threat listing criteria; Hitchmough *et al.*, 2013). The Denniston Plateau holds the largest known population of West Coast green geckos, *Naultinus tuberculatus* (Nationally Vulnerable), as well as abundant forest gecko, *Mokopirirakau granulatus*, and speckled skink, *Oligosoma infrapunctatum*, populations (both 'Declining'; Lettink, 2012; Hitchmough *et al.*, 2013).

18.2.3 Poaching

In recent years, several native reptile populations have been threatened by poaching (Jewell, 2008). Species targeted include at least five species of diurnal, green geckos (genus *Naultinus*) as well as some more common gecko and skink species. The threat is best understood for jewelled geckos, *Naultinus gemmeus*, which have proven highly desirable to the captive market in Europe (Jewell, 2008). Because jewelled geckos are individually identifiable based on unique colour markings and patterns and a large database of known individuals exists (Knox *et al.*, 2013), geckos taken from New Zealand can be tracked when they are sold online. Several individuals and groups have been caught attempting to smuggle jewelled geckos out of New Zealand in the past five years (reported extensively in the New Zealand media), and this has had substantial population-level effects on geckos particularly in the Otago region (Knox, 2010, C. Knox, ecological consultant, personal communication).

18.2.4 Introduced reptiles

Introductions of reptiles to New Zealand present a very real and persistent threat to native reptiles, in their ecological roles as new kinds of predators and competitors, and as hosts of new diseases (Gill *et al.*, 2001; Kikillus *et al.*, 2011). New Zealand reptiles have evolved in the absence of mammalian predators, but are also naïve to the hunting methods of snakes, and in particular have not been constrained in their use of odour (Hoare *et al.*, 2007b). As has occurred in Guam with the invasion of brown tree snakes (*Boiga irregularis*, Wiles *et al.*, 2003), terrestrial snakes would likely be predators of terrestrial native reptiles with devastating consequences for viability of populations. It is illegal to import snakes into New Zealand and there are no snakes in zoos, aquariums or private collections. Sea snakes are occasionally observed in northern New Zealand, but are considered native (Hitchmough *et al.*, 2013).

There are no established populations of exotic snakes in New Zealand, but Boids, Collubrids, Elapids and Vipers have been intercepted entering the country in cargo. Most of the arrivals were single animals from tropical countries, and therefore the risk of establishment is relatively low for most occurrences due to limits on survival through winter and reproductive success in a cooler climate (Gill *et al.*, 2001). However, several occurrences were of species of very high concern due to their invasiveness elsewhere (e.g.

brown tree snake), reproductive mode (e.g. parthenogenic, Brahminy blind snake (*Ramphotyphlops braminus*)), or similarity of climate in their native range to New Zealand (Gill *et al.*, 2001).

The greatest pressure for introduction comes from the lizard fauna. Of reptiles intercepted in cargo (1929–2000), 65% were lizards of which 82% were geckos, predominantly made up of the 'house geckos' *Hemidactylus frenatus*, *Lepidodactylus lugubris* and *Gehyra oceanica* (of Asian and Pacific origins; Gill *et al.*, 2001). All three species are nocturnal and lay eggs, are widespread and/or invasive elsewhere, and *L. lugubris* is parthenogenic, meaning only one egg or individual could start a new population (Gill *et al.*, 2001). Owing to the ongoing and high pressure of these and other species entering the country in cargo, often alive, there is a high risk of establishment in warmer parts of New Zealand especially for subtropical generalists. Tropical species would find it harder to find a suitable climate all year round, but buildings can make excellent warm refuges (Gill *et al.*, 2001). To date, the only lizard to establish is the rainbow skink (Robb, 1986; Gill & Whitaker, 1996), which is now established throughout northern parts of the North Island, with records of occurrence just north of Wellington in Horowhenua (DOC, 2013a). Limited data are available on the direct (e.g. predation) or indirect (e.g. competition for habitat or food, introduction of new diseases or increased disease transmission potential) effects of the rainbow skink on the native reptile fauna (Peace, 2004).

Five species of introduced lizard are commonly kept and bred in captivity, including the dragons *Pogona vitticeps*, *Pogona barbata*, and *Physignathus lesuerii*, blue-tongued skink *Tiliqua scincoides* and leopard gecko *Eublepharis macularius*. None are known to have been released into the natural environment, likely due to their value in the pet trade (Kikillus *et al.*, 2012). There are other species kept in captivity, but they are largely represented by single individuals or single sex groups in zoos, aquariums and private collections, and therefore of low risk for establishment (Kikillus, 2010).

Other exotic reptiles in captivity include red-eared slider (*Trachemys scripta elegans*), Reeves (*Chinemys reevesii*) and snake-neck (*Chelodina longicollis*) turtles, and Greek (*Testudo graeca*) and Hermanns (*T. hermanni*) tortoises (Kikillus *et al.*, 2012). Red-eared sliders have been in New Zealand for decades, and are the only reptiles known to be released or escape regularly into the natural environment (Kikillus *et al.*, 2012). Although they have not established (managed to successfully reproduce) since their introduction, the propagule pressure is high and individuals can survive in a feral state in warmer parts of the North Island. Based on modelling information, the thermal conditions for successful egg incubation of three species of turtles currently held in captivity legally in New Zealand can occasionally be achieved in natural habitats in northern parts of the North Island (Kikillus, 2010). However, these species have temperature-dependent sex determination, so all hatchlings from nests in New Zealand without artificial incubation will be male unless temperatures rise by several degrees. Although Crocodilians have been identified in fossil deposits (Molnar & Pole, 1997), they have not been seen in our territorial waters. However, a crocodile was seen in 1970 not far outside territorial waters, 300 miles off North Cape, swimming strongly (Best, 1988). A few individual imported crocodilians have been kept in New Zealand zoos and public aquariums. Legal importation of new species of reptiles and new individuals of species already in New Zealand is strongly limited by legislation (HSNO, 1996). Neither turtles nor crocodilians are known to enter New Zealand in cargo.

18.2.5 Disease

The prevalence of disease-causing organisms is only just beginning to be evaluated in the native reptilian fauna (Middleton *et al.*, 2010). Little is known of their distribution or pathogenicity, or the risks of transmission among individuals and species. The best studied group of organisms in New Zealand reptiles is *Salmonella*, environmentally stable bacteria with the ability to cause death in reptiles and other species. Eleven *Salmonella* serovars were detected in a study of New Zealand lizards on eight islands, with greater prevalence in skinks than in geckos (Middleton *et al.*, 2010). No *Salmonella* was detected in live tuatara (Gartrell *et al.*, 2006, 2007; Middleton, 2012). *Salmonella* was detected in higher prevalence in exotic reptiles in captivity than in native reptiles in captivity (Kikillus *et al.*, 2011).

Introductions of new reptiles (individuals and species) to New Zealand present a high risk of introduction of new diseases, and new serovars of existing diseases. Strong biosecurity mechanisms are in place for legal importations, but reptiles found in cargo are rarely investigated. For example, even when geckos intercepted in cargo were found carrying ectoparasitic mites, no further investigation was made into the potential for transmission of new diseases (Gill *et al.*, 2001). Disease risks from non-reptilian species have not been investigated.

18.2.6 Climate change

Climate change predictions for New Zealand include increased occurrence and severity of droughts and floods, increased water availability in the west and decreased water availability in the east, sea level rise of up to 0.18–0.59 m by 2100 and temperature increases between 0.5–4 °C by 2080 (Hennessy *et al.*, 2007). Implications for the distributions and population viability of New Zealand reptiles are multi-faceted and uncertain. Where species occur only on offshore islands, restricted opportunities for dispersal may result in local extinctions if droughts become more frequent due to already limited water resources on islands and sensitivities with respect to water balance and reproduction (e.g. Thompson *et al.*, 1996). However, for mainland populations, fewer frost events and warmer minimum temperatures may mean distributions could increase for species currently restricted by temperature, if habitat availability allows and vulnerability to introduced mammalian predators is managed. For example, New Zealand's only oviparous lizard (*Oligosoma suteri*) has a northern distribution proposed to be as a result of poorer fitness of offspring at cooler incubation temperatures (Hare *et al.*, 2002). However, warmer temperatures may also result in increased opportunity for establishment of invasives like red-eared slider turtles (Kikillus *et al.*, 2012).

18.3 Conservation strategies
18.3.1 Eradications

Conservation gains for New Zealand's reptiles have been primarily aimed at securing populations on, and reintroducing species to, offshore islands as their introduced mammalian predators are eradicated (Towns *et al.*, 2001). Huge technological advances have been made in the 50 years since the first (unintended) eradication occurred in New Zealand in 1963, enabling the eradication of rodents from islands up to 11 000 ha in size using aerial poisoning (Towns & Broome, 2003; Russell & Clout, 2006).

Many of New Zealand's reptile species on islands to the north-east of the North Island have benefited from rodent eradications. The most well-documented case of reptile recovery following eradication comes from Korapuki Island and neighbouring islands in the Mercury Island group (e.g. Towns, 2002; Towns et al., 2003; Hoare et al., 2007c). Species that benefited from the eradication include those that survived in the presence of Pacific rats, Rattus exulans, and species reintroduced following rodent eradication that have subsequently established. Other success stories include the eradication of mice from Mana Island, off the south-west coast of the North Island, which enabled a population of large-bodied, nocturnal McGregor's skinks to recover from very low levels to become a large and healthy population (Newman, 1994) that has now recolonised much of the island (Z. Kavas, L. Adams and J. Monks, Department of Conservation, unpublished data).

18.3.2   Predator control on the mainland

Perhaps the biggest breakthrough in conservation of New Zealand's reptiles in the past decade has been the protection of populations of large-bodied, vulnerable lizards through intensive predator management on the mainland (Reardon et al., 2012, see Box 18.2). To date, landscape-level management to eliminate and/or suppress predators over a large enough area to produce recoveries in lizard populations has only been successful for two species and within one region (Reardon et al., 2012). Predator suppression at smaller scales has failed to produce recovery in lizard populations, probably due to insufficient buffering to protect a core area for lizards (Norbury et al., 2013). New Zealand's fenced sanctuaries, in which most mammalian predators are eradicated and smaller species (particularly mice, Mus musculus) are controlled to low levels (Innes et al., 2012), also have potential to protect lizard populations. However, responses of lizards to management at mainland sites have mostly been negligible or slow (e.g. Bell, 2009); at least another decade of monitoring is likely to be required to evaluate benefits for lizards.

A lack of knowledge of the key threats to species hampers conservation efforts for many indigenous lizard populations (Box 18.3). At present, 40 lizard species (22 geckos and 18 skinks) are considered to be 'Data Poor' (Hitchmough et al., 2013) on the basis of a lack of knowledge of number of extant populations, population size, area of occupancy and population growth rate that hampers experts' ability to categorise the species under threat listing criteria (Townsend et al., 2008). A further four species (two geckos and two skinks) are so poorly known that their threat status could not be estimated and these are categorised as 'Data Deficient' (Hitchmough et al., 2013). As such, it is unknown whether the large-scale, multi-species predator management required to protect grand and Otago skinks (Reardon et al., 2012) is also required to protect species in different habitats (e.g. the alpine zone; Bell et al., 2009) or with different habitat use patterns (e.g. arboreal; Hoare et al., 2013). Understanding the suite of threats and their complex interactions that need to be managed to halt lizard population declines is a key challenge for the future conservation of New Zealand's reptiles (Norbury et al., 2013).

Box 18.3 New Zealand's numerous poorly known lizards

Information on habitat use and behaviour, which enables the design and testing of effective monitoring techniques, is lacking for many of New Zealand's cryptic lizard species (Hare *et al.*, 2007). Without this vital basic information it is impossible to design a monitoring programme to evaluate population trends and (if trends are negative) the key threats responsible for declines.

The case of the southern forest gecko, *Mokopirirakau* 'Southern Forest', exemplifies the challenge that lack of basic biological knowledge presents in conserving New Zealand's endemic reptile fauna. New Zealand's forest geckos belong to the recently described genus *Mokopirirakau* which comprises four described species and six putatively new species (Nielsen *et al.*, 2011). The southern forest gecko (also known as the blue-eyed gecko (Jewell, 2011); *M.* 'Southern Forest') is one of the species that is yet to be described and is among the least-known lizard species in New Zealand, being nocturnal and cryptic in both colouration and behaviour (Tocher *et al.*, 2000; Hoare *et al.*, 2013). Until 2010, southern forest geckos were known only from anecdotes, museum specimens and discoveries of individual lizards at three sites during lizard surveys (Tocher *et al.*, 2000; Bell & Jewell, 2007). In 2010–11 dedicated search effort at one of the sites (Tahakopa Valley) in which they have recently been found resulted in sightings of 19 individual geckos (based on photo-identification using natural markings), including juveniles, which confirmed that at least one breeding population of this species exists (Hoare *et al.*, 2013). These geckos were found only through searching for emerged individuals at night ('spotlighting') and in diurnal retreat sites; monitoring devices (funnel traps) placed within known territories failed to capture any geckos (Hoare *et al.*, 2013). Encounter rates of 0.6 geckos per person hour under optimal weather conditions (relatively warm, overcast) and 0.08 geckos per person hour during a cooler spell of weather (Hoare *et al.*, 2013) indicate it would be time-intensive to collect meaningful population trend data or evaluate threats for southern forest geckos.

Devising and validating monitoring techniques for cryptic lizard taxa is an area in which several New Zealand practitioners have invested effort in the past decade. The work uses techniques that were until recently considered novel methods suitable for inventory being incorporated into best practice advice for monitoring lizards within a few years. The best example of this is the use of artificial retreats for reptiles. Designs initially provided an alternative to systematic searching for surveys of terrestrial and arboreal lizards (e.g. Lettink & Cree 2007; Bell, 2009). Testing of these retreats resulted in information about the thermal opportunity they provide (Thierry *et al.*, 2007), optimal design of retreats (Lettink & Cree, 2007; O'Donnell & Hoare, 2012), relationships with microhabitat (Chavel *et al.*, 2012) and optimal sampling conditions (Hoare *et al.*, 2009), which culminated in validation of index counts from retreats against population estimates from another technique (Lettink *et al.*, 2011) and their adoption into the New Zealand Department of Conservation's (DOC's) best practice guidance for herpetofauna (DOC, 2013b). Outcomes from research into new monitoring techniques for New Zealand's cryptic lizards include direct benefits for species conservation. Evaluating the use of funnel traps placed in debris dams to monitor chevron skinks, *Oligosoma homalonotum* (Jamieson & Neilson, 2007; Barr, 2009) resulted in downgrading of the national threat ranking for the species because it was discovered that populations were larger than previously thought (Hitchmough *et al.*, 2010).

18.3.3 Recovery plans

In the 1990s, groups of experts comprising scientists, technical advisors, rangers and academics were convened to collate biological knowledge and establish viable conservation options for several high profile species of reptiles in New Zealand. Resulting recovery plans were compiled for several species of lizards (Whitaker's and robust skinks (Towns, 1992b), chevron skink (Towns & McFadden, 1993), Otago and grand skinks (Whitaker & Loh, 1995) and striped skink (Whitaker, 1998)). Two iterations of recovery plans were prepared for tuatara (Cree & Butler, 1993; Gaze, 2001). Recovery plans identified options and priorities for actions and research for species conservation, and resulted in allocation of resources to species conservation as well as more biological knowledge through research outcomes. All species with recovery plans now have more populations and/or greater security of existing populations (see Boxes 18.1 and 18.2). However, due to the intensive effort of preparing plans, the recovery planning process did not allow for plans to be completed for all species, and so second-generation plans for the lizards were based on genera (*Cyclodina* spp. (Towns, 1999); North Island *Oligosoma* spp. (Towns *et al.*, 2002)). Even so, plans were never completed for South Island *Oligosoma* or for any gecko species, and new editions of plans are no longer being compiled.

Species recovery planning targeted at single species or small groups of species is being phased out by the New Zealand Department of Conservation and replaced by site-based biodiversity conservation (DOC, 2013c). A network of ecologically representative sites will be managed to a healthy, functioning state. In addition, some of these sites or parts of them will be managed more intensively to benefit threatened species which require more concentrated browser or predator control, and other sites will be managed specifically for species not represented in the suite of ecosystem management units. These species-focussed sites will be designed to benefit as many different species as possible at the same place and with a single management regime. Because of their sensitivity to a range of mammalian predators including mice, lizards are likely to be a particular challenge, as they require intensive, broad-spectrum predator control wherever they are managed.

18.3.4 Threat listing

The New Zealand Threat Classification System (NZTCS; Molloy *et al.*, 2002) was introduced to provide a national system similar to the IUCN Red List, but with finer-scale discrimination of the status of a biota with many naturally range-restricted island and montane local endemics. The categories and criteria were reviewed and a revised system manual issued in 2008 (Townsend *et al.*, 2008).

The status of New Zealand's reptiles has been assessed four times since the introduction of the New Zealand Threat Classification System (Molloy *et al.*, 2002; Townsend *et al.*, 2008). Species have been added following new discoveries or new recognition of distinctiveness following genetic studies. There is a general trend of increasing numbers of species in the Nationally Endangered, Nationally Vulnerable and Declining categories. Changes in the assignment of taxa to threat categories, while predominantly driven by gradually improving knowledge rather than measured improvement or deterioration in status, reflect ever-increasing concern about the conservation status of our lizards (Table 18.1).

The history of the reptile species which have been listed as Nationally Critical (NC) at any time over the four threat listing reviews illustrates the importance of both

Table 18.1 Comparison of summary statistics between New Zealand Threat Classification System listing cycles. The 2002–07 categories (Molloy *et al.*, 2002) are not directly comparable to the 2008 ones (Townsend *et al.*, 2008), but are aligned with their closest equivalent. The Nationally Vulnerable and Serious Decline categories of Molloy *et al.* (2002) are now mostly included in Nationally Vulnerable *sensu* Townsend *et al.* (2008). The Sparse and Range Restricted categories of Molloy *et al.* (2002) are mostly equivalent to Naturally Uncommon, with some of the Range Restricted taxa moving to Relict *sensu* Townsend *et al.* (2008). In 2002 and 2007 only extinctions since 1800 were listed, whereas from 2008 all extinctions since human settlement were listed.

Category 2002–2007	No. of species in Hitchmough, 2002	No. of species in Hitchmough *et al.*, 2007	Category 2008 onwards	No. of species in Hitchmough *et al.*, 2010	No. of species in Hitchmough *et al.*, 2013
Extinct	1	1	Extinct	2	2
Data Deficient	8	10	Data Deficient	8	4
Nationally Critical	4	5	Nationally Critical	6	3
Nationally Endangered	4	4	Nationally Endangered	3	10
Nationally Vulnerable	0	1	Nationally Vulnerable	8	19
Serious Decline	1	1			
Gradual Decline	15	22	Declining	27	26
Range Restricted	19	22	Naturally Uncommon	10	9
Sparse	16	12	Relict	11	11
			Recovering	3	4
Not Threatened	14	11	Not Threatened	23	13

changes in knowledge and real changes in status in determining perceived threat of extinction (Box 18.4). No species has remained listed as Nationally Critical across all four listings. Ten species have been in this category at some time. Five initially newly discovered and poorly known species have moved out of this category as better information about their distribution and status became available. Three recently discovered or recognised species have joined the category, and may or may not come to be regarded as less threatened with better understanding in future. The grand and Otago skinks moved into the category initially, and out of it later, as a result of well-documented decline, then recovery as a result of an intensive recovery programme (Box 18.2). Other success stories have involved pest eradication and translocation on offshore islands and led to four previously very threatened species now being listed as Recovering (Box 18.4). However, future conservation benefits will only accrue if predator control on the mainland is targeted at levels that benefit reptiles (not just birds), if more widespread effectiveness of predator control is achieved (as proposed by the pest-free New Zealand concept; DOC, 2013c), if mitigation proposals resulting from development are enforced and evaluated, and if biosecurity risk assessments include a more proactive research component.

Box 18.4 Case study of Nationally Critical lizards

The Roy's Peak and Takitimu geckos have identical listing histories, for the same reasons. Both were listed as Nationally Critical by Hitchmough (2002), but moved to Data Deficient by Hitchmough *et al.* (2007) following the discovery of more small isolated populations scattered over a fairly large area. Hitchmough *et al.* (2010, 2013) placed both in Nationally Vulnerable; by then the extent of their patchy distributions was better understood.

The Open Bay Islands gecko and Open Bay Islands skink also have much in common. Both were thought to be confined to a single island where they were subject to predation by introduced weka (*Gallirallus australis*) when they were listed as Nationally Critical by Hitchmough (2002) and Hitchmough *et al.* (2007). After a survey by Marieke Lettink (personal communication) showed the gecko to be slightly more widespread and abundant on the island than had previously been thought, it was moved to Nationally Endangered (Hitchmough *et al.*, 2010, 2013). Lettink's survey showed the skink to be also found on a second island in the group but very uncommon there as well, so it remained as Nationally Critical (Hitchmough *et al.*, 2010). By the time of the most recent reassessment it had been found to be abundant on two very small islands in a second island group, and was moved to Nationally Endangered (Lettink *et al.* 2013; Hitchmough *et al.*, 2013).

Hitchmough *et al.* (2007) added the Coromandel striped gecko to the Nationally Critical category. In 2002 it had been listed as Data Deficient, but targeted surveys in the intervening period around sites of previous casual observations had failed to locate additional specimens. By the time Hitchmough *et al.* (2010) reviewed its status, there were more scattered casual observations indicating that this gecko was probably too widespread and abundant to be considered Nationally Critical and it went back into the Data Deficient category. Most recently (Hitchmough *et al.*, 2013), with continuing casual reports but still no success from survey, it has been inferred to be Nationally Endangered.

The grand and Otago skinks were also added to the Nationally Critical category by Hitchmough *et al.* (2007), having both been previously listed as Nationally Endangered (Hitchmough, 2002). Both remained as Nationally Critical in the listing of Hitchmough *et al.* (2010) but were moved back to Nationally Endangered by Hitchmough *et al.* (2013). Unlike those for the other species discussed above, these changes were driven by real, measured changes and modelled projected changes in numbers of animals, which were suffering a well-documented decline despite considerable management effort in the early 2000s, but then began to recover strongly when management was further intensified in the second half of the decade (Tocher, 2006; Reardon *et al.*, 2012; see Box 18.2).

Three poorly known and recently discovered skink species were added to the Nationally Critical category by Hitchmough *et al.* (2010), and remain in this category in the latest listing (Hitchmough *et al.*, 2013). The Te Kakahu and Rangitata skinks had not yet been discovered at the time of the listing by Hitchmough (2002), and were recently discovered and poorly known when listed as Data Deficient by Hitchmough *et al.* (2007). The third species, the Chesterfield skink, has suffered from a history of confusion about its taxonomic distinctiveness, and was not

> ### Box 18.4 (continued)
>
> recognised as a distinct taxon in the listings of Hitchmough (2002) and Hitchmough *et al.* (2007). All three species are known from only one or two very small localised populations despite searches in similar habitat in surrounding areas. While it is quite possible that these species will follow the same path as the Roy's Peak, Takitimu and Coromandel striped geckos and Open Bay Island skink and move to less acutely threatened categories as our knowledge of them improves, our current understanding places all three firmly in the Nationally Critical category and their future is of grave concern.

REFERENCES

Anderson, P., T. Bell, S. Chapman & K. Corbett (2012). *New Zealand Lizards Conservation Toolkit – a Resource for Conservation Management of the Lizards of New Zealand.* A SRARNZ Miscellaneous Publication. Society for Research on Amphibians and Reptiles in New Zealand. 69 pages.

Barr, B. P. (2009). Spatial ecology, habitat use, and the impacts of rats on chevron skinks *(Oligosoma homalonotum)* on Great Barrier Island. Unpublished MSc thesis, Massey University, Auckland. 166 pages.

Bauer, A. M. & A. P. Russell (1986). *Hoplodactylus delcourti* n. sp. (Reptilia: Geckonidae), the largest known gecko. *New Zealand Journal of Zoology*, **13**, 141–148.

Bell, T. P. (2009). A novel technique for monitoring highly cryptic lizard species in forests. *Herpetological Conservation and Biology*, **4**, 415–425.

Bell, T. P. & T. J. Jewell (2007). *Lizard Survey in Southland Indigenous Reserves Managed by Southwood Export, January – February 2007.* Manaaki Whenua Landcare Research Contract Report: LC0607/101. Landcare Research, Dunedin. 36 pages.

Bell, T. P., G. Patterson & T. Jewell (2009). *Alpine Lizard Research in Fiordland National Park: February–March 2007.* DOC Research & Development Series 304. New Zealand Department of Conservation, Wellington. 18 pages.

Berry, O. F. & D. M. Gleeson (2005). Distinguishing historical fragmentation from a recent population decline – shrinking or pre-shrunk skink from New Zealand? *Biological Conservation*, **123**, 197–210.

Best, S. (1988). Here be dragons. *Journal of the Polynesian Society*, **97**(3), 239–260.

Böhm, M., B. Collen, J. E. M. Baillie *et al.* (2013). The conservation status of the world's reptiles. *Biological Conservation*, **157**, 372–385.

Chambers, G. K., W. M. Boon, T. R. Buckley & R. A. Hitchmough (2001). Using molecular methods to understand the Gondwanan affinities of the New Zealand biota: three case studies. *Australian Journal of Botany*, **49**, 377–387.

Chapman, S., T. Bell, K. Corbett & P. Anderson (2010). The 'tool-kit' approach to the conservation of New Zealand's lizards (abstract). *New Zealand Journal of Zoology*, **37**, 66.

Chapple, D. G. & M. B. Thompson (2009). Isolation and characterization of microsatellite loci from the invasive delicate skink (*Lampropholis delicata*), with cross-amplification in other Australian Eugongylus group species. *Conservation Genetics Resources*, **1**(1), 55–58.

Chapple, D. G., R. A. Hitchmough & T. Jewell (2009a). Taxonomic instability of reptiles and frogs in New Zealand: information to aid the use of Jewell (2008) for species identification [a comment on King 2009 and further commented by Jewell]. *New Zealand Journal of Zoology*, **36**, 59–71.

Chapple, D. G., P. A. Ritchie & C. H. Daugherty (2009b). Origin, diversification, and systematics of the New Zealand skink fauna (Reptilia: Scincidae). *Molecular Phylogenetics and Evolution*, **52** (2), 470–487.

Chavel, E. E., J. M. Hoare, W. G. Batson & C. F. J. O'Donnell (2012). The effect of microhabitat on skink sightings beneath artificial retreats. *New Zealand Journal of Zoology*, **39**, 71–75.

Corbett, K. (1997). New Zealand's reptiles: protected – or not? *New Zealand Journal of Zoology*, **34**, 261–262.

Cree, A. & D. Butler (1993). *Tuatara Recovery Plan* (Sphenodon *spp*.). Threatened Species Recovery Plan 9. New Zealand Department of Conservation, Wellington. 71 pages.

DOC (2013a). Herpetofauna database: http://www.doc.govt.nz/conservation/native-animals/reptiles-and-frogs/reptiles-and-frogs-distribution-information/atlas-of-the-amphibians-and-reptiles-of-nz/electronic-atlas/. Downloaded on 15 April 2013.

DOC (2013b). DOC Procedures and SOPs: Herpetofauna Inventory and Monitoring. http://www.doc.govt.nz/publications/science-and-technical/doc-procedures-and-sops/biodiversity-inventory-and-monitoring/herpetofauna/. Downloaded on 15 April 2013.

DOC (2013c). *Department of Conservation: Statement of Intent 2013–2017*. Wellington: Department of Conservation, 52 pages.

Frances, S. & P. Warren (1999). Conservation of New Zealand's biodiversity into the new decade–opportunities and challenges. *New Zealand Journal of Environmental Law*, **3**, 169–178.

Gartrell, B. D., E. Jillings, B. A. Adlington, H. Mack & N. J. Nelson (2006). Health screening for a translocation of captive-reared tuatara (*Sphenodon punctatus*) to an island refuge. *New Zealand Veterinary Journal*, **54**, 344–349.

Gartrell, B. D., J. M. Youl, C. M. King *et al.* (2007). Failure to detect Salmonella species in a population of wild tuatara (*Sphenodon punctatus*). *New Zealand Veterinary Journal*, **55**, 134–136.

Gaze, P. (2001). *Tuatara Recovery Plan 2001–2011*. Threatened Species Recovery Plan 47. New Zealand Department of Conservation, Wellington. 36 pp.

Gill, B. J. & A. H. Whitaker (1996). *New Zealand Frogs & Reptiles*. Bateman Fieldguide. 112 pages.

Gill, B. J., D. Bejakovtch & A. H. Whitaker (2001). Records of foreign reptiles and amphibians accidentally imported to New Zealand. *New Zealand Journal of Zoology*, **28**(3), 351–359.

Graham, J. B., I. Rubinoff & M. K. Hecht (1971). Temperature physiology of the sea snake *Pelamis platurus*: an index of its colonization potential in the Atlantic Ocean. *Proceedings of the National Academy of Sciences USA*, **68**(6), 1360–1363.

Hardy, G. S. (1977). The New Zealand Scincidae (Reptilia: Lacertilia); a taxonomic and zoogeographic study. *New Zealand Journal of Zoology*, **4**(3), 221–325.

Hare, K., C. H. Daugherty & A. Cree (2002). Incubation regime affects juvenile morphology and hatching success, but not sex, of the oviparous lizard *Oligosoma suteri* (Lacertilia: Scincidae). *New Zealand Journal of Zoology*, **29**(3), 221–229.

Hare, K. M., J. M. Hoare & R. A. Hitchmough (2007). Investigating natural population dynamics of *Naultinus manukanus* to inform conservation management of New Zealand's cryptic diurnal geckos. *Journal of Herpetology*, **41**, 81–93.

Hay, J. M., S. D. Sarre, D. M. Lambert, F. W. Allendorf & C. H. Daugherty (2010). Genetic diversity and taxonomy: a reassessment of species designation in tuatara (*Sphenodon*: Reptilia). *Conservation Genetics*, **11**, 1063–1081.

Hennessy, K., B. Fitzharris, B. C. Bates *et al.* (2007). *Australia and New Zealand. Climate Change 2007: Impacts, Adaptation and Vulnerability. Contribution of Working Group II to the Fourth Assessment Report of the Intergovernmental Panel on Climate Change.* Parry, M. L., O. F. Canziani, J. P. Palutikof, P. J. van der Linden and C. E. Hanson, Eds. Cambridge University Press, Cambridge, UK, pp. 507–540.

Hitchmough, R. (comp.) (2002). *New Zealand Threat Classification System Lists 2002.* Threatened Species Occasional Publication 23. New Zealand Department of Conservation, Wellington. 210 pages.

Hitchmough, R., L. Bull & P. Cromarty (comps.) (2007). *New Zealand Threat Classification System Lists 2005.* New Zealand Department of Conservation, Wellington. 194 pages.

Hitchmough, R. A., J. M. Hoare, H. Jamieson *et al.* (2010). Conservation status of New Zealand reptiles. *New Zealand Journal of Zoology*, **37**(3), 203–224.

Hitchmough, R., P. Anderson, B. Barr *et al.* (2013). *Conservation Status of New Zealand Reptiles, 2012.* New Zealand Threat Classification Series 2. New Zealand Department of Conservation, Wellington. 9 pages.

Hoare, J. M., L. K. Adams, L. S. Bull & D. R. Towns (2007a). Attempting to manage complex predator-prey interactions fails to avert imminent extinction of a threatened New Zealand skink population. *Journal of Wildlife Management*, **71**, 1576–1584.

Hoare, J. M., S. Pledger & N. J. Nelson (2007b). Chemical discrimination of food, conspecifics and predators by apparently visually-oriented diurnal geckos, *Naultinus manukanus*. *Herpetologica*, **63**(2), 184–192.

Hoare, J. M., S. Pledger, N. J. Nelson & C. H. Daugherty (2007c). Avoiding aliens: behavioural plasticity in habitat use enables large, nocturnal geckos to survive Pacific rat invasions. *Biological Conservation*, **136**, 510–519.

Hoare, J. M., C. F. J. O'Donnell, I. Westbrooke, D. Hodapp & M. Lettink (2009). Optimising the sampling of skinks using artificial retreats based on weather conditions and time of day. *Applied Herpetology*, **6**, 379–390.

Hoare, J. M., P. Melgren & E. E. Chavel (2013). Habitat use by southern forest geckos, *Mokopirirakau* "Southern Forest" in the Catlins, Southland. *New Zealand Journal of Zoology*, **40**(2), 129–136.

HSNO (1996). Hazardous Substances and New Organisms Act 1996. New Zealand Act of Parliament.

Innes, J., W. G. Lee, B. Burns *et al.* (2012). Role of predator-proof fences in restoring New Zealand's biodiversity: a response to Scofield *et al.* (2011). *New Zealand Journal of Ecology*, **36**, 232–238.

IUCN (2012). *IUCN Red List of Threatened Species*. Version 2012.2. www.iucnredlist.org. Downloaded on 3 April 2013.

Jamieson, H. & K. A. Neilson (2007). Detecting the undetectable: developing new ways to catch cryptic reptiles (abstract). *New Zealand Journal of Zoology*, **34**, 264.

Jewell, T. (2008). *A Photographic Guide to Reptiles and Amphibians of New Zealand* [with corrections and comments in Chapple & Hitchmough 2009]. New Holland Publishers (NZ) Ltd, Auckland. 143 pages.

Jewell, T. (2011). *A Photographic Guide to Reptiles and Amphibians in New Zealand. Revised Edition*. New Holland Publishers (NZ) Ltd, Auckland.

Kikillus, K. H. (2010). *Exotic Reptiles in the Pet Trade: Are they a Threat to New Zealand?* PhD thesis, Victoria University of Wellington, New Zealand. 309 pages.

Kikillus, K. H., B. D. Gartrell & E. Motion (2011). Prevalence of *Salmonella* spp., and serovars isolated from captive exotic reptiles in New Zealand. *New Zealand Veterinary Journal*, **59**(4), 174–178.

Kikillus, K. H., K. M. Hare & S. Hartley (2012). Online trading tools as a method of estimating propagule pressure via the pet-release pathway. *Biological Invasions*, **14**(12), 2657–2664.

Knox, C. D. (2010). Habitat requirements of the jewelled gecko *(Naultinus gemmeus)*: effects of grazing, predation and habitat fragmentation. Unpublished MSc thesis, University of Otago, Dunedin. 107 pages.

Knox, C. D., A. Cree & P. J. Seddon (2013). Accurate identification of individual geckos *(Naultinus gemmeus)* through dorsal pattern differentiation. *New Zealand Journal of Ecology*, **37**, 60–66.

Lee, M. S. Y., M. N. Hutchinson, T. H. Worthy *et al.* (2009). Miocene skinks and geckos reveal long-term conservatism of New Zealand's lizard fauna. *Biology Letters*, **5**(6), 833–837.

Lettink, M. (2012). *Lizard Survey of the Footprint of the Proposed Escarpment Mine, Denniston Plateau, North Westland*. Unpublished Report prepared by Fauna Finders for the Royal Forest & Bird Protection Society of New Zealand Inc. 10 pages.

Lettink, M. & A. Cree (2007). Relative use of three types of artificial retreats by terrestrial lizards in grazed coastal shrubland, New Zealand. *Applied Herpetology*, **4**, 227–243.

Lettink, M., C. F. J. O'Donnell & J. M. Hoare (2011). Accuracy and precision of skink counts from artificial retreats. *New Zealand Journal of Ecology*, **35**, 236–243.

Lettink, M., G. Hopkins & R. L. Wilson (2013). A significant range extension and sanctuary for the rare Open Bay Island skink (*Oligosoma taumakae*). *New Zealand Journal of Zoology*, **40**, 160–165.

Middleton, D. M. R. L. (2012). *The Ecological and Immunological Relationships between Salmonella and Tuatara*. PhD Thesis, Victoria University of Wellington. 222 pages.

Middleton, D. M. R. L., E. O. Minot & B. D. Gartrell (2010). *Salmonella enterica* serovars in lizards of New Zealand's offshore islands. *New Zealand Journal of Ecology*, **34**(2), 247–252.

Miller, K. A., D. G. Chapple, D. R. Towns, P. A. Ritchie & N. J. Nelson (2009). Assessing genetic diversity for conservation management: a case study of a threatened reptile. *Animal Conservation*, **12**, 163–171.

Miskelly, C. M. (1997). Whitaker's skink *Cyclodina whitakeri* eaten by a weasel *Mustela nivalis*. New Zealand Department of Conservation, Wellington. 4 pages.

Molloy, J., B. Bell, M. Clout *et al.* (2002). *Classifying Species According to Threat of Extinction; a System for New Zealand*. Threatened Species Occasional Publication 22. New Zealand Department of Conservation, Wellington. 26 pages.

Molnar, R. E. & M. Pole (1997). A Miocene crocodilian from New Zealand. *Alcheringa*, **21**(1–2), 65–70.

Newman, D. G. (1994). Effects of a mouse, *Mus musculus*, eradication programme and habitat change on lizard populations of Mana Island, New Zealand, with special reference to McGregor's skink, *Cyclodina macgregori*. *New Zealand Journal of Zoology*, **21**, 443–456.

Nielsen, S. V., A. M. Bauer, T. R. Jackman, R. A. Hitchmough & C. H. Daugherty (2011). New Zealand geckos (Diplodactylidae): Cryptic diversity in a post-Gondwanan lineage with trans-Tasman affinities. *Molecular Phylogenetics and Evolution*, **59**(1), 1–22.

Norbury, G., A. Byrom, R. Pech *et al.* (2013). Invasive mammals and habitat modification interact to generate unforeseen outcomes for indigenous fauna. *Ecological Applications*. http://dx.doi.org/10.1890/12-1958.1

O'Donnell, C. F. J. & J. M. Hoare (2012). Monitoring trends in skink sightings from artificial retreats: influence of retreat design, placement period and predator abundance. *Herpetological Conservation Biology*, **7**, 58–66.

Peace, J. (2004). Distribution, habitat use, breeding and behavioural ecology of rainbow skinks (*Lampropholis delicata*) in New Zealand. Unpublished MSc Thesis, University of Auckland, New Zealand.

Pyron, R. A., F. T. Burbrink & J. J. Weins (2013). A phylogeny and revised classification of Squamata, including 4161 species of lizards and snakes. *BMC Evolutionary Biology*, **13**, 93.

Reardon, J. T., N. Whitmore, K. M. Holmes *et al.* (2012). Predator control allows critically endangered lizards to recover on mainland New Zealand. *New Zealand Journal of Ecology*, **36**, 141–150.

Robb, J. (1986). *New Zealand Amphibians and Reptiles in Colour*, 2nd edn. Collins, Auckland.

Russell, J. C. & M. N. Clout (2006). The eradication of mammals from New Zealand islands. In: *Assessment and Control of Biological Invasion Risks* (Eds. **Koike, F., M. N. Clout, M. Kawamichi, M. De Poorter & K. Iwatsuki**). World Conservation Union (IUCN), Gland, pp. 127–141.

Sherley, G. H., I. A. N. Stringer & G. R. Parrish (2010). *Summary of Native Bat, Reptile, Amphibian and Terrestrial Invertebrate Translocations in New Zealand*. Science for Conservation 303. New Zealand Department of Conservation, Wellington. 39 pages.

Thierry, A., A. Cree & M. Lettink (2007). Influence of thermal and structural characteristics on use of artificial retreats (ARs) by common geckos (*Hoplodactylus maculatus*) and McCann's skinks (*Oligosoma maccanni*). *New Zealand Journal of Zoology*, **34**, 271.

Thompson, M. B., G. C. Packard, M. J. Packard & B. Rose (1996). Analysis of the nest environment of tuatara *Sphenodon punctatus*. *Journal of Zoology*, **238**(2), 239–251.

Tocher, M. D. (2006). Survival of grand and Otago skinks following predator control. *Journal of Wildlife Management*, **70**, 31–42.

Tocher, M. D. (2009). Life history traits contribute to decline of critically endangered lizards at Macraes Flat, Otago. *New Zealand Journal of Ecology*, **33**, 125–137.

Tocher, M. D., T. Jewell & L. McFarlane (2000). *Survey for Forest Geckos (*Hoplodactylus aff. granulatus*) in the Catlins/Southland District*. Conservation Advisory Science Notes No. 285. New Zealand Department of Conservation, Wellington. 22 pages.

Towns, D. R. (1992a). *Distribution and Abundance of Lizards at Pukerua Bay, Wellington: Implications for Reserve Management*. Science and Research Internal Report 125. New Zealand Department of Conservation, Wellington.

Towns, D. R. (1992b). *Recovery Plan for Whitaker's skink and Robust Skink*. Threatened Species Recovery Plan 3. New Zealand Department of Conservation, Wellington. 48 pages.

Towns, D. R. (1999). Cyclodina *spp. Skink Recovery Plan 1999–2004*. Threatened Species Recovery Plan 27. New Zealand Department of Conservation, Wellington. 69 pages.

Towns, D. R. (2002). Korapuki Island as a case study for restoration of insular ecosystems in New Zealand. *Journal of Biogeography*, **29**, 593–607.

Towns, D. R. & G. P. Elliott (1996). Effects of habitat structure on distribution and abundance of lizards at Pukerua Bay, Wellington, New Zealand. *New Zealand Journal of Ecology*, **20**, 191–206.

Towns, D. R. & I. McFadden (1993). *Chevron Skink Recovery Plan* (Leiolopisma homalonotum). Threatened Species Recovery Plan 5. New Zealand Department of Conservation, Wellington. 36 pages.

Towns, D. R., C. H. Daugherty & A. Cree (2001). Raising the prospects for a forgotten fauna: a review of 10 years of conservation effort for New Zealand reptiles. *Biological Conservation*, **99**, 3–16.

Towns, D. R., K. Neilson & A. H. Whitaker (2002). *North Island* Oligosoma *spp. Skink Recovery Plan 2002–2012*. Threatened Species Recovery Plan 48. New Zealand Department of Conservation, Wellington. 60 pages.

Towns, D. R., G. R. Parrish & I. Westbrooke (2003). Inferring vulnerability to introduced predators without experimental demonstration: case study of Suter's skink in New Zealand. *Conservation Biology*, **17**, 1–11.

Townsend, A. J., P. J. de Lange, C. A. J. Duffy *et al.* (2008). *New Zealand Threat Classification System Manual*. New Zealand Department of Conservation, Wellington. 35 pages.

Vallarino, O. & P. J. Weldon (1996). Reproduction in the yellow-bellied sea snake (*Pelamis platurus*) from Panama: field and laboratory observations. *Zoo Biology*, **15**(3), 309–314.

Whitaker, A. H. (1998). *Striped Skink* Oligosoma striatum *Recovery Plan 1998–2003*. Recovery Plan 24. New Zealand Department of Conservation, Wellington. 43 pages.

Whitaker, A. H. & G. Loh. (1995). *Otago Skink and Grand Skink* (Leiolopisma otagense *and* L. grande) *Recovery Plan*. Threatened Species Recovery Plan 14. New Zealand Department of Conservation, Wellington. 40 pages.

Wildlife Act (1953). *New Zealand Act of Parliament* (and its amendments).

Wiles, G. J., J. Bart, R. E. Beck Jr. & C. F. Aguon (2003). Impacts of the brown tree snake: patterns of decline and species persistence in Guam's avifauna. *Conservation Biology*, **17**(5), 1350–1360.

Worthy, T. H. (1987). Osteological observations on the larger species of the skink *Cyclodina* and the subfossil occurrence of these and the gecko *Hoplodactylus duvaucelii* in the North Island, New Zealand. *New Zealand Journal of Zoology*, **14**, 219–229.

Worthy, T. (1991). Fossil skink bones from Northland, New Zealand, and description of a new species of *Cyclodina*, Scincidae. *Journal of the Royal Society of New Zealand*, **21**, 329–348.

CHAPTER 19

Isolation, invasion and innovation: forces of change in the conservation of New Zealand birds

Sarah Withers

Summary

New Zealand is home to a small but strange set of land-birds and is world-renowned as a hotspot of sea-bird diversity. This unique assemblage of species is the result of the fact that the New Zealand landmass is continental in origin but island-like in nature, being isolated from other countries by a large extent of ocean. A lack of mammalian predators means that many New Zealand birds have evolved remarkable features, including gigantism, flightlessness, ground-nesting habits and physiology which protects them from aerial avian predators. Unfortunately, these features have made New Zealand birds highly susceptible to the impacts of introduced mammalian predators and many species are now threatened. The high levels of extinction and rates of decline have spurred the development of highly innovative and revolutionary conservation techniques. New Zealand conservation efforts have paved the way for international conservation projects, in particular with regard to the use of offshore islands, pest control methodologies, species translocation and breeding manipulation. The use of these interventions has led to the recovery of numerous highly endangered species and the discipline continues to evolve, with New Zealand scientists contributing to ongoing development and discussion of conservation methods. While New Zealand is home to one of the largest public conservation organisations in the world, dwindling governmental and financial support is putting many successful conservation projects at risk and private initiatives are becoming increasingly vital for the ongoing protection of New Zealand's unique species.

Austral Ark: The State of Wildlife in Australia and New Zealand, eds. A. Stow, N. Maclean and G. I. Holwell. Published by Cambridge University Press. © Cambridge University Press 2015.

19.1 The history and origin of New Zealand's avifauna

New Zealand has long been recognised as an environment which hosts a remarkable avian assemblage. During its history New Zealand has contained 245 species in 110 genera representing 46 families. Like Australia, New Zealand has a high level of endemism in its avifauna, with 176 (72%) of the 245 resident species endemic to the archipelago (Holdaway *et al.* 2001). Species diversity is dominated by three groups, the Procellariiformes (53 species), Charadriiformes (30 species) and Passeriformes (44 species) and the country has a global reputation as a hot-spot of sea-bird diversity, with almost a quarter of the 359 sea-bird species worldwide occurring as breeding populations in New Zealand, of which 36 (42%) are endemic (Gaskin and Rayner 2013). The prevalence of sea-birds, shore-birds and forest-birds reflects New Zealand's unique environment as an isolated archipelago with a very recent history of human occupation and consequent destruction.

The unique nature of New Zealand's avian biota has resulted in extensive discussion on its origins. Most scientists consider the New Zealand avifaunal assemblage to be the result of a combination of vicariance, dispersal and human-mediated introductions. The presence of endemic, morphologically unique and apparently archaic avian groups has led some researchers to hypothesise that at least some components of New Zealand's avian assemblage are the result of ancient vicariant events, stemming back to the Cretaceous period when New Zealand was connected to the ancient super-continent of Gondwana. Avian groups that have provided the best support for a vicariant origin of New Zealand taxa include the ratites, the parrots and a basal passerine group, the Acanthisittid wrens, which are characterised by either ancient divergence dates from related groups or a high degree of flightlessness, decreasing the likelihood of colonisation via dispersal. While the survival of archaic vicariant lineages is currently a highly debated topic, it is generally accepted that the vast majority of New Zealand's species are the result of colonisation via dispersal. New Zealand started to separate from Gondwanaland at approximately 82 mya (million years ago), and from approximately 80 mya New Zealand's connection with other Gondwanan continents was severed, rendering it an isolated landmass (Gibbs 2006). Molecular studies on a range of plant and animal populations indicate that the majority of taxa arrived in New Zealand via trans-oceanic dispersal following its isolation from Gondwana (Goldberg *et al.* 2008; Wallis and Trewick 2009; Trewick and Gibb 2010). Australia in particular has been a prolific source of dispersing birds, due to the prevalence of westerly winds and oceanic currents which have brought numerous species across the Tasman Sea (Keast 1971; Conner *et al.* 2004; McDowall 2008). In the past century alone, seven species have successfully colonised New Zealand from Australia (Fleming 1962; Wilson 1997) and an additional 11 species cannot be effectively distinguished from Australian species, indicating recent colonisation (Fleming 1962). Also, the relatively recent history of human occupation in New Zealand has resulted in a number of modern-day introductions of avian groups which are unlikely to have dispersed here via natural means. Following human settlement, 41 species have been introduced to New Zealand (Case 1996) and it has been subjected to 149 attempted introductions (the highest number of attempted introductions in the world, with the exception of Hawaii) (Long 1981; Case 1996; Lever 2005).

19.2   Key characteristics of New Zealand's avian biota

The study of New Zealand's biota has been described as 'as close as we will get to the opportunity to study life on another planet' in recognition of the unique nature of many of the plant and animal groups present (Diamond 1990). The high level of endemism in New Zealand's avifauna (80% of terrestrial and wetland species) (Diamond and Veitch 1981; Wilson 1997) is the result of its long isolation and its current position as one of the most isolated landmasses in the world (Fleming 1962; Cooper and Millener 1993). New Zealand's biogeographical history makes it a unique physical environment, whereby the landmass is continental in origin but has attracted an island-like biota due to its isolation via significant oceanic expanses and its complex geological history (Daugherty *et al.* 1993; Atkinson 2002). This has resulted in the mix of archaic Gondwanan species combined with trans-oceanic colonists, restricted to those that have the flight ability to traverse large oceanic distances. One of the key features of New Zealand's avian fauna is that despite the high level of endemism at the species level, species diversity of land-birds is in fact low, often described as depauperate (Daugherty *et al.* 1990; Atkinson and Cameron 1993; Cooper and Millener 1993; Atkinson 2002). This depauperate land fauna is partly the result of New Zealand's isolation, but is also a function of the high levels of extinction experienced by terrestrial animals in New Zealand (Cooper and Millener 1993).

One of the most notable aspects of New Zealand's fauna is the lack of key animal groups present on other Gondwanan countries. Of greatest significance to the evolution of New Zealand's avifauna is the almost complete absence of terrestrial mammals. Many disparate New Zealand avian taxa have evolved unique morphology and behaviour as a result of the absence of mammalian predators. For example, New Zealand species have a high incidence of gigantism and flightlessness, in the absence of selective pressure for morphology which allows for escape from ground-dwelling, fast-moving predators (Daugherty *et al.* 1993; Holdaway *et al.* 2001). Flightlessness is found in 33% of forest species and a further 9% have reduced flight. Behavioural traits such as ground-nesting habits and 'behavioural flightlessness' are also exhibited by species which have retained flight morphology in the absence of ground-dwelling mammal predators (Diamond 1990; Dowding and Murphy 2001). New Zealand species also have greater longevity, longer incubation and fledging times, and generally low reproductive rates (Daugherty *et al.* 1993; Dowding and Murphy 2001; Franklin and Wilson 2003). New Zealand birds have also evolved to fill niches left available by the absence of mammals. Large flightless birds have assumed the niche space of browsers (e.g. moa) or grazers (e.g. takahe (*Porphyrio hochstetteri*) and kakapo (*Strigops habroptilus*)) (Caughley 1989; Daugherty *et al.* 1993). In the absence of ground-burrowing carnivores, medium-sized flightless birds evolved to probe the soil for invertebrates (e.g. kiwi (*Apteryx*)). Kiwi in particular exhibit traits usually associated with mammals, including nocturnality, burrow-nesting, olfactory dominance and near-sightedness, and the occupation and defence of year-round territories (Daugherty *et al.* 1993). The absence of terrestrial mammals has meant that the predominant predators for New Zealand birds were also avian, including extinct eagles (*Harpagornis moorei*), adzebills (*Aptornis*), and extant falcons (*Falco novaeseelandiae*), harriers (*Circus approximans*), crows (*Corvus frugilegus*), skua (*Stercorarius*) and owls (*Ninox novaeseelandiae*) (Daugherty *et al.* 1993). New Zealand species exhibit several morphological, behavioural and life-history traits reflective of their evolution as prey to avian predators. Many New Zealand species have cryptic dorsal colouring and are nocturnal, traits considered likely to afford additional protection from aerial diurnal predators (King 1990; Daugherty *et al.* 1993). Naivety is also a trait characteristic of New Zealand species, and is observed in both forest-birds and sea-birds (Dowding and Murphy 2001).

19.3 Threats to New Zealand's avifauna

New Zealand's turbulent geological and climatic history combined with the recent impacts of human occupation, have resulted in New Zealand having one of the highest avian extinction rates in the world. At least 76 species (31%) within New Zealand's original avian fauna are now extinct (Atkinson 2001; Holdaway *et al.* 2001), including 36 (40%) of the 89 species of land-birds. The endemic taxa of New Zealand have been especially hard-hit, with 72 of the 174 endemic species (41%) now extinct, and four of seven endemic families (57%) now lost (Atkinson 2001; Holdaway *et al.* 2001; McDowall 2008). This loss of endemic and uniquely adapted species has caused significant disruption to New Zealand ecosystems. The loss of the speciose moa group resulted in a significant loss of both species diversity, and a completely unique guild of grazers. Additionally, almost all of the large frugivores have been lost from New Zealand forests, impacting ecosystem function in terms of plant pollination and dispersal (Diamond and Veitch 1981).

The high level of extinction experienced by New Zealand species reflects a dynamic history of both natural and anthropogenic impacts. Drastic sea level change during the Oligocene resulted in a reduction of New Zealand's emergent land to only 18% of its current size, severely restricting the distribution of all terrestrial species (Cooper and Cooper 1995). Ongoing geological change including periods of uplift, submersion and volcanism throughout most of New Zealand's history has resulted in multiple phases of land alteration, effectively causing the restriction and fragmentation of land taxa. While climate change is likely to have significantly impacted the distribution of land species, the majority of land taxa are thought to have survived the glacial maximum (McGlone 1989).

By far the most powerful extinction force for New Zealand's avian biota has been the colonisation of the landmass by humans. Since the arrival of humans, New Zealand has suffered several pulses of extinction, related to periods of human colonisation and exotic species introductions (Holdaway 1989). The arrival of Polynesian settlers (Māori) in approximately 1300AD brought about drastic changes in New Zealand terrestrial ecosystems through the influence of habitat clearance, human hunting and the introduction of mammalian species including the pacific rat or kiore (*Rattus exulans*) and dog or kuri (*Canis lupus familiaris*). It is estimated that 34 land-bird species were made extinct following Māori settlement (Diamond and Veitch 1981), including 36% of endemic land-birds. The first pulse of extinctions in New Zealand's avifauna largely resulted from the impacts of human hunting and predation by kiore (Holdaway 1989). Flightless, large birds with long generation times were particularly vulnerable to hunting pressure by early humans (Owens and Bennett 2000; Cassey 2001; Duncan and Worthy 2002). Perhaps the best-known example of the impacts of human hunting involved the complete extinction of the moa group (Holdaway and Jacomb 2000).

The second pulse of extinctions during the human history of New Zealand involved species that were less vulnerable to predation pressure from kiore and hunting pressure from humans, but were affected by the ongoing effects of habitat clearance by Polynesian humans. It is estimated that 3000 years ago, 85%–90% of New Zealand was covered in forest. While periods of volcanism are likely to have contributed to the destruction of native forest, deforestation by early humans is considered the primary cause of loss of habitat for terrestrial forest species in New Zealand (McGlone 1989; Atkinson and Cameron 1993). It is estimated that by the time of European settlement in the late 1800s, close to half of the original forest had been destroyed (McGlone 1989; Atkinson and Cameron 1993). This loss of habitat is considered likely to have drastically reduced habitat availability for native forest-birds, contributing to a number of extinctions of New Zealand avifauna (McGlone 1989; Holdaway 1990; Holdaway and Jacomb 2000).

Finally, the arrival of Europeans began a period of rapid change for the New Zealand biota, with further habitat destruction and modification, and the introduction of both mammalian and avian exotic species. Since European settlement, forest cover in New Zealand has been reduced to only 23% and ten additional land-bird species have become extinct (Atkinson and Cameron 1993). European colonists have been responsible for the introduction of a suite of exotic animals. Introduced animals have had a devastating effect on the avifauna of New Zealand (Diamond and Veitch 1981; Holdaway 1989), and remain the most significant threat to forest-birds in New Zealand (Innes *et al.* 2010). Invasive species in New Zealand are listed as the primary threat for 57% of the 359 taxa that are classed as threatened by the New Zealand Department of Conservation (DoC) (Clout 2002). These introduced species have a myriad of effects on native species, including the direct effects of predation and additional impacts of competition for resources, destruction of habitat, disease or hybrisation with local species (Clout 2002). Forest passerines have been particularly hard hit, with six of the 16 native passerines extinct, rare and locally restricted, or isolated on offshore island sanctuaries (Diamond and Veitch 1981).

Ground-dwelling mammalian predators such as rats (*Rattus rattus* and *Rattus norvegicus*), cats (*Felis catus*) and mustelids (Mustelidae) have caused the most drastic and notable declines and extinctions of native New Zealand bird species (King 1990; Dowding and Murphy 2001; Innes *et al.* 2010). Of the 34 established mammalian species introduced to New Zealand, 22 are considered a significant threat to New Zealand's fauna (Atkinson 2001). The unique morphological and behavioural traits that have evolved in New Zealand taxa in the absence of native mammals made them highly vulnerable to predation from these introduced species. Gigantism, flightlessness, ground-nesting habits, loss of defensive behaviours and protracted fledging and incubation times have resulted in numerous New Zealand species falling victim to predation by these species (Daugherty *et al.* 1993; Dowding and Murphy 2001; Innes *et al.* 2010). Introduced rat species including the ship rat and Norway rat are responsible for predation of adults, young and eggs of numerous native species, particularly endangered endemics such as the kiwi and kokako (*Callaeas cinereus*). Indirect effects of rat invasions are also evident through impacts on plant and invertebrate communities (Campbell and Atkinson 2002; Towns *et al.* 2009). Mustelid species are responsible for predation of eggs, young and adult land-birds including kiwi (McLennan *et al.* 1996) and kaka (*Nestor meridionalis*) (Wilson *et al.* 1998). Stoats have contributed to the extinction of at least nine bird species and threaten many more, while cats have been implicated in the extinction of at least six island endemic bird species, and 70 populations of insular birds. While the effects of carnivorous introduced mammalian species are evident, herbivorous browsers and grazers also have significant impacts on native avian species due to the effects of habitat destruction (Nugent *et al.* 2001). Brush-tail possums (*Trichosurus vulpecula*) in particular have had a drastic effect on native New Zealand forests. Possums were introduced for the fur trade between 1858 and 1900, and population densities have since increased to ten times the usual population sizes in Australia due to the lack of competitors, fewer parasites and lack of natural predators (Clout 2002). Possums have multiple effects on native avifauna including habitat damage (Nugent *et al.* 2001), as vectors for disease (Clout 2002) and as predators (Brown *et al.* 1993). In addition to the effects of introduced mammals, the introduction of exotic bird species by Europeans has also influenced New Zealand's native avifauna through the effects of competition and disease. New Zealand was subject to introduction of 41 exotic land-bird species, a number that approaches the number of extant native species (Case 1996).

19.4   Conservation techniques

Due to the diverse and extensive range of threats faced by New Zealand's avifauna, New Zealand has had a long history of conservation and restoration of threatened species. New Zealand is recognised as having contributed substantially to the field of conservation biology. Conservation scientists and managers in New Zealand have paved the way for the conservation biology discipline through the use of intensive adaptive management techniques and the contribution of research to the practice of conservation. Significant contributions include: (1) the use of offshore islands for sanctuaries, (2) ecological restoration, (3) the control of introduced pests, (4) the introduction, re-introduction and transfer of threatened species, and (5) the use of cross-fostering programmes (Diamond 1990).

19.4.1   Use of offshore islands

While New Zealand is made up of three main islands (North, South and Stewart Islands), the New Zealand territory includes over 735 islands over one hectare in size, extending from the subtropics to the sub-Antarctic (Parkes and Murphy 2003). The New Zealand Department of Conservation (DoC) manages approximately 250 of these islands which have been integral to the conservation management of numerous threatened species over New Zealand's history for several key reasons (Parkes and Murphy 2003). First, some of New Zealand's threatened species are island endemics, found only on offshore islands and absent from the mainland (e.g. Chatham Island black robin (*Petroica traverse*) and Forbes parakeet (*Cyanoramphus forbesi*)). Others were historically found on the mainland but native populations have survived solely on offshore islands. Such pseudo-endemics include unique threatened species such as the saddleback (*Philesturnus carunculatus*) and hihi/stitchbird (*Notiomystis cincta*) (Daugherty *et al.* 1990). Offshore islands have been particularly important for the survival of remnant populations of breeding shore-birds, with four of ten extant threatened shore-bird species and nearly all Procellariform sea-birds now breeding exclusively on offshore islands (Robertson 1985; Dowding and Murphy 2001). While the majority of offshore islands were subject to invasions of the same introduced mammalian species as the mainland of New Zealand, 158 are known to have avoided the introduction of mammals (Parkes and Murphy 2003), making them ideal locations for the protection and preservation of threatened species. Additionally, offshore islands which are smaller in size than the mainland and at least somewhat isolated from sources of re-invasion represent a simpler task for pest eradication and restoration (Simberloff 1990). By 2010 at least 147 populations representing 13 species of invasive vertebrates had been eradicated from over 95 offshore islands in New Zealand (Towns, 2011). The largest of these eradication projects involved the complete eradication of Norway rats from 11 200 ha Campbell Island. Offshore islands were recognised as key resources for species recovery plans early in New Zealand's conservation history. Offshore islands, such as Resolution Island, Little Barrier (Hauturu) Island and Kapiti Island were the recipients of some of the earliest translocations initiated to attempt to recover endangered species, and formed the basis for the recovery of several of New Zealand's most threatened avian species, including kakapo, kiwi, hihi, saddleback, the Chatham Island black robin and the takahe. The survival of endangered New Zealand species has therefore relied heavily on offshore islands as sites for both the preservation of threatened populations and the creation of additional populations via translocation (see Table 19.1).

Table 19.1 Island populations of Endangered and Critically Endangered endemic New Zealand species

Species	Conservation status	Islands naturally occurring	Translocated island populations
Anas chlorotis (Brown Teal)	Endangered, Increasing	Great Barrier Is.	Little Barrier Is., Kapiti Is., Mana Is., Mayor, Is. Tiritiri Matangi Is.
Anas nesiotis (Campbell Is. Teal)	Endangered, Increasing.	Campbell Is.	Codfish Is.
Apteryx mantelli (NI Brown Kiwi)	Endangered, Decreasing		Ponui, Is. Kapiti Is., Kawau Is., Little Barrier Is., Motuora Is., Motutapu Is.
Callaeas cinereus (Kokako)	Endangered, Increasing		Secretary Is., Little Barrier Is., Tiritiri Matangi Is., Kapiti Is.
Charadrius obscurus (NZ dotterel)	Endangered, Increasing	Stuart Is., Great Barrier Is., Waiheke Is., Rangitoto Is., Motutapu Is., Ponui Is.	
Cyanoramphus forbesi (Chatham Is. Parakeet)	Endangered, Stable	Chatham Is., Mangere Is.	
Cyanoramphus malherbi (Malherbe's Parakeet)	Critically Endangered		Chalky Is., Maud Is., Tuhua Is., Blumine Is.
Hymenolaimus malacorhynchos (Blue Duck)	Endangered, Decreasing	Secretary Is., Resolution Is., Long Is.	
Eudyptes sclateri (Erect-crested Penguin)	Endangered, Decreasing	Bounty Is. and Antipodes Is.	
Haematopus chathamensis	Endangered, Increasing	Chatham Is.'s	

Table 19.1 (*cont.*)

Species	Conservation status	Islands naturally occurring	Translocated island populations
(Chatham Is. Oystercatcher)			
Megadyptes antipodes (Yellow-eyed Penguin)	Endangered, Decreasing	Stewart Is., Codfish Is., Auckland Is.'s, Campbell Is.'s	
Mohoua ochrocephala (Yellowhead)	Endangered, Decreasing		Codfish Is.
Nestor meridionalis (Kaka)	Endangered, Decreasing	Codfish Is., D'Urville Is., Kapiti Is., Great Barrier Is., Little Barrier Is., Waewaetorea Is.	
Oceanites maorianus (Storm Petrel)	Critically Endangered, Unknown Trend	Little Barrier Is.	
Petroica traversi (Black Robin)	Endangered, Increasing	Chatham Is.	Mangere Is., Rangatira Is. (Chatham Is.)
Phalacrocorax featherstoni (Pitt Is. Shag)	Endangered, Decreasing	Chatham Is.	
Phalacrocorax onslowi (Chatham Is. Shag)	Critically Endangered Decreasing	Chatham Is.	
Porphyrio hochstetteri (Takahe)	Endangered, Increasing		Kapiti Is., Mana Is., Tiritiri Matangi Is., Maud Is.
Pterodroma axillaris (Chatham Petrel)	Endangered, Increasing	Chatham Is	Pitt Is., Main Chatham Is.
Pterodroma magentae (Magenta Petrel)	Critically Endangered, Increasing	Chatham Is.	
Strigops habroptila (Kakapo)	Critically Endangered, Increasing		Codfish Is., Anchor Is., Chalky Is., Maud Is., Little Barrier Is.
Thinornis novaeseelandiae (Shore Plover)	Endangered, Increasing	Chatham Is.	Rangatira Is., Mangere Is., Waikawa Is., Mana Is.

19.4.2   Ecological restoration and the control of introduced pests

The use of offshore islands has been central to New Zealand's attempts at ecological restoration. New Zealand now has over 22 active restoration projects operating on offshore islands, many of which form the core population for particular endangered species (Saunders and Norton 2001). Owing to the impacts of invasive species on New Zealand's biodiversity, the restoration of New Zealand environments requires at least some level of invasive species control to be successful (Atkinson 2001; Saunders and Norton 2001). Invasive species management is the foremost adaptive management methodology employed by New Zealand conservation managers. Attempts at invasive species control have included a combination of techniques. Human hunting of large herbivorous species such as possums, goats (*Capra aegagrus hircus*) and deer (Cervidae) has been combined with poison baiting and trapping methods which target invasive species such as possums, cats, rats and mustelids. Significant achievements in invasive species eradication have been made on offshore islands around New Zealand. Of nearly 300 islands over 5 ha in size, almost half have been colonised by invasive mammals during New Zealand's history of human occupation. As a result of the intensive eradication efforts of New Zealand scientists and managers, 64 of these are now completely free of introduced mammals (Atkinson 2001, 2002). Of particular significance is the complete eradication of all mammals from increasingly large islands such as Little Barrier Island (3083 ha) (Towns and Ballantine 1993), allowing for regeneration and restoration of large forested environments which have become key sites for the recovery of threatened species. These efforts have resulted in increases of at least 26 species of threatened birds, emphasising the importance of invasive species eradications for the recovery of protected species in New Zealand (Bellingham *et al.* 2010).

Historically, the mainland of New Zealand has been subject to less intensive management, with a focus on habitat preservation predominating. With the growing success of offshore island projects, the lessons learned from these sites have been utilised on the mainland in recent years. Ecological restoration on the mainland started as late as the 1970s and has only become prevalent in the past 15 years (Atkinson 2001). By 2001 there were 18 significant projects active on the mainland (Atkinson 2001), including six 'mainland islands' (Clout 2001; Saunders and Norton 2001) and this number continues to grow. Mainland island sites in New Zealand are characterised by the use of intensive pest control regimes involving multi-species eradications, combined with management of multiple native species, and typically address ecosystem recovery goals as opposed to single species management (Saunders and Norton 2001). Because of the prevalence of invasive species on New Zealand's mainland, pest control is a pivotal component of mainland island conservation projects. A total of 17 invasive species are being actively controlled within mainland islands in New Zealand (Saunders and Norton 2001). Owing to the risk of reinvasion by invasive species, eradication within mainland projects involves long-term intensive pest control regimes, and has commonly incorporated the use of poison-baiting and trapping lines which provide widespread coverage of the habitat. Given the early nature of these ongoing invasive-species control regimes, the long-term success of these methods is yet to be recognised, although evidence of ongoing declines of invasive species and corresponding increases in the density and reproductive success of threatened species indicates the efficacy of these methods (Saunders and Norton 2001; Gillies *et al.* 2003; Starling-Windhof *et al.* 2011). Since 1999, the development of predator-proof fencing has added additional security to mainland sites, allowing the most predation-sensitive species to be translocated into mainland islands (Clapperton and Day 2001; Innes *et al.* 2012).

19.4.3 The introduction, re-introduction or transfer of threatened species

While invasive species control and eradication has been a key component of ecological restoration projects around New Zealand, the re-introduction or translocation of managed species has also been utilised extensively to re-establish native assemblages and ecosystems. Translocation is defined by the IUCN as 'the intentional release of animals for the purposes of establishing, re-establishing or augmenting an existing population' (IUCN 2012). It therefore involves the deliberate movement of individuals to sites in order to address multiple conservation goals. The first attempted translocations in 1895 in New Zealand involved two of its most high profile species: the kakapo and the kiwi. While early translocation attempts failed to result in established populations, the lessons learned regarding transfer methodology were invaluable to the future of the practice. Since these initial efforts, the use of translocation has increased and methodologies are continuing to be refined and developed. Translocation is now a central component to most endangered species recovery plans in New Zealand. In the decade between 1981 and 1990, translocations more than doubled from the previous decade (Atkinson 1990). Between 2002 and 2010, over 300 translocation proposals were approved, 74% of which were for birds. Currently, 61% of New Zealand's extant endemic waterfowl, shore-bird and land-bird taxa have been subject to translocation (Miskelly and Powlesland 2013). The importance of translocation in species recovery is exemplified by five taxa which exist solely as translocated populations and 10 further species which would be confined to a single population if not for translocation (Miskelly and Powlesland 2013). Translocation to offshore islands has predominated. The majority of early translocations from 1960 to 1990 involved transfers to offshore islands (Miskelly and Powlesland 2013). To date, 19 species of NZ birds have been translocated to or between offshore islands in New Zealand. Documentation of these early transfers led the way for the translocation of endangered species to the mainland from the 1990s, focussing on sites undergoing intensive pest control. In the 2000s, translocations to fenced mainland sanctuaries dominated the large number of translocation attempts for the first time (Miskelly and Powlesland 2013).

While New Zealand has experienced significant success in the use of translocation in species recovery, translocation has a mixed history of success globally (Griffith *et al.* 1989; Wolf *et al.* 1996; Fischer and Lindenmayer 2000) and the failure of many transfer attempts in New Zealand has resulted in extensive discussion on factors affecting the success or failure of re-introduction and translocation attempts. The documentation of re-introductions by New Zealand researchers and managers has formed the basis for the development of the re-introduction biology field globally and has involved reporting of translocation outcomes, experimental transfers and extensive post-release monitoring and population modelling. New Zealand research has contributed significantly to discussion on capture and release methodology. For example, publication of translocation methodology for rifleman (*Acanthisitta chloris*) (Leech *et al.* 2007), robins (*Petroica australis*) (Lovegrove and Veitch 1994), hihi/stitchbird (Castro 1995; Armstrong *et al.* 1999), kakapo (Lloyd and Powlesland 1994), saddleback (Lovegrove 1996) and burrow-nesting sea-birds (Miskelly *et al.* 2009) has allowed future translocations to refine transfer processes for both forest- and sea-birds, minimising further losses. Research carried out on New Zealand species has also shed light on factors that may influence translocation success post-release, including post-release management of food availability (Armstrong *et al.* 1999), the social composition of founding groups (Armstrong and Craig 1995; Rowe 2007), factors that influence dispersal behaviour (Masuda and Jamieson 2012; Armstrong and Ewen 2002; Bradley *et al.* 2012; Castro *et al.* 1994; Rickett *et al.* 2013), habitat suitability (Griffith *et al.* 1989; Osborne and Seddon 2012), levels of predation (Armstrong and Davidson 2006) and disease (Derraik *et al.* 2008; Ewen *et al.* 2012; Howe *et al.* 2012). The effects of inbreeding on translocation success have received a lot of attention from New Zealand conservation biologists, as translocations often involve the transfer of a small number of founders which will form the basis for the breeding population. Inbreeding depression caused by high levels of relatedness between breeding individuals has been noted in several species of inbred New Zealand birds, including takahe (Jamieson and Ryan 2000), black robin (Butler and Merton 1992), North Island robin (Jamieson *et al.* 2007) and hihi/stitchbird (Brekke *et al.* 2010). However, many New Zealand species show no signs of inbreeding depression despite being clearly inbred (Taylor *et al.* 2005). Documentation of levels of inbreeding and its effects on New Zealand species has contributed to both practical conservation discussions and theories concerning genetic conservation management.

In addition to the publication of translocation and re-introduction research, New Zealand conservation biologists have used experimental re-introduction methods to test key hypotheses concerning the requirements of successful translocation. For example, experiments during saddleback releases have incorporated different holding times to test the effect of holding time on stress levels and loss of mass during transfer (Adams *et al.* 2010), while releases of hihi have tested the benefits of alternative release strategies for survival of transferred birds (Castro *et al.* 1994). Armstrong and Craig (1995) and Armstrong (1995) experimented with social composition of founding groups to determine whether familiarity with co-founders was necessary to reduce aggression and increase translocation success. Trials with food supplementation and feeding regimes of sea-birds (Miskelly *et al.* 2009) and food supplementation for South Island robins (Mackintosh 2005) and brown teal (*Anas chlorotis*) (Rickett *et al.* 2013) have also provided key findings to further develop translocation methods. New Zealand scientists have also pioneered the use of population modelling and simulation using monitoring data to address multiple issues concerning the success of re-introductions (Armstrong and Reynolds 2012). Modelling has been used to address questions such as the need for supplementary translocations in robins (Armstrong and Ewen 2001); levels of required predator control in managed populations of saddleback and robins (Armstrong and Davidson 2006; Armstrong *et al.* 2006) and to assess the future viability of translocated populations of robins (Armstrong and Ewen 2001; Armstrong and Reynolds 2012) and rifleman (Leech *et al.* 2007), among others.

19.4.4   Cross-fostering and captive breeding

Adaptive management methodologies, including pest eradication combined with re-introduction, have resulted in the effective recovery of numerous endangered species in New Zealand and contributed to ecosystem restoration goals. However, in extreme cases, further adaptive management methods have been required to attempt to save endangered species from extinction. Cross-fostering and egg manipulation have been used to assist in the recovery of some of New Zealand's most at-risk species including the takahe, kakapo, black stilt (*Himantopus novaezelandiae*) and black robin (Butler and Merton 1992; Reed *et al.* 1993; Elliott *et al.* 2001; Wickes *et al.* 2009). Captive breeding efforts have also been used to assist and complement other adaptive management methods where wild productivity requires supplementation. Captive rearing of wild-caught eggs and young to re-inforce wild populations has been used in the recovery programmes of four bird species: kiwi, blue duck (*Hymenolaimus malacorhynchos*), takahe and black stilt. Population supplementation via translocation has often used captive bred or captive reared individuals. At least 79 translocations involving at least 50% captive bred or reared individuals have been performed on endangered species, including the kiwi, brown teal, NZ scaup (*Aythya novaeseelandiae*), NZ falcon, kaka, orange fronted parakeet (*Cyanoramphus malherbi*) and blue duck (Miskelly and Powlesland 2013).

19.5 Conclusion

The New Zealand avifauna has long been recognised as unique but faces significant threats, predominantly from mammalian predators introduced by colonising humans. Extreme threat, however, has engendered a high degree of commitment and innovation. In many ways New Zealand conservation biologists and managers lead the world in their adaptive management techniques to recover threatened species, including predator control, ecosystem restoration, translocation and re-introduction, breeding manipulation and captive rearing. The use of these interventions has saved several species from extinction.

REFERENCES

Adams, N. J., K. A. Parker, J. F. Cockrem, D. H. Brunton and E. J. Candy (2010). Corticosterone responses and post-release survival in translocated North Island Saddlebacks (*Philesturnus rufusater*) in New Zealand. *Emu* 110(4): 296–301.

Armstrong, D. P. (1995). Effects of familiarity on the outcome of translocations, II. A test using New Zealand robins. *Biological Conservation* 71: 281–288.

Armstrong, D. P. and J. L. Craig (1995). Effects of familiarity on the outcome of translocations, I. A test using saddlebacks *Philesturnus carunculatus rufusater*. *Biological Conservation* 71: 133–141.

Armstrong, D. P. and R. S. Davidson (2006). Developing population models for guiding reintroductions of extirpated bird species back to the New Zealand mainland. *New Zealand Journal of Ecology* 30(1): 73–85.

Armstrong, D. P. and J. G. Ewen (2001). Assessing the value of follow-up translocations: a case study using New Zealand robins. *Biological Conservation* 101(2): 239–247.

Armstrong, D. P. and J. G. Ewen (2002). Dynamics and viability of a New Zealand robin population reintroduced to regenerating fragmented habitat. *Conservation Biology* 16(4): 1074–1085.

Armstrong, D. P. and M. H. Reynolds (2012). Modelling reintroduced populations: the state of the art and future directions. In *Reintroduction Biology: Integrating Science and Management*, J. G. Ewen, D. P. Armstrong, K. Parker and P. J. Seddon (eds.). Oxford, U.K., Wiley-Blackwell, pp. 165–222.

Armstrong, D. P., I. Castro, J. C. Alley, B. Feenstra and J. K. Perrott (1999). Mortality and behaviour of hihi, an endangered New Zealand honeyeater, in the establishment phase following translocation. *Biological Conservation* 89: 329–339.

Armstrong, D. P., E. H. Raeburn, R. M. Lewis and D. Ravine (2006). Estimating the viability of a reintroduced New Zealand robin population as a function of predator control. *Journal of Wildlife Management* 70(4): 1020–1027.

Atkinson, I. A. E. (1990). Ecological restoration on islands: Prerequisites for success. In *Ecological Restoration of New Zealand Islands*, D. R. Towns, C. H. Daugherty and I. A. E. Atkinson (eds.). Conservation Sciences Publication No.2, Department of Conservation, Wellington.

Atkinson, I. A. E. (2001). Introduced mammals and models for restoration. *Biological Conservation* **99**: 81–96.

Atkinson, I. A. E. (2002). Recovery of wildlife and restoration of habitats in New Zealand. *Pacific Conservation Biology* **8**(1): 27–35.

Atkinson, I. A. E. and E. K. Cameron (1993). Human influence on the terrestrial biota and biotic communities of New Zealand. *Trends in Ecology & Evolution* **8**(12): 447–451.

Bellingham, P. J., D. R. Towns, E. K. Cameron, *et al.* (2010). New Zealand island restoration: sea-birds, predators and the importance of history. *New Zealand Journal of Ecology* **34**(1): 115–136.

Bradley, D. W., L. E. Molles, S. V. Valderrama, S. King and J. R. Waas (2012). Factors affecting post-release dispersal, mortality, and territory settlement of endangered kokako translocated from two distinct song neighborhoods. *Biological Conservation* **147**(1): 79–86.

Brekke, P., P. M. Bennett, J. Wang, N. Pettorelli and J. G. Ewen (2010). Sensitive males: inbreeding depression in an endangered bird. *Proceedings of the Royal Society Biological Sciences Series B* **277**(1700): 3677–3684.

Brown, K., J. Innes and R. Shorten (1993). Evidence that possums prey on and scavenge birds' eggs, birds and mammals. *Notornis* **40**: 169–177.

Butler, D. and D. Merton (1992). *The Black Robin: Saving the World's Most Endangered Bird*. Auckland, New Zealand, Auckland University Press.

Campbell, D. J. and I. A. E. Atkinson (2002). Depression of tree recruitment by the Pacific rat (*Rattus exulans Peale*) on New Zealand's northern offshore islands. *Biological Conservation* **107**: 19–35.

Case, T. J. (1996). Global patterns in the establishment and distribution of exotic birds. *Biological Conservation* **78**: 69–96.

Cassey, P. (2001). Determining variation in the success of New Zealand land-birds. *Global Ecology & Biogeography* **10**: 161–172.

Castro, I. (1995). Behavioural ecology and management of hihi (*Notiomystis cincta*), an endemic New Zealand honeyeater, Massey University.

Castro, I., J. C. Alley, R. A. Empson and E. O. Minot (1994). Translocation of hihi or stitchbird *Notiomystis cincta* to Kapiti Island, New Zealand: transfer techniques and comparison of release strategies. In *Reintroduction Biology of Australian and New Zealand Fauna*. M. Serena, Surrey Beatty and Sons.

Caughley, G. (1989). New Zealand plant–herbivore systems: past and present. *New Zealand Journal of Ecology* **12**: 3–10.

Clapperton, B. K. and T. D. Day (2001). *Cost-effectiveness of Exclusion Fencing for Stoat and Other Pest Control Compared with Conventional Control*. DoC Internal Series 14. Wellington, Department of Conservation.

Clout, M. N. (2001). Where protection is not enough: active conservation in New Zealand. *Trends in Ecology & Evolution* **16**(8): 415–416.

Clout, M. N. (2002). Biodiversity loss caused by invasive alien vertebrates. *Zeitschrift fuer Jagdwissenschaft* **48**(Supplement): 51–58.

Conner, R. N., D. Saenz, R. R. Schaefer, *et al.* (2004). Group size and nest success in Red-cockaded Woodpeckers in the West Gulf Coastal Plain: helpers make a difference. *Journal of Field Ornithology* **75**(1): 74–78.

Cooper, A. and R. A. Cooper (1995). The Oligocene bottleneck and New Zealand biota: genetic record of a past environmental crisis. *Proceedings of the Royal Society of London B* **261**(1362): 293–302.

Cooper, R. A. and P. R. Millener (1993). The New Zealand biota: historical background and new research. *Trends in Ecology and Evolution* **8**(12): 429–433.

Daugherty, C. H., G. W. Gibbs and R. A. Hitchmough (1993). Mega-island or micro-continent? New Zealand and its fauna. *Trends in Ecology & Evolution* **8**(12): 437–442.

Daugherty, C. H., D. R. Towns, I. A. E. Atkinson and G. Gibbs (1990). The significance of the biological resources of New Zealand islands for ecological restoration. In *Ecological Restoration of New Zealand Islands*, D. R. Towns, C. H. Daugherty and I. A. E. Atkinson (eds.). Conservation Sciences Publication No.2, Department of Conservation, Wellington.

Derraik, J. G. B., D. M. Tompkins, M. R. Alley, P. Holder and T. Atkinson (2008). Epidemiology of an avian malaria outbreak in a native bird species (*Mohoua ochrocephala*) in New Zealand. *Journal of the Royal Society of New Zealand* **38**(4): 237–242.

Diamond, J. M. (1990). New Zealand as an archipelago: an international perspective. In *Ecological Restoration of New Zealand Islands*, D. R. Towns, C. H. Daugherty and I. A. E. Atkinson (eds.). Conservation Sciences Publication No.2, Department of Conservation, Wellington.

Diamond, J. M. and C. R. Veitch (1981). Extinctions and introductions in the New Zealand avifauna: cause and effect? *Science* **211**(4481): 499–501.

Dowding, J. E. and E. C. Murphy (2001). The impact of predation by introduced mammals on endemic shore-birds in New Zealand: a conservation perspective. *Biological Conservation* **99**(1): 47–64.

Duncan, R. P. B. and T. H. Worthy (2002). Prehistoric bird extinctions and human hunting. *Proceedings: Biological Sciences* **269**(1490): 517–521.

Elliott, G. P., D. V. Merton and P. W. Jansen (2001). Intensive management of a critically endangered species: the kakapo. *Biological Conservation* **99**: 121–133.

Ewen, J. G., K. Acevedo-Whitehouse and M. R. Alley, Eds. (2012). Empirical consideration of parasites and health in reintroduction. In *Reintroduction Biology: Integrating Science and Management*. Oxford, UK, Wiley-Blackwell, pp. 290–335.

Fischer, J. and D. B. Lindenmayer (2000). An assessment of the published results of animal relocations. *Biological Conservation* **96**: 1–11.

Fleming, C. A. (1962). History of the New Zealand land-bird fauna. *Notornis* **9**(8): 270–275.

Franklin, D. C. and K.-J. Wilson (2003). Are low reproductive rates characteristic of New Zealand's native terrestrial birds? Evidence from the allometry of nesting parameters in altricial species. *New Zealand Journal of Zoology* **30**(3): 185–204.

Gaskin, C. P. and M. J. Rayner (2013). Sea-birds of the Hauraki Gulf: Natural History, Research and Conservation, Hauraki Gulf Forum.

Gibbs, G. (2006). *Ghosts of Gondwana: The History of Life in New Zealand.* Nelson, Craig Potton Publishing.

Gillies, C. A., M. R. Leach, N. B. Coad, *et al.* (2003). Six years of intensive pest mammal control at Trounson Kauri Park, a Department of Conservation 'mainland island', June 1996 – July 2002. *New Zealand Journal of Zoology* **30**(4): 399–420.

Goldberg, J., S. A. Trewick and A. M. Paterson (2008). Evolution of New Zealand's terrestrial fauna: a review of molecular evidence. *Philosophical Transactions: Biological Sciences* **363**(1508): 3319–3334.

Griffith, B., J. Michael Scott, J. W. Carpenter and C. Reed (1989). Translocation as a species conservation tool: status and strategy. *Science* **245**: 477–479.

Holdaway, R. N. (1989). New Zealand's prehuman avifauna and its vulnerability. *New Zealand Journal of Ecology* **12**: 11–25.

Holdaway, R. N. (1990). Changes in the diversity of New Zealand forest-birds. *New Zealand Journal of Zoology* **17**(3): 309–322.

Holdaway, R. N. and C. Jacomb (2000). Rapid extinction of the Moas (Aves: Dinornithiformes): model, test and implications. *Science* **287**(5461): 2250–2254.

Holdaway, R. N., T. H. Worthy and A. J. D. Tennyson (2001). A working list of breeding bird species of the New Zealand region at first human contact. *New Zealand Journal of Zoology* **28**(2): 119–187.

Howe, L., I. C. Castro, E. R. Schoener, *et al.* (2012). Malaria parasites (*Plasmodium* spp.) infecting introduced, native and endemic New Zealand birds. *Parisitology Research* **110**: 913–923.

Innes, J., D. Kelly, J. Overton and C. Gillies (2010). Predation and other factors currently limiting New Zealand forest-birds. *New Zealand Journal of Ecology* **34**(1): 86–114.

Innes, J., W. G. Lee, B. Burns, *et al.* (2012). Role of predator-proof fences in restoring New Zealand's biodiversity: a response to Scofield *et al.* (2011). *New Zealand Journal of Ecology* **36**(2): 232–238.

IUCN (2012). *IUCN Red List of Threatened Species.* Version 2012.1. www.iucnredlist.org: Downloaded on 07 August 2012.

Jamieson, I. G. and C. J. Ryan (2000). Increased egg infertility associated with translocating inbred takahe (*Porphyrio hochstetteri*) to island refuges in New Zealand. *Biological Conservation* **94**: 107–114.

Jamieson, I. G., L. N. Tracy, D. Fletcher and D. P. Armstrong (2007). Moderate inbreeding depression in a reintroduced population of North Island robins. *Animal Conservation* **10**(1): 95–102.

Keast, A. (1971). Continental drift and the evolution of the biota on southern continents. *The Quarterly Review of Biology* **46**(4): 335–378.

King, C. M. (1990). *The Handbook of New Zealand Mammals.* Auckland, New Zealand, Oxford University Press.

Leech, T., E. Craig, B. Beaven, D. K. Mitchell and P. J. Seddon (2007). Reintroduction of rifleman *Acanthisitta chloris* to Ulva Island, New Zealand: evaluation of techniques and population persistence. *Oryx* **41**(3): 369–375.

Lever, C. (2005). *Naturalised Birds of the World*. London, U.K., Poyser.

Lloyd, B. D. and R. G. Powlesland (1994). The decline of kakapo *Strigops habroptilus* and attempts at conservation by translocation. *Biological Conservation* **69**(1): 75–85.

Long, J. L. (1981). *Introduced Birds of the World: the Worldwide History, Distribution and Influence of Birds Introduced to New Environments*. Terrey Hills, NSW, Reed.

Lovegrove, T. G. (1996). Island releases of saddlebacks *Philesturnus carunculatus* in New Zealand. *Biological Conservation* **77**(2–3): 151–157.

Lovegrove, T. G. and C. R Veitch. (1994). Translocating wild forest-birds. *Ecological Management* **2**: 23–35.

Mackintosh, M. A. B. (2005). High levels of hatching failure in an insular population of the South Island robin: a consequence of food limitation? *Biological Conservation* **122**: 409–416.

Masuda, B. M. and I. G. Jamieson (2012). Age-specific differences in settlement rates of saddlebacks (*Philesturnus carunculatus*) reintroduced to a fenced mainland sanctuary. *New Zealand Journal of Ecology* **36**(2): 123–130.

McDowall, R. M. (2008). Process and pattern in the biogeography of New Zealand – a global microcosm? *Journal of Biogeography* **35**: 197–212.

McGlone, M. S. (1989). The polynesian settlement of New Zealand in relation to environmental and biotic changes. *New Zealand Journal of Ecology* **12**: 115–129.

McLennan, J. A., M. A. Potter, H. A. Robertson, *et al.* (1996). Role of predation in the decline of kiwi, *Apteryx* spp., in New Zealand. *New Zealand Journal of Ecology* **20**(1): 27–35.

Miskelly, C. M. and R. G. Powlesland (2013). Conservation translocation of New Zealand birds, 1863–2012. *Notornis* **60**: 3–28.

Miskelly, C. M., G. A. Taylor, H. Gummer and R. Williams (2009). Translocations of eight species of burrow-nesting sea-birds (genera *Pterodroma, Pelecanoides, Pachyptila* and *Puffinus*: Family Procellaridae). *Biological Conservation* **142**: 1965–1980.

Nugent, G., W. Fraser and P. Sweetapple (2001). Top down or bottom up? Comparing the impacts of introduced arboreal possums and 'terrestrial' ruminants on native forests in New Zealand. *Biological Conservation* **99**: 65–79.

Osborne, P. E. and P. J. Seddon, Eds. (2012). Selecting suitable habitats for reintroductions: variation, change and the role of species distribution modelling. In *Reintroduction Biology: Integrating Science and Management*. Oxford, UK, Wiley-Blackwell, pp. 73–104.

Owens, I. P. F. and P. M. Bennett (2000). Ecological basis of extinction risk in birds: habitat loss versus human persecution and introduced predators. *PNAS* **97**(22): 12 144–12 148.

Parkes, J. and E. Murphy (2003). Management of introduced mammals in New Zealand. *New Zealand Journal of Zoology* **30**(4): 335–359.

Reed, C. E. M., D. P. Murray and D. J. Butler (1993). *Black Stilt Recovery Plan (Himantopus novaezealandiae)*. Threatened Species Recovery Plan Series No.4. Wellington, Department of Conservation.

Rickett, J., C. J. Dey, J. Stothart, *et al.* (2013). The influence of supplemental feeding on survival, dispersal and competition in translocated Brown Teal, or Pateke (*Anas chlorotis*). *Emu* **113**(1): 62–68.

Robertson, C. J. R. (1985). *Reader's Digest Complete Book of New Zealand Birds*. Sydney, Reader's Digest Services Pty Ltd.

Rowe, S. J. (2007). The influence of geographic variation in song dialect on post-translocation pair formation in North Island kokako (*Callaeas cinerea wilsoni*). *Notornis* **54**: 28–37.

Saunders, A. and D. A. Norton (2001). Ecological restoration at mainland islands in New Zealand. *Biological Conservation* **99**(1): 109–119.

Simberloff, D. (1990). Reconstructing the ambiguous: can island ecosystems be restored? In *Ecological Restoration of New Zealand Islands*, D. R. Towns, C. H. Daugherty and I. A. E. Atkinson (eds.). Conservation Sciences Publication No.2, Department of Conservation, Wellington.

Starling-Windhof, A., M. Massaro and J. V. Briskie (2011). Differential effects of exotic predator control on nest success of native and introduced birds in New Zealand. *Biological Invasions* **13**: 1021–1028.

Taylor, S. S., I. G. Jamieson and D. P. Armstrong (2005). Successful island reintroductions of New Zealand robins and saddlebacks with small numbers of founders. *Animal Conservation* **8**: 415–420.

Towns, D. R. (2011). Eradications of vertebrate pests from islands around New Zealand: what have we delivered and what have we learned? In *Island Invasives: Eradication and Management*. **Veitch, C. R., M. N. Clout and D. R. Towns** (eds.). IUCN, Gland, Switzerland.

Towns, D. R. and W. J. Ballantine (1993). Conservation and restoration of New Zealand ecosystems. *Trends in Ecology & Evolution* **8**(12): 452–457.

Towns, D. R., D. A. Wardle, C. Mulder, *et al.* (2009). Predation of sea-birds by invasive rats: multiple indirect consequences for invertebrate communities. *Oikos* **118**: 420–430.

Trewick, S. A. and G. C. Gibb (2010). Vicars, tramps and assembly of the New Zealand avifauna: a review of molecular phylogenetic evidence. *Ibis* **152**: 226–253.

Wallis, G. P. and S. A. Trewick (2009). New Zealand phylogeography: evolution on a small continent. *Molecular Ecology* **18**: 3548–3580.

Wickes, C., D. Crouchley and J. Maxwell (2009). *Takahe (Porphyrio hochstetteri) Recovery Plan*. Threatened Species Recovery Plan No. 61. Wellington, Department of Conservation.

Wilson, K.-J. (1997). Extinct and introduced vertebrate species in New Zealand: a loss of biodistinctiveness and gain in biodiversity. *Pacific Conservation Biology* **3**(3): 301–305.

Wilson, P. R., B. J. Karl, R. J. Toft, J. R. Beggs and R. H. Taylor (1998). The role of introduced predators and competitors in the decline of kaka (*Nestor meridionalis*) populations in New Zealand. *Biological Conservation* **83**(2): 175–185.

Wolf, C. M., B. Griffith, C. Reed and S. A. Temple (1996). Avian and mammalian translocations: Update and reanalysis of 1987 survey data. *Conservation Biology* **10**(4): 1142–1154.

CHAPTER 20

Australian birds: current status and future prospects

Stephen T. Garnett, Judit K. Szabo and Donald C. Franklin

Summary

As in much of the world, Australia's birds have suffered greatly from habitat loss, feral predators and direct exploitation. Less universal have been the declines caused by post-colonial changes in fire regime after 40 000 years of Indigenous fire management. Climate change and a disengagement by Australians from nature loom as threats for the future. However, Australia is a country of climatic extremes and many birds are well-adapted to stressful conditions. Given adequate investment, all the major classes of threat have potential solutions, with particular success in recent decades in the removal of feral predators from islands and in reducing the by-catch from fishing. The biggest threat of all is possibly a failure to invest in conservation as modern lifestyles take people further and further away from the natural environment.

20.1 Introduction

Australia's birds are, like those in so much of the world, travelling poorly. Of the 1239 species and subspecies regularly occurring in Australia, 17% are Threatened or Near Threatened on the basis of the IUCN Red List Criteria (Garnett *et al.* 2011). This number has been increasing steadily (Szabo *et al.* 2012a) and, while originally it was taxa of Australia's oceanic islands that were most likely to be threatened, taxa from the mainland are now starting to slip away (Szabo *et al.* 2012b). Sadly some of those most threatened are the most distinctive; birds at the end of long slender branches of the evolutionary tree whose closest relatives are long gone. Other species, however, are thriving under the conditions that have arisen over the past few centuries of intense development.

The modern avifauna of Australia is in substantial part the product of four major events in the Earth's recent geological history. The first was Australia being part of the ancient megacontinent Gondwana, along with Africa, India and South America. The avifauna inherited from Gondwana provided the basis for subsequent evolutionary

Austral Ark: The State of Wildlife in Australia and New Zealand, eds. A. Stow, N. Maclean and G. I. Holwell. Published by Cambridge University Press. © Cambridge University Press 2015.

trajectories, with most endemics having a Gondwanan origin (Christidis and Norman 2010). Second was Australia's long voyage northward after its initial separation from Antarctica *c*. 85 million years ago, a separation that sped up markedly *c*. 35 million years ago (Veevers *et al*. 1991). During this time, this most isolated of continents developed many of the forms not found elsewhere, including some, like the corvid passerines, that went on to colonise the rest of the world (Edwards and Boles 2002). The third event was the approach and ongoing collision of the Australian continental plate with that of Asia. While sea barriers have never quite closed, volant species have been able to colonise Australia from the north during the latter part of the plate's northward migration. Finally, ferocious climatic fluctuations, particularly during the Quaternary, caused the once moist and lush continent to dry out (Fujioka and Chappell 2010) and fluctuate between cool, dry and warmer, wetter weather compared to the climate we experience today. While Australia has had a relatively stable climate for the past 6000 or so years, the current conservation status of Australia's birds cannot be understood without knowledge of this evolutionary history and the changes through which the avifauna has passed.

Then came people. The first were Aboriginal people who travelled across from Africa at least 42 000 years ago (O'Connell and Allen 2004), bringing with them a tradition of using fire as a tool for managing resources. Aboriginal hunting and fire almost certainly caused extinctions of several of the largest bird species, such as *Genyornis newtoni* (Miller *et al*. 1999). Undoubtedly the distribution and abundance of many other species was also affected because the change in fire regime was almost certainly responsible for massive changes in vegetation (Bowman 1998). The introduction of the dingo *c*. 5000 years ago (Savolainen *et al*. 2004) coincided with the rapid decline and extinction of at least one bird species from the Australian mainland, the Tasmanian Native-hen *Gallinula mortierii* (Baird 1993).

The biggest driver of change in avian conservation status, however, was the arrival of European settlers in 1788. As in other places where indigenous, non-agricultural societies have been invaded by agricultural and urban people, the result was devastating for both Indigenous people and the environment. For small islands, the effect was almost instantaneous, many losing species within decades of European arrival (e.g. McAllan *et al*. 2004). For mainland birds, however, the impact took time to gather momentum (Recher and Lim 1990). Yet even by the mid nineteenth century, Wheelwright (1868) was mourning the loss of Magpie Geese *Anseranas semipalmata* from swamps near Melbourne.

It was the nineteenth century that laid the foundation for widespread declines in the twentieth century. While land clearance was relatively limited compared to what was to come, overgrazing by domestic Sheep *Ovis aries* in New South Wales and overcutting of forests in Victoria were both so severe that royal commissions of inquiry were called early in the twentieth century. The introduction and proliferation of the House Cat *Felis catus*, Red Fox *Vulpes vulpes* and European Rabbit *Oryctolagus cuniculus* affected even the remotest parts of the country. While large areas of native vegetation were cleared in eastern and south-western Australia in the twentieth century, there has also been a gradual rise of an environmental movement and protection of large areas of habitat that had not yet been alienated. By the 1990s there was increasing recognition that Indigenous land management might have implications for biodiversity conservation (Luckert *et al*. 2007), and attempts have been made to reinstate, in particular, Indigenous fire regimes (Cook *et al.* 2012). Nevertheless, it is the twentieth century that will be seen as having had the greatest influence on natural environments in Australia, with almost every part of the continent being heavily influenced by the growth of population, technology, urbanisation and both intensive and extensive agriculture.

Towards the end of the twentieth century, conservation biology emerged as a scientific discipline. Initially concentrating on the biological characteristics of species and habitats whose persistence was threatened, the discipline gradually expanded to encompass a whole range of social, economic and political drivers of threats to biological diversity. The environment has also gained far greater political prominence with advocates attaining genuine political power and thus the ability to influence decisions that previously would have been made solely on economic grounds. Environment departments have now been established in most branches of government which have, among other responsibilities, a mandate to maintain biodiversity within their jurisdictions. These departments administer increasingly sophisticated and pervasive environmental legislation, a far cry from their origin – which was regulations to limit hunting. In particular, the Commonwealth of Australia's *Environment Protection and Biodiversity Conservation Act* 1999 established in law concepts that had been developing for decades.

The largest ornithological organisation in Australia was formed in 1901 as the Royal Australasian Ornithologists Union, was re-named Birds Australia in 1996 and, in 2012, became BirdLife Australia. It is very much part of the global network of ornithological conservation societies that make up BirdLife International.

While, in 2014, Australia's diverse avifauna is threatened by both ongoing environmental change and by legacy effects from the past two centuries, scientific understanding of the issues and advocacy on behalf of the birds is also strong and growing.

20.2 Conservation issues
20.2.1 Habitat loss

Almost immediately following their settlement of Australia, Europeans started to clear the native vegetation to make land available for agriculture. Over the next two centuries, most potentially productive land in eastern and south-eastern Australia, Tasmania, south-western Western Australia and Norfolk Island was transformed from woodland or forest into crops or pasture (Department of Environment, Sports and Territories 1996). In other parts of the country there has been only limited development of intensive agriculture. While active clearing is much reduced, the legacy of habitat loss continues to affect birds in the remaining fragments (Martin *et al.* 2012a). These effects are manifold. Small fragments are subject to stochastic processes such as drought. Once, when populations winked out as part of natural fluctuations, they would have been restored through immigration. For some species, there are now no immigrants; for example, the Brown Treecreeper *Climacteris picumnus* will not disperse across habitat gaps wider than 1.5 km (Doerr *et al.* 2011). Another problem is the change in forest structure that occurs with fragmentation. Many fragments are now ideal for Noisy Miners *Manorina melanocephala* (Driscoll and Lindenmayer 2010), which, though native, out-compete other woodland species to the extent that the latter cannot persist (MacNally *et al.* 2012). Nest predation rates also tend to be higher in fragments, especially along edges (reviewed by Ford 2011). Finally, the fragments tend to be those patches of the landscape that are least favourable for agriculture, so have less fertile soils (Watson 2011). Before clearing of vegetation, less productive areas were likely to have been population sinks for some species that bred successfully in the more productive parts of the landscape.

Given the vast area that has been cleared, the number of species threatened by land clearing alone is surprisingly small. Land clearing is listed as one of the threats to 34 Australian taxa (Garnett *et al*. 2011), with the threat mostly associated with past rather than ongoing clearing. The highest number of taxa threatened by land clearing is in the mallee woodlands at the junction of South Australia, Victoria and New South Wales, yet even here large areas of habitat have been conserved and it is not clearing or fragmentation that is the major threat, but uncontrolled wildfire. Through the broad belt of grassy woodlands in the south-east, the brigalow of Queensland and even the areas where rainforest has been replaced by sugarcane and dairy farming in the wetter parts of eastern Queensland, there are only a handful of species that meet the IUCN Red List criteria for being threatened as a result of historical land clearance. There are several reasons for this. One is that the full impacts of clearing are yet to be felt. Thus the Superb Parrot *Polytelis swainsoni*, which is entirely confined to the agricultural belt of inland south-eastern Australia, is very likely to suffer from a shortage of nest hollows within the twenty-first century because the remnant trees that supply breeding hollows are gradually dying and are not being replaced (Manning *et al*. 2004). However, the parrot is not listed as threatened under IUCN criteria because the threatening process is likely to take more than three generations to cause substantial population decline.

The climatic history of the region is also likely to have contributed to the responses of birds to fragmentation. Very few species were confined to the areas that were flat, fertile and moist enough for agriculture. This may be a legacy of the Pleistocene when much of the fertile grassy woodland belt was periodically much drier and there was no analogue to the vegetation that existed when Europeans arrived. Species that live in these woodlands are mostly either arid-adapted and thus also occur in the extensive Australian arid zone in which the vegetation remains substantially intact, or are 'intruders' from the adjacent forested hills and mountains. Thus, many grassy woodland species that have proved particularly susceptible to the impacts of fragmentation, for example Hooded Robin *Melanodryas cucullata*, Jacky Winter *Microeca fascinans*, Brown Treecreeper, Crested Bellbird *Oreoica gutturalis* and Spotted Bowerbird *Ptilonorhynchus maculatus* (e.g. Ford *et al*. 2009), also have extensive distributions in arid Australia and are not threatened nationally.

Woodland populations of birds that also commonly occur in the forests of the Great Dividing Range have proved more resilient to fragmentation. Though heavily exploited for timber, the hills have retained a remarkable amount of native vegetation so that few birds that occupy such forests are threatened. Indeed many of these species may now have a larger and less fragmented area of suitable habitat than they had during the dry periods of the Pleistocene. For example, modelling suggests that, compared to the Pleistocene dry periods, the extent of north Queensland rainforests was greater and less fragmented at the time when European settlers first arrived and possibly even now, after so much has been cleared (Hilbert *et al*. 2007). The same is probably true for other forest patches around the country. Thus, many forest species have already survived a fragmentation filter. Only where vegetation clearing has reduced the mesic islands to a tiny proportion of their former extent, such as in the Mount Lofty Ranges of South

Australia, are local populations of forest birds in trouble (Szabo *et al.* 2011). In Western Australia, the same phenomenon has occurred, in which species of the almost completely cleared western wheatbelt also occur extensively in the arid land that has never been cleared (Judd *et al.* 2008).

Currently the habitat loss that causes most damage to Australian birds is happening outside Australia. Over the past 30 years the loss of intertidal habitats along the flyway of migratory shorebirds that visit Australia in the austral summer has increased enormously (MacKinnon *et al.* 2012), with a commensurate decline in shorebird populations (Szabo *et al.* 2012a). As a result, 16 out of the 39 shorebird taxa that migrate annually to Australia from the Northern Hemisphere are Threatened or Near Threatened.

20.2.2 Fire

Lightning-lit fire has been a feature of Australian ecological systems for millions of years and the flora and fauna have developed a wide range of strategies to cope with it; indeed many species require fire to thrive (Bradstock *et al.* 2012). For at least 42 000 years, Aboriginal people used fire extensively and systematically as a tool to manage vegetation and consequently altered the vegetation profoundly (Bowman 1998). More recently, European settlers have been responsible for major changes in fire regimes.

Birds that live in heath and shrub habitats seem to have been particularly vulnerable to recent changes in fire regimes. Some, like the Noisy Scrub-birds *Atrichornis clamosus* and bristlebirds *Dasyornis* spp., are among the evolutionarily oldest endemic Passerines in Australia. Given their distribution was already restricted when Europeans arrived, they may have suffered from Aboriginal fire regimes. Fossil evidence suggests that these species were more widespread until recently (Baird 1993), but they have certainly retreated further since then, where their habitat has not been protected from fire. Another endemic group, the grasswrens *Amytornis* spp., has many species that are restricted to mature stands of the shrub-like hummock grasslands of spinifex *Triodia* spp., a genus that burns readily and can take years to recover its structure (Allan and Southgate 2002). In total, 73 threatened taxa are thought to be affected detrimentally by current fire regimes (Garnett *et al.* 2011). See Figure 20.1 (Plate 38).

20.2.3 Invasive species

Species have been colonising Australia ever since Gondwana drifted close enough to Asia to receive them from the north, including many bird species of Palaearctic origin. These events must have affected endemic species, including some that went extinct without leaving a record. Some invasive species accompanied the earliest people, of which the dingo has already been noted. The Pacific Rat *Rattus exulans* arrived to Norfolk Island along with Polynesian travellers, and it appears to have caused the extinction of several seabird and rail species (Schodde *et al.* 1983). Maccassans introduced the Tamarind Tree (*Tamarindus indicus*) to northern Australia (Macknight 1976) and possibly other plants too. However, the rate and impact of introductions increased by several orders of magnitude when European colonists arrived (Rolls 1969), and this process is still ongoing.

Figure 20.1 (Plate 38) An alarming number of Australian bird species that survived and even thrived through over 40 000 years of Indigenous settlement and landscape burning have been seriously challenged by the sudden shift in fire regime during the 200 years of European settlement (photo Donald Franklin). A black and white version of this figure will appear in some formats. For the colour version, please refer to the plate section.

Black Rat *Rattus rattus*, other rat species and the House Cat have been common on all Australian offshore island groups except Heard Island, although all appear to have been eradicated from Macquarie Island within the last decade (Tasmanian Parks and Wildlife Service 2014). These mammals predate naïve fauna all over the world with catastrophic results (Szabo *et al.* 2012b). On the mainland, Red Fox and the omnivorous Feral Pig *Sus scrofa* are additional introduced predators, but as the avifauna had prior experience with a wider range of predators, most bird species have coped. Introduced birds may also pose problems for native birds, the Common Myna *Sturnus tristis* being widely suspected of competing with local hollow-nesting species (but see Lowe *et al.* 2011).

For mainland birds, the impact of introduced grazing and browsing animals has been at least as devastating as exotic predators. These introductions include wild or feral species (notably European Rabbit, Horse *Equus ferus caballus*, Donkey *E. africanus asinus*, Water Buffalo *Bubalus bubalis* and Dromedary Camel *Camelus dromedarius*) and domestic species (notably Sheep and Cattle *Bos taurus/indica*). Grazing and browsing removes resources needed by birds mostly by direct consumption. These resources include structural features of habitat, but additional indirect impacts sometimes occur. For example, the consumption of grass that would otherwise allow fire to suppress woody tree growth has triggered a cascade of impacts threatening the Golden-shouldered Parrot *Psephotus chrysopterygius* (Crowley *et al.* 2004).

Some other adverse effects of invasive species have also been subtle. For example, Feral Bees *Apis mellifera* occupy tree hollows in potential competition with nesting Black-Cockatoos *Calyptorhynchus* spp. in south-western Australia, where hollows are scarce as a result of timber cutting (Abbott and Whitford 2001). On Kangaroo Island fertiliser applied to improve pasture has favoured native Brushtail Possums *Trichosurus vulpecula* increasing nest failure of Glossy Black-Cockatoo *C. lathami* (Garnett *et al.* 1999). Exotic diseases (a less obvious group of invasive species) not only had a devastating effect on Aboriginal populations and their land management when European colonists first arrived (Crosby 2004), but also affected populations of numerous native mammals. The evidence among birds is more equivocal, but an introduced air-sac mite *Sternostoma tracheacolum* may have played a major role in the decline of the Gouldian Finch *Erythrura gouldiae* (Tidemann *et al.* 1992; Bell 1996) and south-western Australian heathland, inhabited by a number of threatened birds, is severely affected by an introduced variety of the Cinnamon Fungus *Phytophthora cinnamomi* (Shearer *et al.* 2007).

Some birds, however, have benefited from the availability of new resources. For instance, rabbits are a staple food for many raptors, although they continue to eat native species where rabbits are not present. Even some threatened species have learnt to forage on the flowers, fruit and seeds of exotic species, sometimes to the ire of the people cultivating these plants. However, sometimes adaptation has almost taken too long. Both Long-billed Corella *Cacatua tenuirostris* and Turquoise Parrot *Neophema pulchella* were scarce in the late nineteenth century, but their populations have now recovered, speculatively because they learnt to eat introduced plants. Nevertheless, the number of opportunists is far lower than the number of sensitive taxa that are still having trouble coping with the changed circumstances caused by introduced species. Overall some 125 taxa are threatened by invasive species, 84 of them endemic to islands.

20.2.4 Use of biological resources

The first extinctions after colonisation were the result of settlers hunting the tame wildlife of Norfolk, Lord Howe, King and Kangaroo Islands. Hunting for food and pleasure has continued ever since on the mainland, with many states and territories having licensed hunting of waterfowl and quail. However, far fewer people are hunting birds now than two decades ago and no taxon is threatened by the practice. Similarly, apart from the controlled and carefully monitored traditional Aboriginal harvest of Short-tailed Shearwaters *Puffinus tenuirostris* around Tasmania, the commercial hunting of birds for sale ceased over 80 years ago. The story was very different during the nineteenth century when huge numbers of penguins were killed on Macquarie and Heard Islands and albatrosses on Albatross Islands, all for their oil, while herons were taken on mainland Australia to feed a fashion industry with plumes on ladies' hats.

There has also long been conflict between farmers and birds over the food and habitat that farmers provide. It was to save crops that the Lord Howe subspecies of Tasman Parakeet *Cyanoramphus cookii subflavescens* was exterminated (McAllan *et al.* 2004), and Muir's Corella *Cacatua pastinator pastinator*, the southern subspecies of Western Corella, almost went the same way. Farmers still kill significant numbers of threatened species – Baudin's Black-Cockatoo *Calyptorhynchus baudinii* because they eat apples (Garnett *et al.* 2011) and the eastern subspecies of Regent Parrot *Polytelis anthopeplis anthopeplis* because they have learnt to eat almonds (Baker-Gabb and Hurley 2011).

Many albatrosses and large petrels have been killed at sea in recent years as by-catch arising from modern fishing methods, the birds having learnt to take baits from hooks deployed on long lines (Lebreton and Veran 2013). Trawlers can also net birds like the Flesh-footed Shearwater *Puffinus carneipes* that dive deep for their prey (Lokkeborg 2011). Fishing by-catch is not a problem that can be solved solely within Australian territory. Most seabirds affected by fishing travel around the Southern Hemisphere. Others migrate regularly to the northern Pacific and Indian Oceans they encounter fishing boats wherever they go.

20.2.5 Climate change

Earth's climate is starting to change with increasing rapidity as a result of anthropogenic emissions. Already there is evidence of shifts in the distribution and phenology of species in response (Parmesan and Yohe 2003). The vulnerability of birds to climate change is a function of their exposure to change – i.e. how much the climate is likely to change in the places where they live now – and their sensitivity – i.e. how are they likely to be affected by any change that occurs. A recent analysis of the effects of climate change on Australian birds suggests that the climate space of over 100 Australian terrestrial bird taxa is likely to be entirely gone by 2085, 16 marine taxa will be exposed to marine productivity declines adjacent to their breeding sites of at least 10%, and at least 55 terrestrial taxa are likely to be exposed and sensitive to more frequent or intense fires (Garnett and Franklin, 2014). Modelling of the climate in 2085 based on current rates of greenhouse gas emissions suggests that many birds confined to any one of Cape York Peninsula, the Wet Tropics, the Top End of the Northern Territory (particularly the Tiwi Islands), the arid zone, King Island and southern South Australia (particularly Kangaroo Island) are likely to lose all or most of their current climate space. The rainforest birds of Cape York Peninsula are the largest group likely to be climatically challenged. For marine birds, those nesting on Lord Howe and Norfolk Islands, the Great Barrier Reef and the Houtman Abrolhos and with restricted foraging ranges whilst breeding, are thought the most likely to face declines in local marine productivity. The endemic terrestrial birds of Lord Howe and Norfolk Islands could be adversely affected by projected decreases in summer rainfall (Widlansky *et al.* 2013), but climate projections are not available for other island groups. Sea level rise could also affect a small group of beach-nesting and saltmarsh birds. Based on a set of ecological and morphological metrics (reproductive rate, relative brain size, population size as a proxy for genetic variability, and specialisation with regard to climate niche, diet, foraging and habitat) thought to correspond to an individual's capacity to adapt, seabirds were identified as the most sensitive Australian bird taxa (Garnett and Franklin, in press). Small island taxa were most likely to be both exposed and sensitive to climate change, followed by marine and shoreline taxa. While threatened taxa were more likely to be rated as exposed or be sensitive to climate change than non-threatened taxa, a substantial proportion of both groups was neither.

20.2.6 Governance, commitment and legislation

Australian states and territories have a fine set of laws for protecting biodiversity. However, all of them are relatively weak, like the departments that administer them, compared to other laws and departments. The exception is the EPBC Act referred to above. This law allows the Australian Commonwealth government to intervene on any matters that impinge on issues of national environmental significance. This includes those relating to sites and species listed under Commonwealth law, particularly threatened and migratory species, as well as the conservation of World Heritage areas. Unfortunately, while the law can prevent damage to sites that meet certain criteria, regardless of whether that prevention leads to genuine conservation actions, it is unable to deal with bigger issues relating to land use. Other serious weaknesses under the Act are that, while actions of non-government actors can be proscribed, there is no obligation on the part of any government to allocate resources to protect and manage species – it is quite possible for a species to be given the highest level of protection without any obligation to prevent its extinction. Second, much emphasis is placed on the list of threatened species. This was derived from a previous list when the Act was first passed into law but was already poorly aligned with the IUCN Red List threatened species lists of the time. Since then, the process of modifying the list has been slow and cumbersome, so that the official EPBC Act List has become increasingly divorced from the level of knowledge about what is genuinely threatened. The current EPBC Act list is thus quite different from the most recent assessment of Australian birds that uses the IUCN Red List categories and criteria (Garnett *et al.* 2011). This has serious consequences, because many threatened birds are not listed but should be, meaning they are not protected under the Act and are largely ineligible for Commonwealth funding should it become available.

The erroneous listing of birds we now know to be secure, or which have recovered, is also of concern. Such birds are the ones most likely to be detected when development proposals are considered, often causing lengthy delays to proposals that will inevitably be approved but whose proponents are forced to spend substantial sums undertaking mitigation that may be of no benefit to the listed birds or their environment.

This brings the Act into disrepute and feeds attempts to reform it and remove some of the useful control it exerts over development proposals that may be environmentally harmful. Even worse, it could feed a sentiment that threatened species are an impediment to development. In fact, at the moment, this view is held by very few Australians with by far the majority being willing to invest substantially in threatened bird protection (Zander *et al.* 2014). Such support, however, may be fragile. One of the most frightening threats to Australian birds is one that is rarely raised in standard treatments of conservation management – erosion of societal support. This refers not to any decline in support for environmental political parties but of a societal shift that is gradually separating people from the environment. Indeed there is already evidence that this is occurring (Mccallum and Bury, 2013). This includes both the global phenomenon of urbanisation and lifestyles in which children no longer routinely experience the environment. Ultimately, unless young people are engaged with birds, they will not care if species are lost and nor will they care if environmental legislation is weakened to allow that to happen.

20.3 Management solutions

20.3.1 Corridors, restoration and reservation

The remedy for habitat loss and fragmentation is restoration of the habitat in a way that connects disjunct populations. However, land clearance can be difficult to reverse, particularly given that restored vegetation rarely has the complexity of the original habitat it is intended to replace. While the positive impact of corridors and restoration is widely accepted, this effect is weaker for birds than for some other taxa (Gilbert-Norton *et al.* 2010) and there are occasional negative features of connectivity such as facilitation of the spread of disease and fire (Brudvig *et al.* 2012).

Undoubtedly, restoration can achieve biodiversity benefits. A good example is the conservation management of the Grey-crowned Babbler *Pomatostomus temporalis* in southern Australia where the species had been declining for a century, but has increased locally as a result of revegetation (Robinson 2006). In north Queensland, the creation of corridors between rainforest fragments provides a focussed attempt to restore genetic exchange (Tucker 2000). Across Australia, there are numerous projects at scales ranging from a few hundred metres to thousands of kilometres, all aiming to ensure some level of habitat connectivity across a landscape where native vegetation was once continuous (Whitten *et al.* 2011).

However, the cost of recovering a landscape is orders of magnitude greater than simply retaining as much native vegetation as possible in a natural state. Indeed, the objective of most large corridor projects is to link existing large patches of conserved natural habitat. In Australia, the area reserved for conservation has increased dramatically over the last 50 years from a handful of national parks in attractive parts of the landscape to some level of targeted preservation of key biodiversity assets. Internationally, protection of some of the migratory shorebird stopover sites will be essential if the loss to land reclamation is to be stemmed.

However an extensive protected area estate will never conserve all taxa. A characteristic of Australian birds is that many of them are highly mobile in response to climatic variation. Thus no reserve system is ever going to capture all life-history stages of all species (Woinarski *et al.* 1992). Also, some sedentary species rely heavily on private land, especially the very large areas of Aboriginal and pastoral land in remote and regional Australia where native vegetation remains relatively intact. Even taxa that occur largely in protected areas do not have an assured future. There are now several cases where that confidence has been misplaced because of other threatening processes (Martin *et al.* 2012b). Thus attention needs to be paid to the birds that need additional actions besides conservation of their habitat.

20.3.2   Fire management

In some areas in remote Australia, traditional Aboriginal fire management has persisted into the present day, continuing an ancient tradition. In other parts of northern Australia, attempts are being made to reconstitute at least some elements of Aboriginal burning, particularly patch burning in the early dry season that prevents the spread of large fires when the weather is hotter and the fuel drier. There is evidence that this management is succeeding, with reduced areas burnt late in the season (Price *et al.* 2012).

In much of the rest of Australia, however, the traditions have been lost, the vegetation altered, and the imperative to protect life and property are so high that altogether different fire regimes are in place. For the most part this consists either of attempts to prevent fire for as long as possible, for instance in many areas of heath and mallee. While this is possible where there are natural fire breaks, such as at Mount Gardner where the Noisy Scrub-bird persisted when all other populations had been lost (Danks 1997), such successes are rare. Although the techniques for preventing fire are growing more sophisticated, fire weather alerts have improved, suppression techniques are more effective and we understand the capabilities of fire breaks better, many fires still escape control, particularly on hot windy days. As a result there is less and less long-unburnt vegetation available for birds that require it to prosper.

The other approach has been to undertake large-scale burning in cooler weather, primarily to reduce fuel loads and therefore minimise the likelihood of fires spreading to infrastructure. However, conflicts can arise between conservation and fuel-reduction burning particularly if fires are widespread and set at frequent and/or regular intervals or the canopy is damaged. The forests occupied by the south-eastern subspecies of Red-tailed Black-Cockatoo *Calyptorhynchus banksii graptogyne*, for instance, contain sufficient seed for foraging cockatoos only after extended periods without fire, and fuel-reduction fires scorch the canopy and disrupt the cockatoo's food chain (Koch 2005).

However, some species benefit from fire. As a result of exclusion of fire, habitat of the Golden-shouldered Parrot is changing from grassy woodland to shrubland (Crowley and Garnett 1998), in which the species is likely to suffer from higher rates of predation. Orange-bellied Parrots *Neophema chrysogaster* also appear to benefit from patchy burning of their moorland habitat in Tasmania (Marsden-Smedley and Kirkpatrick 2000). However, some species, once thought dependent on fire to rejuvenate habitat, such as the Eastern Ground Parrot *Pezoporus wallicus*, are now known to persist in relatively high numbers in unburnt habitat, so are not fire-dependent (Baker *et al.* 2010).

20.3.3 Invasive species control

The fauna and flora of New Zealand has been affected more by invasive species than that of most countries around the world, so it is fitting that New Zealanders developed the techniques for controlling invasive species on a large scale. As a consequence, invasive mammals in particular have been eliminated from many islands around the world. In Australia, the most ambitious island eradication project to date (using blanket baiting followed by dog tracking of any animals missed) has been on Macquarie Island (Raymond *et al.* 2011). During this project, it appears that cats, rabbits, rats and House Mice *Mus musculus* have all been eliminated, with an almost instantaneous increase in the numbers of breeding seabirds. Smaller projects have also been successful. For instance, Rat Island in the Houtman Abrolhos was given its name in 1840 but is now rat-free and covered with breeding terns *Sterna* spp. and noddies *Anous* spp. (A.A. Burbidge, personal communication). There are ambitious plans to remove rats and mice from Lord Howe Island (Hutton *et al.* 2007). Rats and cats have also been removed from some of the islands in the Cocos (Keeling) group to allow re-introduction of the local subspecies of Banded Rail *Gallirallus pectoralis andrewsi* (J. C. Z.Woinarski, personal communication). On Christmas Island, and with the long-term aim of eradication, domestic House Cats are now legally required to be de-sexed, tattooed and micro-chipped for registration and no further importation of cats is permitted (Algar *et al.* 2011).

However, control of invasive species at mainland sites and on the largest islands is far more challenging. For the conservation of native mammals, areas have been protected within large enclosures from which all invasive fauna have been removed. Other than this, eradication is rarely possible so perpetual control and management is necessary. Sometimes, invasive species have to be deterred at very local scales: individual nests of the Glossy Black-Cockatoo *Calyptorhynchus lathami* on Kangaroo Island are protected from invasive Brushtail Possums (Harris *et al.* 2009). At other sites, broad-scale baiting has been undertaken at intervals. However, care must be taken that there are not unintended consequences. For instance, baiting for foxes in Western Australia has allowed feral cats to proliferate (de Tores and Marlow 2012), and there are indications that cats are a more effective predator of several threatened bird species there, such as the Western Ground Parrot *Pezoporus flaviventris* and the Western Bristlebird *Dasyornis longirostris*, compared to the Foxes (S. Comer, personal communication).

20.3.4   Direct exploitation

Hunting and persecution of native species perceived as a threat to agriculture is an activity that is ostensibly easy to control. The Western Australian authorities did such a good job of protecting Muir's Corella from agriculturists that the numbers recovered so well that it was possible to remove this subspecies from the threatened list (Garnett *et al.* 2011). However, illegal shooting of the much more threatened Baudin's Black-Cockatoo because of damage caused to orchards has been a problem for decades (Garnett *et al.* 2011), with no immediate prospect of greater vigilance. Legal recreational hunting has been strictly controlled for many years with reasonably good monitoring to set the bag limits, at least for waterfowl. While there is some collateral damage to protected species like Freckled Duck *Stictonetta nervosa* and Blue-billed Duck *Oxyura australis*, none is seriously threatened as a result.

In contrast, the by-catch from fishing would have been extremely difficult to control had it not been in the interests of the fishers themselves to reduce the number of long-line baits taken by albatrosses and petrels. As a result, innovative ways of laying baits and deployment of various forms of discouragement over the past two decades have substantially reduced the by-catch of these species (Lokkeborg 2011). The coordination of activities through an international treaty, the Agreement on the Conservation of Albatrosses and Petrels (ACAP) (Cooper *et al.* 2006), has greatly enhanced international collaboration. While the issue is by no means solved, the signs are good and some of the populations once thought highly vulnerable appear to be recovering, or at least to have stabilised, although others are still suffering.

20.3.5   Climate change adaptation

To minimise the adverse consequences of climate change on bird species and populations, and given the great uncertainty about how and when they will be affected, immediate and ongoing monitoring is needed. Biodiversity monitoring in Australia is generally of a low standard (Lindenmayer *et al.* 2012). However, a substantial body of baseline data in the Atlas of Australian Birds (Blakers *et al.* 1984; Barrett *et al.* 2003) may facilitate assessment of changes to their ranges and abundance. Another no-regrets action is fine-scale modelling of regions identified as having numerous highly exposed bird taxa to identify refugia that can be protected and managed before climate change has a major impact. It is also essential that conservation management of existing threats, e.g. fire management, weed and feral animal control and, for marine taxa, controls on fishing, is ongoing and intensified because the impacts of these threats are often likely to be compounded by climate change (Brook *et al.* 2008).

Even with intensive management, it is almost inevitable that some bird populations will decline irreversibly, rendering captive breeding the only option to prevent extinction. However, most of those for which captive breeding is recommended as a necessary last resort are subspecies of species that are widespread, either in Australia or in New Guinea (Garnett *et al.* 2014). Indeed, most of the bird taxa that are likely to be affected most severely by climate change in the near future are subspecies of more widespread species, and may be judged appropriate to devote resources to them. Surprisingly, the analysis of climate change impacts suggested that there is little urgent requirement for corridors for the maintenance of taxa likely to be threatened with extinction – those few taxa not already living in areas where there are likely to be refugia will require assistance to colonise new climate space and would be unlikely to use corridors. However, maintenance and improvement of corridors and landscape connectivity may be key to avoiding declines and regional extinction in many taxa not necessarily threatened with global extinction by climate change.

For a detailed review of options for the management of Australian birds in the face of climate change, see Franklin *et al.* (2014).

20.3.6 Governance, commitment and legislation

Some aspects of environmental governance are under active review. In particular, the capacity of the Australian law to protect threatened species has recently been examined by a committee of the Australian senate after the extinction of some species that were found only in national parks. Also, the committee responsible for advising the Minister on which species should be listed under the EPBC Act has recognised the problem arising from inaccurate listings and is actively pushing to update it, in the case of birds based primarily on the assessments of Garnett *et al.* (2011).

The retention of societal support for bird conservation is a far larger issue. Advocacy groups have largely assumed that society sympathises with their views and that the public only needs to be informed for political pressure to be applied. However, decreased personal contact with the environment and politicisation of its management are far larger issues that challenge the values on which the conservation movement is based. How advocates for bird conservation manage these issues over the next few decades will be critical to the future of bird conservation.

20.4 Conclusion

While 11% of Australia's bird species and subspecies are Threatened or Near Threatened, the remaining 89% persist despite the massive changes to the environment of the past two centuries. Many of the 89% are well placed to thrive. Many live in places where there has been little development, many occur in national parks, and a sizeable number have adapted to living in farmland or cities and towns. The evolutionary history of Australian birds, particularly their filtering by an often unpredictably variable climate, has selected for traits that make many of them resilient to change. It is not for nothing that of the most widely kept cage birds around the world, the Budgerigar *Melopsittacus undulatus*, Cockatiel *Nymphicus hollandicus* and Zebra Finch *Taenopygia guttata*, have their origins in inland Australia where they cope with temperatures ranging from frost to among the hottest on Earth, and with decade-long droughts and vast floods.

Those that are threatened or declining, however, require the full spectrum of conservation interventions if they are to persist. While some threatened species will benefit from generalised interventions such as habitat protection, and seabirds from better management of the long-line fishery, many of the more threatened species require carefully tailored interventions built on a detailed knowledge of their ecology and how they are affected by threatening processes. Simply targeting the threatening processes in a general manner and working to retain a healthy landscape will not be sufficient.

Tailored interventions are often expensive, but the expense cannot be avoided if the taxa are to persist. Beneath all conservation, however, is the ongoing need for a licence that underpins legislation and legitimises expenditure of public funds.

REFERENCES

Abbott, I., Whitford, K. (2001) Conservation of vertebrate fauna using hollows in forests of south-west Western Australia: strategic risk assessment in relation to ecology, policy, planning, and operations management. *Pacific Conservation Biology* **7**, 240–255.

Algar, D., Hilmer, S., Nickels, D., Nickels, A. (2011) Successful domestic cat neutering: first step towards eradicating cats on Christmas Island for wildlife protection. *Ecological Management and Restoration* **12**, 93–101.

Allan, G. E., Southgate, R. I. (2002) Fire regimes in the spinifex landscapes of Australia. In *Flammable Australia. The Fire Regimes and Biodiversity of a Continent* (Eds. R. A. Bradstock, J. E. Williams and M. A. Gill) pp. 145–176. (Cambridge University Press: Cambridge.)

Baird, R. F. (1993) Pleistocene avian fossils from Pyramids Cave (M-89), eastern Victoria, Australia. *Alcheringa* **17**, 383–404.

Baker-Gabb, D., Hurley, V. G. (2011) Draft National Recovery Plan for the Regent Parrot (eastern subspecies) *Polytelis anthopeplus monarchoides*. Department of Sustainability and Environment, Melbourne.

Baker, J. R., Whelan, R. J., Evans, L., Moore, S., Norton, M. (2010) Managing the Ground Parrot in its fiery habitat in south-eastern Australia. *Emu* **110**, 279–284.

Barrett, G., Silcocks, A., Barry, S., Cunningham, R., Poulter, R. (2003) *The New Atlas of Australian Birds*. (Royal Australasian Ornithologists Union: Hawthorn East.)

Bell, P. J. (1996) Survey of the nasal mite fauna (Rhinonyssidae and Kytoditidae) of the Gouldian finch, *Erythrura gouldiae*, and some co-occurring birds in the Northern Territory. *Wildlife Research* **23**, 675–685.

Blakers, M. S., Davies, J. J. F, Reilly, P. N. (1984) *The Atlas of Australian Birds*. (Melbourne University Press: Melbourne.)

Bowman, D. J. M. S. (1998) The impact of Aboriginal landscape burning on the Australian biota. *New Phytologist* **140**, 385–410.

Bradstock, R. A., Gill, A. M., Williams, R. J. (2012) *Flammable Australia. Fire Regimes, Biodiversity and Ecosystems in a Changing World*. (CSIRO Publishing: Collingwood.)

Brook, B. W., Sodhi, N. S., Bradshaw, C. J. A (2008) Synergies among extinction drivers under global change. *TRENDS in Ecology and Evolution* **23**, 453–460.

Brudvig, L. A., Wagner, S. A., Damschen, E. I. (2012) Corridors promote fire via connectivity and edge effects. *Ecological Applications* **22**, 937–946.

Christidis, L., Norman, J. A. (2010) Evolution of the Australasian songbird fauna. *Emu* **110**, 21–31.

Cook, G. D., Jackson, S., Williams, R. J. (2012) A revolution in northern Australian fire management: recognition of Indigenous knowledge, practice and management. In *Flammable Australia. Fire Regimes, Biodiversity and Ecosystems in a Changing World* (Eds.

R. A. Bradstock, A. M. Gill and R. J. Williams) pp. 293–306. (CSIRO Publishing: Collingwood)

Cooper, J., Baker, G. B., Double, M. C., *et al.* (2006) The agreement on the conservation of albatrosses and petrels: rationale, history, progress and the way forward. *Marine Ornithology* **34**, 1–5.

Crosby, A. W. (2004) *Ecological Imperialism: The Biological Expansion of Europe, 900–1900.* (Cambridge University Press: Cambridge.)

Crowley, G. M., Garnett, S. T. (1998) Vegetation change in the grasslands and grassy woodlands of east-central Cape York Peninsula, Australia. *Pacific Conservation Biology* **4**, 132–148.

Crowley, G. M., Garnett, S. T., Shephard, S. (2004) *Management Guidelines for Golden-shouldered Parrot Conservation.* Queensland Parks and Wildlife Service, Brisbane.

Danks, A. (1997) Conservation of the Noisy Scrub-bird: a review of 35 years of research and management. *Pacific Conservation Biology* **3**, 341–349.

de Tores, P. J., Marlow, N. (2012) The relative merits of predator-exclusion fencing and repeated fox baiting for protection of native fauna: Five case studies from Western Australia. In *Fencing for Conservation: Restriction of Evolutionary Potential or a Riposte to Threatening Processes?* (Eds. M. J. Somers and M. W. Hayward) pp. 21–42. (Springer: New York.)

Department of Environment, Sports and Territories (1996) *Australia: State of the Environment.* CSIRO, Collingwood.

Doerr, V. A. J., Doerr, E. D., Davies, M. J. (2011) Dispersal behaviour of Brown Treecreepers predicts functional connectivity for several other woodland birds. *Emu* **111**, 71–83.

Driscoll, D. A., Lindenmayer, D. B. (2010) Assembly rules are rare in SE Australian bird communities, but sometimes apply in fragmented agricultural landscapes. *Ecography* **33**, 854–865.

Edwards, S. V., Boles, W. E. (2002) Out of Gondwana: the origin of passerine birds. *Trends in Ecology & Evolution* **17**, 347–349.

Ford, H. A. (2011) The causes of decline of birds of eucalypt woodlands: advances in our knowledge over the last 10 years. *Emu* **111**, 1–9.

Ford, H. A., Walters, J. R., Cooper, C. B., Debus, S. J. S., Doerr, V. A. J. (2009) Extinction debt or habitat change? – Ongoing losses of woodland birds in north-eastern New South Wales, Australia. *Biological Conservation* **142**, 3182–3190.

Franklin, D. C., Reside, A. E., Garnett, S. T. (2014) Conserving Australian bird populations in the face of climate change. In *Climate Change Adaptation Plan for Australian Birds* (Eds. S. T. Garnett and D. C. Franklin) pp. 53–78. (CSIRO: Collingwood.)

Fujioka, T., Chappell, J. (2010) History of Australian aridity: chronology in the evolution of arid landscapes. *Geological Society, London, Special Publications* **346**, 121–139.

Garnett, S. T., Franklin, D. C. (in press) *Climate Change Adaptation Plan for Australian Birds.* (CSIRO: Collingwood.)

Garnett, S. T., Pavey, C. R., Ehmke, G., *et al.* (2014) Adaptation outlines for species that are both highly sensitive and highly exposed. In *Climate Change Adaptation Plan for Australian Birds.* (Eds. S. T. Garnett and D. C. Franklin) pp. 79–241. (CSIRO: Collingwood.)

Garnett, S. T., Pedler, L. P., Crowley, G. M. (1999) The breeding biology of the Glossy Black-Cockatoo *Calyptorhynchus lathami* on Kangaroo Island, South Australia. *Emu* **99**, 262–279.

Garnett, S. T., Szabo, J. K., Dutson, G. (2011) *The Action Plan for Australian Birds 2010.* (CSIRO Publishing: Collingwood.)

Gilbert-Norton, L., Wilson, R., Stevens, J. R., Beard, K. H. (2010) A meta-analytic review of corridor effectiveness. *Conservation Biology* **24**, 660–668.

Harris, J. B. C., Damien, A., Fordham, D. A., *et al.* (2009) *Spatially Explicit Population Viability Analysis of the South Australian Subspecies of the Glossy Black-Cockatoo (Calyptorhynchus lathami halmaturinus) Under Climate Change.* Report to the Glossy Black-Cockatoo Recovery Program, Kingscote.

Hilbert, D. W., Graham, A., Hopkins, M. S. (2007) Glacial and interglacial refugia within a long-term rainforest refugium: the Wet Tropics Bioregion of NE Queensland, Australia. *Palaeogeography, Palaeoclimatology, Palaeoecology* **251**, 104–118.

Hutton, I., Parkes, J. P., Sinclair, A. R. E. (2007) Reassembling island ecosystems: the case of Lord Howe Island. *Animal Conservation* **22**, 22–29.

Judd, S., Watson, E. M., Watson, A. W. T. (2008) Diversity of a semi-arid, intact Mediterranean ecosystem in southwest Australia. *Web Ecology* **8**, 84–93.

Koch, P. (2005) Factors influencing food availability for the endangered south-eastern Red-tailed Black Cockatoo *Calyptorhynchus banksii graptogyne* in remnant stringybark woodland, and implications for management. PhD thesis, University of Adelaide, Adelaide.

Lebreton, J.-D., Veran, S. (2013) Direct evidence of the impact of longline fishery on mortality in the Black-footed Albatross *Phoebastria nigripes*. *Bird Conservation International* **23**, 25–35.

Lindenmayer, D. B., Gibbons, P., Bourke, M., *et al.* (2012) Improving biodiversity monitoring. *Austral Ecology* **37**, 285–294.

Lokkeborg, S. (2011) Best practices to mitigate seabird bycatch in longline, trawl and gillnet fisheries-efficiency and practical applicability. *Marine Ecology Progress Series* **435**, 285–303.

Lowe, K. A., Taylor, C. E., Major, R. E. (2011) Do Common Mynas significantly compete with native birds in urban environments? *Journal of Ornithology* **152**, 909–921.

Luckert, M. K., Campbell, B. M., Gorman, J. T., Garnett, S. T. (2007) *Investing in Indigenous Natural Resource Management.* (Charles Darwin University Press: Darwin)

MacKinnon, J., Verkuil, Y. I., Murray, N. (2012) *IUCN Situation Analysis on East and Southeast Asian Intertidal Habitats, with Particular Reference to the Yellow Sea (Including the Bohai Sea).* IUCN, Gland, Switzerland and Cambridge, UK.

Macknight, C. C. (1976) *The Voyage to Marege. Macassan Trepangers in northern Northern Australia.* (Melbourne University Press: Melbourne.)

MacNally, R., Bowen, M. E., Howes, A., McAlpine, C., Maron, M. (2012) Despotic, high-impact species and the subcontinental scale control of avian assemblage structure. *Ecology* **93**, 668–678.

Manning, A. D., Lindenmayer, D. B., Barry, S. C. (2004) The conservation implications of bird reproduction in the agricultural "matrix": a case study of the vulnerable superb parrot of south-eastern Australia. *Biological Conservation* **120**, 363–374.

Marsden-Smedley, J. B., Kirkpatrick, J. B. (2000) Fire management in Tasmania's Wilderness World Heritage Area: ecosystem restoration using Indigenous-style fire regimes? *Ecological Management and Restoration* 1, 195–203.

Martin, T. G., Catterall, C. P., Manning, A. D., Szabo, J. K. (2012a) Australian birds in a changing landscape: 220 years of European colonization. In *Birds and Habitat: Relationships in Changing Landscapes* (Ed. R. J. Fuller) pp. 453–480. (Cambridge University Press.)

Martin, T. G., Nally, S., Burbidge, A. A., *et al.* (2012b) Acting fast helps avoid extinction. *Conservation Letters* 5, 274–280.

McAllan, I. A. W., Curtis, B. R., Hutton, I., Cooper, R. M. (2004) The birds of the Lord Howe Island Group: a review of records. *Australian Field Ornithology* 21, 1–82.

Mccallum, M. L., Bury, G. W. (2013) Google search patterns suggest declining interest in the environment. *Biodiversity and Conservation* 22, 1355–1367.

Miller, G. H., Magee, J. W., Johnson, B. J., *et al.* (1999) Pleistocene extinction of *Genyornis newtoni*: human impact on Australian megafauna. *Science* 283, 205–208.

O'Connell, J. F., Allen, J. (2004) Dating the colonization of Sahul (Pleistocene Australia–New Guinea): a review of recent research. *Journal of Archaeological Science* 31, 835–853.

Parmesan, C., Yohe, G. (2003) A globally coherent fingerprint of climate change impacts across natural systems. *Nature* 421, 37–41.

Price, O. F., Russell-Smith, J., Watt, F. (2012) The influence of prescribed fire on the extent of wildfire in savanna landscapes of western Arnhem Land, Australia. *International Journal of Wildland Fire* 21, 297–305.

Raymond, B., McInnes, J., Dambacher, J. M., Way, S., Bergstrom, D. M. (2011) Qualitative modelling of invasive species eradication on subantarctic Macquarie Island. *Journal of Applied Ecology* 48, 181–191.

Recher, H. F., Lim, L. (1990) A review of current ideas of the extinction, conservation and management of Australia's terrestrial vertebrate fauna. *Proceedings of the Ecological Society of Australia* 16, 287–301.

Robinson, D. (2006) Is revegetation in the Sheep Pen Creek area, Victoria, improving Grey-crowned Babbler habitat? *Ecological Management and Restoration* 7, 93–104.

Rolls, E. C. (1969) *They All Ran Wild: The Story of Pests on the Land in Australia.* (Angus and Robertson: Sydney.)

Savolainen, P., Leitner, T., Wilton, A. N., Matisoo-Smith, E., Lundeberg, J. (2004) A detailed picture of the origin of the Australian dingo, obtained from the study of mitochondrial DNA. *PNAS* 101, 12387–12390.

Schodde, R., Fullagar, P., Hermes, N. (1983) *A Review of the Status of Norfolk Island Birds: Past and Present.* Australian National Parks and Wildlife Service, Canberra.

Shearer, B. L., Crane, C. E., Barrett, S., Cochrane, A. (2007) *Phytophthora cinnamomi* invasion, a major threatening process to conservation of flora diversity in the South-west Botanical Province of Western Australia. *Australian Journal of Botany* 55, 225–238.

Szabo, J. K., Baxter, P. W. J., Vesk, P. A., Possingham, H. P. (2011) Paying the extinction debt: woodland birds in the Mount Lofty Ranges, South Australia. *Emu* 111, 59–70.

Szabo, J. K., Butchart, S. H. M., Possingham, H. P., Garnett, S. T. (2012a) Adapting global biodiversity indicators to the national scale: a Red List Index for Australian birds. *Biological Conservation* **148**, 61–68.

Szabo, J. K., Khwaja, N., Garnett, S. T., Butchart, S. H. M. (2012b) Global patterns and drivers of avian extinctions at the species and subspecies level. *PLoS ONE* **7**, e47080.

Tasmanian Parks and Wildlife Service (2014) Macquarie Island Pest Eradication. http://www.parks.tas.gov.au/?base=13013. Accessed 20 July 2014.

Tidemann, S. C., McOrist, S., Woinarski, J. C. Z., Freeland, W. J. (1992) Parasitism of wild Gouldian finches (*Erythrura gouldiae*) by the air-sac mite *Sternosoma tracheacolum*. *Journal of Wildlife Diseases* **28**, 80–84.

Tucker, N. I. J. (2000) Linkage restoration: Interpreting fragmentation theory for the design of a rainforest linkage in the humid Wet Tropics of north-eastern Queensland. *Ecological Management and Restoration* **1**, 35–41.

Veevers, J., Powell, C., Roots, S. (1991) Review of seafloor spreading around Australia. I. Synthesis of the patterns of spreading. *Australian Journal of Earth Sciences* **38**, 373–389.

Watson, D. M. (2011) A productivity-based explanation for woodland bird declines: poorer soils yield less food. *Emu* **111**, 10–18.

Wheelwright, H. W. (1868) *Bush Wanderings of a Naturalist: or, Notes on the Field Sports and Fauna of Australia Felix*. (Routledge, Warne, & Routledge: London)

Whitten, S. M., Freudenberger, D., Wyborn, C., Doerr, V. A. J., Doerr, E. D. (2011) *A Compendium of Existing and Planned Australian Wildlife Corridor Projects and Initiatives, and Case Study Analysis of Operational Experience*. A report for the Australian Government Department of Sustainability, Environment, Water, Population and Communities. CSIRO Ecosystem Sciences.

Widlansky, M. J., Timmermann, A., Stein, K., *et al.* (2013) Changes in South Pacific rainfall bands in a warming climate. *Nature Climate Change* **3**, 417–423.

Woinarski, J. C. Z., Whitehead, P. J., Bowman, D. M. J., Russell-Smith, J. (1992) Conservation of mobile species in a variable environment: the problem of reserve design in the Northern Territory, Australia. *Global Ecology and Biogeography Letters* **2**, 1–10.

Zander, K. K., Ainsworth, G., Meyerhoff, J., Garnett, S. T. (2014) Threatened bird valuation in Australia. *PLoS ONE* **9**, e100411.

CHAPTER 21

Austral amphibians – Gondwanan relicts in peril

Jean-Marc Hero, J. Dale Roberts, Conrad J. Hoskin, Katrin Lowe,
Edward J. Narayan and Phillip J. Bishop

Summary

Over 30% of Australasian amphibians are currently threatened with extinction. While habitat loss, introduced species and disease have been identified as major threats, the impacts of climate change are understudied. Threatened frogs fall into distinct biogeographical and ecological groupings that can be linked to specific threats (e.g. mountain-top endemics and climate change; stream-dwelling wet forest frogs and disease; and small island endemics and feral pests). The impacts of gradual climate change over millions of years has isolated specific species into climatic refugia (resulting in restricted geographic ranges), which combined with the ecological traits of these species (e.g. small clutch-size) dramatically increases extinction risk. Australasian frogs demonstrate intrinsic links between biogeographic history, species ecology and conservation status. The solutions to most threats are clear at a broad level, stop land clearing, reduce CO_2 emissions and control feral animals; however, declines linked to the disease chytridiomycosis are not easily resolved. Chytridiomycosis is not a universal threat and understanding the causes of variation in impact is critically important. While the threats of land clearing, disease and introduced species are regional and/or species specific, the impacts of climate change must be examined carefully as all species are likely to affected. Here we cover these issues for Australasian frogs, presenting regional examples that highlight threats and avenues for future research and management.

21.1 Introduction
21.1.1 Phylogenetic and biogeographic history

Over 30% of amphibian species are threatened with extinction globally making them the most threatened of the vertebrate groups (Wake and Vredenburg 2008). There are multiple threats to Austral frogs: e.g. disease – critically chytrid fungus for species with more aquatic lifestyles; small clutch size and limited range associated with higher decline or extinction risk; introduced species (*Gambusia* and trout in

Austral Ark: The State of Wildlife in Australia and New Zealand, eds. A. Stow, N. Maclean and G. I. Holwell. Published by Cambridge University Press. © Cambridge University Press 2015.

Australia, Gillespie and Hero, 1999; Murray *et al.*, 2011; *Rattus* in New Zealand, Thurley and Bell, 1994; and mongoose in the Pacific Islands, Pernetta and Watling, 1979) and less specific threats, identified in both Austral and global analyses of amphibian declines: e.g. climate change (Hero *et al.*, 2006, 2008; Hof *et al.*, 2011) and habitat loss and fragmentation (Hero *et al.*, 2008). These factors pose serious threats in many other regions of the world (Stuart, 2008) and their impacts vary among species and genera, depending on their current distribution and habitat use (Table 21.1). Here we examine

(1) the biogeographic history of Austral amphibians to set a historical context for this frog fauna, and a novel perspective on the conservation of phylogenetic diversity,

(2) disease – particularly the amphibian chytrid fungus, and new insights into disease dynamics that might allow frogs to avoid disease impacts,

(3) specific and more generic examples of impacts of habitat loss, case studies and evidence of impacts at landscape scales,

(4) the potential impacts of climate change, and

(5) the potential impacts of introduced predators.

We include case studies on regional faunas that highlight impacts or novel responses to one or more of these threats, or that highlight unique problems for particular species or communities.

21.1.2 Biogeographic history of Austral amphibians

Conservation biology frequently focusses on particular species or species richness but there is a long-standing view that we should also consider phylogenetic diversity (e.g. Tucker and Cadotte, 2013). To consider that approach we discuss the phylogenetic history of Austral frogs and include six well-defined lineages found in Australia, New Guinea, the Pacific islands and New Zealand (Hero *et al.*, 2008).

(1) The Myobatrachidae (*sensu* Pyron and Wiens, 2011) containing 127 species in 20 genera – split into two subfamilies (or families by Frost *et al.*, 2006): the Limnodynastinae and Myobatrachinae (Pyron and Wiens, 2011). The Myobatrachidae, a Gondwanan lineage, are most closely related to the Calyptocephallelidae from Chile (Pyron and Wiens, 2011) with one of the basal genera *Rheobatrachus* now extinct, and another basal lineage *Taudactylus* with most of the species either extinct or seriously threatened.

(2) The subfamily Pelodryadinae with 86 Australian 'tree frogs' in a single genus, *Litoria*, is most closely related to the Phyllomedusinae from South and particularly Central America. Although there has been selective species loss, there are no specific lineages under threat.

(3) Microhylidae: Australia has 23 species all in the subfamily Asterophryinae restricted to northern, tropical regions, with a large number of species with very small geographic ranges, many restricted to mountain tops (Hoskin and Hero, 2008) where they may be at particular risk due to warming climates. The Asterophryinae are also very diverse in New Guinea (184 species: www.amphibiaweb.org May 2013).

Table 21.1 Threatened amphibians of the Austral region (Australia, New Zealand and the Pacific Islands). Conservation status follows the IUCN Red List Categories and Criteria (IUCN, 2012) EX = Extinct; CR = Critically Endangered; EN = Endangered; VU = Vulnerable. The * symbol reflects proposed status following recent revision of the Microhylids in the Wet tropics (see Hoskin, 2004, 2007). Species have been grouped by key elements of their biogeographic history and ecology. Primary threats follow Hero and Morrison 2004 and Hero *et al.*, 2005, 2008. The symbol '?' represents likely threat however empirical evidence is needed.

Species	IUCN status	Location	Primary threats			
			Restricted range	Habitat loss	Chytrid	CC
NZ endemics: ancient Gondwanan relics Small clutch size, restricted distributions						
Leiopelma archeyi	CR	NZ	✓	✓		✓
Leiopelma hamiltoni	EN	NZ	✓	✓	?	✓
Leiopelma hochstetteri	VU	NZ	✓	✓		✓
Leiopelma pakeka	VU	NZ	✓	✓		✓
Isolated endemics of Fiji Small clutch size, restricted distributions						
Platymantis vitianus	EN		✓	✓		
Platymantis vitiensis	NT		✓	✓		
Isolated endemics of Western Australia Small clutch size, restricted distributions						
Geocrinia alba	CR	SW Western Australia	✓	✓		?
Geocrinia vitellina	VU		✓	✓		?
Spicospina flammocaerulea	VU		✓	✓		?
Isolated endemics of Eastern Australia Small clutch size, restricted distributions						
Pseudophryne covacevichae	EN	Wet Tropics North Qld	✓	✓		?
Pseudophryne australis	VU	Sydney sandstone	✓	✓		
Acid frogs in coastal wallum habitats of mid-eastern Australia Restricted distributions						
Litoria cooloolensis	EN	SE Qld	✓			?
Crinia tinnula	VU	mid-eastern Australia		✓		?
Litoria freycineti	VU	mid-eastern Australia		✓		?
Litoria olongburensis	VU	mid-eastern Australia		✓		?
Localized mountain distributions in rainforest isolates of Eastern Australia Small clutch size, restricted distributions						
Cophixalus concinnus	CR	Wet Tropics North Qld	✓			✓
Philoria frosti	CR	Mt Baw Baw Victoria	✓			✓
Pseudophryne corroboree	CR		✓			✓
Cophixalus mcdonaldi	EN	Wet Tropics North Qld	✓			✓
Cophixalus monticola	EN	Wet Tropics North Qld	✓			✓

Table 21.1 (*cont.*)

Species	IUCN status	Location	Primary threats			
			Restricted range	Habitat loss	Chytrid	CC
Cophixalus neglectus	EN	Wet Tropics North Qld	✓			✓
Philoria kundagungan	EN	SE Qld	✓			✓
Philoria loveridgei	EN	SE QldN NSW	✓			✓
Philoria pughi	EN	NE NSW	✓			✓
Philoria richmondensis	EN	NE NSW	✓			✓
Philoria sphagnicolus	EN	Mid-E NSW	✓			✓
Pseudophryne pengilleyi	EN	NSW	✓			✓
Cophixalus aenigma	VU	Wet Tropics North Qld	✓			✓
Cophixalus hosmeri	VU	Wet Tropics North Qld	✓			✓
Cophixalus saxatilis	VU	Wet Tropics North Qld	✓			✓
Stream-dwelling endemics of wet forests in Eastern Australia Small clutch size, restricted distributions						
Rheobatrachus silus	EX	SE Qld	✓	✓	?	?
Rheobatrachus vitellinus	EX	Eungella	✓		?	?
Taudactylus diurnus	EX	SE Qld	✓		?	?
Litoria booroolongensis	CR	NSW	✓	✓	✓	?
Litoria lorica	CR	Wet Tropics North Qld	Rediscovered in 2008		?	?
Litoria myola	CR*	Wet Tropics North Qld	✓	✓		?
Litoria nyakalensis	CR	Wet Tropics North Qld			?	?
Litoria piperata	CR	N NSW	✓		?	
Litoria spenceri	CR	Victoria highlands	✓	✓	✓	?
Taudactylus acutirostris	CR	Wet Tropics North Qld			✓	
Taudactylus pleione	CR	Kroombit Tops SE Qld	✓		?	?
Taudactylus eungellensis	CR	Eungella	✓		✓	
Taudactylus rheophilus	CR	Wet Tropics North Qld	✓		✓	✓
Litoria nannotis	EN	Wet Tropics North Qld			✓	?
Litoria rheocola	EN	Wet Tropics North Qld			✓	?
Mixophyes fleayi	EN	SE Qld			?	?
Mixophyes iteratus	EN	SE Qld		✓	?	?
Nyctimystes dayi	EN	Wet Tropics North Qld			✓	?

Table 21.1 (*cont.*)

Species	IUCN status	Location	Primary threats			
			Restricted range	Habitat loss	Chytrid	CC
Litoria andiirrmalin	VU	Cape Melville, North Queensland	Restricted to 4 streams		?	?
Litoria subglandulosa	VU	N NSW	✓	✓	✓	
Mixophyes balbus	VU	S NSW	✓		?	?
Uncommon species in Eastern Australia Widespread distributions						
Litoria castanea / flavipunctata	CR	Southern Tablelands, NSW	Rediscovered in 2008	✓		✓
Litoria brevipalmata	EN	SE Qld and NSW	✓	✓		
Litoria raniformis	EN	NSW and Victoria			✓	
Heleioporus australiacus	VU	NSW and Victoria	✓	✓		✓
Litoria aurea	VU	Mid NSW – Victoria				

(4) Ranidae: one species in far north Queensland, Australia, and 11 species in Papua New Guinea, in a family that is widespread and diverse from Africa through to SE Asia.

(5) Ceratobatrachidae has 86 species (Genus *Platymantis* 71 species: www. amphibiaweb.org May 2013). This group of frogs is found in The Philippines, New Guinea (Indonesia and Papua) and numerous Pacific Islands. Our focus is on the Pacific Islands, where there is the potential for strong 'island effects' on species persistence.

(6) Leiopelmatidae: the four living species of *Leiopelma* are the only native frogs in New Zealand. They possess ancestral traits such as no vocal sac, the presence of amphicoelous vertebrae and paired caudalipuboischiotibialis 'tail-wagging' muscles. Their closest relatives are 'tailed-frogs' (Family Ascaphidae) found in northwestern North America (Pyron and Wiens, 2011). They are a basal anuran lineage (Pyron and Wiens, 2011) whose loss would exemplify the critical importance of a phylogenetic perspective in conservation goals.

The Gondwanan break-up is likely to account for the relationship of the Myobatrachidae and Pelodryadinae to South and Central American lineages. The Microhylidae and Ranid species are relatively recent arrivals derived from SE Asian (Asterophryninae) or African–Asian lineages (Ranidae) that post-date the collision of the Australian and Eurasian and Philippine Sea continental plates ≈ 30 mybp (Seton *et al.*, 2012). The Cane Toad (family Bufonidae) was deliberately introduced into Australia in 1935 (see Chapter 5 by Shine and Phillips in this book). This long biogeographic history has resulted in a species-rich frog fauna (over 230 species) with most endemic to Australia. Papua New Guinea is also an area of high diversity with over 340 species currently described (amphibiaweb.org), but with only 15 species co-occurring in both Australia and New Guinea.

This is a continuously evolving fauna, while the oldest Australian frog lineages have been in Australia for over 100 million years, many of the current genera are also old, evolving well back in the Tertiary (55–1.8 mybp). Morgan *et al.* (2007) dated the subfamily split in the Myobatrachidae at about 100 mybp (≈80 mybp in Maxson, 1992), with generic divisions dating back 5–70 million years. The origin of currently recognised species varies with well-defined, sympatric, sister species of *Heleioporus* splitting 6 mybp (Morgan *et al.*, 2007) but there is also evidence of relatively recent speciation with relatively low levels of differentiation in mitochondrial DNA between some *Mixophyes* and *Crinia* species (Oza *et al.*, 2012; Read *et al.*, 2001). In several currently recognised species from south-western Australia, there are genetic subdivisions associated with Pleistocene/Pliocene climate fluctuations and geological faults or developing river gorge systems, isolation on mountain ranges and isolation in drainage systems (Edwards *et al.*, 2007). In rainforest systems in north Queensland, Bell *et al.* (2012) reported two phases of differentiation of frog lineages driven by contraction of rainforest blocks: in the Pliocene (5.3–1.8 mybp), and in the mid Pleistocene (1.8 mybp–10 000 years ago). Drying phases in the Pleistocene may have had less impact (Bell *et al.*, 2012).

The frogs of the family Ceratobatrachidae are closely related to the Ranidae (Pyron and Wiens, 2011) and are found in the Philippines, New Guinea and numerous Pacific Islands. This group has many island endemics that are especially vulnerable to island effects (see below). In contrast the native frogs of New Zealand, all in one genus, *Leiopelma*, (Family Leiopelmatidae) are a basal anuran lineage. The Ascaphidae and the Leiopelmatidae are basal lineages with respect to all other frog families (Pyron and Wiens, 2011).

21.2   Primary threats to Austral amphibians

We discuss four well-recognised, major threats: habitat loss and modification, disease, introduced predators, and climate change. These threats are widespread affecting many species/systematic groupings; however, some have critical relevance to only part of the region.

21.2.1   Habitat loss and fragmentation

The impacts of habitat destruction and fragmentation are directly visible and measurable with substantial habitat lost over the past 250 years. These changes are inevitably linked with climate variation and directional climate change. In some cases, e.g. *Geocrinia alba*, declines are clearly associated with range fragmentation and isolation of small populations caused by clearing and this, coupled with low dispersal, has led to loss of many local populations (Roberts *et al.*, 1999). In heavily modified agricultural landscapes in northern Victoria the array of frog species was similar in cleared and forested land surveyed in 2006–2007 following >12 years of extreme drought. But, compared with the late 1950s to 1980s, many rarer species had been lost, and abundance of remaining species was lower (MacNally *et al.*, 2009). Wassens *et al.* (2013) measured breeding activity (evidenced by presence of tadpoles) in rain-fed ponds in agricultural landscapes in southern New South Wales in 2010 following drought from 2002 to 2009. While some common species were still evident, their study found extreme climate events led to a decline in species richness. Similarly, in southern New South Wales, some anuran species occurring in natural water bodies were missing from constructed farm dams but patterns of abundance were similar in natural and man-made or heavily modified natural water bodies (Hazell *et al.*, 2004). In the same region a total shift in vegetation to introduced pine trees radically reduced anuran diversity (Parris and Lindenmayer, 2004). In the West Australian Wheatbelt all frog species ever known were encountered in comprehensive surveys conducted in 1999–2003 despite extensive clearing and secondary salinity resulting from vegetation loss (Burbidge *et al.*, 2004). Salinity limits tadpole occurrence of all anuran species in similar, seasonally arid environments in western Victoria (Smith *et al.*, 2007) but is not a limiting factor for all Australian frog species (Janicke and Roberts, 2010). Habitat loss or modification does not necessarily lead to species extinction or decline, and some species can survive even extreme drought conditions in heavily modified landscapes, although as MacNally *et al.* (2009) note, the fauna may be heavily winnowed by clearing and drought to maintain only tough, resilient species.

In Australia, there are no examples of habitat loss or modification causing extinction but habitat loss or modification may have reduced ranges or caused range shifts: the historical record of frog distributions is too poor to make reliable assessments. Distributions may also be affected by long-term or episodic climate events. Roberts (1993) reported range shifts in *Limnodynastes tasmaniensis* following flood events on the River Murray in South Australia, and MacNally *et al.* (2013) have demonstrated increases in reproductive activity of frogs following drought-breaking rains in north-central Victoria. Modified habitats are dynamic landscapes that can maintain many frog species but are not always a one-way street to decline.

21.2.2 Disease

Chydriomycosis a disease caused by infection by the fungal pathogen (*Batracochytrium dendrobatidis – Bd*) is associated with decline and probable extinction of frogs all over the world (Wake and Vredenburg 2008). A huge volume of research has focussed on this disease (see reviews by Fisher *et al.*, 2009, Kilpatrick *et al.*, 2010, and Blaustein *et al.*, 2012); however, many questions remain unanswered (Muths and Hero, 2010).

Declines of many Austral frog species have been associated with chytridiomycosis (Schloegel *et al.*, 2006, Skerratt *et al.*, 2007; Murray *et al.* 2013). However, the impact of this disease can be heavily moderated by climate, location or life history (Kriger and Hero 2007a,b, 2008; Kriger *et al.*, 2007; Van Sluys and Hero, 2010; Puschendorf *et al.*, 2009). Hauselberger and Alford (2012) reported a very high probability of absence of chytrid in a direct developing rainforest species, *Cophixalus ornatus*, consistent with earlier generalities reported by Kriger and Hero (2007a), suggesting direct-developers were less likely to be infected.

In regions with more seasonal environments or where frogs can exploit thermal niches that expose disease organisms to lethal temperatures or conditions that inhibit chytrid growth, frogs may coexist with chytrid (Kriger and Hero, 2006, 2007b, 2008; Kriger *et al.*, 2007). Habitat and vegetation type influence both the severity and prevalence and intensity of infection in frogs (Van Sluys and Hero, 2010). The capacity to selectively use microhabitats may also explain the persistence of *Litoria lorica*. This species, a stream breeding/living species from the wet tropics thought to be extinct was rediscovered in 2008 in sites that allowed frogs to maintain body temperatures that inhibited chytrid growth (Puschendorf *et al.*, 2011) and recently Rowley and Alford (2013) showed frogs that stayed warmer in the field were less likely to suffer from chytrid infections in three other *Litoria* species.

In New Zealand, chytridiomycosis has tentatively been linked with an 88% decline of *L. archeyi* in the Coromandel Peninsula (Bell *et al.*, 2004) but recent experimental work suggests that under laboratory conditions three out of the four species of *Leiopelma* are not particularly susceptible to chytridiomycosis and can self-cure (Shaw *et al.*, 2010; Ohmer *et al.*, 2013). It is of interest to note that despite many surveys the most aquatic species of *Leiopelma* (*L. hochstetteri*) has never been found to be infected by *Bd* in the wild (Thurley and Haigh 2008; Shaw *et al.*, 2013). However, the susceptibility of leiopelmatids to chytrids in the wild, and population-wide effects of the disease, are yet to be determined.

In south-western Australia, a region with a marked Mediterranean climate, chytrid is widespread having been reported from 15 species: including species with direct development (*Geocrinia* species), and in species breeding in very ephemeral water bodies (*Crinia* spp.) (Riley *et al.*, 2013). Riley *et al.* (2013) discuss several models that might explain persistence of chytrid in south-western Australia without detectable impact on population persistence, including seasonal environments limiting chytrid growth, immunity and strain identity. Riley *et al.*'s (2013) data and discussion suggests very different disease dynamics cautioning against blanket conclusions about chytrid risks and impact of life histories on risk. This caution is particularly important given recent modelling of environmental suitability for chytrid at continental and global scales with an implied associated risk of frog decline (discussed by Riley *et al.*, 2013). Studies on chytrid now critically need data on strain identity given the variation in virulence of chytrid strains identified by Farrer *et al.* (2011).

Global predictions about chytrid impact are currently impossible as disease dynamics and susceptibility vary. Management of chytrid risk requires a better understanding of strain types, strain virulence and the dynamics of disease risk given both the similarities and radical differences between disease dynamics in south-western Australia, eastern and north-eastern Australia and New Zealand. We also need to understand the complex relationships with climate variation as disease and climate change are intricately connected (Pounds *et al.*, 2006; Rohr *et al.*, 2008, 2011; Rohr and Raffel, 2010; Liu *et al.*, 2013; Li *et al.*, 2013; Raffel *et al.*, 2013).

21.2.3 Climate change

Climate change is now looming as the greatest threat to biodiversity (Parmesan and Singer, 2008; Williams *et al.* 2008; Cabrelli *et al.*, Chapter 4 this book). Recent reviews on the impacts of climate change on amphibians (Corn, 2005; Blaustein *et al.*, 2012; Li *et al.*, 2013) are compelling, and the evidence is mounting that impacts are likely to be widespread (Reading, 2007; Parmesan and Singer, 2008; Rohr and Raffel, 2010; Rohr and Palmer, 2013). Shoo *et al.* (2011) have recently suggested management actions to minimise the impacts of climate change on amphibians. However the impacts will be difficult to manage in almost any anuran species, with mountain-top species offering particular challenges.

The degree to which Australia's mountain-top species have been, or will be, impacted by climate change depends on: (1) the exact nature of climate change at these sites, (2) the physiological tolerances of these species, (3) the use of microhabitats that are buffered to some degree, (4) other impacts on the mountain-top systems induced by climate change (e.g. changes in prey, competitors, parasites, predators), and (5) adaptation in physiology and behaviour. Currently there are not enough data on any of these points to reliably assess the likely impact of climate change on Australian montane species. The various mountain tops inhabited by Microhylid species vary greatly in substrate, from earth with no rocks, through earth with rocky patches, to boulder-field. Deeply layered rocky areas provide cooler microhabitats with depth and can be effective at retaining moisture (Couper and Hoskin 2008; Shoo *et al.*, 2009). Just as these areas have acted as lithorefugia for the persistence of rainforest lineages in eastern and northern Australia (Couper and Hoskin, 2008), they may also buffer organisms to some degree from short- and long-term changes in temperature or moisture and enhance persistence in the face of climate change (Couper and Hoskin, 2008; Shoo *et al.*, 2009).

The long history of many anuran lineages (see above) indicates they have survived radical climate variation in the past: there may be retained genetic variation allowing rapid plastic responses or, natural section may select appropriate genotypes (Byrne *et al.*, 2011; Moritz and Agudo 2013). However, as Moritz and Agudo (2013) note, we are dealing with temperature shifts that are faster and larger than global faunas have previously experienced, at least in the Plio-Pleistocene period. Moritz and Agudo (2013) also summarise extensive data sets on documented responses to climate change in the twentieth century and note many counterintuitive shifts in range suggesting even the most sophisticated global models of climate change, habitat loss or changes in disease risk may not be a sufficient predictor of response by anuran or other faunas.

While the threats of habitat loss, disease and climate change directly impact amphibians, identifying the additive or synergistic effects of these threatening processes is challenging. Despite this difficulty, there is mounting evidence that disease and climate change are intricately connected (Pounds *et al.*, 2006; Rohr *et al.*, 2008, 2011; Rohr and Raffel 2010; Liu *et al.*, 2013; Li *et al.*, 2013; Raffel *et al.*, 2013).

21.2.4 Introduced predators

In Australia, critical introduced predators are fish species that primarily affect tadpoles (Gillespie and Hero, 1999). Murray *et al.* (2011) identified the presence of *Gambusia* species as a significant predictor of decline or extinction risk in the Australian frog fauna but did not relate that impact to particular frog species or explain the specific mechanism of impact. Gillespie (2001) showed that introduced trout species were a critical tadpole predator and may have been a strong contributor to declines in populations of *Litoria spenceri* in north-eastern Victoria. In New Zealand and Fiji, there is direct evidence of impacts from a range of introduced mammal predators, e.g. rats and mongoose, eating frogs (Veron *et al.*, 2010). Although an identifiable threat, there are very few quantitative data sets defining the explicit impacts of introduced predators on population dynamics or species persistence.

21.3 Conservation and management of Austral amphibians

We have discussed threats that are common across many areas of Australia, New Zealand and the Pacific Islands. Our focus here is on ecological groups that share a biogeographic history and contain taxa of conservation concern – using examples of species, regions or taxonomic groupings, that exhibit peculiar or unique conservation problems. Some areas are not discussed for one of three reasons:

(a) they may contain species of uncertain status (e.g. several *Uperoleia* species from north-western Australia are listed as data deficient by the IUCN),

(b) regions where there is no evidence of decline: e.g. we make no reference to any primarily desert species because there is no evidence of decline, though we acknowledge these species may be relatively poorly studied,

(c) in some areas, e.g. south-western Australia, we restrict our discussion to species where there is ongoing uncertainty about status to illustrate difficulties for assessment of status and ongoing management.

Decline broadly in Austral frogs is predicted by life-history features, species at risk have small geographic ranges, low fecundity and are bog/soak, and stream breeders, however these characteristics are mediated by their biogeographic history (Table 21.1: Hero *et al.*, 2005, 2006, 2008, 2012; Hero and Morrison 2012; Murray *et al.*, 2011; Morrison and Hero 2012). In this section we discuss the conservation issues facing each specific ecological group of threatened Austral frogs.

21.3.1 NZ endemics: ancient relics

There are four extant species of *Leiopelma* formally recognised; *Leiopelma archeyi*, *L. hamiltoni*, *L. hochstetteri* and *L. pakeka* and all are listed on the IUCN Red Data List. In addition, all four species are listed in the top 100 EDGE amphibians, with *L. archeyi* topping the list as the world's most evolutionarily distinct and globally endangered amphibian species (EDGE, 2013). The mainly terrestrial leiopelmatid frogs were previously widespread and common, including in lowland habitats, throughout New Zealand, but are now much reduced in geographic and altitudinal range (Table 21.1). Remnant populations of *L. hamiltoni* and *L. pakeka* only occur on predator-free islands in the Marlborough Sounds at the northern end of the South Island. *Leiopelma archeyi* and the semi-aquatic *L. hochstetteri* are still found in scattered locations on the mainland of the North Island.

The main causal factors of decline for leiopelmatids have not been conclusively demonstrated, particularly at the population level. The primary threats are considered to be predation by introduced mammals (rats are known predators of frogs in New Zealand (Thurley and Bell, 1994)), habitat loss and modification and climate change. Two introduced frogs from Australia (*Litoria aurea* and *Litoria raniformis*) are known to prey on leiopelmatids, however, the extent of this threat is unclear (Thurley and Bell, 1994). It has also been suggested that other introduced mammals, such as mice, stoats, hedgehogs, possums or pigs, prey on leiopelmatids, but the significance of these predators has yet to be determined (Bell, 1994; Newman, 1996; Bell *et al.*, 2004; Bishop *et al.*, 2013).

All the leiopelmatids are adapted to cool, moist, habitats in native forests between 200 – 800 m above sea level. While climate change is generally considered detrimental to amphibians, recent ecological niche modelling by Fouquet *et al.* (2010) suggests that climate change will cause an increase in the area of climatic suitability for *L. hochstetteri*. However, given the degree of fragmentation of the populations, the very low dispersal ability of leiopelmatids, coupled with low reproductive rates (<20 eggs per season) and high site fidelity, it is highly unlikely that they will respond favourably to an increase in climatically suitable habitat.

Figure 21.1 (Plate 39) Fijian Ground Frog leaves its eggs on land, young froglets hatch directly from large, fertilised eggs. A black and white version of this figure will appear in some formats. For the colour version, please refer to the plate section.

21.3.2 Pacific Island endemics of Fiji

The Platymantid frogs belong to the largest anuran family (Ranidae), and have diverged throughout the Asia Pacific Islands (Allison, 1996, 2009). Platymantines are noted for their unusual geographical distribution, highly variable and unique morphology, direct development and their ability to colonize habitats that lack other ranid frogs (Brown *et al.*, 2006). The Fiji Islands have the eastern-most distributions of frogs in the southwest Pacific, with two Platymantid species: the Fijian ground frog *Platymantis vitiana*, and the Fijian tree frog, *Platymantis vitiensis* (Table 21.1; see also Figure 21.1, Plate 39). Both of these frogs have entirely terrestrial reproduction, where young froglets hatch directly from large, fertilised eggs (Narayan *et al.*, 2011).

Threats to island frog species come from introduced fauna: cane toads, rats, mongooses, feral cats, dogs, cattle, and pigs (Morrison, 2003). Three species of mongoose (*Herpestes auropunctatus*, *Herpestes edwardsi* and *Herpestes javanicus*) have been introduced to Fiji and as habitat generalists they are now widespread throughout Viti Levu and Vanua Levu and many smaller islands (Ryan, 1988). Anecdotal evidence suggests that mongoose prey on Fijian ground frog egg clutches, juveniles and adults. Impacts of these species are poorly known but it has been assumed the mongoose is responsible for the restricted distribution of the Fijian tree frog, and extinction of Fijian Ground Frogs on many islands (Veron *et al.*, 2010; Kuruyawa *et al.*, 2004). Conservation of the endangered Fijian ground frog is now focussed on Mongoose-free islands (Viwa, Ovalau, Taveuni and Gau Islands).

Fijian Ground Frogs are may also threatened by the introduced Cane Toad (*Rhinella marina*) which is widespread in the Fijian Islands. The impacts of Cane Toads on Fijian Ground Frogs are poorly known; however, the exceptionally high densities of Cane Toads coexisting with low densities of Fijian Ground Frogs on Viwa Island suggest there are likely to be direct impacts of competition for food and shelter (Thomas *et al.*, 2011). There is also evidence that even the presence of Cane Toad has a physiological impact on Fijian Ground Frog. Narayan *et al.* (2013) demonstrated experimental exposure to the sight of Cane Toads can induce a corticosterone response and lead to increased fearfulness in Fijian Ground frogs.

21.3.3   Isolated endemics of south-western Australia

South-western Australia has three frog species that are formally listed on the IUCN Red List, and in slightly different categories under the EPBC Act in Australia (Table 21.1). *Geocrinia alba* has a range of about 130 km^2 and is under severe threat from range fragmentation caused by clearing and subsequent loss of isolated, small populations. *Geocrinia vitellina* has a range of about 6 km^2 but, despite that, and despite experiencing both a currently drying climate (but south-western WA has probably been equally dry historically; Cullen and Grierson, 2009) and severe fire in 1997, still has good population densities and has shown good persistence post wild-fire (Bamford and Roberts, 2003). The third listed species, *Spicospina flammocaerulea*, has a limited distribution (about 300 km^2) and is restricted to peat swamps forming in first-order streams relatively high in the landscape (Edwards and Roberts, 2011). *Spicospina* is a monotypic genus. It is spectacularly coloured with ventral colours of blue and orange with the back dark, peat coloured, and is most closely related to the widespread Australian genus *Uperoleia* (Pyron and Wiens, 2011). The majority of its known range is in land managed by the WA Department of Parks and Wildlife giving this species a relatively high degree of security from habitat modification under current climate conditions. This species presents a serious dilemma for conservation managers. Populations can persist for many years with little evidence of breeding activity but then breed explosively, in large numbers, sometimes after fire events. For example, Mountain Road was burnt in 1994 and explosive breeding followed but then numbers of calling frogs declined to almost zero over the next five years (Bamford and Roberts, 2003). Frogs were again abundant at this site by 2002–2003 (collections reported in Edwards and Roberts, 2011). Fire is not a necessary disturbance: in disturbed sites, or permanently wet sites, breeding can occur more frequently (Roberts *et al.*, 1999; Bamford and Roberts, 2003). This makes reliable assessment of population size and population status difficult and has led to frequent media reports in Western Australia of this species being in serious trouble when the reality is either (i) uncertainty about status or (ii) the species is persisting.

21.3.4   Isolated endemics of eastern Australia

Two species of *Pseudophryne* in eastern Australia (Table 21.1) are similar to the Western Australia endemics as they are habitat specialists with low fecundity and restricted geographic range (Hero and Morrison 2004, 2012). *Pseudophryne covacevichae* is geographically restricted to a small area of Eucalypt forest, historically surrounded by rainforest at the southern end of the Wet Tropics. *Pseudophryne australis* (Red-Crowned Toadlet) is restricted to the sandstone habitat surrounding Sydney areas that are heavily impacted by urban development (Hero and Morrison, 2004) which has dramatically fragmented remaining populations.

21.3.5   Acid frogs: coastal wallum frogs of mid-eastern Australia

Acid frogs are a unique group of frogs (Table 21.1) that have the capacity to live and breed in acidic wetlands, typically pH 3.0–5.0, and include four Australian species: *Litoria olongburensis*, *L. freycineti*, *L. cooloolensis*, and *Crinia tinnula* (Hines and Meyer, 2011; Ingram and Corben, 1975; Ingram and McDonald, 1993; Kikkawa *et al.*, 1979). They are restricted to freshwater wallum habitats in coastal lowlands of eastern Australia, characterised by flora-rich heathland, sedgeland and scrubland vegetation communities on low-nutrient, acidic, sandy sediments (Griffith *et al.*, 2003). Wallum is distributed between Shoalwater Bay in Queensland, and Newcastle in New South Wales, however the area of occurrence of wallum was historically much larger, due to a sea level 28 m lower (Griffith *et al.*, 2003; Coaldrake, 1961). Subsequent rises in sea level and changes in climate of the interglacials have isolated current acid frog populations, and increased their vulnerability to future rises in sea level (Ingram and Corben, 1975; Coaldrake, 1961; Hennessy *et al.*, 2007).

Acid frogs are highly sensitive to habitat disturbance and are generally not abundant in areas close to human habitation (Hines and Meyer, 2011; Ingram and Corben, 1975, Shuker *et al.*, 2012, Simpkins *et al.*, 2013) but in protected habitats populations are relatively stable (Hines *et al.*, 1999; Lewis and Goldingay, 2005). Substantial modification, reduction and fragmentation of wallum habitat has occurred since European settlement, and as with other coastal ecosystems, wallum is highly threatened by pressures for coastal development (Hines *et al.*, 1999; Ingram and McDonald, 1993). Changes to hydrological regimes, increased nutrients or sediments (e.g. from storm water, sewage and agricultural run-off), weed invasion, inappropriate fire regimes, and predation from introduced fish all threaten current acid frog populations (Hines *et al.*, 1999; Gillespie and Hero, 1999; Komak and Crossland, 2000; Ingram and Corben, 1975; Lowe *et al.*, 2013). Acid frogs are variously listed as threatened under state and federal legislation, as well as by the International Union for the Conservation of Nature (IUCN 2012 Red List; Meyer *et al.*, 2006).

Fire is a common occurrence in all wallum habitats and plays an integral part in maintaining vegetation communities (Griffith *et al.*, 2003, 2008; Westgate *et al.*, 2012; Specht 1981). Acid frogs have high levels of resilience to wildfire both over the short- and long-term (Westgate *et al.*, 2012; Lowe *et al.*, 2013). More frequent and more severe fires are expected for many ecosystems as a result of climate change and land-use modifications (Williams *et al.*, 2001; Hennessy *et al.*, 2005). Climate change may still pose significant threats to their continued survival, predominantly through changes to hydrological patterns, a critical factor for successful juvenile recruitment (Lowe *et al.*, 2013).

21.3.6 Localized mountain distributions in rainforest isolates of eastern Australia

High-elevation habitats in Australia are largely restricted to the Great Dividing Range from Cape York in the north to Tasmania in the south. Although not particularly high (typically <1300 m above sea level), there is nonetheless a significant area above 1000 m. These upland areas are particularly isolated along the Queensland coast, and in the far north typically consist of isolated mountain tops separated by low valleys. Despite limited altitudinal range, lowland and upland habitats often differ radically: from warm, lowland rainforests to cool temperate rainforests, stunted montane forests and heath. Mountain tops are characteristically moist due to high rainfall and cloud intercept. The most significant mountain-top frog genera are *Cophixalus* and *Philoria* (Table 21.1; Figure 21.2, Plate 40), however the only remaining populations of the once widespread Northern tinker Frog (*Taudactylus rheophilus*) are now restricted to Mt Lewis and Mt Bellenden Ker (Hoskin and Hero, 2008).

Figure 21.2 (Plate 40) Most *Cophixalus* have highly localised distributions with over half restricted to mountain tops. A black and white version of this figure will appear in some formats. For the colour version, please refer to the plate section.

Cophixalus are microhylid frogs from the Wet Tropics and Cape York Peninsula in north-east Queensland (Zweifel, 1985; Hoskin, 2004): 14 species occur in rainforests and four are restricted to boulder-fields (Hoskin and Aland, 2011; Hoskin, 2012). All species have terrestrial breeding, with eggs laid in moist situations (e.g. leaf-litter, under rocks and logs) and fully formed froglets hatching out (Hoskin, 2004). Most *Cophixalus* have highly localised distributions: over half are restricted to a single mountain range with several rainforest species restricted to a single mountain top: e.g. *Cophixalus concinnus* (>1100 m on Thornton Peak), *C. monticola* (>1100 m on the Carbine Tableland) and *C. neglectus* (generally >1200 m on the Bellenden Ker Range; Hoskin, 2004; Shoo and Williams, 2004; Hoskin and Higgie, 2005; Hoskin, 2008; Hoskin and Hero, 2008). These species have altitudinal ranges spanning just 200–400 m of elevation. *Philoria* contains five species restricted to the mountain ranges of north-east New South Wales and south-east Queensland, *P. kundagungan*, *P. loveridgei*, *P. pughi*, *P. richmondensis* and *P. spagnicolus*, and *P. frosti* found on the Baw Baw Plateau Victoria. Most species are restricted to cool upland temperate rainforests (generally >600 m) and all have highly localised distributions. *Philoria* breed in gully headwaters and seepage areas, with all development occurring in a moist nest cavity where tadpoles live in a jelly-like substance nourished by their yolk-sac (Knowles *et al.*, 2004).

The mountain-top *Cophixalus* and *Philoria* are generally listed at the IUCN, national and state level. On IUCN listings, *Cophixalus coninnus* is listed as Critically Endangered, *C. monticola*, *C. neglectus* and *C. mcdonaldi* are listed as Endangered, and *C. aenigma* is listed as Vulnerable. *Philoria frosti* is listed as Critically Endangered and the other four *Philoria* are all listed as Endangered. The threatened species listings for these mountain-top frogs are due to small distributions, projected declines in habitat area and quality due to climate change or for *P. frosti*, demonstrated reductions in range and population size. Additional factors listed for some *Philoria* species include localised impacts from clearing for farmland and logging (Knowles *et al.*, 2004) and chytridiomycosis for *P. frosti* (Hero *et al.*, 2004). For mountain-top species, climate change predictions are severe. Williams *et al.* (2003) modelled climate change impacts in the Wet Tropics region and highlighted *Cophixalus* as a group particularly threatened with suggested loss of core environment and potential extinction of *C. concinnus* in a matter of decades. *Philoria frosti* was surveyed intensively in 1983–1984 and again in 1993–2002 with strong evidence of a down-slope movement to lower altitudes (Hollis, 2004). For *Cophixalus neglectus* there are anecdotes suggesting a recent upward altitudinal shift of its lower range limit over the past few decades. Currently *C. neglectus* is largely restricted to >1200 m above sea level (with small numbers patchily distributed in boulder areas down to about 975 m; Hoskin, 2008), but there are anecdotes of males calling in recent times at significantly lower elevations (e.g., 700 m above sea level), in areas where they are no longer present (C. J. Hoskin, unpublished data).

What limits these species to such high altitudes is not known (e.g. physiology, competition, diet). No data on physiological tolerances have been published for these species but there are anecdotes that some species (e.g. *C. neglectus*, *C. concinnus*) lose righting response when brought to warmer temperatures of the lowlands (Zweifel, 1985; C. J. Hoskin, personal observation). The terrestrial breeding of *Cophixalus* and the use of moist seepage areas by *Philoria* requires extended periods of consistent moisture. Reductions in rainfall or cloud intercept or changes in seasonality of precipitation may have significant impacts on the breeding success of these species.

21.3.7 Stream-dwelling wet-forest species of eastern Australia

Many stream-dwelling frogs in rainforests of eastern Australia have suffered severe declines (Table 21.1) following the disappearance of *Taudactylus diurnus* in 1979 (Czechura and Ingram, 1990; Richards *et al.*, 1993; Hero *et al.*, 2005). These declines have been documented globally and have been described as the the sixth major extinction event in history (Hoffmann *et al.*, 2010; Wake and Vredenburg, 2008). Initially these declines were described as 'enigmatic' as many species disappeared from relatively pristine habitats (Hoffmann *et al.*, 2010). Since the first species disappeared in 1979, several other Austral species have subsequently disappeared including: two species of Gastric Brooding frogs (*Rheobatrachus vitellinus* and *R. silus*), two species of Tinker frogs (*Taudactylus diurnus, T. acutirostris*) the Mountain Mistfrog (*Litoria nyakalensis*) and the Peppered Treefrog (*Litoria piperata*). Many other species have declined dramatically and have disappeared from upland areas throughout most of their historical range. Many of these species are now only found in small isolated populations at lower elevations (Hero *et al.*, 2005, 2008). These include two species of the genus *Taudactylus* (*T. eungellensis* and *T. pleioni*). Enigmatic declines in populations were a mystery until 1998, when the discovery of a global pandemic disease chytridiomycosis was found to be at least partially responsible (Berger *et al.*, 1998; Schloegel *et al.*, 2006).

A species that was once found at several sites in the northern Wet Tropics region (*L. lorica*; Figure 21.3a,b, Plates 41 and 42) was missing for 16 years but was rediscovered at a high-elevation dry sclerophyll site that had little canopy cover, lower annual precipitation, and a more defined dry season than a nearby rainforest (Puschendorf *et al.*, 2011). This population persists despite high prevalence of chytrid infection, and it is presumed the more open and, consequently, warmer habitat, plays an important role in controlling chytrid infection (Puschendorf *et al.*, 2011); as was also suggested for another east Australian frog species (Van Sluys and Hero, 2010).

Figure 21.3a (Plate 41) *Litoria lorica* was not detected for 16 years. A black and white version of this figure will appear in some formats. For the colour version, please refer to the plate section.

Figure 21.3b (Plate 42) The habitat in which *L. lorica* was rediscovered. A black and white version of this figure will appear in some formats. For the colour version, please refer to the plate section.

21.3.8 Generalist and historically widespread species in eastern Australia

This group of high fecundity threatened species (Table 21.1) have declined dramatically with many populations disappearing over large parts of what were relatively large geographic ranges (Hero *et al.*, 2005). This group includes the Bell frogs: *Litoria raniformis,*

L. aurea, and *L. castanea/flavipunctata* (rediscovered in 2008), and *Helioporus australiacus* and *Litoria brevipalmata*. These species are likely to have been affected by disease and habitat loss; however, the impacts of climate change on these species has not been investigated. Two of the Bell frog species (*Litoria aurea* and *L. raniformis*) have been successfully introduced into New Zealand, and despite deaths associated with the chytrid disease they are widespread and common in their non-native habitat.

21.3.9 Non-threatened species – indicators of environmental health

We have discussed the threatened species of Australia; however, they are limited to the frogs we know about, and other species may be at risk. Over 230 species of frogs have been identified in Australia. However, information is often lacking on their distributions and population sizes. Subsequently over 20 species in Australia are currently listed as Data Deficient (Hero and Morrison, 2004; IUCN, 2012). Research on the conservation status of these species is urgently required.

Monitoring rare species is important for maintaining our knowledge on the conservation status of our endangered fauna; however, research on common widespread and abundant species provide much better opportunities for measuring the impacts of climate change (Shoo *et al.*, 2005). The uncertain future and changing environments will result in species that dominate (winners), those that are unable to adapt and whose populations will decline and become locally extinct (losers), and introduced species which continue to expand their geographic range (McKinney and Lockwood, 1999). Amphibian species richness declines following human disturbance, but some species have attributes that allow persistence despite habitat modification (Parris, 2006; Hazell *at al.*, 2004; MacNally *et al.*, 2009). In SE Queensland the highly fecund species that breed in permanent ponds and streams (*Limnodynastes peronii* and *Litoria fallax*) are the dominant species encountered in suburban areas. In the city of Melbourne, *Litoria ewingii*, *Crinia signifera*, *Limnodynastes peronii* and *L. tasmaniensis* dominate the urban ponds (Parris, 2006). In agricultural landscapes of coastal southern New South Wales *Limnodynastes tasmaniensis* and *Uperolia laevigata* were more frequently encountered in farm dams than in adjacent natural pond habitats (Hazell *at al.*, 2004).

Introduced species of the Austral region include Cane Toads in Australia, Papua New Guinea and the Pacific Islands, *Litoria dentata* on Lord Howe Island, and *Litoria ewingii*, *Litoria aurea* and *L. raniformis* in New Zealand. The success of these species is not a direct threat to other amphibians however it is a direct result of their extinction.

21.4 Conclusions

Herein we have identified the strong links between biogeographic history, the ecology of austral frogs, and their vulnerability to extinction processes. Habitat loss, disease and climate change are the primary drivers of recent extinctions, however introduced species (mongoose, rats and *Gambusia*) are also a major threat for some species. The vast majority of the threatened species have small clutch sizes and restricted distributions (relics of biogeographic history), however there are several exceptions.

The endemic frogs of New Zealand are ancient relics of the past and are threatened by habitat loss and climate change. The frogs of the Pacific islands are primarily threatened by introduced species including mongoose, rats and Cane Toads. In Australia the threatened species can be split into habitat specialists (WA endemics, acid frogs, two species of

Pseudophryne, and mountain-top endemics of eastern Australia) that are primarily threatened by habitat loss, disease and climate change. Among the stream-dwelling frogs of eastern Australia, there are many species that are restricted to relatively undisturbed isolated wet-forest fragments. This species group is highly vulnerable to the disease chytridiomycosis which appears to be the primary driver of extinction; however, the synergistic impacts of climate change may also be a major contributor. Interdisciplinary research teams are needed to focus on the potential links between disease as a direct or proximate cause and environmental change as an indirect and ultimate cause of population decline.

The final group of threatened species are widespread species that are likely to be threatened by all of these threatening processes (habitat loss and fragmentation, disease, introduced species and climate change) which are intrinsically linked to each other. Research is urgently needed to understand these synergies using multi-scale, multi-disciplinary approaches.

The solutions to most of these threats are obvious, stop land clearing, reduce CO_2 emissions and control feral animals, however declines linked to the disease chytridiomycosis are not easily resolved. Chytridiomycosis is not a universal threat and understanding the causes of variation in impact is critically important.

Mitigating threats require an extensive range of *in-situ* and *ex-situ* management actions. Mitigate climate change, halt land-clearing, *ex-situ* conservation (captive husbandry) for species on the brink. Hero and Shoo (2003) proposed establishing a global network of 'Amphibian Research and Conservation Centers' to facilitate conservation initiatives: (1) coordinate training for local scientists, (2) document species diversity, abundance and endemism, (3) map the distribution of species at the regional scale and focus conservation efforts in areas of high species richness and endemism, (4) identify threatened and geographically restricted species and set priorities for conservation and protection, (5) study the ecology of threatened species to assist management agencies, (6) monitor threatened species and habitat loss across ecological, latitudinal and altitudinal gradients, (7) restore habitats and fragmented landscapes, and (8) provide information to local communities and government agencies to promote effective amphibian conservation. Since 2003, several international initiatives have been initiated, including the Amphibian Survival Alliance (Mendelson *et al.*, 2006; Gascon *et al.*, 2013), the Amphibian Ark (Zippel *et al.*, 2011) and an international NGO 'Save the Frogs' dedicated to the conservation of frogs. While these initiatives are exemplary they are primarily dependent on public donations and subsequently they are poorly funded. The survival of the Earth depends on frogs: if we save the frogs, we save ourselves.

REFERENCES

Allison, A. (1996). Zoogeography of amphibians and reptiles of New Guinea and the Pacific region. In: *The Origin and Evolution of Pacific Island Biotas, New Guinea to Eastern Polynesia*, (Eds.) Keast, A. and Miller, S. E. SPB Academic Publishing, Amsterdam.

Allison, A. (2009). Patterns of geographical distribution: animals. In: *Oceans and Aquatic Ecosystems*, (Ed.) E. Wolanskiay, Vol. **2**, pp. 358. Encyclopedia of Life Support Systems (EOLSS).

Bamford, M. J., Roberts, J. D. (2003). The impact of fire on frogs and reptiles in south-western Australia. In: *Fire in Ecosystems of South-West Western Australia: Impacts and*

Management, (Eds.) Abbott, I. and Burrows, N. Backhuys Publishers: Leiden, The Netherlands, pp. 349–361.

Bell, B. D. (1994). A review of the status of New Zealand *Leiopelma* species (Anura: Leiopelmatidae), including a summary of demographic studies in Coromandel and on Maud Island. *New Zealand Journal of Zoology*, **21**, 341–349.

Bell, B. D., Carver, S., Mitchell, N. J., Pledger, S. (2004). The recent decline of a New Zealand endemic: how and why did populations of Archey's frog *Leiopelma archeyi* crash over 1996–2001? *Biological Conservation*, **120**, 189–199.

Bell, R. C., MacKenzie, J. B., Hickerson, M. J. *et al.* (2012). Comparative multi-locus phylogeography confirms multiple vicariance events in co-distributed rainforest frogs. *Proceedings of the Royal Society of London B*, **279**, 991–999.

Berger, L., Speare, R., Daszak, P. *et al.* (1998). Chytridiomycosis causes amphibian mortality associated with population declines in the rainforest of Australia and Central America. *Proceedings of the National Academy of Sciences*, **95**, 9031–9036.

Bishop, P. J., Daglish, L. A., Haigh, A. J. M. *et al.* (2013). Native frog (*Leiopelma* species) recovery plan, 2013–2018. DOC, Threatened Species Recovery Plan (in press).

Blaustein, A. R., Gervasi, S. S., Johnson, P. T. J. *et al.* (2012). Ecophysiology meets conservation: understanding the role of disease in amphibian population declines. *Philosophical Transactions of the Royal Society of London B*, **367**, 1596–1688.

Brown, R. M., Foufopoulos, J., Richards. S. J. (2006). New species of *Platymantis* (Amphibia; Anura; Ranidae) from New Britain and redescription of the poorly known *Platymantis nexipus*. *Copeia*, **4**, 674–695.

Burbidge, A. H., Rolfe, J. K., McKenzie N. L., Roberts, J. D. (2004). Biogeographic patterns in small ground-dwelling vertebrates of the Western Australian wheatbelt. In: *A Biodiversity Survey of the Western Australian Agricultural Zone*, (eds.) G. J. Keighery, S. A. Halse, M. S. Harvey and N. L. McKenzie, pp. 139–202. Records of the Western Australian Museum Supplement No. 67, Western Australian Museum, Kewdale, Western Australia.

Byrne, M., Steane, D. A. Joseph, L. *et al.* (2011). Decline of a biome: evolution, contraction, fragmentation, extinction and invasion of the Australian mesic zone biota. *Journal of Biogeography*, **38**, 1635–1656.

Coaldrake, J. E. (1961). The ecosystem of coastal lowlands ('wallum') of southern Queensland. *CSIRO Bulletin, Melbourne*, **283**, 1–138.

Corn, P. S. (2005). Climate change and amphibians. *Animal Biodiversity and Conservation*, **28**, 59–67.

Couper, P. J., Hoskin, C. J. (2008). Litho-refugia: the importance of rock landscapes for the long-term persistence of Australian rainforest fauna. *Australian Zoologist*, **34**, 554–60.

Cullen, L. E., Grierson, P. F. (2009). Multi-decadal scale variability in autumn-winter rainfall in south-western Australia since 1655 AD as reconstructed from tree rings of *Callitris columellaris*. *Climate Dynamics*, **33**, 433–444.

Czechura G. V., Ingram, G. J. (1990). *Taudactylus diurnus* and the case of the disappearing frogs. *Memoirs of the Queensland Museu*, **29**, 361–365.

EDGE website (2013). http://www.edgeofexistence.org/amphibians/top_100.php. Accessed 15 June 2013.

Edwards, D., Roberts, J. D. (2011). Genetic diversity and biogeographic history inform future conservation management strategies for the rare Sunset Frog (*Spicospina flammocaerulea*). *Australian Journal of Zoology*, **59**, 63–72.

Edwards, D., Roberts J. D., Keogh, S. J. (2007). Impact of Plio-Pleistocene arid cycling on the population history of a southwestern Australian frog. *Molecular Ecology*, **16**, 2782–2796.

Farrer, R. A., Weinert, L. A., Bielby, J. *et al*. (2011). Multiple emergences of genetically diverse amphibian-infecting chytrids include a globalized hyper-virulent recombinant lineage. *Proceedings of the National Academy of Sciences*, **108**, 18 732–18 736.

Fisher, M. C., Garner, T. W. J., Walker, S. F. (2009). Global emergence of *Batrachochytrium dendrobatidis* and amphibian chytridiomycosis in space, time and host. *Annual Review of Microbiology*, **63**, 291–310.

Fouquet, A., Ficetola, G. F., Haigh, A., Gemmell, N. (2010). Using ecological niche modelling to infer past, present and future environmental suitability for *Leiopelma hochstetteri*, an endangered New Zealand native frog. *Biological Conservation*, **143**, 1375–1384.

Frost, D. R., Donnellan, S. C., Raxworthy, C. *et al*. (2006). The amphibian tree of life. *Bulletin of the American Museum of Natural History*, **297**, 1–370.

Gascon, C., Collins, J. P., Church, D. R. *et al*. (2013). Scaling a global plan into national strategies for amphibian conservation. *Alytes*, (in press).

Gillespie, G. (2001). The role of introduced trout in the decline of the spotted tree frog (*Litoria spenceri*) in south-eastern Australia. *Biological Conservation*, **100**, 187–198.

Gillespie, G., Hero, J. M. (1999). Potential impacts of introduced fish and fish translocations on Australian amphibians. In: *Declines and Disappearances of Australian Frogs*, Campbell, A. (ed.). Canberra: Environment Australia. Pages 131–144.

Griffith, S. J., Bale, C., Adam, P. (2008). Environmental correlates on coastal heath and allied vegetation. *Australian Journal of Botany*, **56**, 512–526.

Griffith, S. J., Bale, C., Adam P., Wilson, R. (2003). Wallum and related vegetation on the NSW North Coast: description and phytosociological analysis. *Cunninghamia*, **8**, 202–252.

Hauselberger, K. F., Alford, R. A. (2012). Prevalence of *Batrachochytrium dendrobatidis* infection is extremely low in direct-developing Australian microhylids. *Diseases of Aquatic Organisms*, **100**, 191–200.

Hazell, D., Hero, J.-M., Lindenmayer, D., Cunningham, R. (2004). A comparison of constructed and natural habitat for frog conservation in an Australian agricultural landscape. *Biological Conservation*, **119**, 61–71.

Hennessy, K. J., Lucas, C., Nicholls, N. *et al*. (2005). *Climate Change Impacts on Fire-Weather in South-East Australia*. Melbourne, Australia: CSIRO Marine and Atmospheric Research.

Hennessy, K. J., Fitzharris, B., Bates, B. C. *et al*. (2007). Australia and New Zealand. In: *Climate Change 2007: Impacts, Adaptation and Vulnerability*, (Eds.) Parry, M. L., Canziani, O. F., Palutikof, J. P. *et al*. Contribution of Working Group II to the Fourth Assessment Report of the Intergovernmental Panel on Climate Change. Cambridge, UK: Cambridge University Press.

Hero J.-M., Morrison, C. (2004). Frog declines in Australia: global implications. *The Herpetological Journal*, **14**, 175–186.

Hero J.-M., Morrison, C. (2012). Life history correlates of extinction risk in amphibians, Chapter 10. In: *Amphibian Biology, Vol. 10, Conservation and Decline of Amphibians: Ecological Aspects, Effect of Humans, and Management*, (Eds.) H. Heatwole and J. W. Wilkinson. Surrey Beatty and Sons, NSW, pp. 3567–3576.

Hero, J.-M., Shoo, L. (2003). Conservation of amphibians in the Old World tropics: defining unique problems associated with regional fauna, Chapter 6. In: *Amphibian Conservation*, (Ed.) R. D. Semlitsch. Smithsonian Institution Press, Washington, D.C., pp. 70–84.

Hero, J.-M., Williams S. E., Magnusson, W. E. (2005). Ecological traits of declining amphibians in upland areas of eastern Australia. *Journal of Zoology London*, **267**, 221–232.

Hero, J.-M., Morrison, C., Gillespie, G. *et al.* (2006). Overview of the conservation status of Australian Frogs. *Pacific Conservation Biology*, **12**, 313–320.

Hero, J.-M., Richards, S., Alford, R. A. *et al.* (2008). Amphibians of the Australasian Realm. Chapter 6, in *Threatened Amphibians of the World*. Lynx Ediciones, pp. 65–70.

Hero, J.-M., Morrison, C., Chanson, J. *et al.* (2012). Phylogenetic correlates of extinction risk in amphibians, Chapter 8. In: *Amphibian Biology, Vol. 10, Conservation and Decline of Amphibians: Ecological Aspects, Effect of Humans, and Management*, (Eds.) H. Heatwole and J. W. Wilkinson. Surrey Beatty and Sons, NSW, pp. 3539–3551.

Hines, H., Mahony M., Mcdonald, K. (1999). An assessment of frog declines in wet subtropical Australia. In: *Decline and Disappearances of Australian Frogs*, (Ed.) Campbell, A. Environment Australia.

Hines, H. B., Meyer, E. A. (2011). The frog fauna of Bribie Island: an annotated list and comparison with other Queensland dune islands. *Proceedings of the Royal Society of Queensland*, **117**, 261–274.

Hof, C., Araújo, M. B., Jetz, W., Rahbek, K. (2011). Additive threats from pathogens, climate and land-use change for global amphibian diversity. *Nature*, **480**, 516–519.

Hoffmann, M., Hilton-Taylor, C., Angulo, A. *et al.* (2010). The impact of conservation on the status of the world's vertebrates. *Science*, **330**, 1503–1509.

Hollis, G. J. (2004). *Ecology and Conservation Biology of the Baw Baw frog* Philoria frosti *(Anura: Myobatrachidae): Distribution, Abundance, Autoecology and Demography*. PhD Thesis, Department of Zoology, University of Melbourne.

Hoskin, C. J. (2004). Australian microhylid frogs (*Cophixalus* and *Austrochaperina*): phylogeny, taxonomy, calls, distributions and breeding biology. *Australian Journal of Zoology*, **52**, 237–269.

Hoskin, C. J. (2007). Description, biology and conservation of a new species of Australian tree frog (Amphibia: Anura: Hylidae: *Litoria*) and an assessment of the remaining populations of *Litoria genimaculata* Horst, 1883: systematic and conservation implications of an unusual speciation event. *Biological Journal of the Linnean Society*, **91**, 549–563.

Hoskin, C. J. (2008). A key to the microhylid frogs of Australia, and new distributional data. *Memoirs of the Queensland Museum*, **53**, 233–247.

Hoskin, C. J. (2012). Two new frog species (Microhylidae: *Cophixalus*) from the Australian Wet Tropics region, and redescription of *Cophixalus ornatus*. *Zootaxa*, **3271**, 1–16.

Hoskin, C. J., Aland, K. (2011). Two new frog species (Microhylidae: *Cophixalus*) from boulder habitats on Cape York Peninsula, north-east Australia. *Zootaxa*, **3027**, 39–51.

Hoskin C. J., Higgie, M. (2005). Minimum calling altitude of *Cophixalus* frogs on Thornton Peak, northeastern Queensland. *Memoirs of the Queensland Museum*, **51**, 572.

Hoskin, C. J., Hero, J.-M. (2008). *Rainforest Frogs of the Wet Tropics, North-East Australia*. Griffith University, Gold Coast, 96 pages.

Ingram, G. J., Corben, C. J. (1975). The frog fauna of North Stradbroke Island, with comments on the 'acid' frogs of the wallum. *Proceedings of the Royal Society of Queensland*, **86**, 49–54.

Ingram, G. J., Mcdonald, K. R. (1993). An update on the declines of Queensland's frogs. In: *Herpetology in Australia: A Diverse Discipline*, (Eds.) Lunney, D. and Ayers, D. Mosman, NSW: Royal Zoological Society of New South Wales.

IUCN (2012). *IUCN Red List of Threatened Species*. Version 2012.2. www.iucnredlist.org. Downloaded on 12 June 2013.

Janicke, J., Roberts J. D. (2010). *Litoria cyclorhyncha* (Spotted Thighed Frog). Saline water. *Herpetological Review*, **41**, 199–200.

Kikkawa, J., Ingram, G. J., Dwyer P. D. (1979). The vertebrate fauna of Australian heathlands: an evolutionary perspective. In: *Ecosystems of the World 9A. Heathlands and Related Shrublands*, (ed.) Specht, R. L. Amsterdam: Elsevier Scientific Publishing Company.

Kilpatrick, A. M., Briggs, C. J., Daszak, P. (2010). The ecology and impact of chytridiomycosis: an emerging disease of amphibians. *Trends in Ecology and Evolution*, **25**, 109–118.

Knowles, R., Mahony, M. Armstrong, J. *et al.* (2004). Systematics of Sphagnum Frogs of the genus *Philoria* (Anura: Myobatrachidae) in eastern Australia, with the description of two new species. *Records of the Australian Museum*, **56**, 57–74.

Komak, S., Crossland. M. R. (2000). An assessment of the introduced mosquitofish (*Gambusia affinis holbrooki*) as a predator of eggs, hatchlings and tadpoles of native and non-native anurans. *Wildlife Research*, **27**, 185–189.

Kriger, K. M., Hero, J.-M. (2006). Survivorship in wild frogs infected with chytridiomycosis. *EcoHealth*, **3**, 171–177.

Kriger, K. M., Hero, J.-M. (2007a). The chytrid fungus *Batrachochytrium dendrobatidis* is non-randomly distributed across amphibian breeding habitats. *Diversity and Distributions*, **13**, 781–788.

Kriger, K. M., Hero, J.-M. (2007b). Large-scale seasonal fluctuations in the prevalence and severity of chytridiomycosis. *Journal of Zoology, London*, **271**, 352–359.

Kriger, K. M., Hero, J.-M. (2008). Altitudinal distribution of chytrid (*Batrachochytrium dendrobatidis*) infection in 2 subtropical Australian frogs. *Austral Ecology*, **33**, 1022–1032.

Kriger, K. M., Pereoglou, F., Hero J.-M. (2007). Latitudinal variation in the prevalence and severity of chytrid (*Batrachochytrium dendrobatidis*) infection in Eastern Australia. *Conservation Biology*, **21**, 1280–1290.

Kuruyawa, J., Osborne, T., Thomas, N. *et al.* (2004). *Distribution, abundance and conservation status of the Fijian Ground Frog (Platymantis vitianus).* Unpublished Technical Report for the BP Conservation Programme, 16 pages.

Lewis, B. D., Goldingay, R. L. (2005). Population monitoring of the vulnerable wallum sedge frog (*Litoria olongburensis*) in north-eastern New South Wales. *Australian Journal of Zoology*, **53**, 185–194.

Li, Y., Coheb, J. M., Rohr, J. R. (2013). Review and synthesis of the effects of climate change on amphibians. *Integrative Zoology*, **8**, 145–161.

Liu, X., Rohr J. R., Li, Y. (2013). Climate, vegetation, introduced hosts and trade shape a global wildlife pandemic. *Proceedings of the Royal Society of London B*, **280**, 20122506.

Lowe, K., Castley J. G., Hero, J.-M. (2013). Acid frogs can stand the heat: amphibian resilience to wildfire in coastal wetlands of eastern Australia. *International Journal of Wildland Fire.* http://dx.doi.org/10.1071/WF12128.

MacNally R., Horrocks, G., Lada, H. *et al.* (2009). Distribution of anuran amphibians in massively altered landscapes in south-eastern Australia: effects of climate change in an aridifying region. *Global Ecology and Biogeography*, **18**, 575–585.

MacNally, R., Nerenberg, S., Thomson, J. R., Lada, H., Clarke, R. H. (2013). Do frogs bounce, and if so, by how much? Responses to the 'Big Wet' following the 'Big Dry' in south-eastern Australia. *Global Ecology and Biogeography* available on-line, DOI: 10.1111/geb.12104.

Maxson, L. R. (1992). Tempo and pattern in anuran speciation and phylogeny: an albumin perspective. *Herpetology: Current Research on the Biology of Amphibians and Reptiles*, pp. 41–57.

McKinney, M. L., Lockwood, J. L. (1999). Biotic homogenization: a few winners replacing many losers in the next mass extinction. *Trends in Ecology and Evolution*, **14**, 450–453.

Mendelson, J. R., Lips, K. R. Gagliardo, R. W. *et al.* (2006). Biodiversity – confronting amphibian declines and extinctions. *Science*, **313**, 5783.

Meyer, E., Hero, J. M., Shoo, L., Lewis, B. (2006). *National Recovery Plan for the Wallum Sedgefrog and other Wallum-dependent Frog Species.* Report to Department of the Environment and Water Resources, Canberra: Queensland Parks and Wildlife Service, Brisbane.

Morgan, M., Keogh, S., Roberts, J. D. (2007). Molecular phylogenetic dating supports an ancient endemic speciation model in Australia's biodiversity hotspot. *Molecular Phylogenetics and Evolution*, **44**, 371–385.

Moritz, C., Agudo, R. (2013). The future of species under climate change: resilience or decline? *Science*, **341**, 504–508.

Morrison, C. (2003). *A Field Guide to the Herpetofauna of Fiji.* Institute of Applied Sciences, University of the South Pacific.

Morrison, C., Hero, J.-M. (2012). Geographic correlates of extinction risk in amphibians, Chapter 9. In *Amphibian Biology*, Vol. 10, *Conservation and Decline of Amphibians: Ecological Aspects, Effect of Humans, and Management*, (Eds.) H. Heatwole and J. W. Wilkinson. Surrey Beatty and Sons, NSW, pp. 3552–3556.

Murray, K. A., Rosauer, D., McCallum, H., Skerrat, L. F. (2011). Integrating species traits with extrinsic threats: closing the gap between predicting and preventing species declines. *Proceedings of the Royal Society of London, B*, **278**, 1515–1523.

Murray K. A., Skerratt, L. F., Garland, S., Kriticos D., McCallum, H. (2013). Whether the weather drives patterns of endemic amphibian chytridiomycosis: a pathogen proliferation approach. *PLoS ONE*, **8**, e61061.

Muths, E., Hero, J.-M. (2010). Amphibian declines: promising directions in understanding the role of disease. *Animal Conservation*, **13**, 33–35.

Narayan, E., Hero, J.-M., Christi, K., Morley, C. (2011). Early developmental biology of *Platymantis vitiana* including supportive evidence of structural specialization unique to the ceratobatrachidae. *Journal of Zoology, London*, **284**, 68–75.

Narayan, E., Cockrem J. F., Hero, J.-M. (2013). Sight of a predator induces a corticosterone stress response and generates fear in an amphibian. *PLoS One*, **8**, e73564.

Newman, D. (1996). *Native Frog* (Leiopelma *spp.*) *Recovery Plan*. Threatened Species Recovery Plan No. 18, Department of Conservation, Wellington. 35 p.

Ohmer, M. E., Herbert, S. M., Speare, R., Bishop, P. J. (2013). Experimental exposure indicates the amphibian chytrid pathogen poses low risk to New Zealand's threatened endemic frogs. *Animal Conservation*, **16**, 422–429.

Oza A. U., Lovett, K. E., Williams, S. E., Moritz. C. (2012). Recent speciation and limited phylogeographic structure in *Mixophyes* frogs from the Australian Wet Tropics. *Molecular Phylogenetics and Evolution*, **62**, 407–413.

Parmesan, C., Singer, M. C. (2008). Amphibian extinctions: disease not the whole story. *Proceedings of the National Academy of Sciences*, **105**, 17 436–17 441.

Parris, K. M. (2006). Urban amphibian assemblages as metacommunities. *Journal of Animal Ecology*, **75**, 757–764.

Parris, K. M., Lindenmayer, D. B. (2004). Evidence that creation of a *Pinus radiata* plantation in south-eastern Australia has reduced habitat for frogs. *Acta Oecologica-International Journal of Ecology*, **25**, 93–101.

Pernetta, J. C., Watling, D. (1979). The introduced and native terrestrial vertebrates of Fiji. *Pacific Science*, **32**, 223–244.

Pounds, J. A., Bustamante, M. R., Coloma, L. A. *et al.* (2006). Widespread amphibian extinctions from epidemic disease driven by global warming. *Nature*, **439**, 161–167.

Puschendorf, R., Carnaval, A. C., VanDerWal, J. *et al.* (2009). Distribution models for the amphibian Chytrid *Batrachochytrium dendrobatidis* in Costa Rica: proposing climatic refuges as a conservation tool. *Diversity and Distributions*, **15**, 401–408.

Puschendorf, R., Hoskin, C. J., Cashins, S. D. *et al.* (2011). Environmental refuge from disease-driven amphibian extinction. *Conservation Biology*, **25**, 956–964.

Pyron, R. A., Wiens, J. J. (2011). A large-scale phylogeny of Amphibia including over 2800 species, and a revised classification of extant frogs, salamanders, and caecilians. *Molecular Phylogenetics and Evolution*, **61**, 543–583.

Raffel, T. R., Romansic, J. M., Halstead, N. T. *et al.* (2013). Disease and thermal acclimation in a more variable and unpredictable climate. *Nature Climate Change*, **3**, 146–151.

Read, K., Keogh, J. S., Scott, I. A. W., Roberts J. D., Doughty. P. (2001). Molecular phylogeny of the Australian frog genera *Crinia* and *Geocrinia* and allied taxa (Anura: Myobatrachidae). *Molecular Phylogenetics and Evolution*, **21**, 294–308.

Reading, C. J. (2007). Linking global warming to amphibian declines through its effects on female body condition and survivorship. *Oecologia*, **151**, 125–131.

Richards, S. J., McDonald, K. R., Alford, R. A. (1993). Declines in populations in Australia's endemic tropical rainforest frogs. *Pacific Conservation Biology*, **1**, 66–77.

Riley, K., Berry O. F., Roberts, J. D. (2013). Do global models predicting environmental suitability for the amphibian fungus, *Batrachochytrium dendrobatidis*, have local value to conservation managers? *Journal of Applied Ecology*, **50**, 713–720

Roberts, J. D. (1993). Hybridisation between the western and northern call races of the *Limnodynastes tasmaniensis* complex (Anura: Myobatrachidae) on the Murray River in South Australia. *Australian Journal of Zoology*, **41**, 101–122.

Roberts, J. D., Conroy, S., Williams, K. (1999). Conservation status of frogs in Western Australia. In: *Declines and Disappearances of Australian Frogs*, (ed.) A. Campbell. Environment Australia, Canberra, pp. 177–184.

Rohr J. R., Palmer, B. D. (2013). Climate change, multiple stressors, and the decline of ectotherms. *Conservation Biology*, **4**, 741–751.

Rohr, J. R., Raffel, T. R. (2010). Linking global climate and temperature variability to widespread amphibian declines putatively caused by disease. *Proceedings of the National Academy of Sciences*, **107**, 8269–8274.

Rohr, J. R., Raffel, T. R., Romansic, J. M., McCallum H., Hudson, P. J. (2008). Evaluating the links between climate, disease spread, and amphibian declines. *Proceedings of the National Academy of Sciences*, **105**, 17 436–17 441.

Rohr, J. R., Halstead N. T., Raffel, T. R. (2011). Modelling the future distribution of the amphibian chytrid fungus: the influence of climate and human-associated factors. *Journal of Applied Ecology*, **48**, 174–176.

Rowley J. L., Alford, R. A. (2013). Hot bodies protect amphibians against chytrid infection in nature. *Scientific Reports*, **3**, 1515 DOI: 10.1038/srep01515

Ryan, P. (1988). *Fiji's Natural Heritage*. South-Western Publishing Limited, Auckland, New Zealand, pp. 11, 113.

Schloegel, L., Hero, J.-M., Berger, L. *et al.* (2006). The decline of the sharp-snouted day frog (*Taudactylus acutirostris*): the first documented case of extinction by infection in a free-ranging wildlife species? *Ecohealth*, **3**, 35–40.

Seton, M., Müller, R. D., Zahirovic, S. *et al.* (2012). Global continental and ocean basin reconstructions since 200 Ma. *Earth-Science Reviews*, **113**, 212–270.

Shaw, S. D., Bishop, P. J., Berger, L. *et al.* (2010). Experimental infection of self-cured *Leiopelma archeyi* with the amphibian chytrid, *Batrachochytrium dendrobatidis*. *Diseases of Aquatic Organisms*, **92**, 159–163.

Shaw, S. D., Skerratt, L. F., Haigh, A. *et al.* (2013). The distribution and host range of *Batrachochytrium dendrobatidis* in New Zealand spanning surveys from 1930–2010. *New Zealand Journal of Ecology* (in press).

Shoo, L. P., Williams, Y. (2004). Altitudinal distribution and abundance of microhylid frogs (*Cophixalus* and *Austrochaperina*) of north-eastern Australia: baseline data for detecting biological responses to future climate change. *Australian Journal of Zoology*, **52**, 667–676.

Shoo, L. P., Williams S., Hero, J.-M. (2005). Decoupling of trends in distribution area and population size of species with climate change. *Global Change Biology*, **11**, 1469–1476.

Shoo, L. P., Storlie, C., Williams, Y. M., Williams, S. E. (2009). Potential for mountaintop boulder fields to buffer species against extreme heat stress under climate change. *International Journal of Biometeorology*, **54**, 475–478.

Shoo, L. P., Olson, D. H., McMenamin, S. K. *et al.* (2011). Engineering a future for amphibians under climate change. *Journal of Applied Ecology*, **48**, 487–492.

Shuker, J. D., Hero, J. M. (2012). Perch substrate use by the threatened wallum sedge frog (*Litoria olongburensis*) in wetland habitats of mainland eastern Australia. *Australian Journal of Zoology*, **60**, 219–224.

Simpkins, C., Shuker, J. D., Lollback, G. W., Castley J. G., Hero, J.-M. (2013). Environmental variables associated with the distribution and occupancy of habitat specialist tadpoles in naturally acidic, oligotrophic waterbodies. *Austral Ecology* (in press).

Skerratt, L. F., Berger, L., Speare, R., Cashins, S. *et al.* (2007). Spread of chytridiomycosis has caused the rapid global decline and extinction of frogs. *EcoHealth*, **4**, 125–134.

Smith, M. J., Schreiber, E. S. G. Scroggie, M. P. *et al.* (2007). Associations between anuran tadpoles and salinity in a landscape mosaic of wetlands impacted by secondary salinization. *Freshwater Biology*, **52**, 75–84.

Specht, R. L. (1981). Conservation: Australian heathlands. In *Ecosystems of the World* 9B, (Ed.) Specht, R. L. Heathlands and related shrublands: Analytical studies. Amsterdam: Elsevier.

Stuart, S. N. (2008). *Threatened Amphibians of the World*. Lynx Edicions, 758 pages.

Thomas, N., Morrison, C., Winder L., Morley, C. (2011). Spatial distribution and habitat preferences of co-occurring vertebrate species: case study of an endangered frog and an introduced toad in Fiji. *Pacific Conservation Biology*, **17**, 68–77.

Thurley, T., Bell, B. D. (1994). Habitat distribution and predation on a western population of terrestrial *Leiopelma* (Anura: Leiopelmatidae) in the northern King Country, New Zealand. *New Zealand Journal of Zoology*, **27**, 431–436.

Thurley, T., Haigh, A. (2008). *Hochstetter's Frog Amphibian Chytrid Fungus Survey Report*. Waikato Conservancy, Department of Conservation (unpublished).

Tucker, C. M., Cadotte, M. W. (2013). Unifying measures of biodiversity: understanding when richness and phylogenetic diversity should be congruent. *Diversity and Distributions*, **19**, 1472–4642.

Van Sluys, M., Hero, J.-M. (2010). How does chytrid infection vary among habitats? The case of *Litoria wilcoxii* (Anura, Hylidae) in SE Queensland, Australia. *EcoHealth*, **6**, 576–583.

Veron, G., Patou, M.-L., Simberloff, D., McLenachan, P. A., Morley, C. (2010). The Indian brown mongoose, yet another invader in Fiji. *Biological Invasions*, **12**, 1947–1951.

Wake, D. B., Vredenburg, V. T. (2008). Are we in the midst of the sixth mass extinction? A view from the world of amphibians. *Proceedings of the National Academy of Sciences*, **105**, 11 466–11 473.

Wassens, S, Walcott, A., Wilson, A., Freire, R. (2013). Frog breeding in rain-fed wetlands after a period of severe drought: implications for predicting the impacts of climate change. *Hydrobiologia*, **708**, 69–80.

Westgate, M. J., Driscoll, D. A., Lindenmayer, D. B. (2012). Can the intermediate disturbance hypothesis and information on species traits predict anuran responses to fire? *Oikos*, **121**, 1516–1524.

Williams, A. A., Karoly, D. J. Tapper, N. (2001). The sensitivity of Australian fire danger to climate change. *Climatic Change*, **49**, 171–191.

Williams S. E., Bolitho, E. E., Fox, S. (2003). Climate change in Australian tropical rainforests: an impending environmental catastrophe. *Proceedings of the Royal Society of London B*, **270**, 1887–1892.

Williams, S. E., Shoo, L. P., Isaac, J. L. *et al.* (2008). Towards an integrated framework for assessing the vulnerability of species to climate change. *PLoS Biology*, **6**, 2621–2626.

Zippel, K., Johnson, K., Gagliardo, R. *et al.* (2011). The Amphibian Ark: a global community for *ex situ* conservation of amphibians. *Herpetological Conservation and Biology*, **6**, 340–352.

Zweifel, R. G. (1985). Australian frogs of the family Microhylidae. *Bulletin of the American Museum of Natural History*, **182**, 265–388.

CHAPTER 22

Predators in danger: shark conservation and management in Australia, New Zealand and their neighbours

Paolo Momigliano, Vanessa Flora Jaiteh and Conrad Speed

Summary

In this chapter we examine the biodiversity and the status of conservation and management of shark species in Australasia and Indonesia. Almost 17% of shark species in the region are listed by the International Union for the Conservation of Nature (IUCN) as threatened, and approximately 40% are of conservation concern, their future being dependent on the implementation of appropriate management strategies.

Overfishing is a major threat to sharks, as their life-history strategies make them susceptible to even modest levels of fishing mortality. In Australia and New Zealand many shark stocks experienced dramatic declines as a consequence of overfishing; however, in the past few decades substantial improvements in the management of shark fisheries have taken place. On the other hand, shark fishing in Indonesia is largely unreported and unregulated and fishing by Indonesian vessels is likely to have consequences that go beyond the depletion of local populations, affecting shark populations in neighbouring countries such as Australia.

We illustrate examples of overfishing in the region, discuss the potential effects of habitat degradation and climate change in the future and examine current management frameworks for the conservation of shark species in the region with an emphasis on the implementation of Nation Plans of Action for the Conservation and Management of Sharks (NPoAs).

Austral Ark: The State of Wildlife in Australia and New Zealand, eds. A. Stow, N. Maclean and G. I. Holwell. Published by Cambridge University Press. © Cambridge University Press 2015.

22.1   Shark biodiversity in the Indo-Australasian region

The history of shark evolution has been one of enduring success. Since their first appearance in the late Silurian more than 400 MYA (Last & Stevens, 2009), sharks have colonised most of the marine realm, from the clear tropical waters of shallow coral reefs to the freezing depths of the Arctic Circle (Benz *et al.*, 2004). With over 500 extant species representing eight orders and more than 30 families, sharks are an extremely diverse group (Compagno, 2001). A few shark species (approximately 5%) are oceanic species with a circumglobal distribution while the vast majority of taxa are comprised of demersal species, which inhabit the tropical and temperate continental shelves and slopes (Tittensor *et al.*, 2010).

At the global scale, patterns of shark species richness are governed by a small set of environmental variables, with sea surface temperature (SST), coastline length and primary productivity being the main drivers of diversity (Lucifora *et al.*, 2011; Tittensor *et al.*, 2010). Shark species richness is highest at mid latitudes (20°–40°) in coastal areas with high productivity (Lucifora *et al.*, 2011; Tittensor *et al.*, 2010). Indo-Australasia, the geographic region including Indonesia, Australia, New Guinea, New Zealand and New Caledonia supports an extraordinary diversity of shark species. Approximately 240 species are found on the continental shelves and slopes of this region, accounting for nearly 50% of global diversity. Within the region, diversity is highest in Australia (182 species), which alone accounts for 36% of global biodiversity (Last & Stevens, 2009). Indonesia, with approximately 100 species is second, and the remaining subregions (New Zealand, New Guinea and New Caledonia) follow with approximately 60 species (White & Kyne, 2010). The highest species richness is found in the temperate continental shelf of New South Wales, the tropical waters of the Great Barrier Reef on the eastern Australian coast and the coral reefs of Western Australia (Plate 12). A large proportion of the shark species that inhabit the region are found nowhere else in the world, and Australia has the highest proportion of endemic species (39%).

Diversity and endemism are very high for a few families of sharks. One of these families are the Orectolobidae (wobbegongs), which are bottom-dwelling sharks that occur mainly in the western Pacific Ocean. This family includes 12 species, 11 of which occur in Indo-Australasia: seven are endemic to Australia, another four to the Australasian region and only one species (*Orectolobus Japonicus*) is known to occur elsewhere (Last & Stevens, 2009). Of approximately 27 species of dogfishes (Squalidae) recorded globally, 19 occur in Indo-Australasia, and 12 are found in Australia (nine of which are endemic). The Indo-Australasian region also supports 36 species of whaler sharks (Carcharhinidae), approximately 70% of global diversity with at least six endemic species. Requiem sharks are a large family that includes many apex predators of commercial importance such as the blacktip shark (*Carcharinus limbatus*) and the dusky whaler (*Carcharhinus obscurus*)(Last & Stevens, 2009). Sharks belonging to this family tend to favour warm, tropical coastal waters and species richness is highest on the tropical continental shelves of Northern Australia and Indonesia (Plate 13).

Owing to increasing fishing pressure and habitat degradation, shark populations in the region are under threat (Figure 22.1). At the time of writing, in the whole Indo-Australasian region 17% of all shark species are listed as critically endangered (CR), endangered (EN) or vulnerable (VU) by the International Union for the Conservation of Nature (IUCN)

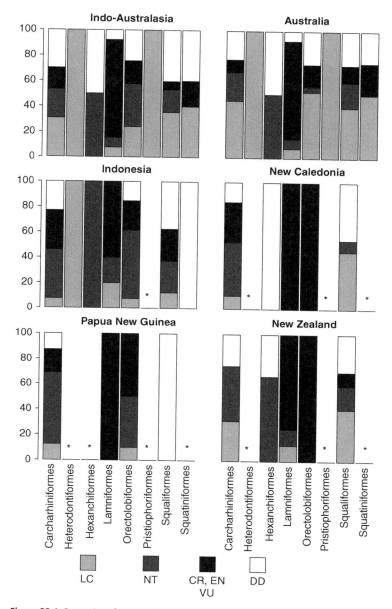

Figure 22.1 Proportion of species Threatened (CR, EN, VU), near Threatened (NT), least concern (LC) and data deficient (DD) in the Indo-Australasian region. Data obtained from the International Union for the Conservation of Nature (IUCN, http://www.iucnredlist.org/).

and approximately 40% are listed as species of conservation concern (CR, EN, VU and near Threatened – NT) (data from IUCN: http://www.iucnredlist.org/). The growing threats to shark biodiversity in the region have potential consequences that go far beyond the extinction of a few shark species. Sharks are an extremely important part of the ecosystem, and fulfil a range of ecological roles which are crucial in maintaining ecosystems. If they disappear, consequences are likely to be far-reaching (Box 22.1).

Box 22.1 The ecological consequences of shark declines

Sharks are generally classified as high trophic level predators (Figure 22.2), which are thought to structure marine communities in two main ways: (1) through the removal of prey, and (2) by altering the behaviour or distribution of prey species through predator avoidance, termed 'risk effects' (Heithaus *et al.*, 2009). The role of sharks and the effects of their removal have been modelled based on dietary studies for some economically important species (Kitchell *et al.*, 2002). In particular, there has been an increase in studies that have addressed the ecosystem effects of declines in shark populations, which are due to overwhelming increases in shark removal, particularly in coastal areas (Myers *et al.*, 2007; Robbins *et al.*, 2006). The effects of predator removal have been examined by quantifying the abundance and distribution of sharks and associated prey or benthic communities (Friedlander & DeMartini, 2002). One example found that fishing pressure on north Australian reefs, by artisanal fishers predominantly from Indonesia (see Box 22.3), led to total absences and decreased numbers of large species of sharks (Field *et al.*, 2009b). Studies of pristine vs. fished reef communities in Hawaii have also noted comparably fewer large predators on fished reefs, and an increase in biomass of primary (herbivores) and secondary consumers (planktivores), and primary producers (macroalgae) (Friedlander and DeMartini, 2002; Sandin *et al.*, 2008). However, the implications of such studies can often be difficult to interpret, due to differences in localised biological and environmental variables among geographic areas.

In some instances, the removal of apex predators has led to an increase in meso-predator populations, known as 'predator release' (Myers *et al.*, 2007), or even resulted in trophic cascades (Heithaus *et al.*, 2008). A study of coastal elasmobranch (sharks and rays) communities in the northwest Atlantic found that there was a dramatic decrease in abundance of large apex predators (e.g. bull, tiger, scalloped hammerhead

Apex
Predators

Meso
Predators

Primary
consumers

Primary
producers

Figure 22.2 Schematic representation of a marine food web, showing sharks as high trophic level predators.

sharks) and a concordant increase in mesopredators that are preyed upon by apex predators (e.g. cownose rays, little skates, and chain catsharks) (Myers *et al.*, 2007). This increase in mesopredators also coincided with a decrease in abundance of bay scallops, which are a known prey item of the cownose ray, leading to the collapse of the bay scallop fishery, which had been operating for a century.

A more recent study found that the lack of reef sharks on remote fished reefs in north-western Australia led to both predator release and trophic cascades (Ruppert *et al.*, 2013), which showed a stark contrast to healthy unfished reefs in the region. Community-level responses on fished reefs were shown to be a result of both top-down effects due to shark removal and bottom up effects from cyclones and coral bleaching. The authors also provide evidence that shark removal not only created mesopredator release, but also reduced the number of primary consumers (herbivores).

The complex behavioural interactions between sharks and their prey can provide further insights into how sharks indirectly structure prey communities and prey resource availability. For example, studies on the west Australian coast have found that the presence of tiger sharks affects community dynamics of their preys (dolphins, dugongs, turtles, seabirds), which respond by moving away from prey-rich shallow habitats when sharks are present, into deeper habitats with lower prey abundance (Heithaus *et al.*, 2009). This shift in behaviour could potentially affect growth or reproduction potential of prey species due to inadequate use of resources, and also result in a reduction of seagrass grazing in dangerous areas and intensification of grazing in safe habitats (Heithaus *et al.*, 2008).

Shifts in prey community structure are possibly a result of both removal of prey as well as risk effects, although there is some controversy surrounding this hypothesis due to non-convergence of dietary links and differences in the distribution of meso- and apex predators (Heithaus *et al.*, 2010). The role sharks play in shaping marine ecosystems is therefore likely related to their abundance, distribution and foraging strategies, as well as prey removal. Our current limited understanding of these processes and their importance to marine systems is of concern given rapidly declining shark populations and increasing human populations within the Indo-Pacific region. Determining the functional role of differing trophic groups (secondary consumers, meso-, and apex predators) is essential to assessing how marine systems will cope with reduced numbers of sharks in the future.

22.2 Sharks' life-history strategies: implications for management and conservation

Sharks are usually long-lived, slow growing and take a long time to reach sexual maturity; however, considerable differences in life-history traits do exist. The spiny dogfish (*Squalus achantias*) for example can live up to 70 years and reaches sexual maturity at more than 20 years, but age at maturity can be as low as one to two years for the Australian sharpnose shark (*Rhizoprionodon taylori*) (Last & Stevens, 2009). In general, age at maturity has a bimodal distribution, with most species reaching maturity at 5–6 years or 15–25 years. Fertilisation is internal in all shark species, and embryonic development follows three distinct patterns (Wourms & Demski, 1993). In some species (most whaler sharks and hammerhead sharks) the embryos develop within one of the uteri and are attached to a yolk-sac placenta from which they receive nutrients (placental viviparity). In ovoviviparous species (such as cow sharks – family Hexanchidae), the embryos develop in the uterus but food is supplied by large yolk-sacs, while oviparous species such as horned sharks (Heterodontidae) lay eggs protected in leathery cases where the embryos develop (Last & Stevens, 2009). These strategies all require considerable energetic investments and result in the production of a small number of well-developed pups with high survival rates. Reproductive rates are variable among species. In some taxa females are able to produce a new litter every year, while other species require one or more years of 'rest' between pregnancies. The school shark (*Galeorhinus galeus*), a species of low productivity and high commercial importance in Australia, reproduces every third year (Lucifora *et al.*, 2004). The vast majority of species produce a small number of pups (5–15) (Wourms & Demski, 1993).

These life-history strategies differ substantially when compared to most commercially harvested teleosts (bony fish), and these differences bear important implications for conservation and fisheries management. Teleosts have extremely high fecundity, producing thousands to millions of eggs per year. Mortality is very high and density-dependent, and the result is that recruitment to adult populations is broadly independent of adult population size (Shepherd & Cushing, 1980). As density decreases due to fishing mortality, resources become more abundant and competition decreases, boosting productivity through density-dependent compensations on recruits' survival. Therefore, teleost adult populations can be significantly depleted without negative effects on recruitment. In sharks, however, stock size and recruitment are closely linked and since fecundity and juvenile mortality are low, any density-dependent compensation in terms of survival rates is greatly constrained (Smith *et al.*, 1998). Recruitment overfishing, the level of population depletion at which recruitment becomes affected, therefore will start to occur early in the history of a shark fishery (Smith *et al.*, 1998).

Owing to their life-history characteristics, many taxa of commercial importance are particularly vulnerable to overexploitation through fisheries. For example, spurdogs (Squalidae) and gulper sharks (Centrophoridae) have been extensively harvested in south-eastern Australia despite their slow growth and low productivity, resulting in severe declines (refer to Box 22.2). Therefore, strict management is recommended for species with low rebound potential, but it is also likely that a certain level of protection is necessary to maintain sustainable fisheries of most coastal species. Management actions should take into account the basic life-history strategies of sharks, and focus on maintaining reproductive potential, taking into account the strong relationship between adult population size and recruitment.

Box 22.2 Fishing the depths: the collapse of deepwater shark populations in south-east Australia

Deepwater sharks inhabit the cold depths over and beyond the continental slope. Historically, they were protected from overfishing until the second half of the twentieth century, when the decline of shallow water fisheries and the advancement of fishing technologies drove a global shift to deeper fishing grounds (Morato *et al.*, 2006). Deepwater fisheries are seldom sustainable; life in the deep has a slow pace and most species have conservative life-history strategies that make them extremely susceptible to overfishing (Kyne & Simpfendorfer, 2007). By-catch is high, and discarded catch has little chance of survival (Gordon, 2001). Detailed life-history data and long-term fishery trends are missing for the vast majority of deepwater sharks, hindering appropriate assessment of the sustainability of current catches. In the few cases for which long-term data are available, the outlook is grim.

A rare case where extensive data are available is the deepwater Commonwealth Trawling Sector (CTS) operating in south-east Australia. The CTS, part of the southern and eastern Shark and Scalefish Fishery (SESSF), is one of the principal threats to deepwater shark populations in Australia. Its fishing grounds extend from Sydney southwards to Tasmania, and westwards to Cape Jervis in South Australia. Commercial trawling of the upper slopes started in 1968, primarily targeting gemfish, but dogfishes (Squalidae) and angel sharks (*Squatina* spp.) were a valuable bycatch of the fishery, particularly as demand for squalene (an oil extracted from deepwater shark livers) increased (Graham *et al.*, 2001). Species caught included slow-growing gulper sharks (*Centrophorus harrisoni*, *C. zeehani* and *C. moluccensi*) as well as other species of dogfishes and the sharpnose sevengill shark *Heptranchias perlo* (Graham *et al.*, 2001). Gulper shark stocks were substantially depleted as the fishery developed, and by the late 1990s stocks of the upper slope of NSW had declined by 98%–99% (Graham *et al.*, 2001). A decline of >90% was recorded also for the greeneye spurdog (*Squalus chloroculus*), and similar (albeit less dramatic) trends have been reported for other species of dogfishes and the sharpnose sevengill shark.

The broad-scale ecological consequences of such declines are largely unknown. Gulper sharks and sevengill sharks are high trophic level organisms, and a depletion of >90% of apex predator biomass is likely to have far-reaching top-down effects (see Box 22.1). Both the Harrison's dogfish (*C. harrisoni*) and the southern dogfish (*C. zeehani*) are endemic species, and their entire range falls within the SESSF-managed area. As a result of these declines the Harrison's dogfish has been listed as Critically Endangered by the IUCN, and southern dogfish as well as the greeneye spurdog have been nominated to be listed as Threatened by the Environmental Protection and Biodiversity Act 1999 (EPBC Act).

The Australian Fisheries Management Authority (AFMA) implemented a management strategy involving a drastic reduction in fishing effort and the total closure to fisheries of approximately 25% of suitable habitat area, with the aim of rebuilding stocks to 25% of the virgin population size (AFMA, 2012). Despite management efforts, a recent review suggests that these measures may not be sufficient to prevent further decline and promote the recovery of these species (Musick, 2011). The conservative life history strategies of these organisms, coupled with the strong stock–recruitment relationship that characterises shark populations means that even if appropriate management strategies are in place, recovery will be extremely slow. Many decades (nearly a century for the Harrison's dogfish) may be required for these species to recover to approximately 25% of pre-exploitation levels.

22.3 The status of shark conservation in the Indo-Australasian region

Fishing pressure is the main anthropogenic process threatening shark populations on a global and local scale. According to the United Nations' Food and Agriculture Organization (FAO), landings in the past six decades have nearly tripled, resulting in a worldwide decline of shark populations and local extinctions. The International Union for the Conservation of Nature (IUCN) carried out extensive assessments worldwide on the conservation status of shark species, and recently considerable effort has been placed on local assessments of shark populations in the Indo-Australasian region (Cavanagh *et al.*, 2003, White & Kyne, 2010). At the regional level, approximately 17% of the total number of species is under threat, being listed by the IUCN as Critically Endangered (CR), Endangered (EN) or Vulnerable (VU) (Figure 22.1). Approximately 40% are of conservation concern, being listed either as Threatened (CR, EN, VU) or Near Threatened (NT), and will be dependent on the establishment of effective management in the future. Only 29% of the species in the region are listed as Least Concern (LC), and the remaining 31% are data deficient.

The conservation status of a species is the product of the threatening processes to which it is subjected, the extent of its distribution, the level of habitat specialisation and the life-history characteristics that determine how populations respond to such processes. Species with limited distributions, subjected to high fishing pressure throughout their range and with low rebound potential are under the highest threat of extinction (Field *et al.*, 2009a). Sharks exhibit a variety of life-history strategies, and fishing pressure varies greatly among shark species and geographic areas within the region. As a result different taxa are inherently more vulnerable than others, and species that may be under imminent threat of extinction at the local scale may be of no conservation concern in other regions.

The proportion of species under threat and of conservation concern is highest in Indonesia and Papua New Guinea, where more than 30% of shark species are listed as Threatened and the vast majority of sharks (66% and 78% respectively) are of conservation concern (White & Kyne, 2010). New Zealand and New Caledonia follow with 21% and 35% of threatened species, and 47% and 59% of species of conservation concern respectively (White & Kyne, 2010). Australia is the country with the fewest threatened species, as considerable efforts have been put in place over previous decades in developing effective management strategies. Some species that are listed as threatened in the rest of the region (such as the zebra-shark *Stegostoma fasciatum*), are of no conservation concern in Australian waters (Cavanagh *et al.*, 2003). A few shark species, however, are under greater threat in Australia than anywhere else. This is the case for Harrison's dogfish (refer to Box 22.2), and for the Critically Endangered eastern population of grey nurse sharks (*Carcharias taurus)* which is demographically isolated and continues to decline despite being protected since 1984 (Ahonen *et al.*, 2009, Otway *et al.*, 2004).

One group, the Lamniformes, have low to moderate rebound potential (Smith *et al.*, 1998) and are under fishing pressure throughout the region, where they are either directly targeted for their fins and meat or caught as by-catch in the billfish and tuna long-line fisheries (Francis *et al.*, 2001, Dharmadi *et al.*, 2007). As a result the vast majority of Lamniformes are currently listed as under threat (Figure 22.1). Within this order are four families for which 100% of the species are listed as under threat throughout the region: Alopiidae (thresher sharks), Lamnidae (mackerel sharks), Odontapsidae (sand tiger sharks) and the monospecific family of the basking shark (Cetorhinidae). Another family of sharks that is of high conservation concern are the hammerhead sharks (Sphyrnidae, order Carcharhiniformes). Four species of hammerhead shark occur in the region, two of which are Endangered (*Sphyrna lewini* and *Sphyrna mokarran*), one Vulnerable (*Sphyrna zygaena*) and one near threatened (*Eusphyra blochii*). Hammerhead sharks are under heavy fishing pressure in the region, particularly in

Indonesian waters where they are directly targeted (particularly *S. lewini*) for their fins and meat (White *et al.*, 2008); as a result, they were recently listed on Appendix 2 of the Convention on International Trade in Endangered Species of Wild Fauna and Flora (CITES). Since the stock of these species is likely shared between Indonesia and neighbouring countries (Ovenden *et al.*, 2009), the effect of overfishing in Indonesian waters may have consequences that go far beyond national boundaries.

Not all sharks are under threat from overfishing. Many species are not under strong fishing pressure, and some have a high rebound potential. Bull-headed sharks for example (Heterodontidae) are relatively fecund (Last & Stevens, 2009) and there seem to be no major threats to these species in the region. Of the 15 species of lantern shark (family: Etmopteridae, order: Squaliformes) present in the region, 11 are not facing any threat of extinction and four are listed as data deficient. While lantern sharks are deep-water species with strongly K-selected life-history traits (Kyne & Simpfendorfer, 2007), they are not directly targeted in the region because of their small size and low commercial value. Some species however are caught as by-catch by deepwater trawl fisheries. While they are usually discarded, survival rates from trawl discards are likely to be low and the current expansion of deepwater trawl fisheries may pose higher risks to these species in the future.

22.4   Major threatening processes

22.4.1   Overfishing in Australian waters

Australia has one of the most extensive Exclusive Economic Zones in the world, extending for more than 8 million km^2 and harbouring a very high diversity of shark species. Shark fisheries, however, make up a small fraction of the country's landings. Over the past 20 years, catches ranged between 6700 and 11 500 t per year. Landings increased in the early 2000s, reaching 11 500 t in 2005 but, as a result of the implementation of several fishery restrictions, declined to less than 7000 t in 2010 (data from FAO). This equates to less than 1% of global catch, and less than 10% of the catch of Indonesia, one of Australia's neighbours. In Australia sharks are caught by commercial, recreational and traditional fisheries. Within commercial fisheries, sharks are both directly targeted and caught as by-catch, which may be retained or discarded. Shark fisheries fall within multiple jurisdictions, with about 50% of shark catches falling within the jurisdiction of Commonwealth Managed Fisheries, while the rest is managed by individual states (Woodhams *et al.*, 2012).

22.4.1.1   Southern and eastern shark fisheries

The Commonwealth managed Southern and Eastern Scalefish and Shark Fishery (SESSF), where sharks are targeted within the Gillnet Hook and Trap and Commonwealth Trawl Sectors (CTS), is responsible for approximately half of the Australian shark catch (Woodhams *et al.*, 2012).

The main target of the SESSF is the gummy shark (*Mustelus lenticulatus*), which is harvested by gillnets. Gummy sharks are not currently overfished and there is no major concern about the sustainability of the fishery in the future, however the same fishery has historically targeted school sharks (*Galeorhinus galeus*), a species more prone to overfishing. Since the 1920s, school sharks were directly targeted using longlines, but declining school shark catches and concerns over high mercury content, combined with the adoption of monofilament gillnets, resulted in a transition towards higher catches of gummy sharks in the 1970s and 1980s (Punt *et al.*, 2000). School sharks are now considered as by-catch of the gummy shark fisheries; however, landings are still substantial (averaged at 325 t between 2001 and 2006). Multiple assessments concluded that school shark stocks have been severely depleted, with estimated biomass in the mid 1990s ranging from 15% to 45% of pre-exploitation equilibrium size (Punt *et al.*, 2000), and recruitment in 1997 estimated to be only 12 to 18% of pre-exploitation levels (Punt *et al.*, 2000). In 2009, the species was listed as conservation dependent under the Environmental Protection and Biodiversity Conservation Act (EPBC Act), and a rebuilding strategy based on by-catch quotas and closure of some fishing grounds has been set in order to rebuild stock biomass to 20% of virgin biomass by 2020. Adult biomass levels have since stabilised; however, the current harvesting level is still too high to allow recovery, and recent biomass estimates range between 8–17% of pristine levels (Woodhams *et al.*, 2012).

The deepwater commonwealth trawling sector (CTS) of the SESSF operating in southeast Australia is a major threat to deepwater shark populations (see Box 22.2). While gemfish are the main target of the fishery, dogfishes and angel sharks are important by-catch. Increased fishing effort in the region in the 1970s, 1980s and 1990s has resulted in a serious depletion of deepwater shark populations, with declines of up to 99% (Graham *et al.*, 2001). Despite management effort in the past decade, there are serious concerns that these measures may not be enough to promote the recovery of affected populations (Box 22.2).

22.4.1.2 Northern shark fisheries

On the northern coast of Australia, direct targeting of sharks using gillnets and longlines started in the early 1980s. The main target group are requiem sharks (Carcharhinidae), including blacktip (*Carcharhinus limbatus* and *C. tilstoni*), spot-tail (*C. sorrah*), sandbar (*C. plumbeus*), and dusky sharks (*C. obscurus*), although hammerhead (*Sphyrna* spp.) and tiger sharks (*Galeocerdo cuvier*) are often caught (Woodhams *et al.*, 2012). There are concerns over the sustainability of the blacktip shark fishery in the Northern Territory, where catches reached a peak of nearly 900 t in 2003 and were still around 800 t in 2010, but dropped to approximately 400 t in 2011 (Northern-Territory-Government, 2011). An assessment of the sustainability of the fishery is complicated by the fact that substantial illegal fishing from foreign boats occurs in the area (see Box 22.3). There is also uncertainty about the sustainability of the blacktip shark fishery in the Gulf of Carpentaria off the Queensland coast (Roelofs, 2012). Furthermore, the fact that Australian blacktip (*C. tilstoni*) and common blacktip (*C. limbatus*) sharks hybridise and can be distinguished only via vertebrae counts or using genetic methods, complicates the assessment of the fisheries for these species (Boomer *et al.*, 2010, Ovenden *et al.*, 2010).

On the north coast of Western Australia there was a tenfold increase in fishing effort in the period 2000–2005, with catches reaching more than 1000 t in 2004–2005. An assessment by the Western Australian Department of Fisheries (DoF WA) concluded that fishing mortality was unsustainable for the sandbar shark (the principal target), and there were serious concerns about the sustainability of several other species (McAuley & Sarginson, 2011).

Box 22.3 Fishing across the border: Indonesian fishing in Australian waters

Illegal, unreported and unregulated (IUU) fishing poses a severe threat to shark populations globally (Bonfil, 1994). Within Australia, illegal fishing for sharks is largely attributed to Indonesian and, in smaller numbers, Taiwanese vessels, which venture into the Australian Fishing Zone (AFZ) to harvest shark products (mainly the highly priced fins) (Marshall, 2011). Fishing activity by Indonesian fishers in the shallow waters along Australia's northern continental shelf began long before European settlement and it was not until the mid twentieth century that it became illegal (Máñez & Ferse, 2010). In 1933, Australia gained sovereignty over Ashmore and Cartier island after the Western Australian Government appealed to the Commonwealth about illegal Indonesian fishing. However, surveillance of the area was infrequent and Indonesian fishing in northern and north-western Australian waters continued relatively unhindered until the late 1960s (Stacey, 2007).

The expansion of the AFZ to 12 nm (1968), and later to 200 nm (1979) with the declaration of Australia's Exclusive Economic Zone (EEZ) (Figure 22.3) brought about the loss of vast fishing grounds that had been frequently fished by Indonesian fishers since at least the 1920s (Stacey, 2007). A Memorandum of Understanding (MoU) between Australia and Indonesia prohibited Indonesian fishing in the AFZ, but allowed traditional fishing with unmotorised vessels on specified offshore islands and reefs under Australian jurisdiction in the Timor Sea, an area known as the 1974 MoU box (Figure 22.3). Shark fishing in the areas open to traditional fishing became more important from the early 1980s onwards, driven by a sharp increase in the demand for shark fins (Stacey, 2007).

The loss of shallow water fishing grounds along the continental shelf and the confinement to the deeper water fishing grounds of the MoU Box led to the development and use of new fishing techniques, particularly longlines in place of traditionally used shark

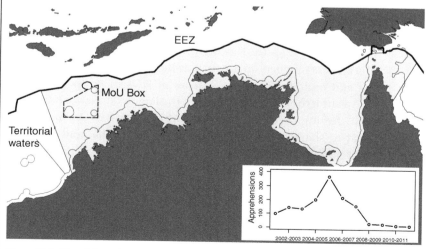

Figure 22.3 Map of the Exclusive Economic Zone off the northern Australian coast. The area mainly affected by IUU from foreign vessels is highlighted in grey. The inset shows trends in apprehensions of foreign fishing vessels in the past decade (data from the Australian Fisheries Management Authority).

Box 22.3 (continued)

rattles (goro-goro) and hand lines, which proved ineffective in deep water (Stacey, 2007). This development of fishing gears demonstrates that the word 'traditional', if used in relation to fishing techniques that are –erroneously – thought to remain consistent over time, is problematic. As conditions have changed, 'traditional' Indonesian shark fishers have adapted their fishing grounds and techniques in response to market demand, catch rates and international regulations (Fox *et al.*, 2009).

Three decades of heavy shark fishing in Eastern Indonesia have resulted in depleted local fishing grounds (Field *et al.*, 2009b) causing a southward movement of fishing effort into Australian waters, where sharks are reportedly larger in size and more abundant (Field *et al.*, 2009b) (personal observation by V. Jaiteh, 2012). A study of catch composition from 13 Indonesian and two Taiwanese foreign fishing vessels (FFVs) that had entered the AFZ between 2006 and 2009 identified 1182 individual sharks from 33 species within eight families (Marshall, 2011). Members of the family Charcharhinidae dominated the seized illegal catch, both in terms of numbers and biomass. Silky sharks (*Carcharhinus falciformis*) and blue sharks (*Prionace glauca*) were the most abundant species while blacktip sharks (*C. limbatus/tilstoni*) were the third most abundant species in terms of numbers, and the less abundant but larger tiger sharks (*Galeocerdo cuvier*) made the third biggest contribution to the overall biomass of the total catch. Certain species, such as the grey reef shark (*C. amblyrhynchos*), spinner shark (*C. brevipinna*), pigeye shark (*C. amboinensis*), bull shark (*C. leucas*) and scalloped hammerhead (*Sphyrna lewini*) were more common on Indonesian vessels, while tiger sharks and sandbar sharks (*C. plumbeus*) were predominantly recorded from Taiwanese vessels (Marshall, 2011). The majority of individuals recorded from the illegal fishing vessels – 59.4% and 44.1% on Indonesian and Taiwanese vessels, respectively – were immature (Marshall, 2011). Based on an estimate of 22 FFVs per day in the year 2006 (Salini, 2007), Marshall (2011) determined an annual shark catch by Indonesian vessels alone of approximately 290–1071 tonnes in 2006, which was comparable to the largest commercial Australian shark fishery operating in northern Australia at the time, the Northern Territory Offshore Net and Line Fishery (Marshall, 2011). These numbers do not include the catches of Taiwanese fishing vessels in that year and indicate that shark catches of FFVs in northern Australian waters were equivalent to or larger than those of the largest domestic shark fisheries.

IUU fishing by Indonesian and other south-east Asian vessels poses a serious threat to shark populations in northern Australia, and also to local Aboriginal livelihoods that depend on coastal marine resources (Field *et al.*, 2009b). Illegal fishing activities also raise customs and quarantine concerns and issues of national security (Vince, 2007). As a result, shark fishing boats entering northern Australian waters are regularly seized and destroyed and their crews arrested (Vince, 2007). The number of apprehensions of foreign vessels in northern Australian waters increased steadily in the early 2000s peaking at nearly 367 vessels between 2005 and 2006. Since then increased surveillance and tightened enforcement measures by Australian authorities have resulted in a reduction of illegal fishing and the number of apprehensions has steadily and sharply decreased since 2007 (Figure 22.3). Nevertheless, Indonesian shark fishers continue to enter the AFZ to fish illegally, often having little choice but to continue fishing for sharks in the absence of alternative livelihoods, due to logistical obstacles that arise from their distance to fresh fish markets, a lack of land rights and the resulting confinement to deriving a livelihood from the sea, or forbidding debt relationships with shark fin traders (Fox *et al.*, 2009; Resosudarmo *et al.*, 2009).

The DoF WA implemented drastic management measures in order to reduce fishing effort, setting a 20 t per year target. No fishing occurred in 2009/2010, however breeding stock levels of sandbar sharks remain low, and given the long life span and slow growth of this species, future recovery will be slow (McAuley & Sarginson, 2011).

22.4.2   New Zealand shark fisheries

New Zealand is one of the top 20 shark fishing nations, with an average of approximately 18 000 t landed every year in the first decade of this century (data from FAO). Shark catches were below 5000 t per year before 1980, but have steadily increased in the two following decades to reach 19 000 t in the late 1990s (Francis, 1998). Commercially important species that are directly targeted by local fisheries include school sharks, gummy sharks and the spiny dogfish (*Squalus acanthias*), as well as various rays and chimaeras. These species are managed under the New Zealand Quota Management System (QMS) through the allocation of Individual Transferable Quotas (ITQs) that are based on estimates of Total Allowable Catch (TAC) to ensure that species are not harvested over the Maximum Sustainable Yield (MSY) levels (Francis, 1998). Species managed under the QMS contribute to more than 80% of the total shark catch.

Direct targeting of other species of sharks is prohibited, however more than 60 other species of sharks, chimaeras and rays are landed as by-catch (Francis, 1998, Francis *et al.*, 2001). Of these by-catch species, only three migratory taxa (blue sharks, porbeagle sharks and mako sharks) that contribute to approximately 4% of total shark catch are subject to the QMS regulations (White & Kyne, 2010). Those species are the major by-catch of the pelagic longline tuna fishery, however it is unlikely that the current levels of by-catch are seriously affecting New Zealand's pelagic shark stocks. Of more concern is the status of species that are not managed through the allocation of ITQs, for which there is a paucity of long-term fishery data from which to estimate the sustainability of current catches.

22.4.3   Shark fisheries in Indonesia, past to present

Indonesia has the largest shark fishery in the world, with a reported average annual catch of 110 528 t between 2000 and 2007, or about 14.09% of global annual shark catches (data from FAO). Elasmobranch catches experienced a boom starting in the mid 1980s in response to a rise in the demand and price for shark products, particularly shark fins (Clarke, 2004). Catches peaked at 118 000 t in 2003 and appear to have experienced a slight but steady decrease since 2006, remaining at or above 100 000 t per year (data from FAO). While shark captures are often reported as by-catch in tuna longline and fish or prawn trawl fisheries (Dharmadi *et al.*, 2007, Tull, 2010), a large target fishery that is for the most part under- or unreported also exists (Dharmadi *et al.*, 2007; Stacey, 2007). Sharks are caught with a variety of gears, predominantly gill nets (Jaring insang), bottom-set longlines (Rawai dasar) and surface longlines (Rawai hanyut) (Dharmadi *et al.*, 2007).

Today, the majority of shark fins are harvested and traded in Eastern Indonesia, where data on catch rates and species composition remain limited (Bonfil, 1994; Resosudarmo *et al.*, 2009). Based on a study of shark landings at Indonesian fish markets with a particular focus on Bali, Java and Lombok, Carcharhinidae make up almost 70% of the total number and *c.* 60% of the biomass of all sharks landed (White, 2007). Blue sharks (*Prionace glauca*) contribute most to overall carcharhinid biomass (16.3%), while the most abundant species in terms of numbers is the spadenose shark (*Scoliodon laticaudus*) (32.5%). Other important target species include silky sharks (*C. falciformis*) (White, 2007) and scalloped hammerhead sharks (*Sphyrna lewini*), the latter representing up to 12% of landed biomass (White *et al.*, 2008). Catches of endangered and protected species are of particular concern, especially where regional information on biological and fishery aspects is insufficient to conduct species assessments and where fishing effort is unreported and unregulated. White *et al.* (2008) found that the vast majority of landed scalloped hammerhead sharks are immature individuals, a clear sign of growth overfishing. These catches are likely to constitute an unsustainable level of fishing mortality and have the potential to cause serious depletions of this species in Indonesian waters (White *et al.*, 2008). Another Endangered species listed under CITES Appendix II, the oceanic whitetip shark *Carcharhinus longimanus*, is also caught in the Indonesian shark fishery, although its contribution in numbers and biomass to the total catch is far smaller than that of *Sphyrna lewini* (White *et al.*, 2008; White, 2007).

Obtaining data on catch rates and composition is more difficult in the more remote provinces of Nusa Tenggara Timur, Maluku, Papua and West Papua, where ice and storage facilities are not normally available and sharks are usually not sold at markets, but finned at sea and the carcasses returned to the ocean (Marshall, 2011). Shark fishers in Eastern Indonesia generally use small wooden boats of less than 20 GT, often powered by sails and without refrigeration (Tull, 2010). As a result, the fishery is often described as artisanal, which is misleading: unlike other artisanal fisheries, the shark fishery is not limited to near-shore environments, and shark fishers often venture far, undertaking regular voyages across national sea boundaries (see Box 22.4) (Resosudarmo *et al.*, 2009; Stacey, 2007). Rarely, if ever, are sharks targeted for local consumption; the fishery is mostly driven by the international demand for shark fins (Dulvy *et al.*, 2008), and only by-products of this fishery may be used locally, such as the salted meat of sharks caught towards the end of a trip. Although there are no available estimates for the number of fishers targeting sharks and rays exclusively and no reliable catch data exist for large parts of the country, the life histories of most commercially valuable species leave little doubt that the Indonesian shark fishery has a serious impact on shark populations and marine ecosystems. Some shark stocks, including mobile species such as blue sharks and the Endangered hammerheads, are shared between Australia and Indonesia, being affected by the harvest of both nations (Ovenden *et al.*, 2009). Overfishing in Indonesian waters, therefore, may have consequences that go beyond the depletion of local stocks, affecting populations in neighbouring countries. Indonesian fishermen from the island of Rote frequently venture into Australian waters, where many have been arrested for fishing illegally in what used to be their traditional fishing grounds (see Box 22.3). This has created forbidding debt relationships between fishers and shark fin traders who sponsor the fishing trips, as well as conflicts among the two nations over shared marine resources and boundaries.

Box 22.4 Reef sharks: predators under threat

Reef sharks exert important top-down control on coral reef ecosystems, and their removal may have dramatic consequences, accelerating reef degradation through trophic cascades (Bascompte *et al.*, 2005). Declines of reef shark populations have been recorded worldwide, from the Chagos Archipelago in the Indian Ocean to the greater Caribbean (Graham *et al.*, 2010; Ward-Paige *et al.*, 2010). Even within the Great Barrier Reef Marine Park (GBR), the most well-managed network of marine reserves in the world, reef shark populations have undergone dramatic declines. The two most abundant and commonly caught reef sharks in Australia are the grey reef shark (*Carcharhinus amblyrhynchos*) and the whitetip reef shark (*Triaenodon obesus*) (Heupel *et al.*, 2009). As habitat specialists, they are found exclusively on, or very near to, coral reefs (Chin *et al.*, 2012) and exhibit a combination of life-history strategies (slow growth, late maturity, low reproductive output) and behavioural characters (site fidelity, habitat specialisation), which make them susceptible to even modest levels of fishing pressure. Within the GBR, reef sharks are caught as by-catch in the Coral Reef Finfish Fishery (Heupel *et al.*, 2009), and grey reef sharks make up to 6% of the shark catch of the Queensland East Coast Shark Fishery (Gribble *et al.*, 2005). Reef sharks are also targeted by Indonesian fishers who hunt for sharks illegally in Australian waters, with grey reef sharks contributing up to 18% of the catch (Box 22.3).

In the absence of specific legislation regulating the harvest of reef sharks, their conservation relies heavily on networks of marine reserves, although there is evidence that present management is only partially effective. Catch rates in the GBR have been relatively stable in the past 20 years, suggesting declines since 1986 have been subtle (Heupel *et al.*, 2009). However, the tendency to under-report discarded sharks, the lack of detailed information prior to the year 2000 (Heupel *et al.*, 2009), and temporal changes in target behaviour may seriously compromise the reliability of catch data to infer population trends (Hisano *et al.*, 2011). Robbins *et al.* (2006) found strictly enforced no-entry zones afford effective protection to reef shark populations in the Northern GBR, although no-take marine reserves are not as effective. Shark density within no-entry zones was an order of magnitude higher when compared not only to areas open to fisheries, but also to less strictly enforced no-take zones. Another study carried out in the central GBR yielded very similar results, with no-entry zones having four times as many sharks as areas open to fisheries (McCook *et al.*, 2010). No-take areas in the central GBR seem to afford some protection to reef shark populations, but densities were still only half of those recorded in no-entry areas. The ineffectiveness of no-take zones in the protection of shark populations is most likely due to direct fishing mortality due to illegal poaching (Hisano *et al.*, 2011; Robbins *et al.*, 2006). Similar patterns were reported for other coral reef predators targeted by fishermen, such as coral trout and snapper, supporting the hypothesis that no-take zones may be substantially depleted by poaching (McCook *et al.*, 2010). While no-take zones make up a substantial part of the GBR marine park (30%), no-entry zones encompass only 1% of the available habitat and are unlikely to play a significant role in the regional conservation of reef sharks. Compliance and effective enforcement of established marine reserve networks will be critical in ensuring the recovery of reef

Box 22.4 (continued)

Figure 22.4 (Plate 44) The grey reef shark, the quintessential coral reef associated shark. Photo by Robert Harcourt. Reproduced with permission. A black and white version of this figure will appear in some formats. For the colour version, please refer to the plate section.

hark populations on the GBR, but this is a challenging task given the scale at which regulations need to be enforced and the remoteness of many areas of the GBR (McCook *et al.*, 2010).

Demographic analyses suggest that populations of reef sharks on the GBR are severely depleted and are undergoing rapid declines (Hisano *et al.*, 2011; Robbins *et al.*, 2006). It is not surprising that shark populations in relatively remote habitats may be threatened by modest levels of poaching. In the Chagos Archipelago, an extremely remote area of the Indian Ocean, reef shark populations collapsed entirely because of illegal shark fishing from foreign fleets (Graham *et al.*, 2010). Since reef sharks are facing imminent threats even in the most well-managed network of marine reserves in the world, it is likely that the situation is even more dramatic in areas where similar management strategies are not in place. In Indonesia, according to one study, grey reef sharks make up 2%–3% of total landed shark biomass, approximately 3000 t per year (which is equivalent to 30% of total yearly shark landings in Australia) (White, 2007). This is likely a gross underestimate of landings, since shallow water coral reef fisheries in eastern Indonesia are largely unreported and unregulated (Varkey *et al.*, 2010). Although data to estimate past and future population trends in Indonesia are not available, given the levels of fishing pressure it seems most likely that reef shark populations would be undergoing even sharper declines than on the GBR.

While direct fishing mortality is at present the highest threat to reef shark populations, habitat degradation is likely to have profound effects on these species in the near future. Coral reefs are declining worldwide at an alarming rate due to the synergistic effects of overfishing, increasing sea surface temperatures, ocean acidification, crown-of-thorns starfish outbreaks, nutrient enrichment and increased tropical storm intensity (De'ath *et al.*, 2012; Wilkinsons, 2008). Within the GBR, coral cover has declined by 50% in the past three decades (De'ath *et al.*, 2012), and the outlook in other regions is even worse. In Indonesia, nearly 90% of coral reefs are under threat, mostly because of overfishing and destructive fishing (Wilkinsons, 2008). With coral reefs shrinking, so is the habitat on which reef sharks rely for their existence.

22.4.4 Habitat loss and climate change

Many shark species rely on a range of different habitats for foraging, mating and reproduction. Mangrove systems, estuaries and seagrass beds function as important nursery grounds, providing habitats sheltered from predators and inclement environmental conditions (Heupel *et al.*, 2007). In addition, some species have strong natal philopatry to specific nursery areas (Hueter *et al.*, 2005). Some sharks may be restricted to a single habitat throughout their life-cycle; this is the case of strictly coral reef associated sharks such as the grey reef shark (*C. amblyrhynchos*) (see Figure 22.4, Plate 44) and the whitetip reef shark (*Triaenodon obesus*) (Chin *et al.*, 2012) and of some deepwater species with restricted distributions. Other species have specific habitat requirements in different stages of their life-cycle, showing ontogenetic shifts in habitat use (Grubbs, 2010). Species showing a high degree of habitat specialisation at any stage of their life-cycle can therefore be highly vulnerable to habitat degradation (Chin *et al.*, 2010). Even for species without strict habitat associations, the degradation of some habitats may imply the loss of important foraging grounds. The potential effect of habitat loss and habitat degradation on shark populations is not as well understood as the effect of direct fishing mortality.

The synergistic effects of climate change, coastal development and the use of destructive fishing methods is leading to the degradation of many habitats that are crucial for shark populations. Eutrophication and increased water turbidity due to inadequately managed land use (Brodie *et al.*, 2011), the rise in sea surface temperatures (Hoegh-Guldberg and Bruno, 2010) and a reduction in the ocean pH due to increased atmospheric CO_2 concentrations (Doney *et al.*, 2009) are likely to have profound effects on ecosystem function, indirectly affecting shark populations that rely on these habitats (Chin *et al.*, 2010). Even in relatively well-managed areas, such as the coasts of Australia, seagrass beds (Walker and McComb, 1992), mangrove forests and coral reefs (De'ath *et al.*, 2012) have been declining at alarming rates. In other areas of the region, such as Papua New Guinea and Indonesia, the outlook is considerably worse (Wilkinsons, 2008). The effects of habitat degradation on local shark populations are difficult to quantify. The degradation of coral reefs may have dramatic effects on sharks that are strictly coral-reef associated (Chin *et al.*, 2010) and the loss of coastal nursery areas may have profound effects on future recruitment of shark populations, exacerbating the effects of already unsustainable harvesting regimes (Field *et al.*, 2009a).

22.5 Current management and conservation strategies

Australia and New Zealand have long histories of shark fishery management. Shark fisheries have been traditionally managed using size and bag limits, restrictions in licensing and fishing gear, along with demographic models that are more often than not better suited to highly productive species of fish (Musick *et al.*, 2000). The first management actions date back to the late 1950s with the introduction of minimum length requirements in the school shark fishery in New South Wales. Starting from the 1980s and 1990s management strategies based on restrictions of licensing and the introduction of individual transferable quotas (ITQs) of total allowable catch were introduced in Australia and New Zealand. These management strategies, however, were usually restricted to a few commercially important species and were only partially effective. It was not until the last decade that coordinated national strategies for the conservation of shark populations were implemented.

22.5.1 National Plans of Action for the Conservation and Management of Sharks

Given growing concerns regarding the over-exploitation of shark stocks, in 1998 the Food and Agriculture Organization of the United Nations (FAO) launched an International Plan of Action for the Conservation and Management of Sharks (IPoA) (FAO, 1998), with the aim of ensuring the long-term conservation of shark populations and the sustainability of shark fisheries. The IPoA called upon nations to prepare National Plans of Action (NPoA) to ensure the sustainability of shark fisheries with a particular focus on threatened and vulnerable shark stocks by: engaging stakeholders in the development of the plan, creating a shark assessment report (SAR), establishing long-term monitoring of species-specific fishery data, minimising incidental catch, and contributing to the protection of biodiversity and ecosystem function. In the Indo-Australasian region NPoAs have been developed by Australia, New Zealand and Indonesia. However, only Australia has followed the recommendations outlined in the IPoA closely.

The first Australian SAR was prepared in 2001, and the first NPoA was developed in 2004 following the guidelines and objectives outlined by the IPoA (NPoA-Australia, 2004). The NPoA played a major role in the development of coordinated management strategies at the national level, including improvements in observer and monitoring programmes, the development of fishery-specific risk assessments for the majority of shark fisheries, and the implementation of specific management responses for threatened and protected species. Most management actions that have been undertaken in the past decades and resulted in improved management of shark fisheries (such as the Commonwealth school shark rebuilding strategy and targeted management of sandbar shark populations in Western Australia) occurred within the framework of the NPoA. In 2009 a new SAR (SAR2) (Bensley *et al.*, 2009) was developed and the following year a review of the NPoA was released. These documents identified improvements in the management of commercial harvest as well as the main shortcomings in terms of the development of effective conservation strategies. Among the most important issues was the inadequacy of data reporting, which resulted in the absence of reliable and validated species-specific catch data, particularly for by-catch species, and consequently the need to implement precautionary measures to prevent future shark declines where uncertainties exist. A new NPoA, which was formulated primarily following the recommendations of SAR2 and the 2010 reviews was developed in 2012 (NPoA-Australia, 2012).

In 2008, an NPoA was approved in New Zealand (NPoA-New-Zealand, 2008); however, the approval of the NPoA did not follow the development of a nation-wide SAR comparable to the one undertaken in Australia in 2001. New Zealand's NPoA is mostly a description and evaluation of current management strategies for shark fisheries, based on the New Zealand Quota Management System. A few key areas where improvement is necessary are highlighted, including the need for actions to eliminate live shark finning, protect threatened and endangered species that are protected under the Convention of Migratory Species (particularly the basking shark), and a review of management strategies for the spiny dogfish. One of the IPoA recommendations was that NPoAs should be reviewed and revised every four years. However, New Zealand has yet to produce a SAR or a formal review of the implementation of the actions suggested in the NPoA developed in 2008.

Indonesia has committed to managing its shark fisheries through the formulation of an NPoA; however, rather than proposing management strategies informed by catch data and fishery-independent assessments, the NPoA released in 2010 outlines the need for increased research effort and data collection to inform management, without providing much guidance on how, where, when and by whom the required data should be collected. Fishery surveys require extended periods of time because in the remote regions of Eastern Indonesia, where there are often no major landing sites, whole sharks are rarely landed and fins are sold directly from fishers to traders, not at fish markets. The lack of reporting requirements and the resulting scarcity of catch data have been an enormous obstacle to species assessments, management strategies and effective conservation initiatives (Blaber *et al.*, 2009; Dharmadi *et al.*, 2007). Given the many obstacles for an accurate assessment of Indonesia's shark fishery, traditional management based on sufficient fishery-, population- and species-specific data is unlikely to be implemented in time to prevent further species declines and local extinctions (Camhi, 1998). Precautionary measures, implemented strategically and efficiently, are therefore needed if the effects of several decades of overfishing are to be mitigated against and the population declines of some stocks are to be slowed down until better data become available to inform management.

22.5.2 The role of Marine Protected Areas

The establishment of areas permanently closed to fishing has been proposed as an essential tool in the protection of shark populations. Networks of Marine Protected Areas (MPAs) have been established in many countries, including Australia, New Zealand and Indonesia. Possibly the most well-known example of an extensive network of MPAs is the Great Barrier Reef Marine Park, extending for more than 2000 km along the Queensland coast. The effectiveness of MPAs, which are usually not designed specifically for shark conservation, in the protection of shark populations has recently sparked debate (see Box 22.4) (Heupel *et al.*, 2009; Robbins *et al.*, 2006). Recent studies have assessed MPAs effectiveness for coastal shark protection, although only a few used empirical evidence that has included long-term movement in and around MPAs (e.g. Knip *et al.*, 2012, Speed *et al.*, in review). One study focussed on the effectiveness of an MPA within Ningaloo Marine Park in Western Australia for three species of reef sharks: grey reef sharks (*Carcharhinus amblyrhynchos*), blacktip reef sharks (*C. melanopterus*), and sicklefin lemon sharks (*Negaprion acutidens*) (Speed *et al.*, in review). The authors found that the largest species (grey reef sharks) spent very little (< 1%) of their time within the MPA, whereas juvenile blacktip and sicklefin lemon sharks spent 84%–99 % of their time within the protected area. The MPA provided a reasonable level of protection for juveniles, although long-distance movements (10–260 km) and larger home ranges by adult grey reef and blacktip reef sharks exposed them to recreational fishing pressure within the Marine Park. These results suggest that the placement and size of the MPA does not incorporate habitats commonly used by adult grey reef and blacktip sharks, such as the reef slope and channels connecting the reef and lagoon.

Dedicated MPAs, designed especially for shark conservation (shark sanctuaries), have been established in many tropical reef areas over the past few years (e.g. Palau, Honduras, Maldives, The Marshall Islands, Tokelau, Raja Ampat) (PEW, 2012). The effectiveness of shark sanctuaries is currently under debate (Chapman *et al.*, 2013; Davidson, 2012), due to scepticism surrounding enforcement. Socio-economic studies of the value of live sharks for tourism compared to fished sharks, have assisted in driving the recent focus on shark sanctuaries as conservation tools (Vianna *et al.*, 2012). One of these studies from Palau found that a population of 100 grey reef sharks is estimated to be worth US $10 800 to the fishing industry over their lifetime, which is only 0.006 % of their value to the tourism industry if kept alive (Vianna *et al.*, 2012).

In some situations the establishment of a network of MPAs may be the best strategy to protect shark populations. For example, given the obstacles for traditional management of Indonesia's shark fishery, shark management should be approached from different angles, such as the protection of fishing grounds through no-take zones or for eco-tourism purposes, alongside the facilitation of alternative livelihoods. In some regions, tourism has already provided an impetus for shark conservation. For example, the development of Raja Ampat as a dive and nature tourism destination has led to initiatives by resort owners, NGOs and communities to protect the region's shark and ray fauna. These efforts have resulted in the country's first shark and manta ray sanctuary, where shark fishing is prohibited by law. While this is a commendable development, shark fishers that have been displaced from their previous fishing grounds are forced to search for different fishing grounds – which will then experience greater fishing pressure – or new livelihoods. The identification of alternative livelihoods and the provision of support in the transition phase must therefore form an integral part of shark conservation initiatives. This requires effective communication between scientists, managers, tourist operators, NGOs and government agencies, as well as improved research capacity and increased allocation of management responsibility to provincial and regency levels.

REFERENCES

AFMA 2012. *Upper-Slope Dogfish Management Strategy*. Consultation Document. Australian Fisheries Management Authority, Canberra.

Ahonen, H., Harcourt, R. & Stow, A. 2009. Nuclear and mitochondrial DNA reveals isolation of imperilled grey nurse shark populations (*Carcharias taurus*). *Molecular Ecology*, **18**, 4409–4421.

Bascompte, J., Melián, C. J. & Sala, E. 2005. Interaction strength combinations and the overfishing of a marine food web. *Proceedings of the National Academy of Sciences of the United States of America*, **102**, 5443–5447.

Bensley, N., Woodhams, J., Patterson, H. *et al.* 2009. *Shark Assessment Report for the Australian National Plan of Action for the Conservation and Management of Sharks*. Department of Agriculture, Fisheries and Forestry, Bureau of Rural Sciences, Canberra.

Benz, G. W., Hocking, R., Kowunna, A., Bullard, S. A. & George, J. C. 2004. A second species of Arctic shark: Pacific sleeper shark *Somniosus pacificus* from Point Hope, Alaska. *Polar Biology*, **27**, 250–252.

Blaber, S., Dichmont, C., White, W. *et al*. 2009. Elasmobranchs in southern Indonesian fisheries: the fisheries, the status of the stocks and management options. *Reviews in Fish Biology and Fisheries*, **19**, 367–391.

Bonfil, R. 1994. *Overview of World Elasmobranch Fisheries*. FAO Rome.

Boomer, J., Peddemors, V. & Stow, A. 2010. Genetic data show that *Carcharhinus tilstoni* is not confined to the tropics, highlighting the importance of a multifaceted approach to species identification. *Journal of Fish Biology*, **77**, 1165–1172.

Brodie, J. E., Devlin, M., Haynes, D. & Waterhouse, J. 2011. Assessment of the eutrophication status of the Great Barrier Reef lagoon (Australia). *Biogeochemistry*, **106**, 281–302.

CAMHI, M. 1998. *Sharks and their Relatives: Ecology and Conservation*. World Conservation Union.

Cavanagh, R. D., Kyne, P. M., Fowler, S. L., Musick, J. A. & Bennett, M. B. 2003. The conservation status of Australasian chondrichthyans. IUCN Shark Specialist Group. Australia and Oceania Regional Red List Workshop, 2003, pp. 7–9.

Chapman, D. D., Frisk, M. J., Abercrombie, D. L. *et al*. 2013. Give shark sanctuaries a chance. *Science (New York, NY)*, **339**, 757.

Chin, A., Kyne, P. M., Walker, T. I. & McAuley, R. 2010. An integrated risk assessment for climate change: analysing the vulnerability of sharks and rays on Australia's Great Barrier Reef. *Global Change Biology*, **16**, 1936–1953.

Chin, A., Tobin, A., Simpfendorfer, C. & Heupel, M. 2012. Reef sharks and inshore habitats: patterns of occurrence and implications for vulnerability. *Marine Ecology Progress Series*, **460**, 115–125.

Clarke, S. 2004. Understanding pressures on fishery resources through trade statistics: a pilot study of four products in the Chinese dried seafood market. *Fish and Fisheries*, **5**, 53–74.

Compagno, L. J. 2001. *Sharks of the World: An Annotated and Illustrated Catalogue of Shark Species Known to Date*. FAO.

Davidson, L. N. K. 2012. Shark sanctuaries: substance or spin? *Science*, **21**, 1538.

De'ath, G., Fabricius, K. E., Sweatman, H. & Puotinen, M. 2012. The 27–year decline of coral cover on the Great Barrier Reef and its causes. *Proceedings of the National Academy of Sciences*, **109**, 17 995–17 999.

Dharmadi, Sumadhiharga, K. & Fahmi 2007. Biodiversity and length frequencies of sharks caught in the Indian Ocean. *Marine Research in Indonesia*, **32**, 139–146.

Doney, S. C., Fabry, V. J., Feely, R. A. & Kleypas, J. A. 2009. Ocean acidification: the other CO_2 problem. *Annual Review of Marine Science*, **1**, 169–192.

Dulvy, N. K., Baum, J. K., Clarke, S. *et al*. 2008. You can swim but you can't hide: the global status and conservation of oceanic pelagic sharks and rays. *Aquatic Conservation: Marine and Freshwater Ecosystems*, **18**, 459–482.

FAO 1998. *International Plan of Action for the Conservation and Management of Sharks*. Document FI:CSS/98/3.

Field, I. C., Meekan, M. G., Buckworth, R. C. & Bradshaw, C. 2009a. Susceptibility of sharks, rays and chimaeras to global extinction. *Advances in Marine Biology*, **56**, 275–363.

Field, I. C., Meekan, M. G., Buckworth, R. C. & Bradshaw, C. J. 2009b. Protein mining the world's oceans: Australasia as an example of illegal expansion-and-displacement fishing. *Fish and Fisheries*, **10**, 323–328.

Fox, J. J., Adhuri, D. S., Therik, T. & Carnegie, M. 2009. Searching for a livelihood: The dilemma of small-boat fishermen in Eastern Indonesia. In: *Working with Nature Against Poverty: Development, Resources and the Environment in Eastern Indonesia*, pp. 201–225.

Francis, M. P. 1998. New Zealand shark fisheries: development, size and management. *Marine and Freshwater Research*, **49**, 579–591.

Francis, M. P., Griggs, L. H. & Baird, S. J. 2001. Pelagic shark bycatch in the New Zealand tuna longline fishery. *Marine and Freshwater Research*, **52**, 165–178.

Friedlander, A. M. & DeMartini, E. E. 2002. Contrasts in density, size, and biomass of reef fishes between the northwestern and the main Hawaiian islands: the effects of fishing down apex predators. *Marine Ecology Progress Series*, **230**, 253–264.

Gordon, J. 2001. Deep-water fisheries at the Atlantic Frontier. *Continental Shelf Research*, **21**, 987–1003.

Graham, K., Andrew, N. & Hodgson, K. 2001. Changes in relative abundance of sharks and rays on Australian South East Fishery trawl grounds after twenty years of fishing. *Marine and Freshwater Research*, **52**, 549–561.

Graham, N. A., Spalding, M. D. & Sheppard, C. R. 2010. Reef shark declines in remote atolls highlight the need for multi-faceted conservation action. *Aquatic Conservation: Marine and Freshwater Ecosystems*, **20**, 543–548.

Gribble, N., Whybird, O., Williams, L. & Garrett, R. 2005. Fishery assessment update 1988–2003: Queensland east coast shark. *Department of Primary Industries and Fisheries, Brisbane, Queensland. Report # QI04070*.

Grubbs, R. D. 2010. Ontogenetic shifts in movements and habitat use. In: Carrier, J.C., Musick, J.A. & Heithaus, M.R. (eds.) *Sharks and their Relatives II*, Boca Raton: CRC Press, pp. 319–351.

Heithaus, M. R., Frid, A., Vaudo, J. J., Worm, W. & Wirsing, A. J. 2010. Unraveling the ecological importance of elasmobranchs. In: Carrier, J. C., Musick, J. A. & Heithaus, M. R. (eds.) *Sharks and Their Relatives II – Biodiversity, Adaptive Physiology, and Conservation*. Boca Raton: CRC Press, pp. 611–638.

Heithaus, M. R., Frid, A., Wirsing, A. J. & Worm, B. 2008. Predicting ecological consequences of marine top predator declines. *Trends in Ecology and Evolution*, **23**, 202–210.

Heithaus, M. R., Wirsing, A. J., Burkholder, D., Thomson, J. & Dill, L. M. 2009. Towards a predictive framework for predator risk effects: the interaction of landscape features and prey escape tactics. *Journal of Animal Ecology*, **78**, 556–562.

Heupel, M., Williams, A., Welch, D. *et al.* 2009. Effects of fishing on tropical reef associated shark populations on the Great Barrier Reef. *Fisheries Research*, **95**, 350–361.

Heupel, M. R., Carlson, J. K. & Simpfendorfer, C. A. 2007. Shark nursery areas: concepts, definition, characterization and assumptions. *Marine Ecology Progress Series*, **337**, 287–297.

Hisano, M., Connolly, S. R. & Robbins, W. D. 2011. Population growth rates of reef sharks with and without fishing on the Great Barrier Reef: robust estimation with multiple models. *PLoS ONE*, **6**, e25028.

Hoegh-Guldberg, O. & Bruno, J. F. 2010. The impact of climate change on the world's marine ecosystems. *Science*, **328**, 1523–1528.

Hueter, R., Heupel, M., Heist, E. & Keeney, D. 2005. Evidence of philopatry in sharks and implications for the management of shark fisheries. *Journal of Northwest Atlantic Fishery Science*, **35**, 239–247.

Kitchell, J. F., Essington, T. E., Boggs, C. H., Schindler, D. E. & Walters, C. J. 2002. The role of sharks and longline fisheries in a pelagic ecosystem of the central Pacific. *Ecosystems*, **5**, 202–216.

Knip, D. M., Heupel, M. R. & Simpfendorfer, C. A. 2012. Evaluating marine protected areas for the conservation of tropical coastal sharks. *Biological Conservation*, **148**, 200–209.

Kyne, P. M. & Simpfendorfer, C. A. 2007. A collation and summarization of available data on deepwater chondrichthyans: biodiversity, life history and fisheries. *IUCN SSC Shark Specialist Group*.

Last, P. R. & Stevens, J. D. 2009. *Sharks and Rays of Australia*. Collingwood, Australia, CSIRO Publishing.

Lucifora, L., Menni, R. & Escalante, A. 2004. Reproductive biology of the school shark, *Galeorhinus galeus*, off Argentina: support for a single south western Atlantic population with synchronized migratory movements. *Environmental Biology of Fishes*, **71**, 199–209.

Lucifora, L. O., García, V. B. & Worm, B. 2011. Global diversity hotspots and conservation priorities for sharks. *PLoS ONE*, **6**, e19356.

Máñez, K. S. & Ferse, S. C. 2010. The history of Makassan trepang fishing and trade. *PLoS ONE*, **5**, e11346.

Marshall, L. 2011. *The Fin Blue Line: Quantifying Fishing Mortality Using Shark Fin Morphology*. University of Tasmania.

McAuley, R. & Sarginson, N. 2011. Northern shark fishery status report 2010/2011. In: Fletcher, W. J. & Santoro, K. (eds.) *State of the Fisheries and Aquatic Resources Report*. Department of Fisheries, Western Australia.

McCook, L. J., Ayling, T., Cappo, M., *et al.* 2010. Adaptive management of the Great Barrier Reef: a globally significant demonstration of the benefits of networks of marine reserves. *Proceedings of the National Academy of Sciences*, **107**, 18 278–18 285.

Morato, T., Watson, R., Pitcher, T. J. & Pauly, D. 2006. Fishing down the deep. *Fish and Fisheries*, **7**, 24–34.

Musick, J., Burgess, G., Cailliet, G., Camhi, M. & Fordham, S. 2000. Management of sharks and their relatives (Elasmobranchii). *Fisheries*, **25**, 9–13.

Musick, J. A. 2011. *An Evaluation of the Australian Fisheries Management Authority (AFMA) Upper-Slope Dogfish Management Strategy – Commonwealth-Managed Fisheries.*

Myers, R. A., Baum, J. K., Shepherd, T. D., Powers, S. P. & Peterson, C. H. 2007. Cascading effects of the loss of apex predatory sharks from a coastal ocean. *Science*, **315**, 1846–1850.

Northern-Territory-Government 2011. Fishery Status Reports 2011. *Northern Territory Government Department of Resources. Fishery Report No. 111.*

Npoa-Australia 2004. *National Plan of Action for the Conservation and Management of Sharks (Shark-plan)*. Department of Agriculture, Fisheries and Forestry (Daff), Australian Government.

Npoa-Australia 2012. *National Plan of Action for the Conservation and Management of Sharks 2012, Shark-plan 2*. Department of Agriculture, Fisheries and Forestry (Daff), Australian Government.

Npoa-New-Zealand 2008. *New Zealand National Plan of Action for the Conservation and Management of Sharks*. Ministry of Fisheries, New Zealand Government.

Otway, N. M., Bradshaw, C. J. & Harcourt, R. G. 2004. Estimating the rate of quasi-extinction of the Australian grey nurse shark *Carcharias taurus* population using deterministic age-and stage-classified models. *Biological Conservation*, **119**, 341–350.

Ovenden, J. R., Kashiwagi, T., Broderick, D., Giles, J. & Salini, J. 2009. The extent of population genetic subdivision differs among four co-distributed shark species in the Indo-Australian archipelago. *BMC Evolutionary Biology*, **9**, 40.

Ovenden, J., Morgan, J., Kashiwagi, T., Broderick, D. & Salini, J. 2010. Towards better management of Australia's shark fishery: genetic analyses reveal unexpected ratios of cryptic blacktip species *Carcharhinus tilstoni* and *C. limbatus*. *Marine and Freshwater Research*, **61**, 253–262.

PEW 2012. *Navigating Global Shark Conservation: Current Measures and Gaps*. Washington DC: PEW Environment Group.

Punt, A. E., Pribac, F., Walker, T. I., Taylor, B. L. & Prince, J. D. 2000. Stock assessment of school shark, *Galeorhinus galeus*, based on a spatially explicit population dynamics model. *Marine and Freshwater Research*, **51**, 205–220.

Resosudarmo, B., Napitupulu, L. & Campbell, D. 2009. Illegal Fishing in the Arafura Sea. *Working With Nature against Poverty Development Resources and the Environment in eastern Indonesia*. Singapore: Institute for Southeast Asian Studies, pp. 178–200.

Robbins, W. D., Hisano, M., Connolly, S. R. & Choat, J. H. 2006. Ongoing collapse of coral-reef shark populations. *Current Biology*, **16**, 2314–2319.

Roelofs 2012. *Annual Status Report 2010, Gulf of Carpentaria Inshore Fin Fish Fishery*. Queensland Department of Primary Industries and Fisheries, Brisbane.

Ruppert, J. L. W., Travers, M. J., Gilmour, J. P. *et al.* 2013. Caught in the middle: combined impacts of shark removal and coral loss on fish communities. *PLoS ONE* **8**(9): e74648.

Salini, J. P. 2007. *Estimating Reliable Foreign Fishing Vessel Fishing Effort from Coastwatch Surveillance and Apprehension Data: Estimating FFV Effort*. CSIRO Marine and Atmospheric Research.

Sandin, S. A., Smith, J. E., DeMartini, E. E. *et al.* 2008. Baselines and degradation of coral reefs in the Northern Line Islands. *PLoS ONE*, **3**, 1–11.

Shepherd, J. & Cushing, D. 1980. A mechanism for density-dependent survival of larval fish as the basis of a stock-recruitment relationship. *Journal du Conseil*, **39**, 160–167.

Smith, S. E., Au, D. W. & Show, C. 1998. Intrinsic rebound potentials of 26 species of Pacific sharks. *Marine and Freshwater Research*, **49**, 663–678.

Speed, C. W., Meekan, M. G., Field, I. C., *et al.* in review. Marine parks for reef sharks: shark residency relative to a marine protected area.

Stacey, N. 2007. *Boats to Burn: Bajo Fishing Activity in the Australian Fishing Zone*. ANU E Press.

Tittensor, D. P., Mora, C., Jetz, W. *et al.* 2010. Global patterns and predictors of marine biodiversity across taxa. *Nature*, **466**, 1098–1101.

Tull, M. 2010. *The History of Shark Fishing in Indonesia*: a HMAP Asia Project Paper.

Varkey, D. A., Ainsworth, C. H., Pitcher, T. J., Goram, Y. & Sumaila, R. 2010. Illegal, unreported and unregulated fisheries catch in Raja Ampat Regency, Eastern Indonesia. *Marine Policy*, **34**, 228–236.

Vianna, G., Meekan, M., Pannell, D., Marsh, S. & Meeuwig, J. 2012. Socio-economic value and community benefits from shark-diving tourism in Palau: a sustainable use of reef shark populations. *Biological Conservation*, **145**, 267–277.

Vince, J. 2007. Policy responses to IUU fishing in Northern Australian waters. *Ocean & Coastal Management*, **50**, 683–698.

Walker, D. & McComb, A. 1992. Seagrass degradation in Australian coastal waters. *Marine Pollution Bulletin*, **25**, 191–195.

Ward-Paige, C. A., Mora, C., Lotze, H. K. *et al.* 2010. Large-scale absence of sharks on reefs in the greater-Caribbean: a footprint of human pressures. *PLoS ONE*, **5**, e11968.

White, W. 2007. Catch composition and reproductive biology of whaler sharks (Carcharhiniformes: Carcharhinidae) caught by fisheries in Indonesia. *Journal of Fish Biology*, **71**, 1512–1540.

White, W., Bartron, C. & Potter, I. 2008. Catch composition and reproductive biology of *Sphyrna lewini* (Griffith & Smith)(Carcharhiniformes, Sphyrnidae) in Indonesian waters. *Journal of Fish Biology*, **72**, 1675–1689.

White, W. & Kyne, P. 2010. The status of chondrichthyan conservation in the Indo-Australasian region. *Journal of Fish Biology*, **76**, 2090–2117.

Wilkinsons, C. 2008. *Status of the Coral Reefs of the World: 2008*. Global Coral Reef Monitoring Network and Reef and Rainforest Research Centre. Townsville, Australia, 296 pages.

Woodhams, J., Vieira, S. & Stobutzki, I. 2012. *Fishery Status Report 2011*, Australian Bureau of Agricultural and Resource Economics and Sciences, Canberra.

Wourms, J. & Demski, L. 1993. The reproduction and development of sharks, skates, rays and ratfishes: introduction, history, overview, and future prospects. *Environmental Biology of Fishes*, **38**, 7–21.

CHAPTER 23

'Ragged mountain ranges, droughts and flooding rains': the evolutionary history and conservation of Australian freshwater fishes

Leanne Faulks, Dean Gilligan and Luciano B. Beheregaray

Summary

Australia hosts a unique assemblage of flora and fauna derived from a combination of Gondwanan relict and more recently evolved endemic taxa and is recognised as one of the world's megadiverse countries. Despite the continent's high species biodiversity, the Australian freshwater fish fauna is relatively depauperate. The conservation of freshwater fishes in Australia is of increasing importance as many species are listed as threatened by the IUCN. The major threatening processes for Australian freshwater fishes are habitat degradation, river regulation, anthropogenic barriers to dispersal, introduced species, disease and climate change. The use of molecular genetic tools to infer evolutionary history and to inform conservation is well recognised and is one way of predicting how fish may respond to these threatening processes. Nonetheless, there are few Australian cases that allow a bigger picture assessment of evolutionary processes across a broad range of environments, yet within a single taxonomic group. The temperate freshwater perches of the genus *Macquaria* provide an exception. This chapter uses this fish group as a case study in phylogeography and population genetics to explore and identify evolutionary processes relevant for aquatic conservation across a large section of eastern and central Australia.

23.1 Australian freshwater fishes: biodiversity and conservation

Australia hosts a unique assemblage of flora and fauna derived from a combination of Gondwanan relict and more recently evolved endemic taxa (Allen *et al.*, 2002;

Austral Ark: The State of Wildlife in Australia and New Zealand, eds. A. Stow, N. Maclean and G. I. Holwell. Published by Cambridge University Press. © Cambridge University Press 2015.

Sanmartin & Ronquist, 2004) and is recognised as one of the world's megadiverse countries (Mittermeier *et al.*, 1997). Despite the continent's high species biodiversity, the Australian freshwater fishes are relatively depauperate with only 59 families and ~ 300 species currently recognised (Humphries & Walker, 2013) (Figure 23.1, Plate 54). However, following recent and ongoing molecular and taxonomic surveys, many cryptic species are being described, meaning that the true level of freshwater species richness may well exceed this estimate. In any case, this restricted diversity is a consequence of relatively low and highly variable freshwater run-off particularly in the central arid area of the continent (Puckridge *et al.*, 1998; Kennard *et al.*, 2010). Given that species can diversify only where there is available niche space, it is no surprise that less water equals less potential for fish diversification. In fact, the majority of the freshwater fish species are found in the moist coastal and tropical regions, with only 28 species endemic to the arid inland basins (Unmack, 2001; Allen *et al.*, 2002). The Australian freshwater fish fauna is characterised by a predominance of species that have invaded and adapted to freshwaters from a marine environment (Unmack *et al.*, 2013a). The majority of these are small-bodied species such as the Gobiidae (gudgeons), Melanotaeniidae (rainbowfishes), Galaxiidae, and Atherinidae (silversides). A small proportion of these species are large bodied and tend to be members of either the Percichthyidae (perch) or Siluriform (catfish) groups. Only two families of fish are truly freshwater, the Osteoglossidae (saratoga) and Ceratodontidae (lungfish). These species are believed to be Gondwanan relicts from up to 150 million years ago (mya) (Allen *et al.*, 2002). Australia also hosts a unique assemblage of desert fishes, many of which are endemic to specific waterways or artesian springs (e.g. the redfinned blue eye (*Scaturiginichthys vermeilipinnis*) from Edgbaston Springs, western Queensland).

The conservation of freshwater resources in Australia is of increasing importance (Dudgeon *et al.*, 2006), as freshwater habitats continue to be significantly altered for agricultural and urban use, and native fish communities are dramatically degraded (Harris & Gehrke, 1997; Davies *et al.*, 2008). Despite growing public awareness, research and management intervention, these ecosystems remain under substantial pressure. Approximately 10% of species are listed as threatened by the IUCN

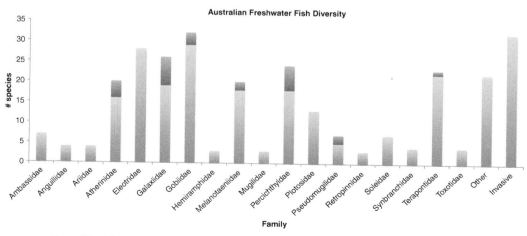

Figure 23.1 (Plate 54) Diversity of Australian freshwater fishes by family. Other: includes families represented by only one or two species. Red indicates the proportion of species listed as threatened by the IUCN Red List. A black and white version of this figure will appear in some formats. For the colour version, please refer to the plate section.

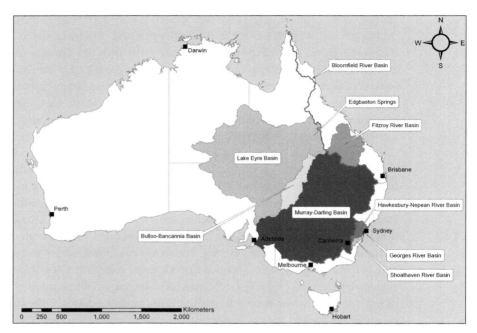

Figure 23.2 (Plate 23) Map of Australia showing the major drainage basins as well as geographical features and locations mentioned throughout the chapter. The great dividing range is indicated by the red line. A black and white version of this figure will appear in some formats. For the colour version, please refer to the plate section.

(Figure 23.1, Plate 54). In addition some populations and ecosystems have been given special protection within the national framework. For example, in the Murray–Darling Basin (MDB) (Figure 23.2, Plate 23) the status of fish communities is estimated to be just 10% of pre-European settlement levels, with five species feared as locally extinct in the lower MDB (MDBC, 2003). As a result, long-term management goals were defined to help restore the basin to its natural state (MDBC, 2003).

The major threatening processes for Australian freshwater fishes are: habitat degradation, river regulation, anthropogenic barriers to dispersal, introduced species and disease (Cadwallader, 1978; Gehrke *et al.*, 1999; Lintermans, 2007). In the future, freshwater ecosystems may also be under additional pressure due to extreme conditions brought on by the uncertain effects of climate change (Beare & Heaney, 2002; Preston & Jones, 2008; Balcombe *et al.*, 2011; Chessman, 2013). Although it is difficult to anticipate the environmental conditions that will be experienced by freshwater taxa, it is predicted that aridification and temperatures in south-eastern Australia will increase, and sea levels will rise (Christensen *et al.*, 2007; Meehl *et al.*, 2007). These changes will alter flow regimes and shift or reduce the availability of suitable habitat. The response of species will be largely dependent upon the geographical region in which they occur, coupled with their biological characteristics (Balcombe *et al.*, 2011). For example, cold-tolerant species and fish assemblages in already disturbed regions like the lower Murray River will probably be most affected (Balcombe *et al.*, 2011). In particular, species that have specific habitat preferences and limited dispersal (e.g. the southern pygmy perch *Nannoperca australis*) are more likely to be affected by habitat degradation than generalist species (e.g. the golden perch *Macquaria ambigua*). For more migratory species, environmental conditions that facilitate dispersal of various life stages may become increasingly important to population persistence. In an attempt to predict the response of fish to climate change, Chessman (2013) compared long-term monitoring data with biological trait analysis, determining a suite of traits that are more likely to deem a species vulnerable to prolonged drought. An additional way of predicting how fish may respond to these changes is by drawing inferences from how palaeoclimatic conditions affected the evolutionary history of species. These types of predictions can guide the conservation of freshwater taxa into the future.

23.2 Australian freshwater fishes: evolutionary origins

The fossil record indicates that freshwater fish, similar to some present-day forms, occurred across both inland and coastal Australia as long as ~30 mya (Longman, 1929; Turner, 1982; Estes, 1984), and although there are some true freshwater groups, e.g. lungfish, the majority of taxa are derived from marine ancestors (Whitley, 1960; Unmack *et al.*, 2013a). Some of these ancient species were morphologically similar to extant Australian species, e.g. the fossil remains of *'Percalates antiquus'* have been compared to contemporary Murray cod (Longman, 1929; Hills, 1934, 1943). The evolution of freshwater fish from marine ancestors could have occurred several times during Australia's history. Fluctuations in sea level and the inundation of large areas of inland Australia occurred throughout the Tertiary (the past 65 my) (Stephenson, 1986), and could have facilitated freshwater incursions and colonisations of drainage networks by marine fish. These environments have experienced episodic changes in habitat, especially in terms of hydrology and water quality. In particular, gradual reductions in salinity following the retreat of inland seas could have selected for euryhaline species that eventually developed tolerance to freshwater. Geographic isolation due to land-locking of marine-derived lineages in multiple freshwater drainages, in association with selection to a riverine life-history, are likely major processes accounting for the diversification and evolution of Australian freshwater fish fauna. This general pattern of marine–freshwater transition associated with both allopatric and ecological diversification in disconnected drainages has been reported in both hemispheres for several freshwater fish groups (Bell & Forster, 1994; Echelle & Echelle, 1984; Lovejoy *et al.*, 1998; Beheregaray *et al.*, 2002; Cooke *et al.*, 2012; Jones *et al.*, 2012).

An understanding of the chronology of events associated with marine incursions and diversification of Australian freshwater fishes is hampered by the paucity of studies reporting on divergence times amongst endemic lineages. Although the major splits within some of the most diverse freshwater fish clades are old and likely of Miocene and Oligocene age (e.g. hardyheads: Unmack & Dowling, 2010; rainbowfishes: Unmack *et al.*, 2013a), several examples exist where diversification along the tips of these clades (both within species complexes and species) were likely influenced by climatic changes during the Pleistocene (past 2 my) (Table 23.1). During the Pleistocene, the Australian continent underwent extensive aridification, morphing from a moist, tropical-like landscape with abundant vegetation and run-off, to a more arid landscape (Bowler, 1990). Throughout this time there were also fluctuations between glacial and interglacial stages. Although glaciation itself was very restricted in Australia, the glacial stages were characterised by cooler, drier conditions, followed by warmer, moister conditions (Kershaw *et al.*, 2003). These changes had a significant impact on the availability and nature of freshwater environments, subsequently influencing the distribution and diversification of freshwater fishes.

Table 23.1   Summary of phylogeographic studies on Australian freshwater taxa. Divergence column indicates drainage basins across which significant levels of genetic divergence have been detected. The timing column shows estimates of the dates of divergence indicated in the previous column. The mechanism column provides suggestions on how taxa were able to breach the drainage divides stated in the divergence column. Abbreviations: BULL = Bulloo basin; Clarence = Clarence River, northern New South Wales; EC = east coast; GC = Gulf of Carpentaria; LEB = Lake Eyre basin; Mary = Mary River, southern Queensland; MDB = Murray–Darling basin; NT = Northern Territory; SC: Southeastern coast; TAS = Tasmania; WA = Western Australia.

Group	Species	Distribution	Divergence	Timing/ Mechanism	References
Perch	*Macquaria ambigua*	MDB, LEB, BULL, EC	MDB–LEB, EC–MDB	Low divides, Pleistocene moist conditions	Musyl & Keenan, 1992; Faulks *et al.*, 2010a, c
	M. australasica	MDB, EC	EC–MDB	Low divides, Pleistocene moist conditions	Dufty, 1986; Faulks *et al.*, 2010b, 2011
	M. novemaculeata	EC			Chenoweth & Hughes, 1997; Jerry, 1997; Jerry & Baverstock, 1998; Shaddick *et al.*, unpublished
	M. colonorum	EC	North–south	Sea level fluctuations	Shaddick *et al.*, 2011a, b
	Nannoperca australis	MDB, EC, TAS	Murray–SE		Hammer, 2001; Cook *et al.*, 2007; Unmack *et al.*, 2013b
	N. obscura	MDB, SC	Murray–SC	Low divides, sea level fluctuations	Hammer *et al.*, 2010; Brauer *et al.*, 2013
	N. oxleyana	EC	North–south	Sea level fluctuations, dunes	Hughes *et al.*, 1999; Knight *et al.*, 2009; Page *et al.*, 2012
	Leiopotherapon unicolor	MDB, LEB, BULL, EC, WA, GC			Bostock *et al.*, 2006
Blackfish	*Gadopsis marmoratus*	MDB, EC, TAS	Northern–southern		Ovenden *et al.*, 1988; Miller *et al.*, 2004
	G. bispinosus	MDB			Ovenden *et al.*, 1988; Miller *et al.*, 2004
Cod	*Maccullochella* spp.	MDB, Mary, Clarence	MDB–EC		Rowland, 1993; Rourke, 2007
Bream	*Nematalosa erebi*	MDB, LEB, EC, GC, WA	MDB–LEB LEB–GC	150 kya 160–350 kya	Hughes & Hillyer, 2006; Masci *et al.*, 2008
Catfish	*Tandanus tandanus*	MDB, EC	EC–EC NEC–SEC SEC–MDB	Miocene 11 mya 2.6 mya Volcanism Aridification?	Musyl & Keenan, 1996; Jerry & Woodland, 1997; Jerry, 2008
	N. hyrtlii	MDB, LEB, BULL, WA, GC, EC	MDB–LEB, LEB–GC	40–72 kya, 196 kya, Gene flow in recent flood events	Huey *et al.*, 2006; Huey *et al.*, 2008
	Porochilus argenteus	LEB, BULL			Huey *et al.*, 2006
Gudgeons	*Mogurnda adspersa*	MDB, EC	EC–EC, MDB–EC	Drainage rearrangement, 1.6 mya	Hurwood & Hughes, 1998; Faulks *et al.*, 2008

Table 23.1   *(cont.)*

Group	Species	Distribution	Divergence	Timing/Mechanism	References
	Gobiomorphus australis	EC			Page *et al.*, 2012
	Philypnodon	MDB, EC	MDB–EC		Thacker *et al.*, 2008
	Hypseleotris spp.	MDB, LEB, BULL, EC	MDB–EC, EC–EC	Low divides, sea level fluctuations, dunes	McGlashan & Hughes, 2001a; Thacker *et al.*, 2007; Sharma & Hughes, 2011; Page *et al.*, 2012
Rainbowfish	*Cairnsichthys rhombosomoides*	EC QLD			Thuesen *et al.*, 2008
	Melanotaenia australis	WA		Lowland wet season dispersal	Phillips *et al.*, 2009
	Melanotaenia duboulayi	EC	EC–EC	650 kya– 1.26 mya, Sea level fluctuations, dunes	Page *et al.*, 2012
	Rhadinocentrus ornatus	EC	EC–EC	50 kya–7 mya, Aridification, Sea level fluctuations, dunes	Page *et al.*, 2004; Sharma & Hughes 2011; Page *et al.*, 2012
Hardyhead	*Craterocephalus* sp., *Craterocephalus stercusmuscarum*	MDB, EC, GC, NT	MDB–EC	1 mya, Drainage rearrangements, aridity, sea level fluctuations	McGlashan & Hughes, 2000; McGlashan & Hughes, 2001b; Unmack *et al.*, 2010
Smelt	*Retropinna semoni*	MDB, LEB, EC	MDB–LEB, EC–EC,	1.5 mya, 5 spp.?	Hughes & Hillyer, 2006; Hammer *et al.*, 2007
Perchlet	*Ambassis* sp.	LEB	LEB–LEB	130 kya–1.5 mya, Independent colonisations over low divides	Cook *et al.*, 2007; Huey *et al.*, 2011
Minnow	*Galaxiella pusilla*	Southern Australia	East–west	5.7 mya, aridification	Unmack *et al.*, 2012
Blue eye	*Pseudomugil mellis*	EC	North–south	560 kya–1.7 mya, Sea level fluctuations, dunes	Page *et al.*, 2012
	Pseudomugil signifier	EC	North–south	420 kya– 1.26 mya, Burdekin gap, sea level fluctuations, dunes	Wong *et al.*, 2004; Page *et al.*, 2012

As a consequence of its geographic setting, the Australian freshwater fauna is high in endemism and a comprehensive understanding of the processes that have driven diversification and influenced the persistence of populations is needed. There is a strong interaction between the environment and evolution (at various spatial and temporal scales) and examining these using phylogeographic (the use of genetic datasets to infer population history and diversification) and landscape genetic (the use of population genetic data to assess environmental correlates of biodiversity) approaches can provide insights into how current and future environmental conditions may affect the evolutionary potential of organisms. For example, Byrne *et al.* (2008) reviewed the influence of climate changes on the diversification and assembly of biota in the Australian arid zone, suggesting that Pleistocene environmental changes can impact upon biota in a range of ways and that further research is required for a comprehensive understanding. In the case of freshwater fauna, the use of molecular tools to infer evolutionary history and inform conservation is well recognised (Hughes *et al.*, 2009). Nonetheless, there are few cases that allow a bigger picture assessment of evolutionary/biodiversity processes across a broad range of environments, yet within a single taxonomic group. The temperate freshwater perches of the genus *Macquaria* provide an exception since they have been used as a case study to identify evolutionary processes across a large section of eastern and central Australia. The remainder of this chapter focusses on how recent studies of *Macquaria* have illuminated our understanding of historical and contemporary factors underpinning patterns of biodiversity in Australian freshwater fishes, as well as how this approach can be valuable for conservation and fisheries management.

23.3 The application of phylogeography and landscape genetics: a case study of the freshwater fishes *Macquaria ambigua* and *Macquaria australasica*

23.3.1 The importance of taxonomy and geology to conservation: the coastal origin of freshwater fishes

An important initial step in conservation is accurate taxonomic classification. In this case, molecular techniques have helped to reveal levels of biodiversity that were previously unrecognised. However, the taxonomy of the *Macquaria* genus examined in this case study is far from being resolved. The genus currently consists of four species, *M. ambigua* (with a proposed but undescribed subspecies *M. ambigua oriens* (Musyl & Keenan, 1992)), *M. australasica*, *M. colonorum* and *M. novemaculeata*. *Macquaria ambigua* and *M. australasica* are strictly freshwater species and are found in both coastal and inland drainages. In contrast, *M. colonorum* and *M. novemaculeata* are found only in coastal drainages and have diadromous life-history strategies. As a result of these differences, the freshwater and diadromous species were previously considered separate genera, *Plectroplites* and *Percalates* respectively, until being assigned to *Macquaria* by MacDonald (1986). In support of the former distinction, the Percichthyidae phylogeny (Jerry *et al.*, 2001) indicates that there are many other strictly freshwater species more closely related to *M. ambigua* and *M. australasica* than their diadromous congeners, e.g. the pygmy perch, *Nannoperca* spp. In addition, recent higher-level molecular phylogenetic studies have found that the diadromous *M. colonorum* and *M. novemaculeata* represent a highly divergent clade that is not part of the Percichthyidae family (Near *et al.*, 2012; Chen *et al.*, 2014). Also of note is the recent discovery of a new Percichthyd species, *Guyu wujalwujalensis* (Pusey *et al.*, 2004), which appears closely related to *M. australasica*. These taxonomic uncertainties can hinder appropriate conservation and management, therefore the ongoing discovery and description of new species should be supported.

Following on from taxonomic studies, intraspecific studies based on genetic data can potentially inform on evolutionary processes leading to diversification and as such provide valuable information for the conservation of these processes and biodiversity in the future. It was with this goal in mind that a comprehensive research programme on the phylogeography and landscape genetics of the *Macquaria* genus was undertaken. Firstly, the *Macquaria* genus was examined in order to investigate one particular biogeographic barrier that is believed to have been influential in shaping the biodiversity of freshwater fishes, the Great Dividing Range (GDR) (Figure 23.2, Plate 23). Many groups of Australian freshwater fishes have disjunct distributions on either side (coastal and inland) of the GDR (Table 23.1). Nonetheless, there is still limited understanding of the processes accounting for this biogeographic pattern, due to differences in molecular, geological and fossil data (Longman, 1929; Ollier, 1978; Wellman, 1979; Turner, 1982; Estes 1984).

Genealogical analyses of *M. ambigua* (Faulks *et al.*, 2010a) and *M. australasica* (Faulks *et al.*, 2010b) revealed a consistent pattern of basal lineages found in coastal populations, to the east of the GDR. Further, phylogeographic analyses indicated demographic and geographic population expansions in the inland basins to the west of the GDR. These results support the hypothesis that both species originated in coastal drainages and subsequently colonised inland basins. It is plausible that, in the past, a common ancestor for the freshwater *Macquaria* species occupied a large section of the east coast of Australia and subsequently diverged into the various present-day forms. In support of this hypothesis, fossil evidence indicates that there was a Percichthyd ancestor, *Percalates antiquus*, in drainages along the coast of Queensland approximately 22–38 mya (Hills, 1934). The following sequence of events could explain this commonly observed phylogeographic pattern: GDR uplift; ancestral freshwater fishes found throughout coastal catchments and in inland lakes and drainages; ancestors become extinct, perhaps due to climate and environmental pressures, or persist and are eventually displaced by present forms; present forms are found along the east coast and then colonise inland drainages. This would account for the current evidence of an ancient and stable GDR, relatively recent inland colonisations and variable divergences among species suggesting that they crossed the divide at different times and in various locations.

Various mechanisms by which fish could be exchanged across drainage basin boundaries have been proposed, e.g. drainage rearrangements (Hurwood & Hughes, 1998), continuous wet divides and episodic tributary connections via axial valleys (Craw *et al.*, 2007), but finding evidence for these processes along an extensive barrier like the GDR is a challenging task. A full-scale survey of the GDR incorporating geomorphological data and a molecular investigation of species with disjunct distributions may help to identify mechanisms of potential 'fish crossings'. Indeed, the role of mountain ranges and headwaters in driving freshwater biodiversity has also been hypothesised in North America, where up to 70% of drainage area is estimated to be in the headwaters and where a range of habitats create niches for species diversification (Lowe & Likens, 2005). This process may be particularly important in Australia where there is little geological activity and relatively low run-off, which otherwise limit diversification opportunities. Thus the headwaters should be conserved as they have the potential to be sources of biodiversity as well as refuge areas under future climate change.

23.3.2 Understanding the processes driving the patterns of biodiversity for more effective conservation management

The effect of environmental conditions on the evolutionary history of fauna and flora has been extensively investigated, with particular focus on the Pleistocene

(Avise & Walker, 1998; Hewitt, 2000; Soltis *et al.*, 2006; Beheregaray, 2008). This recent period in the Earth's history, ~2 mya, was characterised by cyclical fluctuations in climate. These fluctuations are evident in the oxygen isotope record, which is broken up into a series of peaks and troughs that indicate warm interglacials with increased sea level and cold glacial phases with decreased sea level (Imbrie *et al.*, 1984). Although glaciation in Australia was restricted to Tasmania and the NSW/Victorian alpine region (Barrows *et al.*, 2001), evidence from pollen and geomorphological records support the oxygen isotope data, suggesting that the climate fluctuated between cool/dry and warm/moist conditions (Nanson *et al.*, 1992; Kershaw *et al.*, 2003; Byrne *et al.*, 2008). These changes in climate undoubtedly had flow-on effects to environmental conditions and habitat availability. The Australian freshwater fishes have been significantly shaped by both Pleistocene events and contemporary conditions. Australia now experiences some of the most variable climatic and hydrological conditions on Earth, particularly across the inland drainage basins (Poff *et al.*, 2006). An approach combining phylogeography and landscape genetics, in which patterns of gene flow and diversity are assessed in relation to environmental variables across the landscape, is particularly useful for addressing questions about how climatic conditions and species-specific biological characteristics have shaped the demographic history of populations (Manel *et al.*, 2003). These processes need to be recognised and maintained in order to conserve the evolutionary potential of species, particularly under climate change and habitat degradation.

Moister climate conditions, particularly during interglacial phases increased the connectivity of freshwater environments and may have facilitated the expansion of fish populations into and among inland drainages. For example, the excellent dispersal ability of *M. ambigua* (Reynolds, 1983; O'Connor *et al.*, 2005), especially under moister climatic conditions allowed the species to colonise a large area of inland Australia, taking advantage of rare flow/flooding events (Faulks *et al.*, 2010a). Indeed, this process remains important to this species' population dynamics. For example, an increase in the volume of water flow in a river at the start of the spring breeding season was inferred to promote the migration and outbreeding of adults from different catchments, to facilitate larval drift and dispersal, and consequently increase the genetic diversity of populations (Faulks *et al.*, 2010c). In addition, populations in catchments with less variability in flow regime throughout the year (increased perenniality) had higher levels of genetic diversity, a pattern also seen in the catfish, *Neosilurus hyrtlii* (Huey *et al.*, 2008).

The isolation of populations can play an important role in the diversification of freshwater fauna, providing the opportunity for genetic drift and selection to different environments to create divergent and locally adapted lineages. Increasingly dry conditions in Australia during the Pleistocene isolated freshwater environments (Kershaw *et al.*, 2003; Byrne *et al.*, 2008). In the case of *M. ambigua*, these changes had an impact at a large spatial scale with the isolation of neighbouring basins, e.g. Bulloo from the Lake Eyre Basin (Faulks *et al.*, 2010a). In contrast, *M. australasica* populations became isolated in adjacent catchments, e.g. the Lachlan from the remainder of the Murray–Darling Basin (Faulks *et al.*, 2010b). Similar processes have been observed in many other Australian freshwater taxa (Table 23.1). Isolation of freshwater environments can also occur via changes in sea level. Sea level fluctuations are acknowledged to have had an impact on the distribution of taxa living in the estuarine zone and river delta/mouth (Shaddick *et al.*, 2011) and creating alternative patterns of connectivity among basins. However, the influence on strictly freshwater taxa is likely to be subtler. For example, sea level fluctuations may have caused the isolation of *M. australasica* populations in the lower reaches of catchments due to changes in the position of the saline tidal water limit. This isolation could contribute to driving the evolution of divergent lineages within catchments, resulting in the higher levels of genetic diversity observed in coastal populations of *M. australasica* (Faulks *et al.*, 2010b). In addition, anthropogenic structures such as weirs were found to impede fish dispersal and contribute to the genetic differentiation of *M. australasica* populations (Faulks *et al.*, 2011).

These processes of population isolation are predicted to become more common in the future due to climate change and habitat degradation, therefore it is imperative that the information gained from molecular studies is incorporated into conservation and fisheries management strategies. This case study highlights the need to consider historical, contemporary, natural and anthropogenic factors when assessing population genetic structure and diversity within species. The evolutionary and contemporary genetic characteristics of species are also intimately related to aspects of their biology, such as life-history strategy, dispersal capabilities and habitat requirements. These represent a challenge for conservation and fisheries management, where restricted resources often result in an attempt to formulate generalised strategies for a diverse suite of taxa.

23.4 Conservation and management of Australian freshwater fishes

The central aim of conservation management programmes is to maintain species in a manner that enables them to adapt to changes in environmental conditions. This comes with particular challenges for freshwater taxa, which are generally restricted to a specific habitat: a linear and hierarchical riverine network. Thus, as movement to more favourable environmental conditions is likely to be limited for the majority of freshwater taxa, the potential for species to adapt to these changes is particularly important. Perhaps for this reason, many examples of relatively rapid genotypic evolution and phenotypic plasticity have been observed in freshwater taxa (Knudsen *et al.*, 2006; BondurianNky *et al.*, 2012; McCairns & Bernatchez, 2012). Although the role of environmental changes, whether natural or anthropogenic, has been highlighted as an important process in the evolution of freshwater taxa, the role of community-level processes must also be acknowledged. For example, the presence of introduced species, disease and competitors can influence population dynamics and trigger adaptive changes.

The best way to mitigate all of these impacts is to help maintain the natural evolutionary processes experienced by species (Fraser & Bernatchez, 2001; Moritz, 2002; Wishart & Davies, 2003; Davis *et al.*, 2008; Latta, 2008). In the case of freshwater taxa, this can be achieved by protecting and restoring habitat (Rahel, 2007), promoting natural levels of gene flow by constructing fishways and removing redundant fish passage barriers (Raeymaekers *et al.*, 2008; Stuart *et al.*, 2008), deregulating flow regimes by releasing environmental flows (Kingsford & Auld, 2005; Howard 2008) and controlling introduced species (Strayer, 2010). Although these management outcomes may be challenging to achieve, they are likely to provide long-term benefits for species and their communities. Despite this, augmentation of populations by translocation and stocking remains one of the most common management actions for fisheries agencies in Australia and throughout the world (Cowx & Gerdeaux, 2004).

Unfortunately, many stocking activities commenced with little understanding of their potential risks of admixture of lineages, inbreeding and outbreeding depression and the introduction of disease (Hindar *et al.*, 1991; Tallmon *et al.*, 2004; Ward, 2006; Edmands, 2007). The specific risks associated with translocation and stocking practices in Australian freshwater fishes were recently highlighted (Moore *et al.*, 2010) and the genetic risks to species were considered high given that very little is known about historical or contemporary population genetic structure and diversity prior to commencement of stocking activities. In some species, zones for broodstock collection and fingerling release have been designated based on the limited data (Moore *et al.*, 2010). However, unregulated translocation and stocking occurred prior to designation of these stocking management zones and has led to the admixture of genetically and morphologically distinct lineages. For example in *M. australasica*, fish translocated from the MDB into a coastal dam (Cataract) have escaped downstream and have inter-bred with coastal lineages (Faulks *et al.*, 2010b). If introgression continues, this situation could compromise the genetic integrity of the coastal lineages by disrupting locally co-adapted gene complexes and lead to outbreeding depression (Ward, 2006). There has also been widespread stocking of *M. ambigua* across its range since the 1960s (Brumley, 1987). In this species, there is evidence for inter-basin translocation with a single MDB lineage present within our sample of Fitzroy basin fishes, but despite records of MDB fishes being translocated into the LEB, no mixture of lineages was detected (Faulks *et al.*, 2010a). This could indicate that there is poor survival and/or recruitment of translocated fish under different ecological conditions (Woodworth *et al.*, 2002). There is also little evidence of impacts arising from extensive stocking of *M. ambigua* within the MDB (Faulks *et al.*, 2010a, c). Although the level of mitochondrial genetic diversity in *M. ambigua* is fairly low, the levels of microsatellite genetic diversity are comparatively high, indicating that low levels of diversity are more likely due to historical processes than stocking activities. This conclusion was also supported by genetic studies of Murray cod pre- and post-stocking in the Murray–Darling Basin (Rourke *et al.*, 2010). Despite little evidence of stocking activities having reduced genetic diversity in the two species of *Macquaria*, maintaining genetic diversity into the future should be a vital consideration in stocking programmes. It is recommended that the basic conservation breeding and reintroduction principles outlined by Ryman & Utter (1987), Schramm & Piper (1995), Frankham *et al.* (2002) and Hallerman (2003) be adhered to in order to maintain genetic diversity and integrity and to maximise the ability of populations to adapt. In particular, the broodstock used should include as much of the local genetic diversity as possible, especially rare lineages, in order to prevent swamping wild populations with a few lineages, especially those adapted to artificial conditions. This could be achieved by: sampling and crossing broodstock from a variety of locations and regularly replacing these with new wild fish from unstocked locations, and by creating a DNA database where broodfish are 'fingerprinted'.

This case study has demonstrated the valuable role of phylogeographic and landscape genetic surveys to help mitigate fisheries management issues. For example, management units (Moritz, 1994; Crandall *et al.*, 2000) for both *M. ambigua* and *M. australasica* have been designated; given the high degree of population genetic structure in *M. australasica*, catchment-scale management units have been recommended (Faulks *et al.*, 2010b, 2011), whereas in *M. ambigua* basin-scale management may be more appropriate (Faulks *et al.*, 2010a, c). Despite differences in the spatial scale of management in the two species, some common isolating processes have been identified. For example, the Lachlan catchment is distinct from other MDB catchments in *M. ambigua* and *M. australasica*, as well as other freshwater taxa (Rourke, 2007). This example demonstrates how phylogeographic studies of freshwater taxa could be integrated to form a strategy for identifying populations of priority for conservation management. In addition, this information could be further extrapolated to help make management decisions for species where there is an absence of detailed population genetics data and a lack of time to conduct genetic assessments. Species with similar life histories, migratory abilities and habitat preferences could be predicted to display similar patterns of genetic structure and diversity, e.g. *M. ambigua* and silver perch, *Bidyanus bidyanus*. This approach could give fisheries management agencies the potential to act in situations requiring immediate conservation intervention and would help foster the implementation of adaptive management strategies (Sabine *et al.*, 2004; Walters, 2007).

REFERENCES

Allen, G. R., Midgley, S. H., Allen, M. (2002). *Field Guide to the Freshwater Fishes of Australia*. Perth, Australia, Western Australian Museum.

Avise, J. C., Walker, D. (1998). Pleistocene phylogeographic effects on avian populations and the speciation process. *Proceedings of the Royal Society London B*, **265**, 457–463.

Balcombe, S. R., Sheldon, F., Capon, S. J. *et al.* (2011). Climate-change threats to native fish in degraded rivers and flood plains of the Murray–Darling Basin, Australia. *Marine and Freshwater Research*, **62**, 1099–1114.

Barrows, T. T., Stone, J. O., Fifield, L. K., Creswell, R. G. (2001). Late Pleistocene glaciation of the Kosciuszko massif, Snowy Mountains, Australia. *Quaternary Research*, **55**, 179–189.

Beare, S., Heaney, A. (2002). Climate change and water resources in the Murray–Darling Basin, Australia. In: *Proceedings of the 2002 World Congress of Environmental and Resource Economists*. Monterey, California, USA.

Beheregaray, L. B. (2008). Twenty years of phylogeography: the state of the field and the challenges for the Southern Hemisphere. *Molecular Ecology*, **17**, 3754–3774.

Beheregaray, L. B., Sunnucks, P., Briscoe, D. A. (2002). A rapid fish radiation associated with the last sea-level changes in southern Brazil: the silverside *Odontesthes perugiae* complex. *Proceedings of the Royal Society of London B*, **269**, 65–73.

Bell, M. A., Foster, S. A. (1994). *The Evolutionary Biology of the Threespine Stickleback*. Oxford University Press.

Bonduriansky, R., Crean, A. J., Day, T. (2012). The implications of nongenetic inheritance for evolution in changing environments. *Evolutionary Applications*, **5**, 192–201.

Bostock, B. M., Adams, M., Laurenson, L. J. B., Austin, C. M. (2006). The molecular systematics of *Leiopotherapon unicolor* (Gunther, 1859): testing for cryptic speciation in Australia's most widespread freshwater fish. *Biological Journal of the Linnean Society*, **87**, 537–552.

Bowler, J. (1990). The last 500, 000 years. In: *The Murray*, (eds. **Mackay, N, Eastburn, D.**). Canberra, Murray Darling Basin Commission, pp. 95–109.

Brauer, C. J., Unmack, P. J., Hammer, M. P., Adams, M., Beheregaray, L. B. (2013). Catchment-scale conservation units identified for the threatened Yarra pygmy perch (*Nannoperca obscura*) in highly modified river systems. *PLoS ONE*, **8**(12), e82953.

Brumley, A. (1987). Past and present distributions of golden perch *Macquaria ambigua* (Pisces: Percichthyidae) in Victoria, with reference to releases of hatchery produced fry. *Proceedings of the Royal Society of Victoria*, **99**, 111–116.

Byrne, M., Yeates, D. K., Joseph, L. *et al.* (2008). Birth of a biome: insights into the assembly and maintenance of the Australian arid zone biota. *Molecular Ecology*, **17**, 4398–4417.

Cadwallader, P. L. (1978). Some causes of the decline in range and abundance of native fish in the Murray–Darling River system. *Proceedings of the Royal Society of Victoria*, **90**, 211–224.

Chen, W., Lavoué, S., Beheregaray, L. B., Mayden, R. L. (2014). Historical biogeography of a new antitropical clade of temperate freshwater fishes. *Journal of Biogeography*, doi:10.1111/jbi.123333.

Chenoweth, S. F., Hughes, J. M. (1997). Genetic population structure of the catadromous Perciform: *Macquaria novemaculeata* (Percichthyidae). *Journal of Fish Biology*, **50**, 721–733.

Chessman, B. C. (2013). Identifying species at risk from climate change: traits predict the drought vulnerability of freshwater fishes. *Biological Conservation*, **160**, 40–49.

Christensen, J. H., Hewitson, B., Busuioc, A. *et al.* (2007). *Regional Climate Projections. Climate Change 2007: the Physical Science Basis. Contribution of Working Group I to the Fourth Assessment Report of the Intergovernmental Panel for Climate Change* (eds. **Solomon, S., Qin, D., Manning, M.** *et al.*). Cambridge, Cambridge University Press.

Cook, B. D., Bunn, S. E., Hughes, J. M. (2007). Molecular genetic and stable isotope signatures reveal complementary patterns of population connectivity in the regionally vulnerable southern pygmy perch (*Nannoperca australis*). *Biological Conservation*, **138**, 60–72.

Cooke, G. M., Chao, N. L., Beheregaray, L. B. (2012). Marine incursions, cryptic species and ecological diversification in Amazonia: the biogeographic history of the croaker genus *Plagioscion* (Sciaenidae). *Journal of Biogeography*, **39**, 724–738.

Cowx, I. G., Gerdeaux, D. (2004). The effects of fisheries management practises on freshwater ecosystems. *Fisheries Management & Ecology*, **11**, 145–151.

Crandall, K. A., Bininda-Emonds, O. R. P., Mace, G. M., Wayne, R. K. (2000). Considering evolutionary processes in conservation biology. *Trends in Ecology and Evolution*, **15**, 290–295.

Craw, D., Burridge, C. P., Anderson, L., Waters, J. M. (2007). Late Quaternary river drainage and fish evolution, Southland, New Zealand. *Geomorphology*, **84**, 98–110.

Davies, P., Harris, J., Tillman, T., Walker, K. (2008). *SRA Report 1: a Report on the Ecological Health of Rivers in the Murray–Darling Basin, 2004–2007*. Canberra, Australia, Murray–Darling Basin Commission.

Davis, E. B., Koo, M. S., Conroy, C., Patton, J. L., Moritz, C. (2008). The California hotspots project: identifying regions of rapid diversification in mammals. *Molecular Ecology*, **17**, 120–138.

Dudgeon, D., Arthington, A. H., Gessner, M. O. *et al.* (2006). Freshwater biodiversity: importance, threats, status and conservation challenges. *Biological Review*, 81, 163–182.

Dufty, S. (1986). Genetic and morphological divergence between populations of Macquarie perch (*Macquaria australasica*) east and west of the Great Dividing Range. Honours thesis, University of New South Wales, Sydney.

Echelle, A. A., Echelle, A. F. (1984). Evolutionary genetics of a 'species Flock': atherinid fishes on the Mesa Central of Mexico. In: *Evolution of Fish Species Flocks* (eds. Echelle, A. A., Kornfield, I.), pp. 93–110. Orono, University of Maine at Orono Press.

Edmands, S. (2007). Between a rock and a hard place: evaluating the relative risks of inbreeding and outbreeding for conservation and management. *Molecular Ecology*, 16, 463–475.

Estes, R. (1984). Fish, amphibians and reptiles from the Etadunna Formation, Miocene of South Australia. *Australian Zoologist*, 21, 335–343.

Faulks, L. K., Gilligan, D. M., Beheregaray, L. B. (2008). Phylogeography of a threatened freshwater fish (*Mogurnda adspersa*) in eastern Australia: conservation implications. *Marine and Freshwater Research*, 59, 89–96.

Faulks, L. K., Gilligan, D. M., Beheregaray, L. B. (2010a). Clarifying an ambiguous evolutionary history: range-wide phylogeography of an Australian freshwater fish, Golden Perch *(Macquaria ambigua). Journal of Biogeography*, 37, 1329–1340.

Faulks, L. K., Gilligan, D. M., Beheregaray, L. B. (2010b). Evolution and maintenance of divergent lineages in an endangered freshwater fish, *Macquaria australasica*. *Conservation Genetics*, 11, 921–934.

Faulks, L. K., Gilligan, D. M., Beheregaray, L. B. (2010c). Islands of water in a sea of dry land: hydrological regime predicts genetic diversity and dispersal in a widespread fish from Australia's arid zone, the golden perch *Macquaria ambigua. Molecular Ecology*, 19, 4723–4737.

Faulks, L. K., Gilligan, D. M., Beheregaray, L. B. (2011). The role of natural vs. anthropogenic in-stream structures in determining the genetic diversity of an endangered freshwater fish, Macquarie perch (*Macquaria australasica*). *Evolutionary Applications*, 4, 589–601.

Frankham, R., Ballou, J. D., Briscoe, D. A. (2002). *Introduction to Conservation Genetics*. Cambridge, Cambridge University Press.

Fraser, D. J., Bernatchez, L. (2001). Adaptive evolutionary conservation: towards a unified concept for defining conservation units. *Molecular Ecology*, 10, 2741–2752.

Gehrke, P. C., Astles, K. L., Harris, J. H. (1999). Within-catchment effects of flow alteration on fish assemblages in the Hawkesbury–Nepean River system, Australia. *River Research and Applications*, 15, 181–198.

Hallerman, E. M. (2003). *Population Genetics: Principles and Applications for Fisheries Scientists*. Bethesda, American Fisheries Society.

Hammer, M. (2001). *Molecular systematics and conservation biology of the southern pygmy perch Nannoperca australis (Gunther, 1861) (Teleostei: Percichthyidae) in south-eastern Australia*. Adelaide, Australia, Honours Thesis, Adelaide University.

Hammer, M., Adams, M., Unmack, P., Walker, K. F. (2007). A rethink on *Retropinna*: conservation implications of a new taxa and significant genetic sub-structure in Australian smelts (Pisces: Retropinnidae). *Marine and Freshwater Research*, **58**, 327–341.

Hammer, M., Unmack, P., Adams, M., Johnson, J., Walker, K. (2010). Phylogeographic structure in the threatened Yarra pygmy perch *Nannoperca obscura* (Teleostei: Percichthyidae) has major implications for declining populations. *Conservation Genetics*, **11**, 213–223.

Harris, J., Gehrke, P. (1997). *Fish and Rivers in Stress: the NSW Rivers Survey*. Sydney, Australia, New South Wales Fisheries Office of Conservation and Cooperative Research Centre for Freshwater Ecology.

Hewitt, G. (2000). The genetic legacy of the Quarternary Ice Ages. *Nature*, **405**, 907–913.

Hills, E. S. (1934). Tertiary freshwater fishes from southern Queensland. *Memoirs of the Queensland Museum*, **39**, 131–174.

Hills, E. S. (1943). Tertiary freshwater fishes and crocodilian remains from Gladstone and Duaringa, Queensland. *Memoirs of the Queensland Museum*, **12**, 96–101.

Hindar, K., Ryman, N., Utter, F. (1991). Genetic effects of cultured fish on natural fish populations. *Canadian Journal of Fisheries & Aquatic Sciences*, **48**, 945–957.

Howard, J., L. (2008). The future of the Murray River: emerity re-considered? *Geographical Research*, **46**, 291–302.

Huey, J. A., Hughes, J. M., Baker, A. M. (2006). Patterns of gene flow in two species of eel-tailed catfish, *Neosilurus hyrtlii* and *Porochilus argenteus* (Siluriformes: Plotosidae), in western Queensland's dryland rivers. *Biological Journal of the Linnean Society*, **87**, 457–467.

Huey, J. A., Baker, A. M., Hughes, J. M. (2008). The effect of landscape processes upon gene flow and genetic diversity in an Australian freshwater fish, *Neosilurus hyrtlii*. *Freshwater Biology*, **53**, 1393–1408.

Huey, J. A., Baker, A. M., Hughes, J. M. (2011). Evidence for multiple historical colonizations of an endoreic drainage basin by an Australian freshwater fish. *Journal of Fish Biology*, **79**, 1047–1067.

Hughes, J. M., Hillyer, M. (2006). Mitochondrial DNA and allozymes reveal high dispersal abilities and historical movement across drainage boundaries in two species of freshwater fishes from inland rivers in Queensland, Australia. *Journal of Fish Biology*, **68**, 270–291.

Hughes, J., Ponniah, M., Hurwood, D., Chenoweth, S. F., Arthington, A. (1999). Strong genetic structuring in a habitat specialist, the Oxylean Pygmy Perch *Nannoperca oxleyana*. *Heredity*, **83**, 5–14.

Hughes, J. M., Schmidt, D. J., Finn, D. S. (2009). Genes in streams: using DNA to understand the movement of freshwater fauna and their riverine habitat. *BioScience*, **59**, 573–583.

Humphries, P., Walker, K. (2013). *Ecology of Australian Freshwater Fishes*. CSIRO Publishing, Victoria Australia

Hurwood, D. A., Hughes, J. M. (1998). Phylogeography of the freshwater fish, *Mogurnda adspersa*, in streams of north-eastern Queensland, Australia: evidence for altered drainage patterns. *Molecular Ecology*, **7**, 1507–1517.

Imbrie, J., Hays, J., Martinson, D. G. *et al.* (1984). The orbital theory of the Pleistocene climate: support from a revised chronology of the marine $\delta^{18}O$ record. In: *Milankovitch and Climate* (eds. **Berger, A. L., Imbrie, J., Hays, J., Kukla, G., Salzman, B.**), Part 1. Riedel, Dordrecht, pp. 269–305.

Jerry, D. R. (1997). Population genetic structure of the catadromous Australian bass from throughout its range. *Journal of Fish Biology*, **51**, 909–920.

Jerry, D. R. (2008). Phylogeography of the freshwater catfish *Tandanus tandanus* (Plotosidae): a model species to understand evolution of the eastern Australian freshwater fish fauna. *Marine and Freshwater Research*, **59**, 351–360.

Jerry, D. R., Baverstock, P. R. (1998). Consequences of a catadromous life-strategy for levels of mitochondrial DNA differentiation among populations of the Australian bass, *Macquaria novemaculeata*. *Molecular Ecology*, **7**, 1003–1013.

Jerry, D. R., Woodland, D. J. (1997). Electrophoretic evidence for the presence of the undescribed 'Bellinger' catfish (*Tandanus* sp.) (Teleostei : Plotosidae) in four New South Wales mid-northern coastal rivers. *Marine and Freshwater Research*, **48**, 235–240.

Jerry, D. R., Elphinstone, M. S., Baverstock, P. R. (2001). Phylogenetic relationships of Australian members of the family Percichthyidae inferred from mitochondrial 12S rRNA sequence data. *Molecular Phylogenetics and Evolution*, **18**, 335–347.

Jones, F., Grabherr, M., Chan, Y. *et al.* (2012). The genomic basis of adaptive evolution in threespine sticklebacks. *Nature*, **482**, 55–61.

Kennard, M. J., Pusey, B. J., Olden, J. D. *et al.* (2010). Classification of natural flow regimes in Australia to support environmental flow management. *Freshwater Biology*, **55**, 171–193.

Kershaw, P., Moss, P., Van Der Kaars, S. (2003). Causes and consequences of long-term climatic variability on the Australian climate. *Freshwater Biology*, **48**, 1274–1283.

Kingsford, R. T., Auld, K. M. (2005). Waterbird breeding and environmental flow management in the Macquarie Marshes, arid Australia. *River Research and Applications*, **21**, 187–200.

Knight, J. T., Nock, C. J., Elphinstone, M. S., Baverstock, P. R. (2009). Conservation implications of distinct genetic structuring in the endangered freshwater fish *Nannoperca oxleyana* (Percichthyidae). *Marine and Freshwater Research*, **60**, 34–44.

Knudsen, R., Klemetsen, A., Amundsen, P. A., Hermansen, B. (2006). Incipient speciation through niche expansion: an example from the Artic charr in a subarctic lake. *Proceedings of the Royal Society of London B*, **273**, 2291–2298.

Latta, R. G. (2008). Conservation genetics as applied evolution: from genetic pattern to evolutionary process. *Evolutionary Applications*, **1**, 84–94.

Lintermans, M. (2007). *Fishes of the Murray–Darling Basin: An Introductory Guide.* Canberra, Murray Darling Basin Commission.

Longman, H. A. (1929). Specimens from a well at Brigalow. *Memoirs of the Queensland Museum*, **9**, 247.

Lovejoy, N., Bermingham, E., Martin, A. P. (1998). Marine incursion into South America. *Nature*, **396**, 421–422.

Lowe, W. & Likens, G. (2005). Moving headwater streams to the head of the class. *BioScience*, **55**, 196–197.

MacDonald, C. M. (1986). Morhpological and biochemical systematics of Australian freshwater and estuarine Percichthyd fishes. *Australian Journal of Marine & Freshwater Research*, **29**, 667–698.

Manel, S., Schwartz, M. K., Luikart, G., Taberlet, P. (2003). Landscape genetics: combining landscape ecology and population genetics. *Trends in Ecology and Evolution*, **18**, 189–197.

Masci, K. D., Ponniah, M., Hughes, J. M. (2008). Patterns of connectivity between the Lake Eyre and Gulf drainages, Australia: a phylogeographic approach. *Marine and Freshwater Research*, **59**, 751–760.

McCairns, R., Bernatchez, L. (2012). Plasticity and heritability of morphological variation within and between parapatric stickleback demes. *Journal of Evolutionary Biology*, **25**, 1097–1112.

McGlashan, D. J., Hughes, J. M. (2000). Reconciling patterns of genetic variation with stream structure, earth history and biology in the Australian freshwater fish *Craterocephalus stercusmuscarum*. *Molecular Ecology*, **9**, 1737–1751.

McGlashan, D. J., Hughes, J. M. (2001a). Low levels of genetic differentiation among populations of the freshwater fish *Hypseleotris compressa* (Gobiidae: Eleotridinae): implications for its biology, population connectivity and history. *Heredity*, **86**, 222–233.

McGlashan, D. J., Hughes, J. M. (2001b). Genetic evidence for historical continuity between populations of the Australian freshwater fish *Craterocephalus stercusmuscarum* (Atherinidae) east and west of the Great Dividing Range. *Journal of Fish Biology*, **59**, 55–67.

MDBC (2003). *Native Fish Strategy for the Murray Darling Basin 2003–2013.* Available online. http://www.mdba.gov.au/programs/native fish strategy.

Meehl, G. A., Stocker, T. F., Collins, W. D. *et al.* (2007). Global climate projections. In: *Climate Change 2007: the Physical Science Basis. Contribution of Working Group 1 to the Fourth Assessment Report of the Intergovernmental Panel on Climate Change* (eds. **Solomon, S., Qin, D., Manning, M.** *et al.*). Cambridge, Cambridge University Press.

Miller, A., Waggy, G., Ryan, S. G., Austin, C. M. (2004). Mitochondrial 12S rRNA sequences support the existence of a third species of freshwater blackfish (Percichthyidae: *Gadopsis*) from south-eastern Australia. *Memoirs of the Museum of Victoria*, **61**, 121–127.

Mittermeier, R. A., Gil, P. R., Mittermeier, C. G. (1997). *Megadiversity: Earth's Biologically Wealthiest Nations.* Conservation International, Cemex.

Moore, A., Ingram, B. A., Friend, S. *et al.* (2010). *Management of Genetic Resources for Fish and Crustaceans in the Murray–Darling Basin.* Canberra, Bureau of Rural Sciences.

Moritz, C. (1994). Defining evolutionary significant units for conservation. *Trends in Ecology and Evolution*, **9**, 373–375.

Moritz, C. (2002). Strategies to protect diversity and the evolutionary processes that sustain it. *Systematic Biology*, **51**, 238–254.

Musyl, M. K., Keenan, C. P. (1992). Population genetics and zoogeography of Australian freshwater golden perch, *Macquaria ambigua* (Richardson 1845) (Teleostei: Percichthyidae), and electrophoretic identification of a new species from the Lake Eyre basin. *Australian Journal of Marine and Freshwater Research*, **43**, 1585–1601.

Musyl, M. K., Keenan, C. P. (1996). Evidence for cryptic speciation in Australian freshwater eel-tailed catfish, *Tandanus tandanus* (Teleostei: Plotosidae). *Copeia*, **3**, 526–534.

Nanson, G. C., Price, D., Short, S. A. (1992). Wetting and drying in Australia over the last 300 ka years *Geology*, **20**, 791–794.

Near, T. J., Sandel, M., Kuhn, K. L. *et al.* (2012). Nuclear gene-inferred phylogenies resolve the relationships of the enigmatic pygmy sunfishes, elassoma (teleostei: Percomorpha). *Molecular Phylogenetics and Evolution*, **63**, 388–395.

O'Connor, J. P., O'Mahony, D. J., O'Mahony, J. M. (2005). Movements of *Macquaria ambigua*, in the Murray River, south-eastern Australia. *Journal of Fish Biology*, **66**, 392–403.

Ollier, C. D. (1978). Tectonics and geomorphology of the eastern highlands. In: *Landform Evolution in Australasia* (eds. Davies, J. L., Williams, M. A. J.). Canberra, Australian National University Press, pp. 5–48.

Ovenden, J. R., White, R. W. G., Sanger, A. C. (1988). Evolutionary relationships of *Gadopsis* spp. inferred from restriction enzyme analysis of their mitochondrial DNA. *Journal of Fish Biology*, **32**, 137–148.

Page, T. J., Sharma, S., Hughes, J. M. (2004). Deep phylogenetic structure has conservation implications for ornate rainbowfish (Melanotaeniidae: *Rhadinocentrus ornatus*) in Queensland eastern Australia. *Marine and Freshwater Research*, **55**, 165–172.

Page, T. J., Marshall, J. C., Hughes, J. M. (2012). The world in a grain of sand: evolutionary relevant, small-scale freshwater bioregions on subtropical dune islands. *Freshwater Biology*, **57**, 612–627.

Phillips, R. D., Storey, A. W., Johnson, M. S. (2009). Genetic structure of *Melanotaenia australis* at local and regional scales in the east Kimberley, Western Australia. *Journal of Fish Biology*, **74**, 437–451.

Poff, N. L., Olden, J. D., Pepin, D. M., Bledsoe, B. P. (2006). Placing global stream-flow variability in geographic and geomorphic contexts. *River Research and Applications*, **22**, 149–166.

Preston, B. L., Jones, R. N. (2008). Screening climatic and non-climatic risks to Australian catchments. *Geographical Research*, **46**, 258–274.

Puckridge, J. T., Sheldon, F., Walker, K. F., Boulton, A. J. (1998). Flow variability and the ecology of large rivers. *Marine and Freshwater Research*, **49**, 55–72.

Pusey, B., Kennard, M., Arthington, A. (2004). *Freshwater Fishes of North-Eastern Australia*. Australia, CSIRO Publishing.

Raeymaekers, J. A. M., Maes, G. E., Geldof, S. *et al.* (2008). Modelling genetic connectivity in sticklebacks as a guideline for river restoration. *Evolutionary Applications*, **1**, 475–488.

Rahel, F. J. (2007). Biogeographic barriers, connectivity and homogenization of freshwater faunas: it's a small world after all. *Freshwater Biology*, **52**, 696–710.

Reynolds, L. F. (1983). Migration patterns of five fish species in the Murray–Darling river system. *Australian Journal of Marine and Freshwater Research*, **34**, 857–871.

Rourke, M. (2007). *Population genetic structure of Murray cod (Maccullochella peelii peelii) and impacts of stocking in the Murray–Darling Basin*. PhD thesis, Melbourne, Australia, Monash University.

Rourke, M. L., McPartlan, H. C., Ingram, B. A., Taylor, A. C. (2010). Biogeography and life history ameliorate the potentially negative genetic effects of stocking on Murray cod (*Maccullochella peelii peelii*). *Marine and Freshwater Research*, **61**, 918–927.

Rowland, S. J. (1993). *Maccullochella ikei*, an endangered species of freshwater cod (Pisces: Percicthyidae) from the Clarence River system, NSW and *M. peeli mariensis*, a new subspecies from the Mary River system, Qld. *Records of the Australian Museum*, **45**, 121–145.

Ryman, N., Utter, F. (1987). *Population Genetics and Fishery Management*. Washington, USA, University of Washington.

Sabine, E., Schreiber, G., Bearlin, A. R., Nicol, S. J., Todd, C. R. (2004). Adaptive management: a synthesis of current understanding and effective application. *Ecological Restoration and Management*, **5**, 177–182

Sanmartin, I., Ronquist, F. (2004). Southern Hemisphere biogeography inferred by event-based models: plant versus animal patterns. *Systematic Biology*, **53**, 216–243.

Schramm, H. L., Piper, R. G. (1995). *Uses and Effects of Cultured Fishes in Aquatic Ecosystems*. Bethesda, American Fisheries Society.

Shaddick, K., Burridge, C. P., Jerry, D. R. *et al.* (2011a). A hybrid zone and bidirectional introgression between two catadromous species: Australian bass *Macquaria novemaculeata* and estuary perch *Macquaria colonorum*. *Journal of Fish Biology*, **79**, 1214–1235.

Shaddick, K., Burridge, C., Jerry, D. *et al.* (2011b). Historic divergence with contemporary connectivity in a catadromous fish, the estuary perch (*Macquaria colonorum*). *Canadian Journal of Fisheries and Aquatic Sciences*, **68**, 304–318.

Sharma, S., Hughes, J. M. (2011). Genetic structure and phylogeography of two freshwater fishes, *Rhadinocentrus ornatus* and *Hypseleotris compressa*, in southern Queensland, Australia, inferred from allozymes and mitochondrial DNA. *Journal of Fish Biology*, **78**, 57–77.

Soltis, D. E., Morris, A. B., McLachlan, J. S., Manos, P. S., Soltis, P. S. (2006). Comparative phylogeography of unglaciated eastern North America. *Molecular Ecology*, **15**, 4261–4293.

Stephenson, A. E. (1986). Lake Bungunnia – a Plio-Pleistocene mega-lake in southern Australia. *Palaeogeography, Palaeoclimatology, Palaeoecology*, **57**, 137–156.

Strayer, D. (2010). Alien species in freshwater: ecological effects, interaction with other stressors, and prospects for the future. *Freshwater Biology*, **55** (**Suppl. 1**), 152–174.

Stuart, I. G., Zampatti, B. P., Baumgartner, L. J. (2008). Can a low-gradient vertical-slot fishway provide passage for a lowland river fish community? *Marine and Freshwater Research*, **59**, 332–346.

Tallmon, D. A., Luikart, G., Waples, R. S. (2004). The alluring simplicity and complex reality of genetic rescue. *Trends in Ecology and Evolution*, **19**, 489–496.

Thacker, C., Unmack, P., Matsui, L., Rifenbark, N. (2007). Comparative phylogeography of five sympatric *Hypseleotris* species (Teleostei: Eleotridae) in south-eastern Australia reveals a complex pattern of drainage basin exchanges with little congruence across species. *Journal of Biogeography*, **34**, 1518–1533.

Thacker, C., Unmack, P., Matsui, L., Duong, P., Huang, E. (2008). Phylogeography of *Philypnodon* species (Teleostei: Eleotridae) across south-eastern Australia: testing patterns of connectivity across drainage divides and among coastal rivers. *Biological Journal of the Linnean Society*, **95**, 175–192.

Thuesen, P. A., Pusey, B. J., Peck, D. R., Pearson, R., Congdon, B. C. (2008). Genetic differentiation over small spatial scales in the absence of physical barriers in an Australian rain forest stream fish. *Journal of Fish Biology*, **72**, 1174–1187.

Turner, S. (1982). A catalogue of fossil fish in Queensland. *Memoirs of the Queensland Museum*, **20**, 599–611.

Unmack, P. J. (2001). Biogeography of Australian freshwater fishes. *Journal of Biogeography*, **28**, 1053–1089.

Unmack, P. J., Dowling, T. E. (2010). Biogeography of the genus *Craterocephalus* (Teleostei: Atherinidae) in Australia. *Molecular Phylogenetics and Evolution*, **55**, 968–984.

Unmack, P. J., Bagley, J. C., Adams, M., Hammer, M. P., Johnson, J. B. (2012). Molecular phylogeny and phylogeography of the Australian freshwater fish genus *Galaxiella*, with an emphasis on dwarf galaxias (*G. pusilla*). *PLoS ONE*, **7**, e38433.

Unmack, P. J., Allen, G. R., Johnson, J. B. (2013a). Phylogeny and biogeography of rainbowfishes (Melanotaeniidae) from Australia and New Guinea. *Molecular Phylogenetics and Evolution*, **67**, 15–27.

Unmack, P. J., Hammer, M. P., Adams, M., Johnson, J., Dowling, T. E. (2013b). The role of continental shelf width in determining freshwater phylogeographic patterns in south-eastern Australian pygmy perches (Teleostei: Percichthyidae). *Molecular Ecology*, **22**, 1683–1699.

Walters, C. J. (2007). Is adaptive management helping to solve fisheries problems? *Ambio*, **36**, 304–307.

Ward, R. D. (2006). The importance of identifying spatial population structure in restocking and stock enhancement programmes. *Fisheries Research*, **80**, 9–18.

Wellman. P, (1979), On the Cainozoic uplift of the south-eastern Australian highland. *Journal of the Geological Society of Australia*, **26**, 1–9.

Whitley, G. P. (1960). *Native Freshwater Fishes of Australia*. Brisbane, Australia, Jacaranda.

Wishart, M. J., Dawies, B. R. (2003). Beyond catchment considerations in the conservation of lotic biodiversity. *Aquatic Conservation: Marine and Freshwater Ecosystems*, **13**, 429–437.

Wong, B. B. M., Keogh, J. S., McGlashan, D. J. (2004). Current and historical patterns of drainage connectivity in eastern Australia inferred from population genetic structuring in a widespread freshwater fish *Pseudomugil signifer* (Pseudomugilidae). *Molecular Ecology*, **13**, 391–401.

Woodworth, L. M., Montgomery, M. E., Briscoe, D., Frankhan, R. (2002). Rapid genetic deterioration in captive populations: causes and conservation implications. *Conservation Genetics*, **3**, 277–288.

CHAPTER 24

Down under Down Under: Austral groundwater life

Grant C. Hose, Maria G. Asmyhr, Steven J. B. Cooper and William F. Humphreys

Summary

Aquifers of the Austral region are globally significant in terms of their biodiversity. They support a rich and unique fauna, specifically adapted to the harsh subterranean environment. In this chapter we review the nature and diversity of groundwater ecosystems across the Austral region. We consider first the global origins of the Australian groundwater fauna, and their distributions across Gondwana. As the Australian continent evolved, the western shield emerged from the sea during the Proterozoic, which has led to a distinct fauna in those ancient landscapes. In the 'newer' eastern Austral regions there has also emerged a rich groundwater fauna, and here we review the current knowledge of fauna in eastern Australia and New Zealand. Mining and agricultural development threaten groundwater ecosystems across the region, but perhaps the greatest threat is our current lack of knowledge of these unique and important ecosystems and their biota. New approaches for conservation planning provide hope for improved recognition and protection of groundwater ecosystems, but with relatively little surveying of groundwater fauna having been done across the region, much remains undiscovered.

24.1 Introduction

Being the driest inhabited continent on Earth, the availability of water has always been a critical factor shaping the evolution and distribution of species across Australia. So too, the availability of water is critical to the survival and prosperity of human populations across the broader region, from small outback towns to major capital cities. As human pressures increase demand for water, groundwater is increasingly being used to meet water needs of households, industries and farms. Groundwater use accounts for around 20% of the total water used across Australia, it is more than 50% in New Zealand (Fenwick *et al.* 2004), and in many areas it is the only reliable water supply.

Groundwater is essential to maintain a diverse range of groundwater-dependent ecosystems across Australia. These include some iconic and well-known surface

Austral Ark: The State of Wildlife in Australia and New Zealand, eds. A. Stow, N. Maclean and G. I. Holwell. Published by Cambridge University Press. © Cambridge University Press 2015.

landscapes in which groundwater is surface expressed, or where surface biota can access the subterranean water (Eamus *et al.* 2006). Such ecosystems include the red gum forests of the Murray River flood plain in the south, the savanna of the Daly River floodplain in the north, the *Banksia* forests of the Gnangara Mound and swan coastal plain in the west, and the artesian mound springs as oases across the central arid zone. But groundwater itself is also an ecosystem. Water moving through the rocks and sediments at depth creates unique environments, inhabited by a distinct and diverse fauna, containing species not found in surface environments. Only in recent decades has the true biological diversity of aquifers begun to emerge. The diversity of Austral groundwater ecosystems is indeed comparable to that of other regions of the globe, yet we have only begun to scratch the subsurface. Since aquifers cover much of Australia and New Zealand, aquifers are perhaps the most widespread ecosystem type across the Austral region.

Groundwater biota is under threat from a range of pressures, some specific to the region, others of a more cosmopolitan nature. However, unlike other ecosystems, the conservation of groundwater ecosystems is limited and, indeed, difficult to achieve (Boulton 2009). This chapter examines the Australian groundwater fauna and considers first how processes of continental drift and sea level change have shaped the diversity across the Austral region and, second, the threats facing the groundwater ecosystems in the region, and the current and potential conservation strategies that can ensure the protection of aquifers in terms of both the water resource and biodiversity value.

24.2 Overview of groundwater ecosystems

Groundwater ecosystems differ greatly from surface environments, in terms of both biota and the key ecological variables and processes. There is no light underground and, consequently, there are usually no primary producers in groundwater ecosystems. The ecosystem is thus largely dependent on carbon that filters from the surface, through the soil to the groundwater. This carbon is used by bacteria and fungi and provides the basis of the rather simple aquifer food web (Humphreys 2006).

In many aquifers, where pore spaces are large enough, higher order micro- and meio-invertebrates such as Turbellaria, Rotifera, Nematoda and Protozoa (Humphreys 2006) and some larger invertebrates are also present. The meiofauna are predominantly crustaceans, such as Copepoda, Syncarida, Amphipoda, Isopoda and Ostracoda, but may include insects, nematodes, molluscs, oligochaetes and mites. Of the crustaceans, some groups are found exclusively in groundwater (Remipedia, Thermosbanacea and Speleaogriphacea), and it is common for crustaceans to make up more than 50% of the total species abundance and richness of the groundwater invertebrate community (Korbel & Hose 2011). Insects, in particular, are relatively uncommon in groundwater (Humphreys 2006) although diverse coleopteran assemblages are emerging in some areas (Cooper *et al.* 2002; Leys *et al.* 2003; Watts *et al.* 2007) and stream insects can migrate long distances to be found in alluvial aquifers remote from surface waters (e.g. Stanford & Ward 1988).

Overall, vertebrates are rare in groundwater both in Australia and globally, and because of their large size, tend to be limited to karst aquifers where large water-filled voids exist. There are only three species of groundwater-adapted vertebrates known in the region. These are the blind cave eel, *Ophisternon candidum* (Mees 1962)

(Synbranchiformes: Synbranchidae) and two species of cave gudgeon, *Milyeringa species* (Perciformes: Eleotridae). *Milyeringa veritas* (Whitley 1945) co-occurs with *O. candidum* in the karst of Cape Range, north-western Australia (Humphreys 2006), and a second species, *Milyeringa justitia* has been recently described from nearby Barrow Island, WA (Larson *et al.* 2013) .

Most biota found in groundwater ecosystems are highly evolved, obligate groundwater-dwelling animals (stygobites) not found in surface environments. The dark, nutrient- and space-poor aquifer environment has shaped the convergent evolution of groundwater fauna. Groundwater species from different biological groups have independently evolved the common morphological traits of lack of eyes, hardened body parts, lack of body pigments, worm-like body shapes and enhanced non-ocular sensory appendages (Humphreys 2006). As a result, many species are morphologically very similar, requiring specialist taxonomic expertise, or genetic analysis to distinguish different species. The conditions of the groundwater realm favour those organisms so adapted, so healthy groundwater ecosystems will have a relatively high proportion of stygobites in comparison to non-groundwater-adapted surface species (stygoxenes) (Malard 2001; Stein *et al.* 2010).

The supply of external carbon limits the productivity of groundwater ecosystems, usually constraining pristine groundwaters to be low-energy environments, with low biomass and abundance of microbial and invertebrate fauna. With the exception of the occasional chemoautotrophic Prokaryotes, groundwater ecosystems are generally devoid of primary producers (such as plants or algae that cannot grow in the dark). With the additional lack of vertebrates, groundwater ecosystems tend to be relatively simple systems (Humphreys 2006). However, despite the relatively stable temperature and water quality conditions, groundwater ecosystems are characterised by biotic heterogeneity, meaning that there is an uneven distribution of biota over space and time.

Microbial assemblages are the foundation of aquifer ecosystems (Gibert *et al.* 1994; Humphreys 2006). Through heterotrophic or chemotrophic pathways, they capture energy and form the basis of the aquifer food web, providing a food source for invertebrates (Novarino *et al.* 1997; Humphreys 2006). The majority of microbes are distributed sparsely within the aquifer matrix, occurring as single cells or small colonies attached to sediment surfaces; generally less than 1% of available sediment surfaces are colonised by bacteria (Griebler *et al.* 2002; Anneser *et al.* 2010). Healthy, undisturbed aquifers tend to have very low microbial diversity and activity relative to surface waters (Griebler & Lueders 2009), due mainly to naturally low concentrations of nutrients, carbon and oxygen (Gounot 1994). Most microbes inhabiting aquifers are attached rather than being free-living (Gounot 1994; Griebler & Lueders 2009; Anneser *et al.* 2010) although the ratio of attached to free-living bacteria can change with contamination (e.g. Griebler *et al.* 2002). See also Box 24.1.

In terms of meiofauna, groundwater ecosystems are typified by low α diversity (few species at any one locality), with a 'truncated' functional and taxonomic diversity (Gibert & Deharveng 2002). This creates a system with (generally) low horizontal (within trophic level) and vertical (between trophic level) diversity (*sensu* Duffy *et al.* 2007) in a given location, and short food chains. However, isolation has created a fauna dominated by short-range endemic species (Harvey 2002), providing high ß diversity (many species across localities) of invertebrates (Humphreys 2008). This trend is not evident in microbial assemblages which appear to be much more widely distributed (Griebler & Leuders 2009).

> **Box 24.1** Microbial assemblages
>
> Like other elements of the groundwater environment, the microbial assemblages inhabiting aquifers in Australia have been poorly studied, with the exception of those organisms and assemblages used for the bioremediation of contaminated aquifers. The current paradigm of groundwater microbiology is that 'everything is everywhere' meaning that the microbial flora of groundwater is widespread, and comprises species commonly found in soil and surface water environments, with no endemic groundwater microbes yet recorded (Griebler & Leuders 2009).
>
> Recent research in karst regions of the Nullabor and Cape Range (Seymour *et al.* 2007; Humphreys *et al.* 2012) suggest the potential for novel 'groundwater-only' species or at least specific assemblages may exist. The caves of the Nullabor host unique microbial biofilms, or 'cave slimes' that occur as dense 'mantles' or curtains that are suspended in the water column of the large water-filled limestone caverns (Holmes *et al.* 2001). In this environment, carbon for growth and metabolism is extremely limited. Accordingly, the microbes in the biofilm rely heavily on oxidation of nitrogen for energy. Detailed genetic screening of the biofilms identified many microorganisms known from other environments, but also identified several novel species which may well be endemic to these unique microbial communities (Tetu *et al.* 2013). Studies in Cape Range (Seymour *et al.* 2007; Humphreys *et al.* 2012) and in the Ashbourne aquifer, South Australia (Smith *et al.* 2012), have shown strong vertical stratification in microbial assemblages and are further evidence that groundwater assemblages are unique compared to those of other aquatic and terrestrial habitats.
>
> In general, archaea and bacteria are diverse and abundant in groundwater communities, but the micro-eukaryotes, such as fungi, are relatively poorly studied. However, the little information available suggests that fungi are generally common but in lower abundance and diversity than they are in surface water ecosystems (Lategan *et al.* 2012). Indeed, fungi and yeasts made up less than 5% of the total abundance of cultivable heterotrophic microbiota in groundwater samples from aquifers across NSW (Lategan *et al.* 2012). Like bacterial assemblages, there is commonality of most taxa between aquifers at the genus level (Lategan *et al.* 2012), although some taxa were limited to specific wells or aquifers, particularly those that had been contaminated with chlorinated hydrocarbons. Indeed, low level contamination of groundwater by chlorinated hydrocarbons may stimulate the richness of fungi and yeasts. Highly chlorinated sites had few or no cultivable fungi and the uncontaminated reference sites were similar in richness and diversity to other aquifers across the state (Lategan *et al.* 2012).

24.3 Austral diversity

Just as in other ecosystems, and as evident in other chapters of this book, the biota of groundwater ecosystems reflects the unique history and evolution of the Australian continent and the region. Many groundwater species are relics of once common surface species that sought refuge in the stable groundwater environment during times of past climate change. Consequently, the Austral groundwater fauna reflects the various stages of the evolution of the Austral region, starting with the break-up of Pangaea and

Gondwana, the drift north and the changing climate and aridity, the early and sustained emergence of large parts of the Australian continent from the sea during the Proterozoic, and the more recent evolution of the east coast landscapes and New Zealand. The evolution of the Austral landscape differed considerably from that of the Northern Hemisphere, particularly in that it lacked the Pleistocene glaciations that are considered to have been critical to the evolution of groundwater diversity in the Northern Hemisphere (Guzik *et al.* 2010). For this reason, and the extensive areas of saline groundwater and aridity across the region, Austral groundwater ecosystems were long considered to be depauperate of specialised groundwater fauna. However, since the 1990s, it has been established that the subterranean realm in Australia supports an enormous range of obligate subterranean animals in both terrestrial and aquatic habitats. Furthermore, as the species are typically endemic to very small areas, there is a very high biodiversity at both landscape and continental scales, a story that is still unfolding (Humphreys 2008, 2012; Guzik *et al.* 2010).

In 2000, there were, globally, over 7800 known stygofaunal species (Juberthie 2000). However, recent estimates will be far greater, due to large research efforts in Australian (see Guzik *et al.* 2010) and European karst regions. The growing interest in groundwater fauna across the Austral region is reflected in the recent growth in species descriptions, particularly arising from studies in the Yilgarn and Pilbara, but also further afield. In 2010, Guzik *et al.* reported some 770 stygofauna taxa known in the west of Australia. From this they predicted that around 80% of the groundwater fauna are unknown, and the true richness of the region may be as high as 4140 stygobitic species. With this high diversity, and the largely unexplored diversity of eastern Australia, it is clear that the Austral region is a biodiversity hotspot of global significance. In the next section we will consider the global origins of some elements of the groundwater fauna and the differences across the Austral region.

24.4   Global origins of Austral groundwater fauna

24.4.1   Tethyan connections

In a small area of north-western Australia, along the coast at Ningaloo Reef, Barrow Island and the adjacent Pilbara coast, the coastal groundwater in karst, though not directly connected with the sea, is affected by marine tides, is highly stratified with brackish water overlying sea water, and in total darkness (Jaume & Boxshall 2013). This anchialine habitat harbours a remarkable fauna largely comprising crustaceans restricted to anchialine systems the generic composition of which is predictable however far apart in the world they occur. This region of Australia abutted the Tethys, a shallow sea that spread between Africa and Europe into the opening Atlantic during the Jurassic as a result of the fragmentation of the supercontinents Pangaea and subsequently Gondwana. This connection has left a strong Tethyan signature in the fauna of these Australian anchialine habitats which are closely related to those inhabiting caves on either side of the Atlantic, in the Canary Islands and northern Caribbean. The marine part of the anchialine system includes many higher taxa that are new to the Southern Hemisphere including Remipedia, a new class described from the Caribbean as recently as 1981. Also copepods of the order Misophrioida and families Epacteriscidae and Ridgewayiidae, and the order Thermosbaenacea, tiny shrimp-like crustaceans that uniquely carry their eggs and young in a dorsal rather than a ventral pouch (see Figure 24.2, Plate 57). One lineage, atyid shrimps of the genus *Stygiocaris* (see Figure 24.2, Plate 57), has been examined using molecular phylogenetics and found to be only distantly related to the other Australian atyid shrimps but which is the sister taxon to *Typhlatya* that occurs in similar habitats on either side of the North Atlantic (Page *et al.* 2008; von Rintelen *et al.* 2012). It has not yet been resolved whether this vicariant distribution resulted from fragmentation of continents, dispersal within Tethys or both (Bauzà-Ribot *et al.* 2012; Phillips *et al.* 2013).

24.4.2 Gondwanan connections: Phreatoicidea

Phreatoicidean isopods have a Gondwanan distribution and occur widely across southern Australia (and in tropical Arnhem Land and the Kimberley). Their distribution is strongly associated with the areas of the continent not submerged by Cretaceous seas. Five hypogean species are known from the Precambrian western shield where the family Hypsimetopidae is represented by the genera *Pilbarophreatoicus* in the Pilbara and *Hyperoedesipus* in the Yilgarn. These genera are closely related to the hypogean genera found in the Ganges Valley and from caves of Andrah Pradesh (India). These occurrences suggest that Hypsimetopidae were hypogean prior to the separation of Greater Northern India from the western shore of Australia (*c.* 130 My BP) (Wilson 2008).

Crenisopus, which is a stygobiont genus found in aquifers in the Kimberley region of Western Australia (Figure 24.1, Plate 56), is ancestral to most families in the Phreatoicidea, suggesting that the divergence between them occurred after they entered freshwater but prior to the fragmentation of East Gondwana during the Mesozoic era, this providing a link between African and Australasian lineages of phreatoicideans (Wilson & Keable 1999).

24.4.3 Gondwanan connections: Spelaeogriphacea

The crustacean order Spelaeogriphacea, which comprises four species of stygobionts, was described from a cave at an altitude of about 900 m up Table Mountain in South Africa, a feature pre-dating the separation of South America and Africa. A second species occurs in caves in Mato Grosso do Sul, western Brazil, these locations suggesting a split prior to the separation of Africa and South America, 110 Ma in the Early Cretaceous. The finding of two species in the Fortescue Valley, Pilbara, strengthens the Gondwana link. Living spelaeogriphaceans all occur in contact with geological contexts that are earliest Cretaceous or older. The colonisation of Gondwanan freshwater is likely to have occurred after the retreat of the Gondwanan ice sheet (after 320 Ma BP) and prior to the dissolution of Gondwana (142–127 Ma BP) (Jaume 2008).

24.4.4 Gondwanan connections: Bathynellacea

The tiny, vermiform and mostly interstitial stygobionts belonging to the Bathynellacea are another taxon suggesting links with other continents but these have yet to be tested using molecular phylogenetic methods. They are restricted to groundwater systems and occur throughout the world, suggesting that they are of Pangaean origin. However, most species are known only from their type locality and many belong to monotypic genera so that it is not surprising that their geographic distributions contain strong continental signatures. Some Australian parabathynellid genera have apparently clear affinities with

Figure 24.1 (Plate 56) Upper, *Crenisopus* sp. from Koolan Island, WA (photo George Wilson); middle, Phreatoicid isopod from WA (photo George Wilson); lower, *Brevisomabathynella* sp. (Parabathynellidae) from calcrete aquifer in Western Australia (photo Kym Abrams). A black and white version of this figure will appear in some formats. For the colour version, please refer to the plate section.

eastern Gondwana; *Chilibathynella* and *Atopobathynella* are known from Chile and southeastern Australia, the latter also from India. *Notobathynella* also occurs in New Zealand while other Australian genera have a broader distribution with *Hexabathynella*, from eastern Australia, being found in New Zealand, southern Europe, Madagascar and South America. Conversely the genera *Kimberleybathynella*, *Brevisomabathynella*, *Billibathynellla* and *Octobathynella* are endemic to Australia (Schminke 2011).

The Australian fauna has turned up a paradox because the Parabathynellids are widely considered to have a freshwater origin and the species are typically from freshwater systems <500 mg l^{-1}. However, many of the species in Australia occur in brackish water and *Brevisomabathynella clayi*, a large, free-swimming parabathynellid (Figure 24.1, Plate 56), occurs in arid Australia in water close to marine salinity (PSU 35.93) (Cho & Humphreys 2010) in the centre of the Western Shield, where it is associated with a number of maritime copepod lineages such as Ameiridae (Harpacticoida) and *Halicyclops* (Cyclopoida) (Humphreys 2008).

24.5 The arid interior of WA and SA

In the mid 1990s, following discoveries along coastal Western Australia of diverse crustacean fauna with Tethyan and Gondwanan connections, a diverse stygofauna was discovered in the arid Yilgarn and Pilbara regions of Western Australia (Humphreys 2001). These regions together form the Western Shield, a region of the Earth's crust that has remained emergent since the Paleozoic, and consequently containing an ancient history of freshwater habitats. The Western Shield is rich in mineral deposits and the requirement in Western Australia for mineral resource development projects to include subterranean fauna in environmental impact assessments resulted in widespread surveys that have shown that non-karst and non-alluvial substrates, such as pisolites and fractured rocks of the Pilbara region, also support diverse subterranean aquatic and terrestrial faunas. This survey effort has not been applied to many other areas of the Australian arid zone, with the exception of some recent work in South Australia (Flinders Ranges and Eyre Peninsula regions; Remko Leijs, personal communication).

Within the 220 000 km² Pilbara region, ~300 new species of troglobionts have been reported by consultancy companies, many of which appear to be short range endemics (SREs; Harvey 2002) and found in a variety of habitats, including pisolite, fractured rock, limestone karst and calcrete. These include diverse groups of arachnids, such as Schizomida, a tropical forest group that has survived the surface aridity in pesolitic iron ore mesas of the Pilbara region and humid caves in the arid Cape Range, and Palpigradi, with Australian native species only recently discovered in caves or boreholes in karst and fractured rock substrates from the Pilbara and calcretes of the Yilgarn region (Barranco & Harvey 2008). The Pilbara region is also known as a biodiversity hotspot for the crustacean order Ostracoda, with recent finds from the family Candonidae, representing 25% of the known genera in the world (Humphreys 2012; Karanovic 2012). Although the Pilbara and Yilgarn subterranean fauna regions are contiguous, they have radically different subterranean faunas with little overlap, even at the generic level (Table 24.1). The reason for this remarkable disjunction in the composition of the fauna is not understood and is difficult to reconcile with the Western Shield having been a single, continuously emergent land mass since the Paleozoic at least.

Within the Yilgarn region, the fauna, comprising diverse diving beetle species (Dytiscidae) and crustaceans, such as Amphipoda, Isopoda, Bathynellacea and Copepoda, was discovered in groundwater calcretes, shallow (10–20 m thick) carbonates formed by precipitation from groundwater, as a result of near-surface evaporation. Hundreds of isolated calcrete bodies exist in the region, deposited upstream of salt lakes (playas) along ancient palaeovalley systems. Groundwater flow and episodic surface recharge events, and presumably plant root systems, have resulted in numerous minor karst features within the calcretes, providing a suitable habitat for tiny invertebrates (<5 mm) within a groundwater environment that is the equivalent of a

Table 24.1   Overlap of stygal taxa between contiguous northern Yilgarn and Pilbara regions on the Western Shield of Australia. (Abstracted from Humphreys 2008 where the relevant references are given.) Shown are the number of species and genera of Copepoda, Candoninae (Candonidae: Podocopoda), oniscidean and tainisopidean Isopoda, and Dytiscidae (Coleoptera)

Stygal taxon	Unit	Yilgarn	Pilbara	Overlap %
Copepoda	species	30	43	4
Copepoda	genera	15	25	21
Podocopoda: Candoninae: Candonidae	species	5	58	0
Podocopoda: Candoninae: Candonidae	genera	1	13	8
Isopoda: Oniscidea, (stygobionts)	species	c. 20	0	0
Isopoda: Tainisopidea	species	1	5	17
Coleoptera: Dytiscidae	species	97	0	0

subterranean estuary (Humphreys *et al.* 2009). More recently, a diverse troglofauna, comprising taxa, such as oniscidean isopods, collembola, insects, diplurans, myriapods and arachnids (see Barranco & Harvey 2008) that would otherwise be at home in rainforest leaf litter, was also found to be associated with the groundwater calcretes.

Taxonomic treatment of much of the stygofauna and troglofauna is still in its infancy, but studies to date, including molecular phylogenetic and phylogeographic analyses of the stygobitic dytiscid beetles, amphipods and isopods, have shown that numerous species exist, the majority of which are restricted in their distribution to single calcrete bodies, leading to the description of the calcrete system as a subterranean archipelago (Cooper *et al.* 2002, 2007, 2008; Leys *et al.* 2003; Watts & Humphreys 2009). Molecular clock analyses suggest that subterranean lineages evolved 5–8 My ago, placing their evolutionary origins firmly in the period of spreading aridity on the Australian continent, following the drift northwards of the Indo-Australian plate to south-east Asia. This was a period, commencing in the late Miocene (~14 million years ago), of major climatic and environmental changes that resulted in the widespread disappearance of rainforest and permanent sources of water from the interior of Australia. It is likely that the groundwater calcretes acted as refugia for rainforest taxa that were able to adapt to life in darkness and track the water table underground. Notably, it is the groups of animals associated with rainforest leaf litter or capable of living in the hyporheic zone of creeks and rivers that comprise many of the taxa found in subterranean ecosystems. The species have subsequently survived in the calcretes for millions of years, despite considerable fluctuations in surface conditions resulting from ice age climatic changes during the Plio-Pleistocene, when the interior of Australia cycled from wet/warm to arid/cold climates (Byrne *et al.* 2008). They are thus living fossils of the Australian Miocene rainforest environment.

It has been recently discovered that many of the groups of aquatic organisms found in the Western Australian calcretes are also found in other Australian arid zone freshwater 'refugia'. These include the wetlands associated with isolated mound springs of the Great Artesian Basin (GAB: springs that formed by upwelling of groundwater from the GAB) in South Australia, which have provided refugia for a variety of aquatic organisms, including isopod and amphipod crustaceans that are related to species found in the calcretes (Murphy *et al.* 2009). Calcretes of the Ngalia basin, northwest of Alice Springs in the Northern Territory also harbour many of the taxa found in the Western Australian

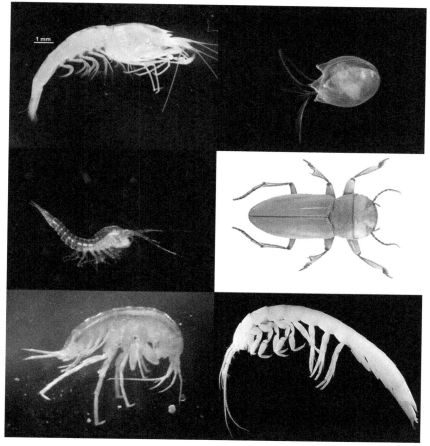

Figure 24.2 (Plate 57) Clockwise from upper left: *Stygiocaris* sp. (Atyidae) from anchialine waters of Ningaloo Coast World Heritage Area (photo Danny Tang); *Danielopolina baltanasi* from anchialine caves on Christmas Island, Indian Ocean (photo Ross Anderson); *Paroster byroensis* from calcrete aquifers, Gascoyne, WA (photo Chris Watts); *Pygolabis* humphreysi (Tainisopodidae) from the Pilbara, a family of isopods, also in the Kimberley, endemic to WA (photo George Wilson); Paramelitid amphipod from Cane River, WA (photo Remko Leijs); *Halosbaena tulki* (Thermosbaenacea), Ningaloo Coast World Heritage Area (photo Douglas Elford, Western Australian Museum). A black and white version of this figure will appear in some formats. For the colour version, please refer to the plate section.

calcretes, including dytiscid beetle species (e.g. Balke *et al.* 2004; Figure 24.2, Plate 57) and a remarkably diverse *Haloniscus* isopod fauna (S. Taiti, unpublished data). These common elements further support the role of aridity in driving relatively widespread rainforest species underground or into isolated groundwater-dependent ecosystems on the surface.

Ongoing research of groundwater systems in the South Australian arid zone, also indicate that a diverse stygofauna exists in alluvial aquifers of the Flinders Ranges and Eyre Peninsula, including a dytiscid beetle (Leys *et al.* 2010) and parabathynellids (Abrams *et al.*, 2013). Amphipoda (family Chiltoniidae) are also particularly diverse in these groundwater systems, and again reveal the ancient connections between now isolated freshwater systems (R. King & R. Leijs, personal communication). Like the alluvial aquifers of eastern Australia there are many more subterranean species in the arid interior of Australia that await discovery.

24.6 The 'new' east

In their recent review, Guzik *et al.* (2010) predicted lower diversity in the newer landscapes of eastern Australia. The east of the continent is geologically younger, having had marine inundations through the Cretaceous period, and overall a much more dynamic geological history than parts of western Australia which have been emergent from the sea since the Proterozoic (BRM Palaoegeographic Group 1990).

Contemporary studies of groundwater biota in the east have focussed on applied ecology, with a focus to understand relationships to land use (Korbel *et al.* 2013), pollution impacts (Sinton 1984; Hartland *et al.* 2011; Stephenson *et al.* 2013), hydrology (Tomlinson 2008; Stumpp & Hose 2013), and aspects of groundwater sampling and assessment (Hancock & Boulton 2009; Hose & Lategan 2012; Korbel & Hose 2011). While some studies have compared the diversity of different aquifers and regions (Hancock & Boulton 2008) they have generally not been detailed, broad studies of phylogeography such as those undertaken in the Yilgarn and Pilbara (Guzik *et al.* 2008; Cooper *et al.* 2008), although some small-scale phylogenetic analyses have been done (Hose 2009; Cook *et al.* 2012).

The structure of groundwater assemblages in eastern Australia is similar to those in the west. Certainly the same higher taxa are present, although, as expected given the geographical separation, taxa collected from eastern Australia often form separate line-ages within continental phylogenies (Leys & Watts 2008; Abrams *et al.*, 2013). Further, there appear differences in the relative abundances of different taxa. While amphipods and isopods are widespread across the west (e.g. Eberhard *et al.* 2009; Schmidt *et al.* 2007) they are encountered less frequently in the east (e.g. Hancock & Boulton 2008, 2009), which may reflect the history of marine inundations in the east, which elsewhere are considered to be responsible for the absence of some of the more ancient crustacean lineages of amphipods and isopods in the groundwater fauna (Bradbury 1999; Wilson & Johnson 1999). Instead, crustacean faunas are dominated by syncarids (Bathynellidae, Parabathynellidae and Psammaspididae) and Harpacticoid and Cyclopiod copepoda. Anaspidacea are only found in eastern Australia (Hobbs 2000), with several families found across the region (Serov 2002). Hancock & Boulton (2008) reported stygobitic elimid and dytiscid beetles from the Hunter and Peel River valleys, NSW (Watts *et al.* 2007, 2008), and these records remain the only examples of stygobitic beetles inhabiting alluvium in Australia. Although diverse dytiscid water beetle assemblages are present in the calcretes of the Yilgarn (Leys *et al.* 2003; Watts *et al.* 2008), such beetles are rare in the east.

Karstic and limestone habitats are traditionally the hotspots of groundwater diversity, and these habitats are somewhat rarer in eastern Australia than in the Western Shield and throughout Europe. However, much of the karst of New South Wales is impounded and as such is geologically isolated, and in some cases is of Devonian origin (Osborne *et al.* 2006) so there is great potential for these environments to support novel taxa through such long isolation. To date 82 stygobitic species have been recorded in the east coast karsts (sum-marised in Thurgate *et al.* 2001). Caves of Tasmania and New South Wales have the most diverse groundwater faunas (Thurgate *et al.* 2001), although this richness may also reflect the greater sampling effort in those states. The species richness of most cave and karst regions in eastern Australia is low by world standards (Thurgate *et al.* 2001) but this may be a consequence of the small, impounded nature of these features.

The deep alluvial deposits of the east coast have yielded a rich fauna. Hancock & Boulton (2008) examined parts of the Peel and Hunter Valley alluviums in NSW, and the Burnett and Pioneer Valley alluviums in Queensland. The study recorded 87 taxa, with seven to 33 taxa per aquifer. Tomlinson (2008) expanded on Hancock & Boulton's (2008) survey of the Peel alluvium and reported 54 taxa of which 33 were obligate groundwater inhabitants. More recent surveys of the alluvial aquifers associated with inland rivers have yielded similar richness, with 20 taxa in the Gwydir (Korbel *et al.* 2013), 15 in the Namoi (Korbel 2013), and ten taxa in a small area of the Macquarie River alluvium (Hancock & Boulton 2009). Similar higher taxa were recorded across all studies, but it was often the Acarina that were the most (morpho)species-rich group. Although detailed species-level taxonomy has not been performed on the specimens collected in these studies, it is likely that each taxon represents a new species, with perhaps the exception of potential hydrologic connectivity and hence species overlap between the Peel and Namoi Rivers.

Upland swamps are distinctive features of low-relief plateau areas in eastern Australia (Young 1986; Dodson *et al.* 1994). They form in shallow depressions in the landscape and, fed by rainfall and regional groundwater, provide a unique, shallow groundwater ecosystem. Sampling of swamps in the Southern Highlands and Blue Mountains of NSW has identified assemblages variably dominated by harpacticoid and cyclopoid copepods, nematodes and ostracods, but also containing syncarids, mites and amphipods. While not all species are likely to be stygobitic, these systems are potentially a diverse source of fauna. Initial studies show particularly fine-scale endemism, with swamps in the Southern Highlands separated by only several hundred metres containing genetically distinct harpacticoid copepod populations (Hose 2009). Similar patterns of endemism are likely in Blue Mountains swamps as seen in other taxa (Dubey & Shine 2010).

The fractured rock systems of eastern Australia appear somewhat less diverse than those of the west (e.g. Eberhard *et al.* 2005), but nevertheless, stygofauna have been recorded in the fractured Triassic Hawkesbury sandstone to the north (11 stygobitic taxa; Hose & Lategan 2012) and south of Sydney (three stygobitic taxa; Hose 2008, 2009), with the likelihood that each area contains locally endemic taxa. The assemblages are dominated by copepods (Harpacticoida and Cyclopoida), but syncarids are also common in some areas. The fauna appears spatially limited within the aquifers, and constrained to shallow water bearing zones (Hose, personal observation). There is a clear need to extend the taxonomy of these surveys to better describe and contextualise the diversity of this region.

The coastal sand aquifers are a common feature of the east coast of Australia. They are an important water source for local communities, but given the high population density on the east coast, the integrity of these aquifers is under threat. For example, the Botany sands aquifer in Sydney is perhaps the most contaminated groundwater site in Australia, with contamination from industrial chlorinated hydrocarbons that began in the 1940s (Acworth 2001). The site is now one of Australia's largest groundwater clean-up projects. While sampling of the aquifer did not reveal meiofauna, diverse prokaryote and eukaryote assemblages were evident, with changes in assemblage structure evident and associated with the contamination gradient (Stephenson *et al.* 2013). The pore space provided by sand aquifers is generally too small to permit stygofauna, although sampling in the coastal sands of the Tomago-Tomaree and Woy Woy and Umina sandbeds on the NSW central coast region of the NSW central coast has revealed three stygobitic taxa (Hose, unpublished data).

Some of the earliest records of stygofauna in the Austral region emanated from the rich alluvial aquifers of the Canterbury plains (Chilton 1882). This region, and indeed the rest of the New Zealand islands have proved a rich source of stygofauna with over 160 known groundwater taxa (Fenwick & Scarsbrook 2004), but the true richness is likely to exceed several hundred taxa (Fenwick *et al.* 2004).

As seen in Western Australia over the past decade, the intensity of mineral resource exploration and extraction has fuelled the search for stygofauna as part of environmental impact assessments and mine approvals. Following the lead of the west, new approvals for coal and coal seam gas extraction are also requiring assessments of groundwater ecosystems. To date this has been largely *ad hoc*, and done by environmental consultants as needed on a site-specific basis. There is a clear need for a comprehensive survey of groundwater ecosystems in eastern Australia, supported by government agencies to examine the diversity of these ecosystems before more regions are developed for mineral exploitation.

24.7 Threats to groundwater ecosystems

The greatest threat facing groundwater ecosystems both in the Austral region and globally, is the current lack of understanding of structure, functioning and diversity of these systems. Indeed, many topics that are well studied in other branches of ecology are essentially unexplored in groundwater (Larned 2012). In particular, little is known of the ecology, distribution and life history of groundwater taxa, which makes conservation efforts particularly challenging (see also Box 24.2).

Box 24.2 Ecosystem services of groundwater ecosystems

Ecosystem goods and services (collectively 'ecosystem services') are natural products and ecological functions that are of benefit to humans (Millennium Ecosystem Assessment 2005). Groundwater ecosystems provide a number of such services (Table 24.2) which include water storage and transport, water purification, and the support of groundwater-dependent ecosystems (Haack & Bekins 2000; Herman *et al.* 2001; Murray *et al.* 2006; Boulton *et al.* 2008).

Table 24.2 Ecosystem services provided by groundwater (from Tomlinson & Boulton 2008)

Type of service	Examples
Provisioning	Water for drinking, irrigation, stock and industrial uses
Supporting	Bioremediation (water purification), ecosystem engineering (flow maintenance), nutrient cycling, sustaining linked ecosystems, providing refugia
Regulating	Flood control and erosion prevention
Cultural	Religious and scientific values, tourism

Perhaps the most critical of these services is the provision of water for human consumption, industry and agriculture. Currently, aquifers provide 600–700 km^3 of water annually, making it the world's most extracted raw material (Zektser & Everrett 2004). In many developing countries it is the most important and safest source of drinking water, yet it is equally important in developed countries, providing nearly 70% of the piped water supply in EU countries (Zektser & Everrett 2004).

The capacity for water quality improvements in aquifers, such as the removal of nitrogen, breakdown of organic contaminants and the assimilation of DOC, is determined largely by the microbial assemblages within an aquifer (Gounot 1994; Chapelle 2001; Griebler 2001; Haack & Bekins 2000, Griebler & Lueders 2009). Under anaerobic conditions, denitrifying bacteria can oxidise nitrate to gaseous nitrogen which is then removed from the groundwater (Gounot 1994). Similar microbial-mediated processes are responsible for the breakdown of organic chemicals; however, the exact mechanisms remain unclear (Chapelle 2001; Goldscheider *et al.* 2006).

Stygofauna contribute to water quality improvement by grazing on microbial biofilms which in turn promotes biofilm activity and hence purification capacity (Chapelle 2001; Gounot 1994). The burrowing activity of macroinvertebrates and the biofilm grazing may help maintain the porosity of the aquifer matrix, thereby maintaining flow and the movement of nutrients through the aquifer (Boulton *et al.* 2008; Danielopol *et al.* 2000; Haack & Bekins 2000; Des Châelliers *et al.* 2009). There is evidence that macroinvertebrate burrowing enhances flow in benthic and hyporheic environments and although likely to occur, it remains untested in aquifers (Boulton *et al.* 2008).

The economies of Australia and New Zealand are both dependent on primary production and, particularly for Australia, mineral exports. Unfortunately, both of these major industries pose a threat to groundwater ecosystems (e.g. Korbel *et al.* 2013). Physical threats to groundwater ecosystems arise through changes in the groundwater regime, such as changes in the availability of groundwater in terms of its flow, depth and timing, and the quality of the water.

With groundwater ecosystems heavily dependent on carbon and oxygen infiltration from surface environments, changes to the rate and volume of groundwater moving through an aquifer can alter the distribution of such nutrients, and consequently impact significantly on fauna. Reductions in flow and the resupply of oxygen can lead to anoxia and limited carbon availability. Natural patterns of decreasing nutrients with depth limit most groundwater fauna to shallow depths (Danielopol *et al.* 2000), and changing the groundwater regime further exacerbates this limitation to fauna distributions. Such changes may occur by way of reduced groundwater recharge, such as interception of run-off by impervious surfaces, or groundwater abstraction leading to changes in the natural hydraulic gradient and flow. Lowering of the watertable in the Jewel Cave Karst system has led to drying of cave root mat habitats and the loss of fauna (Eberhard & Davies 2011).

Lowering of groundwater tables is a common consequence of over-abstraction of groundwater or reduced groundwater recharge, such that the rate at which water is extracted exceeds the rate at which water recharges the aquifer. Fauna are able to tolerate small changes in groundwater levels, but ongoing declines in water levels reduce the volume of accessible habitats for stygofauna and can disconnect groundwater from surface waters (Hancock 2009). Rapid declines in groundwater may lead to stranding of organisms in layers above the water table (Tomlinson 2008; Stumpp & Hose 2013), but the mobility of the taxa is important in enabling their survival. Stygofauna may not survive 48 h in drying sediments (Tomlinson 2008; Stumpp & Hose 2013).

Anthropogenic activities have impacted the quality of groundwater the world over, reducing its suitability for human use and as a habitat for groundwater biota. In the Austral region, increasing concentrations of dissolved ions (salinity), nutrients (particularly carbon and nitrogen) such as from sewage contamination (Sinton 1984; Hartland *et al.* 2011), metals and organic chemicals (Stephenson *et al.* 2013) all threaten groundwater biota and can lead to changes in groundwater assemblages. Unfortunately, changes associated with groundwater quality and contamination are difficult to fix, and if possible, remediation is a long and costly process. Changes to groundwater quality can be directly toxic to biota (e.g. Hose 2005, 2007; Humphreys 2007), or through changes, facilitate the establishment of stygoxenes that may be better able to tolerate the new conditions. Even though direct impacts of stygoxenes on groundwater assemblages may not be evident (Jasinska *et al.* 1993), exotic species pose a considerable threat to fauna (Proudlove 2001), often leading to shifts in ecosystem structure and function.

Dryland and irrigation salinity threatens large areas of productive agricultural lands in Australia and New Zealand. In such cases, the rise of water tables through saline sediments increases the salinity of the groundwater. Similarly, the extraction of fresh groundwater in coastal areas can permit intrusion of sea water into otherwise freshwater aquifers as hydraulic gradients are changed. With many stygofauna having marine ancestry, it is likely that stygofauna have some capacity to tolerate small increases in salinity. Indeed, crustaceans from surface waters tend to be relatively tolerant to increases in salinity relative to other taxa (Kefford *et al.* 2003). Generally stygofauna are rarely found at salinities above 3000 µS/cm (Hancock & Boulton 2008) although in the Yilgarn this condition is commonly and greatly exceeded (Humphreys *et al.* 2009). The threat of saline groundwater is not limited to the aquifer ecosystem but extends to adjoining surface waters (Halse *et al.* 2003).

24.8 Conservation strategies, issues and priorities for groundwater ecosystems

With most groundwater species known from only one or two sites, the majority of them could potentially be classified as vulnerable or endangered taxa (Gibert & Deharveng 2002; Humphreys 2006). Indeed, some of the most defining characteristics of subterranean fauna are their seemingly narrow distributions and extreme endemism over short geographical ranges (short-range endemics, *sensu* Harvey 2002). However, lack of formal species descriptions and reliable data on distribution for the majority of stygofauna in the Austral region would result in them being classified as data deficient in the IUCN Red List of threatened species (International Union for Conservation of Nature). Given this lack of data, the wiser approach is to protect entire habitats and their surface recharge areas (Boulton *et al.* 2003). Two global agreements are in place for protecting subterranean habitats; the Ramsar Convention (Ramsar Convention on Wetlands 1971) a global treaty on conservation and the sustainable use of wetlands including subterranean wetlands, and the UNESCO world heritage list (Culver & Pipan 2009). Although not the main focus of these lists, some stygofauna hotspots have been included and are protected under these agreements (Culver & Pipan 2009).

The field of conservation biology of stygofauna is in its infancy across the world, and concerted surveys of biodiversity on a large scale are not common. An exception, however, is in Europe, where great progress has been made towards elucidating biodiversity and developing conservation strategies as a result of the large-scale subterranean biodiversity survey project PASCALIS (Protocol for the ASsessment and Conservation of Aquatic Life In the Subsurface) (Gibert *et al.* 2005). This project sampled a variety of groundwater habitats across several European countries and the results that emerged from it were used to aid the selection of sites for a groundwater reserve network (Michel *et al.* 2009).

'There are simply too many subterranean species at risk to deal with them one at the time' (Culver *et al.* 2000), thus designing reserve networks to protect all species may be impractical. However, because subterranean species are geographically concentrated in a small percentage of the landscape, the geographical extent of a reserve does not have to take up large areas. Based on grid cells containing stygofauna across Belgium, France, Italy, Portugal, Slovenia and Spain, Michel *et al.* (2009) found that by conserving as little as 10% of the landscape containing groundwater the majority of the species (73.8%) were included in the reserve network. In France, Ferreira *et al.* (2007) found that less than 2% of the landscape was needed to capture 60% of known groundwater species.

In contrast to Europe (PASCALIS project) and parts of the USA (see Culver *et al.* 2000), for Australian aquifers, where very little stygofaunal sampling has occurred, integrated plans for groundwater use and conservation remain non-existent or limited (Boulton 2009). Given the short duration of research targeting these ecosystems in Australia (the past two decades), the amount of biodiversity that remain undiscovered is likely to be enormous. With the majority of subterranean groundwater ecosystems in Australia occurring in areas of high and competing demands for water and mineral resources – such as the Pilbara and Yilgarn regions of Western Australia and the Hunter Valley in NSW – it is urgent to develop efficient conservation strategies for these ecosystems.

So far, the majority of groundwater surveys in Australia have been fuelled by industrial interests, and have not been integrated into groundwater management. Water resource legislation in most Australian states provides for an assessment of the environmental impact of water extraction; however, the focus is generally on groundwater-dependent ecosystems (GDEs) on the surface (Tomlinson *et al.* 2007). Only in Western Australia are there specific requirement for stygofauna surveys before development assessments (Tomlinson *et al.* 2007). Albeit a few Australian subterranean species and several groundwater-dependent communities are listed under biodiversity protection legislation, such listings are only applicable to species and communities under a demonstrable threat (Tomlinson & Boulton 2010).

There is an almost complete lack of information on the basic biology and ecology of the Australian subterranean fauna (Humphreys 2008), but this is not a local phenomenon (Larned 2012). Virtually no data exist on life-cycles, reproduction or population size or genetic structure. As such, groundwater fauna are not applicable for inclusion in endangered species lists (e.g. the Red List or The Endangered Species Act). Moreover, the creation of national parks based on subterranean value seems unlikely in Australia; given that most subterranean ecosystems exist in areas of considerable economic interest (Boulton 2009).

A few authors have put forward possible approaches towards stygofauna conservation (see Tomlinson *et al*. 2007; Tomlison & Boulton 2008; Boulton 2009). In a review, Tomlinson & Boulton (2008) suggest taking an ecohydrological approach for management of groundwater. Groundwater habitat type is a key determinant of biodiversity, and combined with ecologically relevant aspects such as resource supply (via hydrological connectivity) and living space (aquifer void characteristics), and can together with distributional data be used to guide conservation efforts (Tomlinson & Boulton 2008). A DNA-based approach to rapid biodiversity assessment was proposed by Asmyhr & Cooper (2012). DNA sequences can serve as surrogates for species diversity, an approach that is particularly useful for ecosystems comprising a high number of cryptic species in which morphological identification is difficult (Bradford *et al*. 2010; Asmyhr & Cooper 2012). Although not proposed as a way of discouraging traditional taxonomic surveys, molecular methods may be used as an initial screening, allowing systems to be preserved while awaiting formal descriptions.

There is great potential in combining these two approaches for designing groundwater reserve networks in Australia. The level of genetic divergence among stygofauna samples from different sites may serve as the basis for selection of sites comprising the most evolutionary distinct taxa (Asmyhr *et al*. unpublished). A habitat typology, as proposed by Tomlinson & Boulton (2008) can be used to predict possible stygofauna hotspots, and serve as a surrogate for groundwater biodiversity in areas of high anthropogenic pressure. Although still at an early stage, the use of surrogates for groundwater biodiversity has been explored by a few authors (indicator species: Stoch *et al*. 2009; aquifer type: Linke *et al*., unpublished).

Finally, considering that aquifers are not stand-alone ecosystems, rather they are an essential part of a wide range of surface freshwater GDEs, there is a need for a holistic approach towards freshwater conservation. Without a significant increase in reserve cost, aquifers can be included in a broader freshwater conservation network, such as, for example, together with rivers and wetlands (Linke *et al*., unpublished).

REFERENCES

Abrams, K. M., King, R. A., Guzik, M. T., Cooper, S. J. B., Austin, A. D. (2013) Molecular phylogenetic, morphological and biogeographic evidence for a new genus of parabathynellid crustaceans (Syncarida: Bathynellacea) from groundwater in an ancient southern Australian landscape. *Invertebrate Systematics*, **27**, 146–172.

Acworth, R. I. 2001. Physical and chemical properties of a DNAPL contaminated zone in a sand aquifer. *Quarterly Journal of Engineering, Geology and Hydrogeology*, **34**, 85–98.

Anneser, B., Pilloni, G., Bayer, A., *et al.* (2010) High resolution analysis of contaminated aquifer sediments and groundwater – what can be learned in terms of natural attenuation? *Geomicrobiology Journal*, **27**, 130–142.

Asmyhr, M. G., Cooper, S. J. B. (2012). Difficulties barcoding in the dark: the case of crustacean stygofauna from eastern Australia. *Invertebrate Systematics*, **26**(6), 583–591. doi:10.1071/IS12032

Balke, M., Watts, C. H. S., Cooper, S. J. B., Humphreys, W. F., Vogler, A. P. (2004) A highly modified stygobitic diving beetle of the genus *Copelatus* (Coleoptera, Dytiscidae): taxonomy and cladistic analysis based on mtDNA sequences. *Systematic Entomology*, **29**, 59–67.

Barranco, P. Harvey, M. S. (2008) The first indigenous palpigrade from Australia: a new species of *Eukoenenia* (Palpigradi: Eukoeneniidae). *Invertebrate Systematics*, **22**, 227–234.

Bauzà-Ribot, M. M., Juan, C., Nardi, F., *et al.* (2012) Mitogenomic phylogenetic analysis supports continental-scale vicariance in subterranean thalassoid crustaceans. *Current Biology*, **22**, 2069–2074

Boulton, A. J. (2009) Recent progress in the conservation of groundwaters and their dependent ecosystems. *Aquatic Conservation: Marine and Freshwater Ecosystems*, **19**, 731–735.

Boulton, A. J., Fenwick, G., Hancock, P., Harvey, M. (2008) Biodiversity, functional roles and ecosystem services of groundwater invertebrates. *Invertebrate Systematics*, **22**, 103–116.

Boulton, A. J., Humphreys, W. F., Eberhard, S. M. (2003). Imperilled subsurface waters in Australia: biodiversity, threatening processes and conservation. *Aquatic Ecosystem Health and Management*, **6**, 37–41. doi:10.1080/14634980390151565

Bradbury, J. H. (1999). The systematics and distribution of Australian freshwater amphipods: a review. In: *Proceedings of the Fourth International Crustacean Congress, Amsterdam, The Netherlands*, Schram, F. R., von Vaupel Klein, J. C. (eds.). Leiden, The Netherlands, Brill, pp. 533–540.

Bradford, T., Adams, M., Humphreys, W. F., Austin, A. D., Cooper, S. J. B. (2010) DNA barcoding of stygofauna uncovers cryptic amphipod diversity in a calcrete aquifer in Western Australia's arid zone. *Molecular Ecology Resources*, **10**, 41–50. doi:10.1111/j.1755-0998.2009.02706.x

BMR Palaeogeographic Group (1990) *Australia, Evolution of a Continent*. Canberra, Australia: Australian Government Publishing Service.

Byrne, M., Yeates, D. K., Joseph, L., *et al.* (2008) Birth of a biome: synthesizing environmental and molecular studies of the assembly and maintenance of the Australian arid zone biota. *Molecular Ecology*, **17**, 4398–4417.

Chapelle, F. H. (2001) *Groundwater Microbiology and Geochemistry*. New York, John Wiley & Sons.

Chilton, C. (1882) On some subterranean Crustacea. *Transactions and Proceedings of the New Zealand Institute*, **14**, 174–180.

Cho, J.-L., Humphreys, W. F. (2010) Ten new species of the genus *Brevisomabathynella* Cho, Park and Ranga Reddy, 2006 (Malacostraca, Bathynellacea, Parabathynellidae) from Western Australia. *Journal of Natural History*, **44**, 993–1079.

Cook, B. D., Abrams, K. M., Marshall, J. *et al.* (2012) Species diversity and genetic differentiation of stygofauna (Syncarida : Bathynellacea) across an alluvial aquifer in north-eastern Australia. *Australian Journal of Zoology*, **60**, 152–158.

Cooper, S. J. B., Hinze, S., Leys, R., Watts, C. H. S., Humphreys, W. F. (2002) Islands under the desert: molecular systematics and evolutionary origins of stygobitic water beetles (Coleoptera: Dytiscidae) from central Western Australia. *Invertebrate Systematics*, **16**, 589–598.

Cooper, S. J. B., Bradbury, J. H., Saint, K. M., *et al.* (2007) Subterranean archipelago in the Australian arid zone: mitochondrial DNA phylogeography of amphipods from central Western Australia. *Molecular Ecology*, **16**, 1533–1544.

Cooper, S. J. B., Saint, K. M., Taiti, S., Austin, A. D., Humphreys, W. F. (2008) Subterranean archipelago: mitochondrial DNA phylogeography of stygobitic isopods (Oniscidea:Haloniscus) from the Yilgarn region of Western Australia. *Invertebrate Systematics*, **22**, 195.

Culver, D. C., Pipan, T. (2009) *The Biology of Caves and Other Subterranean Habitats*, 1st edn. New York, Oxford University Press.

Culver, D. C., Master, L. L., Christman, M. C., Hobbs III, H. H., (2000). Obligate cave fauna of the 48 Contiguous United States. *Conservation Biology*, **14**, 386–401.

Danielopol, D. L., Pospisil, P., Rouch, R. (2000) Biodiversity in groundwater: a large-scale view. *Trends in Ecology & Evolution*, **15**, 223–224.

Des Châtelliers, M. C., Juget, J., Lafont, M, Martin, P. (2009) Subterranean aquatic Oligochaeta. *Freshwater Biology*, **54**, 678–690. doi: 10.1111/j.1365-2427.2009.02173.x

Dodson, J. R., Roberts, F. K., DeSalis, T. (1994) Palaeoenvironments and human impact at Burraga Swamp in montane rainforest, Barrington Tops National Park, New South Wales, Australia. *Australian Geographer*, **25**, 161–169.

Dubey, S., Shine, R. (2010) Restricted dispersal and genetic diversity in populations of an endangered montane lizard (*Eulamprus leuraensis*, Scincidae). *Molecular Ecology*, **19**, 886–897.

Duffy, J. E., Cardinale, B. J., France, K. E. *et al.* (2007) The functional role of biodiversity in food webs: Incorporating trophic complexity. *Ecology Letters*, **10**, 522–538.

Eamus, D., Froend, R., Murray, B. R., Hose, G. C. (2006) A functional methodology for determining the groundwater regime needed to maintain health of groundwater dependent ecosystems. *Australian Journal of Botany*, **54**, 97–114.

Eberhard, S. M., Davies, S. (2011) Impacts of drying climate on aquatic cave fauna in Jewel Cave and other caves in southwest Western Australia. *Journal of the Australasian Cave & Karst Management Association*, **83**, 6–13.

Eberhard, S. M., Halse, S. A., Humphreys, W. F. (2005) Stygofauna in the Pilbara region, north-west Western Australia: a review. *Journal of the Royal Society of Western Australia*, **88**, 167–176.

Eberhard, S. M., Halse, S. A., Williams, M. *et al.* (2009) Exploring the relationship between sampling efficiency and short-range endemism for groundwater fauna in the Pilbara region, Western Australia. *Freshwater Biology*, **54**, 885–901.

Fenwick, G. D. & Scarsbrook, M. (2004) Lightless, not lifeless: New Zealand's subterranean biodiversity. *Water & Atmosphere*, **12**(3). Accessed 12/04/13, www.niwa.co.nz/sites/default/files/import/attachments/lightless.pdf

Fenwick, G. D., Thorpe, H. R., White, P. A. (2004) Groundwater systems. In *Freshwaters of New Zealand*, Harding, J., Mosley, P., Pearson, C., Sorrell, B. (eds.). Christchurch: New Zealand Hydrological Society and New Zealand Limnological Society, pp. 291–298.

Ferreira, D., Malard, F., Dole-Olivier, M.-J., Gibert, J. (2007). Obligate groundwater fauna of France: diversity patterns and conservation implications. *Biodiversity and Conservation*, 16, 567–596. doi:10.1007/s10531-005-0305-7

Gibert, J. (2001) *Protocols for the Assessment and Conservation of Aquatic Life in the Subsurface (PASCALIS): a European Project (EVK2-CT-2001–00121).* Available at www. pascalis-project.com. Accessed 1/5/13./

Gibert, J., Deharveng, L. (2002) Subterranean ecosystems: a truncated functional biodiversity. *BioScience*, 52, 473–481.

Gibert, J., Stanford, J. A., Dole-Oliver, M. J., Ward, J. (1994). Basic attributes of groundwater ecosystems and prospects for research. In *Groundwater Ecology*, Gibert, J., Danielopol, D., Stanford, J., (eds.). California, Academic Press, pp. 7–40.

Gibert, J., Brancelj, A., Camacho, A. *et al.* (2005) Groundwater biodiversity, Protocols for the ASsessment and Conservation of Aquatic Life In the Subsurface (PASCALIS): overview and main results. In: *Proceedings of an International Symposium on World Subterranean Biodiversity*, Gibert, J. (ed.). University of Lyon, Lyon, Villeurbanne, 8–10 December 2004, pp. 39–52.

Goldscheider, N., Hunkeler, D., Rossi, P. (2006) Review: microbial biocenoses in pristine aquifers and an assessment of investigative methods. *Hydrogeology Journal*, 14, 926–941.

Gounot, A. M. (1994) Microbial ecology of ground waters. In *Groundwater Ecology*, Gibert, J., Danielopol, D., Stanford, J., (eds.). San Diego, California, Academic Press, pp. 189–215.

Griebler, C. (2001) Microbial ecology of subsurface ecosystems. In: *Groundwater Ecology: A Tool for Management of Water Resources*, Griebler, C., Danielopol, D., Gibert, J., Nachtnebel, H. P., Notenboom, J. (eds.). Official Publication of the European Communities, Luxembourg, pp. 81–108.

Griebler, C., Lueders, T. (2009) Microbial biodiversity in groundwater ecosystems. *Freshwater Biology*, 54, 649–677.

Griebler, C., Mindl, B., Slezak, D., Geiger-Kaiser, M. (2002) Distribution patterns of attached and suspended bacteria in pristine and contaminated shallow aquifers studied with an *in situ* sediment exposure microcosm. *Aquatic Microbial Ecology*, 28, 117–129.

Guzik, M. T., Cooper, S. J. B., Humphreys, W. F., Cho, J.-L., Austin, A. (2008). Phylogeography of the ancient Parabathynellidae (Crustacea: Bathynellacea) from the Yilgarn region of Western Australia. *Invertebrate Systematics*, 22, 205–216. doi:10.1071/IS07040

Guzik, M. T., Austin, A. D., Cooper, S. J. B., *et al.* (2010) Is the Australian subterranean fauna uniquely diverse? *Invertebrate Systematics*, 24, 407–418. doi:10.1071/IS10038

Haack, S. K., Bekins, B. A. (2000) Microbial populations in contaminant plumes. *Hydrogeology Journal*, 8, 63–76.

Halse, S. E., Ruprecht, J. K., Pinder, A. M. (2003) Salinisation and prospects for biodiversity in rivers and wetlands of south-west Western Australia. *Australian Journal of Botany*, **51**, 673–688.

Hancock, P. J. (2009) *Alluvial Aquifer Fauna During and Following Drought*. Groundwater in the Sydney Basin Symposium, Sydney, August 4–5.

Hancock, P. J., Boulton, A. J. (2008) Stygofauna biodiversity and endemism in four alluvial aquifers in eastern Australia. *Invertebrate Systematics*, **22**, 117–126.

Hancock, P. J., Boulton, A. J. (2009) Sampling groundwater fauna: efficiency of rapid assessment methods tested in bores in eastern Australia. *Freshwater Biology*, **54**, 902–917.

Hartland, A., Fenwick, G. D., Bury, S. J. (2011) Tracing sewage derived organic matter into a shallow groundwater food web using stable isotope and fluorescence signatures. *Marine and Freshwater Research*, **62**, 119–129.

Harvey, M. S. (2002) Short-range endemism in the Australian fauna: some examples from non-marine environments. *Invertebrate Systematics*, **16**, 555–570.

Herman, J. S., Culver, D. C., Salzman, J. (2001) Groundwater ecosystems and the service of water purification. *Stanford Environmental Law Journal*, **20**, 479–495.

Hobbs III, H. H. (2000) Crustacea. In: *Ecosystems of the World 30. Subterranean Ecosystems*, Wilkens, H., Culver, D. C., Humphreys, W. F. (eds.). Amsterdam, Elsevier.

Holmes, A. J., Tujula, N. A., Holley, M. *et al.* (2001) Phylogenetic structure of unusual aquatic microbial formations in Nullarbor caves, Australia, *Environmental Microbiology*, **3**, 256–264.

Hose, G. C. (2005) Assessing the need for groundwater quality guidelines using the species sensitivity distribution approach. *Human and Ecological Risk Assessment*, **11**, 951–966.

Hose, G. C. (2007) A response to comments on assessing the need for groundwater quality guidelines using the species sensitivity distribution approach. *Human and Ecological Risk Assessment*, **13**, 241–246.

Hose, G. C. (2008) *Stygofauna Baseline Assessment for Kangaloon Borefield Investigations – Southern Highlands, NSW*. Report to Sydney Catchment Authority, Access Macquarie Ltd, North Ryde.

Hose, G. C. (2009) *Stygofauna Baseline Assessment for Kangaloon Borefield Investigations – Southern Highlands, NSW*. Supplementary Report – Stygofauna molecular studies. Report to Sydney Catchment Authority, Access Macquarie Ltd, North Ryde.

Hose, G. C., Lategan, M. J. (2012) *Sampling Strategies for Biological Assessment of Groundwater Ecosystems*. CRC CARE Technical Report no. 21, CRC for Contamination Assessment and Remediation of the Environment, Adelaide, Australia.

Humphreys, W. F. (2001). Groundwater calcrete aquifers in the Australian arid zone: The context to an unfolding plethora of stygal biodiversity. *Records of the Western Australian Museum*, (Supplement 64), 63–83.

Humphreys, W. (2006) Aquifers: the ultimate groundwater dependent ecosystem. *Australian Journal Botany*, **54**, 115–132.

Humphreys, W. F. (2007) Comment on: Assessing the need for groundwater quality guidelines for pesticides using the species sensitivity distribution approach by Hose. *Human and Ecological Risk Assessment*, **13**, 236–240. doi:10.1080/10807030601107551

Humphreys, W. F. (2008) Rising from Down Under; Developments in subterranean biodiversity in Australia from groundwater fauna perspective. *Invertebrate Systematics*, **22**, 85–101.

Humphreys, W. F. (2012) Diversity patterns in Australia. In *Encyclopedia of Caves*, 2nd edn, Culver, D., White, W. (eds.). San Diego, Academic Press, pp 203–219.

Humphreys, W. F., Watts, C. H. S., Cooper, S. J. B., Leijs, R. (2009) Groundwater estuaries of salt lakes: buried pools of endemic biodiversity on the western plateau, Australia. *Hydrobiologia*, **626**, 79–95.

Humphreys, W., Tetu, S., Elbourne, L., *et al.* (2012) Geochemical and microbial diversity of bundera sinkhole, an anchialine system in the eastern Indian ocean. *Natura Croatica*, **21**, 59–63.

Jasinska, E. J., Knott, B., Poulter, N. (1993) Spread of the introduced yabby, *Cherax* sp (Crustacea: Decapoda: Parastacidae), beyond the natural range of freshwater crayfishes in Western Australia. *Journal of the Royal Society of Western Australia*, **76**, 67–69.

Jaume, D. 2008 Global diversity of spelaeogriphaceans & thermosbaenaceans (Crustacea; Spelaeogriphacea & Thermosbaenacea) in freshwater. *Hydrobiologia*, **595**, 219–224

Jaume, D., Boxshall, G. A. (2013) Life in extreme environments: anchialine caves. *Marine Ecology. Encyclopedia of Life Support Systems (EOLSS)*.

Juberthie, C. (2000) The diversity of the karstic and pseudokarstic hypogean habitats in the world. In *Ecosystems of the World 30. Subterranean Ecosystems*, Wilkens, H., Culver, D. C., Humphreys, W. F. (eds.). Amsterdam, Elsevier, pp. 17–39.

Karanovic, I. (2012) *Recent Freshwater Ostracods of the World: Crustacea, Ostracoda, Podocopida*. Berlin, Springer-Verlag.

Kefford, B. J., Papas, P. J., Nugegoda, D. (2003) Relative salinity tolerance of macroinvertebrates from the Barwon River, Victoria, Australia. *Marine and Freshwater Research*, **54**, 755–765.

Korbel, K. (2013) Robust and sensitive indicators of groundwater health and biodiversity. Unpublished PhD Thesis. University of Technology, Sydney, Australia.

Korbel, K. L., Hose, G. C. (2011) A tiered framework for assessing groundwater ecosystem health. *Hydrobiologia*, **661**, 329–349.

Korbel, K. L., Hancock, P. J., Serov, P., Lim, R. P., Hose, G. C. (2013) Groundwater ecosystems change with landuse across a mixed agricultural landscape. *Journal of Environmental Quality*, **42**, 380–390.

Larned, S. T. (2012) Phreatic groundwater ecosystems: research frontiers for freshwater ecology. *Freshwater Biology*, **57**, 885–906.

Larson, H. L., Foster, R., Humphreys, W. F., Stevens, M. I. (2013). A new species of the blind cave gudgeon *Milyeringa* (Gobioidei, Eleotridae, Butinae) from Barrow Island, Western Australia, with a redescription of *M. veritas* Whitley. *Zootaxa*, **3616**(2), 135–150.

Lategan, M. J., Torpy, F., Newby, S., Stephenson, S., Hose, G. C. (2012) Fungal communities vary among aquifers, providing potential indicators of groundwater contamination. *Geomicrobiology Journal*, **29**, 352–361.

Leys, R., Watts, C. H. S. (2008) Systematics and evolution of the Australian subterranean hydroporine diving beetles (Dytiscidae), with notes on *Carabhydrus*. *Invertebrate Systematics*, **22**, 217–225.

Leys, R., Watts, C. H., Cooper, S. J., Humphreys, W. F. (2003) Evolution of subterranean diving beetles (Coleoptera: Dytiscidae: Hydroporini, Bidessini) in the arid zone of Australia. *Evolution*, **57**, 2819–2834.

Leys, R., Roudney, B., Watts, C. H. S. (2010) *Paroster extraordinarius* sp. nov., a new groundwater diving beetle from the Flinders Ranges, with notes on other diving beetles from gravels in South Australia (Coleoptera: Dystiscidae). *Australian Journal of Entomology*, **49**, 66–72.

Malard, F. (2001) Groundwater contamination and ecological monitoring in a Mediterranean karst ecosystem in southern France. In *Groundwater Ecology: A Tool for Management of Water Resources*, Griebler, C., Danielopol, D., Gibert, J., Nachtnebel, H. P., Notenboom, J. (eds.). Official Publication of the European Communities, Luxembourg, pp. 183–194.

Michel, G., Malard, F., Deharveng, L. *et al.* (2009) Reserve selection for conserving groundwater biodiversity. *Freshwater Biology*, **54**, 861–876. doi:10.1111/j.1365–2427.2009.02192.x

Millennium Ecosystem Assessment, (2005) *Ecosystems and Human Well-being: Biodiversity Synthesis*. Washington, DC, World Resources Institute.

Murphy, N. P., Adams, M., Austin, A. D. 2009. Independent colonization and extensive cryptic speciation of freshwater amphipods in the isolated groundwater springs of Australia's Great Artesian Basin. *Molecular Ecology*, **18**, 109–122.

Murray, B. R., Hose, G. C., Lacari, D., Eamus, D. (2006) Valuation of groundwater dependent ecosystems: a functional methodology incorporating ecosystem services. *Australian Journal of Botany*, **54**, 221–229.

Novarino, G., Warren, A., Butler, H. *et al.* (1997) Protozoan communities in aquifers: a review. *FEMS Microbiology Review*, **20**, 261–275.

Osborne, R. A. L., Zwingmann, H., Pogson, R. E., Colchester, D. M. (2006) Carboniferous clay deposits from Jenolan Caves, New South Wales: implications for timing of speleogenesis and regional geology. *Australian Journal of Earth Sciences*, **53**, 377–405. doi:10.1080/ 08120090500507362

Page, T. J., Humphreys, W. F., Hughes, J. M. (2008) Shrimps Down Under: Evolutionary relationships of subterranean crustaceans from Western Australia (Decapoda: Atyidae: *Stygiocaris*). *PLoS ONE*, **3**, e1618, 1–12.

Phillips, M. J., Page, T. J., de Bruyn, M. *et al.* (2013) The linking of plate tectonics and evolutionary divergences (Reply to Bauzà-Ribot et al.). *Current Biology*, **23**, 603–605.

Proudlove, G. S. (2001) The conservation status of hypogean fishes. *Environmental Biology of Fish*, **62**, 239–249.

Schmidt, S., Hahn, H. J., Hatton, T., Humphreys, W. F. (2007) Do faunal assemblages reflect the exchange intensity in groundwater zones? *Hydrobiologia*, **583**, 1–19.

Schminke, H. K. (2011) Arthropoda: Crustacea: Malacostraca: Bathynellacea Parabathynellidae In: *Invertebrate Fauna of the World*, Vol. **21**. Incheon, Republic of Korea: National Institute of Biological Resources.

Serov, P. A. (2002) *A Preliminary Identification of Australian Syncarida (Crustacea)*. CRC Freshwater Ecology, Albury.

Seymour, J., Humphreys, W. F., Mitchell, J. G. (2007). Stratification of the microbial community inhabiting an anchialine sinkhole. *Aquatic Microbial Ecology*, **50**, 11–24.

Sinton, L. W. (1984) The macroinvertebrates of a sewage polluted aquifer. *Hydrobiologia*, **119**, 161–169.

Smith, R. J., Jeffries, T. C., Roudnew, B. *et al.* (2012) Metagenomic comparison of microbial communities inhabiting confined and unconfined aquifer ecosystems. *Environmental Microbiology*, **14**, 240–253.

Stanford, J. A., Ward, J. V. (1988) The hyporheic habitat of river ecosystems. *Nature*, **335**, 64–66.

Stein, H., Kellermann, C., Schmidt, S. I. *et al.* (2010) The potential use of fauna and bacteria as ecological indicators for the assessment of groundwater quality. *Journal of Environmental Monitoring*, **12**, 242–254.

Stephenson, S., Chariton, A., O'Sullivan, M. *et al.* (2013) Changes in microbial assemblages along a gradient of hydrocarbon contamination in a shallow coastal sand aquifer. *Geomicrobiology Journal*, in press.

Stoch, F., Artheau, M., Brancelj, A., Galassi, D. M. P., Malard, F. (2009). Biodiversity indicators in European ground waters: towards a predictive model of stygobiotic species richness. *Freshwater Biology*, **54**, 745–755. doi:10.1111/j.1365–2427.2008.02143.x

Stumpp, C., Hose, G. C. (2013) Impact of water table drawdown and drying on subterranean aquatic fauna in in-vitro experiments. *PLoS ONE*, **8**(11), e78502. doi: 10.1371/journal.pone.0078502

Tetu, S. G., Breakwell, K., Elbourne, L. D. H. *et al.* (2013) Life in the dark: metagenomic evidence that a microbial slime community is driven by inorganic nitrogen metabolism. *ISME Journal*, in press.

Thurgate, M. E., Gough, J. S., Clarke, A. K., Serov, P., Spate, A. (2001) Stygofauna diversity and distribution in eastern Australian caves and karst areas. *Records of the Western Australian Museum Supplement*, **64**, 49–62.

Tomlinson, M. (2008) A framework for determining environmental water requirements for alluvial aquifer ecosystems. PhD thesis. Armidale, Australia: University of New England.

Tomlinson, M., Boulton, A. (2008) Subsurface groundwater dependent ecosystems: a review of their biodiversity, ecological processes and ecosystem services. Waterlines Occasional Paper No. 8, National Water Commission, Canberra, Australia.

Tomlinson, M., Boulton, A. J. (2010) Ecology and management of subsurface groundwater dependent ecosystems in Australia: a review. *Marine and Freshwater Research*, **61**, 936–949.

Tomlinson, M., Boulton, A. J., Hancock, P. J., Cook, P. G. (2007) Deliberate omission or unfortunate oversight: should stygofaunal surveys be included in routine groundwater monitoring programs? *Hydrogeology Journal*, **15**, 1317–1320. doi:10.1007/s10040-007-0211-z

von Rintelen, K., Page, T. J., Cai, Y. *et al.* (2012) Drawn to the dark side: a molecular phylogeny of freshwater shrimps (Crustacea: Decapoda: Caridea: Atyidae) reveals frequent cave invasions and challenges current taxonomic hypotheses. *Molecular Phylogenetics and Evolution*, **63**, 82–96.

Watts, C. H. S., Humphreys, W. F. (2009) Fourteen new Dytiscidae (Coleoptera) of the genera *Limbodessus* Guignot, *Paroster* Sharp and *Exocelina* Broun, from underground waters in Australia. *Transactions of the Royal Society of South Australia*, **133**, 62–107.

Watts, C. H. S., Hancock, P. J., Leys, R. (2007) A stygopitic *Carabhydrus* Watts (Dytiscidae, Coleoptera) from the Hunter Valley in New South Wales, Australia. *Australian Journal of Entomology*, **46**, 56–59.

Watts, C. H. S., Hancock, P. J., Leys, R. (2008) *Paroster peelensis* sp. nov.: a new stygobitic water beetle from alluvial gravels in northern New South Wales (Coleoptera: Dytiscidae). *Australian Journal of Entomology*, **47**, 227–231.

Wilson, G. D. F. (2008) Gondwanan groundwater: subterranean connections of Australian phreatoicidean isopods to India and New Zealand. *Invertebrate Systematics*, **22**, 301–310.

Wilson, G. D. F., Johnson, R. T. (1999) Ancient endemism among freshwater isopods (Crustacea, Phreatoicidea). In *The Other 99%. The Conservation and Biodiversity of Invertebrates*, **Ponder, W., Lunney, D.** (eds.). Transactions of the Royal Zoological Society of New South Wales. Mosman, Australia, Royal Zoological Society of New South Wales, pp. 264–268.

Wilson, G. D. F., Keable, S. J. (1999). A new genus of phreatoicidean isopod (Crustacea) from the North Kimberley Region, Western Australia. *Zoological Journal of the Linnean Society, London*, **126**, 51–79.

Young, A. R. (1986) The geomorphic development of dells (Upland Swamps) on the Woronora Plateau, N. S. W., Australia. *Zeitschrift for Geomorphologie*, **30**, 317–327.

Zektser, I. S. and Everett, L. G. (eds.) 2004. *Groundwater Resources of the World and Their Use*, UNESCO IHP-VI Series on Groundwater No. 6. Paris, France, UNESCO.

CHAPTER 25

Fire and biodiversity in Australia

John C. Z. Woinarski, Allan H. Burbidge, Sarah Comer, Dan Harley, Sarah Legge, David B. Lindenmayer and Thalie B. Partridge

Summary

Fire has a major influence on the management and conservation of Australian biodiversity. Notwithstanding a long history of fire on the continent, inappropriate contemporary fire regimes are a key threatening process for many Australian plant and animal species. Fire regimes vary appreciably across the continent, and different species and taxonomic groups respond in markedly different ways to different regimes. A set of case studies highlights the diversity of wildlife responses to fire, although we acknowledge that this set is inevitably far from a comprehensive assessment of the response of all biodiversity components to all fire regimes. Managing fires for biodiversity remains a challenge, particularly in the more remote parts of the continent or when management is driven mostly by human safety and economic assets. Some notable examples of local- and regional-scale fire management programs for biodiversity conservation are presented. Replicating the conservation benefits of these programmes across other parts of Australia will be difficult and will require improved understanding of the fire regimes required by biodiversity, significant effort in implementation and monitoring of outcomes and better understanding of fire by the Australian community.

25.1 Introduction

Australia is the most fire-prone continent. Fire has long shaped its ecosystem processes, the juxtaposition and extent of its ecological communities, the structure and floristics of its vegetation types, the ecology of many species, and the distribution, abundance or extinction of individual species. Much of this potency relates to Australian climatic regimes and Australia's relative lack of topographic relief (and hence protection from extensive fire). Marked wet–dry (monsoonal) seasonality characterises Australia's north, catalysing frequent (but relatively low-intensity) fire as the annual crop of tall savanna grasses cures during the long dry season. There is marked seasonality also in

Austral Ark: The State of Wildlife in Australia and New Zealand, eds. A. Stow, N. Maclean and G. I. Holwell. Published by Cambridge University Press. © Cambridge University Press 2015.

the Mediterranean and temperate climates of south-eastern and south-western Australia, and their hot summers prompt high-intensity wildfires. Seasonality is less pronounced in the arid inland areas, but recurring but irregular patterns of drought and wet periods drive infrequent but extensive fires as vegetation biomass built up in high-rainfall years dries when the rains disappear. These differences in environmental settings dictate that the frequency and impacts of fires vary very substantially across the Australian continent (Plate 14; Russell-Smith *et al.* 2007).

Inherent fire patterns changed with the settlement of Australia by Aboriginal people about 50 000 years ago. Over the long history since then, fire was their main management tool, and their frequent, purposeful and intricate use of fire re-moulded many Australian ecosystems (Bowman 1998). It is likely (indeed an ecological truism) that this re-shaping would have benefited some species and disadvantaged others, and many of the most distinctive and fantastic of Australian vertebrate species disappeared over this period (Johnson 2006).

Those long-established fire regimes, and their consequential ecological equilibriums, collapsed with the usurpation of traditional Indigenous land management over most of the continent in the relatively brief period since European settlement (1788). Aboriginal land managers were driven off their estates and some areas were almost completely depopulated, ignorance (or misapplied extrapolations from elsewhere) replaced masterful and intimate management of fire, high stocking rates of livestock changed the vegetation (and hence the fire regimes that could be supported), and the new settlers (with their fixed infrastructure) often feared and sought to suppress fire. Those many environments and species keyed into the long-established traditional fire regimes have been collateral victims of the change; some have disappeared entirely, others have endured but in a fragile and much-diminished extent or abundance.

The new Australian society has not yet come to terms with fire. Destructive and intense wildfires haunt our lives, especially on the fringes of the major capital cities. Mitigating measures seem relatively ineffective (Gibbons *et al.* 2012) and contested, and may come at considerable cost to biodiversity; and mechanisms to resolve such conflicts in objectives are not well established (Clarke 2008). In many areas, newly established plants (environmental weeds) now fuel ever more intensive fires, engineering an environmentally destructive vortex. Vegetation fragmentation, pastoralism, extensive surface mining and timber harvesting have also been accompanied by or caused further changes in the purpose and outcomes of fire management. Climate change, with its likely increase in incidence of very hot days and drought across much of the continent, is likely to exacerbate fire impacts (Williams *et al.* 2009), although impacts of fire under changing climate are difficult to predict (Bradstock 2010).

But there are also increasing opportunities for environmentally responsible fire management. Fire is recognised as a major tool for biodiversity conservation, and research has provided much knowledge about the tolerance and associations of many species with particular fire regimes. In remote north and central Australia, Aboriginal people have reclaimed land and renewed some components of traditional fire management practices. In some areas, such traditional expertise has been allied with scientific approaches to craft increasingly effective fire management applications for biodiversity conservation (Russell-Smith *et al.* 2009). And intricate fire management is being increasingly rendered operable by resourcing provided by the emerging carbon economy (Bradshaw *et al.* 2013).

The ecological role and management of fire in Australia is intricate, complex and imperfectly known. This short review cannot encompass the breadth and depth of the issue: readers seeking more information should refer also to the recent comprehensive compilation of Bradstock *et al.* (2012).

25.2 Fire and Australian biodiversity: some examples

It is difficult to encapsulate the impact of fire as a threat to biodiversity conservation in Australia, and perhaps equally as difficult to assess the relative effectiveness of fire as a tool for achieving conservation management. One measure of the relative detrimental impact of fire is its ranking in tallies of threats reported for groups of threatened species. For threatened birds in Australia, Woinarski (1999) reported that inappropriate fire regime was recognised as a threat to 51 threatened bird species, behind only 'habitat clearance and fragmentation' (affecting 52 threatened bird taxa) in ranking of threatening processes. A similar analysis by Garnett & Crowley (2000) reported marginally different tallies (in part because of changes in the composition of the threatened bird list), recognising that inappropriate fire regime was a current threat to 47 of 236 threatened Australian bird taxa, behind only 'domestic herbivores' (affecting 65 bird taxa) and land clearance (affecting 60 bird taxa). For Australia's 95 threatened terrestrial mammal species, inappropriate fire regime was recognised as a threat for 50 species, a tally behind only predation by feral cats as a threatening factor (Woinarski *et al.*, 2014). No comparable analyses have been undertaken for other groups of species, but it is likely that these findings are relatively representative: fire is one of the major threats to Australian biodiversity.

But fire affects biodiversity in many ways, and 'fire' itself encompasses a wide spectrum of ecological facets. In terms of biodiversity conservation, single fires leading to high levels of mortality of any particular species are not in general the conservation problem (although the short-term consequences for some species may be profound). Instead, conservation concern mostly relates to the longer-term impacts of fire regimes on population viability. Fire regimes are defined by the pattern and characteristics of multiple fires over time: their frequency, regularity, seasonality, intensity and extent. In turn, fire regimes may be driven by climate, landscape features and human intervention. Furthermore, the impacts of fire regimes on any species or environment may be influenced also by characteristics of the species (for example its dispersal ability, life-history traits and ecological plasticity) and by the cocktail of other threatening factors that may operate independently or interact with fire (for example with predation pressure typically greater in recently burnt areas: Lindenmayer *et al.* 2009b).

Accordingly, it is difficult and perhaps meaningless to generalise about fire impacts on Australian biodiversity. Instead we provide here a series of case studies that illustrate some of the many ways in which current fire regimes are affecting biodiversity. We acknowledge readily that these few examples are inevitably far from a comprehensive catalogue of the effects of different fire regimes on different components of the Australian biota, or of the rich literature reporting research on fire impacts, but they are illustrative of some of the breadth and impact of these relationships. One notable omission from our set relates to the impacts of fire on fungal communities: such impacts have been comprehensively reviewed recently elsewhere (McMullan-Fisher *et al.* 2011).

25.2.1   Case study 1. Partridge Pigeon

The threatened Partridge Pigeon *Geophaps smithii* is restricted to the monsoonal tropics of north-western Australia. It nests and forages (for seeds) on the ground, rarely flying. Its distribution and abundance has declined since European settlement, thought to be due to inappropriate fire regimes, predation by feral cats and habitat change due to the impacts of livestock (farmed and feral) (Garnett *et al.* 2011). In an intensive three-year study in Kakadu National Park, Fraser (2001) demonstrated that, in the dry season, its foraging efficiency in understorey characterised by a dense grass layer was considerably less than that in recently burnt areas. However, it nests on the ground (typically under shelter provided by grass clumps) in the early dry season, so extensive fires then were likely to result in reduced reproductive success. Fraser (2001) concluded that the optimal fire regime for this species was one characterised by very fine-scale burning (at the scale of individual home ranges), that left enough grass shelter for nesting, but also allowed for sufficient areas of burnt ground to enhance foraging. In subsequent management advice, supported by demonstrations of increased density in such fire-managed areas, Fraser *et al.* (2003) equated this preferred fire regime to traditional Indigenous fire management practice.

A comparable 'Goldilocks-like' fire regime (not too little and not too much fire) was advocated by Griffiths & Christian (1996) for the Frilled Lizard *Chlamydosaurus kingii* and by Pardon *et al.* (2003) for the Northern Brown Bandicoot *Isoodon macrourus* in the same location and environment. In both cases, the species were disadvantaged by absence of fire (because the dense grass layer of unburnt areas inhibited foraging) and by extensive fire (because this led directly or through increased predation to higher rates of mortality).

It is perhaps not surprising that terrestrial granivorous birds such as the Partridge Pigeon should benefit from, or require, such fine-grain patchwork burning, for such practice has long been recognised as beneficial to game birds in temperate areas (Leopold 1948): the interesting conclusion here is that it is also the way that Indigenous people in tropical Australia managed their lands.

25.2.2   Case study 2. Kakadu mammal assemblages

Kakadu National Park in northern Australia is one of the world's great conservation reserves. But recent research has demonstrated a drastic and severe decline in its previously diverse native mammal fauna, with populations of many species falling there by >80% over the past 10–20 years (Woinarski *et al.* 2010). The available evidence suggests that this is not an idiosyncratic feature of Kakadu, but may be characteristic of many other areas in Australia's monsoonal tropics (Firth *et al.* 2010). Monitoring evidence (from 136 fixed sites re-sampled at five-year intervals) suggests that the rate of mammal decline was greater at sites that had been exposed to more frequent fire between sampling events (Figure 25.1), but that fire frequency does not explain all of the decline, with even unburnt sites exhibiting some reduction in mammal abundance and richness.

One possible explanation for these results is an overall increase in predation pressure (by feral cats *Felis catus*), but particularly in burnt areas because fires reduce the availability of shelter sites for native mammals (Oakwood 2000). But fire effects may also be direct, through impacts on habitat structure and floristics. The savanna woodlands are simplified to a dense grass layer and a uniform tree overstorey under frequent fire; whereas a dense floristically rich shrubby understorey may develop when fire is

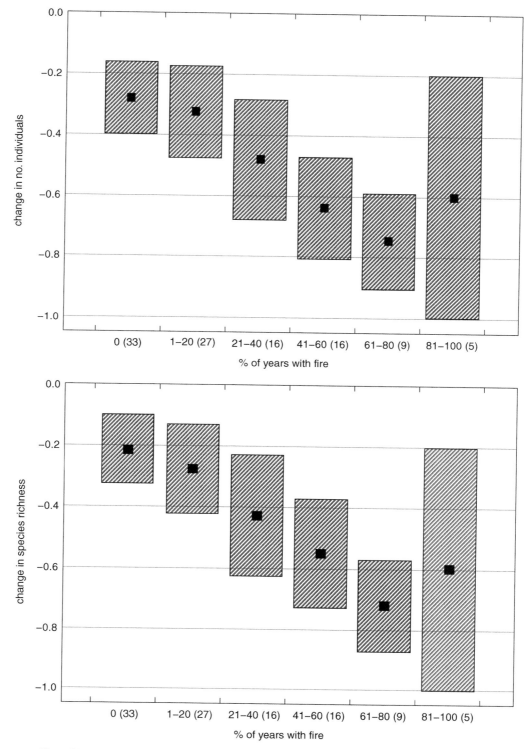

Figure 25.1 Variation in extent of change in the abundance and richness of native mammals across a large set of monitoring plots in Kakadu National Park from baseline sampling between 2001 and 2004 and re-sampling of those

infrequent (Russell-Smith *et al.* 2003; Woinarski *et al.* 2004). This shrub layer is an important component of habitat suitability for many mammal species, particularly for those whose diet includes fruits (Kerle 1985; Friend 1987). Such floristically and structurally diverse savanna woodlands now occupy only a very small proportion of the Kakadu (and broader tropical savanna) landscape (Woinarski 2004). Evidence from archaeology suggests that native fruits were more abundant in these savanna environments before European settlement than now (Atchison *et al.* 2005), indicating that the current fire regime is resulting in simplification of the savannas, to the detriment of the native mammal fauna. Frequent and more intense fires in this environment may also change tree demographic characteristics, notably by reducing the abundance of larger (older-aged) trees that provide the highest abundance and largest of hollows (Williams *et al.* 2003), again to the detriment of the many mammal (and other animal) species that are dependent upon tree hollows.

The lesson from this case study is that the current fire regime is leading to direct and indirect detrimental consequences for key components of biodiversity at broad (regional) scale, with this detriment occurring in conservation reserves at which biodiversity protection is a priority, as well as in lands of other tenure.

25.2.3 Case study 3. Invasive grasses–fire cycle

Fire has direct impacts on biodiversity. But it may also operate interactively (and at times synergistically) with other factors, producing sometimes complex and magnified impacts. In an Australian context, these factors include grazing by livestock and feral herbivores, introduced predators, introduced weeds and habitat fragmentation. Across many regions of Australia, fire regimes are now substantially influenced by the occurrence and abundance of introduced plant species, particularly invasive pasture grasses, mostly deriving from Africa. These were deliberately introduced, mostly in the late twentieth century, to transform natural environments to a more farm-like state that could produce higher plant biomass and hence higher pastoral offtake (Cook & Dias 2006). In many cases, they have matched and exceeded their proponents' expectations; in many cases, they have become almost ineradicable environmental weeds, which spread beyond pastoral properties to become major management problems in conservation reserves.

In a process also widely documented in other countries (see D'Antonio & Vitousek 1992), some of these invasive grasses have catalysed a fire–grass cycle, that iteratively and inexorably drives ecosystem change towards reduced richness of native plant species, simplification of vegetation structure, and increased dominance of non-native species (Setterfield *et al.* 2010). In some cases, they may also dramatically alter nutrient cycling

Figure 25.1 (cont.)

plots between 2007 and 2009, and its association with the fire frequency experienced at those sites in the period between mammal sampling episodes. There was a general trend for decline across all sites, but this trend was most pronounced at sites exposed to most fire. Note that there were relatively few sites in the highest frequency fire category. Values depicted are mean (filled square) and standard error (hatched box); the number of sites sampled is indicated in brackets after the fire frequency class; the index of change varies from a maximum of +1 to a minimum of –1. (Reproduced from Woinarski *et al.* 2010, with permission from CSIRO Publishing.)

(Rossiter-Rachor *et al.* 2009). The most notable examples are Gamba Grass *Andropogon gayanus*, Para Grass *Brachiaria mutica*, Olive Hymenachne *Hymenachne amplexicaulis* and Mission Grasses *Pennisetum* spp. in monsoonal tropical Australia, and Buffel Grass *Cenchrus ciliaris* across most of the lower rainfall rangelands. Collectively, these have the potential to extend over most of the Australian continent.

These invasive grasses produce appreciably higher biomass than native species, and often cure later in the season; accordingly they fuel fires of atypically high intensity and hence detrimental impact to native species: for example, Rossiter *et al.* (2003) reported that fire intensities were eight times greater in areas invaded by gamba grass than in comparable areas supporting native grasses.

Invasion of habitat by these species is enhanced by fire with evidence of a 'Buffel Grass-initiated fire-invasion feedback' occurring in arid parts of Australia, increasing fire frequency and Buffel Grass spread while increasing mortality and reducing recruitment of canopy species leading to a change in the structure of some woodlands, reducing natural firebreaks provided by creeklines and killing hollow-bearing trees (Butler & Fairfax 2003; Pavey & Nano 2009; Miller *et al.* 2010). Even low cover of Buffel Grass can have a detectable influence on community dynamics (Smyth *et al.* 2009). This perennial species is also infesting areas around human assets making fire protection difficult and around the few rocky outcrops and mountain ranges of arid and semi-arid Australia, increasing risk to fire-sensitive species (Paltridge & Latz 2009; Paltridge *et al.* 2011).

25.2.4 Case study 4. Frontier shifts: wet sclerophyll forest and fauna in the Wet Tropics

The boundaries between vegetation types may be rigidly fixed by substrate or topography, but may also be malleable to change associated with more impermanent factors such as fire regimes, grazing regimes or climatic variation. Such boundary changes then result in the expansion or contraction of particular vegetation types. Across much of Australia, vegetation dynamics have long been influenced by climatic variation and fire, notably with contraction of rainforests and other fire-sensitive vegetation types to refugial areas in periods of lower rainfall and/or increasing fire frequency (Bowman 2000). In many parts of tropical Australia, the rainforest estate has been fragmented and reduced by contemporary fire regimes, to the detriment of many narrowly endemic rainforest species (Braby *et al.* 2011). However, against this trend, there has also been evidence in some areas of expansion of rainforest (and other forests) into savanna woodlands and grasslands, possibly associated with plant growth responses to global increases in atmospheric CO_2 (Banfai & Bowman 2006) and with more local-scale reduction in fire frequency and/or intensity (Russell-Smith *et al.* 2004). One notable example of such change is in the Wet Tropics bioregion of north-eastern Australia. In this case, with shift from Aboriginal fire management practice that is inferred to have regularly burnt around rainforest margins to a current regime of fire exclusion, the rainforest boundaries have expanded into a narrow strip of wet sclerophyll (tall eucalypt) forest. In this case, the already small extent of the dwindling habitat means that there may be considerable biodiversity detriment in this fire-mediated change. The most notable impact is for a series of threatened bird and mammal taxa, such as the Northern Bettong *Bettongia tropica*, Mahogany Glider *Petaurus gracilis* and an undescribed subspecies of the Yellow-bellied Glider *P. australis*, that are restricted to the wet sclerophyll forests in this region. For these taxa, the reduction in area of wet sclerophyll forest habitat is the major threatening factor (Harrington & Sanderson 1994; Winter *et al.* 2004; Jackson *et al.* 2011), with changed fire regime the underlying cause of this habitat loss.

This case study illustrates the ecological equilibrium crafted by millennia of fire management by Aboriginal landowners, the susceptibility of that equilibrium to removal of previously consistent management practice, the consequential likely loss of some components of biodiversity, and the conservation imperative to understand and if possible re-institute at least some components of traditional fire management practice.

25.2.5 Case study 5. Leadbeater's Possum and Leichhardt's Grasshopper

The majestic montane forests of south-eastern Australia include some of the tallest hardwood trees in the world. Across extensive areas, these forests may be dominated by a single eucalypt species, typically Mountain Ash *Eucalyptus regnans* or Alpine Ash *E. delegatensis*. Unlike many other eucalypts, these trees are readily killed by fire and regenerate only from seed, released *en masse* from the canopy to germinate in the post-fire ash-beds created by wildfire. Extensive cohorts of similar-aged trees reflect infrequent (i.e. many decades to several centuries) past wildfire events (Ashton 1981). In most circumstances these trees are long-lived, continuing to grow over at least several hundred years (Wood *et al.* 2010). Gradually they develop hollows over this lifespan, with the number and size of hollows increasing with tree age (Lindenmayer *et al.* 1993). Stands of trees under 120 years old are generally unsuitable for the many hollow-dependent fauna species (>40 vertebrate species in Mountain Ash forest) that characterise Australian forest and woodland environments.

Because of their stature, Mountain Ash forests have also been much modified for timber harvesting, and this has substantially reduced the extent and continuity of old-aged forests of highest suitability to hollow-dependent species (Lindenmayer *et al.* 2012). One such species is Leadbeater's Possum, now confined to scattered subpopulations (with a total population size of <2000 individuals) within a small occupied area (*c.* 3500 km^2) of central Victoria. Most populations inhabit Mountain Ash forests, but smaller populations also occur in subalpine woodlands and, at one site, lowland swamp forest. This species requires not only mature forests for the provision of hollows for denning, but also dense midstorey vegetation (typically of *Acacia* or *Leptospermum* species) for the provision of food resources (plant exudates) and to facilitate movement through the forest (Lindenmayer 2009). There is a strong positive relationship between the number of tree hollows at a site and likely occurrence of Leadbeater's Possum (Lindenmayer *et al.* 1991a). Thus, the influence of fire history (and timber harvesting) on tree hollow availability at sites is a major determinant of the distribution and abundance of Leadbeater's Possum populations. Leadbeater's Possum's restricted geographic range makes it highly susceptible to landscape-scale disturbances, such as major fire events. In 1939, a major wildfire burnt approximately two-thirds of the species' range. Post-fire salvage logging continued for a further two decades removing many of the dead and/or fire-scarred but living large old trees (Lindenmayer *et al.* 2008). Owing to diminishing areas and increasing fragmentation of suitable old-age forest, the conservation status of Leadbeater's Possum has become increasingly precarious during recent decades. This situation was exacerbated in a dramatic way in February 2009, when the Black Saturday wildfires burnt about 3300 km^2 throughout the possum's range, destroying within a few days about 45% of the high-quality habitat available to the species (Lindenmayer *et al.* 2012). Post-fire surveys have shown that the possum is absent from burnt sites, regardless of burn severity (Lindenmayer *et al.* 2013). One major subpopulation in sub-alpine woodland on the Lake Mountain plateau, supporting 200–300 individuals before the fire, was reduced to only six individuals after the fire. This single wildfire has had catastrophic impacts on an already endangered species, and highlights that the conservation of Leadbeater's Possum is intimately tied to the influence of fire in space and time.

In recent decades, most Leadbeater's Possum populations have been denning in dead hollow-bearing trees ('stags') created by the 1939 wildfire (Lindenmayer *et al.* 1991b). This situation arose because the 1939 fires burnt significant areas of old forest (i.e. stands that had not experienced major fires for more than a century, possibly much longer). Following 70 years of natural decay in a wet forest environment, the stags created in 1939 are now collapsing at a rapid rate (Lindenmayer *et al.* 2012), and it will be more than 50 years before replacement hollows develop. In contrast to 1939, the 2009 wildfire in the Victorian Central Highlands burnt substantial areas of younger forest, and is unlikely to lead to the stag creation as occurred in 1939. Moreover, the 2009 fires highlighted that stags are extremely susceptible to damage and collapse if subjected to a subsequent fire event (Lindenmayer *et al.* 2012). The contrasting influence of the 1939 and 2009 fires on hollow availability highlights that fire can both create and destroy hollows suitable for Leadbeater's Possum, depending upon the attributes of the forest that is burnt. It can create the species' preferred type of den ('stags') in situations where old living trees are burnt. Yet such trees are particularly susceptible to destruction in subsequent fires. Furthermore, timber harvesting may have additional impacts upon fire regimes and management, as episodes of previous logging may make forests more fire prone for up to 80 years (Lindenmayer *et al.* 2009a), thereby leaving long-lasting detrimental impacts on species such as Leadbeater's Possum.

The example of Leadbeater's Possum and Mountain Ash forests demonstrates that there are Australian vegetation types needing the exclusion of high-intensity fire over long periods (>100 years), and that some individual species are dependent upon such long-unburnt habitat. It also highlights that current management regimes are unlikely to maintain such conditions (Bradstock 2009; Lindenmayer *et al.* 2013). These premises lead to the conclusion that inappropriate fire management may result in substantial numbers of species' extinctions. Hollow-dependent species are particularly sensitive to fire impacts, as it takes more than a century for new hollows to develop in forests regenerating from fire. Furthermore, many animal (and probably plant) species that are associated with fire-sensitive vegetation (or older seral stages) are ancient lineages, with high phylogenetic significance (Woinarski 1999).

Time is relative; the forests of montane south-eastern Australia tick to an ecological rhythm of centuries, and may struggle with fires at multi-decadal intervals. In the sandstone heathlands of Kakadu National Park and Arnhem Land, fire-sensitive heathland plants such as *Pityrodia* species may typically require just three to five years to reach maturity, but – like the Ash forests – may be eliminated if fires recur at intervals shorter than this maturation time. This group of plants provides the sole food source for Leichhardt's Grasshopper *Petasida epiphiggera*, narrowly endemic and with particularly limited dispersal capability. Where frequent fires (such as the annual or biennial fires now typically occurring in this habitat) lead to local losses of *Pityrodia*, the grasshoppers also suffer local extirpation, with such patches of local extirpation compounding haphazardly across the entire range of this species and its host (Barrow 2009). It parallels the story for Leadbeater's Possum: fires that are too frequent will reduce habitat suitability and incrementally drive extinctions. The difference is simply in the contrasting timescale, but such difference serves as a reminder that fire regimes need to be considered and managed at a regional scale and with due recognition of, and objectives relating to, the species and environments with the most specialised fire requirements or those most likely to be disadvantaged by the current regimes.

25.2.6   Case study 6. Fire and germination

In many of the examples given above, fire, or the current fire regime, is detrimental to biodiversity. But much of the Australian biota has persisted in flammable landscapes over evolutionary timescales. Many of the ecological traits of plant or animal species may be adaptations to frequent fire or at least provide some selective advantage to species in landscapes affected by fire. Examples include the epicormic resprouting in many eucalypt species following fire (Gill 1997); the requirement of some plant species for fire to trigger seed shedding (Bradstock & Myerscough 1981), and the beneficial impact of smoke on seed germination for many plants. In one experiment, of 94 plant species from south-western Australia that are normally difficult to germinate in nursery situations, the use of smoke enhanced germination rates for 45 species (Dixon *et al.* 1995). Similar responses were reported for a set of *Grevillea* species in eastern Australia: smoke increased germination rates for all seven species considered, and heat (comparable to bushfire intensity) increased germination rates in four species (Morris 2000). This association is by no means restricted to Australia (van Staden *et al.* 2000), but is particularly pronounced in some fire-prone environments in Australia, particularly in temperate heathlands.

This example indicates that fire is an integral and necessary component for some species and some ecological systems, and that any modification in its timing, severity, scale or frequency may have substantial impacts upon a broad range of species.

25.2.7   Case study 7. Fire and desert reptiles

The arid and semi-arid rangelands of Australia are large (about 70% of the continent's land surface), and sparsely populated, yet the region has seen significant declines in its fauna since European settlement (McKenzie *et al.* 2007). Introduced predators and inappropriate fire regimes have been implicated in this decline. Depending on the region, median annual rainfall varies from <250 mm to 800 mm (Turner *et al.* 2008). Years of above-average rainfall support significant and extensive fires (between 2001 and 2002 over 500 000 km^2 of the arid Northern Territory was burnt: Allan *et al.* 2003) that are more homogeneous than pre-European fire regimes (Latz & Griffin 1978; Wright & Clarke 2007; Edwards *et al.* 2008; Turner *et al.* 2008). Highly flammable spinifex (*Triodia* spp.) hummock grasslands support fires every 7–20+ years, whereas Mulga (*Acacia aneura*) woodlands and scrublands support fire only after several years of consecutive above-average rainfall and subsequent build-up of grassy fuels (Edwards *et al.* 2008). This relationship between fire and rainfall and vegetation structure is particularly important for the rich reptile fauna of the arid zone, many of which are highly specific in their habitat requirements (Pianka 1972).

McDonald *et al.* (2012a) investigated the impact of fire on three snake species in two of the dominant arid vegetation types, hummock grasslands and Mulga woodlands. Changes to the vegetation structure following fire affected the habitat specialists, Desert Death Adder *Acanthophis pyrrhus* and Monk Snake *Parasuta monachus*. Both species were less likely to occur in burnt habitat even eight years post-fire, whereas occurrence of the Stimson's Python *Antaresia stimsoni*, a habitat-generalist, was not influenced by fire. In mulga woodlands the Monk Snake is associated with litter mats which harbour preferred invertebrate and lizard prey (McDonald *et al.* 2012a). Reduction in litter due to fire or grazing reduces the habitat suitability and hence abundance of this and other litter-dwelling reptiles (Caughley 1985; Reid *et al.* 1993; Masters 1996; Pianka 1996; Letnic *et al.* 2004; McDonald *et al.* 2012b). For the Desert Death Adder, this observed fire response may be because large spinifex hummocks occurring only in long-unburnt areas offer a complex microhabitat providing relief from thermal extremes. Habitat preferences for other arid zone reptiles also show an association between thermal preferences and microhabitat structure (Melville & Schulte 2001) and the regeneration stage of vegetation after fire (Masters 1996).

Some arid zone reptile species show a preference for more open vegetation structure. Nocturnal ground-dwelling geckoes not requiring above-ground shade are typically more abundant on recently burnt sites (Letnic *et al.* 2004). The Central Netted Dragon *Ctenophorus nuchalis* was negatively associated with increases in spinifex cover following rain possibly due to its preferred burrow and perching locations as well as large invertebrate prey associated with open microhabitats (Dickman *et al.* 1999). This species is also tolerant of high daytime temperatures. This response was in contrast with the habitat preferences of the related Military Dragon *C. isolepis* which is strongly associated with vegetation cover and rainfall. The relative dominance of these two species switched as vegetation structure changed whether due to rainfall, fire or drought (Masters 1996; Dickman *et al.* 1999; Pianka & Goodyear 2012).

Across its range the Great Desert Skink *Liopholis kintorei* has had conflicting fire history preferences recorded. In sand plains dominated by spinifex and scattered shrubs (particularly *Eremophila* spp.) in the Northern Territory and Western Australia, fire recency may play an important role in creating suitable habitat; burrows were more common in areas burnt within the past 3–15 years (McAlpin 2001; Pearson *et al.* 2001). Patches of vegetation with varying times since fire allow colonies to move into recently burnt habitat as the vegetation changes. This movement of colonies has also been observed in populations recorded near the Watarru community in South Australia (Partridge 2008). In this area, the Great Desert Skink shows a preference for open Mulga habitat where a low frequency of fire may play a role in maintaining the preferred open vegetation structure (Partridge 2008). Further assessment is required to understand the habitat preferences of this threatened species and how potential threatening factors (including inappropriate fire regimes) can be managed.

This strong association with vegetation structure may make Australia's desert reptiles particularly vulnerable to inappropriate fire regimes. Management of landscapes through patch mosaic burning has often been proposed (see Burbidge & McKenzie 1989; Law & Dickman 1998; Letnic 2003), with this recommendation thought to be broadly similar to the historic regime of traditional Indigenous fire management which produced a vegetation mosaic of differing fire histories (Burrows & Christensen 1990; Burrows *et al.* 2006). However, implementing such management is complex and recent work has suggested the scale of this mosaic (coarse- rather than fine-grained) may be particularly important for reptile species with habitat preferences for a particular vegetation structure and small home ranges (Nano *et al.* 2012; Nimmo *et al.* 2012). Some have argued that prescribed fire is of little use for broad-scale conservation of biodiversity as a consequence of taxon-dependent and unpredictable species responses (Pastro *et al.* 2011); however, providing a fire mosaic for multiple taxa at multiple spatial scales and maintaining habitat for fire-sensitive species remains a challenge for fire management and biodiversity conservation across the arid and semi-arid rangelands of Australia.

25.2.8   Case study 8. Fire and birds in south-western Australia

While fire is a prominent natural process in Australia, and has been used purposely by humans for tens of thousands of years, there are still a significant number of bird species that are fire-sensitive – species that are impacted negatively by fire in the short term and which tend to require long intervals between fires. In some areas, older vegetation tends to be important for a greater number of bird species than younger vegetation (Taylor *et al.* 2012).

On the south coast of Western Australia, summer lightning strikes, with resultant intense wildfires in dry fuels, are a regular occurrence, and presumably have been for millions of years. For about the past 40 000 years this has been overlain by Aboriginal fire management, apparently frequent but spatially variable (Abbott 2003; Hassell and Dodson 2003). Interestingly, the Aboriginal fire regimes allowed several rare, fire-sensitive birds – scrub-birds, bristlebirds, ground parrots and whipbirds – to persist in this unfragmented yet fire-prone environment.

Then, about 100 years ago, another major change occurred – Europeans started fragmenting the natural vegetation and made profound changes to existing fire regimes. Generally, fire became less frequent but more severe and more extensive, and recolonisation of burnt areas was hampered by fragmentation of the habitat. This process has been a major factor in the decline in scrub-birds, bristlebirds, ground parrots and whipbirds, and undoubtedly some other species (Burbidge 2003). The Noisy Scrub-bird *Atrichornis clamosus* was thought to be driven to extinction, but was rediscovered on Mt Gardner at Two Peoples Bay, having survived through a severe bottleneck, as the species was reduced to less than 50 territorial males in pockets of long unburnt vegetation protected by large granite outcrops (Smith 1985).

Following the rediscovery of the scrub-bird, a period of intentional fire exclusion (Orr *et al.* 1995) over about 30 years resulted in a slow but steady increase in the number of scrub-birds, peaking in a census of about 180 singing males on Mt Gardner in the mid 1990s (Figure 25.2, Plate 52). This was a logical approach because, as with most other threatened species on the south coast of Western Australia, there is little or no evidence that a juxtaposition of different fire ages is important (with the possible exception of the Western Ground Parrot *Pezoporus flaviventris*: Burbidge *et al.* 2007). Combined with sustained fire management at translocation sites, scrub-bird numbers built up to over 765 singing males by 2001. Then came a harsh (albeit anticipated) lesson – fire exclusion by itself was not enough. The extensive long-unburnt vegetation at the main translocation site, Mt Manypeaks, was at high risk of being burnt in a single uncontrolled bushfire, and plans were being formulated to create a number of fire ages with the intent of making it possible to stop uncontrolled fire in the rough terrain around the peaks.

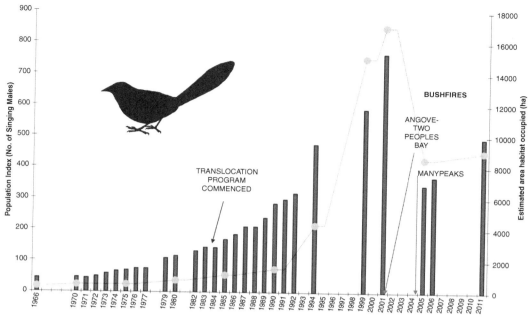

Figure 25.2 (Plate 52) Recovery of the Noisy Scrub-bird population 1966–2011, with territorial males used as an index of population size, and the impact of bushfires since 2001. (Reproduced from S. Comer, unpublished data.) A black and white version of this figure will appear in some formats. For the colour version, please refer to the plate section.

There was also concern for other threatened birds in the area, such as the Western Bristlebird *Dasyornis longirostris* and Western Whipbird *Psophodes nigrogularis* and other threatened animals and plants. Some of these had requirements in relation to fire that were similar to those of scrub-birds, but others were different (Burbidge 2003; Comer & Burbidge 2006; Comer *et al.* 2009). However, before any measures could be put in place safely, a lightning strike in a remote and inaccessible part of the reserve started a fire that burnt almost all the 427 scrub-bird territories known from the Manypeaks area. Combined with other fires around the same time, about two-thirds of the area occupied by scrub-birds was burnt within the space of about three years, and about one-third of the area occupied by the heath subspecies of the Western Whipbird *P. n. nigrogularis*.

Fire management for these species is now attempted at a much finer scale – when there are multiple co-occurring species with stringent but different requirements in relation to fire, management at a regional scale alone, or even the scale of a single reserve, is not acceptable. Nor is it optimal to simply exclude fire for a single species, as in the case of the scrub-bird – this strategy, in the end, had serious detrimental consequences for the scrub-birds at Manypeaks. They are now recolonising the Manypeaks area, though, and it is anticipated that the finer-scale network of slashed breaks and prescription burns currently being maintained will result in a safer environment for all the threatened animals and plants in this area of outstanding biodiversity.

25.3 Policy and management practice

Notwithstanding the substantial extent to which fire regimes affect biodiversity, fire management in Australia is typically based on other factors, especially considerations of risks to human safety and socio-economic impacts. This is the case even in many conservation reserves. Furthermore, the primacy of the financial and safety basis for fire management has been accentuated by a recent series of extensive and uncontrollable wildfires in temperate south-eastern and south-western Australia, judicial inquiries into human casualties caused by these fires, and government responses that aim to provide some reassurance to the community that fire impacts may be alleviated by management that seeks particularly to reduce fuel loads across extensive areas. Attempts to include biodiversity considerations into such policy settings have met limited success, typically resulting in compromise solutions that are not likely to produce desirable conservation outcomes (Clarke 2008).

Management of fire for biodiversity conservation in Australia has been most effective at local scales, particularly at the scale of individual conservation reserves, and for cases where the fire management requirements of a single or small set of threatened species are relatively well-known, and where there are few competing fire management considerations. As described in the last case study above, one notable example is management aimed at the establishment and maintenance of increasing extents of long-unburnt habitat in a series of near-coastal conservation reserves in south-western Australia for the localised and threatened Gilbert's Potoroo *Potorous gilbertii* and several bird species. A notably different example is the frequent application of fire to enhance habitat suitability and reproductive success for the highly localised threatened orchid *Prasophyllum correctum* in south-eastern Australia (Coates *et al.* 2006).

Such closely targeted fire management inevitably becomes more complex and more difficult to achieve when moving from single species at local scale to many species at regional scales. In some cases, the bewildering array of different plant and animal species responses to fire has been synthesised to an array of functional traits (including time to maturity and longevity), that can be interpreted at regional scales to define broad-scale or habitat-specific acceptable fire regimes for multiple species, particularly 'acceptable' intervals between fire recurrence (Bradstock & Kenny 2003; Russell-Smith *et al.* 2010). Such an approach provides a particularly useful broad-scale context to design and implement fire management for biodiversity conservation and to predict the likely consequences for biodiversity of regimes that may be designed for other purposes. However, as noted by Clarke (2008), the approach typically works better for plants than for animals, and needs to consider also the spatial juxtaposition of fires as well as their temporal sequencing.

Compared with the more densely settled temperate southern Australia, in the sparsely populated tropical savannas of northern Australia, fire management is less dominated by community concerns about safety or potential financial loss. As a consequence, it is feasible to design and implement regional-scale fire management purposed primarily or substantially to deliver conservation outcomes. There are two recent notable examples, the Western Arnhem Land Fire Abatement programme in the Northern Territory (Russell-Smith *et al.* 2009) and EcoFire in the Kimberley.

By the early 2000s, most of the Kimberley was being burnt about every two years in relatively high-intensity fires in the mid to late dry season. These fires were mainly lit by people for a variety of intentional and accidental reasons; single fire events covered many hundreds of thousands of hectares. Purposeful fire management was mostly confined to very small scale activities around infrastructure. Some individual properties carried out more extensive prescribed burning activities, but even these efforts were ineffective in the face of the massive wildfires sweeping across entire landscapes.

The EcoFire project arose in response to concerns about the impacts of these fire patterns on a range of issues (conservation, aesthetic, pastoral, cultural) by many sectors of the Kimberley community. The project began in 2007 and covers four million hectares of mixed tenure in the central and north Kimberley, including private conservation land, pastoral land, and indigenous pastoral land supporting small communities; it is managed by the Australian Wildlife Conservancy, funded by State and Commonwealth Governments and guided by a steering group including State government departments and local government.

EcoFire works each year with all property owners and managers to design and deliver a prescribed burning programme that is coordinated across property boundaries whilst respecting tenure-based variation in fire management objectives. The project has clearly defined, and ecologically based, fire pattern 'performance' targets that are measured and reported on each year. These targets include reducing the extent of mid to late season fires and increasing the extent and dispersion of unburnt vegetation (i.e. so that unburnt vegetation is distributed at a finer scale through the landscape). There is a particular focus on increasing the extent and proximity of long-unburnt vegetation (i.e. 3+ years in this context), in recognition of the needs of some declining and threatened taxa that are sensitive to fire frequency (like small mammals, ground-nesting birds, and the threatened Gouldian Finch *Erythrura gouldiae*). A range of biodiversity indicators (riparian specialists, small mammals, grass biomass, ground-dwelling birds, Gouldian Finches) monitored on the AWC properties have shown positive responses to the change in fire patterns (Legge *et al.* 2011).

The Western Arnhem Land Fire Abatement programme is broadly similar in that it seeks to apply considered fire management (including ignition and suppression) at a regional scale collaboratively with landholders with the objective of transforming a previous anarchic fire regime (characterised by unplanned extensive wildfire with severe biodiversity detriment: see Russell-Smith 2006; Russell-Smith *et al.* 2010) to one characterised by more frequent smaller-scale fires and far less extensive, severe and detrimental fire. The programme is notable for its support by a major gas company as a carbon emission offset, for its devolution to management control by Indigenous ranger groups, and for its success to date in achieving its environmental and cultural objectives (Russell-Smith *et al.* 2009).

The EcoFire and Western Arnhem Land Fire Abatement programme represent exemplars that have, to date, succeeded in large-scale changes from fire regimes that had become inappropriate for biodiversity conservation to regimes that deliver good conservation outcomes. But it will long remain difficult to replicate such programmes in regions in which biodiversity conservation is considered by policy-makers to be a secondary consideration to that of human safety or economic productivity.

25.4 Conclusion

Fire shapes most to all of Australian terrestrial environments. Fire regimes have changed markedly in most areas in the *c.* 225 years since European settlement of the continent, and this transformation in fire management has been a major cause of the decline of many species (and complementarily of the increase of others). Fire management varies markedly across the continent, depending in part on intrinsic climatic and topographic factors, but increasingly also on land use, human population density, habitat loss and modification, economic factors and a complex set of policies and legislation. The impacts upon biodiversity of fire may be substantially influenced or compounded by other interacting factors, including weeds, habitat change due to non-native herbivores, habitat fragmentation, and feral predators. As illustrated in the examples included here, the responses of biodiversity to fire may be nuanced and very context-dependent, and the identification of an optimal fire management regime may require much additional research effort.

In some areas, fire management may be a major tool for biodiversity conservation, but across most of Australia, this objective is subordinate to economic and human safety concerns. For fire management to more effectively achieve biodiversity conservation objectives, there needs to be a clearer and evidence-based articulation of the fire regimes required to benefit biodiversity; a more systematic assessment of the costs and benefits of alternative fire management options, including those directed towards goals other than biodiversity conservation; more integrated, long-term, proactive and collaborative regional-scale fire management programmes; linked and adequate monitoring systems; long-term financial support for fire management; sympathetic policy and legislative change; and an increased awareness amongst the Australian community of the ecological importance, role and impacts of fire beyond the prevalent current superficial conception of fire as a destructive force. Research has indicated much of the diversity of fire impacts and identified some of the fire management priorities for many species in many environments, but there remains considerable uncertainty about generalising such management recommendations (Parr & Andersen 2006; Pastro *et al.* 2011). There is much to be gained from further research on the relationships of Australian biota to fire, and much benefit to be gained by seeking to enhance the management of fire for biodiversity conservation.

REFERENCES

Abbott, I. (2003). Aboriginal fire regimes in south-west Western Australia: evidence from historical documents. In *Fire in Ecosystems of South-west Western Australia: Impacts and Management* (eds. I. Abbott & N. Burrows), pp. 119–146. Leiden, The Netherlands: Backhuys.

Allan, G. E., Phillips, N. R. & Hookey, P. (2003). Learning lessons from an exceptional period of fires in central Australia: 1999 to 2002. *Proceedings of the 3rd International Wildland Fire Conference*, p. 126. Sydney.

Ashton, D. H. (1981). Fire in tall open-forests. In *Fire and the Australian Biota* (eds. A. M. Gill, R. H. Groves & I. R. Noble), pp. 339–366. Canberra: Australian Academy of Science.

Atchison, J., Head, L. & Fullager, R. (2005). Archaeobotany of fruit seed processing in a monsoon savanna environment: evidence from the Keep River region, Northern Territory, Australia. *Journal of Archaeological Science*, **32**: 167–181.

Banfai, D. S. & Bowman, D. M. J. S. (2006). Forty years of lowland monsoon rainforest expansion in Kakadu National Park, northern Australia. *Biological Conservation*, **131**: 553–565.

Barrow, P. (2009). Impact of fire regimes on Leichhardt's grasshoppers. In *Kakadu National Park Landscape Symposia Series 2007–2009. Symposium 3: Fire Management, 23–24 April 2008* (eds. S. Atkins & S. Winderlich), pp. 147–149. Supervising Scientist, Darwin.

Bowman, D. M. J. S. (1998). Tansley Review No. 101 – The impact of Aboriginal landscape burning on the Australian biota. *New Phytologist*, **140**: 385–410.

Bowman, D. M. J. S. (2000). *Australian Rainforests: Islands of Green in a Land of Fire*. Cambridge: Cambridge University Press.

Braby, M. F., Willan, R. C., Woinarski, J. C. Z. & Kessner, V. (2011). Land snails associated with limestone outcrops in northern Australia – a potential bioindicator group. *Northern Territory Naturalist*, **23**: 2–17.

Bradshaw, C. J. A., Bowman, D. M. J. S., Bond, N. R., *et al*. (2013). Brave new green world – consequences of a carbon economy for the conservation of Australian biodiversity. *Biological Conservation*, **161**: 71–90.

Bradstock, R. A. (2009). Effects of large fires on biodiversity in south-eastern Australia: disaster or template for diversity? *International Journal of Wildland Fire*, **17**: 809–822.

Bradstock, R. A. (2010). A biogeographic model of fire regimes in Australia: current and future implications. *Global Ecology and Biogeography*, **19**: 145–158.

Bradstock, R. A. & Kenny, B. J. (2003). An application of plant functional types to fire management in a conservation reserve in southeastern Australia. *Journal of Vegetation Science*, **14**: 345–354.

Bradstock, R. A. & Myerscough, P. J. (1981). Fire effects on seed release and the emergence and establishment of seedlings in *Banksia ericifolia*. L. f. *Australian Journal of Botany*, **29**: 521–531.

Bradstock, R. A., Gill, A. M. & Williams, R. J. (2012). *Flammable Australia: Fire Regimes, Biodiversity and Ecosystems in a Changing World*. Melbourne: CSIRO Publishing.

Burbidge, A. A. & McKenzie, N. L. (1989). Patterns in the modern decline of Western Australia's vertebrate fauna: causes and conservation implications. *Biological Conservation*, **50**: 143–198.

Burbidge, A. H. (2003). Birds and fire in the Mediterranean climate of south-west Western Australia. In *Fire in Ecosystems of South-west Western Australia: Impacts and Management* (eds. I. Abbott & N. Burrows), pp. 321–347. Leiden, The Netherlands: Backhuys.

Burbidge, A. H., Rolfe, J. K., McNee, S., Newbey, B. & Williams, M. (2007). Monitoring population change in the cryptic and threatened Western Ground Parrot in relation to fire. *Emu*, **107**: 79–88.

Burrows, N. D. & Christensen, P. E. S. (1990). A survey of Aboriginal fire patterns in the Western Desert of Australia. In *Fire and the Environment: Ecological and Cultural Perspectives. Proceedings of an International Symposium* (eds. S. C. Nodvin & A. W. Thomas), pp. 297–305. USDA Forest Service General Technical Report, Knoxville, Tennessee, USA.

Burrows, N. D., Burbidge, A. A., Fuller, P. J. & Behn, G. (2006). Evidence of altered fire regimes in the Western Desert region of Western Australia. *Conservation Science Western Australia*, **5**: 272–284.

Butler, D. W. & Fairfax, R. J. (2003). Buffel grass and fire in a Gidgee and Brigalow woodland; a case study from central Queensland. *Ecological Management and Restoration*, **4**: 120–125.

Caughley, J. (1985). Effect of fire on the reptile fauna of Mallee. In *Biology of Australasian Frogs and Reptiles*, (eds. G. Grigg, R. Shine & H. Ehmann), pp. 31–34. Australia: Surrey Beatty & Sons Pty Limited.

Clarke, M. F. (2008). Catering for the needs of fauna in fire management: science or just wishful thinking? *Wildlife Research*, **35**: 385–394.

Coates, F., Lunt, I. D. & Tremblay, R. L. (2006). Effects of disturbance on population dynamics of the threatened orchid *Prasophyllum correctum* D. L. Jones and implications for grassland management in south-eastern Australia. *Biological Conservation*, **129**: 59–69.

Comer, S. & Burbidge, A. H. (2006). Manypeaks rising from the ashes. *Landscope*, **22**: 51–55.

Comer, S., Berryman, A. & Burbidge, A. H. (2009). Turning down the heat. The challenges of managing the critically endangered Western Ground Parrot in a wildfire prone environment. *Wingspan*, **19**: 10–13.

Cook, G. D. & Dias, L. (2006). It was no accident: deliberate plant introductions by Australian government agencies during the 20th century. *Australian Journal of Botany*, **54**: 601–625.

D'Antonio, C. & Vitousek, P. (1992). Biological invasions by exotic grasses, the grass/fire cycle and global change. *Annual Review of Ecology and Systematics*, **23**: 63–87.

Dickman, C. R., Letnic, M. & Mahon, P. S. (1999). Population dynamics of two species of dragon lizards in arid Australia: the effects of rainfall. *Oecologia*, **119**: 357–366.

Dixon, K. W., Roches, S. & Pate, J. S. (1995). The promotive effect of smoke derived from burnt native vegetation on seed germination of Western Australian plants. *Oecologia*, **101**: 185–192.

Edwards, G. P., Allan, G. E., Brock, C., *et al.* (2008). Fire and its management in central Australia. *The Rangeland Journal*, **30**: 109–121.

Firth, R. S. C., Brook, B. W., Woinarski, J. C. Z. & Fordham, D. A. (2010). Decline and likely extinction of a northern Australian native rodent, the Brush-tailed Rabbit-rat *Conilurus penicillatus*. *Biological Conservation*, **143**: 1193–1201.

Fraser, F. (2001). *The impacts of fire and grazing on the Partridge Pigeon: the ecological requirements of a declining tropical granivore*. Ph.D., Canberra: Australian National University.

Fraser, F., Lawson, V., Morrison, S., *et al.* (2003). Fire management experiment for the declining Partridge pigeon, Kakadu National Park. *Ecological Management and Restoration*, **4**: 94–102.

Friend, G. R. (1987). Population ecology of *Mesembriomys gouldii* (Rodentia: Muridae) in the wet-dry tropics of the Northern Territory. *Australian Wildlife Research*, **14**: 293–303.

Garnett, S. T. & Crowley, G. M. (2000). *The Action Plan for Australian Birds 2000*. Canberra: Environment Australia.

Garnett, S. T., Szabo, J. K. & Dutson, G. (2011). *The Action Plan for Australian Birds 2010*. Melbourne: CSIRO Publishing.

Gibbons, P., van Bommel, L., Gill, A. M., *et al.* (2012). Land management practices associated with house loss in wildfires. *PLoS ONE*, **7**.

Gill, A. M. (1997). Eucalypts and fire: interdependent or independent? In *Eucalypt Ecology: Individuals to Ecosystems* (eds. J. E. Williams & J. C. Z. Woinarski), pp. 151–167. Cambridge: Cambridge University Press.

Griffiths, A. D. & Christian, K. A. (1996). The effects of fire on the Frillneck Lizard (*Chlamydosaurus kingii*) in Northern Australia. *Australian Journal of Ecology*, **21**: 386–398.

Harrington, G. N. & Sanderson, K. D. (1994). Recent contraction of wet sclerophyll forest in wet tropics of Queensland due to invasion by rainforest. *Pacific Conservation Biology*, **1**: 319–327.

Hassell, C. W. & Dodson, J. R. (2003). The fire history of south-west Western Australia prior to European settlement in 1826–1829. In *Fire in Ecosystems of South-west Western Australia: Impacts and Management* (eds. I. Abbott & N. Burrows), pp. 71–85. Leiden, The Netherlands: Backhuys.

Jackson, S. M., Morgan, G., Kemp, J. E., Maughan, M. & Stafford, C. M. (2011). An accurate assessment of habitat loss and current threats to the Mahogany Glider (*Petaurus gracilis*). *Australian Mammalogy*, **33**: 82–92.

Johnson, C. N. (2006). *Australia's Mammal Extinctions: A 50 000 Year History*. Melbourne: Cambridge University Press.

Kerle, J. A. (1985). Habitat preference and diet of the Northern Brushtail Possum *Trichosurus arnhemensis* in the Alligator Rivers region, N.T. *Proceedings of the Ecological Society of Australia.*, **14**: 161–176.

Latz, P. K. & Griffin, G. F. (1978). Changes in Aboriginal land management in relation to fire and food. In *The Nutrition of Aborigines in Relation to the Ecosystem of Central Australia* (eds. B. S. Hetzel & H. J. Frith), pp. 77–85. Melbourne: CSIRO.

Law, B. S. & Dickman, C. R. (1998). The use of habitat mosaics by terrestrial vertebrate fauna: implications for conservation and management. *Biodiversity & Conservation*, **7**: 323–333.

Legge, S., Murphy, S., Kingswood, R., Maher, B. & Swan, D. (2011). Ecofire: restoring the biodiversity values of the Kimberley region by managing fire. *Ecological Management and Restoration*, **12**: 84–92.

Leopold, A. (1948). *A Sand County Almanac*. New York: Oxford University Press.

Letnic, M. (2003). The effects of experimental patch burning and rainfall on small mammals in the Simpson Desert, Queensland. *Wildlife Research*, **30**: 547–563.

Letnic, M., Dickman, C. R., Tischler, M. K., Tamayo, B. & Beh, C.-L. (2004). The responses of small mammals and lizards to post-fire succession and rainfall in arid Australia. *Journal of Arid Environments*, **59**: 85–114.

Lindenmayer, D. B. (2009). *Forest Pattern and Ecological Process: A Synthesis of 25 Years of Research*. Melbourne: CSIRO Publishing.

Lindenmayer, D. B., Cunningham, R. B., Tanton, M. T., Nix, H. A. & Smith, A. P. (1991a). The conservation of arboreal marsupials in the montane ash forests of the central highlands of Victoria, south-east Australia. III. The habitat requirements of Leadbeater's Possum *Gymnobelideus leadbeateri* and models of the diversity and abundance of arboreal marsupials. *Biological Conservation*, **56**: 295–351.

Lindenmayer, D. B., Cunningham, R. B., Tanton, M. T., Smith, A. P. & Nix, H. A. (1991b). Characteristics of hollow-bearing trees occupied by arboreal marsupials in the montane ash forests of the central highlands of Victoria, south-east Australia. *Forest Ecology and Management*, **40**: 289–308.

Lindenmayer, D. B., Cunningham, R. B., Donnelly, C. F., Tanton, M. T. & Nix, H. A. (1993). The abundance and development of cavities in *Eucalyptus* trees: a case study in the montane forests of Victoria, southeastern Australia. *Forest Ecology and Management*, **60**: 77–104.

Lindenmayer, D. B., Burton, P. J. & Franklin, J. F. (2008). *Salvaging Logging and its Ecological Consequences*. Washington, D. C.: Island Press

Lindenmayer, D. B., Hunter, M. L., Burton, P. J. & Gibbons, P. (2009a). Effects of logging on fire regimes in moist forests. *Conservation Letters*, **2**: 271–277.

Lindenmayer, D. B., MacGregor, C., Wood, J. T., *et al.* (2009b). What factors influence rapid post-fire site re-occupancy? A case study of the endangered Eastern Bristlebird in eastern Australia. *International Journal of Wildland Fire*, **18**: 84–95.

Lindenmayer, D. B., Blanchard, W., McBurney, L., *et al.* (2012). Interacting factors driving a major loss of large trees with cavities in a forest ecosystem. *PLoS ONE*, **7**: 1–16.

Lindenmayer, D. B., Blanchard, W., McBurney, L., *et al.* (2013). Fire severity and landscape context effects on arboreal marsupials. *Biological Conservation*, in review.

Masters, P. (1996). The effects of fire-driven succession on reptiles in spinifex grasslands at Uluru National Park, Northern Territory. *Wildlife Research*, **23**: 39–48.

McAlpin, S. (2001). *A Recovery Plan for the Great Desert Skink* (Egernia kintorei) *2001–2011*. Alice Springs: Arid Lands Environment Centre.

McDonald, P. J., Luck, G. W., Pavey, C. R. & Wassens, S. (2012a). Importance of fire in influencing the occurrence of snakes in an upland region of arid Australia. *Austral Ecology*, **37**: 855–864.

McDonald, P. J., Pavey, C. R. & Fyfe, G. (2012b). The lizard fauna of litter mats in the stony desert of the southern Northern Territory. *Australian Journal of Zoology*, **60**: 166–172.

McKenzie, N. L., Burbidge, A. A., Baynes, A., *et al.* (2007). Analysis of factors implicated in the recent decline of Australia's mammal fauna. *Journal of Biogeography*, **34**: 597–611.

McMullan-Fisher, S. J. M., May, T. W., Robinson, R. M., *et al.* (2011). Fungi and fire in Australian ecosystems: a review of current knowledge, management implications and future directions. *Australian Journal of Botany*, **59**: 70–90.

Melville, J. & Schulte II, J. A. (2001). Correlates of active body temperatures and microhabitat occupation in nine species of central Australian agamid lizards. *Austral Ecology*, **26**: 660–669.

Miller, G., Friedel, M., Adam, P. & Chewings, V. (2010). Ecological impacts of buffel grass (*Cenchrus ciliaris* L.) invasion in central Australia: does field evidence support a fire-invasion feedback? *The Rangeland Journal*, **32**: 353–365.

Morris, E. C. (2000). Germination response of seven east Australian *Grevillea* species (Proteaceae) to smoke, heat exposure and scarification. *Australian Journal of Botany*, **48**: 179–189.

Nano, C. E. M., Clarke, P. J. & Pavey, C. R. (2012). Fire regimes in arid hummock grasslands and *Acacia* shrublands. In *Flammable Australia: Fire Regimes, Biodiversity and Ecosystems in a Changing World* (eds. R. A. Bradstock, A. M. Gill & R. J. Williams). Melbourne: CSIRO Publishing, pp. 195–214.

Nimmo, D. G., Kelly, L. T., Spence-Bailey, L. M., *et al.* (2012). Fire mosaics and reptile conservation in a fire-prone region. *Conservation Biology*, **27**: 345–353.

Oakwood, M. (2000). Reproduction and demography of the northern quoll, *Dasyurus hallucatus*, in the lowland savanna of northern Australia. *Australian Journal of Zoology*, **48**: 519–539.

Orr, K., Danks, A. & Gillen, K. (1995). *Two Peoples Bay Nature Reserve Management Plan 1995–2005*. Perth: Department of Conservation and Land Management.

Paltridge, R. & Latz, P. (2009). *Fire Management Plan for the Mann Ranges and Musgrave Ranges Fire Management Regions of the Anangu Pitjantjatjara Yankunytjatjara Lands*. Umuwa, South Australia: Anangu Pitjantjatjara Yankunytjatjara.

Paltridge, R., Latz, P., Pickburn, A., *et al.* (2011). Working with Anangu to conserve rare plants in the Anangu Pitjantjatjara Yankunytjatjara Lands of South Australia. *Australasian Plant Conservation*, **19**: 10–11.

Pardon, L. G., Brook, B. W., Griffiths, A. D. & Braithwaite, R. W. (2003). Determinants of survival for the Northern Brown Bandicoot under a landscape-scale fire experiment. *Journal of Animal Ecology*, **72**: 106–115.

Parr, C. L. & Andersen, A. (2006). Patch mosaic burning for biodiversity conservation: a critique of the pyrobiodiversity paradigm. *Conservation Biology*, **20**: 1610–1619.

Partridge, T. (2008). *Tjakura Antunymanutjaku: Looking After Tjakura in the Anangu Pitjantjatjara Yankunytjajtara Lands, a Recovery Plan (Draft)*. Umuwa, South Australia: Anangu Pitjantjatjara Yankunytjatjara.

Pastro, L. A., Dickman, C. R. & Letnic, M. (2011). Burning for biodiversity or burning biodiversity? Prescribed burn vs wildfire impacts on plants, lizards and mammals. *Ecological Applications*, **21**: 3238–3253.

Pavey, C. R. & Nano, C. E. M. (2009). Bird assemblages of arid Australia: vegetation patterns have a greater effect than disturbance and resource pulses. *Journal of Arid Environments*, **73**: 634–642.

Pearson, D., Davies, P., Carnegie, N. & Ward, J. (2001). The Great Desert Skink (*Egernia kintorei*) in Western Australia: distribution, reproduction and ethno-zoological observations. *Herpetofauna*, **31**: 64–68.

Pianka, E. R. (1972). Zoogeography and speciation of Australian desert lizards: an ecological perspective. *Copeia*, **1**: 127–145.

Pianka, E. R. (1996). Long-term changes in lizard assemblages in the Great Victoria Desert: dynamic habitat mosaics in response to wildfires. In *Long-term Studies of Vertebrate Communities* (eds. M. L. Cody & J. A. Smallwood), pp. 191–215. California: Academic Press.

Pianka, E. R. & Goodyear, S. E. (2012). Lizard responses to wildfire in arid interior Australia: long-term experimental data and commonalities with other studies. *Austral Ecology*, **37**: 1–11.

Reid, J. R. W., Kerle, J. A. & Baker, L. (1993). *Uluru Fauna. The Distribution and Abundance of Vertebrate Fauna at Uluru (Ayers Rock-Mount Olga) National Park N.T. Kowari.* Canberra: Australian Nature Conservation Agency.

Rossiter, N. A., Setterfield, S. A., Douglas, M. M. & Hutley, L. B. (2003). Testing the grass-fire cycle: alien grass invasion in the tropical savannas of northern Australia. *Diversity and Distributions*, **9**: 169–176.

Rossiter-Rachor, N. A., Setterfield, S. A., Douglas, M. M., *et al.* (2009). Invasive *Andropogon gayanus* (gamba grass) is an ecosystem transformer of nitrogen relations in Australian savanna. *Ecological Applications*, **19**: 1546–1560.

Russell-Smith, J. (2006). Recruitment dynamics of the long-lived obligate seeders *Callitris intratropica* (Cupressaceae) and *Petraeomyrtus punicea* (Myrtaceae). *Australian Journal of Botany*, **54**: 479–485.

Russell-Smith, J., Whitehead, P. J., Cook, G. D. & Hoare, J. L. (2003). Response of *Eucalyptus*-dominated savanna to frequent fires: lessons from Munmarlary, 1973–1996. *Ecological Monographs*, **73**: 349–375.

Russell-Smith, J., Stanton, P. J., Whitehead, P. J. & Edwards, A. (2004). Rain forest invasion of eucalypt-dominated woodland savanna, Iron Range, north-eastern Australia: I. Successional processes. *Journal of Biogeography*, **31**: 1293–1303.

Russell-Smith, J., Yates, C. P., Whitehead, P., *et al.* (2007). Bushfires 'down under': patterns and implications of Australian landscape burning. *International Journal of Wildland Fire*, **16**: 361–377.

Russell-Smith, J., Whitehead, P. J. & Cooke, P. (2009). *Culture, Ecology and Economy of Fire Management in Northern Australia: Rekindling the Wurrki Tradition.* Melbourne: CSIRO Publishing.

Russell-Smith, J., Yates, C. P., Brock, C. & Westcott, V. C. (2010). Fire regimes and interval-sensitive vegetation in semiarid Gregory National Park, northern Australia. *Australian Journal of Botany*, **58**: 300–317.

Setterfield, S. A., Rossiter-Rachor, N. A., Hutley, L. B., Douglas, M. M. & Williams, R. J. (2010). Turning up the heat: the impacts of *Andropogon gayanus* (gamba grass) invasion on fire behaviour in northern Australia. *Diversity and Distributions*, **16**: 854–861.

Smith, G. T. (1985). Population and habitat selection of the Noisy Scrub-bird *Atrichornis clamosus*, 1962–83. *Australian Wildlife Research*, **12**: 479–485.

Smyth, A., Friedel, M. & O'Malley, C. (2009). The influence of buffel grass (*Cenchrus ciliaris*) on biodiversity in an arid Australian landscape. *The Rangeland Journal*, **31**: 307–320.

Taylor, R. S., Watson, S. J., Nimmo, D. G., *et al.* (2012). Landscape-scale effects of fire on bird assemblages: does pyrodiversity beget biodiversity? *Diversity and Distributions*, **18**: 519–529.

Turner, D., Ostendorf, B. & Lewis, M. (2008). An introduction to patterns of fire in arid and semi-arid Australia, 1998–2004. *The Rangeland Journal*, **30**: 95–107.

van Staden, J., Brown, N. A. C., Jager, A. K. & Johnson, T. A. (2000). Smoke as a germination cue. *Plant Species Biology*, **15**: 167–178.

Williams, R. J., Bradstock, R. A., Cary, G. J., *et al.* (2009). *Interactions between Climate Change, Fire Regimes and Biodiversity in Australia – a Preliminary Assessment.* Report to the

Department of Climate Change and Department of the Environment, Water, Heritage and the Arts. Darwin, Canberra: CSIRO.

Williams, R. J., Muller, W. J., Wahren, C., Setterfield, S. A. & Cusack, J. (2003). Vegetation. In *Fire in Tropical Savannas: the Kapalga Experiment* (eds. **A. N. Andersen, G. D. Cook & R. J. Williams**), pp. 79–106. New York: Springer-Verlag.

Winter, J. W., Dillewaard, H. A., Williams, S. E. & Bolitho, E. E. (2004). Possums and gliders of north Queensland: distribution and conservation status. In *The Biology of Australian Possums and Gliders* (eds. R. L. Goldingay & S. M. Jackson). Sydney: Surrey Beatty & Sons.

Woinarski, J. C. Z. (1999). Fire and Australian birds: a review. In *Australia's Biodiversity – Responses to Fire: Plants, Birds and Invertebrates* (eds. A. M. Gill, J. C. Z. Woinarski & A. York), pp. 55–111. Environment Australia, Canberra.

Woinarski, J. C. Z. (2004). The forest fauna of the Northern Territory: knowledge, conservation and management. In *Conservation of Australia's Forest Fauna* (ed. **D. Lunney**), pp. 36–55. Sydney: Royal Zoological Society of New South Wales.

Woinarski, J. C. Z., Risler, J. & Kean, L. (2004). Response of vegetation and vertebrate fauna to 23 years of fire exclusion in a tropical *Eucalyptus* open forest, Northern Territory, Australia. *Austral Ecology*, **29**: 156–176.

Woinarski, J. C. Z., Armstrong, M., Brennan, K., *et al.* (2010). Monitoring indicates rapid and severe decline of native small mammals in Kakadu National Park, northern Australia. *Wildlife Research*, **37**: 116–126, http://www.publish.csiro.au/nid/144/paper/WR09125.htm.

Woinarski, J. C. Z., Burbidge, A. A. & Harrison, P. (2014). *The Action Plan for Australian Mammals 2012*. Melbourne: CSIRO Publishing.

Wood, S. W., Hua, Q., Allen, K. J. & Bowman, D. M. J. S. (2010). Age and growth of a fire prone Tasmanian temperate old-growth forest stand dominated by *Eucalyptus regnans*, the world's tallest angiosperm. *Forest Ecology and Management*, **260**: 438–447.

Wright, B. R. & Clarke, P. J. (2007). Fire regime (recency, interval and season) changes the composition of spinifex (*Triodia* spp.) -dominated desert dunes. *Australian Journal of Botany*, **55**: 709–724.

CHAPTER 26

Terrestrial protected areas of Australia

Ian D. Craigie, Alana Grech, Robert L. Pressey, Vanessa M. Adams, Marc Hockings, Martin Taylor and Megan Barnes

Summary

Australia has a long history of establishing protected areas and they are now the cornerstones of its national and regional conservation strategies, covering over 13% of the country. There are large regional variations in levels of coverage, with most large protected areas placed far from dense human populations and away from productive agricultural land. Most of the recent growth in coverage has been driven by Indigenous Protected Areas and private protected areas, a trend that is likely to increase in the future. It is difficult to say how effective protected areas are in conserving biodiversity due to shortcomings in monitoring and evaluation, but the data that exist show that biodiversity outcomes are variable and that management effectiveness could be substantially improved. Threats to the protected area system are currently increasing with strong government pressure to allow extractive industries, such as mining, logging and grazing, and damaging recreational uses such as hunting to occur on land that is currently protected. If this trend continues, the future holds a great deal of uncertainty for Australia's protected areas.

26.1 Introduction

For centuries people all over the world have set aside places to which they ascribe special values. The reasons for this have been many and various but they are linked by a central purpose – to protect something that humankind perceives as valuable. Over the past century, as human populations have grown and their use of natural resources has increased, so the need to protect the remaining natural areas has also grown. Formally protected areas have become the centrepiece of the global strategy for nature conservation. These are areas where human activities are restricted and that are managed with the primary purpose of nature conservation (Dudley 2008). Australia is no exception in using protected areas as the cornerstones of its national and regional conservation strategies and is a signatory to the Convention on Biological Diversity (CBD). The CBD

Austral Ark: The State of Wildlife in Australia and New Zealand, eds. A. Stow, N. Maclean and G. I. Holwell. Published by Cambridge University Press. © Cambridge University Press 2015.

is an international legally binding treaty that commits Australia to achieving a number of conservation targets. These targets are updated periodically and the most recent version of Target 11, the main target concerning protected areas, reads:

> By 2020, at least 17 per cent of terrestrial and inland water, and 10 per cent of coastal and marine areas, especially areas of particular importance for biodiversity and ecosystem services, are conserved through effectively and equitably managed, ecologically representative and well-connected systems of protected areas and other effective area-based conservation measures, and integrated into the wider landscape and seascapes. (Convention on Biological Diversity 2010.)

Australia was among the earliest nations to dedicate land as national parks, leading the way in protected areas establishment. Wildlife, scenic and cultural conservation, and non-damaging forms of recreation were seen as the primary goals of these lands. In 1879 the National Park, later renamed the Royal National Park, was established mainly as a recreational area for Sydney residents. As the twentieth century progressed, the role of protected areas changed fundamentally, from beautiful places for outdoor recreation, to vital strongholds of the diversity of natural systems and species. Today Australia has 15 natural World Heritage Areas, reflecting both the impressive diversity of habitats within the country and the high importance placed on the many unique Australian biodiversity features by the international community.

Until recent decades, Australia has been generally perceived by its governments as under-populated and under-developed. The opening up of areas for land clearing and agriculture was heavily promoted as recently as the 1970s in some regions (Cresswell and Thomas 1998). The legacy of this viewpoint is the small number of remaining sites outside protected areas that could be considered as pristine wilderness (Bradshaw 2012), with the exception of the arid interior, which offers few opportunities for extractive human use.

The Australian states and territories operate their own systems of terrestrial protected areas under their own legislation, while retaining the term 'National Park' for many of their reserves. Most State and Territory park agencies grew from beginnings in State Forest Services or similar agencies. It was not until the 1960s and 1970s that separate national park services were established. The Federal Government also manages a small number of protected areas originally established on Commonwealth Crown land or in Australian External Territories. The nine separate terrestrial protected area systems (eight States and Territories and the Commonwealth) together with protected areas on private and Indigenous land are collectively known as the National Reserve System (NRS). These multi-jurisdictional arrangements have resulted in a system with more than 40 categories of protected areas nationally. However, in 1994, the jurisdictions agreed to adopt the IUCN 1994 classification[1] of protected areas and to use the IUCN system of management categories. Today, the International Union for the Conservation of Nature (IUCN) considers a protected area as: 'A clearly defined geographical space, recognised, dedicated and managed, through legal or other effective means, to achieve the long-term conservation of nature with associated ecosystem services and cultural values' (Dudley 2008).

This chapter presents an overview of the terrestrial protected areas of Australia. It describes the extent of the current National Reserve System, highlights the major

---

[1] See Table 27.1, p. 584 in Chapter 27 for most recent IUCN classifications of protected areas.

milestones over the last century and discusses some of the failings of the current system in capturing the full range of Australia's biodiversity features. The chapter then looks at the importance of Indigenous Protected Areas (IPAs) in Australia and reviews the recent growth of private protected areas. The evidence for the effectiveness of protected areas in maintaining species abundances and being effectively managed is examined next, and finally, the current and future threats to protected areas in Australia are discussed.

26.2 Number, extent and distribution of Australian protected areas

The most recent statistics show that there are over 9700 individual protected areas covering over 13% of Australia's land area (CAPAD 2010). However, these numbers hide large variations in the amount of land protected in each jurisdiction, with less than 7% protected in Queensland and 28% protected in South Australia (Table 26.1). This level of coverage is comparable to those of the whole globe which has around 13% of the land protected, also with large variations from country to country (Jenkins and Joppa 2009). Information on Australian protected areas are collated by the Commonwealth (Australian) Government and stored in the Collaborative Australian Protected Areas Database[2] (CAPAD), which is updated every few years.

The high level of protected area coverage in some jurisdictions represents a considerable achievement over more than a century. There have been several distinct phases in the history of protected area establishment, which are summarised in Figure 26.1. The first protected areas were small and established primarily for recreational purposes,

Table 26.1   Area (km^2) and percentage terrestrial coverage of declared protected areas in Australia; NSW = New South Wales; NT = Northern Territory; QLD = Queensland; SA = South Australia; TAS = Tasmania; VIC = Victoria; WA = Western Australia; ACT = Australian Capital Territory; JBT = Jervis Bay Territory (Commonwealth land within NSW); *Australia* = total for all jurisdictions. Data sources: CAPAD 2012 and Indigenous Protected Areas August 2013 (Commonwealth of Australia). Note: Protected areas that were classified in CAPAD as not being in the NRS were removed.

Jurisdiction	Protected area (km^2)	Jurisdiction area (km^2)	Jurisdiction within protected area (%)	Area contribution to National Reserve System (%)
ACT	1295	2358	54.9	0.1
NSW	72 015	800 642	9.0	5.2
NT	356 359	1 349 129	26.4	25.7
QLD	116 250	1 730 648	6.7	8.4
SA	305 576	983 482	31.1	22.0
TAS	26 918	68 401	39.4	1.9
VIC	37 735	227 416	16.6	2.7
WA	469 654	2 529 875	18.6	33.9
JBT	*56*	73	76.7	0.0
Australia	1 385 858	7 692 024	18.0	100.0

[2] http://www.environment.gov.au/parks/nrs/science/capad/

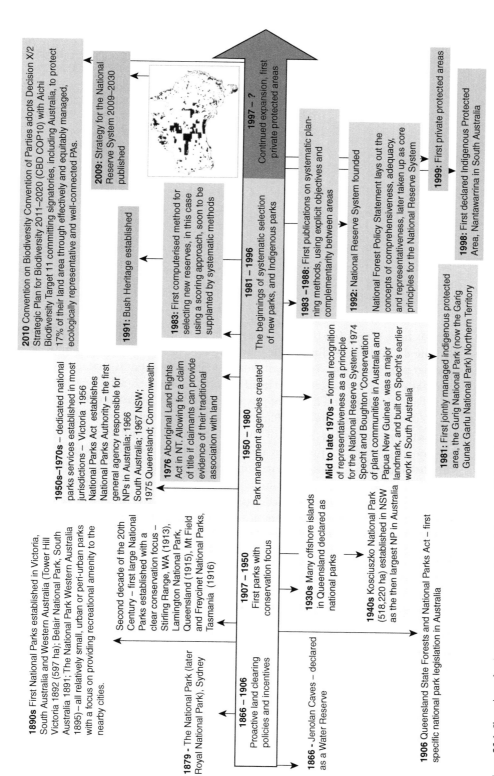

Figure 26.1 Chronology of major events and phases of protected area establishment in Australia.

1890s First National Parks established in Victoria, South Australia and Western Australia (Tower Hill Victoria 1892 (597 ha); Belair National Park, South Australia 1891; The National Park Western Australia 1895) – all relatively small, urban or peri-urban parks with a focus on providing recreational amenity to the nearby cities.

Second decade of the 20th Century – first large National Parks established with a clear conservation focus – Stirling Range, WA (1913), Lamington National Park, Queensland (1915), Mt Field and Freycinet National Parks, Tasmania (1916)

1950s–1970s – dedicated national parks services established in most jurisdictions – Victoria 1956 National Parks Act establishes National Parks Authority – the first general agency responsible for NPs in Australia; 1966 South Australia; 1967 NSW; 1975 Queensland; Commonwealth

1976 Aboriginal Land Rights Act in NT. Allowing for a claim of title if claimants can provide evidence of their traditional association with land

2010 Convention on Biodiversity Convention of Parties adopts Decision X/2 Strategic Plan for Biodiversity 2011–2020 (CBD COP10) with Aichi Biodiversity Target 11 committing signatories, including Australia, to protect 17% of their land area through effectively and equitably managed, ecologically representative and well-connected PAs.

2009: Strategy for the National Reserve System 2009–2030 published

1991: Bush Heritage established

1983: First computerised method for selecting new reserves, in this case using a scoring approach, soon to be supplanted by systematic methods

1879 - The National Park (later Royal National Park), Sydney

1866 – 1906
Proactive land clearing policies and incentives

1866 - Jenolan Caves – declared as a Water Reserve

1906 Queensland State Forests and National Parks Act – first specific national park legislation in Australia

1907 – 1950
First parks with conservation focus

1930s Many offshore islands in Queensland declared as national parks

1940s Kosciuszko National Park (518,220 ha) established in NSW as the then largest NP in Australia

1950 – 1980
Park management agencies created

Mid to late 1970s – formal recognition of representativeness as a principle for the National Reserve System; 1974 Specht and Boughton 'Conservation of plant communities in Australia and Papua New Guinea' was a major landmark, and built on Specht's earlier work in South Australia

1981 – 1996
The beginnings of systematic selection of new parks, and Indigenous parks

1981: First jointly managed indigenous protected area, the Gurig National Park (now the Garig Gunak Garlu National Park) Northern Territory

1983 –1988: First publications on systematic planning methods, using explicit objectives and complementarity between areas

1992: National Reserve System founded

National Forest Policy Statement lays out the concepts of comprehensiveness, adequacy, and representativeness, later taken up as core principles for the National Reserve System

1997 – ?
Continued expansion, first private protected areas

1999: First private protected areas

1998: First declared Indigenous Protected Area, Nantawarrina in South Australia

Table 26.2   Percentages of each jurisdictions' protected areas within five population density categories derived from the remoteness index (Plate 15C, Australian Bureau of Statistics). NSW = New South Wales; NT = Northern Territory; QLD = Queensland; SA = South Australia; TAS = Tasmania; VIC = Victoria; WA = Western Australia; ACT = Australian Capital Territory; JBT = Jervis Bay Territory (Commonwealth land within NSW); *Australia* = total percentage coverage of that population density category.

Jurisdiction	Major city	Inner regional	Outer regional	Remote	Very remote
ACT	5.2	94.8	0.0	0.0	0.0
NSW	0.7	21.8	55.3	7.8	14.4
NT	0.0	0.0	0.2	9.3	90.5
QLD	0.1	3.0	10.7	14.1	72.0
SA	0.0	0.1	1.6	8.9	89.3
TAS	0.0	0.9	41.6	54.3	3.2
VIC	0.3	12.2	67.2	20.3	0.0
WA	0.0	0.4	2.2	5.2	92.1
JBT	0.0	100.0	0.0	0.0	0.0
Australia	0.1	2.3	8.5	9.5	79.6

then as time progressed the need to protect areas with iconic and unique biodiversity became recognised, leading to the establishment of some of today's most famous and popular protected areas. The recent pattern of establishment is characterised by links to international policy targets, an agreed regionalisation for spatial selection of new protected areas and the emergence of private and Indigenous Protected Areas.

The distribution of the NRS (Plate 15A) shows the largest protected areas are in the arid interior, across the North of the country and on indigenous land. There is an uneven distribution of protected areas in relation to bioregions (Plate 15B), reflecting the distribution of environments suitable for human uses. Protection is generally lower in regions with higher suitability for agricultural and pastoral industries. While all of the largest protected areas are far removed from areas of dense human population (see Plate 15C) there are still many protected areas in accessible areas surrounding major cities (Table 26.2), allowing the public to easily enjoy intact natural areas.

26.3 Indigenous Australians and protected area management

Australia is considered a world leader in joint management of protected areas with Indigenous (traditional) owners (Chape *et al.* 2008). The Indigenous Protected Areas (IPAs) program was established by the Commonwealth Government in the mid 1990s to support Indigenous Australians in managing their land for conservation as part of the National Reserve System.

The identity of Aboriginal and Torres Strait Islanders is underpinned by their land and sea country. Indigenous people have cultural and traditional responsibilities to protect and manage their country. Indigenous communities have expressed their customary responsibilities for country via the establishment of dedicated Natural and Cultural Resource Management (NCRM) organisations and ranger groups. Having begun with

Box 26.1 Case study: Queensland's National Park expansion 1908–2010, by Marc Hockings and Peter Ogilvie

The early history of national parks in Queensland began auspiciously with the passage of the first legislation in Australia to generically provide for the declaration of national parks, the *State Forests and National Parks Act* of 1906. After a lengthy campaign by Robert Collins and Romeo Lahey, Lamington National Park was declared in 1915, one of the first national parks in Australia to be declared remote from the capital cities and for a specific nature conservation intent. The growth of the national park estate from the early 1900s to the 1960s was episodic but with some significant and far-sighted additions. Many of these are now iconic parks within the Queensland system such as Carnarvon, Hinchinbrook Island and Eungella National Parks. The declaration of the majority of islands in the Great Barrier Reef region in the 1940s is an example of such a far-sighted decision which has meant that the majority of these islands remain undeveloped to this day. Significant additions of the Simpson Desert (now called Munga Thirri) and Daintree in the mid 1960s significantly boosted the area of national park and more regular, although still gradual, addition of areas was undertaken under the direction of a small but knowledgeable group of foresters and scientists in the National Parks Branch of the Forestry Department. The Queensland National Parks and Wildlife Service was formed in 1975 by combining the then National Parks Branch of the Forestry Department with the Fauna Conservation Branch of the Department of Primary Industries. Subsequently, a program of more rapid expansion of the park estate was begun. From very early on, this was guided by one of the first exercises in conservation planning aimed explicitly at representiveness of the State's ecosystems. In 1977, Peter Stanton and Gethin Morgan published their report *Rapid Selection and Appraisal of Key and Endangered Sites: the Queensland Case Study* which was to become the blueprint for park additions over the coming decades (Stanton and Morgan 1977). In this seminal work, they identified 13 biogeographic regions in Queensland as a basis for conservation planning and established the forerunner of the Interim Biogeographic Regionalisation of Australia that now underpins conservation efforts across the country.

The increasing declaration of protected areas, including parks in many of the unrepresented or poorly represented bioregions, was boosted by the election of the Goss Government in late 1989 with a mandate to double the national park estate. The next six years saw a rapid increase in both the extent and representativeness of protected areas. In 1999, the publication of *The Conservation Status of Queensland's Bioregional Ecosystems* (Sattler and Williams, 1999) provided information on the delineation and status of regional ecosystems across the state and has underpinned subsequent conservation planning and priority setting. The next and most recent expansions of the national park estate have come about through the transfers of former forestry lands to national parks following the Southeast Queensland Forest Agreement, although this process was suspended by the new State Government in 2012. Despite these recent advances, the 2010 Australian Cooperative Protected Areas Database (CAPAD 2010) indicates that Queensland remains the Australian state or territory with the lowest percentage of its land area contained in protected areas (6.65% or half the national average). Ten of the 17 bioregions now recognised in

Box 26.1 (continued)

Queensland have less than 5% of their area conserved (CAPAD 2010). Despite auspicious beginnings, much remains to be done to secure a comprehensive, adequate and representative system of protected areas in Queensland (see Figure 26.2).

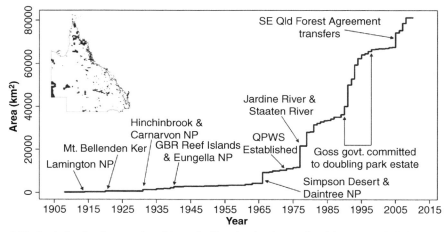

Figure 26.2 Graph showing the expansion of Queensland's national park network with key events labelled. QPWS: Queensland Parks and Wildlife Service. Forest Agreement transfers: a process where previously logged State Forests became managed as National Parks. Inset: Queensland's protected areas in 2010.

limited government support or recognition, in some locations these community-based initiatives have developed into well-established organisations with expertise in knowledge integration, planning, research, training, and environmental management. Tasks performed by Indigenous NCRM organisations and ranger groups include the management of protected areas, heritage sites, surveillance, monitoring, fire management, feral animal control, and the recording of Indigenous ecological and cultural knowledge. Australian governments and non-government organisations are increasingly investing in programmes such as 'Working on Country'[3] to provide education, training and employment opportunities for Indigenous rangers and community members (Altman 2007; Preuss and Dixon 2012) whilst also ensuring the protection and conservation of Australia's natural and cultural heritage (May 2010).

Indigenous communities, NCRM organisations and ranger groups engage in protected area management through cooperatively (jointly) managed protected areas and the 'Caring for Country'[4] Indigenous Protected Areas (IPA) Program. The joint management of national parks emerged in response to the legal recognition of Indigenous rights to traditional lands and the passing of the Commonwealth *Aboriginal Land Rights (Northern Territory) Act* in 1976. Jointly managed protected areas are government-designated

---

[3] http://www.environment.gov.au/indigenous/workingoncountry/index.html
[4] http://www.environment.gov.au/indigenous/ipa.html

protected areas where decision-making power, responsibility and accountability are shared between Indigenous communities and the relevant government agency. The first jointly managed protected area, the Gurig National Park (now the Garig Gunak Garlu National Park) on the Cobourg Peninsula, Northern Territory, was declared in 1981. Most jointly managed protected areas are within the Northern Territory, including the world famous Uluru-Kata Tjuta and Kakadu National Parks. Variations on joint management arrangements have emerged in other regions, notably New South Wales and Queensland.

The Australian Government would have made less progress towards its commitment to the establishment of a comprehensive, adequate and representative (CAR – see Section 26.5) system of protected areas without the IPA Program. IPAs are a class of protected areas on Indigenous-held land where Indigenous communities have entered into an agreement with the Australian Government. The primary goal of the IPA program is to provide Indigenous communities with guidance on how to 'achieve Management Plans that recognise the connections between Indigenous people, country, traditional law and culture, while also meeting national and international standards for protected area management' (Hill et al. 2011). The first IPA (Nantawarrina) was proclaimed in 1998 in the northern Flinders Ranges of South Australia, with several more proclaimed in other States in 1999. IPAs fall into a number of IUCN protected area categories (i.e. II–VI) depending on the type of management arrangements in place. An important feature of the IPA Program is the integration of contemporary protected area management practices with Indigenous ecological and cultural knowledge (Preuss and Dixon 2012). IPAs are actively managed by Indigenous communities, often with the assistance of State/Territory government agencies. As of August 2013 there were 60 IPAs across Australia, covering 478 931 km^2 of terrestrial Australia, making up over 40% of the total extent of the National Reserve System (Plate 16).

26.4 Private protected areas

Recently there has been a rapid growth in conservation covenants on private lands. These contractual protected areas have restrictions on use attached to the title of freehold lands, and special conditions on leasehold lands, to enable their management as private protected areas. These arrangements are an attempt to overcome the challenge of expanding the National Reserve System with 60% of Australia's land lying in private ownership, either as freehold (20%) or Crown leasehold (40%), effectively preventing the establishment of protected areas without expensive land purchases.

Because of the perceived low cost to government, enrolling land in private protected areas is becoming an increasingly prevalent mechanism to support public reserve networks. All Australian states have conservation covenant programs, with approximately 30 000 km^2 enrolled in covenants (see Table 26.3). In addition, non-government organizations such as Australian Wildlife Conservancy and Bush Heritage Australia own significant private reserves (~30 000km^2 and 10 000km^2 as of 2013, respectively).

The bulk of land in conservation covenants is in Queensland, with 20 780 km^2 enrolled through the State's Nature Refuge program. Queensland's Nature Refuge

Table 26.3   Predominant state-based government conservation covenant programs in Australia and total area enrolled, as of June 2011, adapted from Adams and Moon (2013). Note: not all of these covenants may be accepted as part of the National Reserve System

State	Name of program	Total area enrolled in km^2
New South Wales	New South Wales Conservation Agreements Program	1328
Queensland	Queensland Nature Refuge Program	20 780
South Australia	South Australia Heritage Agreement Program	6208
Tasmania	Tasmania Private Land Conservation Covenant Program	787
Victoria	Victoria Trust For Nature Conservation Covenant Program	427
Western Australia	Western Australia Conservation Covenant Program (National Trust of Australia (WA))	620
	The Nature Conservation Covenant Program (Department of Environment and Conservation)	127
Total		30 277

program allows landholders to voluntarily place portions of their properties under conservation covenants. The covenants are attached to the land title in perpetuity, stipulate nature conservation as the primary use, and constitute IUCN category VI protected areas ('protected area with sustainable used of natural resources'). Nature Refuges therefore contribute to the National Reserve System and to meeting global commitments such as targets agreed to under the Convention on Biological Diversity (Convention on Biological Diversity 2010). To encourage enrolment in the Nature Refuge program, Queensland has developed a number of financial and legislative incentives. For example, NatureAssist provides financial assistance to landholders to support management of Nature Refuges, including weed removal and fence building, through a competitive bidding process. Similar incentives are used across the Australian state-based conservation covenant programs. Alternatively, Queensland's Delbessie Agreement (DERM 2007), until recently abandoned by the state government, provided a framework of legislation, policies and guidelines supporting the environmentally sustainable, productive use of rural leasehold land, allowing lessees with properties identified as having conservation value to be awarded a ten-year lease extension if they place a portion of the property under a conservation covenant (DERM 2007; Adams *et al.* 2011).

Differences in personal circumstances (land use, income, education, health) and landholders' norms and attitudes may affect their willingness to participate in different types of private protected area programs. For example, Moon and Cocklin (2011) found that production landholders were more likely to participate in short-term programs requiring smaller percentages of land to be protected with large financial incentives. Conversely, they found that non-production landholders were more likely to participate in long-term programs that are voluntary or offer small financial incentives and apply to larger percentages of properties. Financial motivations, conservation attitudes and production values are common factors determining whether landholders participate in private conservation programs (Kabii and Horwitz 2006; Moon and Cocklin 2011).

26.5 Comprehensiveness, adequacy and representativeness

An important set of principles has long underpinned policy behind the National Reserve System. The principles have their origin more than 20 years ago in the National Forest Policy Statement (Anon. 1992). They are listed here as originally defined.

- Comprehensiveness: 'the reserve system includes the full range of forest communities recognised by an agreed national scientific classification at appropriate hierarchical levels'.

- Adequacy: 'the maintenance of the ecological viability and integrity of populations, species and communities'.

- Representativeness: 'those sample areas of the forest that are selected for inclusion in reserves should reasonably reflect the biotic diversity of communities'.

The importance of these principles in shaping conservation policy in Australia cannot be over-emphasised. They reflect the advanced stage, in a global context, of Australian thinking on conservation of the environment in the late 1980s and early 1990s. The principles quickly became established as an acronym – CAR – and were the foundation of discussion in the 1990s about how they should be operationalised for the Regional Forest Agreement processes in states with extensive native forests (Anon. 1995; Joint ANZECC/ MCFFA National Forest Policy Statement Implementation Sub-committee 1997).

The names of two of the CAR principles are potentially confusing. Much of the work on systematic conservation planning in Australia was based on the idea of representativeness or sampling of biodiversity (Austin and Margules 1986; Margules and Pressey 2000). The CAR principles refer to two levels of sampling, with comprehensiveness referring to sampling of extensive, high-level entities and representativeness referring to sampling of the range of physical and biological variation within those extensive entities. In this sense, representativeness, as a very common term in the scientific literature, applies to both the 'C' and the 'R' in CAR, but at different spatial extents and levels of classification.

More recently, the CAR principles have been defined for the National Reserve System, with comprehensiveness referring to Australia's 89 biogeographic regions[5] and representativeness referring to the 419 subregions nested within them. The most recent data on the National Reserve System indicate that we are some way from a comprehensive system (and see Plate 15B).

Targets for comprehensiveness and representativeness, to be achieved by 2015, have been identified nationally (National Reserve System Task Group 2009).

- For comprehensiveness, include examples of at least 80% of the number of regional ecosystems in each IBRA region (priority will be given to under-represented IBRA bioregions with less than 10% protected in the National Reserve System).

- For representativeness, include examples of at least 80% of the number of regional ecosystems in each IBRA subregion.

---

[5] http://www.environment.gov.au/topics/land/national-reserve-system/science-maps-and-data/australias-bioregions-ibra

The World Wildlife Fund (WWF) in Australia has undertaken a number of analyses of Australia's protected area system and published the results in a series of reports under the title of Building Nature's Safety Net (Sattler and Glanznig 2006; Sattler and Taylor 2008; Taylor *et al.* 2011). Despite progress since the first report, they emphasise that '… significant gaps for protection of both ecosystem and species diversity remain in every bioregion' (Taylor *et al.*, 2011, p. 9). Taylor *et al.* (2011) found that, for comprehensiveness, only five of the (then) 85 biogeographic regions had reasonable protection of at least 80% of their regional ecosystems and that, for representativeness, only 20 of the (then) 403 subregions had reasonable protection of at least 80% of their regional ecosystems. Given the slow development of the National Reserve System so far, the targets for 2015 seem impossible to achieve.

Just as importantly, there is a serious limitation in the formulation of the 2015 targets proposed by the National Reserve System Task Group (2009). The targets rely too heavily on sampling of ecosystems and ignore priority of protection in relation to threat. A pervasive characteristic of Australia's National Reserve System is that it dominantly samples those ecosystems that are 'residual' to extractive land uses, or in least need of protection. This conclusion is supported by more than 50 publications referring to parts or all of Australia, varying from qualitative, authoritative observations to careful quantitative analyses. Residual patterns of protected areas have been demonstrated for Western Australia (Pouliquen-Young 1997) South Australia (Bryan 2002), Victoria (Sharafi *et al.* 2012), Tasmania (Mendel and Kirkpatrick 2002), the Australian Capital Territory (Landsberg *et al.* 2000), New South Wales (Pressey *et al.* 2000), and Queensland (Vanclay 1996). There appear to be no comparable analyses for protected areas in the Northern Territory.

The residual tendencies of protected areas mean that gaps in protection are not random, but systematic in terms of avoiding protection of ecosystems with significant commercial potential (and this is part of a worldwide tendency: Jenkins and Joppa 2009). These are, of course, the ecosystems most threatened by human activities and most in need of protection. This bias in protection is also reflected in the capacity of Australia's protected area system to represent threatened species. It has been found that 12.6% (166) of threatened species occurred entirely outside protected areas and only 19.6% (259) had sufficient areas of their range protected.

In the context of residual protection, the main limitation of the targets for the National Reserve System (National Reserve System Task Group 2009) is that sampling (comprehensiveness and representativeness) is not complemented by prioritisation with respect to irreplaceability and threat, recognised as a strategy for minimising the extent to which objectives for ecosystems will be compromised by ongoing loss and degradation due to extractive activities (Margules and Pressey 2000; Pressey *et al.* 2004). Even if the protected area system achieved the targets of sampling 80% of the number of regional ecosystems in biogeographic regions and subregions, would they be the ecosystems most in need of protection? Australia's track record of protecting ecosystems suggests otherwise: the likelihood is, even if 80% sampling of ecosystems were achieved by some date beyond the target of 2015, the ecosystems in the unsampled 20% would be those with most urgent need for protection due to continuing threats. This risk is supported by chronological studies showing that increases in representation of ecosystems are paralleled by strengthening biases in protection away from the ecosystems that most need protection (Pressey and Taffs 2001; Pressey *et al.* 2002). One of the drivers of this increasing bias is the rapid growth of Indigenous Protected Areas which are almost exclusively established on economically unproductive land. In the past few years this increasing tendency to select residual land for protection may be on the decline with the Caring for Country scheme actively targeting protection of land and biodiversity under immediate threat. Though generally, the national policy needs to be improved to focus on more threatened regional ecosystems before further options for protection are foreclosed.

If comprehensiveness and representativeness of the National Reserve System are not well developed, what about adequacy? Adequacy is defined (National Reserve System Task Group 2009) in terms of how much of each ecosystem should be sampled to provide ecological viability and integrity of populations, species and ecological communities. With few regional ecosystems in Australia having even relatively low levels of protection (Taylor *et al.* 2011), the adequacy of protected areas generally is unlikely to be high. An operational constraint here is that we do not have consensus on the amount of protection needed to confer viability and integrity, although it is likely that the answer will vary among ecosystems according to their composition and exposure to threats arising from extractive uses and climate change.

26.6 Evaluating the effectiveness of protected areas

Protected areas are widely considered the most important tool available to conservationists and other stakeholders to maintain habitat integrity and species diversity in the face of increasing anthropogenic threats worldwide (Brooks *et al.* 2004; Rodrigues 2006; Andam *et al.* 2008; Coad *et al.* 2013). Additionally, protected areas are amongst the most cost-effective conservation methods available (Balmford *et al.* 2002). Increasing percentages of land under formal protection are often used to indicate conservation success (e.g. UNEP and WCMC 2007; Bertzky *et al.* 2012). However, percentages of regions under formal protection, though a useful guide for creating representative networks, cannot be used alone as meaningful measures of conservation progress.

Once a protected area has been established it is important that appropriate management occurs to maintain the desired biodiversity outcomes. Coverage percentages provide no information on whether protected areas are able to retain their biodiversity values, or whether they are able to do so better than alternate conservation mechanisms. Coverage also provides no information on how effectively protected areas have been located to avert loss of natural values. This last point is particularly important given the residual nature of many protected areas (see Section 26.5). Coverage metrics implicitly assume that either implementation *per se* or the management applied to protected areas is appropriate and will result in the conservation of biodiversity values. However, this is often not the case, in particular where management effectiveness is poor, or placement is inappropriate. Given the level of investment in protected areas, it is critical that we understand the mechanisms underlying protected area effectiveness in retaining biological values.

Two types of measures of protected area effectiveness are commonly available: the first is management effectiveness, and the second are a suite of measures that directly assess biodiversity outcomes. Management Effectiveness Evaluation (MEE) is defined as the assessment of how well-protected areas are being managed – primarily, the extent to which management processes are being implemented to protect values, and achieve management goals and objectives (Hockings *et al.* 2006). Biodiversity outcomes can be measured in a number of ways but typically include natural habitat protection (e.g. Andam *et al.* 2008) and whether the abundances of key species are stable or increasing (e.g. Craigie *et al.* 2010).

26.6.1 Management effectiveness

Assessments of management of protected areas in Australia have been conducted through a variety of mechanisms including Auditor-General performance audits and Parliamentary Enquiries, internal agency processes of evaluation and NGO-driven assessments of progress against national and international targets. In 2007 a parliamentary enquiry was held to assess the state of the management of Australia's protected areas. It was titled: Senate Standing Committee on Environment, Communications, Information Technology and the Arts Enquiry into *Conserving Australia: Australia's national parks, conservation reserves and marine protected areas*. Reflecting the primarily State-based responsibility for protected area management in Australia, the Committee concluded, 'The effectiveness of Australia's reserve system relies on a landscape based approach to nature conservation, on good inter-agency and inter-jurisdictional coordination, and on adequate planning and resources for management of parks'. A common theme in submissions to the enquiry was the lack of adequate staff and financial resources to manage the expanding protected area estate. The Committee recommended both increased funding in line with protected area expansion and more transparent reporting of funding allocations by agencies. Despite this call for greater openness in providing data on resourcing of protected area management, comprehensive data across the jurisdictions in Australia remains difficult to obtain.

The most recent available data on investment in protected area management by State and Territory governments are provided by Taylor *et al.* (2011). These figures indicate that while budgets have increased in absolute terms, only in New South Wales have the increases kept pace with the increasing size of the protected area estate. In all other jurisdictions, including the Commonwealth, the management budgets have declined in $/ha terms, adjusted for inflation (Taylor *et al.* 2011). The level of investment in 2008/9 varied greatly from $46.88/ha in New South Wales to $2.45/ha in South Australia with a national average of just $9.56/ha (Taylor *et al.* 2011). They concluded that a seven-fold increase in the funding for the National Reserve System would be needed to fill the gaps identified in the report (Taylor *et al.* 2011). Despite this, the Commonwealth Government announced that there would be no specific allocation to the program in 2013/14 (DSEWPC 2012).

26.6.1.1 Auditor-General performance audits

A number of performance audits of protected area agencies have been conducted over the years, including:

- 1995 Auditor-General Victoria – Managing Parks for Life: The National Parks Service;
- 2001–2 Commonwealth Auditor General – The Management of Commonwealth National Parks and Reserves;
- 2004 NSW Audit Office. Performance audit: managing natural and cultural heritage in parks and reserves: National Parks and Wildlife Service;
- 2010 Queensland Audit Office – Sustainable management of national parks and protected areas; and
- 2013 NSW Audit Office – Management of historic heritage in national parks and reserves.

A common finding across all these audits is that the agencies responsible for management of protected areas lack measurable targets and data to assess their effectiveness in meeting key natural and cultural heritage conservation objectives. Often only a small proportion of protected areas have comprehensive and up-to-date management plans. The reports also comment on a lack of capacity to demonstrate value for money and baselines to measure change in reserve condition and reporting on outcomes. Reflecting this lack of specific and measurable outcomes in those management plans that do exist, the 2007 Senate enquiry recommended that 'management plans clearly identify practical on-ground outcomes and that protected area agencies have in place comprehensive monitoring and evaluation programs to continually assess management effectiveness and the extent to which protected area values are being maintained'.

26.6.1.2   Agency-based assessments of management effectiveness

All state national park agencies in Australia have developed systems for assessing the effectiveness of their management of protected areas, although they vary in breadth of coverage and depth of assessment across the reserves systems (Jacobson *et al.* 2008). Some, like New South Wales, Victoria, Queensland and Tasmania, have developed, or are in the process of developing, comprehensive, system-wide assessment systems. Most of these systems have been based on the IUCN framework for assessing management effectiveness of protected areas (Hockings *et al.* 2000, 2006). Public reporting on the results of the assessments varies across jurisdictions with some agencies providing published reports and others using the assessments only for internal reporting and decision-making. Across all agencies, there has been an increasing focus on linking assessment results to improved planning, priority setting and decision-making within the agency with the development of more sophisticated online systems for timely provision of data to decision makers.

The Australian Strategy for the National Reserve System (Natural Resource Management Ministerial Council (NRMMC) 2010) calls on all protected area agencies to adopt and implement systems for assessing the effectiveness of management of their reserves according to a set of principles for management evaluation (Table 26.4). Jacobson *et al.* (2008) concluded that state and territory agencies had made significant progress towards the NRS target of having management effectiveness evaluation systems in place in each management jurisdiction.

26.6.2   Biodiversity outcomes of protected areas

In Australia, protected area performance in maintaining populations of species remains poorly studied and therefore poorly understood. Long-term systematic population monitoring data is exceptionally rare, but critical for determining species- and community-level changes in natural values. Both population recoveries, e.g. Northern Hairy Nosed Wombats (Horsup 2004), and declines, e.g. mammals in Kakadu National Park (Woinarski *et al.* 2001, 2010, 2011, 2013) have been recorded.

Other than these limited examples, there is evidence, anecdotally, that not all protected areas are equally effective and that some have declining natural values. Assessments of protected area effectiveness that compare changes in natural values

Table 26.4 Principles for assessing management effectiveness (from Natural Resource Management Ministerial Council (NRMMC) 2010)

Principle	Description
Values and threats	Assessments should address the effectiveness of management in protecting and enhancing values and reducing threats to those values.
Landscape context	Assessments must take account of the relationship between protected areas and their biophysical and social landscape.
Internationally recognised framework	Assessments should be customised to local circumstances, but apply an internationally recognised system such as the IUCN World Commission on protected areas management effectiveness evaluation framework.
Clear objectives and assessment criteria	Objectives and criteria for assessing management effectiveness must be clearly defined and understood by assessors, managers and stakeholders.
Clear and cost-effective indicators	Indicators used in assessments must be cost effective and meaningful and capable of integration with broader natural resource management indicators.
Comprehensive engagement and capacity building	Managers, key stakeholders and those with expert knowledge about the environment should be engaged in the assessment process, where possible. Develop the capacity of traditional owners for the long-term effective management of protected areas.
Qualitative and quantitative information	Assessments should make use of all relevant available information rather than deferring assessment pending finalisation of precise quantitative data sets.
Adaptive management	Assessments should be part of adaptive management processes responsive to climate change and other threats.
Peer review	Internal assessments should be subjected to meaningful external peer review.
Transparency	Assessments should be publicly reported and routinely repeated to track trends.

inside vs. outside typically have not accounted for the bias in the current protected area network locations when evaluating impact and so are of limited use in assessing effectiveness. Protected areas in Australia are generally good at retaining habitat due to legislative protection for national parks but also because they have been placed in areas under less threat from clearing (see Section 26.5). However, recent state government decisions permitting livestock grazing and other damaging activities in national parks and the lack of protection against mining in private protected areas, raises the concern that this may not be the case in perpetuity. Nonetheless, there is an urgent need for well-designed studies of the impact of protected areas in Australia, using at least before–after or conservation–intervention comparisons (Geldmann *et al.* 2013) but preferably using robust matching designs for their analysis (Nolte *et al.* 2013).

Without monitoring in protected areas, negative outcomes, including large population declines, can go undetected. For instance, despite intensive management, small mammals in northern Australia and Kakadu National Park have been decimated by a perfect storm of habitat conversion, predation (by feral cats), poisoning (by invading cane toads), novel disease, and poorly managed fire regimes (Woinarski *et al.* 2011). Such examples demonstrate the critical importance of monitoring. Australia-wide, however, monitoring in protected areas is rare.

In the USA, Newmark (1987) illustrated that even extremely large protected areas were losing species over time. Recent research indicates that protected areas are unlikely to retain species unless external threats are actively managed (Laurance *et al.* 2012). It is difficult to say whether protected areas in Australia are effective at retaining biological values and/or mitigating threats overall, but there are numerous case studies demonstrating that they are the last strongholds of once-abundant and extensively distributed species and habitats e.g. Cassowaries (Moore 2007), Hairy-nosed wombats (Horsup 2004), Leadbeaters possum (Lindenmayer and Possingham 2013) and Temperate Rainforests (Lamington National Park), so it is clear they are providing substantial benefits for at least some key biodiversity features.

There is an urgent need to address the current shortfalls in protected area monitoring. Without better evidence on the effectiveness of protected areas as a conservation action, it is not possible to demonstrate the impact of our efforts and investment in protected areas on the conservation bottom line – outcomes, or to improve our performance and inform better decisions in the future (Bottrill and Pressey 2012). Monitoring biological outcomes in protected areas, and in comparable areas outside protected areas, must become routine if we are to achieve the goal of demonstrating the benefits of protected areas to biodiversity across Australia.

26.7   Threats to the Australian protected areas system

The Australian protected area system has grown to over a million square kilometres since its inception with continued government commitments to add to the system to ensure global targets are met (Adams *et al.* 2011). Despite this growth in the system, protected areas are not immune to calls for increased commercial and extractive access to these lands. The practice of protected area *downgrading, downsizing* and *degazettement* (PADDD, Mascia and Pailler 2011) is a globally documented trend with *downgrading* being the reduction in legal restrictions on human activities within a protected area, *downsizing* being the decrease in size of a protected area, and *degazettement* being the loss of legal protection for an entire protected area. PADDD events globally have been driven primarily by industrial-scale petroleum and mineral extraction, but other causal factors include tourism development and local land claims (Mascia and Pailler 2011). Some PADDD events have been argued to be part of protected area reconfigurations (Mascia and Pailler 2011).

There have been many examples of PADDD in Australia, most involving downsizing or degazettement of conservation reserves to allow resource extraction. At least some of these examples precede more recent legislation in some states to ensure the security of protected areas. In Tasmania, parts of reserves have been alienated for mining and quarrying, water impoundment, logging and grazing (Mosley 1969; Kirkpatrick 1987). Mercer and Peterson (1986) documented 23 revocations of parts of reserves, 11 of which were significant either in extent or environmental implications. One revocation for forestry involved the exchange of commercially valuable timber in a national park for a remote and less valuable area. In 1965, the Tasmanian Government announced that Lake Pedder National Park would be 'modified' by extensive inundation for a hydro-electric scheme (Crowley 1999). Similar examples come from other States. In the 1950s, dedicated reserves in South Australia were resumed for commercial uses (Harris 1977). In New South Wales, excisions from Royal National Park were made for commercial and other purposes (Pettigrew and Lyons 1979). Conservation reserves in Western Australia have been used for timber, agriculture, and mining (Ride 1975; Rundle 1996; Pouliquen-Young 1997).

Some scientists have suggested that degazettement followed by new protected area purchases could result in better protection for Australian ecosystems (Fuller *et al.* 2010). A serious problem with this argument is that, if adopted, it would undermine the deliberate legislative security now given to many protected areas, regardless of how effectively they have been located and designed. With this security removed, a host of motivations for removing protected status could be ushered into public debate, not all of them directed at better protection of ecosystems.

PADDD remains a real phenomenon in Australia, particularly in relation to mining. Individual State and Territory legislation specifies where mineral exploration can occur, allowing for exploration in most protected areas with the exception of National Parks, which represent ~30% of the current protected area system. Adams and Moon (2013) detailed the continued mineral exploration of protected areas after gazettement of these areas for protection in Queensland with a clear directive from the state government that downgrading or degazettement would be considered if exploration led to a proposal for mineral extraction. This has certainly been the result in the Bimblebox Nature Refuge. Exposing private protected areas to mineral exploration and extraction can result in an imbalance between the costs and benefits, to landholders and the government, from conservation and mining activities on private properties. Landholders bear the financial costs of a private protected area while, in the event of mining, the government receives economic benefits from the mineral exploration and extraction (Adams and Moon 2013). Among the social consequences of mining private protected areas is diminished legitimacy of the conservation covenant program which might ultimately result in reduced participation and reduced capacity of the State and Federal Governments to meet their policy commitments. Aside from mining, legislation in several Australian states has sought recently to downgrade protected areas for activities such as grazing and hunting (Adams and Moon 2013; Ritchie 2013). These trends challenge the image of Australia's protected areas as permanent vestiges of unique ecosystems and biodiversity.

26.8 Conclusions

Australia's protected areas are the cornerstone of its national- and State-level conservation strategies. Today's network represents a considerable achievement and a major contribution to the nation's biodiversity protection efforts. However, there are several ways in which Australia's protected areas can be improved in the future.

- Target new protected areas towards regional ecosystems and species that are currently inadequately protected and are at risk of loss and degradation, of which there remain many; and expand the application of protected areas as tools for landscape restoration in addition to the preservation of pristine environments.

- Make monitoring of management effectiveness and biodiversity outcomes routine so that examples of best-practice and successful outcomes can be shared to all protected areas.

- Prevent existing protected areas from being damaged by narrow and short-term commercial interests.

In future it appears that much of the progress in increasing coverage of currently under-represented regional ecosystems and species will lie in the increasing protection of private lands and Indigenous Protected Areas. It should be remembered that protected areas are intimately linked to their surroundings and management of areas adjacent to protected areas will often influence biodiversity outcomes within protected areas. This applies especially to threatening processes, such as invasive species, that occur across whole landscapes. The lesson from this is that protected areas should not be viewed and managed in isolation but as parts of broader landscapes that are managed also for their natural values and to strengthen protected areas.

Over the past few decades there has been a worrying trend for governments to focus on economic growth as their main objective, to the detriment of protected areas specifically and conservation in general. This weakening of protections is not new: it has been noted since the 1990s (Figgis 1997). Conservative governments in Australia are typically strongly pro-development and favourably disposed to timber, mining, tourism and rural industry lobbies. This leads to a preference for economic use (in the narrow, cash-generating meaning of the word) of land over conservation use. If this trend continues, there is the potential for shortsighted decisions to undermine the considerable broad social and economic benefits provided by protected areas and to wind back decades of conservation gains. Today, the future holds a great deal of uncertainty for Australia's protected areas (Ritchie 2013).

REFERENCES

Adams, V. M. and K. Moon (2013). Security and equity of conservation covenants: Contradictions of private protected area policies in Australia. *Land Use Policy* **30**(1): 114–119.

Adams, V. M., D. B. Segan and R. L. Pressey (2011). How much does it cost to expand a protected area system? Some critical determining factors and ranges of costs for Queensland. *PLoS ONE* **6**(9): e25447.

Altman, J. C. (2007). *Alleviating Poverty in Remote Indigenous Australia: The Role of the Hybrid Economy*, Australian National University, Centre for Aboriginal Economic Policy Research.

Andam, K. S., P. J. Ferraro, A. Pfaff, G. A. Sanchez-Azofeifa and J. A. Robalino (2008). Measuring the effectiveness of protected area networks in reducing deforestation. *Proceedings of the National Academy of Sciences* **105**(42): 16 089–16 094.

Anon. (1992). *National Forest Policy Statement*. Canberra, Commonwealth of Australia.

Anon. (1995). *National Forest Conservation Reserves: Commonwealth Proposed Criteria*. Canberra, Australian Government Publishing Service.

Austin, M. P. and C. R. Margules (1986). Assessing representativeness. In *Wildlife Conservation Evaluation*, M. B. Usher (ed.). London., Chapman and Hall: pp. 45–67.

Balmford, A., A. Bruner, P. Cooper, *et al.* (2002). Economic reasons for conserving wild nature. *Science* **297**(5583): 950–953.

Bertzky, B., C. Corrigan, J. Kemsey, *et al.* (2012). *Protected Planet Report 2012: Tracking Progress Towards Global Targets for Protected Areas*. IUCN, Gland, Switzerland and UNEP-WCMC, Cambridge, UK.

Bottrill, M. C. and R. L. Pressey (2012). The effectiveness and evaluation of conservation planning. *Conservation Letters* 5(6): 407–420.

Bradshaw, C. J. (2012). Little left to lose: deforestation and forest degradation in Australia since European colonization. *Journal of Plant Ecology* 5(1): 109–120.

Brooks, T., M. Bakarr, T. Boucher, *et al.* (2004). Coverage of the existing global protected area system. *BioScience* 54: 1081–1091.

Bryan, B. A. (2002). Reserve selection for nature conservation in South Australia: past, present and future. *Australian Geographical Studies* 40(2): 196–209.

CAPAD (2010). Collaborative Australian Protected Area Database (CAPAD). Canberra, Department of Sustainability, Environment, Water, Populations and Communities. Australian Goverment.

Chape, S., M. Spalding and M. Jenkins (2008). *The World's Protected Areas: Status, Values and Prospects in the Twenty-First Century.* University of California Press, Berkeley.

Coad, L., F. Leverington and N. Burgess (2013). Progress towards the CBD protected area management effectiveness targets. *Parks* 19: 13–24.

Convention on Biological Diversity (2010). Decision adopted by the conference of the parties to the convention on biological diversity at its tenth meeting [Decision X/2] Nagoya, Japan, Secretariat to the Convention on Biological Diversity, Montreal, Canada.

Craigie, I. D., J. E. M. Baillie, A. Balmford, *et al.* (2010). Large mammal population declines in Africa's protected areas. *Biological Conservation* 143(9): 2221–2228.

Cresswell, I. and G. Thomas (1998). *Terrestrial and Marine Protected Areas in Australia (1997),* Environment Australia, Biodiversity Group, Canberra.

Crowley, K. (1999). Lake Pedder's loss and failed restoration: ecological politics meets liberal democracy in Tasmania. *Australian Journal of Political Science* 34: 409–424.

DERM (2007). Delbessie Agreement (State Rural Leasehold Land Strategy), Available from http://www.nrw.qld.gov.au/land/state/rural_leasehold/strategy.html

DSEWPC (2012). *One Land – Many Stories: Prospectus of Investment 2013–2014,* Commonwealth of Australia.

Dudley, N. (2008). *Guidelines for Applying Protected Area Management Categories.* Gland, Switzerland, IUCN.

Figgis, P. (1997). *Australia's National Parks and Protected Areas: Future Directions.* Canberra, Australia.

Fuller, R. A., E. McDonald-Madden, K. A. Wilson, *et al.* (2010). Replacing underperforming protected areas achieves better conservation outcomes. *Nature* 466 (7304): 365–367.

Geldmann, J., M. Barnes, L. Coad, *et al.* (2013). Effectiveness of terrestrial protected areas in reducing habitat loss and population declines. *Biological Conservation* 161(0): 230–238.

Harris, C. R. (1977). Towards a historical perspective. *Proceedings of the Royal Geographical Society of Australasia South Australian Branch* 78: 55–71.

Hill, R., F. Walsh, J. Davies and M. Sandford (2011). *Our Country Our Way: Guidelines for Australian Indigenous Protected Area Management Plans*. CSIRO Ecosystem Sciences and Australian Government Department of Sustainability, Environment, Water, Population and Communities.

Hockings, M., S. Stolton and N. Dudley (2000). *Evaluating Effectiveness: a Framework for Assessing Management of Protected Areas*. IUCN Cardiff University Best Practice Series, IUCN, Cambridge, UK.

Hockings, M., S. Stolton, F. Leverington, N. Dudley and J. Courrau (2006). *Evaluating Effectiveness: a Framework for Assessing Management Effectiveness of Protected Areas*, 2nd edition. IUCN, Gland, Switzerland, and Cambridge, UK.

Horsup, A. (2004). *Recovery Plan for the Northern Hairy-nosed Wombat* Lasiorhinus krefftii *2004–2008*. Report to the Department of Environment and Heritage, Canberra. Environmental Protection Agency. Queensland Parks and Wildlife Service, Brisbane, Australia.

Jacobson, C., R. Carter and M. Hockings (2008). The status of protected area management evaluation in Australia and implications for its future. *Australasian Journal of Environmental Management* 15(4): 202–210.

Jenkins, C. N. and L. Joppa (2009). Expansion of the global terrestrial protected area system. *Biological Conservation* 142(10): 2166–2174.

Joint ANZECC/MCFFA National Forest Policy Statement Implementation Subcommittee (1997). *Nationally Agreed Criteria for the Establishment of a Comprehensive, Adequate and Representative Reserve System for Forests in Australia*. Canberra, Commonwealth of Australia.

Kabii, T. and P. Horwitz (2006). A review of landholder motivations and determinants for participation in conservation covenanting programmes. *Environmental Conservation* 33: 11–20.

Kirkpatrick, J. B. (1987). Forest reservation in Tasmania. *Search* 18: 138–142.

Landsberg, J., R. J. Hobbs and C. J. Yates (2000). Status of temperate woodlands in the Australian Capital Territory Region. In *Temperate Eucalypt Woodlands in Australia: Biology, Conservation, Management and Restoration*: pp. 32–44.

Laurance, W. F., D. Carolina Useche, J. Rendeiro, *et al.* (2012). Averting biodiversity collapse in tropical forest protected areas. *Nature* 489(7415): 290–294.

Lindenmayer, D. B. and H. P. Possingham (2013). No excuse for habitat destruction. *Science* 340(6133): 680.

Margules, C. R. and R. L. Pressey (2000). Systematic conservation planning. *Nature* 405: 243–253.

Mascia, M. B. and S. Pailler (2011). Protected area downgrading, downsizing, and degazettement (PADDD) and its conservation implications. *Conservation Letters* 4(1): 9–20.

May, K. (2010). *Indigenous Cultural and Natural Resource Management and the Emerging Role of the Working on Country Program*. Centre for Aboriginal Economic Policy Research, Australian National University.

Mendel, L. C. and J. Kirkpatrick (2002). Historical progress of biodiversity conservation in the protected-area system of Tasmania, Australia. *Conservation Biology* **16**(6): 1520–1529.

Mercer, D. and J. Peterson (1986). The revocation of national parks and equivalent reserves in Tasmania. *Search* **17**: 134–140.

Moon, K. and C. Cocklin (2011). A landholder-based approach to the design of private-land conservation programs. *Conservation Biology* **25**(3): 493–503.

Moore, L. A. (2007). Population ecology of the southern cassowary *Casuarius casuarius johnsonii*, Mission Beach north Queensland. *Journal of Ornithology* **148**(3): 357–366.

Mosley, J. G. (1969). Scenic reserve and fauna sanctuary systems of Tasmania. In *The Last of Lands*, L. J. Webb, D. Whitelock and J. L. G. Brereton (eds.). Brisbane, Jacaranda Press: pp. 160–169.

National Reserve System Task Group (2009). *Australia's Strategy for the National Reserve System 2009–2030*. Canberra, Commonwealth of Australia.

Natural Resource Management Ministerial Council (NRMMC) (2010). *Australia's Strategy for the National Reserve System 2009 – 2030*, Australian Government.

Newmark, W. D. (1987). A land-bridge island perspective on mammalian extinctions in western North American Parks. *Nature* **325**: 430–432.

Nolte, C., A. Agrawal, K. M. Silvius and B. S. Soares-Filho (2013). Governance regime and location influence avoided deforestation success of protected areas in the Brazilian Amazon. *Proceedings of the National Academy of Sciences* **110**(13): 4956–4961.

Pettigrew, C. and M. Lyons (1979). Royal National Park: a history. *Parks and Wildlife* **2** (3–4): 15–30.

Pouliquen-Young, O. (1997). Evolution of the system of protected areas in Western Australia. *Environmental Conservation* **24**: 168–181.

Pressey, R. L. and K. H. Taffs (2001). Sampling of land types by protected areas: three measures of effectiveness applied to western New South Wales. *Biological Conservation* **101**: 105–117.

Pressey, R., T. Hager, K. Ryan, *et al.* (2000). Using abiotic data for conservation assessments over extensive regions: quantitative methods applied across New South Wales, Australia. *Biological Conservation* **96**(1): 55–82.

Pressey, R. L., G. L. Whish, T. W. Barrett and M. E. Watts (2002). Effectiveness of protected areas in north-eastern New South Wales: recent trends in six measures. *Biological Conservation* **106**: 57–69.

Pressey, R. L., M. E. Watts and T. W. Barrett (2004). Is maximizing protection the same as minimizing loss? Efficiency and retention as alternative measures of the effectiveness of proposed reserves. *Ecology Letters* **7**: 1035–1046.

Preuss, K. and M. Dixon (2012). 'Looking after country two-ways': insights into Indigenous community-based conservation from the Southern Tanami. *Ecological Management & Restoration* **13**(1): 2–15.

Ride, W. L. D. (1975). Towards an integrated system: a study of selection and acquisition of national parks and nature reserves in Western Australia. In *A National System of*

Ecological Reserves in Australia, F. Fenner (ed.). Canberra, Australian Academy of Science: pp. 64–85.

Ritchie, E. G. (2013). Conservation: relaxed laws imperil Australian wildlife. *Nature* **498** (7455): 434.

Rodrigues, A. S. L. (2006). Are global conservation efforts successful? *Science* **313**: 1051–1052.

Rundle, G. E. (1996). History of conservation reserves in the south-west of Western Australia. *Journal of the Royal Society of Western Australia* **79**: 225–240.

Sattler, P. S. and A. Glanznig (2006). *Building Nature's Safety Net: A Review of Australia's Terrestrial Protected Area System, 1991–2004*. WWF-Australia.

Sattler, P. and M. Taylor (2008). *Building Nature's Safety Net 2008: Progress on the Directions for the National Reserve System*. WWF-Australia.

Sattler, P. S. and R. Williams (1999). *The Conservation Status of Queensland's Bioregional Ecosystems*. Environmental Protection Agency, Queensland Government. Brisbane.

Sharafi, S. M., M. White and M. Burgman (2012). Implementing comprehensiveness, adequacy and representativeness criteria (CAR) to indicate gaps in an existing reserve system: a case study from Victoria, Australia. *Ecological Indicators* **18**: 342–352.

Stanton, J. P. and M. G. Morgan (1977). The rapid selection and appraisal of key and endangered sites: the Queensland case study. Project RAKES – a rapid appraisal of key and endangered sites. University of New England. Report No. 1. Armidale.

Taylor, M., P. Sattler, C. Curnow, *et al.* (2011). *Building Natures Safety Net 2011 – The State of Protected Areas for Australia's Ecosystems and Wildlife*. WWF Australia.

UNEP and WCMC (2007). Indicator 26: Protected Areas Report. *Millenium Development Goals*.

Vanclay, J. K. (1996). Lessons from the Queensland rainforests: steps towards sustainability. *Journal of Sustainable Forestry* **3**(2–3): 1–27.

Woinarski, J. C. Z., D. J. Milne and G. Wanganeen (2001). Changes in mammal populations in relatively intact landscapes of Kakadu National Park, Northern Territory, Australia. *Austral Ecology* **26**(4): 360–370.

Woinarski, J. C. Z., M. Armstrong, K. Brennan, *et al.* (2010). Monitoring indicates rapid and severe decline of native small mammals in Kakadu National Park, northern Australia. *Wildlife Research* **37**(2): 116–126.

Woinarski, J. C. Z., S. Legge, J. A. Fitzsimons, *et al.* (2011). The disappearing mammal fauna of northern Australia: context, cause, and response. *Conservation Letters* **4**(3): 192–201.

Woinarski, J., J. Green, A. Fisher, M. Ensbey and B. Mackey (2013). The effectiveness of conservation reserves: land tenure impacts upon biodiversity across extensive natural landscapes in the tropical savannahs of the Northern Territory, Australia. *Land* **2**(1): 20–36.

CHAPTER 27

Australian marine protected areas

Alana Grech, Graham J. Edgar, Peter Fairweather, Robert L. Pressey and Trevor J. Ward

Summary

Marine protected areas (MPAs) are sites in the ocean and coastal sea that are dedicated to the conservation of biodiversity, fisheries, ecosystem services and cultural values. MPAs range from small, highly protected marine reserves through to large, multiple-use marine parks, such as the Great Barrier Reef Marine Park of Queensland, Australia. This chapter identifies the major policy events and phases of MPA development in Australia, and explores the role and effectiveness of MPAs in conserving Australia's marine environment. The governance of Australian MPAs is complex; the responsibility for their declaration and management is shared between the Australian (Commonwealth), State and Territory Governments. Progress in the declaration and management of MPAs is not uniform across Australia, with some jurisdictions performing better than others. Australia is considered a world leader in the science and implementation of MPAs. However, there are serious weaknesses in the design of MPAs in Commonwealth waters due to the locating of new MPAs where they are least controversial and least costly. Considerable further effort is needed to create an effective national programme for delivering biodiversity conservation in Australia waters. This is particularly important because Australia's oceans face an unprecedented set of pressures from accelerating climate change and coastal development.

27.1 Introduction

Australia is responsible for one of the largest marine jurisdictions in the world, covering an area of more than 13.86 million km^2. This domain stretches across about 45° of latitude from the tropical waters of the north to the sub-Antarctic waters of the Southern Ocean, and encompasses seabed, open ocean and shoreline ecosystems, and near-shore marine and estuarine waters. The marine environment is rich in biodiversity. Over 33 000 identified marine species live in Australian waters, including a large number of endemics.

Austral Ark: The State of Wildlife in Australia and New Zealand, eds. A. Stow, N. Maclean and G. I. Holwell. Published by Cambridge University Press. © Cambridge University Press 2015.

Australia's marine ecosystems provide a wide range of services that benefit people both directly and indirectly. The trade of marine natural resources, including seafood products, oil, natural gas and minerals, is important to Australia's economy, and 99% of Australia's exports are transported by sea. Recreational activities, such as fishing, swimming, boating and diving, contribute to the well-being of many Australians. Parts of the marine environment are also some of Australia's most significant cultural and heritage icons, such as the Great Barrier Reef in Queensland and Shark Bay in Western Australia. Australia's coastal waters are Sea Country for many Aboriginal and Torres Strait Island communities.

More than 85% of the Australian population lives within 50 km of the coast, to capitalise on the many values of the marine environment. But with an increasing coastal population comes increasing pressure from both land-based and marine human activities. Agricultural and urban developments may affect marine ecosystems by modifying sediment regimes and the inflows of nutrients, pesticides and other chemicals, and by replacing or damaging habitats with infrastructure. Marine-based activities such as fishing and mining may deplete fisheries resources, pollute waters, alter habitats and change species composition. Anthropogenic climate change has profound implications for marine ecosystems across Australia. As a result, human activities are adversely affecting marine biodiversity and compromising the delivery of ecosystem services that are crucial to the well-being of many Australians along the coast.

Marine protected areas (MPAs) are intended to mitigate the impacts of human activities on the marine environment. MPAs alone are not sufficient, of course. They will always need to be complemented with other approaches to marine conservation, including education, regulation of extractive activities in the sea, and management of land-based activities to reduce their impacts on marine waters. Nonetheless, MPAs are a key management tool for the marine environment (Gaines *et al.*, 2010; Veitch *et al.*, 2012), so their effectiveness is of considerable interest. This chapter explores the role and effectiveness of MPAs in conserving Australia's marine environment.

27.2 Benefits of marine protected areas

The International Union for the Conservation of Nature (IUCN) considers an MPA as 'A clearly defined geographical space, recognised, dedicated and managed, through legal or other effective means, to achieve the long-term conservation of nature with associated ecosystem services and cultural values' (Day *et al.*, 2012). Most Australian government authorities recognise this IUCN definition, which is widely encompassing and covers areas subject to a great variety of aims, regulations and governance regimes. For example, MPAs can be declared for conservation and for fisheries management, research, education, heritage, tourism and recreational use, or a combination of these. To provide a common currency when designating MPAs, most authorities follow the six categories developed by the IUCN (Dudley, 2008; Table 27.1).

MPAs range from small, highly protected 'no-go' marine reserves where all marine life is fully protected from human use (i.e. IUCN category Ia), through 'no-take' areas where all forms of extraction are prohibited, to large, multiple-use marine parks, where different human activities are zoned and managed in an integrated way (e.g. the Great Barrier Reef Marine Park; Box 27.1). MPAs can therefore provide protection at various levels that range from restrictions on one or more human activities, to the comprehensive protection of an area from all extractive effects. This chapter's focus is on 'no-take' IUCN

Table 27.1   IUCN protected area categories and their definitions (Dudley, 2008). In this chapter, MPAs cover all six categories, while marine reserves are a subset of MPAs being 'no-take' areas that meet the criteria for categories Ia, Ib or II. Note that the appropriateness with which these categories have been applied to Australian MPAs has been questioned (Fitzsimons, 2011).

IUCN category	Definition
Ia (strict nature reserve) and Ib (wilderness area)	Strictly protected areas where human visitation, use and impacts are controlled and limited to ensure protection of conservation values.
II (national park)	Protected areas where large-scale ecological processes are protected, while allowing for environmentally and culturally compatible visitor opportunities.
III (natural monument or feature)	Protected areas that are set aside to protect a specific natural monument. They are generally quite small protected areas and often have high visitor value.
IV (habitat/species management area)	Protected areas that aim to protect particular species or habitats, with management reflecting this priority.
V (protected landscape/seascape)	Protected areas where safeguarding the integrity of a valuable interaction between people and nature is vital to protecting and sustaining natural and other values.
VI (protected area with sustainable use of natural resources)	Protected areas where ecosystems, together with associated cultural values and traditional natural resource management systems, are conserved.

Box 27.1 Re-zoning of the Great Barrier Reef Marine Park

The Great Barrier Reef (GBR) Region is ~346 000 km^2 and stretches along ~2000 km of Australia's north-east coast. The region supports a variety of tropical biota including the world's largest coral reef ecosystem, seagrasses, 1500 species of fishes, turtles, sea snakes, sea birds, whales, dolphins and dugongs. The region is also an important area for recreational and commercial activities, especially marine tourism, shipping and fishing.

The GBR Region is jointly managed by the State of Queensland and Australian (Commonwealth) Governments. The Australian Government's *Great Barrier Reef Marine Park Act* 1975 established both the Great Barrier Reef Marine Park (GBR Marine Park) and the Great Barrier Reef Marine Park Authority (GBRMPA). The GBRMPA is a Commonwealth Statutory Authority and principal adviser to the Australian Government on the GBR Marine Park. The GBR Marine Park is a Commonwealth MPA that protects all the waters below mean low water; most coastal and island waters above mean low water mark are within Queensland (State) waters.

Between 1981 and the late 1990s, approximately 15 800 km^2 (~4.5%) of the GBR Marine Park was zoned as 'no-take' areas (marine reserves); a further 450 km^2 was set aside for scientific research. A review of one section of the GBR Marine Park found that the amount and distribution of 'no-take' areas were inadequate to ensure the protection of biodiversity (Fernandes *et al.*, 2005). The 'Representative Area Program' (RAP) was initiated to re-zone the GBR Marine Park to improve biodiversity protection through a comprehensive and representative multiple-use zoning regime (Day *et al.*, 2000). The objective of the RAP was to maintain biological diversity by optimising the design of a network of marine reserves, covering the range of habitats and communities found within the region while minimising the adverse impacts on users of the area (Day *et al.*, 2000).

The GBRMPA worked with an independent expert Scientific Steering Committee to establish 11 Biophysical Operational Principles and quantitative targets to guide the agency in developing the 'no-take' network (see Fernandes *et al.*, 2005). The quantitative targets, as well as expert opinion, stakeholder involvement and analytical approaches such as the marine reserve design software Marxan (Ball & Possingham, 2000) informed the GBRMPA in the design of the new zoning plan, shown in Plate 17. Stages in the development of the 2004 zoning plan, from raw software outputs to refined boundaries following public participation, are illustrated by Pressey *et al.* (2013).

The Australian Government's *Great Barrier Reef Marine Park Zoning Plan* 2003 and Queensland's *Great Barrier Reef Coast Marine Park Zoning Plan* 2004 were implemented in July 2004. The Plans provide 'no-take' protection for specific areas (i.e. Marine National Park Zones), prohibit public access in very small 'no-go' areas (i.e. Preservation Zones) whilst also allowing a variety of ecologically sustainable activities, such as commercial fishing, to occur in other zones (i.e. General Use and Habitat Protection Zones). The Plans protect approximately 33% of the GBRWHA in 'no-take' and 'no-go' zones.

categories Ia and Ib (I) and II MPAs (Table 27.1), referred to here as marine reserves, because they are the most effective tool for biodiversity conservation. Investigations of ecological conditions in partially protected MPAs (i.e. IUCN categories III, IV, V and VI; Table 27.1) typically reveal no or very little differences from the conditions in fished areas outside MPAs (Lester & Halpern, 2008; Di Franco *et al.*, 2009; Edgar *et al.*, 2011). The benefits that arise from MPAs are highly variable, depending on configuration, regulatory enforcement and level of community engagement. We review the potential benefits of MPAs below.

Direct benefits   MPAs can preserve and/or restore populations of exploited species (e.g. fished species), creating higher densities of such species living inside marine reserves relative to outside (e.g. Williamson *et al.*, 2004; Edgar & Stuart-Smith, 2009; McCook *et al.*, 2010; Edgar & Barrett, 2012). Higher populations of exploited species inside marine reserves can benefit adjacent fisheries via the spillover of individuals that swim across boundaries (Januchowski-Hartley *et al.*, 2013) and by drift of larvae and eggs that reseed the wider region (Jones *et al.*, 2009; Harrison *et al.*, 2013). However, numerous exceptions also exist where densities of exploited species are similar or lower inside marine reserves relative to outside (Edgar *et al.*, 2009), mainly because of poor MPA design, management and compliance.

Indirect benefits   Marine reserves can contain community types not present in fished areas as a consequence of indirect effects of fishing that modify food-web structure (Edgar *et al.*, 2009). For example, the keystone role of large predatory animals, such as sharks and groupers, in controlling prey numbers has been lost in most unprotected areas due to fishing. Long-term studies in marine reserves indicate that recovery of large predators is generally followed at decadal scales by second- and third-order changes to food webs (Babcock *et al.*, 2010) and an overall improvement in ecosystem functioning and ability to cope with change (i.e. 'ecosystem resilience').

Direct and indirect benefits manifest in several contributions of MPAs to conservation and society.

Resource allocation	MPAs can separate biodiversity features from activities that directly threaten them. Through zoning of high-value areas, well-planned MPAs can also assist managers reduce conflict between different user groups (Wood & Dragicevic, 2007).
Insurance	Complete knowledge of marine ecosystems is unattainable because of the massive complexity of species' interactions and their nonlinear population dynamics (Glaser *et al.*, 2013). MPAs assist precautionary approaches to management by providing 'an insurance option' in the absence of complete knowledge (Murray *et al.*, 1999). Such insurance is increasingly important in an era of climate change, when environmental parameters in ecosystem models exceed known bounds, limiting the reliability of predictions of impacts.
Knowledge	MPAs can be regarded as a vast opportunity for providing innumerable improvements in our understanding of how human activities interact and affect natural systems. Marine reserves also provide opportunities for research on species that are normally rare, directly contributing to their effective management outside of MPAs (Dayton *et al.*, 2000). Field estimates of life-history attributes of large predatory lobsters in Tasmania, for example, can only be made in marine reserves because of the near absence of these animals in fished areas (Barrett *et al.*, 2009).
Custodianship	Community interest in MPAs and stewardship of local sites leads to increased awareness of changes to the marine environment and the need to counter potential threats. For example, following local community interest, eradication programmes have been undertaken for invasive black urchins (*Centrostephanus rodgersii*) in Beware Reef Nature Reserve (Victoria) and for wakame kelp (*Undaria pinnatifida*) in the Tinderbox Marine Reserve (Tasmania).
Enhanced human experience	Many users of the marine environment benefit from values additional to exploitable resources (Agardy, 2003). These include diving, photography and other non-consumptive recreational and tourism-related activities, or educational, aesthetic or indigenous cultural values. These benefits are typically enhanced within MPAs compared to unprotected sections of coast or ocean.

While MPAs can generate substantial benefits, costs are rarely trivial and also need to be considered (e.g. Ban *et al.*, 2011; Hunt, 2013). These can include direct planning and ongoing management costs, such as policing and monitoring of conditions, and also costs that affect the marine environment outside MPAs, such as local depletion in fish stocks when fishers move from closed fishing grounds to adjacent areas ('displaced effort'). An additional important cost, which is rarely considered, is ongoing uncontrolled deterioration in the marine environment that accompanies a false sense of security amongst policymakers and the public about environment protection when ineffective MPA networks are declared (Edgar, 2011).

Most reports on the application of MPAs emphasise that they should not be regarded as a panacea for achieving conservation of all marine biodiversity (Kaiser, 2005). MPAs represent an important tool for managers attempting to reduce direct and proximal threats to the marine environment, albeit one that has limited value in countering pervasive broad-scale threats such as climate change, introduced invasive species and chronic pollution associated with terrestrial run-off. MPAs do not occur in isolation and their realised benefits are dependent on the health of surrounding waters. Broad-scale threats are generally best dealt with through other targeted mechanisms. Regardless, MPAs can contribute to some reduction in threats by allowing focussed local control of human activities. Public and managerial attention is typically enhanced in MPAs, and more stringent regulatory controls are applied when dealing with potential threats, for example, new foreshore development, ballast water release and storm-water discharge. The most important consideration is that the net sum of all threats is reduced. While marine ecosystems are generally resilient to individual stresses, they can change, sometimes irreversibly, when tipping points associated with cumulative impacts are exceeded (Hughes *et al.*, 2012).

27.3 Marine protected areas in Australian waters

The declaration and management of MPAs in Australian waters is the responsibility of the Australian (Commonwealth), State and Territory Governments. The Australian Government is responsible for MPAs in Commonwealth waters and the State and Northern Territory Governments are responsible for MPAs in their waters (Plate 11). The States and Northern Territory usually have jurisdiction over near-shore waters out to 3 nautical miles (~ 5.5 kilometres) from the mainland coast and around inshore islands, while the Commonwealth jurisdiction extends from the State and Territory waters boundary to the outer edge of Australia's Exclusive Economic Zone, generally 200 nautical miles from the coast.

The Commonwealth and State Governments both have regulations and legislation dedicated to the design, declaration and management of MPAs (e.g. *South Australian Marine Parks Act* 2007), and agencies to administer that legislation (e.g. the Australian Government Department of Environment (DOE)). The Great Barrier Reef has its own Commonwealth legislation and an agency whose primary purpose is to protect the values of the Great Barrier Reef Region (Box 27.1). The Commonwealth has no statutory role in MPAs in State waters and vice versa. High-level coordination of the design, declaration and reporting of MPAs is conducted by the National Representative System of Marine Protected Areas (NRSMPA) programme in DOE. A chronology of the major events and phases of MPA development in Australia, including the NRSMPA programme, is in Figure 27.1. The primary goal of the NRSMPA was to build a comprehensive, adequate and representative (CAR) system of MPAs to contribute to the long-term conservation of Australia's marine environment. The *Strategic Plan of Action for the NRSMPA* defines the CAR principles as follows (ANZECC TFMPA, 1999, pp. 15–16):

Comprehensiveness	The NRSMPA will include the full range of ecosystems recognised at an appropriate scale within and across each bioregion.
Adequacy	The NRSMPA will have the required level of reservation to ensure the ecological viability and integrity of populations, species and communities.
Representativeness	Those marine areas that are selected for inclusion in MPAs should reasonably reflect the biotic diversity of the marine ecosystems from which they derive.

27.3.1 State and Northern Territory marine protected areas

Governments of all States and the Northern Territory are signatories to the *Strategic Plan of Action for the NRSMPA* (Figure 27.1). However, progress in the declaration and management of MPAs is not uniform across Australia (SOEC, 2011). South Australia, Victoria and Queensland have active MPA programmes that are well advanced. In 2012, South Australia announced the finalisation of management plans and zoning for 19 multiple-use MPAs and 3600 km^2 of marine reserves[1] (Table 27.2). Victoria's marine waters encompass 13 Marine National Parks, 11 smaller Marine Sanctuaries and four Marine/ Marine and Coastal Parks.[2] Together, these parks and sanctuaries protect 5.3 % of Victorian waters in marine reserves. Australia's first MPA was declared in Queensland waters in 1937 (Figure 27.1) and the second declared over Heron and Wistari Reefs in 1974. Queensland now has three large multiple-use MPAs (Great Barrier Reef Coast Marine Park (Box 27.1), Great Sandy Marine Park and Moreton Bay Marine Park), covering most of the nearshore waters along the east-coast of Queensland.[3]

The performance of other States and the Northern Territory in relation to declaration of MPAs can be described as patchy and mediocre at best (Wescott, 2006). The worst performing jurisdictions, in terms of percentage of waters with high protection, are the Northern Territory and Tasmania,[4] where ~1 % of coastal waters are fully protected in marine reserves (Table 27.2), although Tasmania has also protected an additional 747 km^2 of state waters at remote Macquarie Island. The New South Wales system of MPAs encompasses six multiple-use MPAs, 12 aquatic reserves and 62 national parks and reserves with marine components.[5] However, there is currently a moratorium on declaring new MPAs, and on zoning plan reviews and alteration of sanctuary zones in New South Wales waters. Western Australia has declared 13 multiple-use MPAs covering about 12% of coastal waters[6] including about 3% of its waters in reserves (SOEC, 2011). The Western Australian Government has committed to seven additional MPAs in their waters, which are all currently in the planning process.

---

[1] http://www.environment.sa.gov.au/marineparks/About
[2] http://www.coastlinks.vic.gov.au/marineparks.htm   [3] http://www.nprsr.qld.gov.au/marine-parks/
[4] http://www.parks.tas.gov.au/?base=397   [5] http://www.mpa.nsw.gov.au/
[6] http://www.dec.wa.gov.au/management-and-protection/marine-environment/marine-parks-and-reserves.html

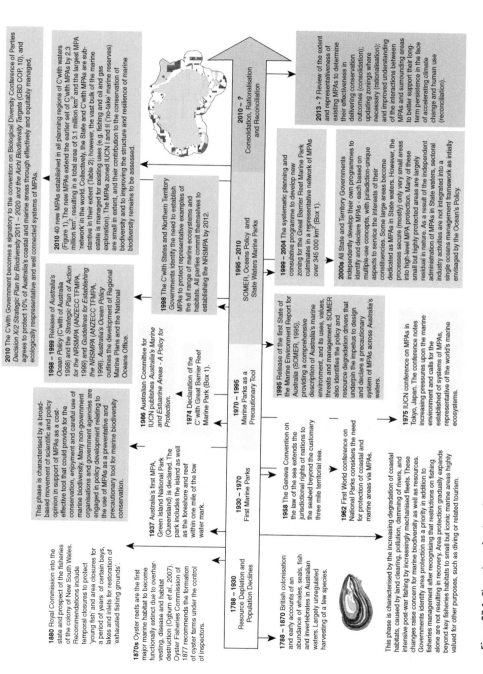

Figure 27.1 Chronology of major policy events and phases of marine protected area development in Australia. Kelleher & Kenchington (1991) discussed three main approaches to resource management and biodiversity conservation in Australian waters, consistent with the first three phases of the chronology. NRSMPA: National Representative System of Marine Protected Areas; C'with = Commonwealth; IUCN = International Union for the Conservation of Nature.

Table 27.2   Area (km^2) of declared marine reserves in 'no-take' marine reserves (IUCN categories I and II; Table 27.1) and other IUCN categories in Commonwealth waters. C'wlth = Commonwealth; NSW = New South Wales; NT = Northern Territory; Qld = Queensland; SA = South Australia; Tas = Tasmania; Vic = Victoria; WA = Western Australia; Australia = total for all jurisdictions; and, IUCN = International Union for Conservation of Nature.

	C'wlth[a]	NSW	NT	Qld	SA	Tas	Vic	WA[e]	Australia
IUCN I	219 968	665	0	412	689	0	0	2960	224 694
IUCN II[b]	977 134	0	740	16 607	2934	1 001[c]	535	0	998 951
Other IUCN	1 974 846	–	–	–	–	–	–	–	–
Sum I and II	1 197 102	665	740	17 019	3623	1001	535	2960	1 223 645
Total waters	8 528 214[d]	8802	71 839	121 994	60 032	23 560	10 213	126 332	8 950 986
% in IUCN I and II	14.04	7.56	1.03	13.95	6.04	4.25	5.24	2.34	13.67

[a] Commonwealth reserves include the Great Barrier Reef Marine Park (Box 27.1) and Commonwealth MPAs declared in 2007 and 2012 (Plate 11).
[b] These figures do not include Recreational Use Zones in the North-west and South-east regions.
[c] Includes 815 km^2 (3.5% of state waters) off sub-Antarctic Macquarie Island.
[d] Total waters of the Commonwealth do not include Antarctic waters.
[e] Five new marine parks in WA waters are well advanced in planning, and will include significant additional IUCN I and II zones; a review of zoning allocations in existing parks is underway to more accurately identify areas of high-protection (IUCN I and II categories).

27.3.2   Marine protected areas in Commonwealth waters

The Australian Government fulfilled its commitment to the NRSMPA by declaring 40 new MPAs in Commonwealth waters in 2012 (Plate 11; Tables 27.2 and 27.3). The new MPAs were added to existing Commonwealth MPAs, including the Great Barrier Reef Marine Park (Box 27.1) and MPAs established in 2007 in the South-east marine region (Plate 11), and expanded the overall size of the Commonwealth MPA estate by 2.3 million km^2. The process of designing the Commonwealth MPAs ('marine bioregional planning') was implemented in six regions covering most of Australia's Exclusive Economic Zone: North, North-west, South-west, South-east (completed in 2007), East and Coral Sea (Plate 11). The primary objective of establishing the Commonwealth MPAs was stated as the conservation of marine biodiversity at minimal cost to users of the marine environment.[7] The Australian Government developed a set of goals and principles to provide a consistent framework for identifying new MPAs in Commonwealth waters.[8] The goals focussed on creating representative MPAs in Australian waters, and the principles ensured that the selection and design of MPAs was done in a way that minimised socio-economic impacts. This enabled the Australian Government to avoid designating MPAs in areas highly valued by industry groups and recreational users.

[7] http://www.environment.gov.au/marinereserves/background.html#nrsmpa
[8] http://www.environment.gov.au/coasts/mbp/publications/general/goals-nrsmpa.html

Table 27.3   Percentage of Commonwealth waters in each of six marine bioregional planning regions within IUCN categories I and II ('no-take' marine reserves), IV and VI. IUCN = International Union for Conservation of Nature. The Heard Island and McDonald Islands (HIMI) Marine Reserve and the Great Barrier Reef Marine Park (Box 27.1) are not included in this table because they were not part of the Commonwealth's recent bioregional planning exercise to establish new MPAs.

	North	North-west[a]	South-west	South-east[a]	East	Coral Sea	All regions
IUCN I	–	0.12	–	9.38	–	–	2.18
IUCN II	2.71	9.65	13.90	–	4.11	50.78	12.19
IUCN IV	–	1.89	9.11	6.46	9.55	29.16	9.50
IUCN VI	22.46	19.76	16.35	7.95	12.48	20.06	15.18
Total waters (km^2)	625 690	1 067 731	1 292 014	1 632 402	1 466 831	989 842	7 074 510

[a] Recreational Use Zones are categorised as IUCN IV in this table.

The Commonwealth MPAs (not including the Great Barrier Reef Marine Park, see Box 27.1) include five types of zones that fall into the IUCN categories of I, II, IV and VI (Plate 11; Tables 27.2 and 27.3). Zones classified as IUCN categories I and II are 'no-take' marine reserves (i.e. Sanctuary and Marine National Park Zones). However, three sites identified by the Commonwealth as category II 'Recreational Use Zones' located at Ashmore Reef and Ningaloo (North-west region) and Freycinet (South-east region), misapply the IUCN categories by allowing recreational fishing activities (Table 27.2; Fitzsimons, 2011). Category IV 'Habitat Protection Zones' allow extractive uses, including some types of commercial fishing (e.g. trolling, purse-seining, pelagic gillnetting and long-lining), pearling, aquaculture and recreational fishing. Category VI 'Multiple Use Zones' allow a much broader range of commercial fishing and other activities, including oil and gas extraction and exploration (mining). The percentage of each bioregional planning region within each IUCN category is shown in Table 27.3.

The Australian Government fulfilled its goal to represent each of the provincial bioregions in its 2007 and 2012 declarations of Commonwealth MPAs, except the Cocos and Christmas Islands provinces (outside of the scope of the NRSMPA planning process) and the Northeast Shelf Province (already protected in the Great Barrier Reef Marine Park). However, the extent of protection differs widely between provincial bioregions. For example, seven of the 38 provincial bioregions have less than 10% of their total areas within MPAs, while bioregions within the Coral Sea are almost fully protected (Devillers *et al.*, 2014). The percentages of provincial bioregions within highly protected marine reserves (IUCN categories I and II) are much smaller (Plate 11). Six provincial bioregions have no marine reserves and only seven have more than 20% coverage (Devillers *et al.*, 2014).

Coastal areas that are exposed to greater intensities of human use are not well represented in the Commonwealth MPAs. Plate 11C shows the percentages of meso-scale (coastal and fine spatial scale) and provincial bioregions covered by marine reserves in Commonwealth waters. Twenty-eight of the 49 meso-scale bioregions in Commonwealth waters (not including the Great Barrier Reef Marine Park) have no marine reserves, five have less than 1% coverage, and the remaining 16 have between 1% and 19% coverage. Overall, only 2.88% of the area of coastal meso-scale bioregions is protected by marine reserves compared to 17.23% of the area of offshore provincial bioregions (see Barr & Possingham, 2013). The bias of marine reserves away from coastal areas is also demonstrated by the coverage of marine reserves within the four geomorphic provinces of Heap & Harris (2008). Less than 3% of the area of Commonwealth waters on the continental shelf are within marine reserves compared to over 20% of the abyssal plain (>4000 m depth; Devillers *et al.*, 2014).

There are many examples of human activities, such as fishing and oil and gas development, being prioritised over marine reserves in Commonwealth waters. In the Coral Sea (Plate 11), marine reserves and category IV zones that precluded pelagic long-lining were configured around areas with high intensities of pelagic long-lining (Hunt, 2013). Devillers *et al.* (2014) found that average fish catches were lower by factors of 5.6–13.9 within Commonwealth MPAs than outside, and that marine reserves avoided oil and gas titles, release areas and wells. Williams *et al.* (2009) found that the 'zone of importance' (Australian waters <1500 m where human uses coincide with the greatest mega-faunal biodiversity) received the least amount of protection by marine reserves in the South east planning process in 2007. The systematic designation of protected areas at sites of least value for extractive uses is known as 'residual protection' (Margules & Pressey, 2000). There are two key reasons why the residual protection of Australia's marine environment is detrimental to biodiversity conservation. First, species and ecosystems exposed to high levels of human use are also those most vulnerable to negative effects and therefore most in need of protection; but residual reservation affords these features least protection. Second, selecting areas for protection that have low levels of human use cannot improve the condition of those areas and creates a false sense of security. A false sense of security can lead to reserve fatigue, where Government, stakeholders and communities use up the limited supply of 'conservation capital', reducing the willingness to extend MPAs in the future, even into areas that most need protection. The primary objective of the Australian Government to minimise the socio-economic impacts or 'opportunity costs' of new MPAs has overridden concerns for biodiversity and therefore jeopardised the effective protection of Australia's marine biodiversity now and into the future (Devillers *et al.*, 2014).

27.4 Improving marine protected areas in Australian waters

The NRSMPA programme has been in operation for over 20 years. During that time, the program has been unable to achieve a cooperative and integrated approach to the planning and management of MPAs in Australian waters outside of the Great Barrier Reef Region (SOEC, 2011). The goals and principles for the establishment of the NRSMPA are intended to be underpinned by the CAR principles described previously. However, a lack of clear and nationally consistent guidelines on implementing the CAR principles has led to several shortcomings in the implementation of the NRSMPA in Australian waters (SOEC, 2011):

(1) the use of MPA designations is not consistent across jurisdictions and do not follow the *Guidelines for Applying the IUCN Protected Area Management Categories to Marine Protected Areas* (Day *et al.*, 2012; Table 27.1), making it difficult to report on and compare the adequacy of MPAs across State, Territory and Commonwealth waters;

(2) minimal cooperation and integrated planning and management across jurisdictions has prevented the designation of MPAs in representative areas of Australia's marine environment;

(3) MPA designations reflect piecemeal or *ad hoc* decision-making, which is all the more difficult to understand given Australia's long-standing leadership in systematic methods for conservation planning; and,

(4) the lack of quantitative goals and targets associated with the CAR principles has constrained the achievement of representative marine protection in Australian waters, in turn allowing economic considerations around extractive uses to dominate MPA planning.

We believe there are four management areas that need to be improved to secure future declarations and the ongoing management of MPAs: governance, measurement of effectiveness of MPAs and conservation outcomes, restoration and stewardship.

27.4.1 Governance

The *Strategic Plan of Action for the National Representative System of Marine Protected Areas* (ANZECC TFMPA, 1999) and *Guidelines for Establishing the National Representative System of Marine Protected Areas* (ANZECC TFMPA, 1998) were developed cooperatively by the Australian and New Zealand Environment and Conservation Council (ANZECC) Task Force on Marine Protected Areas (TFMPA). Both documents highlight the importance of a national and systematic process for the identification of conservation priorities, and encourage cross-jurisdictional cooperation and collaboration. In addition, the *Guidelines* outline a systematic process for the identification and selection of MPA sites based on the principles of comprehensiveness, adequacy and representativeness, but these were defined only in general terms, and the appropriate application was left to each jurisdiction. The ability of each jurisdiction to apply the *Guidelines* also depended partly on how advanced it was in the process of implementing MPAs. For example, South Australia used the *Guidelines* to derive Design Principles for the South Australian Representative System of MPAs, whereas New South Wales used the *Guidelines* only for the few parks that were declared after 1998.

An obvious improvement in the governance arrangements for MPAs would be for the States, Northern Territory and Commonwealth to adopt a systematic and integrated approach to the design, implementation and ongoing management (especially monitoring) of MPAs. This could be achieved through the reinvigoration of the outputs of the ANZECC TFMPA via the development of clear and nationally consistent guidelines for applying the CAR principles. A requirement of the new guidelines should be that any planning exercise that draws on qualitative principles should also interpret these into explicit, defensible quantitative objectives, as seen in the re-zoning of the Great Barrier Reef Marine Park (Box 27.1).

An unresolved question is whether the resources sector (i.e. extractive industries such as fishing, energy production, and mining) should be involved throughout the entire MPA planning process. A common situation in the past has been for parallel, separate processes for planning conservation and resource development to duplicate governance procedures and only to merge at the final decision. This usually involves the dominance of resource extraction in non-transparent negotiations, drawing out expectations from some stakeholders that in the end are not fulfilled. A more open attempt to trade off options across all interested parties in one process would be fairer and more transparent.

27.4.2 Measuring conservation outcomes

Assessing the effectiveness of MPAs requires goals, objectives and sub-objectives of MPAs that are measurable (Barr & Possingham, 2013), and information on the distribution and condition of features (e.g. species and habitats) and processes (e.g. connectivity) both within and outside MPAs. However, information on features and processes is scarce because of the challenges associated with measuring and monitoring the marine environment. Large cetaceans, for example, might occupy an MPA for only part of the year for specific purposes like feeding or breeding, and so effective monitoring needs to tailor sampling to those times and behaviours. In contrast, data on features that are more static, such as reef-attached fishes, can be less sensitive to the timing of surveys. Improving monitoring in MPAs requires some combination of the following.

(1) Better definitions of the metrics of biodiversity to be used for different species, habitats and processes (e.g. is it more appropriate to monitor a marine mammal species by estimating the number of individuals in a population or by survivorship?).

(2) Development of cost-effective procedures that allow monitoring across the broad scales of time and space most relevant to managers (e.g. tapping into the skill-sets and commitment of trained recreational citizen scientists, as in the Reef Life Survey programme[9]).

(3) Improved understanding of cumulative impacts and the many synergies that apply in the marine environment, including environmental research on the cause-and-effect relationships between threats and biodiversity features.

(4) Better links to actions arising from monitoring outcomes, including triggers for feedback loops to management interventions at critical points (Gray & Jensen, 1993) and adaptive management approaches; and,

(5) Improvements in compilation, analysis and presentation of monitoring data for more effective communication to the public and managers, including managers of areas outside MPAs (e.g. GBRMPA 2009).

A critical improvement to measuring the conservation outcomes of Australia's MPAs requires a shift from measuring management responses (e.g. km^2 of MPAs established) to measuring actual benefits in the form of avoided loss of marine biodiversity or averted threats. Although measures of management responses, such as km^2 of MPAs or percentages of jurisdictions under formal protection, are commonly used, they convey no information about the real purpose of MPAs which is to separate marine biodiversity from processes that threaten its persistence. This is all the more true when jurisdictions such as the Commonwealth Government establish MPAs that are strongly residual to the extractive uses that threaten biodiversity (Devillers et al., 2014).

27.4.3 Restoration

MPAs can lead to restoration if they remove human activities such as fishing. However, restoration more generally is seen as an attempt to reverse biodiversity loss by improving the status of a degraded site or feature. Restoration techniques can also be used to address specific problems that limit the values of MPAs. Restoring the marine environment

---

[9] www.reeflifesurvey.com

includes activities such as the assisted translocation of species or populations under climate change and more active remediation such as transplanting juvenile mangroves or seagrasses. Marine restoration is potentially more achievable than in the terrestrial realm due to the greater extent of marine connectivity providing sources of recruitment.

There are many factors to consider when designing restoration programmes, including biodiversity potential, level of degradation, community engagement, and cost. Cost-effective programmes require the prioritisation of potential sites for restoration based on the site's ability to recover. Recovery potential is measured by locating a site along a continuum of alteration from pre-extractive (pristine) condition to its current degraded state. However, there are often high levels of uncertainty about the pre-extractive states of sites and about the ability of sites to recover to their former states and compositions. Most sites are subjected to multiple human uses simultaneously, so untangling which threats are currently limiting the desired states and compositions of sites can be difficult.

There is still much to learn about marine restoration, but MPAs could be the best places to derive information because they exclude at least some of the activities that lead to loss of species. Some species and areas will need active restoration to recover, and MPAs are therefore not the only solution. In many places, restoration might be best approached through effective management of land areas adjacent to the coast, thus reducing pressures from outside MPAs. For example, the Reef Water Quality Protection Plan[10] is a collaborative programme that is improving the quality of water in the Great Barrier Reef region via changes to land management practices (Brodie et al., 2012).

27.4.4 Stewardship

The successful implementation of Government MPA policy requires public support and community buy-in. Public education and participatory programmes facilitate community buy-in by increasing awareness and creating a sense of ownership and responsibility over the implementation of Government policy. There are several Australian examples of education and participatory programmes that have successfully engaged the community in MPA policy. The Port Noarlunga Reef and Onkaparinga Estuary Aquatic Reserve in South Australia established an underwater interpretation trail at Port Noarlunga Reef in the 1990s to raise community awareness of the importance of Aquatic Reserves and protecting the marine environment.[11] The Reef Guardian Schools programme established in 2003 by the Great Barrier Reef Marine Park Authority currently engages more than 293 schools and 114 900 students.[12] The programme's objective is to create awareness, understanding and appreciation for the Great Barrier Reef and its surrounding and connected ecosystems. Students actively participate in activities such as native habitat rehabilitation, beach clean-ups and recycling. These activities, along with education material supplied by the Reef Guardian Schools programme, foster stewardship and promote custodianship for the protection of the Great Barrier Reef. The Reef Guardian Schools programme has been so successful in the GBR that it has been expanded to include Reef Guardian Councils, Reef Guardian Fishers and Reef Guardian Farmers. The implementation of education and participatory programmes such as Reef Guardian Schools would greatly improve public support and community buy-in of MPAs across Australia.

[10] http://www.reefplan.qld.gov.au/index.aspx
[11] http://www.sardi.sa.gov.au/research_sectors/aquatic_sciences/education__and__extension/
[12] http://www.reefed.edu.au/home/guardians

27.5 Conclusions

MPAs, and specifically marine reserves, are an important part of every marine manager's toolkit. However, if they are to be successful, they must be well designed to address biodiversity conservation, and not simply located where they are least controversial and least costly. Residual MPAs are wasteful of resources, consume community goodwill, establish false expectations, convey a false sense of security, and perhaps most importantly, protect areas and features in least need of protection.

While Australia has made domestic and international commitments to develop MPAs across at least 10% of its jurisdiction (e.g. CBD COP, 2010), the inference is that these MPAs will be effective, and non-residual. At the current rate of progress, it seems unlikely that this target will be met for non-residual MPAs. As we write this chapter, the status of even the existing 'residual' Commonwealth MPAs is threatened (Vidot, 2013).

While the many values of MPAs are well demonstrated and widely recognised, the attempts to create a coherent national system of MPAs for Australia have not been successful. Considerable further effort is needed to make the NRSMPA an effective national programme for delivering biodiversity conservation that complements off-reserve initiatives and provides a sound basis for maintaining ocean health and resilience. This is particularly important because Australia's oceans face an unprecedented set of pressures from accelerating climate change and coastal development.

REFERENCES

Agardy, T. (2003). An environmentalist's perspective on responsible fisheries: The need for holistic approaches. In *Responsible Fisheries in the Marine Ecosystem*, eds. **Sinclair, M. & Valdimarsson, G.** Rome, Italy: Food and Agricultural Organisation of the United Nations (FAO), pp. 65–85.

Australian and New Zealand Environment and Conservation Council [ANZECC] Task Force on Marine Protected Areas [TFMPA]. (1998). *Guidelines for Establishing the National Representative System of Marine Protected Areas*, Australian and New Zealand Environment and Conservation Council Task Force on Marine Protected Areas, Canberra, Australia: Environment Australia.

Australian and New Zealand Environment and Conservation Council [ANZECC] Task Force on Marine Protected Areas [TFMPA]. (1999). *Strategic Plan of Action for the National Representative System of Marine Protected Areas: A Guide for Action by Australian Governments*, Australian and New Zealand Environment and Conservation Council Task Force on Marine Protected Areas, Canberra, Australia: Environment Australia.

Babcock, R. C., Shears, N. T., Alcala, A., *et al.* (2010). Decadal trends in marine reserves reveal differential rates of change in direct and indirect effects. *Proceedings of the National Academy of Sciences of the United States of America*, **107**: 18 251–18 255.

Ball, I. & Possingham, H. (2000). MarXan (v1.2) *Marine Reserve Design using Spatially Explicit Annealing: A Manual Prepared for the Great Barrier Reef Marine Park Authority*, Townsville, Australia: Great Barrier Reef Marine Park Authority.

Ban, N., Adams, V., Pressey, R. L. & Hicks, J. (2011). Promise and problems for estimating management costs of marine protected areas. *Conservation Letters*, **4**: 241–252.

Barr, L. & Possingham, H. (2013). Are outcomes matching policy commitments in Australian marine conservation planning? *Marine Policy*, **42**: 39–48.

Barrett, N., Buxton, C. & Gardner, C. (2009). Rock lobster movement patterns and population structure within a Tasmanian Marine Protected Area inform fishery and conservation management. *Marine and Freshwater Research*, **60**: 417–425.

Brodie, J. E., Kroon, F. J., Schaffelke, B., *et al*. (2012). Terrestrial pollutant runoff to the Great Barrier Reef: an update of issues, priorities and management responses. *Marine Pollution Bulletin*, **65**: 81–100.

Commonwealth of Australia (1998). *Australia's Oceans Policy*. Canberra, Australia: Environment Australia.

Convention on Biological Diversity (CBD) Conference of Parties (COP) 10th Meeting. (2010). *Decision X/2 Strategic Plan for Biodiversity 2011–2020 and the Aichi Biodiversity Targets*, available online: http://www.cbd.int/decision/cop/?id=12268

Day, J, Fernandes, L., Lewis, A., *et al*. (2000). The Representative Areas Program for protecting biodiversity in the Great Barrier Reef World Heritage Area. In *Proceedings of the Ninth International Coral Reef Symposium 2000*, ed. **Moosa, M.** Jakarta, Indonesia: Ministry of Environment and Indonesian Institute of Sciences and International Society for Reef Studies, pp. 687–696.

Day, J., Dudley, N., Hockings, M., *et al*. (2012). *Guidelines for Applying the IUCN Protected Area Management Categories to Marine Protected Areas*, Gland, Switzerland: International Union for the Conservation of Nature.

Dayton, P. K., Sala, E., Tegner, M. J. & Thrush, S. (2000). Marine reserves: parks, baselines, and fishery enhancement. *Bulletin of Marine Science*, **66**: 617–634.

Devillers, R., Pressey, R. L., Grech, A., *et al*. (2014). Reinventing residual reserves in the sea: are we favoring ease of establishment over need for protection? *Aquatic Conservation*. DOI: 10.1002/aqc.2445

Di Franco, A., Bussotti, S., Navone, A., Panzalis, P. & Guidetti, R. (2009). Evaluating effects of total and partial restrictions to fishing on Mediterranean rocky-reef fish assemblages. *Marine Ecology Progress Series*, **387**: 275–285.

Dudley, N. 2008. *Guidelines for Applying Protected Area Management Categories*, Gland, Switzerland: International Union for the Conservation of Nature.

Edgar, G. J. (2011). Does the global network of marine protected areas provide an adequate safety net for marine biodiversity? *Aquatic Conservation: Marine and Freshwater Ecosystems*, **21**: 313–316.

Edgar, G. J. & Barrett, N. S. (2012). An assessment of population responses of common inshore fishes and invertebrates following declaration of five Australian marine protected areas. *Environmental Conservation*, **39**: 271–281.

Edgar, G. J. & Stuart-Smith, R. D. (2009). Ecological effects of marine protected areas on rocky reef communities: a continental-scale analysis. *Marine Ecology Progress Series*, **388**: 51–62.

Edgar, G. J., Barrett, N. S. & Stuart-Smith, R. D. (2009). Exploited reefs protected from fishing transform over decades into conservation features otherwise absent from seascapes. *Ecological Applications*, **19**: 1967–1974.

Edgar, G. J., Banks, S. A., Bessudo, S., *et al*. (2011). Variation in reef fish and invertebrate communities with level of protection from fishing across the Eastern Tropical Pacific seascape. *Global Ecology and Biogeography*, **20**: 730–743.

Fernandes, L., Day, J., Lewis, A., *et al*. (2005). Establishing representative 'no-take' areas in the Great Barrier Reef: large-scale implementation of theory on marine protected areas. *Conservation Biology*, **19**(6): 1733–1744.

Fitzsimons, J. (2011). Mislabeling marine protected areas and why it matters – a case study of Australia. *Conservation Letters*, **4**: 340–345.

Gaines, S. D., Lester, S. E., Grorud-Colvert, K., Costello, C. & Pollnac, R. (2010). Evolving science of marine reserves: New developments and emerging research frontiers. *Proceedings of the National Academy of Sciences of the United States of America*, **107**: 18 251–18 255.

Glaser, S. M., Fogarty, M. J., Liu, H., *et al*. (2013). Complex dynamics may limit prediction in marine fisheries. *Fish and Fisheries*, doi:10.1111/faf.12037.

Gray, J. S. & Jensen, K. (1993). Feedback monitoring: a new way of protecting the environment. *Trends in Ecology and Evoloution*, **8**: 267–268.

Great Barrier Reef Marine Park Authority (GBRMPA). (2009). *Outlook Report 2009*. Townsville, Australia: Great Barrier Reef Marine Park Authority.

Harrison, H. B., Williamson, D. H., Evans, R. D., *et al*. (2013). Larval export from marine reserves and the recruitment benefit for fish and fisheries. *Current Biology*, **22**: 1023–1028.

Heap, A. D. & Harris, P. T. (2008). Geomorphology of the Australian margin and adjacent seafloor. *Australian Journal of Earth Sciences*, **55**: 555–585.

Hughes, T. P., Linares, C., Dakos, V., van de Leemput, I. A. & van Nes, E. H. (2012). Living dangerously on borrowed time during slow, unrecognized regime shifts. *Trends in Ecology & Evolution*, **28**: 149–155.

Hunt, C. (2013). Benefits and opportunity costs of Australia's Coral Sea marine protected area: a precautionary tale. *Marine Policy*, **39**: 352–360.

Januchowski-Hartley, F. A., Graham, N. A. J., Cinner, J. E., & Russ, G. R. (2013). Spillover of fish naïveté from marine reserves. *Ecology Letters*, **16**: 191–197.

Jones, G. P., Almany, G. R., Russ, G. R., *et al*. (2009). Larval retention and connectivity among populations of corals and reef fishes: History, advances and challenges. *Coral Reefs*, **28**: 307–325.

Kaiser, M. (2005). Are marine protected areas a red herring or fisheries panacea? *Canadian Journal of Fisheries and Aquatic Science*, **62**:1194–1199.

Kelleher, G. & Kenchington, R. (1991). *Guidelines for Establishing Marine Protected Areas*, Townsville, Australia: Great Barrier Reef Marine Park Authority.

Lester, S. E. & Halpern, B. S. (2008). Biological responses in marine 'no-take' reserves versus partially protected areas. *Marine Ecology Progress Series*, **367**: 49–56.

Margules, C. R. & Pressey, R. L. (2000). Systematic conservation planning. *Nature*, **405**: 243–253.

McCook, L. J., Ayling, T., Cappo, M., *et al*. (2010). Adaptive management of the Great Barrier Reef: A globally significant demonstration of the benefits of networks of marine reserves. *Proceedings of the National Academy of Sciences of the United States of America*, **107**: 18 278–18 285.

Murray, S. N., Ambrose, R. F., Bohnsack, J. A., *et al*. (1999). 'No-take' reserve networks: Sustaining fishery populations and marine ecosystems. *Fisheries*, **24**: 11–25.

Ogburn, D. M., White, I. & McPhee, D. P. (2007). The disappearance of oyster reefs from eastern Australian estuaries – impact of colonial settlement or mudworm invasion? *Coastal Management*, **35**: 271–287.

Pressey, R. L., Mills, M., Weeks, R. & Day, J. C. (2013). The plan of the day: managing the dynamic transition from regional conservation designs to local conservation actions. *Biological Conservation*, **166**: 115–169.

State of the Environment Committee (SOEC). (2011). Marine Environment. In *Australian State of the Environment Report 2011*, Canberra, Australia: Department of Sustainability, Environment, Water, Population and Communities, pp. 372–463.

State of the Marine Environment [SOMER] (1995). *State of the Marine Environment Report for Australia*. Canberra, Australia: Department of the Environment, Sport and Territories, and Townsville, Australia: Great Barrier Reef Marine Park Authority.

Veitch, L., Dulvy, N. K., Koldewey, H., *et al*. (2012) Avoiding empty ocean commitments at Rio+20. *Science* **336**: 1383–1385.

Vidot, A. (2013). June 4. *Opposition Moves to Kill Marine Park Plans*. ABC Rural News. Retrieved June 5, 2013, from http://www.abc.net.au/news/2013-06-04/coalition-moves-marine-park-disallowance/4731478

Westcott, G. (2006). The long and winding road: the development of a comprehensive, adequate and representative system of highly protected marine protected areas in Victoria, Australia. *Ocean and Coastal Management*, **49**: 905–922.

Williams, A., Bax, N., Kloser, R., *et al*. (2009). Australia's deep-water reserve network: implications of false homogeneity for classifying abiotic surrogates of biodiversity. *ICES Journal of Marine Science*, **66**: 214–224.

Williamson, D. H., Russ, G. R. & Ayling, A. M. (2004). No-take marine reserves increase abundance and biomass of reef fish on inshore fringing reefs of the Great Barrier Reef. *Environmental Conservation*, **31**: 149–159.

Wood, L. J. & Dragicevic, S. (2007). GIS-based multicriteria evaluation and fuzzy sets to identify priority sites for marine protection. *Biodiversity and Conservation*, **16**: 2539–2558.

Marine reserves in New Zealand: ecological responses to protection and network design

Nick Shears and Hannah L. Thomas

Summary

Marine reserves are simple management tools that exclude extractive and destructive human activities from areas of the ocean. Given that fishing is one of the greatest impacts in most coastal ecosystems, networks of marine reserves are recognised as a core part of implementing ecosystem-based management in marine systems. Research in New Zealand marine reserves has contributed disproportionately to the global understanding of how species and ecosystems respond to marine reserve protection. We use examples from New Zealand to demonstrate the unequivocal role that marine reserves play in protecting exploited species within their boundaries, and how the recovery of exploited species can have wider conservation and fisheries value through indirect mechanisms and the movement of individuals from reserves. Progress towards developing a comprehensive and representative network of marine reserves in New Zealand has been slow because of a lack of political will, marine protected area legislation, and clear scientific guidance on marine reserve network design. Based on progress in designing networks of marine reserves internationally, and their demonstrated role in protecting biodiversity, we recommend a set of scientific guidelines to aid future development of marine reserves networks in New Zealand, and recommend that such networks be at the core of future marine spatial planning processes.

28.1 Introduction

Marine reserves are areas of the ocean that are protected from all extractive and destructive human activities (Lubchenco *et al.* 2003). They are often referred to as 'no-take' marine reserves as fishing is the main activity that is typically eliminated from a

---

Austral Ark: The State of Wildlife in Australia and New Zealand, eds. A. Stow, N. Maclean and G. I. Holwell. Published by Cambridge University Press. © Cambridge University Press 2015.

particular stretch of coast when a marine reserve is established. Given that fishing is the most widespread and historic human impact in coastal environments worldwide (Jackson *et al.* 2001) marine reserves provide a simple management tool to protect defined areas of the ocean from the impacts of fishing. While management of many fisheries is improving (Worm *et al.* 2009), there have been widespread calls to increase the level of protection for marine species through the implementation of networks of marine reserves worldwide (Wood *et al.* 2008). Fishing has a myriad of impacts on species as well as ecosystems through habitat disturbance and changes to food webs (Dayton *et al.* 2003). While marine reserves are not a panacea, as humans have a wide variety of impacts on marine ecosystems, they can protect the species and ecosystems within their boundaries from the effects of fishing. Accordingly, networks of marine reserves and more broadly marine protected areas (MPAs), that may allow restricted fishing activities, are a core part of marine spatial planning initiatives worldwide and an essential component of ecosystem-based approaches to managing our oceans (Halpern *et al.* 2010).

Marine reserves are most commonly established to meet conservation goals of biodiversity protection. However, they may also be established to help rebuild depleted fish stocks, thus potentially playing a role in fisheries management. However, the value of marine reserves to fisheries remain uncertain and clear demonstrations are scarce (Hilborn *et al.* 2004). Over the past 10–15 years there has been an almost exponential increase in the number of empirical studies on the effects of marine reserves (Caveen *et al.* 2012). A vast literature exists on the efficacy of marine reserves, as well as on the design of marine reserve networks. Early studies espoused the potential of reserves and gave local-scale examples of reserve effects. As these studies increased, meta-analyses showed that positive effects of protection on targeted species generally occur in reserves worldwide (Lester *et al.* 2009). Now, research is focussing more on the finer details of why some reserves are 'successful' and others are not (Freeman *et al.* 2012b) and the wider effects of marine reserve networks (Harrison *et al.* 2012). There is also increasing recognition that marine reserves alone are insufficient to conserve ecosystems where multiple stressors are acting (Agardy *et al.* 2011).

New Zealand is a truly maritime nation with an exclusive economic zone approximately 15 times the land area of the country. The marine environment is of immense commercial, social and cultural value to New Zealanders and, accordingly, there has been a long legacy of conserving and managing marine resources and biodiversity. New Zealand demonstrated its early commitment to marine conservation through the establishment of the Marine Reserves Act in 1971, followed shortly afterwards by one of the world's first no-take marine reserves, the Cape Rodney-Okakari Point (or Leigh) Marine Reserve, in 1977 (Ballantine & Gordon 1979). A total of 38 no-take marine reserves are now currently in place around New Zealand (Figure 28.1, Plate 45). Research carried out in New Zealand marine reserves, particularly in the Leigh Marine Reserve, has contributed greatly to the global understanding of the ecological responses to marine reserve protection and also the potential role of marine reserves as tools for conservation and management (Babcock 2013). Although New Zealand's long history of marine life protection is impressive compared to many other places in the world, and New Zealand's reserves have been of great value for marine science, the existing reserves were mostly established individually and independently, rather than systematically as a coherent network designed to protect biodiversity and ecosystem services at a national scale for the long term. Overall, progress towards establishing a systematic and comprehensive network of marine reserves in New Zealand has been slow. In total, 7% of New Zealand's territorial

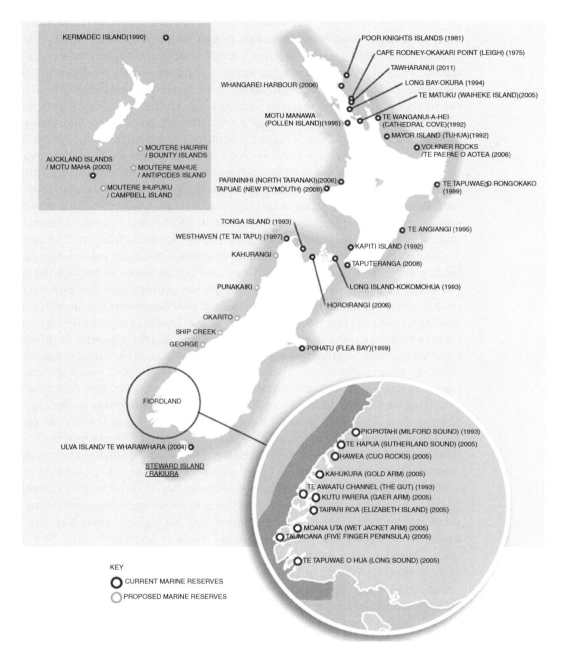

Figure 28.1 (Plate 45) New Zealand's no-take marine reserves. A further five marine reserves have been approved on the West Coast of the South Island (as at 6/8/2014 from www.forestandbird.org.nz). Image courtesy of The Royal Forest & Bird Protection Society of New Zealand Inc. A black and white version of this figure will appear in some formats. For the colour version, please refer to the plate section.

waters are protected within marine reserves, but most of this area is situated in offshore waters (Kermadec Islands and the sub-Antarctic islands) where human influences are minimal. Less than 1% of the waters around mainland New Zealand are currently protected within reserves (Department of Conservation & Ministry of Fisheries 2011).

In this chapter we summarise the ecological responses to marine reserve protection using examples from research primarily carried out in New Zealand marine reserves. We then review current marine reserve network developments in New Zealand, and based on recent progress in designing networks of marine reserves worldwide, we recommend some scientific guidelines to aid future development of marine reserve networks in New Zealand.

28.2 Ecological responses to marine reserve protection

The ecological response to marine reserve protection can be defined as changes in populations, communities, habitats or ecosystem processes that are attributable to the presence of a marine reserve. Given that the primary function of a marine reserve is protecting against fishing, unsurprisingly the greatest response usually observed following protection is the recovery of species that were previously targeted in the area. These changes can be considered the **direct responses** to marine reserve protection and are restricted to target species. When the recovery of targeted species within the reserve results in changes in other species or habitats, these are referred to as **indirect responses** to marine reserve protection. Both the direct and indirect responses to protection are confined to species and habitats within the marine reserve itself (Figure 28.2, Plate 48) and these responses are typically investigated using the **large-scale experimental framework** provided by marine reserves (Box 28.1). When the recovery of targeted species within reserves results in the export of either adults or young (larvae) to areas outside the reserve these can be considered as **external responses**. It is through this 'spillover' of adults or export of larvae to surrounding populations that marine reserves may benefit fisheries.

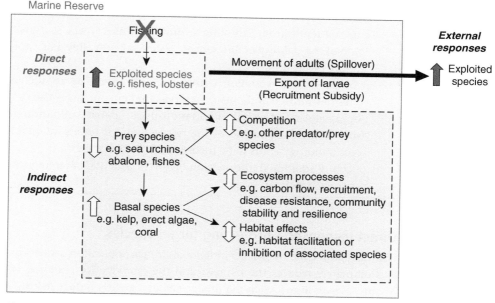

Figure 28.2 (Plate 48) The ecological responses of species and ecosystems to marine reserve protection. Direct and indirect responses occur as a result of reduced fishing mortality inside reserves, whereas external responses are changes in populations outside of reserves due to the spillover of adults or export of larvae beyond reserve boundaries. A black and white version of this figure will appear in some formats. For the colour version, please refer to the plate section.

> ## Box 28.1 Marine reserves as a large-scale experimental framework
>
> By excluding extractive activities such as fishing from areas of the ocean, marine reserves provide a **large-scale experimental framework** to investigate how species and ecosystems respond to protection, and conversely, how populations outside of the reserves have been affected by fishing. The typical approach used to study the direct and indirect response of species to protection includes sampling sites inside and outside the marine reserve and comparing abundance, body size and total biomass between these (for example, see Case study 1). Ideally surveys are conducted at multiple times and if possible both before and after establishment of the reserve (Willis *et al.* 2003b). This before-after control-impact (BACI) design allows researchers to more confidently separate the response to protection from inherent spatial variability between the reserve and fished areas as well as temporal variation. In general, BACI designs remain rare in the marine reserve literature (Shears *et al.* 2006), primarily due to the financial commitments required for monitoring before a reserve is implemented and uncertainty over whether a reserve will actually be approved and the eventual location and boundaries of reserves. Consequently, the majority of studies rely on spatial comparisons between reserve and fished sites following certain periods of protection. While there may be large differences in the abundance of particular species (or extent of habitats) between reserves and adjacent fished areas, care must be taken in attributing differences to protection as political considerations can cause marine reserves to be unrepresentative of wider areas (Edgar *et al.* 2009).
>
> A number of approaches can be used to strengthen the inference of reserve effects from such spatial comparisons.
>
> Multiple-reserves: Replication of reserves to increase the generality of reserve effects.
>
> Temporal-replication: Sampling at multiple times (years, seasons) gives information of the stability of differences. Also, where possible data should be collected before protection.
>
> Encompass-spatial-variation within reserves: Sampling an adequate number of sites both inside and outside reserves that encompasses the range of variation in a given area. Also it is important to avoid spatial confounding (i.e. all control sites located at one end of the reserve), and if necessary quantify potentially confounding factors to be incorporated into analyses.
>
> Collecting additional information on environmental parameters that can be used as covariates in analyses.

28.2.1 Direct responses – protecting target species

Fishing typically removes larger individuals from a population, reduces overall abundance, and ultimately reduces the biomass and the reproductive potential of the population. Marine reserves immediately eliminate or at least **greatly reduce fishing mortality** within an area; therefore the direct response to protection is simply a reversal of the effects of fishing on the species that occur within a marine reserve. It has been widely demonstrated that targeted species recover in marine reserves and these responses are typically evident in the first five years following marine reserve establishment (Babcock *et al.* 2010), although rebuilding of populations can continue for decades (Edgar *et al.* 2009). On average, populations of targeted species are more abundant, larger, and consequently have **higher biomass and reproductive output** within marine reserves compared to fished areas (Lester *et al.* 2009). The snapper *Pagrus auratus* in northern New Zealand provides an example of a highly important commercially and recreationally targeted species that has widely benefited from protection in northern New Zealand marine reserves (Case study 1; Figure 28.3, Plate 46). In addition, populations of targeted species within marine reserves

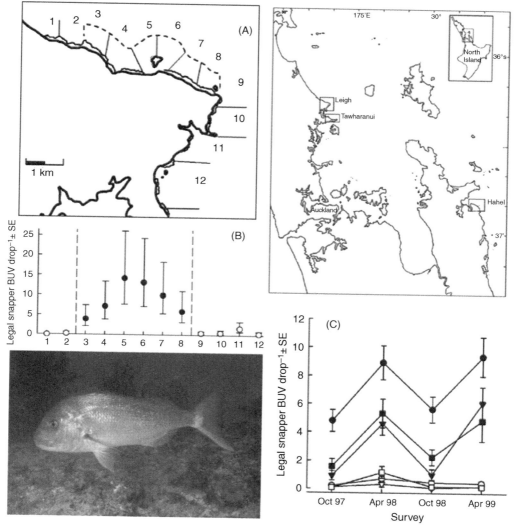

Figure 28.3 (Plate 46) Direct response of snapper *Pagrus auratus* to marine reserve protection in northern New Zealand marine reserves. Willis *et al.* (2003a) used baited underwater video to monitor snapper density and size seasonally inside and outside three marine reserves over two years. In each location-reserve and non-reserve areas were divided into segments (see (A) for design at Leigh) and randomly selected stations were surveyed within each segment using baited underwater video. Densities of legal-sized snapper across the Leigh reserve are shown in (B); clearly demonstrating how densities decline towards boundaries. Despite seasonal variation in snapper, due to immigration and emigration, densities were consistently higher in all three reserves (shaded symbols, (C) compared to fished areas (closed symbols). A black and white version of this figure will appear in some formats. For the colour version, please refer to the plate section.

may be **healthier, grow faster and have greater resilience to climatic impacts** (Freeman & MacDiarmid 2009; Freeman *et al*. 2012a; Micheli *et al*. 2012).

Research worldwide has typically revealed benefits for targeted species within marine reserves. However, the magnitude, and in some cases the direction, of the response can be highly variable among studies (Lester *et al*. 2009). Many factors are known to influence the magnitude of direct responses (Edgar & Stuart-Smith 2009). Key to understanding and predicting the direct response of target species to marine reserve protection is knowledge of what species are fished in a region and how hard they are fished, as well as knowledge of the biology of the species (e.g. mobility) and characteristics of the reserve itself (e.g. size and age). In particular, to maximise the potential benefits for exploited species reserves need to be large enough (Claudet *et al*. 2008), contain suitable habitat (Jack & Wing 2013) and be well enforced (Guidetti *et al*. 2008). Targeted species that are small, sedentary or site-faithful are considered most likely to benefit from protection (Murawski *et al*. 2000), whereas long-lived, slow-growing and highly mobile species are likely to respond more slowly or not at all (Taylor *et al*. 2011). However, larger, more mobile species, such as dugongs, turtles and sharks are benefiting from larger reserve networks like the Great Barrier Reef Marine Park (McCook *et al*. 2010). The level of exploitation on a species in areas surrounding reserves may also influence a species ability to recover (Hilborn *et al*. 2006) as adult movement and larval supply into reserves may be limited when stocks are overfished.

Case study 1. Direct response of snapper to protection in northern NZ marine reserves

Snapper (*Pagrus auratus*) support one of New Zealand's largest inshore commercial fisheries and recreationally is northern New Zealand's most sought after saltwater fish (Parsons *et al*. 2009). Snapper are generally considered to be fairly wide-ranging (tens of kilometres) but studies of movement patterns demonstrate that a subset of the population takes up residence on inshore reefs and restrict their movements to within a few kilometres (Parsons *et al*. 2010). Within the Leigh Marine Reserve the size of snapper has increased considerably and overall biomasses are now ~10 times higher than they were when the reserve was established (Babcock *et al*. 2010). The opportunity to see large snapper (up to a metre in length) in their natural environment has become one of the major attractions for the many tourists that visit this marine reserve. Abundance of legal-sized snapper is highest in the centre of the reserve and declines towards the reserve boundaries as a combined result of fish movement and sustained fishing pressure at the reserve boundaries (Willis *et al*. 2000) (Figure 28.3, Plate 46).

Similar direct responses of snapper have occurred in most northern New Zealand reserves (Willis *et al*. 2003a; Shears *et al*. 2008). On average the abundance of snapper greater than minimum legal size in reserves is estimated to be 14 times higher than in fished areas, and the relative egg production ~18 times higher (Willis *et al*. 2003a). Seasonal monitoring has also provided new insights into the onshore–offshore movement patterns of snapper and suggests recovery of snapper populations within reserves is due to seasonal immigration of individuals from fished areas that take up residence (Figure 28.3, Plate 46). For example, the rapid recovery of snapper at the Poor Knights Islands Marine Reserve was driven by immigration of large fish into the reserve, rather than within-reserve recruitment of juveniles (Denny *et al*. 2004). In a few reserves there has been little to no recovery of snapper, which is thought to be due to illegal fishing or lack of suitable reef habitat (Shears *et al*. 2008).

28.2.2 Indirect responses – protecting ecosystems

Throughout history fishing has targeted large fishes, which typically occur at high trophic levels (Jackson *et al.* 2001). In some cases this has led to cascading effects on lower trophic levels and resulted in shifts in entire ecosystems (Salomon *et al.* 2010). Consequently, predatory species typically show the greatest response to protection in marine reserves (Babcock *et al.* 2010) and their recovery can potentially reverse the ecosystem effects of fishing. Changes in species, habitats or ecosystem processes that stem from **trophic or competitive interactions** with harvested species are **indirect responses to marine reserve** protection (Figure 28.2, Plate 48). A wide variety of indirect responses may occur and understanding these is necessary to predict how non-target species and ecosystem processes may be affected by protection, and to assess whether marine reserves are achieving their conservation goals.

Comparing populations of non-target species and habitats between reserve and non-reserve sites (Box 28.1) can provide insights into the indirect responses of protection. However, in order to confirm the mechanisms responsible for such indirect responses other techniques and experiments are often necessary. For example, to better understand how predators influence prey populations and ecosystem processes a variety of methods have been used including observations of predation events, tethering experiments, gut contents and isotopic analysis (See Shears & Babcock 2002; Salomon *et al.* 2008). Carrying out such studies and measurements inside and outside reserves can also provide greater understanding of marine food webs and the role of predators. Ecosystem modelling may also be used to inform how recovery or removal of certain components of the food web in marine reserves might indirectly influence other species (Pinkerton *et al.* 2008).

The most commonly reported indirect responses to marine reserve protection are driven through predator–prey interactions and these typically involve declines in sea urchins in response to a recovery of large predatory species in reserves (Babcock *et al.* 2010). In some cases, indirect changes in prey populations within marine reserves can lead to further flow-on effects on lower trophic levels and even habitats. For example, the recovery of predators (snapper and lobster) within the Leigh Marine Reserve has resulted in a suite of indirect responses that have cascaded through the entire ecosystem (Case study 2). Such trophic cascades can result in changes in species composition, habitat distributions, ecosystem processes and overall ecosystem resilience. For example, no-take marine reserves can enhance coral reef resilience by allowing recovery of functionally important populations of herbivorous fishes (Stockwell *et al.* 2009) as well as by reducing the occurrence of outbreaks of coral-eating starfish (McCook *et al.* 2010). However, protection within marine reserves may not increase resilience to all external stressors (e.g., sedimentation; Halpern *et al.* 2013).

Indirect responses to protection are harder to predict than direct responses and typically take much longer to manifest (Babcock *et al.* 2010; Case study 2). Predicting what indirect responses might occur requires an understanding of the key trophic linkages within an interaction web and how fishing has impacted on species that may interact with targeted species. Indirect responses are most likely to occur in systems where there are strong interactions among target and non-target species. In more speciose systems where there is greater functional redundancy, such as coral reefs, these indirect effects are likely to be more diffuse and harder to detect. Likewise in ecosystems where bottom-up or environmental influences play a dominant role in structuring biological communities the indirect responses are not likely to be apparent. Conversely, in such systems the effects of fishing are primarily restricted to target species, and the removal of exploited species does not have cascading effects on the entire ecosystem.

Case study 2. Indirect responses to protection in the Leigh Marine Reserve

When the Leigh marine reserve was established in 1977 a large-scale survey of the underwater environment was carried out (Figure 28.4) (Ayling 1978). This information provided an **invaluable baseline** for subsequent research into both the direct and indirect responses to marine reserve protection. When the reserve was established, approximately one-third of subtidal reef was classified as 'urchin barrens' habitat, i.e. areas where grazing by sea urchins had eliminated large seaweeds. Urchin barrens dominated between depths of approximately 3–8 m and had a suite of reef fish species and herbivorous gastropods associated with it. Following the recovery of snapper and lobster in the reserve there was a gradual decline in the abundance of sea urchins, and by the late 1990s most areas of urchin barrens had changed to kelp forest (Babcock *et al.* 1999; Shears & Babcock 2003). Now <1% of the subtidal reef in the reserve is urchin barrens (Figure 28.4; Leleu *et al.* 2012). The theory that the recovery of predators was responsible for these large-scale changes in habitats (a 'trophic cascade') was further supported by the fact that urchin barrens persisted at sites outside the reserve (Babcock *et al.* 1999; Shears & Babcock 2003) and predation rates on sea urchins inside the reserve were approximately seven times higher than outside (Shears & Babcock 2002).

The effects of increased predators in the reserve have been found to extend beyond sea urchins and the reef. Lower densities of small cryptic fishes inside the Leigh Reserve compared to sites outside are thought to be due to higher predation pressure (Willis & Anderson 2003). Similarly, the densities of large bivalves and heart urchins in sandy habitats are lower in the reserve compared to fished areas due to predation by lobster, which forage beyond the reef (Langlois *et al.* 2006).

The increase in kelp forests and decline in sea urchins has resulted in a net increase in primary productivity and also led to habitat-mediated changes in species composition. The increased biomass of kelp in the reserve influences detrital pathways, with kelp making a greater contribution to the diets of filter feeders transplanted into the reserve compared to outside (Salomon *et al.* 2008). Some species of herbivorous gastropods that were once abundant in the barrens have now declined, whereas others have increased (Shears & Babcock 2003). Similarly, habitat-associated changes in fish communities have also been hypothesised (Jones 2013).

The ecological changes within the Leigh Reserve are clear and the evidence for protection from fishing being the cause is strong given the availability of early baseline data, comparisons with non-reserve sites and experimental studies. However, similar habitat-level changes have only been observed in one other reserve in New Zealand; the Tawharanui Marine Reserve, established in 1982 and located approximately 10 km to the south of the Leigh Marine Reserve (Babcock *et al.* 1999; Shears & Babcock 2002). Comparisons of reserve and fished areas throughout New Zealand, spanning a variety of environmental conditions, suggest that these trophic cascade effects are not general to reserves throughout the country (Shears & Babcock 2004; Shears *et al.* 2008). In some cases, reserves examined in these studies had not been protected for long enough to exhibit such habitat-level effects. However, it was found that on many of the reefs examined, both inside and outside of reserves, urchin barrens were rare or absent

(Shears & Babcock 2007). This demonstrates that in many parts of New Zealand sea urchins do not have the same pervasive effects on seaweeds as they do in northern New Zealand, and therefore trophic cascades involving sea urchins would not be expected to occur (Schiel 2013).

28.2.3 External responses – benefits beyond boundaries?

The potential benefits from reserves for populations outside the reserve, and therefore fisheries, can arise through the recovery and subsequent spillover of adults and juveniles into adjacent fisheries, or through the dispersal of larvae that results in successful recruitment elsewhere (recruitment subsidy) (Gell & Roberts 2003). In order for these external responses to occur, **marine reserves must first be effective in protecting exploited species** and promote the recovery of such species within the reserve boundaries. How these changes within reserves may benefit adjacent populations and fisheries are less well understood and the evidence for positive effects on adjacent fisheries is limited.

28.2.3.1 Spillover

The build-up of exploited species within marine reserves can lead to the net movement or **'spillover' of individuals across reserve boundaries**, where they can potentially be caught (Kramer & Chapman 1999). The occurrence and importance of spillover can be investigated by quantifying densities and biomass of exploited species, or catch rates, in relation to distance from reserve boundaries (Halpern *et al.* 2009). However, such studies only provide circumstantial evidence for spillover, and investigating how catch rates change over time since protection (Russ *et al.* 2004) or using tagging studies (Case study 3) provide a more direct means of investigating whether exploited species are moving across boundaries and being caught in the fishery.

The level of spillover from a marine reserve will depend on the size and shape of the reserve, the mobility of exploited species within it, and the distribution of habitats and/or barriers that may influence movement (Freeman *et al.* 2009). Some studies have concluded that marine reserves can benefit local fisheries through the spillover of adults of target species (McClanahan & Mangi 2000; Alcala *et al.* 2005). For example, the marine reserve at Apo Island in the Philippines has been shown to benefit local fisheries at the island through higher catch rates near to the reserve boundary (Russ *et al.* 2004). Increased catch of larger-sized fish near to reserve boundaries has also been attributed to spillover from reserves (Gell & Roberts 2003). However, in general, the effects of spillover from reserves on fisheries are localised and any benefit to fisheries is only evident very close to the boundary. Furthermore, spillover is typically only considered to **compensate for the loss of fishery catch** caused by the loss of fishing area when the reserve is set up, rather than increasing overall catch in a region (Kelly *et al.* 2002; Goñi *et al.* 2010).

Case study 3. Lobster movement in New Zealand marine reserves

Movement studies of the spiny lobster *Jasus edwardsii* inside New Zealand marine reserves have shown that adult individuals carry out seasonal movements onto deeper patch reef and sand for reproduction and foraging (Kelly 2001; Freeman *et al.* 2009). Within the Leigh Marine Reserve this means that large lobsters frequently move beyond the offshore boundary, which is only 800 m from shore (Figure 28.4) (Kelly & MacDiarmid 2003). Consequently, a large amount of commercial and recreational fishing effort is concentrated along the offshore and long-shore boundaries. Trapping studies have found that lobster catch along the offshore boundary was highly variable due to the seasonal nature of offshore movements, that the catch was made up

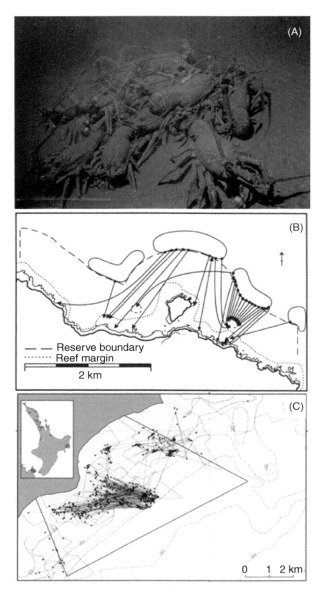

Figure 28.4 Spillover of lobster beyond marine reserve boundaries. Large spiny lobster *Jasus edwardii* in the Leigh Marine Reserve carry out seasonal movements off the reef to forage in deeper sandy habitats (A) and (B); Kelly & MacDiarmid 2003. These movements take them beyond the offshore boundary where they form large aggregations (A) and are vulnerable to fishing (B). Lobster carry out similar inshore–offshore movements in the larger Te Tapuwae O Rongokako Marine Reserve (C) (Freeman *et al.* 2009), but do not cross the offshore boundary due to the greater distance and presence of muddy substrates. Lobsters do, however, cross the long-shore boundaries where the reef (in grey) also crosses the boundary (C).

of large lobsters, and at certain times of the year extremely large numbers of lobsters were caught (Kelly *et al.* 2002). On average, CPUE was similar to unprotected fishing areas nearby, suggesting that spillover maintains catch rates at similar levels to those in the surrounding fishery (Kelly *et al.* 2002). However, large declines in lobster numbers inside the reserve have since been recorded leading to the suggestion that increased fishing on the boundary has impacted on population size in the reserve (Babcock *et al.* 2010). While Kelly *et al.* (2002) indicated that the reserve did not have a negative effect on the livelihood of local fishers, the subsequent declines suggested that the offshore extent of the reserve is not sufficient to adequately protect populations of large individuals within the reserve.

The offshore boundary of the Te Tapuwae o Rongokako Marine Reserve is approximately 5 km offshore (Figure 28.4) and tagging studies have revealed that lobster do not move beyond the offshore boundary (Freeman *et al.* 2009). The deep soft-sediment areas in this reserve are muddy, which is also thought to minimise offshore movements and movement between reefs. Generally, lobster only move beyond the long-shore boundaries where the reef also crosses the boundary (Figure 28.4). These studies clearly demonstrate that populations of target species within marine reserves can contribute to adjacent fisheries through adult movement beyond boundaries and the level of spillover will depend on species mobility and characteristics of the reserve. While spillover will not compromise conservation values if populations in reserves are at carrying capacity, reserves need to be large enough to ensure that populations are adequately protected and spawning biomass is maximised.

28.2.3.2   Recruitment subsidy

It has widely been demonstrated that, due to the accumulation of large individuals in reserves, the reproductive output of harvested species can be considerably greater from reserves than adjacent areas. Therefore, if there is substantial larval export from reserves to fished areas, in theory, reserves can provide benefits to adjacent fisheries through a 'recruitment subsidy' (Gaines *et al.* 2010). However, there is little empirical evidence that reserves provide recruitment benefits and enhance fisheries beyond their immediate boundaries (Steneck *et al.* 2009). This is largely because larvae or eggs are often microscopic and are difficult to tag or track, and recruitment patterns are inherently variable in space and time. Also, until recently there has been an absence of marine reserve networks to examine the wider fishery-scale impacts of reserves. The potential for recruitment effects have largely been studied theoretically using hydrodynamic or dispersal models (see Botsford *et al.* 2001). The importance of reserves in supplying larvae to adjacent fished populations has also been inferred by higher recruitment close to reserve boundaries (Pelc *et al.* 2009) and long-term increases in catch-per-unit-effort in the vicinity of a marine reserve (Kerwath *et al.* 2013). However, recent advances in genetic techniques are providing **direct measurements of the role that networks of marine reserves can play in exporting larvae to populations outside of reserves** (Case study 4).

No direct evidence has been published from New Zealand on the potential importance of larval export from marine reserves. However, such effects have been inferred to occur. For example, Willis *et al.* (2003b) estimated that a reserve the size of Leigh (5 km of coastline) produces a quantity of snapper eggs equivalent to that produced by ~90 km of unprotected coastline. These studies demonstrate how protection of harvested species even within small reserves can contribute disproportionately to the greater larval pool.

Case study 4. Larval export from a network of marine reserves on the Great Barrier Reef

In a recent study on the Great Barrier Reef, Harrison *et al.* (2012) used genetic techniques to resolve patterns of larval dispersal from marine reserves for two species of exploited coral reef fish (Figure 28.5, Plate 47). Mean biomass of reproductive adults was approximately twice as high in no-take reserves compared to surrounding fished reefs. Through DNA parentage analysis of juveniles they estimated that no-take reserves, which accounted for just 28% of the local reef area, produced approximately half of all juvenile recruitment of coral trout to both reserve and fished reefs within 30 km. Furthermore, populations

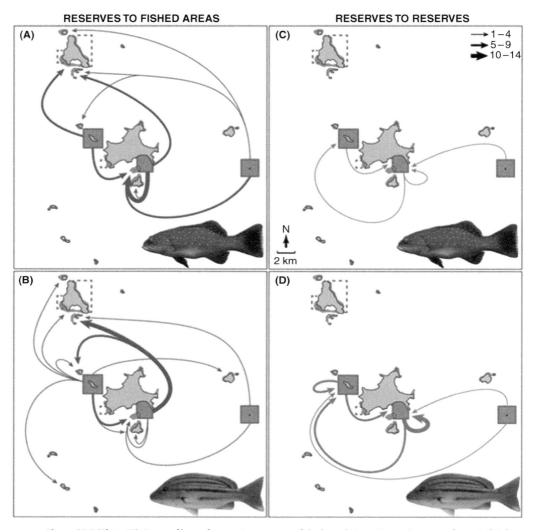

Figure 28.5 (Plate 47) Export of larvae from marine reserves to fished populations. Dispersal patterns of juvenile fish from a network of marine reserves on the Great Barrier Reef Australia (Harrison *et al.* 2012). The three focal marine reserves (green boxes) were an important source of juvenile recruitment for two species of reef fish for local fished areas. For both species, coral trout *Plectropomus maculatus* (A) and stripey snapper *Lutjanus carponotatus* (B), recruits collected in fished areas were assigned to adults from one of three focal reserves. Local retention within focal reserves and connectivity between reserves (dotted green boxes) also made an important contribution to juvenile recruitment in reserves ((C) and (D)). A black and white version of this figure will appear in some formats. For the colour version, please refer to the plate section.

resident in reserves exported 83% (coral trout) and 55% (stripey snapper) of offspring to fished reefs. This study provides the strongest support yet that exploited populations within reserves can provide a significant source of larvae to maintain and replenish populations outside reserves. However, Harrison and colleagues state that the magnitude of larval supply from reserves may not be sufficient to offset a substantial increase in fishing pressure outside reserves, and that reserves need to continue to be coupled with traditional harvest restrictions to ensure that current fishery yields are sustainable.

28.3 Designing marine reserve networks

Although individual, independent marine reserves are a valuable tool for biodiversity conservation and potentially for fisheries management, networks of marine reserves can potentially deliver far greater benefits due to the ecological linkages between reserves and the greater spatial scales involved (Gaines *et al.* 2010). Achieving long-term biodiversity persistence within these networks is more likely if their design is shaped by the best available scientific guidance on conservation planning for both population sustainability and ecosystem protection. Greater awareness of the need for increased marine protection and sustainable resource management is increasingly resulting in the development of marine reserve or MPA networks around the world.

28.3.1 Marine reserve network development in New Zealand

In New Zealand, the Marine Protected Areas Policy and Implementation Plan (MPA Policy) was developed in 2005 to 'protect marine biodiversity by establishing a network of marine protected areas that is comprehensive and representative of New Zealand's marine habitats and ecosystems' (DOC & MFISH 2005). This policy was designed to help achieve the New Zealand Biodiversity Strategy target of protecting 10% of the marine environment by 2010 and meet New Zealand's commitments to the International Convention on Biological Diversity. Currently, the overall level of marine reserve protection for biogeographic regions around mainland New Zealand ranges from between 0% and ~1% (Department of Conservation & Ministry of Fisheries 2011). This clearly demonstrates that New Zealand still has a long way to go towards meeting stated targets and establishing effective networks of marine reserves.

One of the core problems limiting the establishment of networks of marine reserves in New Zealand, and elsewhere, is the vehement and vocal opposition to marine reserves by extractive stakeholder groups. The idea of no longer being able to fish in given areas of the coast is inconceivable to many users, particularly when there are already a variety of rules and regulations in place to manage fish stocks. While the majority of New Zealanders support the idea of marine reserves and most believe the current level of marine reserve protection is far greater than it actually is (Colmar-Brunton 2011), government politicians are not willing to stand up to the likely backlash from extractive user groups, despite international agreements to increase marine reserve protection. This problem is not unique to New Zealand. Progress in developing marine reserve networks in highly populated coastal regions such as California does however demonstrate that this problem is not insurmountable, provided there is **a clear mandate on marine protection, clear scientific guidelines and a strong stakeholder-driven participatory-process** (Gleason *et al.* 2010). These enabling factors were considered key to developing the state-wide network in California, which now includes 124 MPAs protecting approximately 9% of state waters in marine reserves and 6% as partially protected MPAs.

A recent review of New Zealand's current MPA planning framework and a comparison with international developments by Mulcahy *et al.* (2012) has revealed that the current framework is well below international best practice and there is an urgent **need to develop new MPA legislation.** Furthermore, the MPA Policy and MPA classification, protection standard and implementation guidelines (DOC & MFISH 2008) **lack clear guidance on designing marine reserve networks** and how much should be protected. These shortcomings are evident in the recent outcome of the West Coast South Island MPA planning process, where after extensive stakeholder engagement over many years only 1.3% of the entire region will be protected in five marine reserves. This level of protection falls well short of the target of 10% set out in New Zealand's Biodiversity Strategy, which is at the bottom end of global targets that recommend between 10% and 30% protection (Wood *et al.* 2008).

28.3.2 The need for clear scientific guidelines

Development and application of scientifically based guidelines or '**rules of thumb**' have proven to be an effective means of incorporating the best available science into decision-making, design and evaluation of marine reserve networks (Carr *et al.* 2010). Extensive MPA planning processes have recently been carried out in California (Saarman *et al.* 2013), on Australia's Great Barrier Reef (Fernandes *et al.* 2009) and in the United Kingdom (NE & JNCC 2010). These have all used **the best available science** (usually based around a scientific literature review and expert advice) to develop **strong science guidelines** for MPA network development that are often accompanied by clear numerical targets as to the amount of protection required, the size and spacing of reserves, and the level of protection required (e.g. no-take vs. partial protection).

The NZ MPA Policy does have a range of principles and guidelines, and many of these are consistent with international recommendations. However, unlike international examples these principles and guidelines are not prescriptive, numeric or particularly specific. When network design processes involve stakeholder groups, often with differing objectives, the use of specific, science-based numeric targets are an efficient and effective way to produce MPA recommendations, as they can be justified and measured. Simple, yet **highly specific and scientifically justifiable guidelines** contributed to the successful delivery of MPA planning processes in California and the Great Barrier Reef, primarily because stakeholders had clear 'instructions' with which to design reserves and managers had clear measures with which to evaluate the conservation value of proposed networks. Numerical modelling approaches were also used to complement this approach by evaluating the proposals and generating alternative network designs (Hilborn 2012).

28.3.3 Recommendations for future design of marine reserve networks in New Zealand

New Zealand marine ecosystems have a number of functionally analogous species and ecosystems to those found in California. Therefore, we can be confident that recommendations based on the guidelines used in California should support the New Zealand process in delivering an ecologically comprehensive and coherent network of marine reserves.

Consequently, we provide a series of simple recommendations below that are based largely on those used in California (CDFG 2010), but also consistent with guidelines used in other processes worldwide. Although many of these aspects are included to some extent in the existing NZ MPA Policy and Implementation guidelines (DOC & MFISH 2008) (e.g., 2 and 3), we provide more specific recommendations around focussing MPA planning efforts on establishing no-take marine reserves (1), and the size and spacing of marine reserves (4 and 5), to bring New Zealand's efforts more in line with international design processes.

The following are recommended scientific guidelines for marine reserve network design in New Zealand.

(1) **No-take marine reserves make up the core of any MPA network**. MPAs that restrict certain activities areas may be used to supplement networks of marine reserves.

(2) **All habitats are represented in the network**. The appropriate habitat classification should match the spatial scale of the conservation planning efforts.

(3) **Within each biogeographic region several examples of each habitat should be included within separated marine reserves**. A precautionary number of replicates would be three, with two replicates being the minimum.

(4) **Marine reserves should be large enough to cover the majority of species' adult movement distances**. Based on international case studies and extensive studies from existing New Zealand reserves, we would recommend that marine reserves have a minimum coastline length of 5–10 km, preferably 10–20 km, and should extend along the depth gradient from intertidal to deeper offshore waters, preferably to the 12 nautical mile limit.

(5) **The spacing between marine reserves should allow larval dispersal to occur**. We recommend that marine reserves, with similar habitats where possible, should be placed within 50–100 km of each other.

We recommend that MPA network design processes **focus on no-take marine reserves** as these provide the maximum conservation benefits for marine ecosystems. Research in northern New Zealand MPAs that allow recreational fishing shows there is no conservation value in protecting against commercial fishing alone when recreational effort is high (Denny & Babcock 2004; Shears *et al.* 2006). Since the direct and indirect impacts of fishing on ecosystems are often complex and uncertain, if the aim is to ensure biodiversity protection then this requires the precautionary approach through establishing marine reserves that eliminate all extractive activities. Furthermore, marine reserves are a simple and cost-effective tool from a management perspective that removes doubt for users over what activities are allowed and gives no preference to particular extractive stakeholder groups. While our recommendation here is for no-take marine reserves, other forms of no-take MPAs may be established through fisheries closures or traditional approaches such as Taiapure. The design of such MPAs should also follow the scientific guidelines recommended here to ensure that they are effectively integrated into and contribute to a wider network of protection.

The size of individual reserves should be **large enough to encompass the home ranges of desired species** and therefore maximise the direct responses of species to protection. Research from New Zealand marine reserves suggests that the minimum size requirements of no-take zones used in international case studies (e.g. minimum coastline lengths of 5km and preferred of 10–20 km) would be appropriate in the New Zealand context. Babcock *et al.* (2012) found that marine reserves with sizes of approximately 5 km^2 were too small to fully protect resident reserve snapper populations. Therefore, while reserves spanning ~5 km of coast are known to still have conservation benefits and result in increased biomasses of snapper (Case study 1), this size should be considered a minimum. Similarly, Jack & Wing (2013) suggest from research on lobster in Fiordland marine reserves that managers should establish networks of large marine reserves. **Extending reserves offshore** to jurisdictional limits also ensures that species that undergo offshore movements are protected (Case study 3), that deep water habitats are represented, and that individual marine reserves contribute maximally to the overall area protected within a region.

Achieving **connectivity between marine reserves** is necessary to ensure that the overall benefits of the marine reserve network far exceed the summed benefits of the individual MPAs. This fundamentally requires the exchange of eggs, larvae, juveniles and adults among MPAs with the aim of maintaining species populations in the long term across all parts of their range. In reality, larval dispersal patterns and adult movement distances vary hugely between species and will be affected by changing ocean conditions, so any connectivity reserve design must attempt to maximise the possibilities for larval and individual exchange between the widest range of species. To achieve this, MPAs need to be spaced in a fashion (i.e. close enough to each other) that larvae from the large majority of species can disperse from one MPA to the next. This connectivity principle can be incorporated into MPA planning with a simple **spacing guideline** that gives recommended maximum distances between MPAs. In California, this required **reserves to be sited within 50–100 km of each other**, in order to be within the dispersal range of most commercial or recreational fish or invertebrate species (CDFG 2010). We recommend the same guideline be applied in New Zealand. However, spacing could be adjusted depending on regionally important species; for example, for short-dispersing species such as abalone reserves may need to be more closely spaced (Stephens *et al.* 2006).

It is important to note that these recommended science guidelines **do not include numerical targets for the overall amount of protection**. While specific targets provide measurable indicators of success and clear objectives for stakeholder-driven processes, fixed percentages are often controversial (Gleason *et al.* 2010). For example, the World Parks Congress set a target in 2003 to include 20–30% of each habitat within strictly protected areas (Wood *et al.* 2008). An alternative to setting specific targets is the use of other specific network design principles (e.g. 2–5) that automatically establish a proportion of marine reserves within their respective networks. This approach worked well in the California process (Gleason *et al.* 2010) and is recommended here. However, based on recommendations worldwide 10% coverage of marine reserves should generally be considered a minimum. Networks of marine reserves may be complemented, and the overall area protected extended, by broader networks of MPAs that allow restricted activities. For example, on the Great Barrier Reef, in addition to the 33% no-take areas, a large proportion of area (32%) are Habitat Protection Zones where trawling is prohibited.

28.3.4   The management role of marine reserve networks

The development of marine reserve networks is still largely in its infancy worldwide and scientific research is only just starting to provide answers to some of the big questions regarding the fishery benefits of marine reserve networks. The signs from GBRMP are very promising (Harrison *et al.* 2012) and over time the potential benefits of this marine reserve network, and others, to fisheries will be borne out by further study. With continual improvements in our knowledge and useful lessons learned, the suggestion is that designing reserves specifically for both fisheries and conservation benefits is technically possible and highly desirable. However, most current marine reserves are too small to deliver the biological conditions necessary to promote fishery recoveries and the vast majority are still single, isolated reserves, struggling in a sea of increasing pressures without the multiplicative benefits provided by large networks of no-take and multiple-use MPAs (Gaines *et al.* 2010). Analysis of existing reserve performance has led to clearer science guidelines in future network design to achieve improved fisheries benefits without the implicit trade-off against conservation goals (Gaines *et al.* 2010). Any potential 'win-win' scenario must therefore be accompanied by a far greater collaboration between fisheries and conservation to achieve successful ecosystem management (Worm *et al.* 2009).

Although large marine reserve networks are now being designed or implemented to hopefully fulfil this dual expectation (e.g. California, Great Barrier Reef), the debate is by no means concluded. Overfishing is only one of the numerous negative pressures facing ecosystem persistence, and in marine environments, pollution, invasive species, climate change and cumulative impacts present increasingly worrying threats. However, it is important to recognise that unlike these other impacts, the impacts of fishing can be easily controlled and effectively eliminated spatially with the use of marine reserves. Although marine reserves should not be expected to achieve long-term conservation success on their own (Agardy *et al.* 2011), large, well-designed marine reserve networks offer insurance policies against current and future risks to both fisheries and biodiversity (Allison *et al.* 2003).

28.4 Conclusions

Fishing has a widespread impact on many coastal species and ecosystems. Reducing fishing mortality inside marine reserves typically results in a recovery of target species; with higher abundances of larger and sometimes healthier individuals in reserves compared to fished areas. The higher biomass of target species in reserves results in greater reproductive output per unit area compared to fished areas. While these direct responses have been observed in many reserves, the magnitude of the response varies depending on a range of factors related to the species in question, fishing intensity both before protection and outside the marine reserve, the attributes of the reserve itself and other management aspects such as level of enforcement.

Through a variety of indirect responses marine reserves can have conservation value beyond exploited species. Fishing can have cascading effects on ecosystems and via indirect responses marine reserves have the potential to reverse these impacts and restore trophic interactions, ecological processes and habitats. These indirect responses are generally more difficult to predict, take longer to manifest and are often not as obvious as direct responses to protection. They typically require strong direct responses and strong trophic linkages between exploited species and non-target species.

How marine reserves impact on populations of exploited species outside of their boundaries is less well understood and considerably harder to investigate. The potential value of marine reserves to fisheries stems through the external responses of marine reserve; either through the movement of adult individuals across reserve boundaries (spillover) or the export of larvae to adjacent populations (recruitment subsidy). While spillover may compensate for reduced available areas for fishing, emerging evidence demonstrates that the role of reserves as a potential source of larvae for fished populations is of far greater consequence to fisheries management. Maximising the potential for such recruitment effects requires that networks of reserves are individually of sufficient size to protect spawning populations, and sufficiently connected to facilitate transfer between reserves.

New Zealand has a vast, unique and productive marine environment that is of huge importance to New Zealanders. Current levels of marine reserve protection are well below globally recognised targets and levels of protection being achieved in other coastal nations. While management of the variety of human impacts and uses of the marine environment requires a much broader and integrated management approach, we recommend that strategically designed networks of marine reserves, using the scientific guidelines provided here, should be at the core of any marine spatial planning or ecosystem-based management process in New Zealand.

REFERENCES

Agardy T., Di Sciara G. N. & Christie P. (2011). Mind the gap: addressing the shortcomings of marine protected areas through large scale marine spatial planning. *Marine Policy*, **35**, 226–232.

Alcala A. C., Russ G. R., Maypa A. P. & Calumpong H. P. (2005). A long-term, spatially replicated experimental test of the effect of marine reserves on local fish yields. *Canadian Journal of Fish and Aquatic Science*, **62**, 98–108.

Allison G. W., Gaines S. D., Lubchenco J. & Possingham H. P. (2003). Ensuring persistence of marine reserves: catastrophes require adopting an insurance factor. *Ecological Applications*, **13**, S8–S24.

Ayling A. M. (1978). *Cape Rodney to Okakari Point Marine Reserve Survey*. Leigh Laboratory Bulletin 1.

Babcock R. (2013). Leigh Marine Laboratory contributions to marine conservation. *New Zealand Journal of Marine Freshwater Research*, **47**(3), 360–373.

Babcock R. C., Egli D. P. & Attwood C. G. (2012). Incorporating behavioural variation in individual-based simulation models of marine reserve effectiveness. *Environmental Conservation*, **39**, 282–294.

Babcock R. C., Kelly S., Shears N. T., Walker J. W. & Willis T. J. (1999). Changes in community structure in temperate marine reserves. *Marine Ecology Progress Series*, **189**, 125–134.

Babcock R. C., Shears N. T., Alcala A. C., et al. (2010). Decadal trends in marine reserves reveal differential rates of change in direct and indirect effects. *Proceedings of the National Academy of Sciences USA*, **107**, 18 256–18 261.

Ballantine W. J. & Gordon D. P. (1979). New-Zealand's 1st Marine Reserve, Cape Rodney to Okakari Point, Leigh. *Biological Conservation*, **15**, 273–280.

Botsford L. W., Hastings A. & Gaines S. D. (2001). Dependence of sustainability on the configuration of marine reserves and larval dispersal distance. *Ecology Letters*, **4**, 144–150.

Carr M. H., Saarman E. & Caldwell M. R. (2010). The role of "rules of thumb" in science-based environmental policy: California's Marine Life Protection Act as a case study. *Stanford Journal of Law, Science and Policy*, **2**, 1–17.

Caveen A., Sweeting C., Willis T. & Polunin N. (2012). Are the scientific foundations of temperate marine reserves too warm and hard? *Environmental Conservation*, **39**, 199–203.

CDFG (2010). *California Department of Fish and Game. The Marine Life Protection Act Master Plan for Marine Protected Areas.* http://www.dfg.ca.gov/mlpa/pdfs/revisedmp0108.pdf

Claudet J., Osenberg C. W., Benedetti-Cecchi L., *et al.* (2008). Marine reserves: size and age do matter. *Ecology Letters*, **11**, 481–489.

Colmar-Brunton (2011). *Measuring New Zealanders' Attitudes Towards their Oceans and Marine Reserves.* A Colmar Brunton report for WWF-New Zealand Published 26 May 2011. wwf.org.nz

Dayton P. K., Thrush S. & Coleman F. C. (2003). *Ecological Effects of Fishing.* Report to the Pew Oceans Commission, Arlington, Virginia (USA).

Denny C. M. & Babcock R. C. (2004). Do partial marine reserves protect reef fish assemblages? *Biological Conservation*, **116**, 119–129.

Denny C. M., Willis T. J. & Babcock R. C. (2004). Rapid recolonisation of snapper *Pagrus auratus*: Sparidae within an offshore island marine reserve after implementation of no-take status. *Marine Ecology Progress Series*, **272**, 183–190.

Department of Conservation & Ministry of Fisheries (2011). *Coastal Marine Habitats and Marine Protected Areas in the New Zealand Territorial Sea: a Broad scale Gap Analysis.* Department of Conservation and Ministry of Fisheries. Wellington, New Zealand.

DOC & MFISH (2005). *Marine Protected Areas Policy and Implementation Plan.* Ministry of Fisheries and Department of Conservation, Wellington, New Zealand. 54 pages.

DOC & MFISH (2008). *Marine Protected Areas: Classification, Protection Standard and Implementation Guidelines.* Ministry of Fisheries and Department of Conservation, Wellington, New Zealand. 54 pages.

Edgar G. J., Barrett N. S. & Stuart-Smith R. D. (2009). Exploited reefs protected from fishing transform over decades into conservation features otherwise absent from seascapes. *Ecological Applications*, **19**, 1967–1974.

Edgar G. J. & Stuart-Smith R. D. (2009). Ecological effects of marine protected areas on rocky reef communities-a continental-scale analysis. *Marine Ecology Progress Series*, **388**, 51–62.

Fernandes L., Day J., Kerrigan B., *et al.* (2009). A process to design a network of marine no-take areas: Lessons from the Great Barrier Reef. *Ocean & Coastal Management*, **52**, 439–447.

Freeman D. J. & MacDiarmid A. B. (2009). Healthier lobsters in a marine reserve: effects of fishing on disease incidence in the spiny lobster, *Jasus edwardsii*. *Marine and Freshwater Research*, **60**, 140–145.

Freeman D. J., Breen P. A. & MacDiarmid A. B. (2012a). Use of a marine reserve to determine the direct and indirect effects of fishing on growth in a New Zealand fishery for the spiny lobster *Jasus edwardsii*. *Canadian Journal of Fish and Aquatic Science*, **69**, 894–905.

Freeman D. J., MacDiarmid A. B. & Taylor R. B. (2009). Habitat patches that cross marine reserve boundaries: consequences for the lobster *Jasus edwardsii*. *Marine Ecology Progress Series*, **388**, 159–167.

Freeman D. T., Macdiarmid A. B., Taylor R. B., *et al.* (2012b). Trajectories of spiny lobster *Jasus edwardsii* recovery in New Zealand marine reserves: is settlement a driver? *Environmental Conservation*, **39**, 295–304.

Gaines S. D., White C., Carr M. H. & Palumbi S. R. (2010). Designing marine reserve networks for both conservation and fisheries management. *Proceedings of the National Academy of Sciences*, **107**, 18 286–18 293.

Gell F. R. & Roberts C. M. (2003). Benefits beyond boundaries: the fishery effects of marine reserves. *Trends in Ecology & Evolution*, **18**, 448–455.

Gleason M., McCreary S., Miller-Henson M., *et al.* (2010). Science-based and stakeholder-driven marine protected area network planning: a successful case study from north central California. *Ocean & Coastal Management*, **53**, 52–68.

Goñi R., Hilborn R., Díaz D., Mallol S. & Adlerstein S. (2010). Net contribution of spillover from a marine reserve to fishery catches. *Marine Ecology Progress Series*, **400**, 233–243.

Guidetti P., Milazzo M., Bussotti S., *et al.* (2008). Italian marine reserve effectiveness: does enforcement matter? *Biology Conservation*, **141**, 699–709.

Halpern B., Selkoe K., White C., *et al.* (2013). Marine protected areas and resilience to sedimentation in the Solomon Islands. *Coral Reefs*, **32**, 61–69.

Halpern B. S., Lester S. E. & Kellner J. B. (2009). Spillover from marine reserves and the replenishment of fished stocks. *Environmental Conservation*, **36**, 268–276.

Halpern B. S., Lester S. E. & McLeod K. L. (2010). Placing marine protected areas onto the ecosystem-based management seascape. *Proceedings of the National Academy of Sciences*, **107**, 18 312–18 317.

Harrison H. B., Williamson D. H., Evans R. D., *et al.* (2012). Larval export from marine reserves and the recruitment benefit for fish and fisheries. *Current Biology*, **22**, 1023–1028.

Hilborn R. (2012). The role of science in MPA establishment in California: a personal perspective. *Environmental Conservation*, **39**, 195–198.

Hilborn R., Micheli F. & De Leo G. A. (2006). Integrating marine protected areas with catch regulation. *Canadian Jounral of Fish and Aquatic Science*, **63**, 642–649.

Hilborn R., Stokes K., Maguire J. J., *et al.* (2004). When can marine reserves improve fisheries management? *Ocean & Coastal Management*, **47**, 197–205.

Jack L. & Wing S. R. (2013). A safety network against regional population collapse: mature subpopulations in refuges distributed across the landscape. *Ecosphere*, **4**, art57.

Jackson J. B. C., Kirby M. X., Berger W. H., *et al.* (2001). Historical overfishing and the recent collapse of coastal ecosystems. *Science*, **293**, 629–638.

Jones G. (2013). Ecology of rocky reef fish of northeastern New Zealand: 50 years on. *New Zealand Journal of Marine and Freshwater Research*, **47**, 334–359.

Kelly S. (2001). Temporal variation in the movement of the spiny lobster *Jasus edwardsii*. *Marine and Freshwater Research*, **52**, 323–331.

Kelly S. & MacDiarmid A. B. (2003). Movement patterns of mature spiny lobsters, *Jasus edwardsii*, from a marine reserve. *New Zealand Journal of Marine and Freshwater Research*, **37**, 149–158.

Kelly S., Scott D. & MacDiarmid A. B. (2002). The value of a spillover fishery for spiny lobsters around a marine reserve in Northern New Zealand. *Coastal Management*, **30**, 153–166.

Kerwath S. E., Winker H., Götz A. & Attwood C. G. (2013). Marine protected area improves yield without disadvantaging fishers. *Nature Communications*, **4**.

Kramer D. L. & Chapman M. R. (1999). Implications of fish home range size and relocation for marine reserve function. *Environmental Biology of Fishes*, **55**, 65–79.

Langlois T. J., Anderson M. J., Babcock R. C. & Kato S. (2006). Marine reserves demonstrate trophic interactions across habitats. *Oecologia*, **147**, 134–140.

Leleu K., Remy-Zephir B., Grace R. & Costello M. J. (2012). Mapping habitats in a marine reserve showed how a 30-year trophic cascade altered ecosystem structure. *Biology Conservation*, **155**, 193–201.

Lester S. E., Halpern B. S., Grorud-Colvert K., *et al.* (2009). Biological effects within no-take marine reserves: a global synthesis. *Marine Ecology Progress Series*, **384**, 33–46.

Lubchenco J., Palumbi S. R., Gaines S. D. & Andelman S. (2003). Plugging a hole in the ocean: the emerging science of marine reserves. *Ecological Applications*, **13**, S3–S7.

McClanahan T. R. & Mangi S. (2000). Spillover of exploitable fishes from a marine park and its effect on the adjacent fishery. *Ecology Applications*, **10**, 1792–1805.

McCook L. J., Ayling T., Cappo M., *et al.* (2010). Adaptive management of the Great Barrier Reef: A globally significant demonstration of the benefits of networks of marine reserves. *Proceedings of the National Academy of Sciences*, **107**, 18 278–18 285.

Micheli F., Saenz-Arroyo A., Greenley A., *et al.* (2012). Evidence that marine reserves enhance resilience to climatic impacts. *PLoS ONE*, **7**, e40832.

Mulcahy K., Peart R. & Bull A. (2012). *Safeguarding our Oceans: Strengthening Marine Protection in New Zealand*. Environmental Defence Society.

Murawski S. A., Brown R., Lai H. L., Rago P. J. & Hendrickson L. (2000). Large-scale closed areas as a fishery-management tool in temperate marine systems: the Georges Bank experience. *Bulletin of Marine Science*, **66**, 775–798.

NE & JNCC (2010). The Marine Conservation Zone Project: Ecological Network Guidance.

Parsons D. M., Morrison M. A., MacDiarmid A. B., *et al.* (2009). Risks of shifting baselines highlighted by anecdotal accounts of New Zealand's snapper (*Pagrus auratus*) fishery. *New Zealand Journal Marine and Freshwater Research*, **43**, 965–983.

Parsons D. M., Morrison M. A. & Slater M. J. (2010). Responses to marine reserves: Decreased dispersion of the sparid *Pagrus auratus* (snapper). *Biology Conservation*, **143**, 2039–2048.

Pelc R., Baskett M., Tanci T., Gaines S. & Warner R. (2009). Quantifying larval export from South African marine reserves. *Marine Ecology Progress Series*, **394**, 65–78.

Pinkerton M. H., Lundquist C. J., Duffy C. A. J. & Freeman D. J. (2008). Trophic modelling of a New Zealand rocky reef ecosystem using simultaneous adjustment of diet, biomass and energetic parameters. *Journal of Experimental Marine Biology and Ecology*, **367**, 189–203.

Russ G. R., Alcala A. C., Maypa A. P., Calumpong H. P. & White A. T. (2004). Marine reserve benefits local fisheries. *Ecological Applications*, **14**, 597–606.

Saarman E., Gleason M., Ugoretz J., *et al.* (2013). The role of science in supporting marine protected area network planning and design in California. *Ocean & Coastal Management*, **74**, 45–56.

Salomon A. K., Gaichas S. K., Shears N. T., *et al.* (2010). Key features and context-dependence of fishery-induced trophic cascades. *Conservation Biology*, **24**, 382–394.

Salomon A. K., Shears N. T., Langlois T. J. & Babcock R. C. (2008). Cascading effects of fishing can alter carbon flow through a temperate coastal ecosystem. *Ecological Applications*, **18**, 1874–1887.

Schiel D. (2013). The other 93%: trophic cascades, stressors and managing coastlines in non-marine protected areas. *New Zealand Journal of Marine and Freshwater Research*, **47**, 1–18.

Shears N. T. & Babcock R. C. (2002). Marine reserves demonstrate top-down control of community structure on temperate reefs. *Oecologia*, **132**, 131–142.

Shears N. T. & Babcock R. C. (2003). Continuing trophic cascade effects after 25 years of no-take marine reserve protection. *Marine Ecology Progress Series*, **246**, 1–16.

Shears N. T. & Babcock R. C. (2004). Indirect effects of marine reserves on New Zealand's rocky coastal communities. In, Department of Conservation Science Internal Series No. 192. Department of Conservation, Wellington, New Zealand. 49p. Available from http://www.doc.govt.nz/templates/page.aspx?id=39167 (accessed July 2007).

Shears N. T. & Babcock R. C. (2007). Quantitative description of mainland New Zealand's shallow subtidal reef communities. *Science for Conservation*, **280**, 126 pages. http://www.doc.govt.nz/upload/documents/science-and-technical/sfc280entire.pdf

Shears N. T., Babcock R. C. & Salomon A. K. (2008). Context-dependent effects of fishing: variation in trophic cascades across environmental gradients. *Ecology Applications*, **18**, 1860–1873.

Shears N. T., Grace R. V., Usmar N. R., Kerr V. & Babcock R. C. (2006). Long-term trends in lobster populations in a partially protected vs. no-take Marine Park. *Biological Conservation*, **132**, 222–231.

Steneck R., Paris C., Arnold S., *et al.* (2009). Thinking and managing outside the box: coalescing connectivity networks to build region-wide resilience in coral reef ecosystems. *Coral Reefs*, **28**, 367–378.

Stephens S., Broekhuizen N., Macdiarmid A., *et al.* (2006). Modelling transport of larval New Zealand abalone (*Haliotis iris*) along an open coast. *Marine and Freshwater Research*, **57**, 519–532.

Stockwell B., Jadloc C. R. L., Abesamis R. A., Alcala A. C. & Russ G. R. (2009). Trophic and benthic responses to no-take marine reserve protection in the Philippines. *Marine Ecology Progress Series*, **389**, 1–15.

Taylor R. B., Morrison M. A. & Shears N. T. (2011). Establishing baselines for recovery in a marine reserve (Poor Knights Islands, New Zealand) using local ecological knowledge. *Biology Conservation*, **144**, 3038–3046.

Willis T. J. & Anderson M. J. (2003). Structure of cryptic reef fish assemblages: relationships with habitat characteristics and predator density. *Marine Ecology Progress Series*, **257**, 209–221.

Willis T. J., Millar R. B. & Babcock R. C. (2000). Detection of spatial variability in relative density of fishes: comparison of visual census, angling, and baited underwater video. *Marine Ecology Progress Series*, **198**, 249–260.

Willis T. J., Millar R. B. & Babcock R. C. (2003a). Protection of exploited fish in temperate regions: high density and biomass of snapper *Pagrus auratus* (Sparidae) in northern New Zealand marine reserves. *Journal of Applied Ecology*, **40**, 214–227.

Willis T. J., Millar R. B., Babcock R. C. & Tolimieri N. (2003b). Burdens of evidence and the benefits of marine reserves: putting Descartes before des horse? *Environmental Conservation*, **30**, 97–103.

Wood L. J., Fish L., Laughren J. & Pauly D. (2008). Assessing progress towards global marine protection targets: shortfalls in information and action. *Oryx*, **42**, 340–351.

Worm B., Hilborn R., Baum J. K., *et al.* (2009). Rebuilding global fisheries. *Science*, **325**, 578–585.

CHAPTER 29

Conclusion: conservation onboard Austral Ark needs all hands on deck

Adam Stow

The wildlife of the Australian and New Zealand landmasses house a truly unusual component of global biodiversity, a fact now held with a degree of national pride. By global standards, Australia and New Zealand have relatively few people, the majority of whom can trace their ancestry back to other continents in fairly recent time. Nonetheless, one can sense that a frontier mentality of conquer and survive is being replaced with a sense of being custodians for something special. And there is good reason for this. Australian and New Zealand wildlife, uniquely sculpted by long isolation, have been especially vulnerable to the impacts of the anthropocene. This book presents some of the vital research that has been conducted to describe what wildlife is present, what the conservation issues are, and how best to conserve it.

Threats to Australian and New Zealand wildlife are varied and include introduced plants and animals, changes in human land use, pollution, disease and the looming issue of climate change. The loss of native biota has been rapid. A haunting reminder of this can be gained in parts of Central Australia where one can stand on earth raised above the desert plains by the strenuous efforts of the burrowing bettong, a species now removed from the landscape, and their presence still remembered by those living today. Similarly, when listening to the dawn chorus on one of New Zealand's predator-free offshore islands, one is reminded of the relative silence of most mainland forests, sadly devoid of many native songbirds that once were widespread. While this book serves to remind us of the natural heritage that is under threat in Australia and New Zealand, it has also offered hope, describing the work that is underway to conserve it. There are solutions to many of the issues and a community of passionate individuals who are willing to work on them. The issues and solutions facing wildlife on Austral Ark will not be reiterated in any detail here, instead, the focus is on how we might ensure that all of those that are motivated to help conserve wildlife are harnessed for best effect.

The very low population sizes of Australia and New Zealand is a double-edged sword. Australia has among the lowest of population densities for an inhabited continent, is the sixth largest country by area (CIA 2014), and has the largest portion of marine

Austral Ark: The State of Wildlife in Australia and New Zealand, eds. A. Stow, N. Maclean and G. I. Holwell. Published by Cambridge University Press. © Cambridge University Press 2015.

environment as part of its jurisdiction (Chapter 27). New Zealand has a higher population density than Australia (people per km^2; 17 versus 2.7 respectively; CIA 2014), but the total population size is small, just shy of five million. While a larger population would heighten many threatening processes, smaller nations do possess limited resources available for conservation efforts. Given the magnitude of many conservation problems, the resources available, and the rapid pace at which we are losing biodiversity, it is imperative that there is clear communication and coordination among all those who are motivated to help.

It is essential that research biologists, land managers and policy makers continue to talk to each other so the right questions can be asked and answered. However, professional conservation biologists, land managers and policy makers form only a small, though critically important, component of the community. Often inspired by the great interpretative work of institutions like zoos, botanic gardens, museums and non-government conservation organisations, there are many willing volunteers. Through various volunteer schemes much time and effort have been dedicated to removing weeds, cleaning up unwanted material, fencing out introduced predators, and a whole gamut of other conservation-related activities. And this is just as well, as we have seen there are a multitude of issues and much work that needs to be done in these sparsely populated countries.

Well-thought-out conservation strategies and coordinating those willing to work towards protecting our natural heritage will best harness this enthusiasm. Broadly, this can be achieved through direct action to remove threats, such as weed removal, habitat creation or to collect data for science. Collecting scientific data by interested volunteers, 'Citizen Science' has recently been gaining popularity (see Silvertown 2009). While the sorts of projects suitable for Citizen Science need to be carefully considered, and data validated, there are some great examples of where it has been applied to better understand species distributions and for monitoring biodiversity (Silvertown 2009). Recent technological advances in communication and computing allow for rapid and formulated onsite data entry, and also provide a means of lodging data for those who are not literate. These advances are coming up with solutions to efficiently validate data and standardise its entry into databases, thereby strengthening the scientific value of these data.

Especially in the Australian context, there are vast areas in the centre and north that are far from civilisation and difficult to access (Chapter 26, Plate 15C). Many locations are rarely if ever surveyed, and even fewer have been subject to long-term monitoring. When they have, the result has been discovering astonishing levels of diversity, especially with respect to lizards, and many of these might be at low abundance and thereby require long-term search effort to locate them (Pianka 2013). Visitation to remote areas of Australia does occur recreationally, especially among retired members of the community. Furthermore, Indigenous Protected Areas occupy 40% of reserved areas (Chapter 26), and traditional land managers are often more or less permanently present in areas that are rarely subject to western-style data collection. The opportunity to obtain species record data from these sources has been observed by others, and great collaborative work is being done (for collaboration with indigenous groups see Ems 2012), but there is still room for further engagement. Special interest groups such as ornithological and herpetological societies or those with interests in nature-based exploration, such as SCUBA diving groups are often well suited to, and interested in, these sorts of activities. There are many established and developing programmes for community involvement in conservation science now available.

Figure 29.1 (Plate 25) Individual Grey Nurse shark have a unique arrangement of dots that can be used for identification and monitoring this critically endangered species. Image kindly provided by Rob Harcourt. A black and white version of this figure will appear in some formats. For the colour version, please refer to the plate section.

To provide a marine example, the Grey Nurse shark is critically endangered on the east coast of Australia (Chapter 22). A Grey Nurse Shark Community engagement programme has been established by academic staff and the REEF Check program. With support from both government and non-government sources, recreational SCUBA divers survey and photograph Grey Nurse sharks. Spot patterns on the Grey Nurse sharks skin allow for individual recognition (Figure 29.1, Plate 25) and these photographic data are used to observe individual movements and monitor for population recovery. On land, the National Parks Association of NSW carried out a Citizen Science programme called the Great Koala Count that provided information on the land-use types where koalas were sighted. In New Zealand programs such as Nature Watch are coordinating a range of Citizen Science projects that includes recording the locations of weed species and even plant pathogens. This small sample of a collection of many good programmes to which information can be volunteered is provided simply to illustrate the sorts of ways community observations and efforts can be organised to provide conservation-relevant data.

The overview provided in *Austral Ark* of the state of the wildlife in Australia and New Zealand, and its conservation will hopefully not only inform but inspire action. In particular it is hoped that it will stimulate further engagement with sectors of community beyond dedicated conservation organisations. Although the challenges are large, the oddball diversity on the land and in the water of these far-flung southern lands is sure to fascinate and provide for future generations. It's well worth the effort.

REFERENCES

Ems, E. J., Towler, G. M., Daniels, R. *et al.* (2012) Looking back to move forward: collaborative ecological monitoring in remote Arnhem Land. *Ecological Management and Restoration*, **13**, 26–35.

Pianka E. 2013 Rarity in Australian desert lizards. DOI:10/1111/aec.12061

Silvertown, J. 2009 A new dawn for citizen science. *Trends in Ecology & Evolution*, **24**, 467–471.

The World Fact Book https://www.cia.gov/library/publications/the-world-factbook/geos/as.html

INDEX